Sustainable Engineering

Drivers, Metrics, Tools, and Applications

Krishna R. Reddy, Claudio Cameselle, and Jeffrey A. Adams

This edition first published 2019
© 2019 John Wiley & Sons, Inc.

All rights reserved. No part of this publication may be reproduced, stored in a retrieval system, or transmitted, in any form or by any means, electronic, mechanical, photocopying, recording or otherwise, except as permitted by law. Advice on how to obtain permission to reuse material from this title is available at http://www.wiley.com/go/permissions.

The right of Krishna R. Reddy, Claudio Cameselle, and Jeffrey A. Adams to be identified as the authors of this work has been asserted in accordance with law.

Registered Office
John Wiley & Sons, Inc., 111 River Street, Hoboken, NJ 07030, USA

Editorial Office
111 River Street, Hoboken, NJ 07030, USA

For details of our global editorial offices, customer services, and more information about Wiley products visit us at www.wiley.com.

Wiley also publishes its books in a variety of electronic formats and by print-on-demand. Some content that appears in standard print versions of this book may not be available in other formats.

Limit of Liability/Disclaimer of Warranty
In view of ongoing research, equipment modifications, changes in governmental regulations, and the constant flow of information relating to the use of experimental reagents, equipment, and devices, the reader is urged to review and evaluate the information provided in the package insert or instructions for each chemical, piece of equipment, reagent, or device for, among other things, any changes in the instructions or indication of usage and for added warnings and precautions. While the publisher and authors have used their best efforts in preparing this work, they make no representations or warranties with respect to the accuracy or completeness of the contents of this work and specifically disclaim all warranties, including without limitation any implied warranties of merchantability or fitness for a particular purpose. No warranty may be created or extended by sales representatives, written sales materials or promotional statements for this work. The fact that an organization, website, or product is referred to in this work as a citation and/or potential source of further information does not mean that the publisher and authors endorse the information or services the organization, website, or product may provide or recommendations it may make. This work is sold with the understanding that the publisher is not engaged in rendering professional services. The advice and strategies contained herein may not be suitable for your situation. You should consult with a specialist where appropriate. Further, readers should be aware that websites listed in this work may have changed or disappeared between when this work was written and when it is read. Neither the publisher nor authors shall be liable for any loss of profit or any other commercial damages, including but not limited to special, incidental, consequential, or other damages.

Library of Congress Cataloging-in-Publication Data is Applied For

Hardback ISBN: 9781119493938

Cover design: Wiley
Cover image: © Pobytov/Getty Images

Set in 10/12pt WarnockPro by SPi Global, Chennai, India

Printed in the United States of America

V10008852_031419

This book is dedicated to our families for their unwavering support, love, and sacrifice:

- *Krishna Reddy – Hema (wife), Nishith (son), and Navya (daughter)*
- *Claudio Cameselle – Claudio and Rosa (parents), Susana (wife), and Lucia and Guilherme (children)*
- *Jeffrey Adams – Amy (wife) and Lizzie (daughter)*

Contents

Preface *xvii*

Section I   Drivers, Environmental, Economic and Social Impacts, and Resiliency *1*

1	**Emerging Challenges, Sustainability, and Sustainable Engineering** *3*	
1.1	Introduction *3*	
1.2	Emerging Challenges *3*	
1.2.1	Increased Consumption and Depletion of Natural Resources *3*	
1.2.1.1	Easter Island Example *4*	
1.2.1.2	Metallic Ores Consumption Example *5*	
1.2.2	Growing Environmental Pollution *6*	
1.2.3	Increasing Population *7*	
1.2.4	Increasing Waste Generation *8*	
1.2.5	Increasing Greenhouse Gas Emissions *10*	
1.2.6	Decline of Ecosystems *13*	
1.2.7	Loss of Biodiversity *13*	
1.2.8	Social Injustice *14*	
1.2.9	Urban Sprawl *16*	
1.3	The Master Equation or IPAT Equation *17*	
1.4	What Is Sustainability? *17*	
1.5	What Is Sustainable Engineering? *21*	
1.6	Summary *25*	
1.7	Questions *26*	
	References *26*	
2	**Environmental Concerns** *31*	
2.1	Introduction *31*	
2.2	Global Warming and Climate Change *32*	
2.3	Desertification *40*	
2.4	Deforestation *40*	
2.5	Loss of Habitat and Biodiversity *41*	
2.6	Ozone Layer Depletion *43*	
2.7	Air Pollution *44*	
2.8	Smog *46*	
2.9	Acid Rain *47*	
2.10	Water Usage and Pollution *48*	

2.11	Eutrophication	*51*
2.12	Salinity	*52*
2.13	Wastes and Disposal	*52*
2.14	Land Contamination	*59*
2.15	Visibility	*60*
2.16	Odors	*60*
2.17	Aesthetic Degradation	*61*
2.18	Land Use Patterns	*61*
2.19	Thermal Pollution	*61*
2.20	Noise Pollution	*62*
2.21	Summary	*62*
2.22	Questions	*63*
	References	*64*
3	**Social, Economic, and Legal Issues**	***69***
3.1	Introduction	*69*
3.2	Social Issues	*69*
3.2.1	Society	*69*
3.2.2	Developed and Developing Societies	*70*
3.2.3	Social Sustainability Concept	*71*
3.2.4	Social Indicators	*72*
3.2.5	Social Impact Assessment	*73*
3.2.6	Social Sustainability Implementation	*77*
3.3	Economic Issues	*77*
3.3.1	Economic Assessment Framework	*78*
3.3.2	Life Cycle Costing	*79*
3.3.3	True-cost Accounting	*79*
3.4	Legal Issues	*80*
3.5	Summary	*81*
3.6	Questions	*81*
	References	*82*
4	**Availability and Depletion of Natural Resources**	***85***
4.1	Introduction	*85*
4.2	Types and Availability of Resources	*85*
4.2.1	Fossil Fuels	*85*
4.2.2	Radioactive Fuels	*87*
4.2.3	Mineral Resources	*88*
4.2.4	Water Resources	*89*
4.2.5	Other Elemental Cycles	*91*
4.3	Resource Depletion	*94*
4.3.1	Causes of Resource Depletion	*95*
4.3.2	Effects of Resource Depletion	*95*
4.3.3	Overshooting	*98*
4.3.4	Urban Metabolism	*98*
4.4	Summary	*99*
4.5	Questions	*100*
	References	*101*

5	**Disaster Resiliency** *103*	
5.1	Introduction *103*	
5.2	Climate Change and Extreme Events *104*	
5.3	Impacts of Extreme Events *105*	
5.3.1	The 2012 Hurricane Sandy in New York City *105*	
5.3.2	The 2016 Chile's Wildfires by Drought and Record Heat *106*	
5.3.3	The 2017 Worst South Asian Monsoon Floods *106*	
5.4	What Is Resiliency? *106*	
5.5	Initiatives and Policies on Resiliency *109*	
5.6	Resiliency Framework *112*	
5.7	Resilient Infrastructure *115*	
5.8	Resilient Infrastructure Examples *117*	
5.8.1	San Francisco Firehouse Resilient Design *117*	
5.8.2	San Francisco Resilient CSD Design *117*	
5.8.3	Resilient Environmental Remediation *119*	
5.9	Challenges *126*	
5.10	Summary *126*	
5.11	Questions *127*	
	References *127*	

Section II Sustainability Metrics and Assessment Tools *131*

6	**Sustainability Indicators, Metrics, and Assessment Tools** *133*	
6.1	Introduction *133*	
6.2	Sustainability Indicators *133*	
6.3	Sustainability Metrics *136*	
6.4	Sustainability Assessment Tools *137*	
6.5	Summary *139*	
6.6	Questions *139*	
	References *140*	

7	**Material Flow Analysis and Material Budget** *143*	
7.1	Introduction *143*	
7.2	Budget of Natural Resources *143*	
7.3	Constructing a Budget *145*	
7.4	Material Flow Analysis *145*	
7.5	Material Flow Analysis: Wastes *148*	
7.6	National Material Account *151*	
7.7	Summary *155*	
7.8	Questions *156*	
	References *156*	

8	**Carbon Footprint Analysis** *159*	
8.1	Introduction *159*	
8.2	Global Warming Potential and Carbon Footprint *159*	
8.3	Measuring Carbon Footprint *161*	
8.3.1	Define the Scope of Your Inventory *161*	

8.3.2	Measure Emissions and Establish a Baseline *161*
8.3.3	Develop Targets and Strategies to Reduce Emissions *164*
8.3.4	Off-set Unavoidable Emissions *164*
8.3.5	Independent Verification *164*
8.4	Standards for Calculating the Carbon Footprint *164*
8.5	GHG Inventory: Developments in the United States *165*
8.6	USEPA: Greenhouse Gas Reporting Program *166*
8.7	Tools for GHG Inventory *166*
8.8	UIC Carbon Footprint Case Study *167*
8.9	Programs to Mitigate GHG Emissions *171*
8.10	Summary *172*
8.11	Questions *172*
	References *172*

9	**Life Cycle Assessment** *175*
9.1	Introduction *175*
9.2	Life Cycle Assessment *176*
9.2.1	Definition and Objective *176*
9.2.2	Procedure *176*
9.2.3	History *178*
9.3	LCA Methodology *179*
9.3.1	Goal and Scope Definition *180*
9.3.2	Life Cycle Inventory (LCI) *181*
9.3.3	Life Cycle Impact Assessment (LCIA) *184*
9.3.4	Interpretation *188*
9.4	LCA Tools and Applications *189*
9.5	Summary *190*
9.6	Questions *191*
	References *191*

10	**Streamlined Life Cycle Assessment** *193*
10.1	Introduction *193*
10.2	Streamlined LCA (SLCA) *194*
10.3	Expanded SLCA *197*
10.4	Simple Example of SLCA *200*
10.5	Applications of SLCA *202*
10.6	Summary *206*
10.7	Questions *206*
	References *207*

11	**Economic Input–Output Life Cycle Assessment** *209*
11.1	Introduction *209*
11.2	EIO Model *209*
11.3	EIO-LCA *211*
11.4	EIO-LCA Model Results *213*
11.4.1	Interpretation of Results *213*
11.4.2	Uncertainty *213*
11.4.3	Other Issues and Considerations *214*
11.5	Example of EIO-LCA Model *214*

11.6	Conventional LCA versus EIO-LCA	*216*
11.7	EIO versus Physical Input–Output (PIO) Analysis	*218*
11.8	Summary	*221*
11.9	Questions	*221*
	References	*222*

12	**Environmental Health Risk Assessment**	*223*
12.1	Introduction	*223*
12.2	Emergence of the Risk Era	*223*
12.3	Risk Assessment and Management	*224*
12.3.1	Hazard Identification	*225*
12.3.2	Dose–Response Assessment	*225*
12.3.3	Exposure Assessment	*227*
12.3.4	Risk Characterization	*228*
12.4	Ecological Risk Assessment	*230*
12.5	Summary	*231*
12.6	Questions	*232*
	References	*232*

13	**Other Emerging Assessment Tools**	*233*
13.1	Introduction	*233*
13.2	Environmental Assessment Tools/Indicators	*233*
13.3	Economic Assessment Tools	*235*
13.3.1	Life-Cycle Costing	*236*
13.3.2	Cost–Benefit Analysis	*237*
13.4	Ecosystem Services Valuation Tools	*237*
13.5	Environmental Justice Tools	*238*
13.6	Integrated Sustainability Assessment Tools	*239*
13.7	Summary	*241*
13.8	Questions	*241*
	References	*242*

	Section III Sustainable Engineering Practices	*243*

14	**Sustainable Energy Engineering**	*245*
14.1	Introduction	*245*
14.2	Environmental Impacts of Energy Generation	*246*
14.2.1	Air Emissions	*246*
14.2.2	Solid Waste Generation	*250*
14.2.3	Water Resource Use	*250*
14.2.4	Land Resource Use	*250*
14.3	Nuclear Energy	*251*
14.4	Strategies for Clean Energy	*252*
14.5	Renewable Energy	*254*
14.5.1	Solar Energy	*254*
14.5.2	Wind Energy	*255*
14.5.3	Water Energy	*257*
14.5.4	Geothermal Energy	*259*

14.5.5	Biomass Energy	262
14.6	Summary	265
14.7	Questions	266
	References	266

15 Sustainable Waste Management 269

15.1	Introduction	269
15.2	Types of Waste	269
15.2.1	Nonhazardous Waste	270
15.2.2	Hazardous Waste	270
15.3	Effects and Impacts of Waste	270
15.4	Waste Management	271
15.4.1	Pollution Prevention	272
15.4.2	Green Chemistry	272
15.4.3	Waste Minimization	274
15.4.4	Reuse/Recycling	274
15.4.5	Energy Recovery	276
15.4.6	Landfilling	276
15.5	Integrated Waste Management	278
15.6	Sustainable Waste Management	281
15.7	Circular Economy	282
15.8	Summary	283
15.9	Questions	283
	References	284

16 Green and Sustainable Buildings 287

16.1	Introduction	287
16.2	Green Building History	288
16.3	Why Build Green?	288
16.4	Green Building Concepts	289
16.5	Components of Green Building	290
16.6	Green Building Rating – LEED	293
16.7	Summary	297
16.8	Questions	297
	References	298

17 Sustainable Civil Infrastructure 299

17.1	Introduction	299
17.2	Principles of Sustainable Infrastructure	300
17.3	Civil Infrastructure	300
17.4	Envision™: Sustainability Rating of Civil Infrastructure	302
17.5	Sustainable Infrastructure Practices: Example of Water Infrastructure	305
17.5.1	Green Roofs	306
17.5.2	Permeable Pavements	306
17.5.3	Rainwater Harvesting	307
17.5.4	Rain Gardens and Planter Boxes	309
17.5.5	Bioswales	309

17.5.6	Constructed Wetlands and Tree Canopies	*309*
17.6	Summary	*313*
17.7	Questions	*313*
	References	*314*

18 Sustainable Remediation of Contaminated Sites *315*

18.1	Introduction	*315*
18.2	Contaminated Site Remediation Approach	*317*
18.3	Green and Sustainable Remediation Technologies	*318*
18.4	Sustainable Remediation Framework	*323*
18.5	Sustainable Remediation Indicators, Metrics, and Tools	*326*
18.6	Case Studies	*328*
18.7	Challenges and Opportunities	*329*
18.8	Summary	*330*
18.9	Questions	*331*
	References	*332*

19 Climate Geoengineering *333*

19.1	Introduction	*333*
19.2	Climate Geoengineering	*336*
19.3	Carbon Dioxide Removal (CDR) Methods	*336*
19.3.1	Subsurface Sequestration	*336*
19.3.2	Surface Sequestration	*338*
19.3.3	Marine Organism Sequestration	*338*
19.3.4	Direct Engineered Capture	*339*
19.4	Solar Radiation Management (SRM) Methods	*340*
19.4.1	Sulfur Injection	*342*
19.4.2	Reflectors and Mirrors	*343*
19.5	Applicability of CDR and SRM	*344*
19.6	Climate Geoengineering – A Theoretical Framework	*345*
19.7	Risks and Challenges	*345*
19.8	Summary	*347*
19.9	Questions	*348*
	References	*348*

Section IV Sustainable Engineering Applications *351*

20 Environmental and Chemical Engineering Projects *353*

20.1	Introduction	*353*
20.2	Food Scrap Landfilling Versus Composting	*353*
20.2.1	Background	*353*
20.2.2	Methodology	*355*
20.2.2.1	Goal and Scope	*355*
20.2.2.2	Study Area	*355*
20.2.2.3	Technical Design	*355*
20.2.3	Environmental Sustainability	*358*

20.2.4	Life Cycle Assessment	*359*
20.2.5	Economic Sustainability	*359*
20.2.6	Social Sustainability	*365*
20.2.7	ENVISION™	*365*
20.2.8	Conclusions	*368*
20.3	Adsorbent for the Removal of Arsenic from Groundwater	*368*
20.3.1	Background	*368*
20.3.2	Methodology	*369*
20.3.2.1	Goal and Scope	*369*
20.3.2.2	Site Location	*370*
20.3.2.3	Technical Design	*370*
20.3.3	Environmental Sustainability	*372*
20.3.4	Economic Sustainability	*373*
20.3.5	Social Sustainability	*375*
20.3.6	Streamline Life Cycle Assessment (SLCA)	*375*
20.3.7	Envision	*378*
20.3.8	Conclusions	*380*
20.4	Conventional Versus Biocover Landfill Cover System	*381*
20.4.1	Background	*382*
20.4.2	Methodology	*383*
20.4.2.1	Goal and Scope	*383*
20.4.2.2	Landfill Location	*383*
20.4.2.3	Technical Design of Landfill Covers	*383*
20.4.3	Environmental Sustainability	*386*
20.4.4	Economic Sustainability	*391*
20.4.5	Social Sustainability	*393*
20.4.6	Conclusions	*394*
20.5	Algae Biomass Deep Well Reactors Versus Open Pond Systems	*394*
20.5.1	Background	*394*
20.5.2	Methodology	*396*
20.5.2.1	Goal and Scope	*396*
20.5.2.2	Site Location	*396*
20.5.2.3	Technical Design	*396*
20.5.2.4	Sustainability Assessment	*396*
20.5.3	Environmental Sustainability	*400*
20.5.4	Economic Sustainability	*402*
20.5.5	Social Sustainability	*402*
20.5.6	Conclusions	*405*
20.6	Remedial Alternatives for PCB- and Pesticide-Contaminated Sediment	*405*
20.6.1	Background	*405*
20.6.2	Methodology	*406*
20.6.2.1	Goal and Scope	*406*
20.6.2.2	Study Area	*406*
20.6.2.3	Technical Design	*406*
20.6.2.4	Sustainability Assessment Methodology	*409*
20.6.3	Environmental Sustainability	*410*
20.6.4	Economic Sustainability	*411*
20.6.5	Social Sustainability	*412*
20.6.6	Overall Sustainability	*414*

20.6.7	Conclusions	*416*
20.7	Summary	*416*
	References	*417*

21 Civil and Materials Engineering Sustainability Projects *419*
21.1	Introduction	*419*
21.2	Sustainable Translucent Composite Panels	*419*
21.2.1	Background	*419*
21.2.2	Methodology	*420*
21.2.2.1	Goal and Scope	*420*
21.2.2.2	Technical Design	*420*
21.2.3	Environmental Sustainability	*423*
21.2.4	Economic Sustainability	*423*
21.2.5	Social Sustainability	*427*
21.2.6	Conclusions	*430*
21.3	Sustainability Assessment of Concrete Mixtures for Pavements and Bridge Decks	*430*
21.3.1	Background	*430*
21.3.2	Methodology	*432*
21.3.2.1	Goal and Scope	*432*
21.3.2.2	Materials	*432*
21.3.2.3	Technical Design	*433*
21.3.2.4	Sustainability Assessment	*436*
21.3.3	Environmental Sustainability	*439*
21.3.4	Economic Sustainability	*445*
21.3.5	Social Sustainability	*447*
21.3.6	Conclusions	*448*
21.4	Sustainability Assessment of Parking Lot Design Alternatives	*449*
21.4.1	Background	*449*
21.4.2	Methodology	*450*
21.4.2.1	Goal and Scope	*450*
21.4.2.2	Study Area	*450*
21.4.2.3	Technical Design	*450*
21.4.2.4	Sustainability Assessment	*451*
21.4.3	Environmental Sustainability	*452*
21.4.4	Economic Sustainability	*455*
21.4.5	Social Sustainability	*456*
21.4.6	Overall Sustainability	*457*
21.4.7	Conclusions	*457*
21.5	Summary	*458*
	References	*458*

22 Infrastructure Engineering Sustainability Projects *461*
22.1	Introduction	*461*
22.2	Comparison of Two Building Designs for an Electric Bus Substation	*461*
22.2.1	Background	*461*
22.2.2	Methodology	*462*
22.2.2.1	Goal and Scope	*462*
22.2.2.2	Subsurface Soil Profile and Design Requirements	*462*

22.2.2.3	Technical Design	*462*
22.2.2.4	Sustainability Assessment	*463*
22.2.3	Environmental Sustainability	*463*
22.2.4	Economic Sustainability	*467*
22.2.5	Social Sustainability	*469*
22.2.6	Conclusion	*472*
22.3	Prefabricated Cantilever Retaining Wall versus Conventional Cantilever Cast-in Place Retaining Wall	*472*
22.3.1	Background	*473*
22.3.2	Methodology	*473*
22.3.2.1	Goal and Scope	*473*
22.3.2.2	Study Area	*473*
22.3.2.3	Technical Design	*473*
22.3.2.4	Sustainability Assessment	*476*
22.3.3	Environmental Sustainability	*477*
22.3.4	Economic Sustainability	*477*
22.3.5	Social Sustainability	*478*
22.3.6	Conclusion	*483*
22.4	Sustainability Assessment of Two Alternate Water Pipelines	*483*
22.4.1	Background	*483*
22.4.2	Methodology	*484*
22.4.2.1	Goal and Scope	*484*
22.4.2.2	Site Background	*484*
22.4.2.3	Technical Design	*484*
22.4.2.4	Sustainability Assessment	*484*
22.4.3	Environmental Sustainability	*486*
22.4.4	Economic Sustainability	*487*
22.4.5	Social Sustainability	*488*
22.4.6	Conclusion	*489*
22.5	Sustainable Rural Electrification	*491*
22.5.1	Background	*491*
22.5.2	Methodology	*491*
22.5.2.1	Goal and scope	*491*
22.5.2.2	Study Area	*491*
22.5.2.3	Technical Design	*491*
22.5.2.4	Sustainability Assessment	*492*
22.5.3	Environmental Sustainability	*493*
22.5.4	Economic Sustainability	*493*
22.5.4.1	Solar PV Power Generation System Proposal (CAPEX Costs)	*493*
22.5.4.2	Diesel Power Generation System Proposal (OPEX and CAPEX Costs)	*497*
22.5.5	Social Sustainability	*497*
22.5.6	Conclusion	*498*
22.6	Sustainability Assessment of Shear Wall Retrofitting Techniques	*499*
22.6.1	Background	*499*
22.6.2	Methodology	*500*
22.6.2.1	Goal and Scope	*500*
22.6.2.2	Technical Design	*501*

22.6.2.3	Sustainability Assessment	*502*
22.6.3	Environmental Sustainability	*503*
22.6.4	Economic Sustainability	*505*
22.6.5	Social Sustainability	*507*
22.6.6	Overall Sustainability	*507*
22.6.7	Conclusion	*508*
22.7	Summary	*510*
	References	*510*

Index *513*

Preface

The concept of *sustainability* is embedded everywhere. Over the past couple of decades, an unprecedented awareness has arisen among private citizens, governmental bodies, and the business world that human activities continue to impart an enormous effect on the natural environment. Human activities are not only affecting the natural environment, but they are also affecting human health, well-being, and mankind as a whole. Among the most striking environmental impacts are the rapid depletion of natural resources, pollution of environment (air, soil, and water), and global warming and climate change. As the natural environment is impacted, strains are placed on society and the economy.

Consider the manifestation of sea-level rise resulting from climate change. With increases in mean sea level, coastline habitats can become inundated, and salt water intrudes into estuary environments, greatly disrupting flora and fauna in these ecosystems. The built environment is also significantly affected; ports, roads, bridges, and other shoreline structures and improvements are diminished in serviceability or lose function altogether. In many cases, these result in adverse economic effects, either through loss of jobs, associated profits and tax revenues, or required large expenditures to maintain operation of these systems. The lost economic value can have a direct adverse effect on society, leading to increased unemployment, decreased fiscal resources for ancillary social services, and an overall diminished quality of life.

In meeting these increasing challenges faced by humankind and planet Earth, there is an increased focus among engineers, technical and design professionals, government regulators, the business community, and the general public to seek alternatives of development that meet the needs of the present without compromising the ability of future generations to meet their own needs. This precise definition of sustainability is one that serves as the guiding premise of this book. Furthermore, as we look for sustainable solutions to these global challenges, the aforementioned stakeholders increasingly look for solutions that maximize the benefit when considered among environmental, economic, and societal dimensions. The maximization of these benefits or the minimization of the adverse impacts in these dimensions, commonly described as the "triple bottom line," is also a recurring theme of this book. The dimensions or parameters of the triple bottom line may be further described as the following:

- *Environmental* – diversity and interdependence within living systems, the goods, and services produced by the world's ecosystems, and the impacts of human-generated wastes (the planet)
- *Economic* – flow of financial capital and the facilitation of commerce (prosperity)
- *Social* – interactions between institutions/firms and people, functions expressive of human values, aspirations, and well-being, and ethical issues (people)

Sustainable engineering involves the application of sustainability principles and concepts to design environmentally friendly, economically viable, and socially equitable projects, products and systems. It offers a means to develop designs that may positively affect social, economic, and environmental considerations across the entire life cycle of a project, product, or system. Currently, a common goal of sustainable engineering is to engineer systems that are flexible enough to adapt to a range of dynamic, external stressors imposed by the natural environment while prudently utilizing resources and minimizing waste generation and emissions.

This book is organized into four sections. Section I focuses on key drivers that identify the need to incorporate sustainability principles into engineering practices. The major concerns and challenges to minimize the impacts on the environment, the economy, and society – the three dimensions of the triple bottom line – are discussed. This section also highlights the importance of resiliency in engineering design and the key approaches to incorporate it in conjunction with sustainability in engineering project designs. Section II provides a comprehensive presentation of metrics and tools for the sustainability assessment of alternative designs for a project, product, or activity. Detailed explanations on several topics, including material flow analyses and material budgets, carbon footprint analysis, life cycle assessment and streamlined life cycle assessment, economic input and output models, environmental health risk assessments, and other emerging assessment tools are provided in this section. Section III details several sustainable engineering practices, including topics such as sustainable energy engineering, sustainable waste management, green and sustainable buildings and civil infrastructure, green and sustainable remediation, and climate geoengineering. Section IV provides a range of case studies demonstrating the application of sustainability assessment tools to assess sustainability in engineering project designs. These are grouped into several categories, including environmental and chemical engineering projects, civil and materials engineering projects, and infrastructure engineering projects. Each chapter concludes with questions that may be used for review, contemplation, or as coursework exercises.

This book is primarily intended as a textbook for graduate students interested in sustainability and sustainable engineering. However, the structure and the contents of the book make it a valuable reference source for engineers, technicians, administrators, and members of the public interested in the topic of sustainability. The authors have collectively amassed decades of research, teaching, and consulting experience and have strived to write this book so that it can be valuable to practitioners and academic professionals alike.

The authors offer their deepest gratitude to a number of individuals and organizations who have provided assistance, support, patience, brainstorming, and inspiration toward the completion of this book. Many students who have taken the Sustainable Engineering graduate course offered by Krishna Reddy at the University of Illinois at Chicago (UIC) provided motivation to undertake this book project, added to the content, and developed the example projects presented in Chapters 20–22. Their hard work is gratefully acknowledged. The authors are grateful to Girish Kumar, a doctoral student at UIC, for his immensely valuable assistance in preparing, reviewing, and editing of the chapters. The authors are also thankful to other graduate students at UIC, particularly Jyoti Chetri, who also reviewed the chapters. Claudio Cameselle is thankful for the Fulbright Fellowship that allowed him to visit UIC and undertake this book project. The support of researchers and staff at the University of Vigo, especially Susana Gouveia, has been vital in the completion of this book. Jeff Adams is immensely thankful to ENGEO Incorporated, particularly Uri Eliahu, Shawn Munger, Joe Tootle, Scott Johns, and Divya Bhargava, all of whom have thoughtfully provided love and encouragement. Finally, the authors thank Bob Esposito, Michael Leventhal and the entire Wiley editorial group (Beryl Mesiadhas and Grace Paulin) for their cooperation during this book project. We hope that this book will help students and professionals alike to learn and advance the concepts of sustainability and develop sustainable engineering projects.

25 January 2019

Krishna R. Reddy
Chicago, USA
Claudio Cameselle
Vigo, Spain
Jeffrey A. Adams
San Ramon, USA

Section I

Drivers, Environmental, Economic and Social Impacts, and Resiliency

1

Emerging Challenges, Sustainability, and Sustainable Engineering

1.1 Introduction

The concept of *sustainability* is everywhere. In recent years, ever-growing numbers of people around the world have become more aware of strains being placed on the Earth. These strains have been manifested in a variety of ways – accelerated exploitation of natural resources, pollution of air, soil, and water, and climate change. Not only have private citizens taken notice, but governments and the business world have also taken steps to address sustainability. Numerous intra- and intergovernmental initiatives and agreements have been developed to address these strains on the environment and to identify measures that encourage more sustainable practices. Businesses, too, have realized that sustainability is a good practice for a variety of reasons. New systems and products have been developed that are more protective and less wasteful of resources, and the pursuit of the so-called "triple bottom line" of sustainability has been increasingly applied in the new projects and products. The triple bottom line is the reference framework in sustainability that accounts for financial as well as social and environmental metrics.

Several key questions have emerged that necessitate contemplation. What emerging challenges are forcing us to think about sustainability? What is sustainability? How do we take action to further this concept? What is sustainable engineering (SE), and what role can it play in sustainable development? Of equal importance, how do we determine success in pursuit of these initiatives, and how do we measure our progress toward these goals? These evolving and increasingly significant concepts are the focus of this book.

In this chapter, the broader emerging challenges that are forcing us to think about sustainability are described. Next, the general definitions and interpretations of the meaning of sustainability are presented. Finally, sustainable engineering and the role it can play in achieving sustainable development are described.

1.2 Emerging Challenges

Before we delve into the concepts, applications, methods, and measures related to sustainability, let us examine several acute problems and related examples that are faced worldwide and are increasingly having a measurable, detrimental effect on the planet.

1.2.1 Increased Consumption and Depletion of Natural Resources

A key consideration of sustainability focuses on our ability to preserve resources for future generations. This is extremely important, as many essential resources (e.g. precious metals,

Sustainable Engineering: Drivers, Metrics, Tools, and Applications, First Edition.
Krishna R. Reddy, Claudio Cameselle, and Jeffrey A. Adams.
© 2019 John Wiley & Sons, Inc. Published 2019 by John Wiley & Sons, Inc.

fossil fuels) are nonrenewable and are limited in quantity. For many of these resources, we are on a current trajectory of utilization/exploitation in which near-total depletion of economically viable reserves is a very realistic possibility. The alarming rates of consumption of a number of resources not only spell trouble for the availability of these resources for future generations, in many cases, there can be unintended secondary, but catastrophic, side effects on the environment.

1.2.1.1 Easter Island Example

A classic example of the catastrophic consequences of the primary and secondary effects of natural resource depletion is the collapse of a civilization on *Easter Island*. Located in the southeastern Pacific Ocean and arguably the most remote habitable region on the planet, Easter Island gained its name from the sighting/discovery of the island on Easter Sunday, 1722 by Dutch sailors. It was subsequently annexed by Chile in 1888. Large stone statuary called moai, created by early Rapa Nui peoples, were important monuments to a sophisticated culture and civilization that had once flourished on Easter Island but had devolved into a small, primitive culture at the time of European discovery (Fischer 2005).

Although Easter Island is subject to a cold and dry climate, it was at one time heavily forested with palms, conifers, and sandalwood. The first Polynesians arrived at Easter Island in the fifth century and numbered no more than 20 or 30. The harsh climate and nutrient poor soils restricted agricultural activity to the cultivation of sweet potatoes. Nevertheless, a sophisticated and advanced society flourished among the Rapa Nui. To allow for agricultural activity, much of the land was deforested (Stevenson et al. 1999). Trees were also harvested to provide structural materials for housing and boat fabrication and for use as fuel. However, a significant number of trees were also harvested to create a track system to maneuver the large moai from quarry locations to sites where they were erected. As the population grew upwards of 7000 persons, these resources were further utilized to meet increasing demand (Mulrooney 2013).

By the 1600s, the entire island had been deforested. The lack of timber resources eliminated the ability to construct fishing boats and wooden structures. The elimination of bark materials prevented the fabrication of cloth materials. Furthermore, the deforestation greatly accelerated soil erosion, and agricultural capacity of the already nutrient-poor soils was again severely reduced (Rull et al. 2015). Elaborate rituals centered around the moai statuary diminished, placing even greater strain on the social fabric of the declining society, including the breakdown of social and religious conventions. Because boats could no longer be fabricated without timber, the Rapa Nui were trapped on the remote island. Eventually, they were forced to resort to primitive cultural practices where available shelter had been reduced to available caves (Mann et al. 2008). Ongoing turbulent conditions fueled conflict/warfare, slavery, and even reports of cannibalism.

Ultimately, the once great civilization had collapsed. At the time of European discovery, the population had declined precipitously. The collapse of island's agricultural capacity and activity resulted in widespread, ongoing starvation. Subsequent contacts from seafaring groups, such as whalers, introduced sexually transmitted diseases and smallpox. Peruvian slave parties also captured numerous Rapa Nui for use in the slave trade. By 1877, only 111 Rapa Nui remained on Easter Island, and at the time of Chilean annexation, the Chilean government confined the remaining inhabitants to one village.

Rapa Nui remain on Easter Island to this day. Archaeologists brought attention to the island in the mid-twentieth century, which in turn has stimulated tourism and led to the restoration of some of the moai statuary. Conditions have improved for the Rapa Nui, although unemployment remains high and alcoholism and related social strains are quite prevalent. Additionally, they are still dependent on imported food.

Nevertheless, the story of the Rapa Nui has been repeated with other peoples – the Mayan and Inca peoples of Central and South America as well as the ancient Greeks and Romans. All of these peoples offer a cautionary tale – when a society disregards the health of its environment, excessive strains on vital resources, such as soil and water, can lead to a collapse of agricultural activity and other aspects of economy and culture. When basic necessities such as food, clothing, and shelter become scarce, a disparity between "haves" and "have-nots" is often exacerbated, leading to mistrust and resentment between classes. As the problem grows, conflict is inevitable, and collapse of the underlying civilization will occur.

1.2.1.2 Metallic Ores Consumption Example

Several examples can be presented with respect to unsustainable utilization of natural resources. Let us take the example of the usage of metallic ores, and as an example, let us examine the use of zinc (Graedel and Allenby 2010). Consider zinc use over a sustainability design period of 50 years and a global population of 7.5 billion people. Estimated global zinc reserves consist of 330×10^{12} g (330 Tg) of zinc. Considering a 50-year period (after which all resources will be depleted), 6.6 Tg may be used per year. Assuming an even allocation among 7.5 billion people, each person would be entitled to 0.9 kg/year. Allowing for recycling (assuming a 30% increase in zinc supply), the available per capita allotment may be increased to 1.2 kg/person/year.

Consider these industrial/consumer uses of zinc: zinc alloys – 38% (0.5 kg/person/year); construction coatings – 31% (0.4 kg/person/year); zinc chemicals – 15% (0.15 kg/person/year); vehicle plating – 15% (0.15 kg/person/year); cast, rolled zinc – 4% (0.05 kg/person/year).

However, current automobiles contain approximately 32 kg of zinc. Assuming a 1.2 kg/year allocation, this would result in a new car purchase once every 27 years – only if no other uses of zinc were allocated. When compared to the uses listed above, zinc would need to be removed from all other product streams and uses to meet automobile use demand at this rate. Of course, it is a valid argument that not every person on the planet will consume an equal amount of zinc, allowing for some to consume vastly more than others in this example. Nevertheless, when finite resources are available for use, it is necessary to consider the limits of the resource. A famous example takes this further. Imagine a resource that is finite and is open to use to a finite number of users. Even in this case, the users may be motivated to use more of the resource because the side effects do not offset the gain. This famous example is detailed in a seminal paper, *the Tragedy of the Commons* (Hardin 1968, 2009).

Here is the scenario – 20 farmers use a communal pasture to graze 10 cows each, or 200 cows. Each farmer is utilizing 1/20 of the resource – or 5%. Now, let us say one farmer adds an 11th cow. He is increasing his benefit by 10%, but the consequences of the additional cow (let us assume the additional loading begins a process of overgrazing) are borne by all equally, so he suffers 5% of the consequences. He has just increased his marginal benefit of adding one cow by this difference. The farmer has an incentive to act in this manner, but so do all of the farmers. It is reasonable to expect that other farmers will also choose to add cows, increasing their respective marginal benefit, but all of the farmers ultimately will suffer the consequences of overuse together.

The Tragedy of the Commons remains a controversial study, and many have pointed out shortcomings or exceptions to the model (Parslow 2010), but it is a simple illustration that individual parties often have incentive to exploit a resource, even if the consequences will eventually spoil that resource (Weitz et al. 2016). In other words, it remains difficult to get multiple parties to act together to preserve resources (Prentice and Sheldon 2015). This is vividly illustrated in Figure 1.1, which shows the effect of overfishing (Ostrom 2015; Hutter 2015) on the population of sardine near the Pacific coast of North America. Sardine populations have dramatically

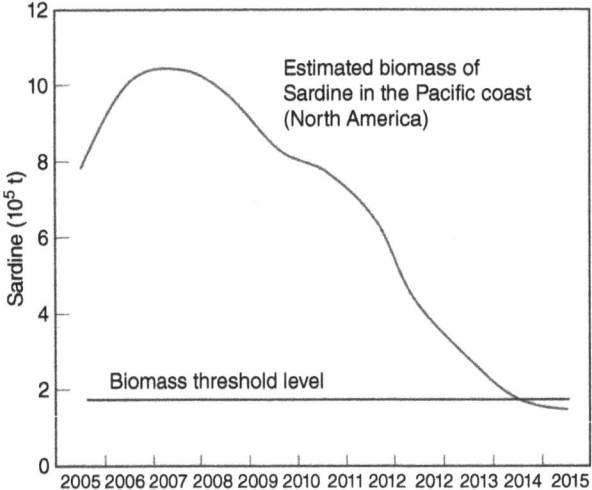

Figure 1.1 Sardine population declining up to below the threshold level. Source: Data from National Marine Fisheries Service, CA, USA.

declined since their peak in 2007. This led to a fishing ban enacted to protect the collapsing population. The fishing ban is not expected to be lifted in the short or mid-term, until the sardine population reaches acceptable levels to support fishing again.

1.2.2 Growing Environmental Pollution

The growth in consumption of natural resources has led to increasingly problematic side effects – one in particular is increased environmental pollution. Air, soil, and water are being increasingly polluted from human development and activities. Control and remediation of the effects of pollution are crucial in order to protect the environment and public health, and a movement toward sustainability is not possible without control or prevention (Trevors 2010).

Water pollution can have many effects. Pollutant loading can directly affect domestic water sources, such as surface water or groundwater resources. Sediment runoff can affect water flow in navigable waterways. In addition to direct deposition of toxic materials to surface water, pollution sources can also have a long-term detrimental effect on surface water bodies.

Soil impacts are also increasing. In the form of mining, waste products are often landfilled at or near the extraction point. Noxious by-product chemicals capable of leaching are often disposed as well, and when mixed with tailings and rainwater, the resulting toxic "soup" can significantly pollute soil as well as groundwater. Not only can mining practices affect soil but also other practices such as deforestation, overgrazing, timber harvesting, and agricultural management can lead to releases of potentially toxic materials or otherwise exacerbate environmental impacts such as erosion of topsoil (Stoate et al. 2001). Further, landfilling practices used in waste management can also affect soil. Older landfills often do not have protective measures in place, such as liners or leachate collection systems, and authorized or unauthorized hazardous waste disposal facilities accept acutely toxic or recalcitrant compounds, such as aromatic and aliphatic hydrocarbons, chlorinated solvents, PCBs, pesticides, and heavy metals, which can be released into the environment (Han et al. 2016).

In addition to water and soil, air is also increasingly polluted by human activities. Combustion of wood and fossil fuels leads to emissions of gases and particulate matter, many of which can affect health or even climate. Evaporated volatile organic compounds (VOCs), carbon monoxide, heavy metals, and particulates can lead to impacts to public health, including elevated asthma or cancer rates. Some emission constituents are subject to secondary reactions in the atmosphere, leading to increased smog, near-surface ozone, or depletion of higher atmospheric ozone (Kampa and Castanas 2008).

1.2.3 Increasing Population

Much of the strain placed on the planet is associated with ever-increasing uses of natural resources, the unintended side effects, and pollution that results from their use to land, air, and water. However, even if measures were taken to curtail these uses and mitigate the effects of pollution, a third problem is leading to strains that may not be as easy to curtail – rapid increases in total human population as well as increasing acceleration in population growth.

The statistics are staggering – nearly 7 billion humans are currently alive on Earth (Goudie 2013). This number has increased by 2 billion in only 25 years and has doubled from approximately 3.5 billion in the past 50 years (Figure 1.2). Currently, world population grows by 75 million on an annual basis. If current growth rate predictions prove to be true, the world population will exceed 9.1 billion by the year 2050. Figure 1.2 demonstrates the rapid acceleration in population growth over the past millennium. It also demonstrates how population totals may grow, level off, or contract by the year 2200 assuming worldwide average female fertility rates over this period of time (Janaskie 2013; Basten et al. 2013).

With additional people added at such an accelerating pace, goods and resources become scarcer. Of course, the growing population will need to use resources for food, clothing, and shelter, as well as participate in a meaningful way in local, national, and/or international economies. Yet, challenges are already faced by a significant portion of the world's population in even trying to meet the most basic human needs. Approximately 985 million people experience extreme poverty (defined as income of less than or equal to US $1 per day) (Kamruzzaman 2015). Over 800 million people face severe malnutrition. A startling number of preschool children, over 6 million, die from hunger/malnutrition-related maladies per year (Dasandi 2014).

With the rapid increase in world population, similar appreciable growth in world economic output has occurred. Global economic production has doubled in the past 25 years, a rate that

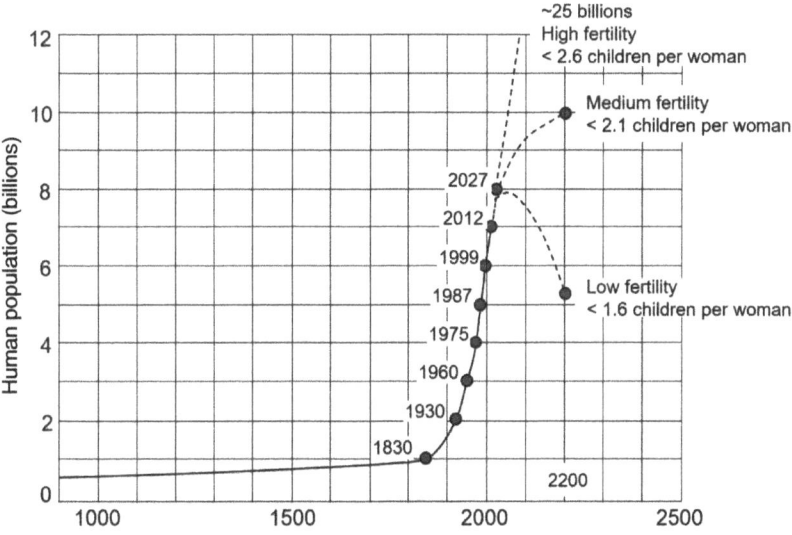

Figure 1.2 World's population explosion and projection to year 2200. Source: Data from United Nations.

even exceeds the rapid rate in population growth (Patnaik 2003). This development has led to a great benefit for many millions of people, as the growth has indeed lifted millions out of poverty. Gross domestic products in numerous developing countries have grown significantly over this time period, and scores of nations throughout the world have grown into world economic powers from modest economic positions within a generation. Nevertheless, economic development in many of these countries has proven to be uneven at best; large inequalities in wealth have developed within these countries as well as compared to the positions of larger, more developed countries, such as those in Europe or North America. As a result, real income in many developing countries is actually falling. Countries with high birth rates are especially susceptible to this growing inequity; stabilization of birth rates will be essential in assuring more equitable economic growth for a given population in the future.

1.2.4 Increasing Waste Generation

As the population continues to grow at an accelerating rate and is coupled with increasing use of natural resources, an additional by-product of these factors is also placing strain on the environment of the planet – increased waste generation. Enormous quantities of waste are being generated every year, and this growth continues unabated. It is especially troubling and detrimental to the environment because much of this waste generation growth is occurring in the developing world where waste management and disposal practices are very basic and not protective of the environment. Much of the developed world has placed an increased focus on waste minimization and diversion, such as reuse and/or recycling; however, large quantities of waste materials cannot be practicably repurposed and therefore require disposal or incineration. As a result, the detrimental environmental effects of both proper and improper waste management continue to grow, further spoiling the environment. Collectively, countries around the world must find ways to reduce this increasing trend of waste generation, as these practices cannot be considered sustainable.

The framework to reduce waste generation and the pollution effects that lead to fouling of the environment are straightforward, and it is a necessary undertaking such that we do not jeopardize the quality of life for present or future generations. Ultimately, we need to identify waste products that have the greatest impact on the environment. Once these materials have been identified, the activities acting as a source of the waste materials can be identified, and steps can be taken to reduce or minimize the activity through alterations in generation practices or by identifying more protective alternative activities. Measures may also be taken to prevent the introduction of unavoidable waste products into the environment in such a way that environmental impacts may occur. Further, pollution impacts that have already occurred may be identified and remedied through a variety of traditional or innovative environmental remediation techniques. Ultimately, in order to achieve sustainability, civilization will need to identify technologically feasible means to transition from polluting activities to environmentally protective activities.

Before techniques and alternatives to reduce waste generation can be identified, it is important to discuss types of waste. In 1976, the Resource Conservation and Recovery Act (RCRA) was enacted into law in the United States. While this was not the first attempt to regulate waste disposal practices, it was the first comprehensive legal framework that reached nearly all facets of waste generation, storage, and disposal for a wide range of waste classifications. Generally speaking, solid waste is generated in a variety of residential, commercial, and industrial settings. Much of this material is not considered hazardous and thus falls into a general category of municipal solid waste (MSW). Figure 1.3 shows a percentage breakdown by waste type of MSW generated in the United States. Much of this material can be recycled and reused or

1.2 Emerging Challenges | 9

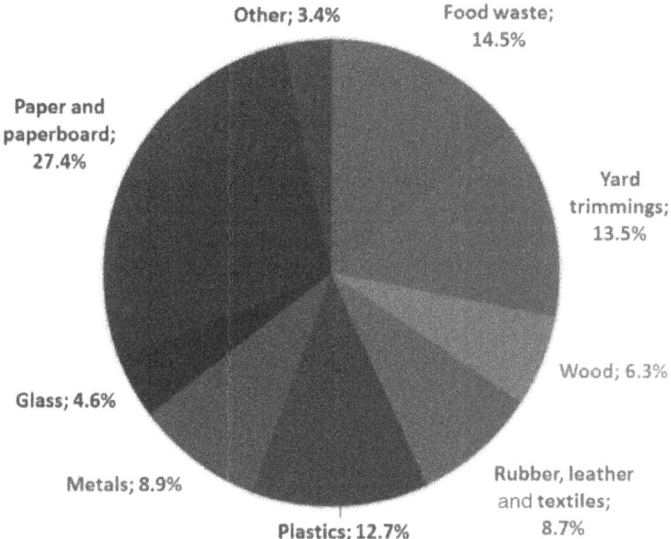

Figure 1.3 Total municipal solid waste (MSW) generated in the United States. Source: USEPA (2014).

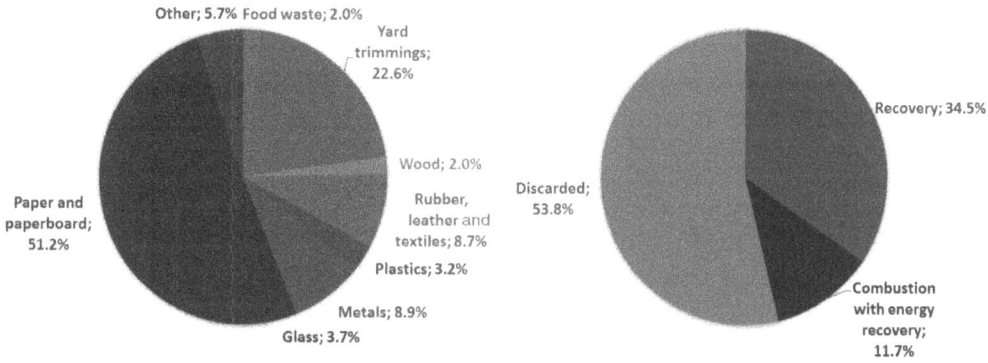

Figure 1.4 Municipal solid waste (MSW) recovery by material and MSW management in the United States. Source: USEPA (2014).

repurposed. Figure 1.4 shows a percentage breakdown of the end fate of MSW waste stream in the United States. Notable progress in recycling programs has greatly expanded the percentage of wastes that are recycled. However, two-thirds of MSW is still not being repurposed and is either being landfilled or incinerated.

As an example of MSW management, consider landfills. Landfills have been used for countless generations in waste management; however, only in the past century have advances been made to help mitigate the environmental side effects of landfilling. Sanitary landfills have been constructed to manage odors, protect underlying soil and groundwater, and facilitate the collection and management of generated landfill gas and leachate. Leachate is a liquid by-product of landfill operation. It consists of infiltrated water (often from rainfall) as well as liquid and "leached" solid wastes from waste materials. Leachate often contains elevated concentrations of a variety of organic and inorganic chemical compounds, heavy metals, and other potentially deleterious materials. If not controlled properly, it can enter groundwater, where in can adversely affect aquifer sources (Baker et al. 2015). Landfill-induced groundwater contamination is a serious problem throughout all 50 states, not only from recently constructed landfills, but more often from old uncontrolled or abandoned landfill sites.

Landfills also generate copious amounts of gases that need to be controlled. Buried wastes commonly undergo anaerobic decomposition, resulting in the production of biogas. This biogas primarily contains methane, carbon dioxide, and hydrogen-containing compounds. If not controlled, this can be a significant health and explosion hazard, both at the landfill site as well as off-site due to horizontal and vertical migration into nearby structures. However, when properly captured and treated, biogas is a valuable resource that can be used as a fuel for power generation (Sadasivam and Reddy 2014). As an example, in 2008, commercial landfill gas produced electricity and natural gas for 1.4 million homes (Themelis and Ulloa 2007). Because it burns relatively clean, its capture and use can displace the use of dirtier fossil fuels, leading to a relative reduction in greenhouse gas emissions and air pollutants.

1.2.5 Increasing Greenhouse Gas Emissions

Increased resource generation and utilization not only leads to increased solid/liquid waste generation, it also leads to increased gas emissions. Gas emissions, particularly those from agriculture and combustion processes in the industrial, transportation, and energy generation sectors, can lead to serious air pollution consequences, but they can also facilitate climate change. Although climate change occurs due to a variety of natural physical processes, overwhelming evidence suggests that anthropogenic sources also contribute to climate change. The magnitude of this contribution and the resulting effects of climate change are still being debated, but very little scientific basis exists that objectively refutes contributions to these effects from anthropogenic sources.

Much of the anthropogenic portion comes from the emission of greenhouse gases. Greenhouse gases, such as carbon dioxide, methane, nitrous oxide, and water vapor, as well as a range of fluorocarbons, act to trap heat from the earth's surface that would normally radiate to space harmlessly (Figure 1.5). This effect, similar to the trapping of heat generated by sunlight within a greenhouse (commonly known as the Greenhouse Effect), warms the lower atmosphere. Higher temperatures result in increased generation of water vapor, which can act to trap more heat. However, higher temperatures also accelerate plant metabolism, which leads to utilization (and removal) of carbon dioxide from the atmosphere. These phenomena are still being studied (Oreskes 2004).

Nevertheless, substantial increases in greenhouse gases have been generated during the industrial era, contributing to a rapid increase in global temperatures (Figure 1.6). Figure 1.7 depicts the correlation between global temperature and atmospheric carbon dioxide concentrations over the past 130 years.

Predictions of further impact are difficult to make and offer wide ranges due to variations in mathematical models. By 2100, global temperatures are predicted to increase between 2.5 and 10.4 °F (1.4–5.8 °C), with temperatures expected to increase even more in the United States. Increased global temperatures can lead to drastic shifts in weather patterns, with more precipitation in some areas, less in others, and more intense storm events. These effects can drastically affect agricultural activities and induce strains on ecosystems and biodiversity. Furthermore, increased temperatures are expected to accelerate melting of polar ice and alpine glaciers, leading to increases in sea elevation, which in turn can inundate coastal areas.

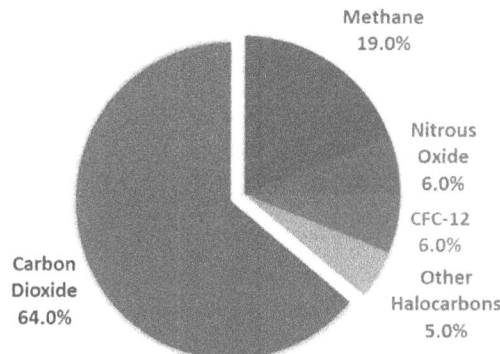

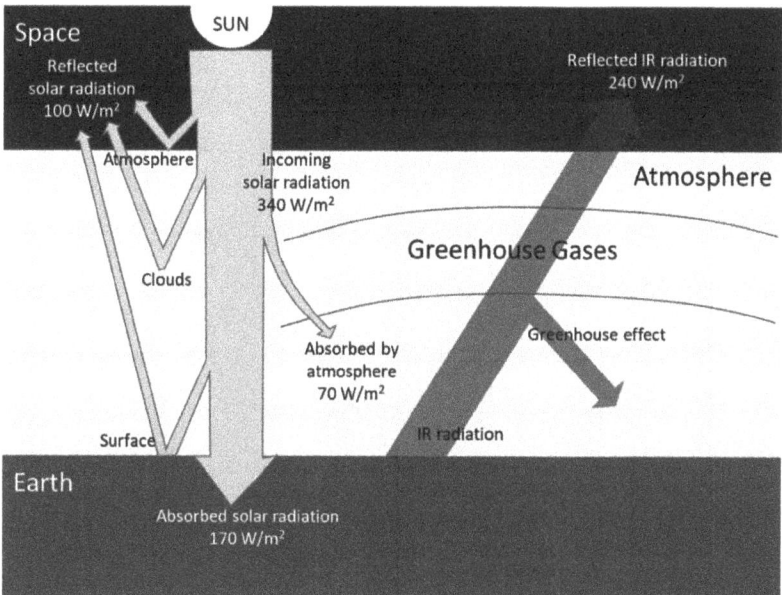

Figure 1.5 Greenhouse gas effect in the atmosphere and the estimated relative contribution of main greenhouse gases.

As a short example of the current impact and sustainability of greenhouse gas emissions, consider carbon dioxide and a sustainability period of 50 years as well as a representative world population of 7.5 billion people. If global emissions were limited to a doubling of atmospheric carbon dioxide concentrations, this would be 7–8 pg (picograms) of carbon (7–8 × 10^{-15} kg). Based on an allocation to 7.5 billion people, this would result in 1 Mg (megagram) of carbon per year per person. Assuming a 40/25/35% breakdown to transportation, energy, and general society emissions, this would translate to 400, 250, and 350 kg of carbon dioxide per year, respectively.

With respect to generation, a typical automobile emits 62 g of carbon/km, and coal-fired power generates 89 kg of carbon/kJ. This translates to 6450 km/year of automobile transport, and 2.8 kJ/year/person in terms of allowable carbon emissions. However, in Germany, the current per capita distance of auto travel is 15 000 km/year, and per capita use of energy is

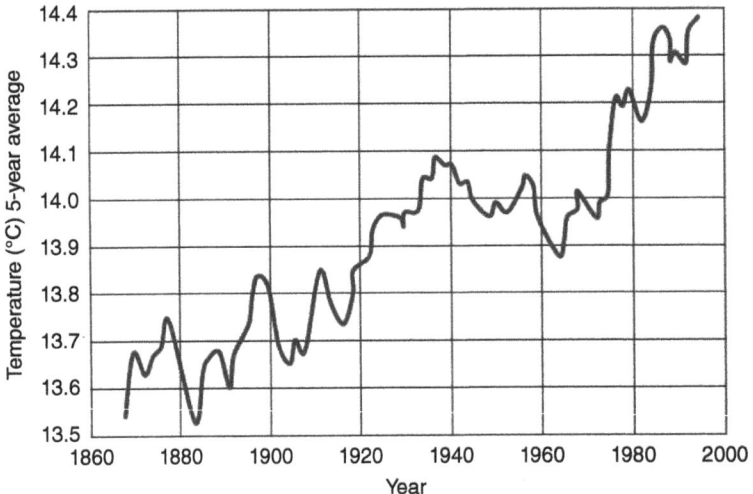

Figure 1.6 Average temperature on Earth at ground level. Source: Data from NASA Goddard Institute for Space Studies, NY, USA.

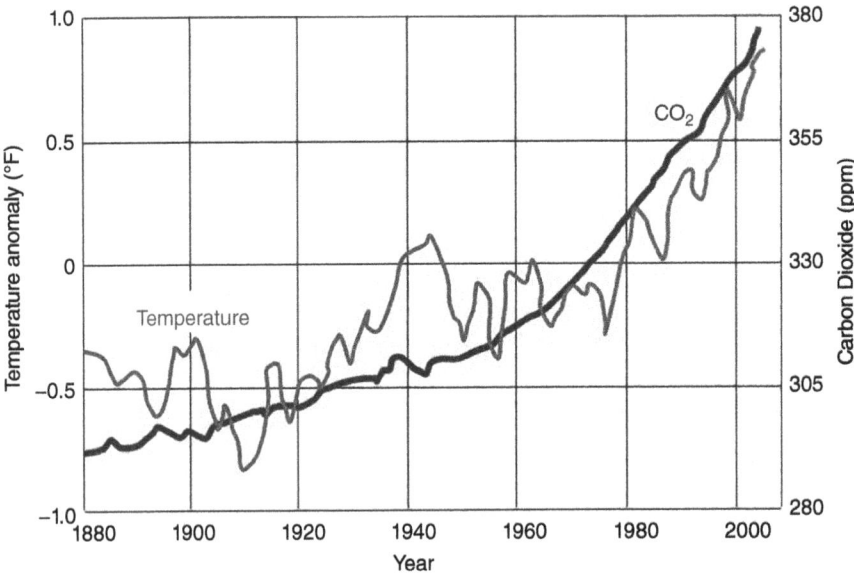

Figure 1.7 Average global temperature change and atmospheric carbon dioxide. Source: Data from Climate Change Impacts in the United States: The Third National Climate Assessment.

270 GJ/year. The current rates are way above an allocated level of emissions that would only lead to a doubling of carbon atmospheric concentrations.

If these levels of consumption and emission generation continue, current predictions for these inputs are for a 3–5 °C increase in global temperature (Knutti and Sedláček 2013). To put it in perspective, a 5 °C difference in global temperature represents the difference between a warm period and an ice age (Marcott et al. 2013). As indicated above, such a temperature will have a significant effect on sea level, agricultural practices, and numerous other activities that can adversely affect coastal and inland populations alike.

1.2.6 Decline of Ecosystems

The ongoing harm and destruction of natural environments resulting from waste generation and emissions lead to an additional problem – the decline of ecosystems. Healthy ecosystems are critical to human civilization – of primary importance; they support human life and economies by providing a supply of materials to be used in goods and services. However, these resources are not being managed in a healthy or sustainable manner. Human populations and consumption, commercial activity, and industrial activity are leading to the depletion and contamination of groundwater, topsoil, forests, and fisheries.

The global economy is highly dependent on natural, renewable resources that are provided by the wide range of ecosystems on the planet. The primary, direct products available in these ecosystems include fresh water, food, fuel, wood/timber, leather, wool, furs, and other products. Secondary and/or refined/synthesized products include raw materials for fabrics, oils, chemicals, and many others. Furthermore, the utilization of these materials in industries such as agriculture, forestry, and fishing are responsible for 50% of jobs worldwide (De Ferranti et al. 2002).

Ecosystems are also a primary source for service-based economic roles. Some services that derive directly from ecosystem management include waste management, climate regulation, erosion control, water treatment, and vector management. These combined goods and services form "ecosystem capital" – economic development, income, and overall human well-being are strongly based on the effective management of this capital. These goods and services are available and feasible as long as the ecosystems from which they are derived are healthy and protected. Therefore, the ecosystem capital in a particular nation and its income-generating capacity represent a critical component of a particular nation's wealth (De Ferranti et al. 2002).

As the new millennium dawned, an assessment was performed to evaluate the overall ecosystem health. The assessment focused on the services that are provided by ecosystems and provided an agenda for ecosystem restoration. Further, the *Millennium Ecosystem Assessment* gathered information on the world's ecosystems. This study, which included the work of 1360 scientists in 95 countries, gathered, analyzed, and synthesized ecosystem-related information to focus on the links between ecosystem goods/services and the well-being of human civilization (Irwin et al. 2007).

The conclusions of the *Millennium Ecosystem Assessment* were sobering: it found that humans have altered the world's ecosystems more rapidly and profoundly over the past 50 years than at any time in the human history. Further, over 60% of ecosystem goods and services are being degraded or not being utilized in a sustainable manner. In the absence of a reversal of these trends, there will soon be critical consequences to human civilization. Ideally, the *Assessment* will serve as a knowledge base for sound decisions and management that can be undertaken by policy-makers and mangers in both the private and public sector.

1.2.7 Loss of Biodiversity

With the ongoing decline of ecosystems comes another related problem – the loss of biodiversity. Biodiversity is generally defined as the variability among living organisms and the greater ecological complexes to which they belong. Millions of species exist on the planet – and many have not been studied or assessed. Of an estimated 5–30 million organisms, only approximately 2 million have been described. Many of these have been adversely affected. Since 1970, vertebrate species have reportedly declined by 27% (Butchart et al. 2010).

Stresses and impacts to biodiversity come from several causes. Many are linked to strains on host ecosystems, including pollution, introduction of invasive species that "crowd out" native species or are predatory on native species, and conversion of natural lands into a developed land use, such as agriculture, timber harvesting, commercial/industrial uses, or residential uses. Additionally, species may be exploited for commercial values, either legally or illegally. For instance, many species continue to be hunted and killed and marketed even after national or international prohibitions have been enacted.

The loss of biodiversity can have substantial negative effects on the environment and human civilization. Diverse species play a critical role in agriculture, the overall food chain, as well as in advances in pharmaceuticals and medicine. Further, a robust, diverse range of flora and fauna are critical in an ecosystem's self-healing mechanisms after major natural or artificial disruptions, such as hurricanes, forest fires, or flooding.

As a counterpoint to overexploitation, the protection of biodiversity can provide short-term and long-term economic benefits. Land and resource stewardship performed in a sustainable manner can have significant economic benefits. These resources can be a source of economic wealth through their incorporation into goods and services, especially among the poor. Further, many argue there are aesthetic, and even more powerfully, moral obligations to protect biodiversity and the range of flora and fauna present on the planet (Vellend 2017).

1.2.8 Social Injustice

Effective stewardship of resources, ecosystems, and their side effects on biodiversity requires some relative form of intervention, both economically and politically. Yet the means and degree of intervention are continuously subject to debate, both locally and globally. Figure 1.8 shows a matrix outlining political and economic systems. On the vertical axis is a depiction of political systems – moving upward along the axis indicates an increasing libertarian view, which espouses that justice is based on the equality of opportunity toward inputs or events. Moving downward along the axis indicates a more egalitarian view, which defines justice as the equality of an outcome based on inputs or events. The horizontal axis depicts different types of economic values. To the left indicates a greater degree of communitarianism, in which welfare is optimized by the absorption of individual economic activity into the community. A position to the right of the access indicates a greater focus on corporatism, in which welfare is optimized by free economic activity of individuals.

With respect to many other nations, the economic and political system of the United States has long been positioned toward the upper right-hand corner of the matrix – indicating a greater focus on libertarian political and corporate economic policies. At first glance, some would argue that this is at the opposite end of the matrix from where a sustainable-inducing position would be located. However, there are many points to refute this. First, many of the

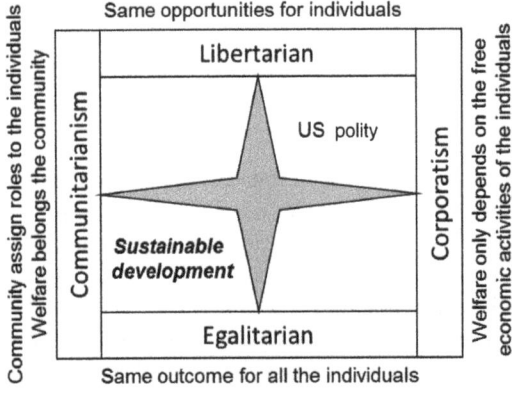

Figure 1.8 Relations between political and economic systems.

worst environmental atrocities and catastrophes have occurred in nations of totalitarian government regimes and communal economic principles – no individual or group could challenge the will of these governments, and a lack of economic competition diminished innovations or better alternatives to practices that led to these outcomes. Second, less regulation and more enterprise can lead to market-based solutions to address sustainability. When financial gain is to be made, the "animal spirits" of economic participants will pursue these gains. The goal should be to find a way to align sustainability goals and financial goals. This is increasingly being applied with the "triple bottom line" – where emphasis is placed on financial, social, and environmental metrics for a particular economic activity.

When it comes to environmental and resource stewardship, it is also important to facilitate just relationships within both economic and political systems. An increasing focus is being placed on environmental justice – a concept that addresses increased notoriety of environmental racism. The concept of environmental racism is better defined in terms of socioeconomic status, although sadly, this range in status often follows along racial or ethnic lines. Environmental racism postulates that waste sites and harmful industries are often places in areas of lower economic status residents – which often include substantial portions of minority or historically disadvantaged residents. This is further exacerbated by the perception of greater infrastructure expenditures and facility improvements, including institutional buildings and facilities, roads, and utility conveyance projects, in areas of greater political activity or economic clout, which often includes greater percentages of white residents. Whether the racial component in specific cases is coincidental or a real influencing factor is irrelevant – the perception is that these decisions are made along racial lines, and this leads to greater perceptions of environmental racism (Pulido 2000).

The concept of environmental justice has received greater focus in recent years throughout the developed world. Citizens of all racial/ethnic and economic backgrounds have coalesced into groups to pursue environmental justice issues. Citizen and "watchdog" groups increasingly bring effective action toward environmental justice goals and to monitor progress toward effective, desirable outcomes (Holifield 2001).

Increased focus on social and environmental justice is also critical in the developing world. In many instances, this has not occurred in areas with rapid economic growth, leading to an alarming disparity between the wealthy and poor. The economic developments in developing countries are often driven by natural resource utilization or low-tech industries. In this process, people often lose their land or their ability to utilize it, leading to strains on their individual economic prospects. Decreased access to food, shelter, and other basic needs in turn often leads to increased health problems. These factors tend to repeat, and more people are trapped in poverty. In fact, over a billion people face extreme poverty from economic and environmental injustice that stems from this pattern. This is exacerbated in developing countries, as they often have weak or immature governing institutions. Corruption between politicians, wealthy individuals, and other officials that are intermediaries between these groups leads to the inappropriate diversion or outright theft of vast sums of money. Reforms, strengthened institutions, and accountability for governments, corporations, and their respective individuals are necessary to reverse these trends (Ako 2013).

Wealthy, industrialized nations also have a role to play to combat these forms of injustice. Many industrialized nations or large developing countries engage in unfair trade practices that prevent others from participating fairly. Tax and tariff systems and quotas often restrict imports that may emanate from other developing countries. Some countries (especially those with heavy mass production-focused manufacturers) will flood markets with products at depressed prices (often for less than the cost to manufacture the products) with the aim to damage competition from other competing countries. These practices destroy competitors and create too many barriers in developing countries from competing in a fair environment. In addition to providing forms of political, advisory, and economic aid to developing countries, wealthy, developed nations can play a role in eliminating these unfair practices. By forming and enforcing free trade agreements as well as educating the marketplace, these unethical and often illegal destructive trade practices can be limited when meaningful sanctions are put in place toward offending nations.

1.2.9 Urban Sprawl

Governments and their institutions not only have a role to facilitate economic and social justice, they also play a greater role in advancing humans preserve their built environment. Urban sprawl is causing an adverse impact on the environment in many developed and developing worlds. Urban sprawl is difficult to define, but it can be described as the pursuit of a built environment that focuses more on low-density residential areas, retail areas (such as low-rise shopping malls), industrial facilities, and other facilities. These features often are polynucleic in nature (areas of intense activities are often situated at relatively large distances from each other) and need to be connected by transportation linkages, such as multilane highways (Ioppolo et al. 2014).

Urban sprawl leads to an ever-expanding boundary for cities into the surrounding countryside. In areas without or with weak planning codes/regulations or traditions, sprawl can occur without due consideration of side effects or a defined limit of where growth will be fostered or controlled. Expansion of the built environment with respect to land uses (residential, commercial, and industrial) and transportation linkages often leads to the destruction of open areas and natural lands. Explosive development around traditional city centers often happened when cheap land was available for development without any forethought to planning principles. Eventually, this development extended out toward existing entities (towns and villages) that had their own jurisdictional authority, or in cases where planning regulations existed, their own land use plans and codes; in many cases, keeping residential and commercial uses separate in manners that make sense within the specific jurisdiction but are incongruous when considered from a regional standpoint. In the absence of land use regulatory frameworks or where conflicting approaches existed, growth has often occurred without undertaking a regional approach to planning or infrastructure development, leading to inefficiency, as well as areas overserved or underserved by infrastructure.

Uncontrolled urban sprawl can lead to significant side effects that can affect sustainability principles. A polynucleic land use approach emanating from unchecked sprawl that requires relatively intensive transportation systems drives up the energy requirements for transportation. Increased needs from transportation affect the environment – approximately 28% of greenhouse gas emissions are related to transportation-based activity (Rode et al. 2017). Much of this is due to intensive automobile use – cars are responsible for approximately 80% of these emissions – and numerous cities in the United States (many lacking a robust mass transit infrastructure) cannot meet air quality standards.

Greater sprawl takes up a greater footprint of the landscape. Impervious materials – paving for roads and parking, roofs, etc. – increasingly replace natural pervious ground surfaces, leading to increased water runoff, erosion, and flooding. This runoff also picks up pollutants – sediments, oils, agricultural chemicals – leading to a general degradation in water quality.

Increased land development resulting from low-density sprawl also takes already productive land out of use. Sprawl leads to the removal of an estimated 1.3 million acres of productive agricultural land per year. As a result, there is a greater reliance of food products from increasingly distant locales – local sources have been replaced by sources often hundreds or thousands of miles away. This, in turn, leads to greater emissions from the requisite transportation activity to deliver these products. Further, 1.8 million acres per year of natural lands are lost due to development. This can affect habitats of species, either through direct reduction or fragmentation of wild areas that can affect species which need large areas to thrive (Lawler et al. 2014).

Sprawl also can generate several secondary effects that can affect the quality of life of people and communities as a whole. A greater reliance on automobile trips in lower-density areas results in higher mileage trips, leading to more energy consumption and wear and tear on roads. A greater number of trips per capita also lead to a greater absolute number of accidents, injuries, and fatalities. When infrastructure cannot keep up with population growth, or when constraints are in place (e.g. existing land uses or natural landforms like water bodies or mountains) to limit expansion of existing systems, traffic congestion is inevitable, leading to lost productivity, increased transit times, and a greater waste of energy to power idling vehicles. Building additional roadways to increase transportation capacity requires greater expenditure, which may lead to increased rate of taxes. The greater trip distances that result from sprawl lead to greater reliance on automobile transportation and a like reduction in physically active transportation, such as biking or walking. As a result, reduced physical activity, increased obesity, and related health effects tend to rise.

1.3 The Master Equation or IPAT Equation

To assess the combined impacts of resource use, environmental impact, and population, the "*Master Equation*" may be used to assess the overall impact of these inputs. The equation is defined as follows:

$$\text{Overall impact} = \text{Population} \times (\text{Resource use/person}) \times (\text{Environmental impact/unit of resource use}) \quad (1.1)$$

This equation is also known as the IPAT equation (Chertow 2000) first proposed by Ehrlich and Holdren (1971), where I = Impact, P = Population, A = Affluence, and T = Technology. As for measurements of inputs, resource use can be set to GDP, and converted to a per capita value to account for a person-by-person basis. As can be seen, with increased population and resource use (which are relatively easy to measure), a reduction in environmental impact per unit of resource use is necessary and must be of sufficient magnitude if overall impact is to be minimized.

1.4 What Is Sustainability?

Several problems faced on a global scale that could severely affect the environment and ultimately human civilizations have been presented in preceding sections. There is a growing desire to pursue sustainable solutions to address these growing problems, but what exactly is sustainability? There is no universally accepted definition, but several common definitions convey similar concepts. The most commonly used definition comes from one presented by the 1987 United Nations World Commission on Environment and Development (commonly known as the Brundtland Commission, named after the commission chair, Gro Harlem Brundtland, former Prime Minister of Norway). In the Commission's report, *Our Common Future*, sustainability was defined as, "Development that meets the needs of the present without compromising the ability of future generations to meet their own needs" (UN 1987).

The United Nations (UN) adopted the 2030 Agenda (Transforming our World: the 2030 Agenda for Sustainable Development) in September 2015 with effective date January 1, 2016 (UN 2015). The Agenda is a commitment to achieve sustainable development by 2030 world-wide. It consists of 17 global goals (known as Sustainability Development Goals or simply SDGs) with associated 169 specific targets. The SDGs cover the following social, economic and environmental development issues: No poverty; Zero hunger; Good health and well-being for people; Quality education; Gender equality; Clean water and sanitation; Affordable and clean energy; Decent work and economic growth; Industry, Innovation, and Infrastructure; Reducing inequalities; Sustainable cities and communities; Responsible consumption and production; Climate action; Life below water; Life on land; Peace, justice and strong institutions; and Partnerships for the goals. These goals are vital and ambitious but necessary to shift the world onto a sustainable and resilient path.

The concept of sustainability and sustainable development supersedes the concept of "environmentalism" that has been in existence for several years. In addition to protection of the environment, sustainability incorporates the concept of equity – the needs of the present shall be met, but of equal importance is the preservation of resources for equally deserving future generations. The incorporation of sustainability means different things to different people, too. Economists typically consider sustainability and its aspects with respect to growth, efficiency, and resource use. Sociologists focus on human needs, including equity, empowerment, social cohesion, and cultural identity. Environmentalists tend to focus on the preservation of the integrity of the environment, living within the limits, means, and carrying capacity of the planet, and exploring ways to minimize environmental impacts while remediating existing impacts from past and present pollution occurrences.

The pursuit and achievement of sustainable solutions presents several challenges, not the least of which is that development and economic activity often have goals that, at first glance, can be in direct conflict with environmental stewardship or resource preservation. However, sustainable solutions are increasingly desired, and the concept of the triple bottom line has been a valuable framework in which to consider the relative efficacy of these solutions. The triple bottom line (Figure 1.9) consists of three measurement schemes to consider these solutions (Halpern et al. 2013):

- *Environmental* – Diversity and interdependence within living systems, the goods and services produced by the world's ecosystems, and the impacts of human-generated wastes (the planet).
- *Economic* – Flow of financial capital and the facilitation of commerce (prosperity).
- *Social* – The interactions between institutions/firms and people, functions expressive of human values, aspirations, and well-being, and ethical issues (people).

Sustainable solutions are those that maximize the relative return to these three categories and their representative metrics. More sustainable solutions lead to a greater intersection of these metrics.

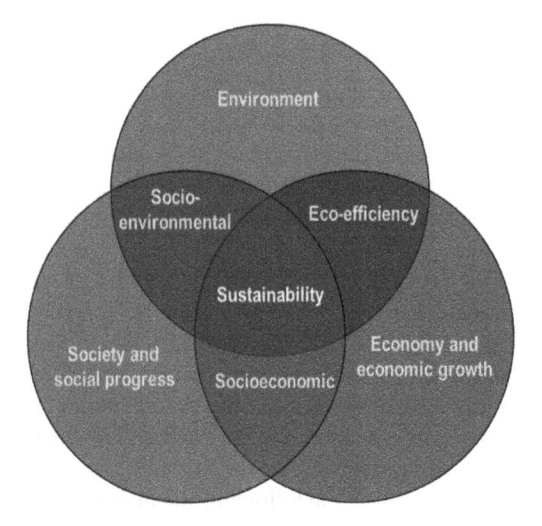

Figure 1.9 The triple bottom line: environment, economic, and social aspects, in sustainability.

A collective transition to more sustainable practices may be hard to imagine, but it is necessary. A stable human population must recognize that the earth has a finite set or resources as well as a finite ability to accept and absorb waste products. If sustainable practices are not incorporated, catastrophic consequences may result. A pursuit of greater sustainability requires care and dedication to the preservation of the natural world and a commitment of justice and equity among various peoples of the planet (Costanza et al. 2014).

Several actions or characteristics are vital for the transformation to more sustainable practices. First, population growth must slow down to a more stable, sustained pace. Political and social institutions need to increasingly embrace environmental stewardship and approaches to reduce poverty and inequity. Resource utilization needs to progress from a sole focus on short-term economic benefit and output to one that incorporates a long-term view with due consideration for the protection of ecosystems and natural resources. Technology needs to play a role in the reduction of greenhouse gases and transition from pollution and waste-intensive processes to more environmentally protective activities. Finally, land use and planning should increasingly focus on community development practices that discourage sprawl-causing development and encourage higher-density uses, including more livable cities.

In summary, the sustainability concept, if not the specific definitions, is relatively easy to comprehend. However, the ability to measure sustainability remains a work in progress. Nevertheless, efforts have been made, and sustainability measuring tools are beginning to evolve. As an example, in 2007, the United Nations Commission on Sustainable Development listed 96 indicators under 14 themes to measure sustainability or sustainability in a particular country (Huang et al. 2015). Some examples are presented in Table 1.1. However, sustainability at project level requires new set of sustainability indicators, which are addressed in Section II of this book.

Table 1.1 Indicators for measuring sustainability (UN 2007).

Theme	Subtheme	Indicator (example)
Poverty	Income poverty	Proportion of population living below national poverty line
	Income inequality	Ratio of share in national income of highest to lowest quintile
	Sanitation	Proportion of population using an improved sanitation facility
	Drinking water	Proportion of population using an improved water source
	Access to energy	Share of households without electricity or other modern energy services
	Living conditions	Proportion of urban population living in slums
Governance	Corruption	Percentage of population having paid bribes
	Crime	Number of intentional homicides per 100 000 population
Health	Mortality	Under-five mortality rate
	Health care Delivery	Percentage of population with access to primary health-care facilities
	Nutritional status	Nutritional status of children
	Health status and risks	Morbidity of major diseases such as HIV/AIDS, malaria, tuberculosis
Education	Education level	Net enrolment rate in primary education
	Literacy	Adult literacy rate

(Continued)

Table 1.1 (Continued)

Theme	Subtheme	Indicator (example)
Demographics	Population	Population growth rate
	Tourism	Ratio of local residents to tourists in major tourist regions and destinations
Natural hazards	Vulnerability to natural hazards	Percentage of population living in hazard prone areas
	Disaster preparedness and response	Human and economic loss due to natural disasters
Atmosphere	Climate change	Carbon dioxide emissions
	Ozone layer depletion	Consumption of ozone depleting substances
	Air quality	Ambient concentration of air pollutants in urban areas
Land	Land use and status	Land degradation
	Desertification	Land affected by desertification
	Agriculture	Arable and permanent cropland area
	Forests	Proportion of land area covered by forests
Oceans, seas and coasts	Coastal zone	Percentage of total population living in coastal areas
	Fisheries	Proportion of fish stocks within safe biological limits
	Marine environment	Proportion of marine area protected
Freshwater	Water quantity	Water use intensity by economic activity
	Water quality	Presence of fecal coliforms in freshwater
Biodiversity	Ecosystem	Proportion of terrestrial area protected, total and by ecological region
	Species	Change in threat status of species
Economic development	Macroeconomic performance	Gross domestic product (GDP) per capita
	Sustainable public finance	Debt to GNI (Gross National Income) ratio
	Employment	Labor productivity and unit labor costs
	Information and communication technologies	Internet users per 100 population
	Research and development	Gross domestic expenditure on R&D as a percentage of GDP
	Tourism	Tourism contribution to GDP
Global economic partnership	Trade	Current account deficit as percentage of GDP
	External financing	Net Official Development Assistance (ODA) given or received as a percentage of GNI
Consumption and production patterns	Material consumption	Material intensity of the economy
	Energy use	Annual energy consumption, total and by main user category
	Waste generation and management	Generation of hazardous waste
	Transportation	Modal split of passenger transportation

1.5 What Is Sustainable Engineering?

With a definition, or at least a general idea of the sustainability in place, it is time to discuss how sustainability principles and goals can be applied to the world around us. Of course, for centuries, engineers have applied scientific principles to devise solutions in nearly all aspects of our lives, including water, sanitation, mobility, energy, information management and propagation, food, health care, shelter, and communications. For much of this time, engineers have developed solutions with an implicit notion of limitless resources and little concern with respect to management of waste or by-products, or at least solutions independent of these factors. Additionally, engineers have commonly sought solutions that have not fully considered the social implications of their solutions.

With an understanding that, in fact, resources, waste management, and social implications are indeed critical factors, the concepts of sustainable engineering have begun to emerge. Sustainable engineering is the development of engineering solutions to advance human life to maximize benefits and minimize adverse impacts to the environment, the economy, and the society ("the triple bottom line") throughout the life cycle of a project. Sustainable engineering can be applied regardless of value chain position or the magnitude of a project – it is scalable and applied across molecular, product, process, and system design.

The goals of sustainable engineering are shared among many specific disciplines. As presented in the American Academy of Environmental Engineers (AAEE) Environmental Engineering Body of Knowledge (2008), a joint declaration was issued with the American Association of Engineering Societies (AAES), American Institute of Chemical Engineers (AIChE), the American Society of Mechanical Engineers (ASME), and the National Academy of Engineering (NAE) that stated the following: "Creating a sustainable world that provides a safe, secure, healthy life for all peoples is a priority for the US engineering community… Engineers must deliver solutions that are technically viable, commercially feasible, and environmentally and socially sustainable." The American Society of Civil Engineers (ASCE) also addressed sustainability and sustainable engineering in 2008 by stating, "Sustainability is the ability to meet human needs for natural resources, industrial products, energy, food, transportation, shelter, and effective waste management while conserving and enhancing environmental quality and the natural resource base essential for the future. Sustainable engineering meets these human needs" (ASCE 2008).

Lately, the concept of green engineering (GE) has begun to emerge, but there are important, clear distinctions between green engineering and sustainable engineering. Generally stated, green engineering is the development of engineering solutions that are protective of human and environmental health throughout the life cycle of a project. Sustainable engineering is broader than green engineering in that it aims to develop engineering solutions that are not only protective of human and environmental health but are also economically and socially responsible throughout the life cycle of a project. In other words, green engineering is a subset of sustainable engineering with a focus to minimize environmental impacts. Table 1.2 and Figure 1.10 summarize key principles of green engineering and sustainable engineering, respectively.

Of equal importance to environmentally protective, green concepts, and financial considerations are several socially focused dimensions. In helping to maximize the benefit within the context of these dimensions, it is important to identify to the extent possible the social issues and values that a particular activity or project will affect, and it is equally important to consider both the positive and the negative social consequences. For many reasons, perhaps due to practice, cultural, or social traditions, it is important that the social context even be considered, as it has not been incorporated in the past for countless projects or activities. To the extent possible, it is important to provide a culture/environment around a project or activity

Table 1.2 Principles of green engineering.

Inherent rather than circumstantial	Designers need to strive to ensure that all materials and energy inputs and outputs are as inherently nonhazardous as possible
Prevention instead of treatment	It is better to prevent waste than to treat or clean up waste after it is formed
Design for separation	Separation and purification operations should be designed to minimize energy consumption and materials use
Maximize efficiency	Products, processes, and systems should be designed to maximize mass, energy, space, and time efficiency
Output-pulled versus input-pushed	Products, processes, and systems should be "output pulled" rather than "input pushed" through the use of energy and materials
Conserve complexity	Embedded entropy and complexity must be viewed as an investment when making design choices on recycle, reuse, or beneficial disposition
Durability rather than immortality	Targeted durability, not immortality, should be a design goal
Meet need, minimize excess	Design for unnecessary capacity or capability (e.g. "one size fits all") solutions should be considered a design flaw
Minimize material diversity	Material diversity in multicomponent products should be minimized to promote disassembly and value retention
Integrate material and energy flows	Design of products, processes, and systems must include integration and interconnectivity with available energy and materials flows
Design for commercial "afterlife"	Products, processes, and systems should be designed for performance in a commercial "afterlife"
Renewable rather than depleting	Material and energy inputs should be renewable rather than depleting

Source: Anastas and Zimmerman (2003). Reproduced with permission of American Chemical Society.

such that dialogue with respect to socially related aspects can occur. In such an environment, social values that may not have been seriously considered or even identified can be integrated into the project-related, decision-making process. Once this culture and dialogue are in place, decisions with respect to important social dimensions may be thoughtfully considered. Some important dimensions that may be considered include effects on population factors, such as birth rate, literacy, life span, cultural acceptance/openness (art, religion, etc.), and gender roles and degree of equity.

Aspects of a project can be designed or re-designed to maximize social or cultural benefits to the maximum extent feasible, while at the same time maximizing relative benefit with respect to traditional financially driven metrics, including costs, rate of return, consumer demand, etc. Of course, this is an iterative process, and it is rare, if not impossible, to maximize return or benefit to all three bottom line aspects. However, with iterations and thoughtful consideration of these three frames of reference, the goal is to maximize the relative benefit of each frame, incorporate into design, execution, and management, and carry out the action to attain the desired result.

Although the framework and the nomenclature of sustainable engineering are relatively new, some concepts have been around for a long time, and there has been a steady progression for decades toward its current state. Figure 1.11 shows how this trend has evolved over the past 40–50 years. For instance, prior to 1970, very little thought was given to social or environmental aspects. However, with significant environmental issues and related government regulation

1.5 What Is Sustainable Engineering?

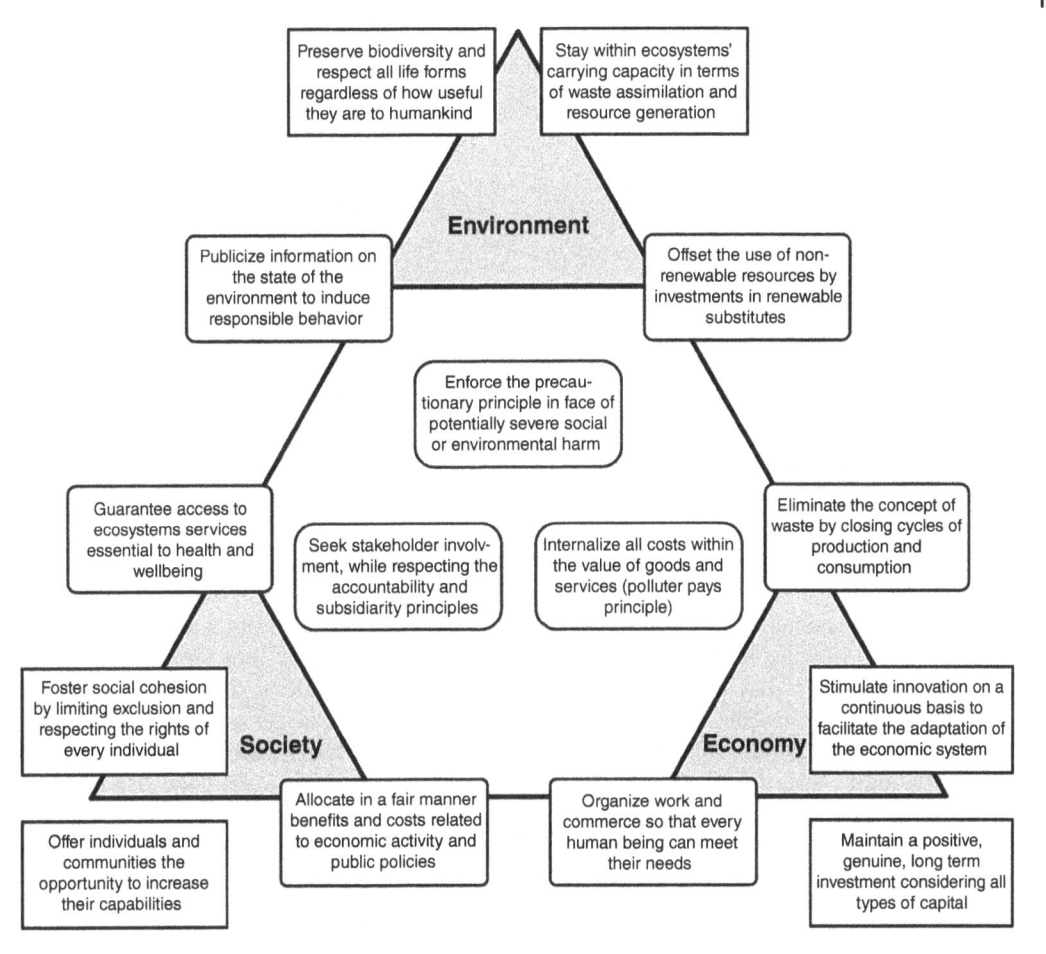

Figure 1.10 Evolution of concepts in sustainable engineering. (Gagnon et al. 2008).

increasingly becoming part of the public consciousness in the 1970s, projects began to react and comply, then anticipate, and later even guide evolving environmental attitudes. As this evolution progressed, increased attention was paid to socially responsible and culturally aware practices, with an increased incorporation into design phases of projects and activities.

As we have introduced and discussed the aspects of sustainable engineering, there are several emerging general strategies or goals that are increasingly being considered core to sustainable engineering. Ultimately, there is a goal to pursue a full integration of engineered systems within the overall natural system of which it is a part. Additionally, it is important to seek flexible, scalable solutions that may be adapted from a micro- to macrolevel, or a local to global scale. Further, thoughtful designs are based on a system approach that incorporates green chemistry, processes, and materials use with comprehensive considerations of life cycle impacts with respect to energy and materials inputs as well as emissions and other waste outputs. Finally, these designs should consider socially responsible outcomes as well as traditional economic metrics, including revenue, profit, profit margin, return on investment (ROI), and return on equity (ROE), among many others.

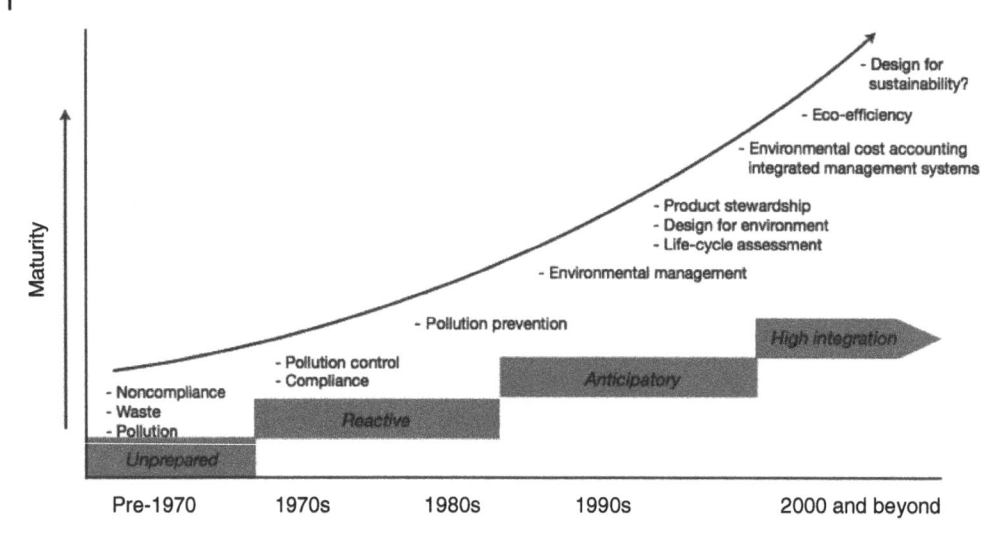

Figure 1.11 Evolution of concepts in sustainable engineering.

As mentioned, sustainable engineering operates at various scales. From a local standpoint, concepts and principles may be incorporated by individual firms as they attempt to design projects or products that are protective of the environment, aim to reduce pollution or waste generation, and utilize resources efficiently. On the next larger scale, firms or other entities may cooperate on larger initiatives, such as intra or intersector initiatives, standards of practice, joint ventures, collaborative studies, or facility development. Finally, on a regional or global scale, larger multinational firms, entire industries, or government entities may participate on even large initiatives, such as budget and cycle studies, materials and energy flow studies, research initiatives, and modeling scenarios that can ultimately lead to practice regulations.

An example of this scalability comes from the automotive industry. Imagine a single firm that fabricates a component of an automobile, such as axles. At the firm level, they can find sustainable designs or implement systems that eliminate waste or add efficiency to the design. At the next scale, that can participate with their client (one of the large automobile companies, such as General Motors or Toyota) as well as with the company's other subcontractors to develop eco-friendly integrated systems for overall design or for manufacturing principles. At the next larger scale, they may work within the industry to find larger-scale sustainable solutions, such as those that may affect the design of roads and infrastructure, refining of traditional motor fuels as well as alternative fuels, etc. Finally, at the global scale, they may work with a number of intra-industry groups as well as those from other industries or from governments to look for other initiatives that directly look to solve society's needs, such as alternative transportation modes, or land use that may optimize trip length and density.

A concrete example of this concept is the design and manufacture of Sertraline by Pfizer, a large pharmaceutical company (USEPA 2002; Nameroff et al. 2004). Sertraline is the active ingredient in Zoloft, a market-leading drug for the treatment of depression. Hundreds of millions of prescriptions have been written for Zoloft in the United States alone. By applying some of the sustainable engineering principles described in this chapter, Pfizer was able to eliminate the use of 140 metric tons per year of titanium tetrachloride and eliminate the generation of 440 metric tons of titanium dioxide waste per year. In addition, 150 metric tons of hydrochloric acid waste and 100 metric tons of sodium hydroxide needed for production were eliminated on a yearly basis. Other raw materials were also eliminated. These reductions saved Pfizer substantial money with respect to raw materials purchases and waste management. Correspondingly resources were preserved, and Pfizer's financial metrics improved.

When designing products or projects that aim to be sustainable, it is important to measure the degree of sustainability involved in the fabrication, use, and disposal of the product. Not only should this study span the project or product itself, it should also consider all inputs, and activities associated with final disposition of the product or project. Such a study may include a life cycle assessment (LCA). An LCA includes an inventory of inputs (e.g. raw materials and energy) and outputs (e.g. products, emissions, and solid wastes) of the product over its entire life. The inventories that are assembled are converted into categories of impact; some of these may include emissions-related effects on climate change, acidification, aquatic toxicity, or human health. The LCA further assigns a quantitative weighting system of the individual categories to compute a single impact number.

Life cycle assessments can be incredibly powerful and useful in assessing sustainability. They consider the entire cycle of a project or product and related effects. It allows for the consideration and incorporation of several value-based assessment criteria, as well as latitude with respect to weighting of these criteria. Their use also promotes comparison between project/product alternatives as well as individual components. Unfortunately, there are some limitations associated with life cycle assessments. They are costly and time-consuming to perform. There is an inherent imprecision of some data associated with the attempt to quantify qualitative aspects or virtues. Additionally, it is difficult to incorporate and/or measure all relevant data, and some impacts may be inadvertently overlooked. Nevertheless, they are useful tools in working to achieve more sustainable designs with respect to material and energy inputs as well waste management/reduction.

A final concept of this chapter, whether with respect to degree of sustainability, the triple bottom line, or applications of life cycle assessments, is the effect of globalization. Increasingly, virtually all aspects of economic activity are moving toward a global approach and participation. Very little economic activity of any importance occurs without some form of global impact. The global aspects of economic activity need to be carefully considered when assessing sustainability. Trade policies and regulations, the interaction of design and manufacturing in distant locations, transportation, and the flow of inputs, outputs, and waste by-products (such as emissions) are important to consider and factor when assessing sustainable alternatives.

1.6 Summary

When defining sustainability, it is important to remember that it is a concept that transcends and supersedes familiar topics associated with environmentalism. Sustainability entails that the human development or activity currently should not jeopardize the needs and the development of future generations. Sustainability is measured across several factors, and the concept of the triple bottom line assesses sustainability within a framework that considers overall effects of a project or product with respect to economics, the environment, and the society. Sustainable practices, especially to be applied in the future, must be applied in such a way that several emerging challenges (as described in this chapter) are addressed and incorporated.

Sustainable engineering is the means by which sustainable concepts can be applied to systems, projects, and products. Sustainable engineering is defined as a tool for engineering responsibly and professionally by integrating the environmental, social, and economic considerations during the life cycle of a project. Finally, the strategies of sustainable engineering aim to engineer systems that can be integrated into natural systems, are flexible enough to adapt to a range of scales from a local to global basis, incorporate design based on a systems approach, and incorporate judicious resources use while minimizing waste generation.

1.7 Questions

1.1 Describe briefly case of Easter Island, Chile. Based on this case, explain how and why sustainability is important.

1.2 Explain environmental impacts and other emerging challenges in the context of sustainability.

1.3 Outline three strategies you would recommend to lessen/abate waste generation.

1.4 Discuss how impacts to rainforest ecosystems could affect human populations that live in other areas.

1.5 Explain the "Master Equation." Discuss the validity of this equation.

1.6 Explain the "Tragedy of Commons." How is it relevant in a sustainability context?

1.7 Define "Sustainability" and "Triple Bottom Line." How does it fit into the current regulatory framework?

1.8 Name three specific metrics that can be used to assess the degree of sustainability from within the environmental dimension.

1.9 Name three specific metrics that can be used to assess the degree of sustainability from within the economic dimension.

1.10 Name three specific metrics that can be used to assess the degree of sustainability from within the social dimension.

1.11 Explain a simple approach to quantify sustainability? Give some examples.

1.12 How can one evaluate sustainability under complex and dynamic conditions?

1.13 Explain the "grand objectives" of sustainability. How can they be related to environmental concerns and targeted activities and specific actions?

1.14 What is sustainability (or sustainable) engineering? Why is it becoming more critical in current engineering practice?

1.15 What are the general strategies for sustainable engineering? Explain them.

1.16 Discuss a consumer product of your choosing, listing five raw material inputs and five emissions/outputs that could be included in an LCA analysis.

References

Ako, R. (2013). *Environmental Justice in Developing Countries: Perspectives from Africa and Asia-Pacific*. Routledge.

Anastas, P.T. and Zimmerman, J.B. (2003). Design through the twelve principles of green engineering. *Environmental Science & Technology* 37 (5): 94A–101A.

ASCE (2008). *Civil Engineering Body of Knowledge for the 21st Century. Preparing the Civil Engineer for the Future*, 2e. American Society of Civil Engineers. ISBN: 978-0-7844-0965-7 Stock # 40965.

Baker, R.J., Reilly, T.J., Lopez, A. et al. (2015). Screening tool to evaluate the vulnerability of down-gradient receptors to groundwater contaminants from uncapped landfills. *Waste Management* 43: 363–375.

Basten, S., Lutz, W., and Scherbov, S. (2013). Very long range global population scenarios to 2300 and the implications of sustained low fertility. *Demographic Research* 28 (39): 1145–1166.

Butchart, S.H., Walpole, M., Collen, B. et al. (2010). Global biodiversity: indicators of recent declines. *Science* 328 (5982): 1164–1168.

Chertow, M.R. (2000). The IPAT equation and its variants. *Journal of Industrial Ecology* 4 (4): 13–29.

Costanza, R., Cumberland, J.H., Daly, H. et al. (2014). *An Introduction to Ecological Economics*. CRC Press.

Dasandi, N. (2014). International inequality and world poverty: a quantitative structural analysis. *New Political Economy* 19 (2): 201–226.

De Ferranti, D., Perry, G.E., Lederman, D., and Maloney, W.E. (2002). *From Natural Resources to the Knowledge Economy: Trade and Job Quality*. Washington, DC: World Bank.

Ehrlich, P.R. and Holdren, J.P. (1971). Impact of population growth. *Science. American Association for the Advancement of Science.* 171 (3977): 1212–1217.

Fischer, S.R. (2005). *Island at the End of the World: The Turbulent History of Easter Island*. Reaktion Books.

Graedel, T.E. and Allenby, B.R. (2010). *Industrial Ecology and Sustainable Engineering*. NJ: Prentice Hall.

Gagnon, B., Leduc, R., and Savard, L. (2008). Sustainable development in engineering: a review of principles and definition of a conceptual framework. Cahier de recherche/Working Paper 08-18.

Goudie, A.S. (2013). *The Human Impact on the Natural Environment: Past, Present, and Future*. Wiley.

Halpern, B.S., Klein, C.J., Brown, C.J. et al. (2013). Achieving the triple bottom line in the face of inherent trade-offs among social equity, economic return, and conservation. *Proceedings of the National Academy of Sciences* 110 (15): 6229–6234.

Han, Z., Ma, H., Shi, G. et al. (2016). A review of groundwater contamination near municipal solid waste landfill sites in China. *Science of the Total Environment* 569: 1255–1264.

Hardin, G. (1968). The tragedy of the commons. *Science* 162 (3859): 1243–1248.

Hardin, G. (2009). The tragedy of the commons*. *Journal of Natural Resource Policy Research* 1 (3): 243–253.

Holifield, R. (2001). Defining environmental justice and environmental racism. *Urban Geography* 22 (1): 78–90.

Huang, L., Wu, J., and Yan, L. (2015). Defining and measuring urban sustainability: a review of indicators. *Landscape Ecology* 30 (7): 1175–1193.

Hutter, J. (2015). Overfishing in Canada and the United States: A Comparative Study of Policy and Legislation. Doctoral dissertation. Faculty of Graduate Studies.

Ioppolo, G., Heijungs, R., Cucurachi, S. et al. (2014). Urban metabolism: many open questions for future answers. In: *Pathways to Environmental Sustainability*, 23–32. Springer International Publishing.

Irwin, F.H., Ranganathan, J., and Bateman, M. (2007). *Restoring Nature's Capital: An Action Agenda to Sustain Ecosystem Services*. World Resources Inst.

Janaskie, S. (2013). *The Human Population Explosion*. Yale Scientific http://www.yalescientific.org/2013/05/the-human-population-explosion/.

Kampa, M. and Castanas, E. (2008). Human health effects of air pollution. *Environmental Pollution* 151 (2): 362–367.

Kamruzzaman, P. (2015). *Dollarisation of Poverty: Rethinking Poverty Beyond 2015*. London, UK: Palgrave Macmillan https://doi.org/10.1057/9781137541437.

Knutti, R. and Sedláček, J. (2013). Robustness and uncertainties in the new CMIP5 climate model projections. *Nature Climate Change* 3 (4): 369–373. https://doi.org/10.1038/nclimate1716.

Lawler, J.J., Lewis, D.J., Nelson, E. et al. (2014). Projected land-use change impacts on ecosystem services in the United States. *Proceedings of the National Academy of Sciences* 111 (20): 7492–7497.

Mann, D., Edwards, J., Chase, J. et al. (2008). Drought, vegetation change, and human history on Rapa Nui (Isla de Pascua, Easter Island). *Quaternary Research* 69 (1): 16–28. http://dx.doi.org/10.1016/j.yqres.2007.10.009.

Marcott, S.A., Shakun, J.D., Clark, P.U., and Mix, A.C. (2013). A reconstruction of regional and global temperature for the past 11,300 years. *Science* 339 (6124): 1198–1201.

Mulrooney, M.A. (2013). An island-wide assessment of the chronology of settlement and land use on Rapa Nui (Easter Island) based on radiocarbon data. *Journal of Archaeological Science* 40 (12): 4377–4399. ISSN 0305-4403, http://dx.doi.org/10.1016/j.jas.2013.06.020.

Nameroff, T.J., Garant, R.J., and Albert, M.B. (2004). Adoption of green chemistry: an analysis based on US patents. *Research Policy* 33 (6–7): 959–974. https://doi.org/10.1016/j.respol.2004.03.001.

Oreskes, N. (2004). The scientific consensus on climate change. *Science* 306 (5702): 1686–1686.

Ostrom, E. (2015). *Governing the Commons*. Cambridge university press.

Parslow, J. (2010). Individual transferable quotas and the "tragedy of the commons". *Canadian Journal of Fisheries and Aquatic Sciences* 67 (11): 1889–1896. https://doi.org/10.1139/F10-104.

Patnaik, U. (2003). Global capitalism, deflation and agrarian crisis in developing countries. *Journal of Agrarian Change* 3: 33–66. https://doi.org/10.1111/1471-0366.00050.

Prentice, M. and Sheldon, K.M. (2015). Evolutionary and social psychological perspectives on human cooperation. In: *Evolutionary Perspectives on Social Psychology*, 267–277. Springer International Publishing.

Pulido, L. (2000). Rethinking environmental racism: white privilege and urban development in Southern California. *Annals of the Association of American Geographers* 90 (1): 12–40.

Rode, P., Floater, G., Thomopoulos, N. et al. (2017). Accessibility in cities: transport and urban form. In: *Disrupting Mobility*, 239–273. Springer International Publishing.

Rull, V., Cañellas-Boltà, N., Margalef, O. et al. (2015). Late Holocene vegetation dynamics and deforestation in Rano Aroi: implications for Easter Island's ecological and cultural history. *Quaternary Science Reviews* 126: 219–226. http://dx.doi.org/10.1016/j.quascirev.2015.09.008.

Sadasivam, B.Y. and Reddy, K.R. (2014). Landfill methane oxidation in soil and bio-based cover systems: a review. *Reviews in Environmental Science and Biotechnology* 13 (1): 79–107. https://doi.org/10.1007/s11157-013-9325-z.

Stevenson, C., Wozniak, J., and Haoa, S. (1999). Prehistoric agricultural production on Easter Island (Rapa Nui), Chile. *Antiquity* 73 (282): 801–812. https://doi.org/10.1017/S0003598X00065546.

Stoate, C., Boatman, N.D., Borralho, R.J. et al. (2001). Ecological impacts of arable intensification in Europe. *Journal of Environmental Management* 63 (4): 337–365.

Themelis, N.J. and Ulloa, P.A. (2007). Methane generation in landfills. *Renewable Energy* 32 (7): 1243–1257.

Trevors, J.T. (2010). What is a global environmental pollution problem? *Water, Air, and Soil Pollution* 210 (1): 1–2. https://doi.org/10.1007/s11270-010-0337-9.

UN (1987). *Report of the World Commission on Environment and Development: Our Common Future*. Oslo, Norway: United Nations General Assembly, Development and International Co-operation: Environment.

UN (2007). *Indicators of Sustainable Development: Guidelines and Methodologies*, 3e. New York: Economic and Social Affairs. United Nations. ISBN: 978-92-1-104577-2.

UN (2015). *Transforming our World: The 2030 Agenda for Sustainable Development*. United Nations. https://sustainabledevelopment.un.org/post2015/transformingourworld/publication (accessed January 24, 2019).

USEPA (2002). *Presidential Green Chemistry Challenge Award Recipients*. Document Number EPA-744-K-02-002. Washington, DC: EPA.

USEPA (2014). *Municipal Solid Waste Generation, Recycling, and Disposal in the United States: Facts and Figures for 2012*. Environmental Protection Agency Report number EPA-530-F-14-001.

Vellend, M. (2017). The biodiversity conservation paradox. *American Scientist* 105 (2): 94.

Weitz, J.S., Eksin, C., Paarporn, K. et al. (2016). An oscillating tragedy of the commons in replicator dynamics with game-environment feedback. *Proceedings of the National Academy of Sciences of the United States of America* 113 (47): E7518–E7525. https://doi.org/10.1073/pnas.1604096113.

2

Environmental Concerns

2.1 Introduction

When considering a range of projects, which may include a tangible product, a real estate development, or a productive service/activity, it is often beneficial to assess and quantify the environmental impacts associated with a range of alternative projects/designs for the purposes of a comparative assessment. In order to measure the degree of sustainability of a project, all potential environmental impacts, or as many as practicable, should be considered and incorporated into an environmental impact assessment, especially when those impacts are quantifiable. These may include both direct and indirect environmental impacts. However, it can often be difficult to quantify some impacts resulting from preproject issues, such as raw materials extraction/fabrication, project activities, outputs (e.g. emissions and financial implications), and post-use issues (e.g. waste management and disposition). Such an environmental impact assessment will identify and allow for the selection of the project/design that minimizes overall environmental impact. Of course, in addition to environmental impacts, various economic and social implications should be considered to select a project as the most sustainable option.

As discussed in Chapter 1, environmental impacts constitute one of the dimensions (pillars) of sustainability. In order to assess environmental concerns, it is useful to utilize a reference framework, such as "the triple bottom line." As previously discussed, the triple bottom line sustainability framework is measured using three criteria – environmental, economic, and social.

Environmental concerns are present during all phases of a project, and they occur over a range of scope and magnitudes. These concerns may occur on a local scale, a regional scale, or a global scale. It is important to remember that these concerns are not universal – many potential environmental concerns that warrant attention may not occur for a specific project. Additionally, virtually every activity will have one or more environmental impact. Several aspects of a project may result in side effects that pertain to a specific or a range of environmental concerns. However, it is also important to point out that many of these may be disregarded, depending on their magnitude or when considered with the benefits and drawbacks of moving forward with a specific project. The key point is to determine the degree of potential impact that may occur and determine the degree to which human health or the environment may be affected. Separating important effects from "background noise" is a key skill in assessing the sustainability of a project, and it is important that these environmental concerns be assessed on a qualitative or quantitative basis.

To perform these assessments, it is critical to understand a range of environmental concerns that can occur from the projects or activities under assessment. This chapter provides background on various environmental issues and impacts related to sustainability. This background is essential to serve as a context for sustainable engineering practices.

Sustainable Engineering: Drivers, Metrics, Tools, and Applications, First Edition.
Krishna R. Reddy, Claudio Cameselle, and Jeffrey A. Adams.
© 2019 John Wiley & Sons, Inc. Published 2019 by John Wiley & Sons, Inc.

2.2 Global Warming and Climate Change

There is an overwhelming scientific consensus that the Earth is warming and undergoing climate change. Although climate change results from a variety of natural physical processes, and global climate appears to follow a natural oscillatory cycle, the scientific evidence suggests that anthropogenic sources, such as greenhouse gases (e.g. methane, carbon dioxide) are contributing to the climate change. The magnitude of this contribution and the resulting effects on climate change in the biosphere is still being debated, but there is a consensus among the scientific community that these anthropogenic sources are contributing in a meaningful way to the climate change. Global warming and climate change can have a range of adverse effects, including increased extreme weather events, effects on natural habitat and species, and adverse effects on the ability of ecosystems to naturally adapt and cope with extreme weather and geophysical events. Collectively, much of the anthropogenic contribution to global warming and climate change is occurring through the "Greenhouse Effect."

The Greenhouse Effect is a beneficial process that helps to maintain the Earth's average temperature around 15 °C/58 °F, making life on Earth possible. Without the Greenhouse Effect, the average temperature in the planet would be −18 °C/0 °F, a 33 °C/58 °F difference (Archer 2016). This effect is not exclusive to Earth; the temperature of other terrestrial planets is also influenced by the Greenhouse Effect (Figure 2.1). Simply stated, the Greenhouse Effect occurs when the energy from the sun, in the form of electromagnetic radiation, heats the surface of the Earth, which in turn regulates Earth's weather and climate. Much of this energy radiates back into space. As this radiated energy travels through the atmosphere, several "greenhouse gases" (mainly water vapor and carbon dioxide) trap some of this energy, much in the same way a greenhouse traps heat. The trapped energy serves to warm the Earth. Figures 2.2 and 2.3 depict the mechanisms of the Greenhouse Effect.

Some of the greenhouse gases, such as carbon dioxide, methane, nitrous oxide, and water vapor, are generated by both natural and anthropogenic processes. Other gases, such as fluorocarbons, are solely generated from anthropogenic processes. Specific gases exhibit a relative "strength" or contribution to the Greenhouse Effect. The global warming potential (GWP) is an index that assesses the heat trapping ability of a similar mass of a particular gas. This dimensionless value is expressed as a factor of carbon dioxide's heat-trapping ability (carbon dioxide's GWP is standardized to 1). Table 2.1 presents the GWP for several greenhouse gases based on the Intergovernmental Panel on Climate Change (IPCC) Fifth Assessment Report (IPCC 2013).

Thick atmosphere. It contains most of CO_2 (96%).
Average temperature: +460 °C

The atmosphere contains 0.03% of CO_2.
Average temperature: +15 °C

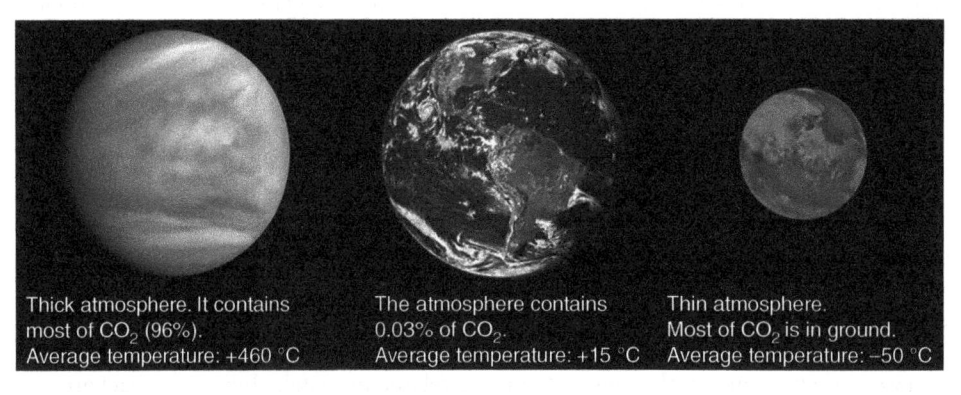

Thin atmosphere.
Most of CO_2 is in ground.
Average temperature: −50 °C

Figure 2.1 Greenhouse Effect on three planets.

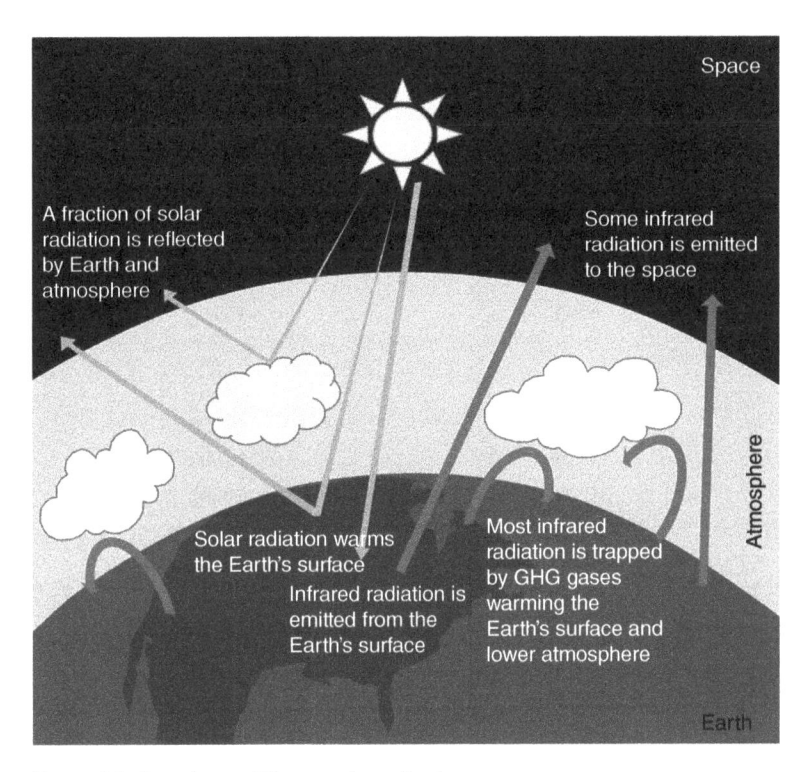

Figure 2.2 Greenhouse Effect on planet Earth.

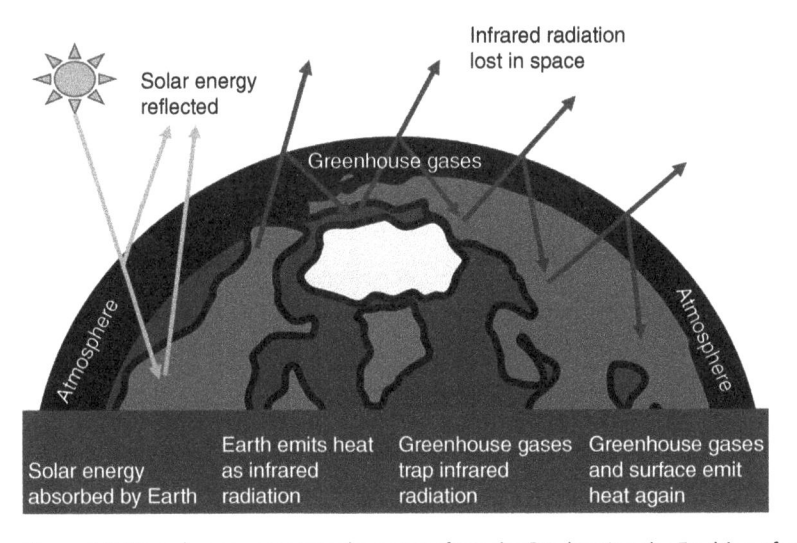

Figure 2.3 Greenhouse gases trap the energy form the Sun heating the Earth's surface.

Anthropogenic contributions of greenhouse gases come from a variety of sources. Carbon dioxide emanates from the combustion of fossil fuels, mainly for energy production and transportation. From a mass balance perspective, 1 kg of fuel-based carbon becomes 3 kg of atmospheric carbon dioxide upon combustion. Increased carbon dioxide also results from deforestation since plants are a sink for carbon dioxide, and the removal of plants via mass

Table 2.1 Global Warming Potential (GWP) for some greenhouse gases in the atmosphere (IPCC 2014; for a 100 year time horizon).

Industrial designation/ common name of the greenhouse gas	Chemical formula	GWP based on IPCC fifth assessment report 2014
Carbon dioxide	CO_2	1
Methane	CH_4	28
Nitrous oxide	N_2O	265
Substances controlled by Montreal protocol		
CFC-11	CCl_3F	4 660
CFC-12	CCl_2F_2	10 200
CFC-13	$CClF_3$	13 900
CFC-113	CCl_2FCClF_2	5 820
CFC-114	$CClF_2CClF_2$	8 590
CFC-115	$CClF_2CF_3$	7 670
Halon-1301	$CBrF_3$	6 290
Halon-1211	$CBrClF_2$	1 750
Halon-2402	$CBrF_2CBrF_2$	1 470
Carbon tetrachloride	CCl_4	1 730
Methyl bromide	CH_3Br	2
Methyl chloroform	CH_3CCl_3	160
HCFC-21	$CHCl_2F$	148
HCFC-22	$CHClF_2$	1 760
HCFC-123	$CHCl_2CF_3$	79
HCFC-124	$CHClFCF_3$	527
HCFC-141b	CH_3CCl_2F	782
HCFC-142b	CH_3CClF_2	1 980
HCFC-225ca	$CHCl_2CF_2CF_3$	127
HCFC-225cb	$CHClFCF_2CClF_2$	525
Hydrofluorocarbons (HFCs)		
HFC-23	CHF_3	12 400
HFC-32	CH_2F_2	677
HFC-41	CH_3F_2	116
HFC-125	CHF_2CF_3	3 170
HFC-134	CHF_2CHF_2	1 120
HFC-134a	CH_2FCF_3	1 300
HFC-143	CH_2FCHF_2	328
HFC-143a	CH_3CF_3	4 800
HFC-152	CH_2FCH_2F	16
HFC-152a	CH_3CHF_2	138
HFC-161	CH_3CH_2F	4
HFC-227ea	CF_3CHFCF_3	3 350
HFC-236cb	$CH_2FCF_2CF_3$	1 210

(Continued)

Table 2.1 (Continued)

Industrial designation/ common name of the greenhouse gas	Chemical formula	GWP based on IPCC fifth assessment report 2014
HFC-236ea	CHF_2CHFCF_3	1 330
HFC-236fa	$CF_3CH_2CF_3$	8 060
HFC-245ca	$CH_2FCF_2CHF_2$	716
HFC-245fa	$CHF_2CH_2CF_3$	858
HFC-365mfc	$CH_3CF_2CH_2CF_3$	804
HFC-43-10mee	$CF_3CHFCHFCF_2CF_3$	1 650
Perfluorinated compounds		
Sulfur hexafluoride	SF_6	23 500
Nitrogen trifluoride	NF_3	16 100
PFC-14	CF_4	6 630
PFC-116	C_2F_6	11 100
PFC-218	C_3F_8	8 900
PFC-318	$c-C_4F_8$	9 540
PFC-31-10	C_4F_{10}	9 200
PFC-41-12	C_5F_{12}	8 550
PFC-51-14	C_6F_{14}	7 910
PCF-91-18	$C_{10}F_{18}$	7 190
Trifluoromethyl sulfur pentafluoride	SF_5CF_3	17 400
Perfluorocyclopropane	$c-C_3F_6$	9 200
Fluorinated ethers		
HFE-125	CHF_2OCF_3	12 400
HFE-134	CHF_2OCHF_2	5 560
HFE-143a	CH_3OCF_3	523
HCFE-235da2	$CHF_2OCHClCF_3$	491
HFE-245cb2	$CH_3OCF_2CF_3$	654
HFE-245fa2	$CHF_2OCH_2CF_3$	812
HFE-347mcc3	$CH_3OCF_2CF_2CF_3$	530
HFE-347pcf2	$CHF_2CF_2OCH_2CF_3$	889
HFE-356pcc3	$CH_3OCF_2CF_2CHF_2$	413
HFE-449sl (HFE-7100)	$C_4F_9OCH_3$	421
HFE-569sf2 (HFE-7200)	$C_4F_9OC_2H_5$	57
HFE-43-10pccc124 (H-Galden 1040×)	$CHF_2OCF_2OC_2F_4OCHF_2$	2 820
HFE-236ca12 (HG-10)	$CHF_2OCF_2OCHF_2$	5 350
HFE-338pcc13 (HG-01)	$CHF_2OCF_2CF_2OCHF_2$	2 910
HFE-227ea	$CF_3CHFOCF_3$	6 450
HFE-236ea2	$CHF_2OCHFCF_3$	1 790
HFE-236fa	$CF_3CH_2OCF_3$	979
HFE-245fa1	$CHF_2CH_2OCF_3$	828
HFE 263fb2	$CF_3CH_2OCH_3$	1

(Continued)

Table 2.1 (Continued)

Industrial designation/ common name of the greenhouse gas	Chemical formula	GWP based on IPCC fifth assessment report 2014
HFE-329mcc2	$CHF_2CF_2OCF_2CF_3$	3070
HFE-338mcf2	$CF_3CH_2OCF_2CF_3$	929
HFE-347mcf2	$CHF_2CH_2OCF_2CF_3$	854
HFE-356mec3	$CH_3OCF_2CHFCF_3$	387
HFE-356pcf2	$CHF_2CH_2OCF_2CHF_2$	719
HFE-356pcf3	$CHF_2OCH_2CF_2CHF_2$	446
HFE 365mcf3	$CF_3CF_2CH_2OCH_3$	<1
HFE-374pc2	$CHF_2CF_2OCH_2CH_3$	627
Perfluoropolyethers		
PFPMIE	$CF_3OCF(CF_3)CF_2OCF_2OCF_3$	9710
Hydrocarbons and other compounds – direct effects		
Chloroform	$CHCl_3$	16
Methylene chloride	CH_2Cl_2	9
Methyl chloride	CH_3Cl	12
Halon-1201	$CHBrF_2$	376

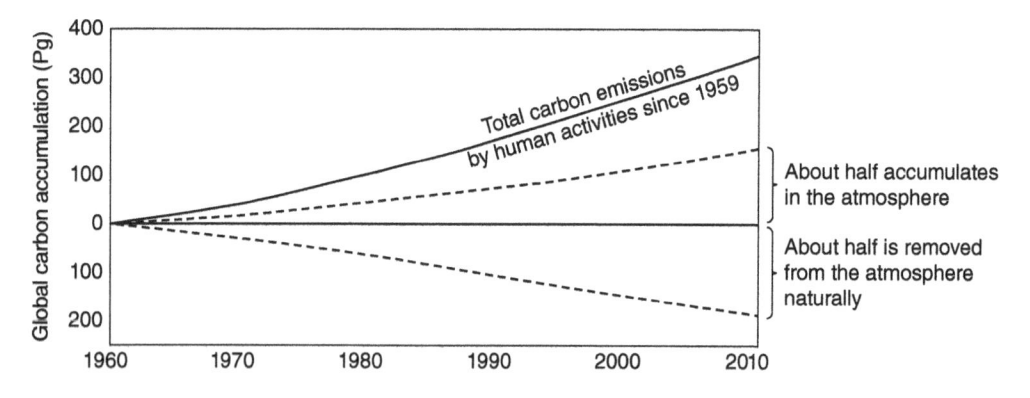

Figure 2.4 Earth's capacity for CO_2 absorption. Source: NOAA's Earth System Research Laboratory.

deforestation reduces this carbon dioxide absorption ability. Methane emitted from landfills, livestock and farming activities, natural gas pipeline losses, and some land use or wetlands alterations considerably adds to the greenhouse emissions into the atmosphere.

It is important to note that the Earth has a natural absorption capacity for carbon dioxide and other greenhouse gases; Figure 2.4 shows this relationship, and Figure 2.5 depicts the planetary carbon cycle. Nevertheless, a substantial amount of greenhouse gases has been generated during the industrial era, contributing to a measured rapid increase in global temperatures. Figure 2.6 depicts the average temperature on Earth over thousands of years, and it highlights

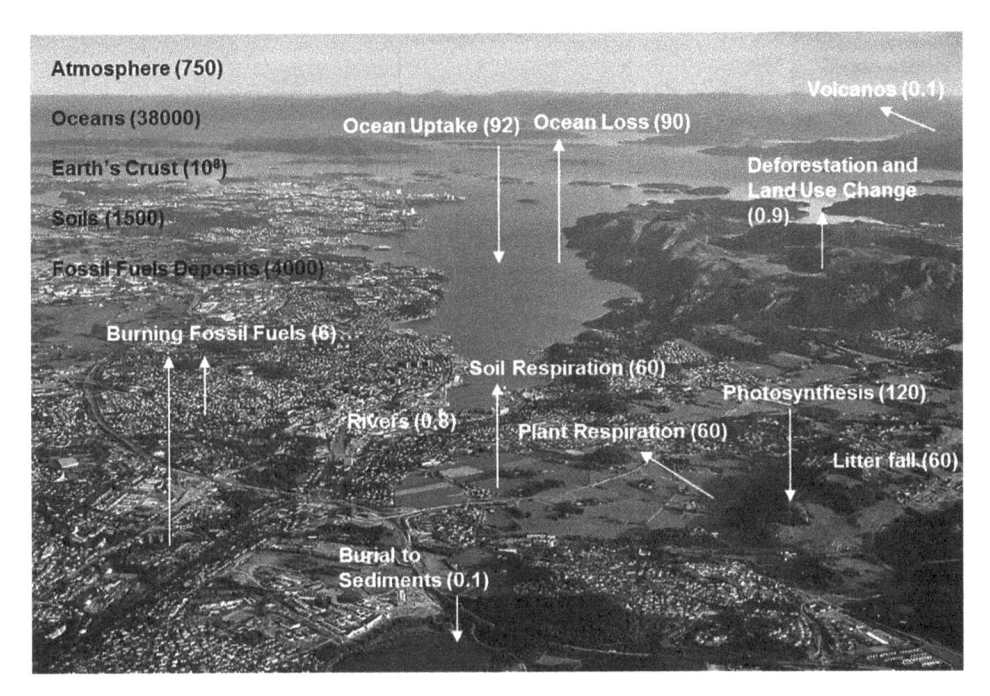

Figure 2.5 Global carbon cycle. Pools in black (Pg). Fluxes in white (Pg/year).

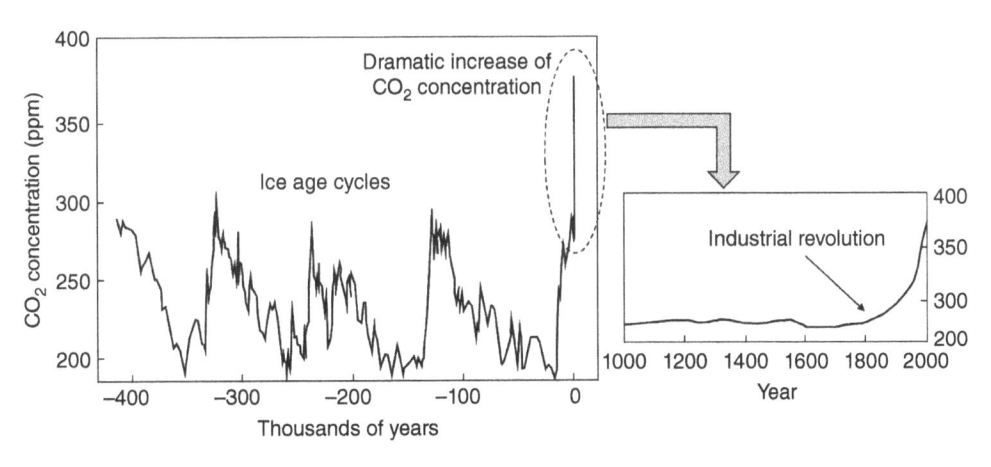

Figure 2.6 Influence of CO_2 emissions in the average temperature of Earth. Source: IPCC (2014).

the increase of temperature because of the massive CO_2 emissions to the atmosphere dating from the onset of the industrial revolution.

Many resources have been devoted to ongoing study of greenhouse gases and resulting climate change. The IPCC was established by the United Nations in 1988. The IPCC provides an assessment and report on global climate change from 1990 when the first report was published. In 2014, the Fifth Assessment Report (FAR) was issued (IPCC 2014). Utilizing input from thousands of scientific experts and hundreds of authors, the FAR provided convincing evidence that emissions-related human activity is already having a severe impact on global climate and sea

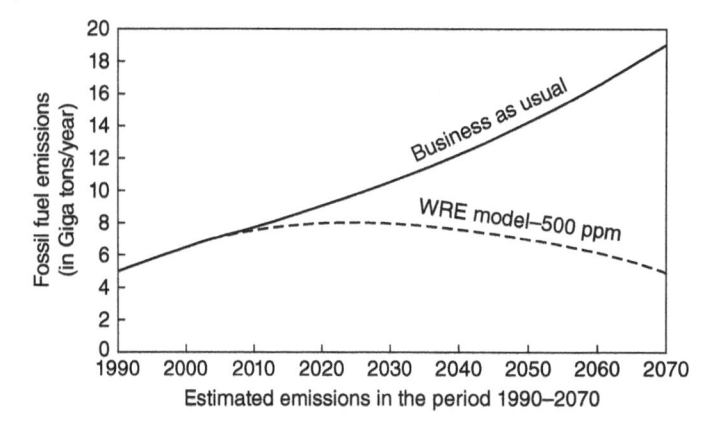

Figure 2.7 Emissions of greenhouse gases due to the utilization of fossil fuels. (WRE: Wigley, Richels, and Edmonds model for "stabilization at 500 ppm.") Source: Pacala and Socolow (2004).

level. Specifically, the IPCC concluded that "An increasing body of observations gives a collective picture of a warming world and other changes in the climate system." Also, "There is new and stronger evidence that most of the warming observed over the past 50 years is attributable to human activities."

Predictions of future impacts are difficult to make and involve wide ranges due to variations in the mathematical models. Figure 2.7 depicts two models and their predicted fossil fuel emissions for the next 50 years. Regarding the effects of these continued emissions, by 2100, global temperatures are predicted to increase between 2.5 and 10.4 °F, with temperatures expected to increase even more in the United States (NCA 2014).

Increased global temperatures can lead to drastic shifts in weather patterns, with intense precipitation and storm events in some areas and less in others (Figure 2.8). These events can drastically affect agricultural activities and induce strains on ecosystems and biodiversity. Furthermore, increased temperatures will likely result in accelerated melting of polar ice cap and alpine glaciers, leading to increases in sea levels, which in turn can inundate coastal areas. These climatic effects may have drastic impacts on several economic and resource sectors, including agriculture, water resources, energy, transportation, public health, and infrastructure. Specific climate effects include the following:

- *Precipitation* – Many areas will experience increased precipitation, including more days with precipitation, more intense events, and numerous storm events.
- *Snow cover* – With increases in temperature, the snow pack and water holding capacity decreases. Snow seasons are shorter, precipitation events consist of rain as opposed to snow, and runoff patterns and timings are altered.
- *Tropical storms* – Increased number of more intense tropical storms are expected to make landfall, as depicted in Figure 2.8c. The Atlantic and Gulf Coasts as well as Hawaii are susceptible to hurricanes. Additionally, one-fourth of US crude oil production is situated in the hurricane-prone Gulf of Mexico (EIA 2017a).
- *Flooding/drought* – Altered precipitation patterns will trigger more frequent and severe flooding in some regions and droughts in others. Approximately 3800 towns and cities of over 2500 inhabitants are located in floodplains, and increases in flooding will lead to increased damage to infrastructure and property, which is already averaging approximately to be $6 billion in annual losses (Hallegatte et al. 2013).

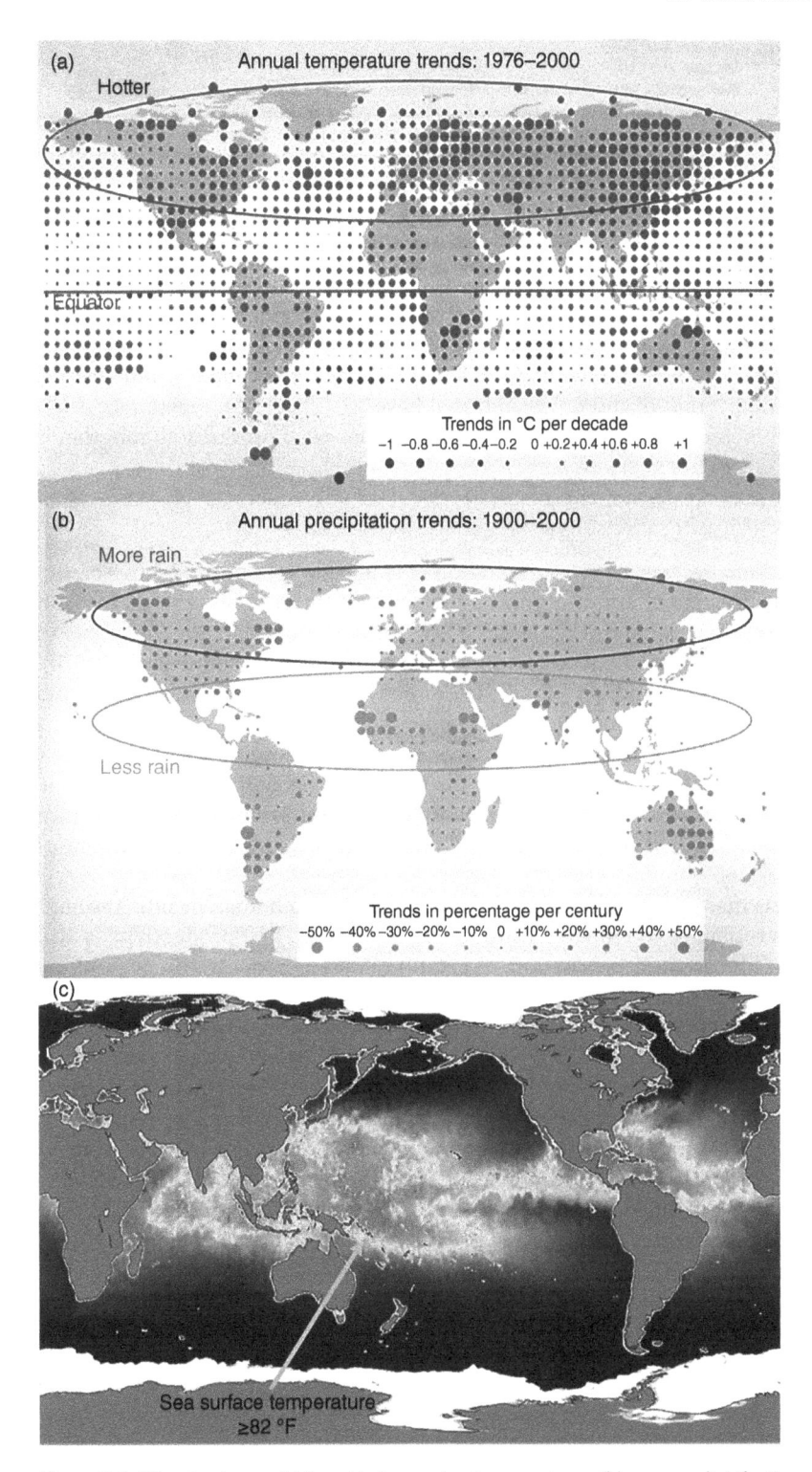

Figure 2.8 Climate change (a) trend in increasing temperatures, (b) uneven distribution of rains, (c) sea-level temperature that supports the formation of tropical storms. Source: IPCC (2014).

2.3 Desertification

Dry lands – defined as areas with arid, semiarid, and dry subhumid climates – constitute approximately 40% of the Earth's land surface. These regions play a vital role for civilization in that they provide much of the world's arable land and rangeland for livestock. However, nearly 70% of these regions are under stress through land degradation, or desertification, which reduces productive use of these regions and ultimately decreases food security. The degree of land degradation falls into one of the following four categories (Lal et al. 2003):

- *Light* – Lands that are mildly salinized or eroded; rangelands that contain approximately 70% native vegetation.
- *Moderate* – Lands that have been moderately eroded or salinized, soil chemical and physical integrity has been compromised, and rangelands that have 30–70% native vegetation.
- *Severe* – Lands that include frequent gullies or hollows; poor crop production; rangelands that have less than 30% native vegetation.
- *Extreme* – No crop growth occurs, and restoration is impossible; damage may have been caused by extreme erosion or salinization.

Land degradation occurs because of overexploitation of land resources or productive capacity. Overcultivation, deforestation, and overgrazing are some of the common causes for land degradation. Soils susceptible to land degradation in dry lands are particularly vulnerable to erosion, chemical and physical deterioration of the soil structure (Blanco-Canqui and Lal 2010).

Unfortunately, more than 250 million people are directly affected by the effects of land degradation, and 1 billion people in more than 100 countries are at risk. Many of these regions include some of the world's poorest and most economically disadvantaged peoples. For instance, in Africa, land degradation is a constant threat to physical and economic survival. Recurrent drought conditions exacerbate ongoing soil degradation problems, which in turn magnify the effects of subsequent droughts, both of which magnify the effect and extent of famines (Reynolds et al. 2007).

Land degradation has many consequences to the environment and human health. The ability of the land and its resilience to climatic disturbances is reduced. Soil productivity is also reduced, resulting in reduced food production. Vegetation cover is reduced, often resulting in the replacement of edible plants with nonedible plants. Further, the detrimental effects of reduced vegetation cover increase susceptibility to erosion, which often results in flooding, increased surface runoff, reduced water quality, sedimentation of water bodies, and siltation of reservoirs and navigable channels (Reynolds et al. 2007). Health problems may also be exacerbated; wind-blown dust can result in chronic or acute respiratory problems, eye infections, allergies, and mental stress.

The UN Convention to Combat Desertification (UNCCD) was enacted by the United Nations in 1996 to address the issue of land degradation (UNCCD 2017). The Convention established at the national level addresses the underlying causes of desertification and drought. The convention also seeks to identify and implement preventative measures that helps to stop and even to invert the general trend of degradation of productive soil and desertification of land.

2.4 Deforestation

Deforestation is a form of land degradation that is occurring globally at an alarming rate. During the 1990s, more than 110 million hectares (or 11 million hectares/year) of forested lands were lost. Much of this loss occurred in developing countries. Unfortunately, about 45% of the Earth's original forests are now gone (Thiele 2016), and deforestation continues to occur mainly in tropical and subtropical forest.

When reviewing these statistics, it would seem logical that the cause of this deforestation is unsustainable timber harvesting. However, this is not the case – the primary causes of deforestation are related to attempts to create agricultural land by clearing forests as well as harvesting to gain wood for fuel. Approximately half of the wood harvested in the world is used as fuel or converted to charcoal, mostly in developing countries (Chidumayo and Gumbo 2013). On the other hand, harvested timber in developed countries is primarily used for industrial uses and construction materials.

Ironically, deforestation to create agricultural land in developing countries is often counterproductive. For instance, soils in tropical or coniferous forests are relatively nutrient-poor and often do not provide prime agricultural land when cleared. The original forests often provided better economic capacity as well as better overall food security than the resulting clear land. Deforestation also reduces biodiversity and eliminates habitat for fauna that often could serve as a sustainable economic resource.

The effect of deforestation goes beyond the destruction of wildlife habitat. When trees are removed, the water absorption and retention capacity of soils are often reduced, contributing to increased surface water runoff and reduced water infiltration. In turn, groundwater recharge is reduced, and aquifer depletion may result. Because groundwater is the primary freshwater source for one-third of the world's population, this reduction in capacity can have disastrous consequences for agriculture, economy, and human well-being (Thiele 2016). Increased surface runoff also increases land degradation effects, including erosion of soils, which further degrades the agricultural capacity of these lands leading to decreased water quality and increased sedimentation in nearby waterways. This, in turn, has a significant secondary effect on the host ecosystem as well as economic prospects, since aquatic life can be adversely affected, leading to reduced hatcheries and reefs that are supportive of aquatic life and related commerce. Further, when water infiltration is reduced, areas of standing water may increase. In tropical areas, this can lead to increased malaria transmission, which annually claims the lives of 750 000 children under the age of 5 (Austin et al. 2017). As with land degradation, strategies to combat deforestation have been enacted. Efforts have been taken on an international level with the formation of the Intergovernmental Panel on Forests (IPF) and its successor, the Intergovernmental Forum on Forests (IFF).

2.5 Loss of Habitat and Biodiversity

The previous section outlined several primary and secondary effects resulting from deforestation. Land degradation and deforestation not only result in the loss of economically valuable natural resources within an ecosystem, they also have a significant adverse effect on other plant and animal species. With the elimination of forests, supportive habitats for a rich diversity of plant and animal species can be irrevocably harmed. As an example, current data suggests that the Earth's primary tropical forests can be essentially eliminated within four to five decades. It is well known that these forests support the widest range of species diversity on the planet. With increased habitat reduction or outright destruction, recent studies have indicated that rates of extinction will continue to accelerate in the future (Ceballos et al. 2015). As these forests continue to be eliminated, it is estimated that 0.2–0.3% of all species within these forests become extinct on an annual basis (Barlow et al. 2016). These percentages are believed to translate to the loss of 4000–6000 species annually, which is approximately 10 000 times the natural rate of species extinction.

These reductions in biodiversity are occurring because of increased conversion of open space to agricultural land or urban use, overharvesting of plants and animals, draining and filling of wetlands, overapplication of agricultural chemicals, overfishing, and air pollution. In 2005, the Millennium Ecosystem Assessment was released, which highlighted a substantial and irreversible loss of biodiversity on Earth. The assessment concluded that approximately 10–30% of bird, mammal, and amphibian species are threatened with extinction due to human activities (MEA 2005). Broken down by species classification, this includes the following:

- One out of eight bird species
- One out of four mammal species
- One out of four conifer species
- One out of three amphibians
- Six out of seven marine turtles.

As a species becomes endangered, its breeding capacity is greatly diminished, and its available gene pool is significantly reduced. This, in turn, reduces its capacity to adapt to environmental changes. If left unchecked, this cycle continues until the species becomes extinct. When a species is lost completely, its genetic material is lost as well. The elimination of the species from an ecosystem can have a significant effect, as its elimination from the food chain/food web can significantly upset the balance of other species, either due its elimination as a predator or prey. Aggressive or invasive species can quickly evolve into a dominant role as these natural checks and balances are removed. This can result in a chain reaction of effects that can be of a global magnitude.

While the reduction of biodiversity has an obvious impact on ecosystem and environmental health, it also results in significant adverse impacts to human health. Disruptions to the food chain can directly or indirectly affect plant-based and animal-based food sources. Additionally, potential or existing sources for medicines and other pharmaceutical products can be lessened or eliminated. In response, communities and governments worldwide have increasingly focused on the protection of habitats and biodiversity by focusing on protection of air and water, maintenance of soil fertility, surface water management, waste treatment, and waste reduction (Clark et al. 2014).

The benefits of biodiversity are many. With respect to the ecosystems, healthy biodiversity and habitat are directly linked to water resource and soil protection, pollution breakdown and absorption, contributions to climate stability, and the ability of an ecosystem to recover from natural disaster or other rare/unforeseen events. Biodiversity also leads directly to beneficial economic output in the forms of food, medicinal sources, timber, ornamental plants, and a diverse gene pool that can benefit wild and domesticated species. Finally, these factors also contribute positively to culture and society, with increased opportunities for research and education, recreation and tourism, and a reinforcement of cultural values (Clark et al. 2014).

Biodiversity is becoming increasingly important at an intergovernmental level. The United Nations Convention on Biological Diversity (UNCBD) was established in 1992 and has been ratified by more than 175 countries (CBD 2014). The convention set forth three main goals: the conservation of biodiversity, the sustainable use of its components, and the fair and equitable sharing of the benefits from the use of genetic resources. The convention also established in 2000, the Cartagena Protocol with the goal of protecting the planet's species and ecosystems from potential risks, from modified organisms, and to establish an advanced informed agreement procedure. Thus, countries are provided with the necessary information to make informed decisions before agreeing to the import of these organisms. The protocol was a landmark agreement because it was a first use of a formal "precautionary approach" as set in the 1992 Rio Declaration on Environment and Development (Rio Summit) (UNCED 1992).

2.6 Ozone Layer Depletion

As discussed in Section 2.7, the presence of ozone near the ground surface is a form of air pollution. The presence of ozone in the stratosphere, however, protects lifeforms from the deleterious effects of ultraviolet (UV) light. UV light, adjacent to the visible light spectrum, is invisible to the naked human eye and is highly energetic, especially UV-C light. Stratospheric ozone blocks the penetration of UV-C, and reducing the incidence of UV-B by 95%, and UV-A by 5%.

Ozone is formed in the stratosphere when UV light splits O_2 molecules, resulting in free oxygen (O). The free oxygen may quickly react with O_2 molecules to form O_3 (ozone) molecules. Figure 2.9 depicts the occurrence of ozone in the atmosphere as a function of altitude above the Earth's surface.

Ozone levels fluctuate with the seasons; concentrations are highest in the northern hemisphere during summer due to the trapping of these molecules by NO_2 and methane, and high concentrations are present near the equator (Harris et al. 2015). This "thinning" is further exacerbated by the presence of chlorofluorocarbons (CFCs) in the atmosphere. Rowland and Molina (1974) concluded that CFCs could damage ozone due to the release of chlorine atoms. CFCs are degraded by UV light, resulting in the generation of free chlorine. The free chlorine, in turn, reacts with ozone to form chlorine monoxide and molecular oxygen. In 1985, noticeable thinning (a 50% reduction) of ozone was detected over the South Pole. Cold temperatures during the Antarctic winter and the vortex in the atmosphere induced by the Earth's rotation facilitate the accumulation of chlorine over Antarctica. Fortunately, summer air replenishes ozone, somewhat reversing the cycle (Abbasi and Abbasi 2017). Nevertheless, the ozone hole has grown to the size of North America, and temperate regions in the southern hemisphere are being affected.

Several human health impacts have measurably increased with impacts to the ozone layer. Some of these include the increased occurrence of melanoma and nonmelanoma skin cancers, acceleration of eye cataracts, and reduced immune system effectiveness. As an example, two out of three people in Queensland, Australia are expected to develop skin cancer (Lane 2015) due to exposure to UV light not blocked by the absent ozone. Additionally, increased UV-B

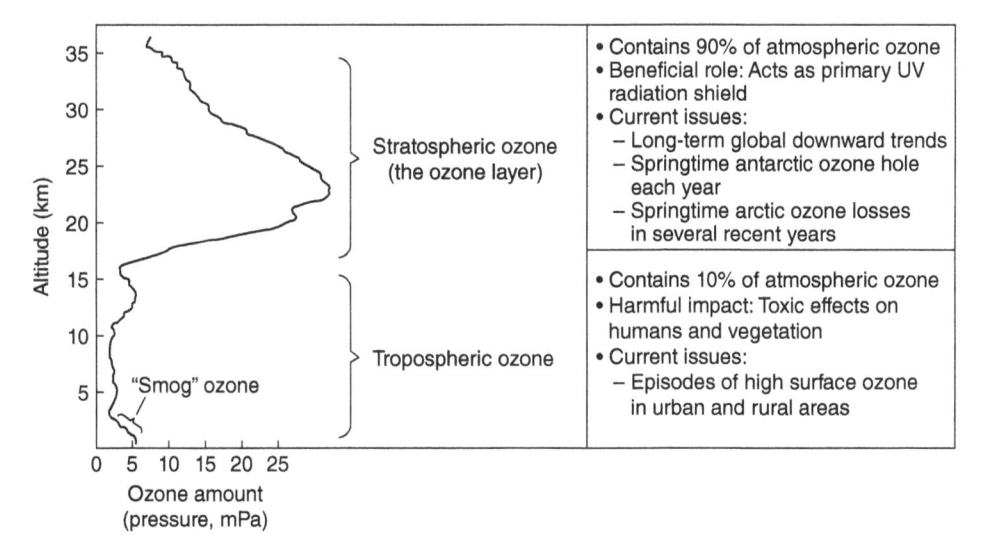

Figure 2.9 Occurrence of ozone in the atmosphere. Source: http://www.ozonelayer.noaa.gov/science/basics.htm, March 30, 2017.

exposure can affect plant growth and crop yields as well as damage the aquatic ecosystems. Although ozone-depleting substances (ODS) concentrations in the atmosphere peaked in 1994, the concentrations have been gradually decreasing due to worldwide efforts to curb the use of CFCs. However, the risks to the ozone layer and associated health impacts from ozone layer depletion persist (Vingarzan 2004).

Because of their stability, halogenated hydrocarbon compounds have been often used as coolant compounds in refrigerators, air conditioners, and heat pumps. They have also been used in the manufacture of plastic foams. Due to their nonreactive nature, they were also used as propellants in aerosol applications. Because of their destructive effect on stratospheric ozone, CFC use in aerosols was banned in 1978. While manufacturers of these products switched to alternative compounds, CFC use has continued in other products. In 1987, UN member countries met in Montreal, leading to the formation of the Montreal Protocol. A total of 194 nations agreed to scale back the use of CFCs by 50% by 2000 (UNEP 2017). Subsequent amendments moved the total phase-out of CFCs to 1996 (Prather et al. 1996). Additionally, other similar compounds have dangerous effects on the ozone layer. For instance, bromine compounds, commonly used in soil fumigants and pesticides, are 60 times as potent as chlorine with respect to ozone depletion (Burkholder et al. 2015).

Although CFCs are still in use to some extent, some substitutes have been developed. For instance, hydrofluorocarbons (HFCs) do not contain chlorine, and fluorine does not pose a threat to the ozone layer. These measures have positive effects; although ozone-depleting substances (ODS) concentrations in the atmosphere peaked in 1994, the concentrations have been gradually decreasing due to worldwide efforts to curb the use of CFCs, and it is expected that pre-1980 levels will be reached by 2050 (Ennis 2017). However, the risks to the ozone layer and associated health impacts from ozone layer depletion persist (Vingarzan 2004).

2.7 Air Pollution

The atmosphere contains a range of naturally occurring gases – nitrogen, oxygen, carbon dioxide, argon, and water vapor – plus numerous trace gases, including ozone, helium, hydrogen, and nitrogen oxides. Naturally occurring aerosols – microscopic liquid or soil particles resulting from the entry of dust, pollen, sea salts, and other sources – are also present in the atmosphere. However, numerous pollutants, many of these from anthropogenic sources, have been emitted to the atmosphere, resulting in a range of potential harmful effects.

Air pollutants can be classified as primary or secondary pollutants. Primary pollutants are the direct by-products of combustion or evaporation emitted to the atmosphere; some include volatile organic compounds (VOCs), particulates, carbon monoxide, nitrogen oxides (NO_x), sulfur dioxide, and lead. Secondary pollutants are the result of the reactions of primary pollutants with other compounds within the atmosphere. Some of these include ozone, pero-oxyacetyl, nitrates, sulfuric acids, and nitric acids.

The presence of air pollutants can result in a range of adverse effects to human health. Acutely elevated concentrations of air pollutants can be lethal to people suffering from cardio-pulmonary diseases. Even moderate levels can affect cardiac functions, affecting those with heart disease. Several common pollutants can also have chronic effects. For example, sulfur dioxide and ozone can lead to irritation and/or damage to the lungs and/or breathing passages. Particulate matter can lead to respiratory and cardiovascular diseases. Nitrogen oxides can impair lung function and affect the immune system. Carbon monoxide reduces the oxygen-carrying capacity of blood. Acute concentrations of carbon monoxide can be fatal, but even chronic lower concentrations can lead to heart disease.

Some air pollutants are also carcinogenic. Benzene, a common VOC present in motor fuels, solvents, other industrial chemicals, explosives, and smoke, is clearly correlated to several types of cancers, including leukemia, blood disorders, and immune system disorders. Some diesel fuel-related emissions are also considered likely carcinogens.

Lead is an example of an anthropogenic air pollutant with significant human health consequences. Lead was historically used as a common additive to gasoline for anti-knock purposes. With its presence in leaded gasoline as a primary source, potential exposure pathways include inhalation of lead-containing emissions. Additionally, subsequent deposition of airborne lead (aerially deposited lead) resulting from combustion provides an exposure pathway through ingestion or dermal contact. Lead poisoning can have serious health consequences, including learning disabilities in children and hypertension in adults. Although lead was gradually phased out of gasoline in the 1970s and 1980s, elevated lead concentrations in blood persisted in children and adults through the 1980s (Annest et al. 1983). Fortunately, the EPA mandated the elimination of leaded gasoline by 1996, leading to dramatic reduction of lead concentrations in the environment (Dapul and Laraque 2014).

Air pollutants can directly affect vegetation. Large-scale sulfur dioxide emissions have been linked to widespread destruction of vegetated areas. Because of its ability to affect carbon and nutrient plant metabolism, ground ozone emissions can reduce or destroy orchards, cereal crops, or forests. Areas in California and Appalachia that are susceptible to increased ozone concentrations have experienced direct tree damage or secondary damage as weakened trees become more susceptible to pests and wildfires.

Common air pollutants can also have a detrimental effect on structures and the built environment. Particulate matter can discolor building façades and windows. Ozone emissions react with rubber, leading to degradation and deterioration. Sulfur oxides and nitrogen oxides corrode metals and deteriorate stonework and masonry. While many materials can be cleaned or repaired, the effect on many historic statues and architecture/artworks are beyond repair.

With increased deterioration of air quality following the end of World War II, by the 1960s, it was becoming more obvious that increased emissions were having a deleterious effect on the atmosphere. Increased public awareness ultimately led to governmental regulation of air pollution. In 1970, Congress passed the Clean Air Act. This landmark legislation (amended on several occasions) serves as the foundation of US anti-air pollution efforts. Ambient standards were set for criteria pollutants, and control measures were established. Some criteria pollutants include sulfur dioxide, carbon monoxide, nitrogen oxides, ozone, and lead. The National Ambient Air Quality Standards (NAAQS) set primary standards for short-term and long-term conditions such that these pollutants would not harm human health (Brunekreef and Holgate 2002). Table 2.2 provides a summary of these standards.

The Clean Air Act requires the EPA to review pollutants every five years, and adjustments are made with advances in scientific knowledge. Additionally, pollutants have been added to the regulatory environment. For instance, in 1980, National Emission Standards for Hazardous Air Pollutants (NESHAPs) were established, requiring tracking and regulation of 187 toxic pollutants (Gerber 2015). The National Emission Inventory was established, which consists of a database of criteria and pollutants that are regulated.

The Clean Air Act has been very useful in addressing air pollution. The regulatory philosophy is to keep criteria pollutant emissions below primary standard levels, leading to an overall improvement in public health. Emitting industries are required to implement controls to address their polluting emissions. Additionally, regions in violation are required to meet compliance. However, enforcement can be difficult, as cities and states are often unwilling to enforce or control emissions to achieve compliance with the air quality standards.

Table 2.2 EPA has set National Ambient Air Quality Standards for six principal pollutants (USEPA 2017f).

Pollutant	Primary/secondary	Averaging time	Level	Form
Carbon monoxide (CO)	Primary	8 h	9 ppm	Not to be exceeded more than once per year
		1 h	35 ppm	
Lead (Pb)	Primary and secondary	3-month rolling average	0.15 µg/m^3	Not to be exceeded
Nitrogen dioxide (NO$_2$)	Primary	1 h	100 ppb	98th percentile of 1-h daily maximum concentrations, averaged over 3 yrs
	primary and secondary	1 yr	53 ppb	Annual mean
Ozone (O$_3$)	Primary and secondary	8 h	0.070 ppm	Annual fourth-highest daily maximum 8-h concentration, averaged over 3 yrs
Particle pollution (PM) PM$_{2.5}$	Primary	1 yr	12.0 µg/m^3	Annual mean, averaged over 3 yrs
	Secondary	1 yr	15.0 µg/m^3	Annual mean, averaged over 3 yrs
	Primary and secondary	24 h	35 µg/m^3	98th percentile, averaged over 3 yrs
PM$_{10}$	Primary and secondary	24 h	150 µg/m^3	Not to be exceeded more than once per year on average over 3 yrs
Sulfur dioxide (SO$_2$)	Primary	1 h	75 ppb	99th percentile of 1-h daily maximum concentrations, averaged over 3 yrs
	Secondary	3 h	0.5 ppm	Not to be exceeded more than once per year

2.8 Smog

Several of the air pollutants discussed in Section 2.7 may form smog. Smog often takes one of two common forms. The first form is industrial smog, which is a combination of smoke and fog. Industrial smog consists of a grayish mix of soot, sulfur compounds, and water vapor. Commonly the result of coal burning, it is most prevalent in cold climates with industrialized areas. London was historically infamous for "London Fog" (actually industrial smog), but it is now common in China, India, Korea, and eastern European countries (Elsom 2014). The second common form of smog, photochemical smog, is primarily associated with vehicle exhaust in warm, sunny areas. Pollutants, such as VOCs and nitrogen oxides, interact with sunlight in warm climates to form a brown, irritating haze. The creation of photochemical smog from automobile exhaust as identified by Arie Jan Haagen-Smit of the California Institute of Technology is presented in Eq. (2.1):

$$\text{NMHC (nonmethane hydrocarbons)} + \text{NO} + h\upsilon \rightarrow \text{NO}_2 + \text{other products} \tag{2.1}$$

Nitrogen dioxide further reacts as shown in Eqs. (2.2) and (2.3):

$$NO_2 + h\upsilon \, (\upsilon < 410\,\text{nm}) \rightarrow NO + O^{\cdot} \quad (2.2)$$

$$O_2 + O^{\cdot} \rightarrow O_3 \quad (2.3)$$

The free oxygen reacts with molecular oxygen to produce ground ozone. Exposure to 0.1–1 ppm of ground ozone is an irritant as described above, leading to headaches, burning eyes, and irritation of breathing passages (Ebi and McGregor 2008).

As demonstrated by the reactions outlined above, photochemical smog can be controlled by limiting the concentrations of NMHCs and nitrogen oxides in the atmosphere. One successful method has been the widespread adoption of automobile catalytic converters. These devices, introduced in the United States in the 1970s, are designed to capture NO, CO, and unburned hydrocarbons from exhaust.

Several large metropolitan areas, including Los Angeles, Beijing, New Delhi, and Mexico City, have been afflicted with chronic smog issues for decades. On some occasions, conditions can exist where acute smog can form, resulting in significant health effects. A famous example occurred in Donora, Pennsylvania, in October 1948. A temperature inversion trapped industrial smog from coal-fired and sulfur-emitting industries. The smog outbreak lasted for several days. Within a month, at least 70 people died, and an estimated 6000 people were struck with a variety of illnesses (Magoc 2014). While local industries denied responsibility, public outrage at this event as well as others eventually leads to governmental intervention through legislation.

2.9 Acid Rain

Air pollution can also have secondary effects on public health and the environment. With the emission of sulfur dioxide and nitrogen oxides, sulfuric and nitric acids can form due to interaction with atmospheric moisture. This, in turn, can result in the deposition of these acids via acid rain, a phenomenon that affects broad areas of industrialized regions of the world.

Rainfall is normally slightly acidic (on the order of pH = 5.6). However, acid rain (defined as precipitation with pH less than 5.5) can be 10–1000 times more acidic. Rain and snow in eastern North America have been measured with pH ranging from 3.0 to 4.6. Mountain forests near Los Angeles have exhibited measurable pH as low as 2.8 (McNeil and Culcasi 2015).

Several natural sources of acid deposition occur. An estimated 50–70 million tons of sulfur dioxide are generated per year from volcanoes, sea spray, and microbial processes. Between 30 and 40 million tons of nitrogen oxides are generated per year from lightning, natural combustion of biomass, and microbial processes. However, significant anthropogenic sources also exist. Between 100 and 130 million tons of sulfur dioxide are emitted primarily from the combustion of coal. Between 60 and 70 million tons of nitrogen oxides are emitted yearly from vehicular fuel combustion sources. Much of these emissions occur in industrialized areas, but downwind areas are also affected (Vet et al. 2014). Fortunately, rates of emission are decreasing due to increased regulations. Although emissions have increased fourfold since 1900, deposition has decreased 33–35% in the past 15 years, as new technologies have been implemented and obsolete coal burning plants have been decommissioned. Nevertheless, these compounds are still generated, and acid rain generation will continue.

The deposition of these acids can have long-term impacts. For instance, consequences are still observed from emission/deposition that occurred decades ago. Acid precipitation can result in the leaching of toxic heavy metals from soil and rock, where they can be mobilized and become available for biological uptake in the environment. Acid deposition can also have a detrimental effect on aquatic environments, as lowered pH can adversely affect enzymes, hormones, and proteins of organisms. Because natural river and lake environments commonly have pH ranging from 6 to 8, deviations can impact the regulatory systems of organisms. Future generations may also be affected; small changes in pH can adversely affect egg and sperm cells (ApSimon et al. 2014).

It is important to note that natural systems have the ability to buffer, or neutralize acid deposition, and not all areas are harmed by deposition. Limestone as well as derived soils have an appreciable buffering capacity. Additionally, the Clean Air Act and related regulations have led to the reductions in atmospheric NO_x and SO_x concentrations. Specifically, Title IV of the Clean Air Act has led to significant reductions, which have occurred faster than expected. This approach has included mandates to introduce emissions-curbing technologies (e.g. scrubbers, use of low-sulfur coal) as well as free-market incentives that utilize emissions allowances. These market-based approaches have fostered technological innovations and provided incentives to reduce emissions while saving costs related to emissions allowances.

2.10 Water Usage and Pollution

Access to clean, fresh water sources is essential to human life. Protection of these resources is equally important. Only 3% of the Earth's water is fresh; the remainder is saline and unusable without some type of desalination measure. Of the available fresh water, about 30% comes from groundwater sources. Nearly 69% of fresh water is trapped in ice caps and glaciers. Only approximately 0.3% exists as surface water. Further, approximately 87% of surface fresh water is present in lakes, 11% exists in fresh wetlands and swamps, and approximately 2% is present in river environments (Figure 2.10) (Shiklomanov 2000).

Human fresh water usage has increased over the past century, as demonstrated in Figure 2.11. Fresh water is used for a variety of key human activities, including agriculture, industrial processes, and domestic uses. Consequently, sources of fresh water are under stress. For instance, drought-induced surface water depletion results in habitat loss, decreased aquifer recharge, impacts on municipal water supplies, and concentration of pollutants. Groundwater depletion due to unsustainable pumping rates and/or reduced recharge can affect private or public water systems, the drying of wells, deterioration of water quality, and land subsidence. Groundwater depletion is especially problematic in semiarid and arid regions of the midwestern

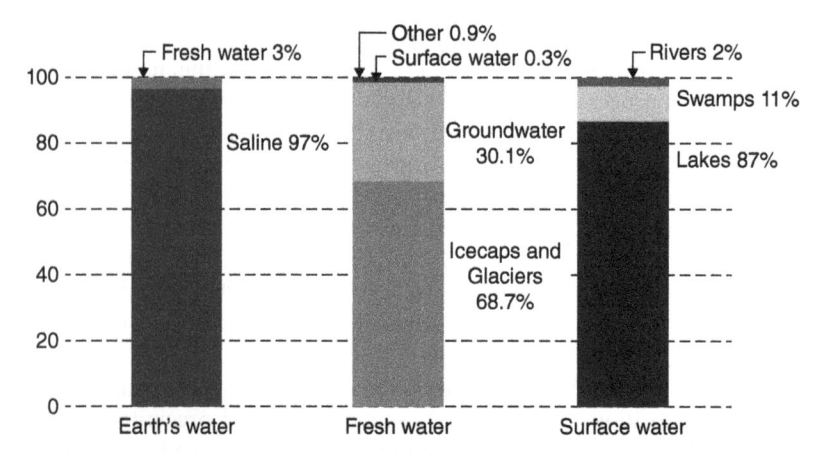

Figure 2.10 Distribution of water on the planet Earth.

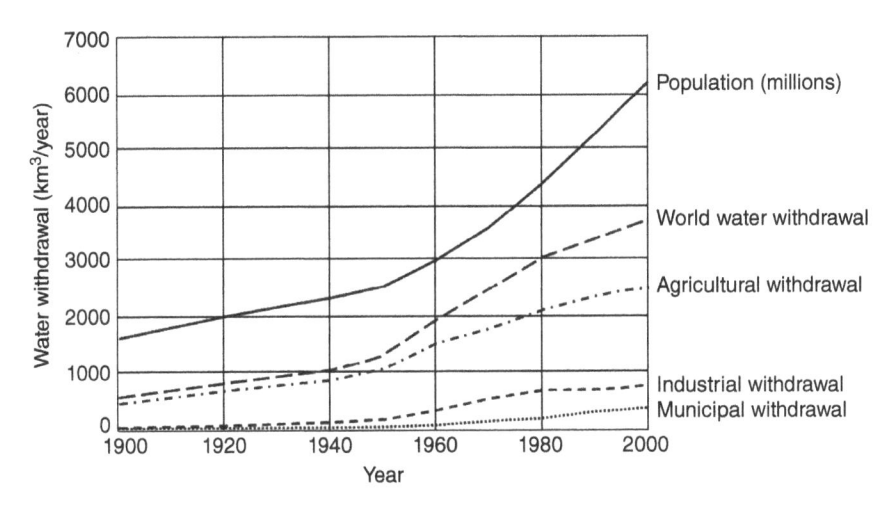

Figure 2.11 Freshwater use and population growing. Source: Data obtained from Shiklomanov (1998).

and western United States, where lowering of localized and regional water tables has been especially acute.

Water pollution can take many forms, and it can have significant environmental effects. The sources of pollution can be point-source and nonpoint source. Point-source pollution comes from a single location or system, such as industrial facilities, wastewater treatment facilities, and mines or other points of resource extraction/refinement. Because of their specific location, point sources are easy to identify, and therefore are relatively straightforward to monitor and regulate. Nonpoint sources consist of sources that do not have a specific point of discharge; rather, they often consist of discharge resulting from areal or regional runoff. Some examples include agricultural runoff, stormwater runoff from vegetated or paved surfaces, and runoff from construction sites. Nonpoint sources rarely feature the acute loading or toxicity of point sources; however, the volume from nonpoint sources can be substantial, and these discharges may be difficult to control and regulate. When dealing with point sources, it can be efficient to treat water before releasing into the environment, whereas nonpoint sources are most efficiently managed by reducing the runoff by some means.

Pathogens are one class of water pollutants. Bacteria, viruses, and parasites often found in human and animal excrement can carry disease. Infections from pathogens can result in widespread symptoms and health problems. Public health departments play an active role in minimizing exposure to pathogens, using measures such as purification and disinfection of public water supplies, sanitary collection and treatment of wastes, regulations pertaining to sanitary food preparation practices, and domestic and personal hygiene practices (Shannon et al. 2008).

Organic wastes are a second type of water pollutants. Organic matter can come from human or animal wastes and by-products, remnant food materials, and decaying plant matter. With the exception of plastic and other select synthetic materials, much of this matter is biodegradable. Bacteria and other organisms will utilize and consume these materials, but oxygen is utilized as well. Biological oxygen demand (BOD) is a quantitative measure of the relative amount of organic matter present in a given water body. Because the saturation limit of dissolved oxygen is relatively low (8–10 ppm), the presence of and subsequent utilization of excessive organic matter can lead to rapid depletion of dissolved oxygen.

Chemical compounds constitute a third category of water pollutants. These can be classified as inorganic chemicals, which include heavy metals, acids, or other compounds (including salts). Likewise, organic chemicals pose a threat, including industrial chemicals, such as solvents, polychlorinated biphenyls (PCBs), detergents, or pesticides. Many chemicals can be toxic to aquatic life at very low concentrations; furthermore, these compounds, if considered "bioaccumulative," can become concentrated and pose a threat to organisms in a more advanced position on the food chain. Additionally, higher concentrations can alter water chemistry, leading to secondary effects on aquatic life.

An acute example of chemical pollution is oil spills. These typically occur in marine or near-shore environments due to accidental releases from oil platforms, tankers, or pipelines. The spills may involve crude oil or lighter-phase refined products, such as gasoline or diesel fuel, as well as remnant heavier phase products, such as bunker oil. Cleanup and recovery can be difficult, expensive, and time-consuming; these efforts can be affected by temperature, weather conditions, currents, and shore topography/conditions. Spills can take months or years to clean up, and as a result drastic effects to aquatic life are quite common. Oil spills can be equally lethal to birds, especially seabirds, which are closer to the source of oil. Oil causes matting and separation of bird's feathers and exposure of their delicate skin to the extreme temperatures. This not only affects their ability to fly but can also be toxic when ingested during preening. Oil spills can also have a range of physiologic and toxic effects to fish and mammals.

Excessive nutrient loading, which commonly includes nitrogen, phosphorus, and potassium, can have significant effects. While these inorganic nutrients are essential for plant life, excessive nutrients in runoff, either from point sources such as wastewater plants with insufficient treatment) or nonpoint sources, including agricultural runoff, lawns, golf courses, can lead to algal blooms. Specifically, when excess nitrogen is introduced into a water body, growth of phytoplankton accelerates. Excess phytoplankton can lead to excessive growth of zooplankton as well. The consumption of phytoplankton by zooplankton, as well as decay of these microorganisms, can lead to an acceleration of oxygen consumption, depleting oxygen concentrations in water bodies. The resulting condition, hypoxia, can severely affect the balance of life in water bodies, and if unchecked, "dead zones" may develop. Dead zones have occurred in water bodies on a seasonal or ongoing basis (Friedrich et al. 2014).

A well-known example of dead-zone formation has occurred in the Gulf of Mexico near the Mississippi River Delta. The Mississippi River collects 40% of the continental United States watershed runoff. Much of this watershed is situated in agricultural intensive areas; so most of the runoff contains agricultural chemicals (including fertilizers) and biomass from animal wastes (Beaty 2015). Much of the runoff water and sediments are devoid of oxygen, and the Louisiana Gulf Coast has become susceptible to dead zone formation during the summer months. Because this area includes an abundant fishing ground valued at approximately $2.8 billion, the formation of dead zones can have a significant economic effect (de Mutsert et al. 2016).

To combat the potential for hypoxia occurrences, in 1998, Congress passed the 1998 Harmful Algal Bloom and Hypoxia Research and Control Act (NCOOS 2017). Further, following confirmation of the relationship of dead zone occurrences and elevated nitrogen concentrations, an interagency taskforce in 2000 identified options to reduce the occurrence and size of dead zones. Options for controlling this include, reduced fertilizer applications and means to restore natural denitrification processes. The latest actions recommend a 45% reduction in nitrogen and phosphorus runoff, but an action plan to reduce the hypoxic area by 50% by 2015 has not led to a reduction in the size of the dead zone (USEPA 2015).

Water pollution has affected many countries and civilizations over the generations. As an example, untreated chemicals and sewage were commonly dumped into waterways during the early stages of the Industrial Revolution. Not surprisingly, public health issues commonly arose as drinking and domestic water supplies were affected. Once sewage-borne pathogens were demonstrated to pose significant health risks, cities began to install rudimentary wastewater disposal systems. Unfortunately, these systems often only served to concentrate wastes, and many receiving bodies evolved into little more than cesspools. Although the public began to take notice, early regulations, such as the Federal Water Pollution Control Act of 1948, did little to combat growing water pollution. Following World War II and a rapid acceleration of industrial and chemical facility prowess, water pollution conditions greatly worsened. Many water bodies throughout the United States were greatly impacted. An acute example, and one that galvanized public outrage, occurred in 1969 when chemical pollution within the Cuyahoga River in Cleveland caught on fire (an event that had occurred several times in the past).

Partially inspired by this event, and with public outcry increasing, Congress enacted the Clean Water Act in 1972. By charging the USEPA with restoring and maintaining waters from a chemical, physical, and biological standpoint, the Clean Water Act has proven to be very effective.

Several other regulations and criteria have either been attributable to or have evolved since the passage of the Clean Water Act. A total of 167 substances have been identified as criteria pollutants based on some characteristics (e.g. toxicity, pH), and maximum allowable concentrations have been assigned and enforced. The concentrations may include Criteria Maximum Concentration (CMC), defined as the highest single or "acute" concentration beyond which impacts may occur; or as Criteria Continuous Concentration (CCC), the highest sustained or "chronic" concentration beyond which impacts may occur (USEPA 2012).

These criteria have been used to develop other standards. For instance, the Safe Drinking Water Act (SDWA) set forth maximum contaminant levels (MCLs) for 94 substances that may affect drinking water USEPA (2017a). Another example is the National Pollution Discharge Elimination System (NPDES), which addresses point-source pollution (USEPA 2017b). Permits are required for wastewater and industrial discharges. Total Maximum Daily Loads (TMDLs) are promulgated to address mainly nonpoint sources (USEPA 2017c).

2.11 Eutrophication

Hypoxia, as described above, is a serious form of water pollution. It is also one manifestation of eutrophication and is defined as an ecosystem response to the addition of natural or artificial substances, such as nutrients, to an aquatic system (Farley 2012). The additions may come from substances associated with anthropogenic activities, such as wastewater or fertilizer-laden runoff, due to runoff from urban or agricultural activities. It also can occur as a natural process, called natural eutrophication, due to accumulation of materials in a depositional environment or from ephemeral flows. Natural eutrophication is a natural succession process within water bodies, such as lakes and ponds.

Typically, eutrophication results in the excessive growth of plants. Although this leads to an increase in oxygen levels, subsequent decay of dead plant matter results in decreases in oxygen levels. Additionally, the stimulation of phytoplankton and zooplankton results in a rapid reduction in water aesthetics and water quality. An increase in these organisms leads to an increase in turbidity, which in turn decreases the depth of sunlight penetration into a given water body. As a result, benthic (bottom-rooted) plants as well as submerged aquatic vegetation cannot receive the necessary sunlight to survive. Therefore, oligotrophic lakes, which are home for a wide range of aquatic life and exhibit several attractive qualities with respect to aesthetics, can be overtaken by phytoplankton and zooplankton. With the growth and decay cycles of these materials, oxygen levels decrease, turbidity increases, and plants and animals are "choked out" as well as poisoned by secreted toxins. A eutrophic condition occurs in the lake, in which it becomes an unappealing body of algae and devoid of other forms of life.

Eutrophication can be slowed or reversed through a range of intervening activities. Active treatments can be applied to specific bodies of water. Herbicides can be applied to control phytoplankton or submerged aquatic vegetation to manageable levels. Of course, these can be toxic to other life forms, and the decay of the resulting dead matter will also consume oxygen. Plant matter may be harvested to maintain a balance, or aerators may be installed to increase dissolved oxygen levels. Water levels may also be drawn down to control plant matter. However, these are temporary measures, as plants will inevitably grow back.

Eutrophication may be controlled on a larger scale by limiting the amount of nutrients that can migrate into water bodies. The NPDES permit process has proven to be effective in reducing point-source pollution. Additionally, controls placed on some materials, such as phosphates, as well as monitoring of nitrogen and phosphorus, have proven effective in limiting nutrient runoff. The TMDLs described in earlier sections have been established to limit pollutant loading into water bodies. Best management practices (BMPs), often consisting of policies or intervention technologies, can also be applied to intercept, treat, and minimize loading from runoff.

With an effective intervention of various authorities, the effects of eutrophication on water bodies can be limited or recovery may be induced. Lake Washington, situated to the east of Seattle, had become an unpleasant water body covered with blue-green algae during the 1940s and 1950s due to the discharge of 20 million gallons per day of treated wastewater. Following the efforts of concerned citizens, wastewater flow was diverted away from the lake, and gradually, the lake was fully recovered by 1975 (Cooke et al. 2016). Citizens all around the United States have acted to protect similar water bodies.

2.12 Salinity

Salinity is defined as the degree of dissolved salt (such as sodium chloride, magnesium and calcium sulfates, and bicarbonates) present in water or soil. It is often reported in units of parts per thousand (grams per kilogram) as well as milligrams per liter (parts per million). Plants (called halophytes) and microbes such as bacteria (called halophiles) can adapt to live in saline conditions. Organisms that can tolerate a wide range of salinity conditions are referred to as euryhalines.

The degree of salinity is believed to contribute to global concentrations of carbon dioxide. Carbon dioxide is less soluble in highly saline water (Sunda and Cai 2012). Additionally, saline intrusion is a critical problem in estuary. As global sea levels rise, more saline waters flow back into estuaries, which can affect the salinity of surface water and groundwater that reside in these locations. This, in turn, can greatly affect aquatic life, especially those not adaptable to saline conditions.

2.13 Wastes and Disposal

Waste generation and disposal is a constant concern with respect to the environment. The proper handling and disposal of waste, as well as its potential negative effects on the environment, is highly dependent on the specific type of waste. Several categories of waste exist based on its specific origin and characteristics, including municipal solid waste (MSW), hazardous waste, e-waste, medical waste, and radioactive waste.

MSW consists of the wastes generated by households and small businesses. It differs from hazardous waste and nonhazardous industrial waste (e.g. construction/demolition waste, agricultural waste, mining waste, and sewage sludge) in that these materials are regulated more stringently. MSW generation in the United States is generally increasing, both on a total basis and a per capita basis. For instance, in 1960, the average person generated 2.7 lb of waste per day. By 2007, this had increased to 4.6 lb of waste per day (USEPA 2016). These wastes vary in composition by region and season. Figure 2.12 shows the relative contribution of materials to the MSW waste stream. An increasing portion of this waste stream is diverted to recycling applications. The remainder of the waste stream is landfilled or incinerated. Unfortunately, for several reasons, it is getting progressively more difficult to safely manage and dispose of these wastes in an appropriate manner.

MSW is collected by local government agencies and/or their agents. In some cases, local governments directly collect and transport wastes using their own workers and equipment. In other cases, they outsource the work to private firms. One of the three fee models may be used to pay for waste collection – taxes, direct service billing to households/businesses, or "pay as you go" tipping fees.

In the past, waste disposal was commonly performed in an uncontrolled or unregulated manner with little regard for effects to the environment. MSW was commonly burned or buried in uncontrolled dumps. Uncontrolled burning contributed to air pollution, and open dumps had serious effects on soil, air, and water quality. Because of growing public pressure, these uncontrolled methods were phased out, with engineered landfills becoming a more preferred solution.

Landfilling of wastes has been a common waste disposal method for generations. In the past, uncontrolled dumping was commonly performed outside of populated areas within natural depressions (wetlands, gullies, and abandoned quarries), where land was cheap and where nuisance conditions (odors, vectors, etc.) would not affect nearby residents. Techniques were gradually developed to deal with these and other serious environmental side effects. As an example, soil cover was placed on waste dumps on a daily basis to help suppress odors and

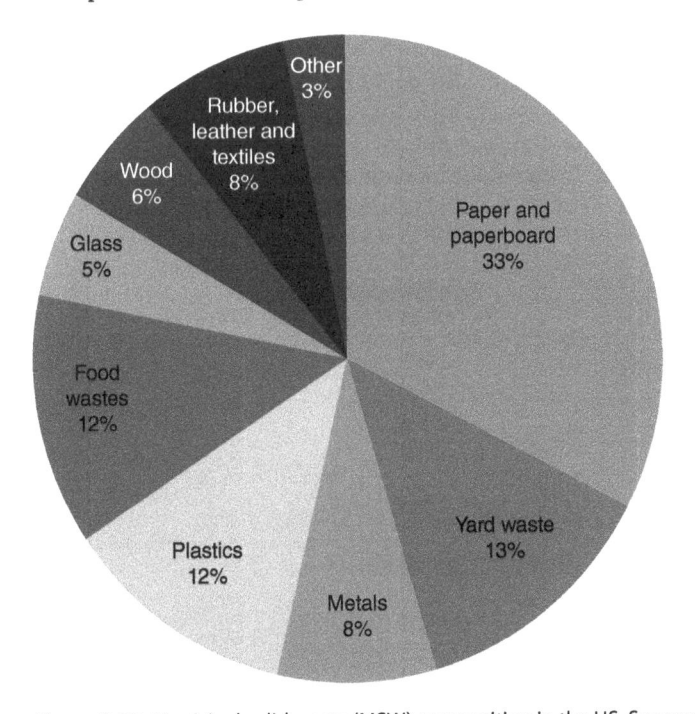

Figure 2.12 Municipal solid waste (MSW) composition in the US. Source: Data from USEPA (2016).

for vector control. As landfilling practices continued to evolve, engineered covers and liners were developed to control infiltration as well as exfiltration of liquids. Gas capture systems were also developed to control nuisance and/or dangerous gases developed within the landfill. Figure 2.13 depicts a typical engineered landfill cross-section.

Many of the engineering advances in landfill design, construction, and operation came about because of the side effects and "by-products" that are generated during the life of a landfill. One by-product is leachate generation. Leachate is a liquid formed within landfills from infiltrated rainwater combined with waste-derived liquids present within the landfill. Because it is commonly acidic, it can mobilize pollutants from solid waste materials. Leachate commonly includes a wide range of pollutants, including heavy metals, organic chemicals, and organic matter. If left uncontrolled, leachate can enter and migrate into groundwater, polluting domestic water sources. Leachate-based groundwater pollution is a common problem throughout the United States and is commonly associated with old uncontrolled and/or abandoned landfill/dump sites without leachate collection systems or liners in place. Fortunately, leachate collection systems, consisting of permeable drainage layers, polymer liners, and drainage tiles, can effectively be used to prevent unwanted leachate migration. Monitoring wells are also typically installed to confirm that groundwater has not been polluted.

In addition to noxious liquid generation, decomposition of refuse within landfills results in the generation of prodigious amounts of biogas, which primarily consists of methane, carbon dioxide, and other nonmethanogenic organic compounds (Themelis and Ulloa 2007). Biogas is highly flammable and mobile – if not controlled; it can migrate and enter structures, presenting an explosion hazard. Additionally, it can be harmful to plant life and can kill vegetation if it reaches the surface. Fortunately, biogas is also potentially valuable – if captured and purified, can be used as a fuel. As an example, in 2008, commercial landfill gases were used to provide natural gas and electricity for 1.4 million homes (DOE 1999). Because this landfill operation by-product can be used to offset the use of fossil fuels, resources may be preserved, and greenhouse gas emissions may be reduced.

Biogas is generated during decomposition of wastes. However, the process of decomposition of waste is not always uniform, some wastes are readily degradable, some may take a longer time, and some may not be degradable at all. For example, plastics resist decomposition, and even biodegradable polymers (such as those derived from cornstarch, cellulose, lactic acid, or soybeans) can take decades to decompose. Even readily biodegradable materials may take years

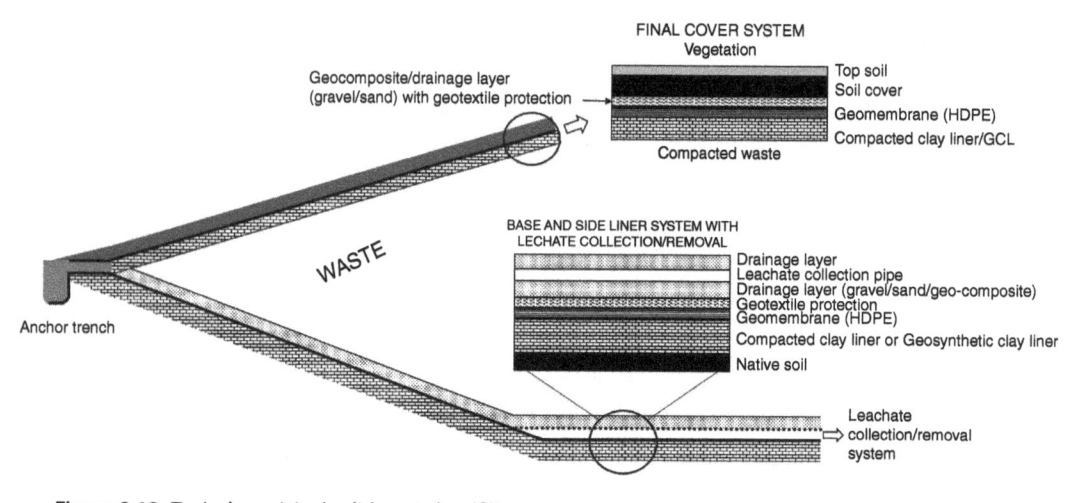

Figure 2.13 Typical municipal solid waste landfill.

or decades to degrade. Decades-old newspapers are commonly exhumed in readable condition from landfills. In fact, newspapers are used to date the age of decades-old trash disposal when exhumed for engineering or forensic purposes.

Because of uneven rates of decomposition, settlement rates associated with waste degradation can be problematic. Settlements are typically large, and differential settlements can occur due to the differing rates of decomposition. This can place significant stresses on landfill covers and monitoring/collection equipment, although design allowances are typically included into these devices. Because of the settlement and compressibility of waste, buildings are not commonly sited on landfill deposits. However, closed landfills have been converted to sites for recreational uses, including golf courses or playgrounds.

With advances in technologies, landfills have evolved into well-engineered systems. Ideally, landfills are constructed on areas at higher elevation, stable ground, groundwater tables at deeper depths, and away from airports (given their propensity to attract birds, which can have a detrimental effect on aircraft). However, because of both the real and perceived environmental effects associated with landfills, it has been increasingly difficult to approve new landfills or expansions of existing facilities. In fact, between 1988 and 2007, the number of active landfills dropped from approximately 8000 to 1754 facilities (Vaughn 2011). Although the USEPA does not believe there is a capacity issue, it presumably will get harder to approve facilities in the future as anti-growth "NIMBY" ("Not In My Backyard") or "BANANA" ("Build Absolutely Nothing Anywhere Near Anything") sentiments persist near suitable landfill locations. Other alternatives will need to be considered if landfill capacity cannot meet the demand. Additionally, with increasing costs, illegal dumping is also a growing concern.

One alternative to this is outsourcing, i.e. the transfer of wastes across state or national boundaries. Currently, 11 states export over one million tons of waste per year, while 13 states import over one million tons of waste per year. This, understandably, can create economic benefits for the importing states but vociferous opposition and resentment toward the exporting states.

Incineration is another option increasingly being considered. A total of 89 facilities in the United States burn 32 million tons of waste per year (Vaughn 2011). This process is considered as a waste reduction, and not disposal method. Although combustion reduces waste weight by 70% and volume by 90%, it results in the generation of ash, which is again a waste product. In some cases, fly ash and/or bottom ash are incorporated into building materials, including concrete.

Combustion does offer several additional benefits. First, neither households nor small businesses (the generators of MSW) need to alter waste generation activities. Second, combustion can be incorporated into waste-to-energy (WTE) facilities. Figure 2.14 depicts the operation of a hypothetical WTE facility. Two-thirds of such facilities are able to capture this energy; MSW combustion releases 35% as much energy when burned, and currently such facilities provide electricity for 2.3 million homes, which in turn eliminates the need for 9.4 billion gallons of diesel per year (EIA 2017b). Further, to achieve more efficient performance, many facilities separate recyclable materials before burning, enhancing resource recovery.

Of course, there are drawbacks associated with combustion. Air pollution is a constant concern, and even small volumes of potentially hazardous household materials can lead to toxic emissions as well as the release of nuisance-causing odorous compounds. Facilities are difficult to site and expensive to build as well. Finally, to be economically feasible with respect to WTE, these facilities need a steady stream of MSW to generate energy. To provide an example, consider an MSW facility receiving 3000 of MSW a day from a service area of one million people. Assuming typical efficiency levels, over the course of the year, 80% of 1 million tons of MSW is burned, producing electricity for 65 000 homes. Additionally, with 12% recovery due to recycling, 40 000 tons of metals are recycled. Of course, the concerns

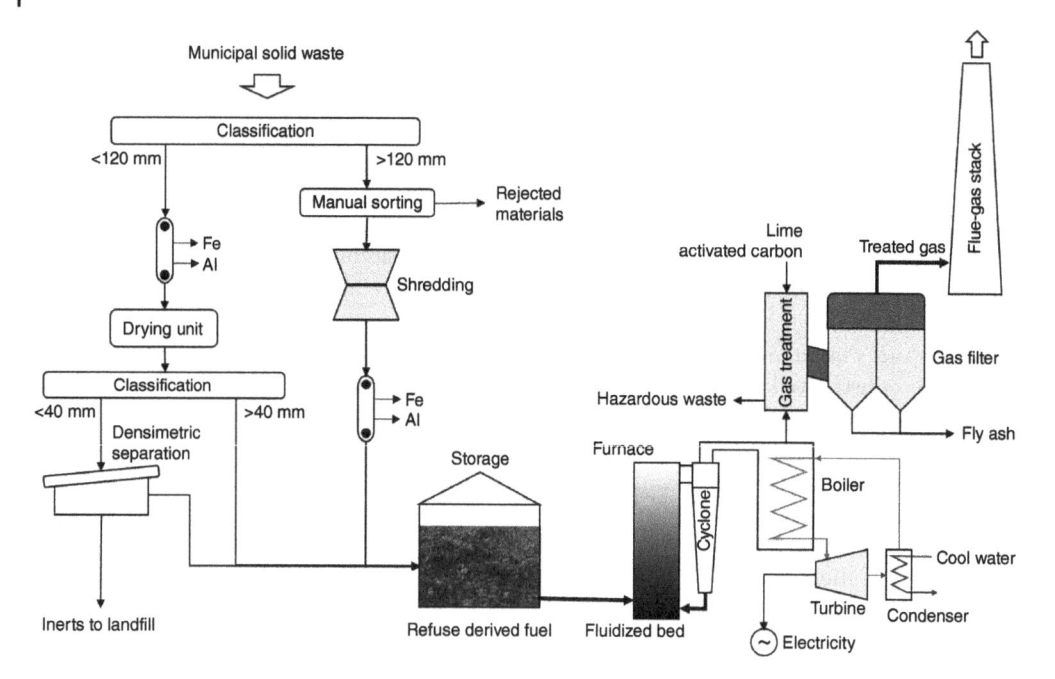

Figure 2.14 Waste to energy (WTE) process. Wastes are classified and shredded before incinerated. Heat is used to produce electric energy.

associated with such facilities are cited as drawbacks, including emissions, odors, and traffic.

In the past decades, federal regulations have been enacted to enforce local control or consolidate such control to the federal level. For instance, in 1965, the Solid Waste Disposal Act gave the Bureau of Solid Waste Management jurisdiction over MSW, although it was limited to financial and technical issues. The Resource Conservation and Recovery Act (RCRA) was passed in 1976 to manage and regulate both hazardous and nonhazardous wastes as well as underground storage tanks (USTs). RCRA also laid emphasis on the recovery and reuse of materials through recycling.

While several alternatives exist for handling MSW, an integrated approach of source reduction, WTE, recycling, composting, and "traditional" landfilling will likely be needed to address growing challenges associated with MSW. Several local, state, and federal programs have been enacted nationwide to encourage a smarter, integrated approach to MSW.

In contrast to nonhazardous wastes, hazardous wastes exhibit, by definition, at least one characteristic that warrants special care (USEPA 2009a):

- *Ignitability* – Fire-related properties, including those materials spontaneously combustible or with flash points less than 140 °F.
- *Corrosivity* – Materials with pH less than 2 or greater than 12.5, including those capable of corroding metallic containers.
- *Reactivity* – Materials susceptible to unstable conditions such as explosions, toxic fumes, gases, or vapors when heated, compressed, or mixed with water under normal conditions.
- *Toxicity* – Substances that can induce harmful or fatal effects if ingested, absorbed, or inhaled.

Electronic waste, also known as e-waste, refers to discarded electronics such as televisions, computerized and peripherals, telephones, batteries, and audio and video equipment. It is estimated that 3 million tons are generated in the United States per year, including 27 million televisions, 205 million computer products, and 140 million cellular phones (Kumar et al. 2017). Less than 15–20% of these materials are refurbished, even though they often include valuable resources as well as potentially hazardous materials (Veit and Bernardes 2015). Some of these materials include heavy metals, organic compounds, and flame retardants. The improper disposal of these materials is a growing concern to the environment as well as the waste of potentially valuable materials. Recently, programs have been developed to encourage special waste diversion, source reduction, and recovery of these materials.

Medical waste is also a specific environmental concern. Because of its inherent dangers, it is regulated by the USEPA, OSHA, CDC, and ATSDR (Windfeld and Brooks 2015). Following well-publicized events in the 1980s when medical waste washed up on the New Jersey shore, Congress enacted the Medical Waste Tracking Act (MWTA) to evaluate risks from disposal. Seven wastes under MWTA include microbiological wastes, human blood products, pathological human wastes, contaminated animal wastes, isolation wastes, contaminated sharps, and uncontaminated sharps. Additionally, the USEPA considered including wastes that had been in contact with infectious agents or blood, including sponges, dressings, surgical gloves, drapes, slides, and laboratory coats.

Radioactive waste, unlike nonhazardous and hazardous waste, is not regulated by RCRA. Instead, it is under the jurisdiction of the Nuclear Regulatory Commission (NRC) (USNRC 2017). Radioactive wastes are characterized into to four categories: high-level wastes (HLW), transuranic waste (TRU), low-level waste (LLW), and mill tailings. The health and environmental dangers due to radiation may persist for hundreds of thousands of years since various radioactive wastes decay at different rates.

The Clean Air and Clean Water Acts prevented disposal of hazardous wastes to air or water; unfortunately, this induced unregulated land disposal. It also led to three common disposal methods.

Deep well injection consisted of injection of materials (commonly pesticides, explosives, fuels, and volatile organics) into deep sealed boreholes thousands of feet below the surface. These materials react with natural deposits, rendering them less hazardous. The EPA oversees this practice through the Underground Injection Control Program. A total of 121 wells operate in the United States; most are located in the Gulf Coast region (USEPA 2017d).

A second method consists of surface impoundments. These consist of depressions (such as ponds) in which materials are placed to induce settlement and evaporation of water, consolidating the hazardous material. Assuming the bottom is well sealed and loading is equal to the rate of evaporation, these facilities can receive wastes indefinitely, or until solids fill the impoundment or are removed. However, surface water infiltration, such as that generated by stormwater capture, can affect operation (Johnson et al. 2003). In 2007, it was reported that 781 million pounds of toxics were released to 18 000 surface impoundments at 7500 facilities. Two-thirds of the impoundments contained materials deemed carcinogenic or posing other health risks. Many of the impoundments were within close proximity to groundwater, and many were not lined. Furthermore, many were located near population centers; over 20 million people live within 1.2 miles of an impoundment. Further, 24% were considered a threat to the environment.

Hazardous materials may also be placed in landfills. The landfills meet specific RCRA criteria: they typically include sophisticated leachate removal and monitoring systems, engineered caps, and surveillance and monitoring programs. Concentrated liquids or solids are placed in drums. Best-demonstrated available technologies (BDATs) are also used to reduce mobility or toxicity, including stabilization, incineration, and chemical oxidation. Currently, only 23 landfills receive off-site hazardous waste (403 million pounds in 2007) (Blackman Jr 2016).

Because of their acute nature, it is essential that hazardous wastes are properly handled. RCRA Subtitle C was developed to manage hazardous wastes, including a regulatory framework for the generation, transportation, storage, and disposal of hazardous waste as well as technical standards for the design and operation of treatment, storage, and disposal facilities (TSDFs). Additionally, hazardous wastes are subjected to "cradle-to-grave" tracking, requiring detailed records for generation, transport, and disposal. The copies of these records must go to the EPA as well.

Although RCRA covered nonhazardous and hazardous waste management for ongoing facilities, it did not address abandoned sites. Several well-publicized incidents highlighted cases where hazardous wastes were negligently handled. One instance is known as Love Canal. In the Niagara Falls, New York area, development occurred over the previously abandoned Love Canal. The resulting development and infrastructure construction pierced the clay-lined canal. Over time, noxious odors were observed, and significant acute and chronic health problems were reported by the citizens. Eventually, follow-up testing and analysis determined the presence of widespread soil and groundwater contamination, and the United States Federal government paid for the relocation of hundreds from the Love Canal area. A second site was the Valley of the Drums. This site, a 23-acre toxic waste site in Bullitt County, Kentucky, was named after the waste-containing drums strewn across the area. In 1966, the site caught fire and burned for over a week (Reddy 2014). Ultimately, it caught the attention of state officials, leading to its fencing off and capping in the 1980s.

As a result, in 1980, the Comprehensive Environmental Response, Compensation, and Liabilities Act (CERCLA), or "Superfund" was passed to address cleanup of these hazardous sites, specifically addressing funding, liability, and prioritization of hazardous and/or abandoned waste sites. A $1.6 billion fund was created from taxes levied on chemical and petroleum industries to finance the cleanup of hazardous waste sites and litigation brought against potentially responsible parties (PRPs). Additional funds ($8.5 billion) were appropriated in 1986 with the passage of the Superfund Amendments and Reauthorization Act (SARA). A $500 million fund was also appropriated for the remediation of leaking USTs. Additionally, community right-to-know provisions were adopted. Controversially, SARA established provisions for cleanup-related legal and financial liability. Disclosure requirements related to annual releases of hazardous substances were also included. Explicit liability provisions directed at current landowners and related innocent landowner provisions became a paramount concern for all entities associated with land transactions.

Both CERCLA and SARA significantly underestimated the potential costs and timing associated with environmental cleanups. When CERCLA was first enacted, approximately 36 000 contaminated sites were identified; of these, 1200 were placed on the national priority list (NPL) OSWER 2010). At the end of Fiscal Year 2010, 1627 sites remained on the NPL, and 475 sites had been closed. However, these closures consumed a significant amount of resources; on average, $40 million was expended per site, requiring an average of 11 years to achieve closure. Further, $6 billion held in trust in 1996 had been exhausted by 2003 (Gamper-Rabindran and Timmins 2011).

Because of fears associated with CERCLA liability, impacted properties with significant reuse potential remained idle and contaminated for long periods of time. Many of these sites became known as Brownfields. A Brownfield is an abandoned, idled, or underutilized industrial or commercial site where expansion or redevelopment is complicated by actual or perceived environmental contamination. The real or perceived contamination can range from minor aesthetic issues to significant subsurface contamination. In the early 1990s, the federal government took action to provide inducements to encourage Brownfield redevelopment. The USEPA also took measures to clarify liability provisions as well as provide for indemnity for prospective purchasers. In 2002, the United States Congress passed the Small Business Liability Relief and Brownfields Revitalization Act to provide certain relief for small businesses from liability imposed under the CERCLA, promote the cleanup and reuse of brownfields, provide financial assistance for brownfields revitalization, and enhance State response programs (USEPA 2009b).

In 2005, the USEPA established the All Appropriate Inquiries (AAI) requirements, which became law on 1 November 2006 (USEPA 2017e). The purpose of AAI was to establish liability protection under CERCLA for innocent landowners, contiguous property owners, or bonafide prospective purchasers provided certain criteria were met. AAI has been a very important milestone in encouraging land acquisition and development. By establishing a framework, prospective land purchasers have a discrete set of actions they must perform to avoid open-ended liability and costs. In this manner, they can help eliminate the unknowns associated with a potential redevelopment project, which facilitates a return to productive use for many impacted properties.

Finally, another source of soil contamination is related to the USTs, which are common at service stations, corporation yards, and a variety of commercial and industrial settings. The underground placement of tanks reduces the danger of explosion and fire, however, because tanks eventually deteriorate, USTs are a very common source for soil and groundwater contamination. Leaks, especially in older tanks, can result in substantial contamination before they are detected. Subtitle I of RCRA regulates USTs and includes provisions for notification, methods for release detection, design and construction standards, and reporting, recordkeeping, and financial responsibility. Corrective actions pertaining to releases from USTs are also regulated under Subtitle I. USTs containing hazardous wastes are regulated under Subtitle C. Furthermore, a leaking UST (LUST) trust fund receives 0.1 cents per gallon of fuel purchased to finance federal oversight activities (Tiemann 2006).

2.14 Land Contamination

Hazardous waste sites, those with contamination from a range of toxic materials, are a serious threat to public health. Both active and inactive sites pose a concern. Whereas active sites can be monitored, and site activities can be regulated or changed to be more protective. More invasive intervention techniques can be implemented at inactive sites, such as capping of large-scale cleanups. Inactive sites were often operated without protective measures in place; for instance, old landfills and dumps were often constructed without liners, which could lead to more widespread soil and groundwater contamination.

Toxic chemicals fall into several categories. Some can be easily degraded and assimilated into the environment, where they may be diluted from stabilized to safe levels; however, others, such as heavy metals and synthetic organics, are resistant to degradation and persist in the environment. Many heavy metals and products (e.g. lead, mercury, arsenic, cadmium, tin, chromium (VI), zinc, copper, and products containing these substances) are persistent, toxic compounds. They have been commonly used in many industrial applications, including metalwork, plating, batteries, and electronics. They were also once commonly used in paints, glazing, inks, and dyes. Some can be soluble in water, and if absorbed, they are commonly bioaccumulative within the body. They can interfere with enzyme function, birth defects, learning disabilities, and severe mental illness.

Synthetic and derived organic compounds can also be toxic to human health and the environment. These compounds are often used as the basis for plastics, fibers, coatings, solvents, pesticides, and preservatives. They are resistant to degradation, allowing them to persist in toxic form within the environment. Upon absorption into the human body, these compounds can interfere with enzyme action. Acute exposures can lead to poisoning or death, and chronic exposure can result in mutagenic, carcinogenic, and teratogenic effects. Several compounds pose acute dangers to health and the environment. Some of these, collectively labeled as persistent organic pollutants (POPs), include halogenated hydrocarbons (those containing chlorine, bromine, fluorine, and iodine), and chlorinated hydrocarbons, which include plastics, pesticides, solvents, and insulation. In 2004, the Stockholm Convention banned or restricted use of 12 of these compounds (nicknamed the "dirty dozen") because of their toxic, carcinogenic, or endocrine disruption effects at low concentrations (Batool et al. 2016). Additional compounds (including the "nasty nine") have since been added (UNEP 2008).

A very common compound which is used in a variety of industrial and consumer applications but exhibits significant impact to health and the environment is tetrachloroethylene (PCE, or "PERC"). PCE is a colorless, nonflammable, halogenated hydrocarbon used as a dry-cleaning solvent, and industrial solvent, and home products. As a dense, nonaqueous phase liquid (DNAPL), it is heavier that water; when it exposed to environment, it is mobile, sinks through groundwater, and is highly resistant to degradation. Acute exposure can lead to dizziness, fatigue, and headaches, and because it has proven to be carcinogenic to rats and mice, it is suspected as a human carcinogen. Dry cleaning accounts for 10% of the United States' annual PCE consumption of 370 million pounds per year, and the remainder is primarily used to make CFCs (ICIS 2017). The use of PCE is being phased out; several European Union are considering ban of its use, and California has mandated its ban in dry cleaning and within related equipment by 1 January 2023 (CARB 2018). Other common organic compounds that have been increasingly linked to health effects include phthalates (used to soften plastics) and perchlorate (used in rocket fuels and other flammable products). These compounds have entered and impacted the environment (soil sediments, water) as well as affected human health due to their presence in a range of products.

2.15 Visibility

Visibility is defined as the greatest distance over which one can see and identify familiar objects with the naked eye (Hyslop 2009). Visibility is based on two factors – first is the perception abilities of the individual, second is related to the degree to which light coming from the object is absorbed or scattered. Two principal causes of decreased visibility include the emission of small particles on the same order of visible light, about 0.2–0.7 µm as well as the emission of reactive gases that are subsequently converted into small particles (Delucchi et al. 2002). Visibility impacts can affect urban or rural areas. The EPA passed the Regional Haze Rule in 1999 with the goal of improving visibility and reducing emissions in parks and wilderness areas.

2.16 Odors

Odors, resulting from both organic and inorganic compounds, can have different degree of unpleasantness. Regardless of the aesthetics and thresholds, some odors can be indicative of a potential health hazard; others may result in a perceived danger or affect quality of life. From a public health standpoint, odors may have a direct adverse effect on health or may induce other symptoms associated with illness.

Odors can emanate from a variety of sources. Some industrial sources include wood treatment facilities, paper mills, landfills, refineries, solvent processing facilities, and wastewater treatment plants. Nonindustrial sources may include pesticide and fertilizer applications, feedlots, and facilities with a large number of idling trucks. Odors from these sources elicit far more community complaints than odorless air pollutants.

In some cases, odors do not represent an overexposure to a specific compound. In other cases, such as VOC exposures, odors can be indicative of environmental quality problems as well as a sign of unhealthy overexposure. The resulting health effects depend on several factors, including sensitivity and response to odors on an individual basis. Health effects may include irritation and adverse effects to mucous membranes, air passages, blood vessels, heart, stomach and intestines, brain, and psychological well-being (Bernstein et al. 2008).

2.17 Aesthetic Degradation

Aesthetic degradation is any reduction in environmental quality that people interpret as a reduction in attractiveness (Verburg 2014). This may include siting of unsightly industrial or institutional facilities, or other cases where the built environment affects the natural environment. These may also include visual degradation to soil, air, and water. Specific sources can include the placement of debris, unnatural colors or textures, turbidity or haziness to water/air, or scum and floating materials on water. With respect to water, natural or anthropogenic substances can degrade water quality, such as water color, odor, or clarity. The monitoring of aesthetics can be relatively inexpensive. It is primarily based on subjective criteria, as visual appreciation will vary from person to person. It can also be a source of controversy, as the ranges of personal senses of aesthetics can make it difficult to reach a consensus of these matters.

2.18 Land Use Patterns

Land use pertains to how the physical environment is utilized for resources and/or converted to the built environment. It can include specific resource utilization, such as mining, forestry, or agricultural uses, or it can pertain to consumption, such as residential, commercial, industrial, or recreational activities. Economic and social factors are paramount to land use decisions and patterns. As the built environment expands, there is often stress or conflict with the natural environment. Sprawl of metropolitan areas can often lead to the loss of natural vegetation, open space, wetlands, and other habitats. Agricultural lands are also often lost to growing communities. As these land use conversions continue, more public attention is drawn to the environmental impacts that result from a range of side effects, including loss of vegetation and habitat, traffic, and utilization of other resources.

2.19 Thermal Pollution

Thermal pollution is the degradation of the environment due to processes that change ambient temperature. These effects can have significant consequences and are commonly observed due to the discharge of heated or cooled water resulting from human activity. A typical occurrence is the release of cooling water from power plants, refineries, paper mills, chemical plants, and steel mills. These cooling waters are often warmer than the receiving water, and the resulting elevation in temperature can affect the ecosystem. Thermal effects can also occur from urban runoff and stormwater runoff that may be generated from changes in vegetation. As impervious surfaces replace natural vegetation, runoff temperature may be elevated as compared to a natural condition, leading to similar effects as observed within cooling water discharge. When warmer water is discharged, dissolved oxygen rates decrease because oxygen solubility decreases. Also, the increased temperature can enhance metabolism in organisms, leading to greater utilization of oxygen and decreased dissolved oxygen levels, ultimately resulting in eutrophic conditions.

Thermal pollution can also be directly toxic to aquatic life since enzyme systems can be suddenly disrupted or shocked. Organisms can be killed by sudden thermal changes that are beyond the tolerance limits of their metabolic systems. Ironically, in some instances, thermal pollution can be viewed as beneficial ("thermal enrichment"). Heated water can result in longer commercial fishing seasons and the reduction of winter ice cover in cold areas.

Thermal pollution effects can be controlled with several strategies. Heat transfer can be safely induced using cooling towers, cooling ponds, or cogeneration applications where the heat may be safely extracted or dissipated. Cooling water can also be retained and recirculated through the heat generating system, thereby reducing direct discharge into vulnerable water bodies. Stormwater runoff can be treated by increasing residence times using bioretention and infiltration systems, thereby allowing a gradual reduction in temperature.

2.20 Noise Pollution

Noise pollution is excessive environmental noise that can disrupt the quality of life. Noise pollution may emanate from human, animal, or machine-related sources. Typical sources of noise may include construction or transportation activities as well as industrial activities. It may also result from undesirable activities and practices as well as inappropriate urban planning in which incompatible activities are placed near to one another. Noise pollution can have physical health effects, such as hearing loss, cardiovascular effects, such as hypertension, as well as psychological effects, and diminished well-being. Animals can be affected, as noise can diminish or eliminate natural predator or prey behaviors.

Federal regulations have been enacted to abate noise pollution. Title IV of the Clean Air Act focuses on noise pollution (USEPA 2018). Additionally, the Noise Control Act of 1972 and the Quiet Communities Act of 1978 provide a regulatory framework for noise pollution reduction (Robinson 2016). These regulations as well as other mandated or industry-led activities have identified methods to reduce noise, such as design improvement in roads and vehicles, quieter jet engines, and refinements in industrial equipment design.

2.21 Summary

Environmental concerns of varying scope and magnitude may occur during all phases of a project. These concerns may occur on a local scale, a regional scale, or a global scale, but regardless of the scope and scale, virtually every activity will have one or more environmental impact.

It is also important to point out that the environmental impacts can have a significant detrimental effect on human health and the ecosystem. Separating important effects from "background noise" is a key skill in assessing the sustainable impacts of a project, and it is important that these environmental concerns are assessed on a qualitative or quantitative basis.

When assessing a project or an activity, it is essential to understand a range of environmental concerns that can occur. This chapter summarizes a wide range of environmental concerns that can directly affect human health or the greater environment. When assessing the sustainability of a project, all potential environmental impacts, or as many as practicable, should be considered and incorporated into the assessment, especially the impacts that are quantifiable. These may include both direct and indirect environmental impacts. It is also often beneficial to assess and quantify the environmental impacts associated with a range of alternative projects/designs for the purposes of a comparative assessment. Such an assessment will identify and allow for the selection of the project/design that causes the least overall environmental impact.

2.22 Questions

2.1 What are the environmental concerns from sustainability perspective? Explain in detail.

2.2 What are the global scale environmental concerns? Explain them briefly.

2.3 What are the regional scale environmental concerns? Explain them briefly.

2.4 What are the local scale environmental concerns? Explain them briefly.

2.5 What are greenhouse gases? How are they helpful in maintaining favorable temperatures on the Earth for the survival of most of the living organisms? What are the consequences of high amounts of greenhouse gases in the atmosphere?

2.6 Explain global climate change and list the main greenhouse gases of anthropogenic origin. What are the sources and impacts?

2.7 Describe how the GWP assesses the relative strength of greenhouse gases?

2.8 Name some key desert areas, and identify two large metropolitan areas near each location?

2.9 Why is deforestation to create agricultural land often counter-productive?

2.10 Discuss in detail why losses in genetic diversity can be harmful to ecosystems?

2.11 Describe the difference between stratospheric ozone and ground-level ozone.

2.12 Which are the compounds that abet ozone layer depletion? Describe in detail the process of how CFCs result in ozone-layer depletion.

2.13 Explain the physiological and psychological effects of lead poisoning.

2.14 Discuss meteorological and topographical conditions that may exacerbate industrial or photochemical smog.

2.15 What are the two types of smog? Explain briefly.

2.16 What is acid rain and how does it occur? Explain briefly.

2.17 What is eutrophication? Explain briefly.

2.18 Provide specific examples of point source and nonpoint source water pollution.

2.19 Explain how climate change and water cycle are related.

2.20 Explain how climate change and water hazards are related.

2.21 Explain how climate change and water quality are related.

2.22 Discuss the different types of solid wastes and characteristics of each.

2.23 What are the characteristics of hazardous wastes? Explain briefly.

2.24 What is land degradation and mention the different categories of land degradation.

2.25 What are the different sources of odor in the environment, and how does it impact the human health?

2.26 What is thermal pollution and what are its effects on the environment? Explain briefly.

2.27 List examples of health effects caused by noise pollution.

References

Abbasi, S.A. and Abbasi, T. (2017). Strategies to contain the ozone hole. In: *Ozone Hole*, 113–120. New York: Springer.

Annest, J.L., Pirkle, J.L., Makuc, D. et al. (1983). Chronological trend in blood lead levels between 1976 and 1980. *The New England Journal of Medicine* 308 (23): 1373–1377.

ApSimon, H., Pearce, D., and Ozdemiroglu, E. (2014). *Acid Rain in Europe: Counting the Cost*. Routledge.

Archer, D. (2016). *The Long Thaw: How Humans are Changing the Next 100,000 Years of Earth's Climate*. Princeton University Press.

Austin, K.F., Bellinger, M.O., and Rana, P. (2017). Anthropogenic forest loss and malaria prevalence: a comparative examination of the causes and disease consequences of deforestation in developing nations. *AIMS Environmental Science* 4 (2): 217–231.

Barlow, J., Lennox, G.D., Ferreira, J. et al. (2016). Anthropogenic disturbance in tropical forests can double biodiversity loss from deforestation. *Nature* 535 (7610): 144–147.

Batool, S., Ab Rashid, S., Maah, M.J. et al. (2016). Geographical distribution of persistent organic pollutants in the environment: a review. *Journal of Environmental Biology* 37 (5): 1125.

Beaty, T.A. (2015). Life on the Mississippi: reducing the harmful effects of agricultural runoff in the Mississippi River Basin. *Ohio Northern University Law Review* 41: 819–965.

Bernstein, J.A., Alexis, N., Bacchus, H. et al. (2008). The health effects of nonindustrial indoor air pollution. *The Journal of Allergy and Clinical Immunology* 121 (3): 585–591.

Blackman, W.C. Jr., (2016). *Basic Hazardous Waste Management*. CRC Press.

Blanco-Canqui, H. and Lal, R. (2010). Soil and water conservation. In: *Principles of Soil Conservation and Management*, 1–19. Netherlands: Springer.

Brunekreef, B. and Holgate, S.T. (2002). Air pollution and health. *The Lancet* 360 (9341): 1233–1242. https://doi.org/10.1016/S0140-6736(02)11274-8.

Burkholder, J.B., Cox, R.A., and Ravishankara, A.R. (2015). Atmospheric degradation of ozone depleting substances, their substitutes, and related species. *Chemical Reviews* 115 (10): 3704–3759.

CARB (California Air Resources Board) (2018). Dry cleaning program website. https://www.arb.ca.gov/toxics/dryclean/dryclean.htm.

CBD (2014). Global Biodiversity Outlook 4. Secretariat of the Convention on Biological Diversity. Montréal, p. 155. ISBN- 92-9225-540-1.

Ceballos, G., Ehrlich, P.R., Barnosky, A.D. et al. (2015). Accelerated modern human-induced species losses: entering the sixth mass extinction. *Science Advances* 1 (5): e1400253.

Chidumayo, E.N. and Gumbo, D.J. (2013). The environmental impacts of charcoal production in tropical ecosystems of the world: a synthesis. *Energy for Sustainable Development* 17 (2): 86–94.

Clark, N.E., Lovell, R., Wheeler, B.W. et al. (2014). Biodiversity, cultural pathways, and human health: a framework. *Trends in Ecology & Evolution* 29 (4): 198–204.

Cooke, G.D., Welch, E.B., Peterson, S., and Nichols, S.A. (2016). *Restoration and Management of Lakes and Reservoirs*. CRC Press.

Dapul, H. and Laraque, D. (2014). Lead poisoning in children. *Advances in Pediatrics* 61 (1): 313–333.

Delucchi, M.A., Murphy, J.J., and McCubbin, D.R. (2002). The health and visibility cost of air pollution: a comparison of estimation methods. *Journal of Environmental Management* 64 (2): 139–152.

DOE (1999). Natural Gas 1998: Issues and Trends. Energy Information Administration. Office of Oil and Gas. U.S. Department of Energy. *Report no. DOE/EIA-0560(98)*. Washington, DC.

Ebi, K.L. and McGregor, G. (2008). Climate change, tropospheric ozone and particulate matter, and health impacts. *Environmental Health Perspectives* 116 (11): 1449.

EIA (2017a). Petroleum and other liquids: crude oil production. US Energy Information Administration. Independent Statistics and Analysis. https://www.eia.gov/dnav/pet/pet_crd_crpdn_adc_mbblpd_a.htm (accessed 16 October 2018).

EIA (2017b). Landfill gas and biogas. US Energy Information Administration. Independent Statistics and Analysis. https://www.eia.gov/energyexplained (accessed 10 April 2017).

Veit, H.M. and Bernardes, A.M. (2015). *Electronic Waste. Recycling Techniques*, Topics in Mining, Metallurgy and Materials Engineering. Switzerland: Springer International Publishing. ISBN: 978-3-319-15714-6 https://doi.org/10.1007/978-3-319-15714-6;

Elsom, D. (2014). *Smog Alert: Managing Urban Air Quality*. Routledge.

Ennis, C.A. (2017). Scientific assessment of ozone depletion: 2006. *World Meteorological Organization Global Ozone Research and Monitoring Project—Report No. 50*. Geneva, Switzerland.

Farley, M. (2012). Eutrophication in fresh waters: an international review. In: *Encyclopedia of Lakes and Reservoirs* (ed. L. Bengtsson, R.W. Herschy and R.W. Fairbridge), 258–270. Dordrecht: Springer Netherlands https://doi.org/10.1007/978-1-4020-4410-6_79.

Friedrich, J., Janssen, F., Aleynik, D. et al. (2014). Investigating hypoxia in aquatic environments: diverse approaches to addressing a complex phenomenon. *Biogeosciences (BG)* 11: 1215–1259.

Gamper-Rabindran, S. and Timmins, C. (2011). Valuing the Benefits of Superfund Site Remediation: Three Approaches to Measuring Localized Externalitites. Triangle RDC.

Gerber, V. (2015). The policy and politics of pollution: exploring regulation, environmental justice, and toxic air emissions dispersion trends. Doctoral dissertation. University of California, Berkeley.

Hallegatte, S., Green, C., Nicholls, R.J., and Corfee-Morlot, J. (2013). Future flood losses in major coastal cities. *Nature Climate Change* 3 (9): 802–806.

Harris, N.R., Hassler, B., Tummon, F. et al. (2015). Past changes in the vertical distribution of ozone–Part 3: analysis and interpretation of trends. *Atmospheric Chemistry and Physics* 15 (17): 9965–9982.

Hyslop, N.P. (2009). Impaired visibility: the air pollution people see. *Atmospheric Environment* 43 (1): 182–195. https://doi.org/10.1016/j.atmosenv.2008.09.067.

ICIS (2017). Perchloroethylene uses and market data. ICIS web site. https://www.icis.com/resources/news/2007/11/06/9076131/perchloroethylene-uses-and-market-data (accessed 25 July 2017).

IPCC (2013). Climate Change 2013: The Physical Science Basis. In: *Contribution of Working Group I to the Fifth Assessment Report of the Intergovernmental Panel on Climate Change* (eds T.F. Stocker, D. Qin, G.-K. Plattner, M. Tignor, S.K. Allen, J. Boschung, A. Nauels, Y. Xia, V. Bex, and P.M. Midgley). 1535 p. Cambridge: Cambridge University Press.

IPCC (2014). Climate change 2014: synthesis report. In: *Contribution of Working Groups I, II and III to the Fifth Assessment Report of the Intergovernmental Panel on Climate Change*. Geneva, Switzerland: IPCC 151 pp.

Johnson, B., Balserak, P., Beaulieu, S. et al. (2003). Industrial surface impoundments: environmental settings, release and exposure potential and risk characterization. *Science of the Total Environment* 317 (1): 1–22.

Kumar, A., Holuszko, M., and Espinosa, D.C.R. (2017). E-waste: an overview on generation, collection, legislation and recycling practices. *Resources, Conservation and Recycling* 122: 32–42.

Lal, R., Iivari, T., and Kimble, J.M. (2003). *Soil Degradation in the United States: Extent, Severity, and Trends*. CRC Press.

Lane, J.L. (2015). Stratospheric ozone depletion. In: *Life Cycle Impact Assessment*, 51–73. Netherlands: Springer.

Magoc, C.J. (2014). Reflections on the public interpretation of regional environmental history in western Pennsylvania. *The Public Historian* 36 (3): 50–69.

McNeil, B.E. and Culcasi, K.L. (2015). Maps on acid: cartographically constructing the acid rain environmental issue, 1972–1980. *The Professional Geographer* 67 (2): 242–254.

MEA (2005). *Millennium Ecosystem Assessment – Information to Conserve Ecosystems and Enhance Human-Well Being*. Washington, DC: New Island.

de Mutsert, K., Steenbeek, J., Lewis, K. et al. (2016). Exploring effects of hypoxia on fish and fisheries in the northern Gulf of Mexico using a dynamic spatially explicit ecosystem model. *Ecological Modelling* 331: 142–150.

NCA (2014). *The U.S. Global Change Research Program's Third Assessment Report*. National Climate Assessment. U.S. Global Change Research Program. ISBN: 9780160924026. nca2014.globalchange.gov.

NCOOS (2017). Harmful algal bloom and hypoxia research and control act. National Centers for Coastal Ocean Science. https://coastalscience.noaa.gov/research/habs/habhrca (accessed 10 April 2017).

OSWER (2010). *Fiscal Year 2010 End of Year Report*. Office of Solid Waste and Emergency Response. United States Environmental Protection Agency, Washington, DC.

Pacala, S. and Socolow, R. (2004). Stabilization wedges: solving the climate problem for the next 50 years with current technologies. *Science* 305 (5686): 968–972.

Prather, M., Midgley, P., Rowland, F.S., and Stolarski, R. (1996). The ozone layer: the road not taken. *Nature* 381 (6583): 551.

Reddy, K.R. (2014). Evolution of geoenvironmental engineering. *Environmental Geotechnics* 1 (3): 136–141.

Reynolds, J.F., Smith, D.M.S., Lambin, E.F. et al. (2007). Global desertification: building a science for dryland development. *Science* 316 (5826): 847–851.

Robinson, N.A. (2016). *Environmental Regulation of Real Property*. Law Journal Press.

References

Rowland, F.S. and Molina, M.J. (1974). Stratospheric ozone destruction catalyzed by chlorine atoms from photolysis of chlorofluoromethanes. *Transactions of the American Geophysical Union* 55 (1974): 1153–1153.

Shannon, M.A., Bohn, P.W., Elimelech, M. et al. (2008). Science and technology for water purification in the coming decades. *Nature* 452 (7185): 301–310.

Shiklomanov, I.A. (1998). *World Water Resources. A New Appraisal and Assessment for the 21st Century*. Paris: United Nations, Educational, Scientific and Cultural Organization (UNESCO).

Shiklomanov, I.A. (2000). Appraisal and assessment of world water resources. *Water International* 25 (1): 11–32.

Sunda, W.G. and Cai, W.J. (2012). Eutrophication induced CO_2-acidification of subsurface coastal waters: interactive effects of temperature, salinity, and atmospheric CO_2. *Environmental Science & Technology* 46 (19): 10651–10659.

Themelis, N.J. and Ulloa, P.A. (2007). Methane generation in landfills. *Renewable Energy* 32 (7): 1243–1257.

Thiele, L.P. (2016). *Sustainability*. Wiley.

Tiemann, M. (2006). Leaking underground storage tanks: program status and issues. *CRS Report for Congress*. Order Code RS21201.

UNCCD (2017). United Nations convention to combat desertification. http://www2.unccd.int (accessed 7 April 2017).

UNCED (1992). Rio declaration on environment and development. United Nations Conference on Environment and Development, Rio de Janeiro, Brazil.

UNEP (2008). Stockholm Convention. Protecting human health and environment from persistent organic pollutants. http://chm.pops.int/TheConvention/ThePOPs (accessed 10 April 2017).

UNEP (2017). The Montreal protocol on substances that deplete the ozone layer. United Nations Environment Programme. Ozone Secretariat. http://ozone.unep.org (accessed 9 April 2017).

USEPA (2009a). Hazardous waste characteristics. *US Environmental Protection Agency. Report*. 30 pages.

USEPA (2009b). Brownfields fact sheet – EPA brownfields grants, CERCLA liability, and all appropriate inquiries. EPA 560-F-09-026, Washington, DC.

USEPA (2012). Water quality standards handbook. Chapter 3: Water Quality Criteria. *Report No. EPA-823-B-12-002*.

USEPA (2015). Mississippi River/Gulf of Mexico watershed nutrient task force. US Environmental Protection Agency. *HTF 2015 Report to Congress*.

USEPA (2016). Municipal solid waste generation, recycling, and disposal in the United States: facts and figures. https://archive.epa.gov/epawaste/nonhaz/municipal/web/html/msw99.html (accessed 9 April 2017).

USEPA (2017a). Safe drinking water act (SDWA). https://www.epa.gov/sdwa (accessed 10 April 2017).

USEPA (2017b). National pollutant discharge elimination system (NPDES). https://www.epa.gov/npdes (accessed 10 April 2017).

USEPA (2017c). Program overview: total maximum daily loads (TMDL). https://www.epa.gov/tmdl/program-overview-total-maximum-daily-loads-tmdl (accessed 10 April 2017).

USEPA (2017d). Underground injection control regulations and safe drinking water act provisions. Environmental Protection Agency. https://www.epa.gov/uic.

USEPA (2017e). Brownfields all appropriate inquiries. US Environmental Protection Agency. https://www.epa.gov/brownfields/brownfields-all-appropriate-inquiries (accessed 10 April 2017).

USEPA (2017f). National ambient air quality standards (NAAQS). US Environmental Protection Agency. https://www.epa.gov/criteria-air-pollutants (accessed 10 April 2017).

USEPA (2018). Clean air act title IV – noise pollution. US Environmental Protection Agency. https://www.epa.gov/clean-air-act-overview/clean-air-act-title-iv-noise-pollution (accessed 25 July 2018).

USNRC (2017). Radioactive waste. United States Nuclear Regulatory Commission. https://www.nrc.gov/waste.html (accessed 10 April 2017).

Vaughn, J. (2011). *Environmental Politics: Domestic and Global Dimensions*, 6e. Boston, MA, USA: Wadsworth, Cengage Learning.

Verburg, P.H. (2014). The representation of human-environment interactions in land change research and modelling. Chapter 8. In: *Understanding Society and Natural Resources* (ed. M.J. Manfredo, J.J. Vaske and E.A. Duke), 161–177. Dordrecht: Springer. ISBN: 978-94-017-8958-5.

Vet, R., Artz, R.S., Carou, S. et al. (2014). A global assessment of precipitation chemistry and deposition of sulfur, nitrogen, sea salt, base cations, organic acids, acidity and pH, and phosphorus. *Atmospheric Environment* 93: 3–100.

Vingarzan, R. (2004). A review of surface ozone background levels and trends. *Atmospheric Environment* 38 (21): 3431–3442.

Windfeld, E.S. and Brooks, M.S.L. (2015). Medical waste management – a review. *Journal of Environmental Management* 163: 98–108.

3

Social, Economic, and Legal Issues

3.1 Introduction

As presented in Chapter 1, the concepts of sustainability and sustainable development transcend the idea of "environmentalism." In addition to environmental protection and stewardship, the concept of sustainability incorporates the notion of equity, including social, economic, and legal aspects. From an economic standpoint, sustainability is usually framed with respect to growth, efficiency, and resource use; whereas the social framework often focuses on human needs, including social equity, empowerment, social cohesion, and cultural identity.

The global concept of sustainability defines sustainable solutions as those that maximize the positive impacts of the three dimensions of the triple bottom line – environmental, social, and economic. The concept of the triple bottom line has continued to evolve into a valuable framework for the assessment of a project or an activity. More sustainable solutions lead to a greater intersection of these metrics, and as sustainable solutions are increasingly desired, it is important to understand the social, economic, and legal issues that become relevant when considering a project or an activity. In this chapter, the social and economic issues associated with sustainability are presented and discussed. Furthermore, the importance of legal issues cannot be underestimated due to their influence on economic activities and social organization. A brief outline of these issues is presented in this chapter.

3.2 Social Issues

Regardless of the precise definition that one may choose or prefer for sustainability, each definition, in some way, includes the necessity to consider the needs of future generations. Further, while considering the needs of future generations, it is important to consider the effects of actions on local society as well as how these actions may affect people across the globe. The effects of these actions, both positive and negative, are reflected in the social aspects of the triple bottom line. Let us now examine the social dimension of the triple bottom line.

3.2.1 Society

Society is a social system functioning to reproduce the biological and cultural aspects of humankind within a territory (Castells 2011). In addition to the reproduction of a human population, societies aim to create biological and physical entities that provide utility, are assigned some form of ownership, and may be produced or reproduced. Crop cultivation and the raising of livestock, man-made infrastructure, and industrial output are three examples of these entities.

Sustainable Engineering: Drivers, Metrics, Tools, and Applications, First Edition.
Krishna R. Reddy, Claudio Cameselle, and Jeffrey A. Adams.
© 2019 John Wiley & Sons, Inc. Published 2019 by John Wiley & Sons, Inc.

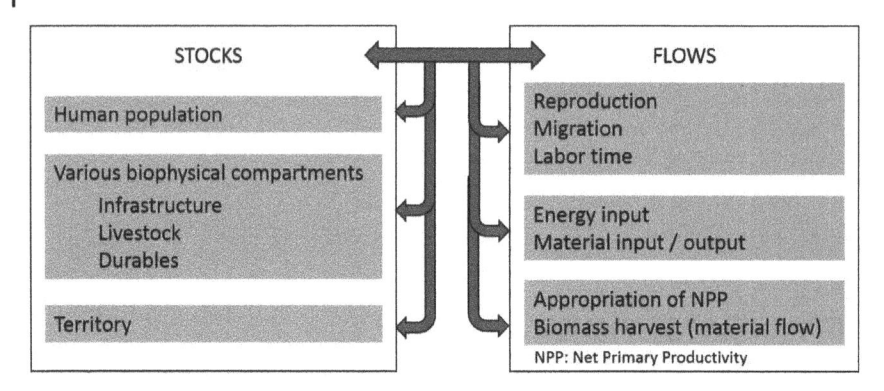

Figure 3.1 Relevant biophysical dimensions of social systems.

Considered on a larger scale, we can divide social systems into these entities (stocks) and actions undertaken by these entities or stimuli imparted to these entities (denoted as flows in Figure 3.1). Some examples of stocks include humans themselves, the output of human-directed activity (crops/livestock, industrial output, infrastructure), and physical territory. Flows include reproduction, migration, labor activity, energy inputs and outputs, material inputs and outputs, appropriation of NPP (net primary productivity) and agricultural harvesting/culling.

The structure that a particular society creates in order to govern itself (i.e. creates rules and norms, enacts and enforces rules and norms, and interprets rules and norms) is an important aspect of a society. Assuming a particular society is able to impose its collective free will and choose a system of its own particular liking (something that, unfortunately, the majority of peoples over the millennia have not been able to do), the respective system is often a reflection of the importance of a range of attributes emphasized by that society. As discussed in Chapter 1, different nation-states throughout history have placed a wide range of emphasis on economic systems, from centrally planned to free markets; the type or inspiration of a legal framework, often based on a particular religion, a series of cultural norms, or historic precedent; and the degree central government strength/control, from libertarian to authoritarian approaches. These different economic and governance models have an enormous influence on the society development and structure, as well as in the linking and relationship among society members.

3.2.2 Developed and Developing Societies

The characteristics of a society (i.e. organization, equality, and liberty) may vary depending on the wealth and standard of living of each country. The access to the resources and opportunities, public services, and well-being can be very different depending on the level of development of the country and the available resources and services for their citizens. Within the frame of reference of social sustainability, a compelling aspect is how societies evolve and deteriorate under stress. In general, wealthier societies are more stable societies, whereas poorer societies struggle to provide the necessary inputs to ensure a stable society. Further, the degree of stability may be influenced by the range of wealth distribution equity or inequity.

In discussing these differences, it is important to distinguish between more (or most) developed countries (MDCs) versus less (or least) developed countries (LDCs). An MDC is typically associated with highly developed economies, advanced technological infrastructure, and relatively high standards of living. LDCs, on the other hand, are typically associated with low standards of living, undeveloped or underdeveloped economies, and underdeveloped industrial bases.

LDCs often serve as lower-cost sources of labor for international companies (often based in MDCs) and attract lower-value industries because of their underdeveloped economies and relatively less skilled workforce. Unfortunately, these industries are often significant contributors to environmental pollution. LDCs harbor an increasing percentage of the world's industrial facilities, leading to increased pollution (McMichael 2000). An example of this is the textile industry in Tirupur, India. This region had traditionally served as a center of textile fabrication, which included 4000 independent small business units (Tewari 2006). Until the 1980s, many of these businesses manufactured white cotton undershirts ("banians") commonly worn in India. However, a shift in production occurred in response to lucrative international markets associated with colored T-shirts. The focus of textile industry in Tirupur shifted, and industry participants began to incorporate significant dying operations. This, combined with an appreciable increase in emissions associated with the increase of production to meet the demand of the external market, resulted in a significant increase in pollution related to this industry. Salt used in the dying process resulted in the generation of highly saline wastewater laden with excess dyes and other chemicals. The disposal of this wastewater further impacted water sources and the environment in general (Dhanya et al. 2005; Arumugam et al. 2014). Specifically, the uncontrolled discharge of wastewaters contaminated groundwater and made it unusable for the local community for domestic use.

When analyzing sustainable engineering-related issues associated with LDCs, it is important to remember that the relative degree of achieving sustainable outcomes will be largely based on the activities undertaken in the developing world. LDCs hold large percentages of nonrenewable resources but often consume relatively small amounts. However, these resources are increasingly utilized, especially in growing industries located in LDCs.

Fortunately, LDCs do hold significant advantages when compared to the humble industrialization beginnings of most MDCs. For instance, LDCs are starting at a far more advanced technological "starting point" as compared to where many MDCs started. LDCs can learn from the development experiences, both good and bad, exhibited by MDCs during the past. For instance, with respect to energy utilization and infrastructure, LDCs may opt to avoid incorporation and utilization of fossil fuels, instead choosing to utilize low-emission or emission-free energy sources, such as wind or solar energy sources. Since LDCs have not developed an underlying infrastructure to an appreciable extent, they have the opportunity to "do it right the first time" and avoid the past mistakes or regrets experienced by MDCs (Archibugi and Pietrobelli 2003).

3.2.3 Social Sustainability Concept

Sustainability is associated with the preservation of resources for the benefit of future generations; thus, it is common to refer to sustainability primarily in terms of natural resources, their economic and environmental benefits and drawbacks, and the impact of present actions for future generations. However, there is also an increased focus on the concept of social resources – as well as access to these within the same generation or across generational lines. Some of these social resources include culture and cultural resources, human rights, labor rights, and corporate governance. The concept of available access to these within and among generations is referred to as "social sustainability" (Dillard et al. 2009). With respect to human development, the environment and natural resources may be used to achieve better standards of living. Income and social resources are also important factors in achieving quality-of-life and well-being.

As the key issues that affect society and social sustainability can vary significantly, there is no straightforward definition of social sustainability. The Western Australia Council of Social Services (WACOSS) defines social sustainability as the following (McKenzie 2004):

Social sustainability occurs when the formal and informal processes, systems, structures, and relationships actively support the capacity of current and future generations to create healthy and livable communities. Socially sustainable communities are equitable, diverse, connected and democratic and provide a good quality of life.

WACOSS has outlined several dimensions associated with social sustainability. These include the following (WACOSS 2013):

- *Equity* – The community provides equitable opportunities and outcomes for all its members, especially the poor and the vulnerable.
- *Diversity* – The community promotes and encourages diversity.
- *Social cohesion/interconnection* – The community provides processes, systems, and structures that promote connectedness within and outside of the community at the formal, informal, and institutional level.
- *Quality of life* – The community ensures that basic needs are met and fosters a good quality of life for all its members (e.g. health, housing, education, employment, and safety).
- *Democracy and governance* – The community provides democratic processes and open and accountable governance structures.
- *Maturity* – The individual accepts responsibility of consistent growth and improvement through broader social attributes (e.g. communication styles, behavioral patterns, indirect education, and philosophical explorations).

3.2.4 Social Indicators

The evolution of a society toward more sustainable conditions, or the level of social sustainability, can be measured through several important social issues that can be considered as the indicators of the status of a society. These issues or indicators can give a quantitative or qualitative measure of the detrimental effect on society and the ability of societies to achieve social well-being (Table 3.1).

Poverty is defined as the lack of sufficient wealth to meet basic human needs. It is caused by a lack of essential resources, lack of education, political and societal oppression, and similar factors. Often, poverty results in physical and mental health deterioration because of a lack of resources to acquire proper or enough nutrition and health resources. Ultimately, poverty can result in death, either directly or indirectly, and can exacerbate civil unrest and crime.

Table 3.1 Global social issues that affect the sustainability of societies.

Poverty
Exploitation
Deteriorated physical and mental health
Lack of essential services
Civil unrest
Crime
Depression

In its most basic definition, exploitation is the unfair treatment of people, often to achieve a physical, economic, or social benefit. It occurs when social, financial, or physical power is used to force less empowered members of society to serve the needs or desires of more powerful members of society to the detriment of themselves. Often, it exacerbates the differences between these classes, as the exploiting classes amass more wealth and power and the exploited classes become weaker and poorer. It is often manifested by the physical exploitation of people as labor or through financial exploitation. If left unchecked or unabated, it can lead to civil unrest.

Another key issue is mental/physical health. When a society is in a stressed or depressed condition, it can lead to widespread deterioration of individual mental and physical health. Some of the causes include inadequate nutrition, exercise, and/or recreation, increased pollution, physical strain and injury, stress, lack of mental stimulation, and lack of purpose or meaningful contribution opportunities. Both physical health and mental health are directly connected to the physical environment, and effects to the physical environment can transcend and affect the overall health. When mental and physical health is adversely affected, it can lead to reduced productivity, a strong effect on quality-of-life, and a strong enhancement of mortality factors.

One manifestation of diminished mental health is depression. This is typically associated with feelings of unhappiness, hopelessness, powerlessness, and lack of self-worth. It is often rooted in poor treatment, poor health, circumstances of poverty, exploitation, and isolation. It can result in reduced societal contributions and diminished quality-of-life.

Mental and physical health can also drastically suffer when there is a lack of essential services or a lack of access to such services. In many cases, this access is adversely affected when there are insufficient systems in place to provide these services. This diminished access can result from poverty, ineffective or corrupt governance, lack of community empathy or responsibility, crime, natural disasters, or other natural or man-made barriers. If the basic needs of the community are not met, significant stress will be placed on society, which could lead to reduced economic output, diminished physical and mental health, and an acceleration of crime, poverty, and other ill effects to the society.

When poverty, corruption, and a lack of essential services exist, it can lead to civil unrest. These reactions are manifested in rebellion and mob/mass action of citizens against the systems and organizations that are in place. Civil unrest is often rooted in feelings of resentment toward poor management, neglect, and exploitation, often at the hands of government institutions. Civil unrest can lead to disruptions of already stressed services and infrastructure and can lead to significant changes or replacement (sometimes for the better; other times for the worse) of the organizations that are perceived to be inadequately meeting the needs of the society.

3.2.5 Social Impact Assessment

Projects and other interventions, usually associated with infrastructure and industrial activities, may have an important influence on the surrounding communities. A Social Impact Assessment (SIA) may analyze and measure the impact of such interventions in the community. The International Association of Impact Assessment (IAIA) states that a SIA includes the processes of analyzing, monitoring, and managing the intended and unintended social consequences, both positive and negative, of planned interventions (policies, programs, plans, and projects) and any social change processes invoked by those interventions. Its primary purpose is to bring about a more sustainable and equitable biophysical and human environment.

The SIA is a method to evaluate the impact of a project or activity in the social community, in a way similar to the Environmental Impact Assessment (EIA). The EIA was first developed in USA in 1970s, and the SIA has been developed to give a response to the increasing demands of society about the impact of different interventions in the social and biophysical media (Dendena and Corsi 2015).

SIAs are performed by professionals with diverse backgrounds, including expertise in sociology, anthropology, geography, development studies, planning, and other fields. They are often carried out with EIAs, but they have not been as widely adopted, and often play a minor role in EIAs. Nevertheless, the present tendency is to incorporate the SIA with the same weight of EIA in an Environmental and Social Impact Assessment (ESIA) (Esteves et al. 2012). Several frameworks have been developed on a national basis; we will take a brief look at the framework commonly used in the United States. The US framework includes several key aspects as follows:

- Achieve extensive understanding of local and regional settings to be affected by the action or policy
    o Identify interested/affected stakeholders and other parties
    o Develop baseline profiles of local and regional communities
- Focus on key elements of the human environment
    o Identify the key social and cultural issues that could be affected or related to the action/policy
    o Select social and cultural variables that measure and explain the issues identified
- Identify research methods, assumptions, and significance
    o Research methods should be holistic in scope
    o Research methods must describe cumulative social effects
    o Ensure methods and assumptions are transparent and replicable
    o Select appropriate forms and levels of data collection and analysis
- Provide quality information for use in decision-making
    o Collect sufficient and useful quantitative and qualitative data
    o Ensure methods are scientifically robust
    o Ensure the integrity of the collected data
- Ensure that any environmental justice issues are fully described and analyzed
    o Ensure methods consider underrepresented and vulnerable stakeholders and populations
    o Consider the distribution of all impacts to different social groups
- Undertake evaluation, monitoring, and mitigation
    o Establish mechanisms for evaluation and monitoring of the action, policy, or program
    o Provide an effective mitigation plan where needed
    o Identify data gaps and plan for filling these needs

Figure 3.2 and Table 3.2 illustrate a flowchart and a matrix, respectively, associated with this SIA framework. Examples of key SIA indicators are as follows:

- *Population characteristics* – Present and expected population, ethnic/racial diversity, demographic mix, and seasonal variations.
- *Community and institutional structures* – Local government and connections to the larger system, patterns of employment and economic diversification, voluntary organizations, and religious and other interest organizations.
- *Political and social resources* – Distribution of power/authority, public participation, income/wealth distribution, and legal and civil rights.
- *Individual and family* – Health, education, safety, and friendship networks.
- *Community resources* – Natural resources/land use, physical environment, recreation, availability of housing and services, viability of community life, and historical/cultural resources.

When compiling and quantifying these indicators, a ranking system may be applied to normalize the quantification process. Scaled rankings may be applied to both positive and negative

3.2 Social Issues

Activities	Actions
1. Develop public involvement program	Public involvement
2. Describe proposed action and alternatives	Identification
3. Describe relevant human environment and zones of influence	Community profile
4. Identify probable impacts	Scoping
5. Investigate probable impacts	Projection of estimated effects
6. Determine probable response of affected parties	
7. Estimate secondary and accumulative impacts	
8. Recommend changes in proposed actions or alternatives	Formulation of alternatives
9. Mitigation, remediation and enhancement plant	Mitigation
10. Develop and implement monitoring program	Monitoring

Include interested and affected parties in all steps of the SIA process

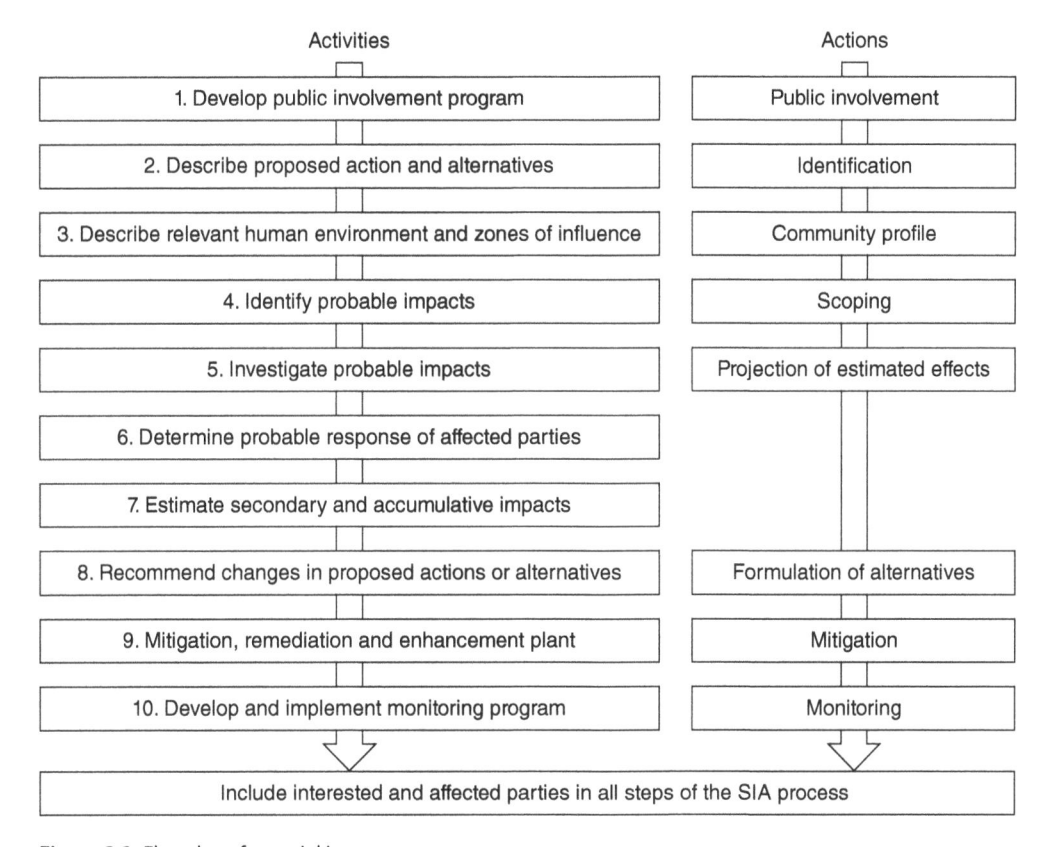

Figure 3.2 Flowchart for social impact assessment process.

impacts; these are often based on the number of people or percentage of the population that may be affected. The rankings can be applied to individual subgroups or the community and the scores may be summed as appropriate.

The quantification and measurement of positive and negative attributes remain to be a work in progress; however, it is important to incorporate it on an objective basis to the maximum degree practicable. Moreover, when negative, substantive impacts are identified, and mitigation efforts should be made to minimize or avoid the impact.

It is important to note that quantification is not a goal in and of itself – it is meant to be a process where a comparison and assessment can be made to allow for informed decisions about project design, implementation, and mitigation as necessary.

The quantification of the variables identified in the SIA can be based on a ranking system considering the amount or fraction of people that it is affected, positively or negatively, by a specific action or activity. Thus, for example, when evaluating social costs and benefits of a specific project or action, a value from −4 to 0 or 0 to 4 can be assigned to the variables based on the number of people affected by that action (Table 3.3).

The social impacts can be assessed and quantified for individuals or small groups and the whole community separately. Thus, the SIA may identify what groups or individuals are more affected from an action or a project, and in turn, analyze its influence on the whole community. The social benefits for the individuals and the community can be added and then the costs subtracted to get an overall score for each action, project, or alternative. Thus, the SIA may decide between different alternatives based on the benefits/costs for the community; or the SIA may

Table 3.2 Matrix of social impact assessment variables.

Social impact assessment variables	General planning, policy development preliminary assessment	Detailed planning, funding, and impact assessment	Construction implement	Operation/ maintenance	Decommission/ abandonment
Population change					
Population size, density, and change					
Ethnic and racial composition and change					
Relocating people					
Inflow/outflow of temporaries					
Presence of seasonal residents					
Community and institutional structures					
Voluntary associations					
Interest group activity					
Size and structure of local government					
Historical experience with change					
Employment/income characteristics					
Employment equity of disadvantaged groups					
Local/regional/national linkages					
Industrial/commercial diversity					
Presence of planning and zoning					
Political and social resources					
Distribution of power and authority					
Conflict newcomers and old-timers					
Identification of stakeholders					
Interested and affected parties					
Leadership capability and characteristics					
Interorganizational cooperation					
Community and family changes					
Perception of risk, health and safety					
Displacement/relocation concerns					
Trust in political and social institutions					
Residential stability					
Density and acquaintanceships					
Attitudes toward proposed action					
Family and friendship networks					
Concerns about social well-being					
Community resources					
Change in community infrastructure					
Indigenous populations					
Changing land use pattern					
Effects on cultural, historical, sacred and archaeological resources					

Table 3.3 Ranking system for quantification of variables in SIA based on the fraction of people affected.

To assess social costs: rank −4–0	
0	No or minimal disadvantage to individuals or communities
−1	Significant disadvantage to a small group of individuals or minimal disadvantage to <5% of population
−2	Disadvantage to 5–20% of population
−3	Disadvantage to 20–50% of population
−4	Disadvantage to >50% of population

To assess social benefits: rank 0–4	
0	Minimal or no advantage to individuals or communities
1	Significant advantage to a small group of individuals or minimal advantage to <5% of population
2	Advantage to 5–20% of population
3	Advantage to 20–50% of population
4	Advantage to >50% of population

even be used to propose modifications to a project or an action to reduce those negative impacts in the community or individuals. In the case of a social impact that is irreversible or of substantial significance, no matter how small the affected group is, those negative impacts must be highlighted in the final assessment, and such impacts should be avoided with alternative solutions or modifications to the project. The main limitation of this quantification and assessment method is associated with the subjectivity of many social variables or impacts that may change the final scoring of the SIA over a wide range. Regardless, it is necessary to investigate social sustainability for completeness of a triple bottom line assessment.

3.2.6 Social Sustainability Implementation

Social sustainability attributes are necessarily implemented at the project/activity level. In designing and implementing projects, it is important to incorporate and respond to the input of the stakeholders in the local community, consider local opinions and contexts in design aspects, and facilitate designs that promote social inclusion, cohesion, and accountability. Citizen participation, engagement, and empowerment are key facets since the populace is always a key stakeholder, and acceptance of a project requires their emotional investment. To achieve this, it is important to perform adequate social analyses, which in turn allow for identification of social opportunities as well as mitigation of social impacts and risks. When designing projects, it is important to understand the social environment as much as the physical environment and pursue the goals of social sustainability along with economic and environmental sustainability.

3.3 Economic Issues

Many economic activities, such as mining, industry, transportation, rely upon the utilization of natural resources, and they have an enormous impact on the environment and on the future availability of those resources. Thus, any discussion about sustainability must include an

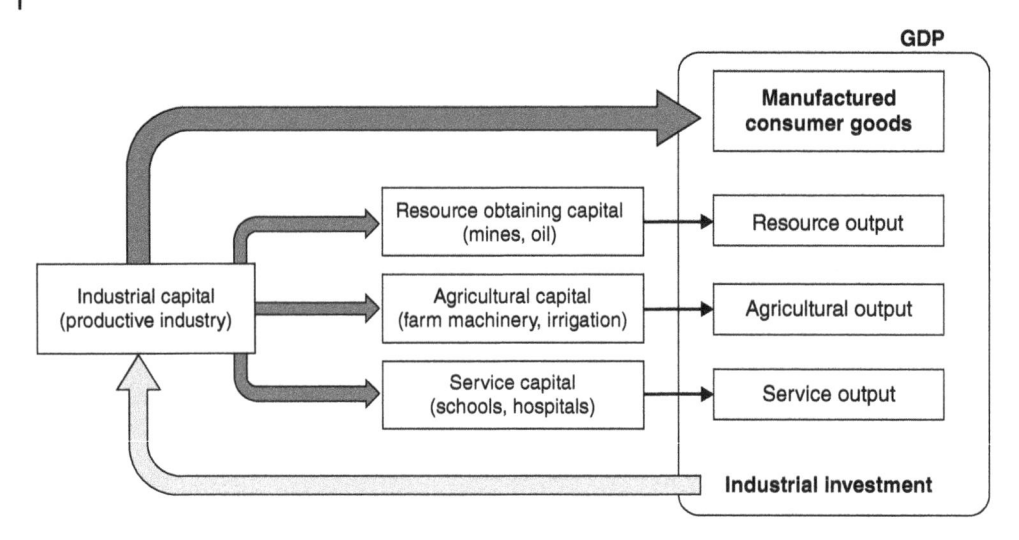

Figure 3.3 Capital flow in productive economy.

economic context. When considering the economic aspects of sustainability, it is important to consider a multidimensional economic analysis of a particular activity or set of activities that accounts for ecological and environmental effects. Additionally, social, cultural, and public health related aspects should also be incorporated.

There are different ways in which the economic context can be defined. One model offers two distinctive economic disciplines: the *physical economy*, defined as activities in which tangible natural resources act as a limit to activity; and the *money economy*, in which activities are defined in monetary means and are not constrained by resources or physical laws (Robertson 2005). Figure 3.3 depicts a schematic of capital flow within an economy, in which investments are made in productive capacity, which in turn yield beneficial output. However, the important question to consider is how to measure economic output from a sustainability perspective? There are traditional methods – cost–benefit analyses, either considering a payback method or using a discount rate. However, it is difficult to incorporate these approaches in environmental or social issues. Alternatively, other methods, such as the human capital method, the cost of illness method, the preventive/mitigative expenditure method, the wage differential method, the contingent valuation method, and the surrogate actions method may be able to incorporate different human effects with varying degrees of precision (Dixon et al. 2013). While all of these methods try to assign dollar values to the aforementioned issues, they can be complicated, have inherent drawbacks, and should be applied with caution.

3.3.1 Economic Assessment Framework

To provide a robust assessment framework for environmental issues, "green accounting" and "environmental accounting" methods have been developed. The purpose is to accurately estimate environmental costs associated with economic activity. Some additional factors may include the cost of raw materials, manufacturing costs, costs associated with resource use, waste and scrap generation, and product design. Environmental costs are assigned to the associated activity or activities with the purpose of managing these costs using traditional cost-benefit analyses (Jasch 2006).

A multi-tier environmental costing framework can be used – it consists of five tiers associated with the costs and benefits of economic activity (Langfield-Smith et al. 2015).

- Tier I costs are those that are normally captured in traditional economic engineering applications and include the following: capital equipment, materials, labor, supplies, utilities, structures, and salvage value.
- Tier II costs incorporate environmental costs that are commonly applied to overhead, including the following: off-site waste management charges; waste treatment equipment and operating expenses; permit costs; sampling and analysis of waste streams; waste/emissions inventory assessments; documentation and inspection of waste storage and handling; emergency response plans; documentation of chemical usage; development, implementation, and reporting associated with pollution prevention plans and related activities.
- Tier III incorporate environmental liability costs, including the following: compliance obligations; remediation obligations; fines and penalties; obligations to compensate private parties for personal injury, property damage, and economic loss; punitive damages; and natural resource damages. Tier III costs depend on the probability that an event will occur, the costs associated with the event, and when the event occurs.
- Tier IV consists of costs and benefits (internal to a company) associated with improved environmental performance.
- Tier V consists of costs and benefits (external to a company) associated with improved environmental performance.

3.3.2 Life Cycle Costing

Life cycle analysis (LCA), which will be presented in greater detail in Chapter 9, is a rapidly evolving framework in which the costs and benefits associated with an economic activity are assessed over the full life of a product or an activity. According to the International Organization for Standardization (ISO), LCA is the "compilation and evaluation of inputs, outputs, and the potential environmental impacts of a product system throughout its life cycle" (ISO 2006). Further, economic and social aspects are, typically, outside the scope of the LCA. Other tools may be combined with LCA for more extensive assessments (Klöpffer 2003; Lacasa et al. 2016).

Two extensions of LCA include life cycle costing (LCC) to account for the economic dimension, and accounting for the work environment and employment to incorporate the social dimension (Dhillon 2009). For LCC, the costs induced by a product, both indirect and direct, in public and private activities, are accounted. It is important to determine proper cost categories, cost bearers, cost models, and cost aggregations (Steen 2005).

Cost categories are grouped according to economics (budget costs, market costs, social costs, and alternative costs); life cycle stages (research and development, primary production, manufacturing, use, and disposal); and activity types (design, transport, sales, and manufacturing) (Dhillon 2009). Cost bearers include producers, supply chain participants/activities, owners, users, regional/national society, and global society.

Several traditional cost models may be used. Some include net present value (NPV), average yearly cost, steady-state cost, annuity, payback time, and cost–benefit ratio. The discount rate used in these analyses is also key – it may be statutorily assigned, market-based (CPI, CCI, Libor) or a specific company's internal preference. Additionally, the appropriate time horizon needs to be determined.

3.3.3 True-cost Accounting

Another approach gaining popularity is true-cost accounting (TCA), a method that seeks to calculate the social, environmental, and economic costs and benefits of a product over its lifespan instead of focusing solely on the cost of its initial purchase (Burritt and Schaltegger 2000). TCA tries to accomplish the following (Bainbridge 2009):

- *Goal definition and scoping* – Identifies and defines the project and purpose of the assessment.
- *Streamlining the analysis* – Refines the first step by connecting the objectives and other elements to sustainability metrics and impact categories.
- *Identify potential risks* – Evaluates the relative importance of the impact categories and the current feasibility of expressing the costs for each attribute of the project.

To perform the TCA, the Tier I through V costs are assigned, and an impact assessment is conducted to determine the largest contributory costs for each category and the information is then used for the decision-making.

3.4 Legal Issues

Table 3.4 shows a matrix that highlights the evolution of environmental regulations. In the past, evolutions in the framework were directed at remediation – the handling and remedy of waste and the polluted environment. Today, framework changes are primarily focused on compliance – the regulation and mitigation of effects associated with emissions. In the future, the framework will likely again shift focus toward more sustainable economic output, with a greater focus on life cycles.

The interaction of national and international frameworks poses an interesting dilemma. A national statutory framework is primarily limited to within a nation's borders. However, environmental issues – acid precipitation, GHG emissions, ozone depletion, and watershed pollution – often extend across borders and have continental or global effects. The scale of these effects and the political bodies and related legal frameworks do not match; one nation's laws cannot solely be applied to battle environmental issues of such extent. Therefore, it is critical to develop international frameworks or political bodies that can act on a global basis.

Legal frameworks should be cooperative in nature. They should be established to pursue both intragenerational and intergenerational equity. They should also focus less on sanctions and

Table 3.4 Evolution of the environmental regulations.

Time	Activity	Focus	Geographical temporary scale	Regulation	Leader
Past	Remediation	Waste	Local Immediate	Command and control	USA
Present	Compliance	Emissions	Point source Immediate	Command and control, end of pipe	More developed countries
Future	IE/DfE	Products and services life cycle	Regional and global systems All time scales	Establish boundary conditions	EU, Netherlands and Germany

DfE: Design for Environment; IE: Industrial Ecology.

punitive activity and find a means to facilitate cooperation thorough incentives or rewards for positive actions. Additionally, they should be established where legal jurisdiction can match the reach of the particular environmental issue. To put it another way, legal frameworks should attempt to invoke the following (Rogers et al. 2008):

- *Motivation for initiators* – Quality of living standards should be, in the very least, maintained at current levels, but behaviors that lead to an improvement should be encouraged.
- *Intergenerational equity* – The condition of the environment should be preserved "as-is" or better for future generations.
- *Intragenerational equity* – Efforts should be made to provide a decent standard of living for all through "humane globalization."

3.5 Summary

Although it is difficult to separate environmental issues and concerns from considerations of sustainability, it is equally important to consider issues and concerns associated within social, economic, and legal frameworks. An understanding of these concepts is essential when considering and assessing projects or activities from a triple bottom line framework. Sustainable solutions should strive to maximize the overall triple bottom line and their representative metrics. More sustainable solutions lead to a greater intersection of these metrics.

When considering social metrics, projects and activities should be assessed on how they may exacerbate or ameliorate poverty, exploitation, physical and mental health, social services, civil unrest, or crime. Economic metrics are of critical importance as both for-profit and not-for-profit entities seek to maximize rates of return on investment measured in numerous ways, including cost–benefit analyses, discounted rate models, or simple payback models. Additionally, accounting and financial models are evolving in which direct or indirect costs are assigned to a range of sustainable-related factors. Legal frameworks are also important, as a keen understanding of national legal models are key to any successful venture within that specific country. However, international and intergovernmental legal frameworks will likely become more essential in the future to deal with environmental and sustainability issues that span across national jurisdictions and shared borders.

The concepts associated with the social, economic, and legal issues presented in this chapter are essential in providing an appropriate context of sustainability presented and applied throughout this book.

3.6 Questions

3.1 Mention some of the examples for SIA indicators.

3.2 Select a regional society with which you are familiar and discuss indicators of social sustainability associated with it.

3.3 What are the economic issues related to sustainability perspective?

3.4 Discuss the difference between the physical economy and the money economy.

3.5 What is environmental accounting? How do you evaluate environmental cost?

3.6 Explain briefly the multitier environmental costing framework.

3.7 Explain life cycle costing.

3.8 Discuss the differences between NPV and payback financial assessments.

3.9 Define CCI, CPI, and LIBOR. How may these be used in financial assessments?

3.10 Explain true cost accounting.

3.11 Select a product or a project, and list economic metrics in each of the five tiers of environmental costing that may be applied.

3.12 What are legal issues related to sustainability?

3.13 List examples where extranational legal issues between nations may arise with respect to environmental issues.

References

Archibugi, D. and Pietrobelli, C. (2003). The globalisation of technology and its implications for developing countries: windows of opportunity or further burden? *Technological Forecasting and Social Change* 70 (9): 861–883.

Arumugam, K., Rajesh Kumar, A., and Elangovan, K. (2014). Assessment of groundwater quality using water quality index in Avinashi–Tirupur–Palladam Region, Tamil Nadu, India. *International Journal of Applied Engineering Research* 9 (22): 12177–12191.

Bainbridge, D.A. (2009). *Rebuilding the American Economy with True Cost Accounting*. San Diego, CA: Rio Redondo Press.

Burritt, R. and Schaltegger, S. (2000). *Contemporary Environmental Accounting: Issues, Concepts and Practice*. Sheffield, UK: Greenleaf Publishing Limited.

Castells, M. (2011). *The Rise of the Network Society: The Information Age: Economy, Society, and Culture*, vol. 1. Wiley.

Dendena, B. and Corsi, S. (2015). The environmental and social impact assessment: a further step towards an integrated assessment process. *Journal of Cleaner Production* 108: 965–977. https://doi.org/10.1016/j.jclepro.2015.07.110.

Dhanya, D., Tamilarasi, S., Subashkumar, R., and Lakshmanaperumalsamy, P. (2005). Impact of dyeing industrial effluent on the ground water quality and soil microorganisms in Tirupur. *Indian Journal of Environmental Protection* 25 (6): 495.

Dhillon, B.S. (2009). *Life Cycle Costing for Engineers*. CRC Press.

Dillard, J., Dujon, V., and King, M.C. (2009). *Understanding the Social Dimension of Sustainability*, 1–300. New York: Routledge. ISBN: 9781135924935.

Dixon, J., Scura, L., Carpenter, R., and Sherman, P. (2013). *Economic Analysis of Environmental Impacts*. Routledge.

Esteves, A.M., Franks, D., and Vanclay, F. (2012). Social impact assessment: the state of the art. *Impact Assessment and Project Appraisal* 30 (1): 34–42. https://doi.org/10.1080/14615517.2012.660356.

ISO, E. (2006). *14040: 2006. Environmental Management-Life Cycle Assessment-Principles and Framework*. European Committee for Standardization.

Jasch, C. (2006). How to perform an environmental management cost assessment in one day. *Journal of Cleaner Production* 14 (14): 1194–1213. https://doi.org/10.1016/j.jclepro.2005.08.005.

Klöpffer, W. (2003). Life-cycle based methods for sustainable product development. *International Journal of Life Cycle Assessment* 8: 157. https://doi.org/10.1007/BF02978462.

Lacasa, E., Santolaya, J.L., and Biedermann, A. (2016). Obtaining sustainable production from the product design analysis. *Journal of Cleaner Production* 139: 706–716.

Langfield-Smith, K., Thorne, H., Smith, D.A., and Hilton, R.W. (2015). *Management Accounting: Information for Creating and Managing Value*, 7e. North Ryde: NSW McGraw-Hill Education Australia Pty Ltd.

McKenzie, S. (2004). *Social Sustainability: Towards Some Definitions*. Hawke Research Institute, University of South Australia Magill 31 pages.

McMichael, A.J. (2000). The urban environment and health in a world of increasing globalization: issues for developing countries. *Bulletin of the World Health Organization* 78 (9): 1117–1126.

Robertson, J. (2005). *The New Economics of Sustainable Development*, vol. 97, 1–9. EC, Briefing for Policy Makers.

Rogers, P.P., Jalal, K.F., and Boyd, J.A. (2008). *An Introduction to Sustainable Development*. London, UK: Glen Educational Foundation, Inc. ISBN: 978-1-84407-520-6 (hardback).

Steen, B. (2005). Environmental costs and benefits in life cycle costing. *Management of Environmental Quality: An International Journal* 16 (2): 107–118.

Tewari, M. (2006). Adjustment in India's textile and apparel industry: reworking historical legacies in a post-MFA world. *Environment and Planning A* 38 (12): 2325–2344.

WACOSS (2013). Sustainable Community Services. WA State Election 2013 Party Policies. wacoss.org.au/StateElection2013/PartyPolicies/SustainableCommunityServices.aspx.

4

Availability and Depletion of Natural Resources

4.1 Introduction

Since the primary focus of sustainability principles is the preservation of natural resources for future generations, it is critical to consider the prudent utilization of natural resources. When we discuss the availability and utilization of resources, it is important to make a distinction between renewable and nonrenewable resources. A nonrenewable resource cannot be reproduced or regenerated to any appreciable extent within the timeframe of human life, and any material rate of consumption of a nonrenewable resource will result in its eventual depletion. In contrast, renewable resources can be replaced by natural processes and are a result of natural forces/phenomena that are persistent and virtually inexhaustible in the environment. This may also include recyclable materials, which when used for a certain period can be reformulated and/or recaptured for future beneficial use. Examples include timber, solar power, wind power, wave/tidal power, and geothermal energy (Bilgili et al. 2015).

In order to achieve sustainability, the long-term protection and preservation of natural resources is critical. In particular, it is important to preserve nonrenewable resources, as they are finite. This chapter presents information on the availability, use, and depletion of natural resources, which are essential aspects of sustainability assessments.

4.2 Types and Availability of Resources

4.2.1 Fossil Fuels

Many developed nations are dependent on the use of nonrenewable fossil fuel for energy. These fuel sources are a result of the decomposition of plant and animal matter over the course of millions of years due to environmental changes. Primary nonrenewable sources include coal, petroleum, and natural gas. Oil shales and tar sands are regarded as secondary sources. These fuel sources are relatively easily refined and used in a variety of energy-consuming applications but result in the generation of a high amounts of carbon dioxide emissions.

Coal is considered the most abundant fossil fuel, with an estimated reserve of one trillion metric tons (Riazi and Gupta 2015). Coal is most abundant in Asia, Eastern Europe, and the United States. As of 2005, China was the global leader in extraction (44%), followed by the United States (20%), India (8%), and Australia (7%). It is commonly burned as a primary source of heat, and it is used to convert water to steam for electrical generation.

An important concept regarding coal use (or use of another fuel source) is "peak coal." It is defined as the point in time at which a maximum global production peak is reached, after which the rate of production will enter a period of terminal decline (Milici et al. 2013). It may

Sustainable Engineering: Drivers, Metrics, Tools, and Applications, First Edition.
Krishna R. Reddy, Claudio Cameselle, and Jeffrey A. Adams.
© 2019 John Wiley & Sons, Inc. Published 2019 by John Wiley & Sons, Inc.

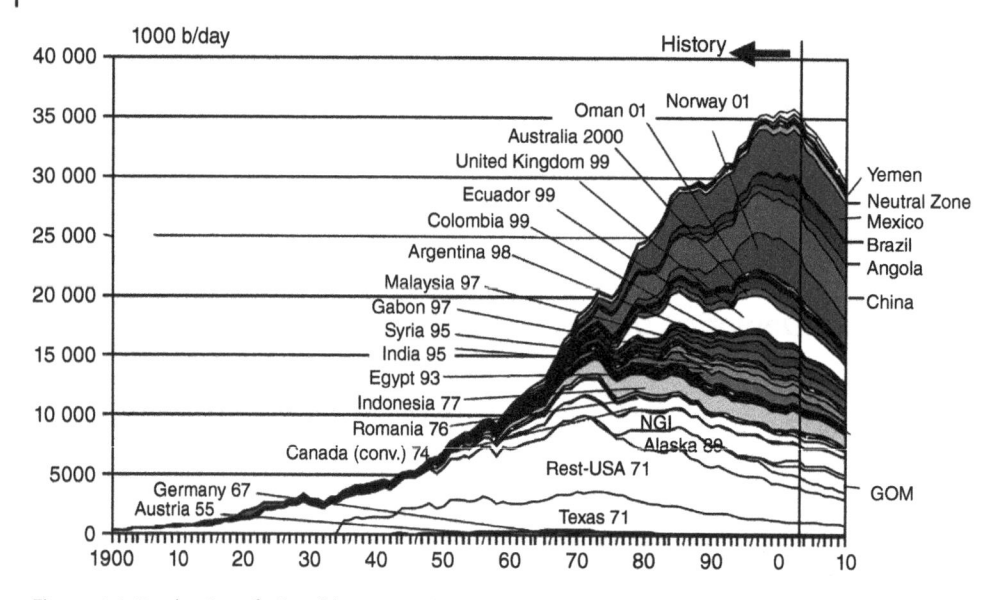

Figure 4.1 Production of oil and future predictions for a scenario in 2004. Source: Wikipedia 2018.

be measured in two ways – by mass of coal used or by energy output. For coal, the energy output per mass has dropped since 2000, indicating that the energetic peak will come sooner than the mass peak. Estimates for the mass peak vary widely, but research by the University of Newcastle (Australia) estimates that the peak may occur between now and 2048.

Petroleum, or crude oil, is a second primary fossil fuel. It is versatile and may be easily refined into a range of products, including motor fuels, material feedstock, lubricants, and heating oils. More than 50% of proven reserves are present in the Middle East (Chapman 2014). However, resources are limited, and unless consumption is controlled, resources could be depleted within 30 years (Miller and Sorrell 2014). From a US peak oil perspective, domestic oil production peaked in about 1970. Further, global production has stabilized close to 2005 highs (approximately 74 million barrels per day). Figures 4.1 and 4.2 depict actual and predicted sources of petroleum.

Oil can be difficult to locate and extract. Current technology can extract about 40% of oil from most wells, and much of the oil that has been considered relatively easy to withdraw from the Earth has been extracted and utilized. However, new extraction technologies are allowing for the extraction of additional reserves, and oil fields long considered exploited are again being returned to an economically productive state. Further, higher prices for crude oil can render oil reserves requiring more difficult and costly extraction to become economically feasible.

Natural Gas is a mixture of gaseous hydrocarbons (methane, ethane, propane, and butane) where methane is the most abundant. It can also be refined and reconstituted, and it is cheap. It can be stored and transported easier than coal or oil and is relatively environmentally friendly. Nearly one-quarter of world energy is derived from natural gas, and its consumption has nearly doubled in the past 30 years (Cronshaw 2015). Significant growth in consumption is coming from developing countries. Figures 4.3 and 4.4 depict natural gas reserves and production.

Oil Shales and Tar Sands are used less frequently but are considered a secondary source of fossil fuels. Oil shale contains kerogen, a waxy form of carbon that can be vaporized at very high temperatures and condensed into shale oil. Tar sand contains low-quality thick, crude oil. Both oil shales and tar sands require significant energy inputs for extraction; however, when global petroleum prices are high, their extraction and use become cost-effective.

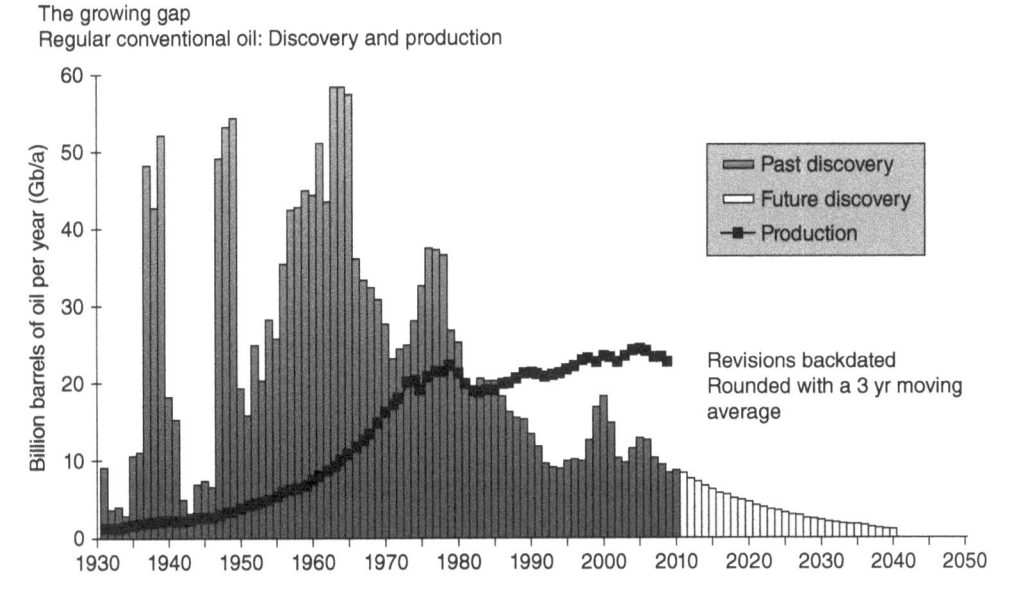

Figure 4.2 Oil production and reserves. Source: From http://www.resilience.org/primer/.

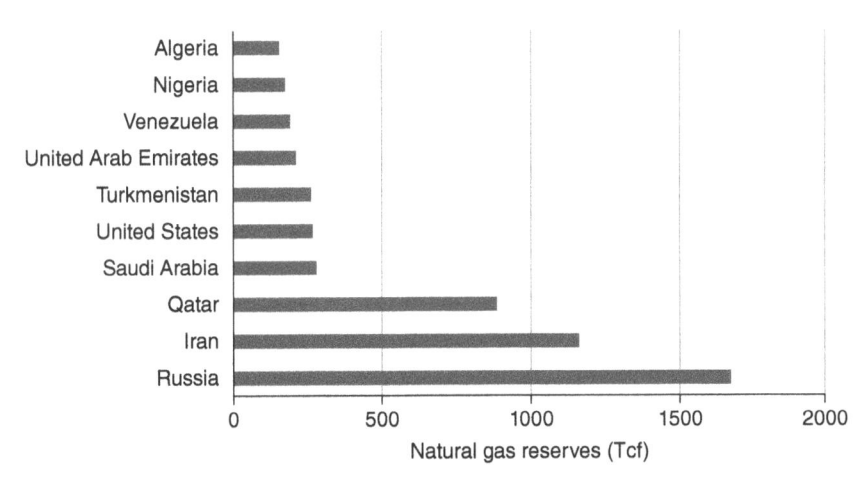

Figure 4.3 Natural gas reserves. Source: U.S. EIA from Oil and Gas Journal data (2012).

4.2.2 Radioactive Fuels

Nuclear power relies upon the fission of fuel atoms (typically uranium) to produce heat, which in turn generates steam for electricity production. Uranium can also be bred into other compounds, such as plutonium, allowing for more efficient fuel use. Uranium may also be refined for use in nuclear explosives/weapons. Nuclear power accounts for approximately 6% of the world's energy and 13–14% of the world's electricity (NEI 2017). It is a clean source of energy since no emissions are generated; however, spent fuel is acutely dangerous for a long period of time. It is estimated that every year 200 000 metric tons of low/intermediate level radioactive waste (LILW) are generated, whereas high-level waste (HLW) accounts for 10 000 metric tons.

When considering peak use of uranium, there is debate as to know if the peak has occurred. Some argue that the peak happened in the 1980s, whereas others claim that a second peak could

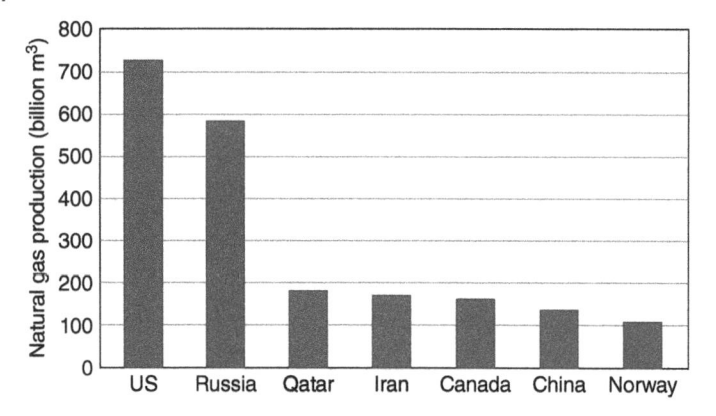

Figure 4.4 Natural gas production in 2014. Source: Data from BP Statistical Review of World Energy (2015).

happen around 2035 (WNA 2016). World demand is expected to continue to increase – it is expected that between 80 and 100 kilotons of uranium will be needed per year by 2025. Uranium is present in many rock formations as well as in seawater. As many other metals, it is found in relatively low concentrations, which limits its economical extraction. Typically, anything containing over 0.075% uranium is considered ore and can be mined.

4.2.3 Mineral Resources

A mineral resource is "a concentration or occurrence of material of intrinsic economic interest in or on the Earth's crust in such form, quantity, or quality that there are reasonable prospects for eventual economic extraction" (Dill 2010). These may be further divided in three categories: inferred, indicated, and measured; depending on the degree of confidence of their amount, quality, and concentration. Figure 4.5 depicts the elemental resources in the Earth's crust.

Inferred mineral resources are those whose concentration, quality, and amount can be estimated with only a low level of confidence. Information derived from geologic evidence (e.g. trenches, outcrops) is used to classify a possible mineral resource in this group. Indicated mineral resources are those whose characteristics: tonnage, grade and density; can be estimated to a reasonable level of confidence based on sampling. If the sampling goes further, a "indicated resource" may be classified as "measured mineral resource" when the estimate regarding content, density, tonnage, and quality can be made with a high degree of confidence. Mineral resources may also be classified as "reserve" when they are known to be economically feasible for extraction (Meinert et al. 2016). Probable reserves are those with similar characteristics than a proven reserve but with a lower level of confidence.

Metal ores include various minerals such as oxides, sulfides, silicates, or native metals. They can be found in the Earth's crust at various concentrations. The rocks and minerals in the deposit must be processed to separate first the metal ore from the waste rock. Then, the metal is extracted from the ore by reducing the metal to the elemental state. With increased demands for the mineral of interest, there is an increased demand for the parent ore. With increased demand comes a greater value and more exploitation. However, this results in higher extraction and a decrease in ore supply as reserves are depleted. Additionally, greater mining efforts often result in detrimental environmental effects.

One example of a mineral resource is copper. Copper, due to its ability to be recycled, has been reused many times over. An estimated 80% of all copper ever mined is still available for use (Boryczko et al. 2014). Chief copper producers include Chile, the United States, and Peru.

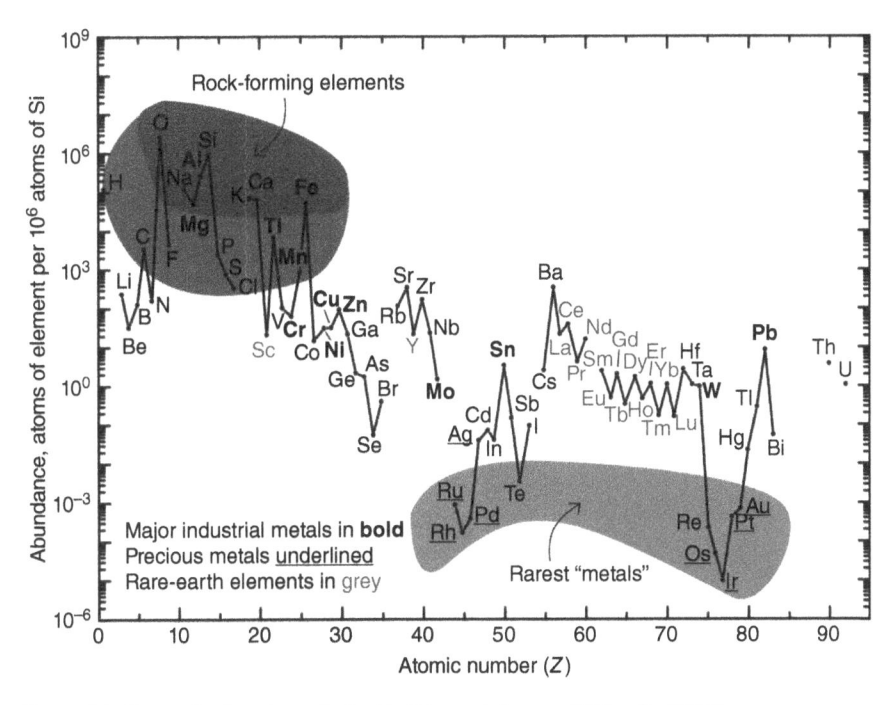

Figure 4.5 Elements abundance in the Earth's crust. Source: Wikipedia (2018).

Copper is a very important industrial metal and is commonly used in power and transmission cables, electrical and automotive equipment, ammunition, and jewelry. Copper has been in use for over 10 000 years, but 95% in use has been extracted since 1900. As developing countries demand more copper, supplies are becoming more strained, leading to increased prices and increased copper theft.

Total world copper production is approximately 15 million tons per year. The demand is growing by approximately 575 000 tons per year and continues to accelerate (Mikesell 2013). Some recent estimates have determined that the demand in 2100 will exceed the quantity available for extraction; others have indicated that copper may run out within 25 years using extrapolation of reasonable growth demand estimates (2% per year). Approximately 21 of the 28 large copper mines cannot be expanded, and many of these mines will have been exhausted between 2010 and 2015 (Sverdrup et al. 2014).

4.2.4 Water Resources

Water is a vital natural resource. Over 70% of the Earth's surface is covered by water. Of course, the majority of this is salt water (97.5%) (Oki and Kanae 2006). Table 4.1 presents the major salt compounds present in seawater, which can be effectively extracted and used for useful purposes. Only 2.5% of the water on Earth is usable fresh water. Unfortunately, of this fresh water, nearly 70% is trapped in icecaps in Antarctica and Greenland; much of the remainder is present as soil moisture or deep groundwater not accessible for human use. Ultimately, the percentage of water available for human use is only about 0.007%. This water is in lakes and reservoirs, rivers and shallow groundwater. The water consumed from natural resources is replenished by precipitations (rainfall and snowfall), and these sources are therefore considered renewable. Figure 4.6 depicts the water cycle responsible for movement and replenishment of water.

Table 4.1 Composition of major constituents in sea water (Salinity: 35) (DOE 1994).

Constituent	Concentration (mol/kg-soln)
Na^+	0.469 11
Mg^{2+}	0.052 83
Ca^{2+}	0.010 36
K^+	0.010 21
Cl^-	0.549 22
SO_4^{2-}	0.028 24

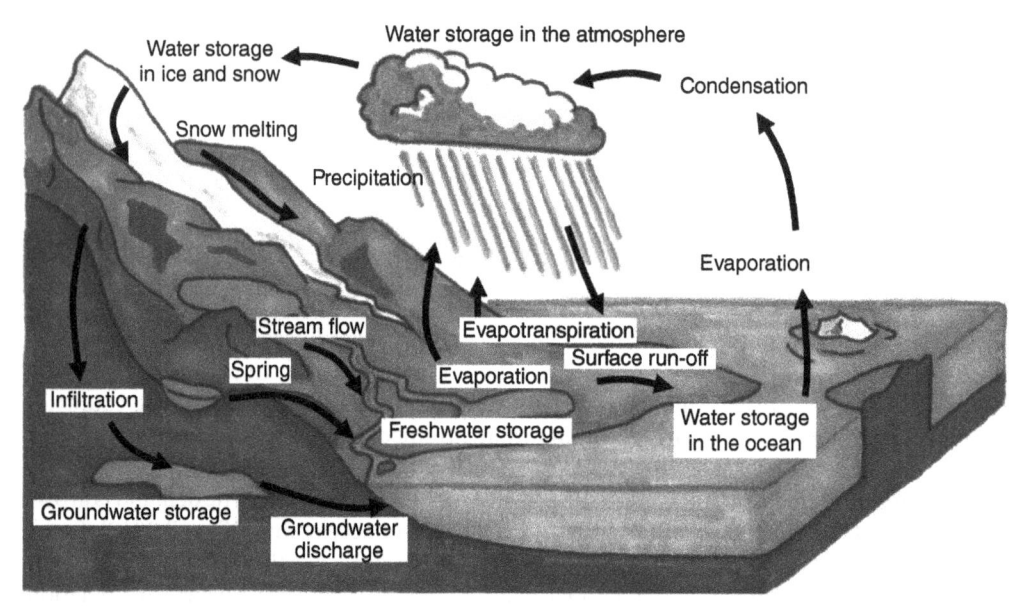

Figure 4.6 Natural cycle of water.

Fresh water is essential for several uses, including domestic use (approximately 8%), irrigation (70%), and industrial uses (22%). Most supplies are derived from surface runoff, although groundwater is also a primary source in many locations. Water use for these sources has increased dramatically over recent decades, as depicted in Figure 4.7. With increased population projected in the future, the demand for water will only increase. Unfortunately, the supply of usable fresh water is decreasing (due to impacts from contamination). As an example, 2 million tons of wastewater and other effluents drain into the world's water. Further, in developing countries, 70% of untreated industrial wastes are dumped into surface waters, where they contribute to contamination of the useable water supply. Therefore, water scarcity will only increase in the future, compounded by the fact that water use has been growing at more than twice the growth of population over the past century (Haddeland et al. 2014). Further, much of the increase in fresh water withdrawals is occurring and will continue to occur in developing countries for the next decade. Currently, 30% of accessible fresh water meets this need; however, it is projected that by 2025 this will increase to 70%.

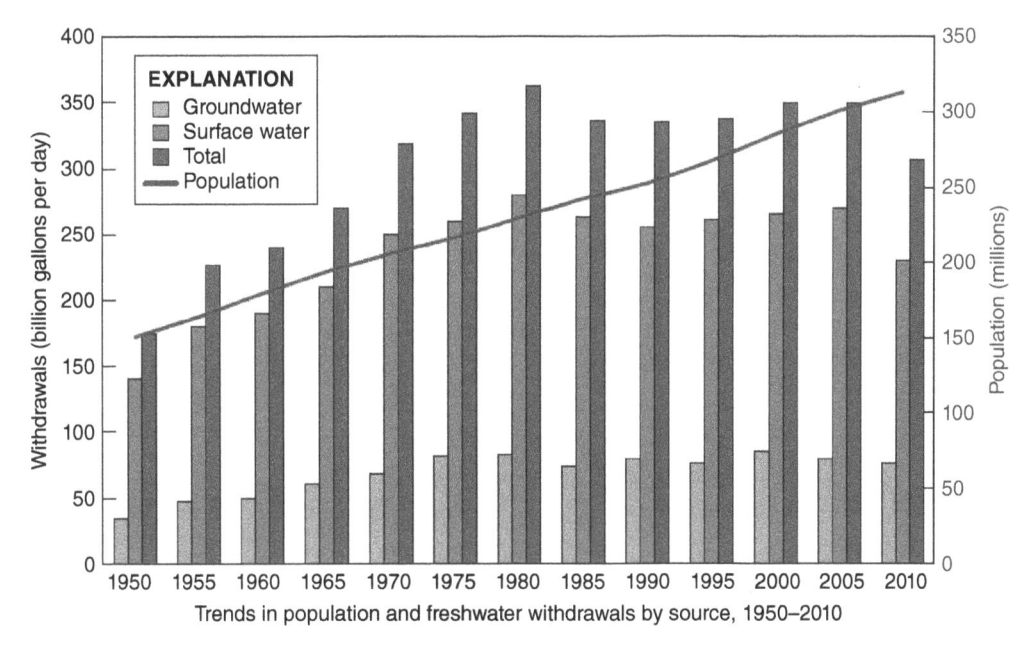

Figure 4.7 Increasing demand of water resources by population in the USA. Source: https://water.usgs.gov/edu/wateruse-trends.html.

Fresh water supplies are essential to health, economic activity, and overall quality of life. Unfortunately, approximately 1 of every 6 people in the world lack basic water services, and over 20% of people do not have access to sanitation services (Gleick 2003). As a result, millions of people die of water-borne diseases every year. Water is also the basis of international conflict. It is essential to provide access to clean fresh water universally for adequate health, welfare, and security for people worldwide. To accomplish this, it is important to provide fresh water sources for domestic use while finding efficient use and allocation for irrigation and industrial purposes.

With respect to water availability, it is helpful to assess the "water footprint." The water footprint assesses the amount of water necessary in all aspects of the production, delivery, use, and disposal of goods and services. The water footprint illustrates how daily consumption/utilization and the problems of water depletion and pollution exist in the respective regions where consumption/utilization occurs. The water footprint of nearly every type of good or service can be comparatively assessed.

4.2.5 Other Elemental Cycles

About 15–20 elements are commonly utilized and deposited or redistributed throughout the global ecosystem in a manner and timeframe where the cycle can be meaningfully analyzed. From a biological standpoint, the four most important elements and related cycles include carbon, nitrogen, sulfur, and phosphorus (Stevenson and Cole 1999).

Figure 4.8 presents a schematic of the global carbon cycle with the sinks and sources (as well as their relative magnitudes). Carbon is provided to the atmosphere through decomposition and respiration in the form of carbon dioxide (Eq. (4.1)):

$$C_6H_{10}O_4 + 6.5O_2 = 6CO_2 + 5H_2O + \text{energy} \qquad (4.1)$$

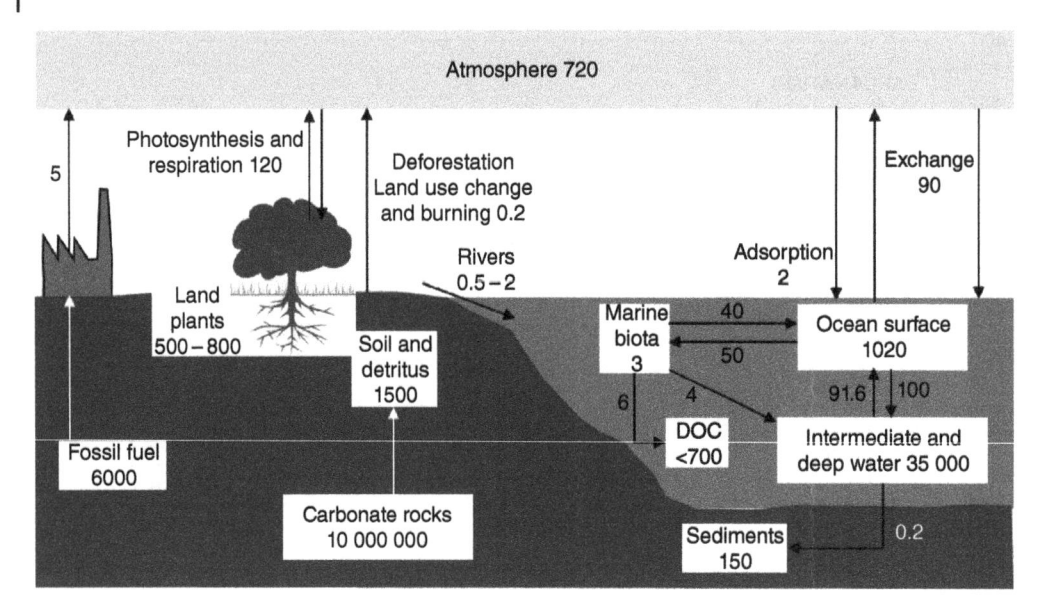

Figure 4.8 Global carbon cycle. (Reservoirs in Pg and fluxes in Pg/yr. 1 Pg = 10^{15} g).

Carbon is brought back to the land and oceans via photosynthesis (Eq. (4.2)):

$$6CO_2 + 5H_2O + \text{sunlight} = C_6H_{10}O_4 + 6.5O_2 \qquad (4.2)$$

Approximately 6000 Pg of carbon are present within reserves of fossil fuels (Trabalka and Reichle 1986). Considering approximately 5 Pg/year of carbon are liberated from the combustion of fossil fuels on an annual basis, this results in approximately 1200 years' worth of total fossil fuel reserves.

Sulfur is present in the lithosphere, hydrosphere, and atmosphere. In the lithosphere, sulfur is present in elemental form as well as in the mineral form (e.g. gypsum). Approximately 7800 trillion tons are present in sedimentary rocks, while 2375 trillion tons are present in oceanic rocks. Approximately 1280 trillion tons of dissolved sulfate ions are present in the hydrosphere, while 6 million tons of sulfur (mostly carbonyl sulfide) is present in the atmosphere (Brimblecombe et al. 1989). Figure 4.9 presents the sulfur cycle. Prior to industrial activity, approximately 55 million tons of reduced sulfide or dusts and 20 million tons of SO_2 from volcanic eruptions would be transported annually between the atmosphere and land. Additionally, 140 million tons would be transported as ocean spray between the atmosphere and the oceans. However, global human activities now release 93 million tons of

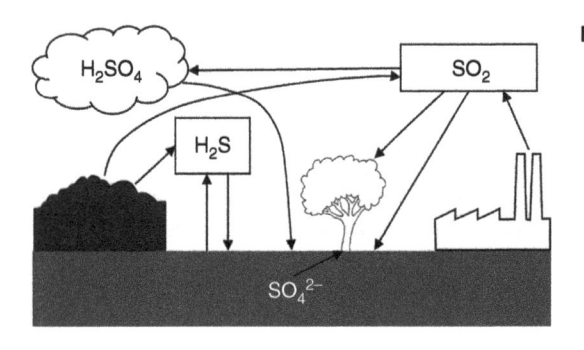

Figure 4.9 Global sulfur cycle.

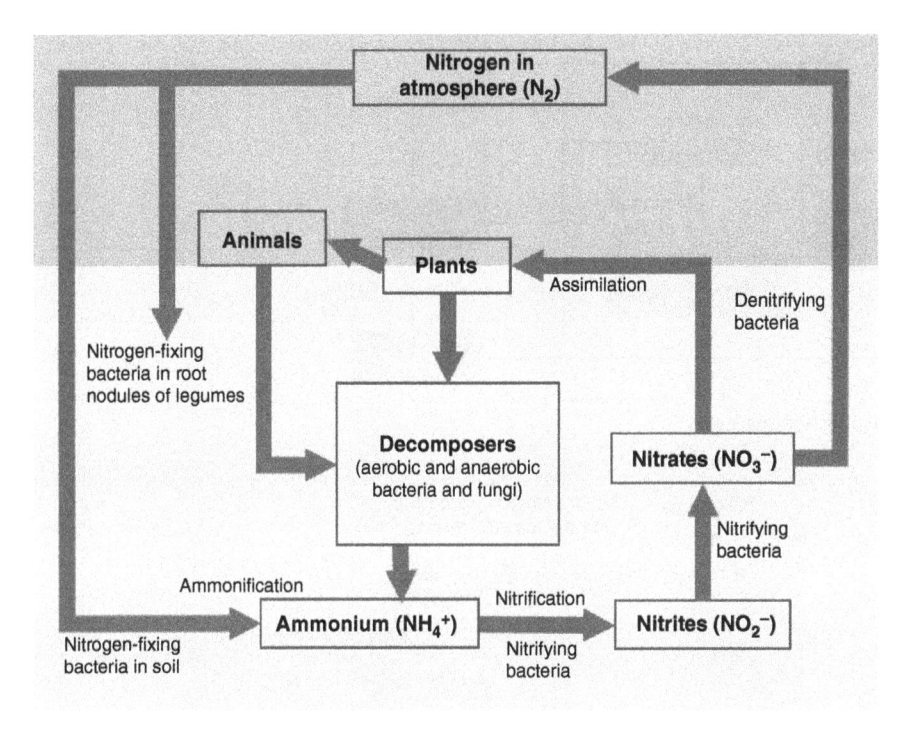

Figure 4.10 Sources and sinks of nitrogen in the environment.

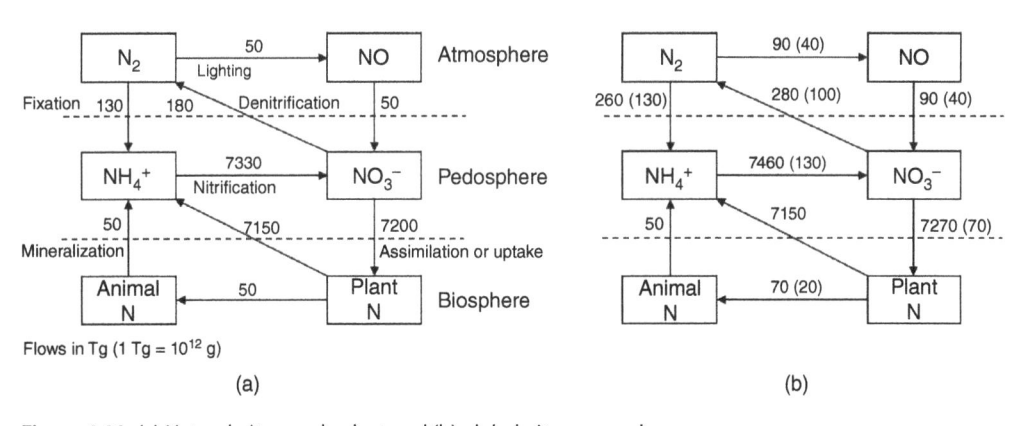

Figure 4.11 (a) Natural nitrogen budget and (b) global nitrogen cycle.

SO_2, 10 million additional tons of sulfur-containing dust, 29 million tons via fertilizers, and 13 million tons via wastewater. The human impact is estimated at 150 million tons annually (Erisman et al. 2013).

Nitrogen is the most prevalent gas in the atmosphere, constituting approximately 78%. It is also ubiquitous in the lithosphere. Figure 4.10 shows the sources and sinks of nitrogen in the environment, and Figure 4.11 illustrates the natural nitrogen budget and global nitrogen cycle, respectively. Nitrogen is a fundamental element for plant growth, and it is intensively used in agriculture to increase land productivity. As a consequence, there are increasing amounts of nitrogen in water that lead to eutrophication of the rivers, lakes, and sea estuaries (Erisman et al. 2013).

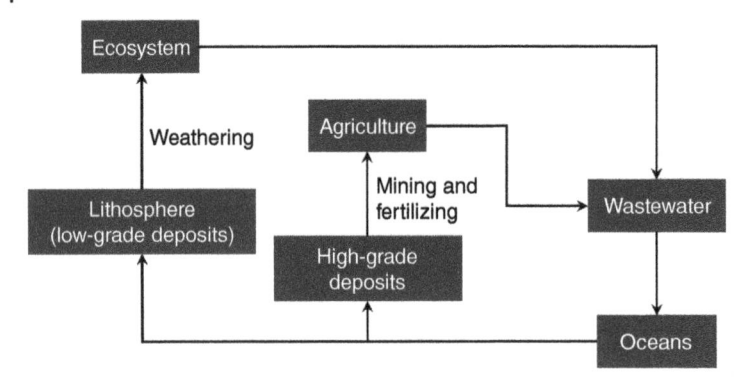

Figure 4.12 Global phosphorous cycle.

Lithosphere	Billion tons
Rocks and sediments, continents	19
ocean	8.4×10^5
DETRITUS	
Soil, peat, litter	96–120
BIOSPHERE	
Continents	2.6
Oceans	0.05–0.12
HYDROSPHERE	
Fresh water	0.09
Oceans	80

(a)

(b)

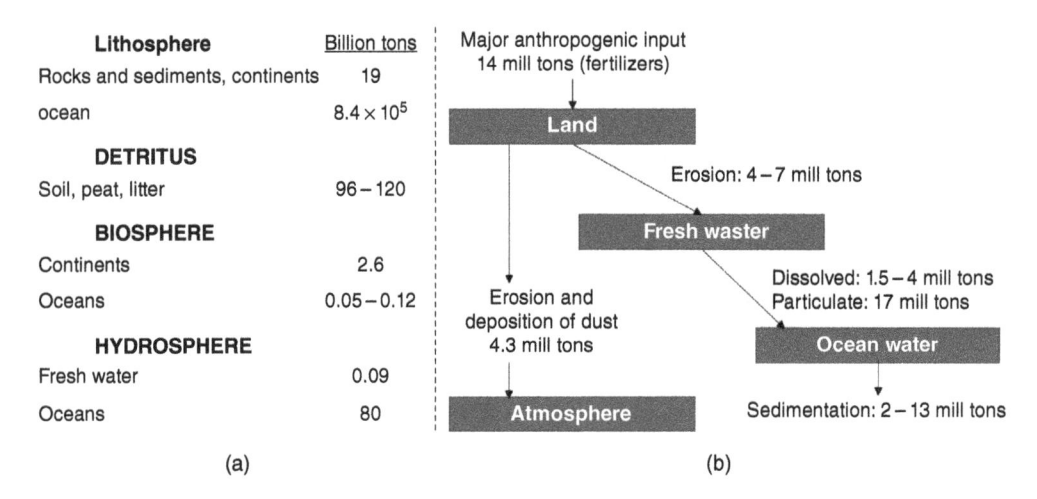

Figure 4.13 The phosphorous cycle (a) global inventory and (b) present flow of P (tons per year).

Phosphorus in rocks and sediments is estimated to be 19 billion tons and 8.4×10^5 billion tons, respectively. In addition, approximately 100 billion tons are present in soil, peat, and litter. The major anthropogenic source of phosphorus is fertilizers, constituting approximately 14 million tons annually. Erosion, deposition, and dust emissions transport phosphorus to the hydrosphere and atmosphere. Figures 4.12 and 4.13 depicts the phosphorus cycle, and their flows and inventory. Similar to nitrogen, phosphorous is also used in agriculture and is also responsible for the eutrophication of waterbodies (Penuelas et al. 2013).

4.3 Resource Depletion

A sustained economic development based on the consumption of nonrenewable resources is impossible within a closed system, such as Earth. The concept of "resource depletion" refers to the exhaustion of raw materials in a region beyond the rate of restoration. Resource depletion is most commonly referred to various human activities such as farming, fishing, mining exploitation, and fossil fuels consumption. It is surprising how little attention is directed toward resource depletion while much attention is focused on climate change. Interestingly, these two challenges are linked because many measures to combat climate change will also reduce resource utilization.

4.3.1 Causes of Resource Depletion

The causes of resource depletion are many, including the following:

- Overconsumption/unnecessary use of resources
- Nonequitable distribution of resources
- Overpopulation
- Slash and burn agricultural practices
- Technological development
- Erosion
- Habitat and biodiversity destruction
- Irrigation
- Mining/extraction
- Aquifer depletion
- Inadequate forestry stewardship
- Pollution or contamination of resources.

"The Limits to Growth" was a relevant study published in 1972. This study alerted the governments, corporations, and society that the human activities would start to exceed the capacity of the planet if the humanity did not change their rate of resource consumption (Meadows et al. 1972). Just within one generation, various precious metals and oil will likely be scarce and much more expensive due to the increasing demand and relatively lower availability. Since that study was published, the resource use and waste generation, instead of decrease, have continue to grow at an accelerated pace due to population growth and the development of the economies of many countries. Thus, it is estimated that the humanity's ecological footprint is at least 20% beyond what the Earth can sustain (Bravo 2014). The United Nations Economic Program, reported in 2011, estimated that the demand of minerals, ores, fossil fuels, and biomass will be 140 billion tons by 2050 (Lacy and Rutqvist 2015). This is three times the current consumption, and much more than the Earth can supply. To combat this, there needs to be a collective commitment to improve recycling and repurposing of materials as well as investments into renewable energy.

4.3.2 Effects of Resource Depletion

Utilization and future resource depletion is illustrated in Figures 4.14–4.17. Figures 4.14 and 4.15 show the utilization of metals in the twentieth century as well as the history of Australian ores, plotted as the percent of metal mined in rock. The trend of mining poorer and poorer metal ores is obvious (Mudd 2007). Figures 4.16 and 4.17 demonstrate the energy required to produce or extract a range of materials. The energy required to extract and process copper ore at different ore grades is roughly linear on a log–log scale. It indicates that mining of an ore with a grade ten times poorer than that being mined today results in energy requirements roughly ten times as much as those required today, mostly to separate the ore mineral grains from their waste rock host. Hence, energy limitations are readily imaginable if we continue extracting copper from ore of rapidly diminishing quality. Finally, Figure 4.18 depicts the environmental disturbance and toxicity associated with specific mining activities. The toxicity of metals differs greatly; however, this chart (Figure 4.18) indicates a much higher risk associated with the mining of arsenic, cadmium, or mercury compared with copper, iron, or zinc.

96 | *4 Availability and Depletion of Natural Resources*

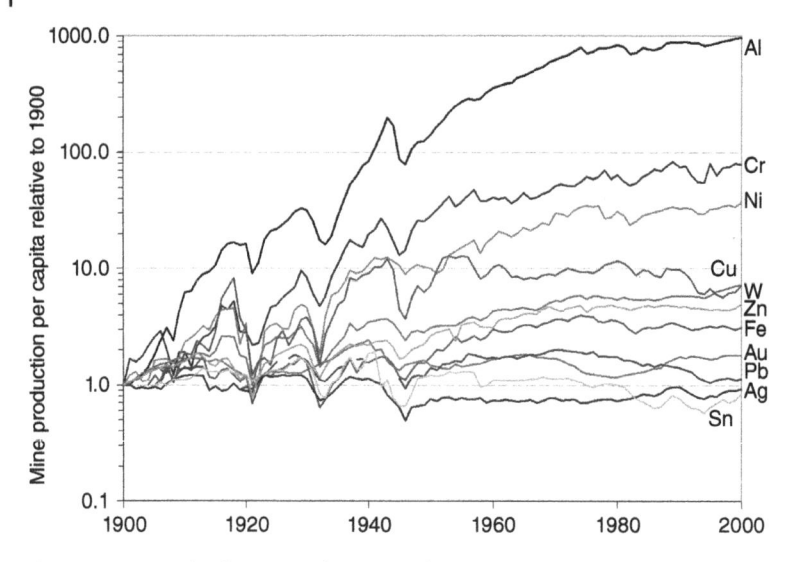

Figure 4.14 Metal utilization in the twentieth century.

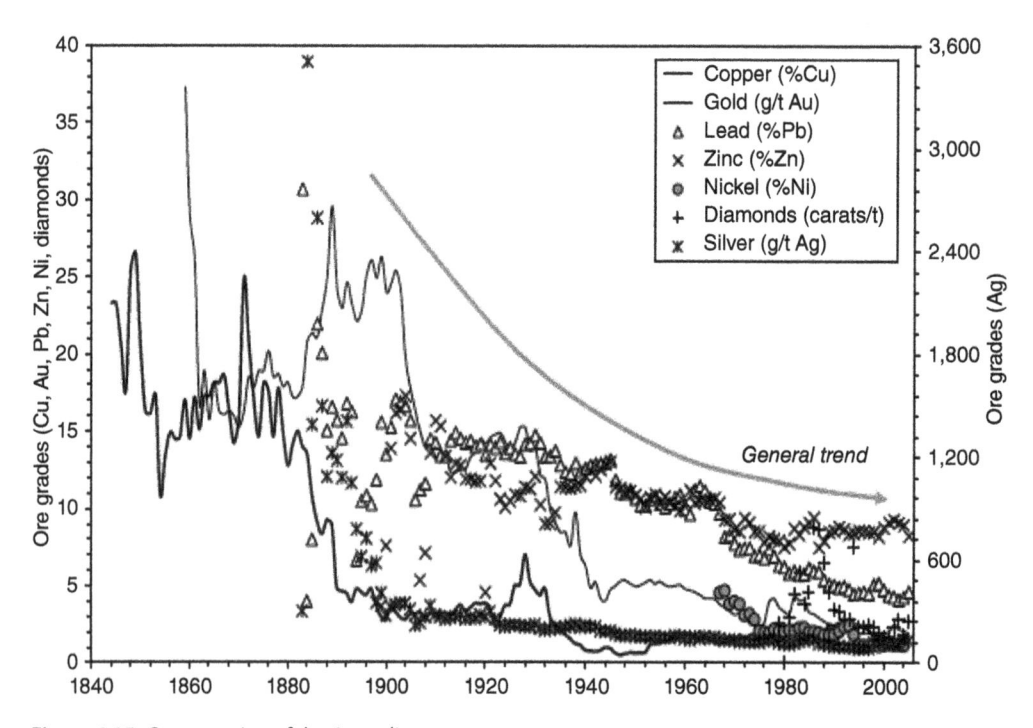

Figure 4.15 Consumption of the Australian ores.

Ultimately, resources are being used at increasing rates, which in turn requires increased energy for extraction and transportation. The uncontrolled mining and acquisition of these materials can have significant adverse effects on the environment. Oddly, materials are further being utilized in a manner that are increasingly unsustainable as compared to other alternatives. The continued scarcity of these materials and the prices associated with them will be subjected to amplified pressures from increased demand and decreased supply.

4.3 Resource Depletion | 97

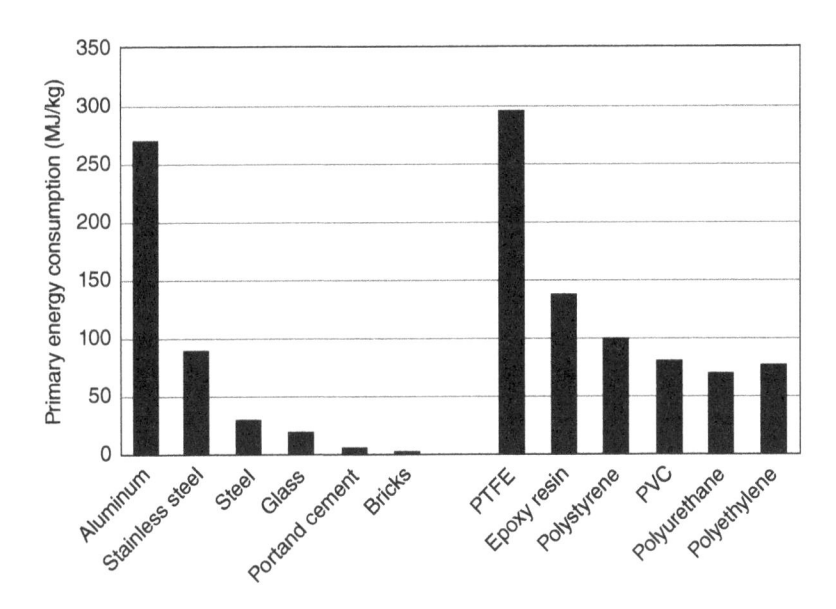

Figure 4.16 Energy requirements per unit of material. Source: Schuckert et al. (1997).

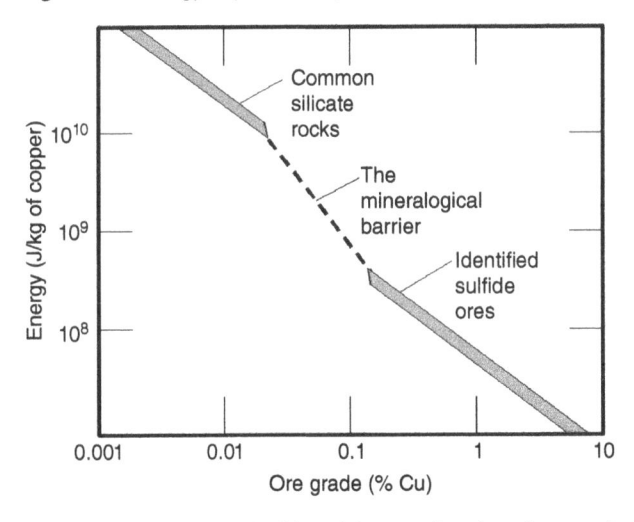

Figure 4.17 Energy required for mining as a function of ore grade. Source: Gordon et al. (1987). Reproduced with permission of Harvard University Press.

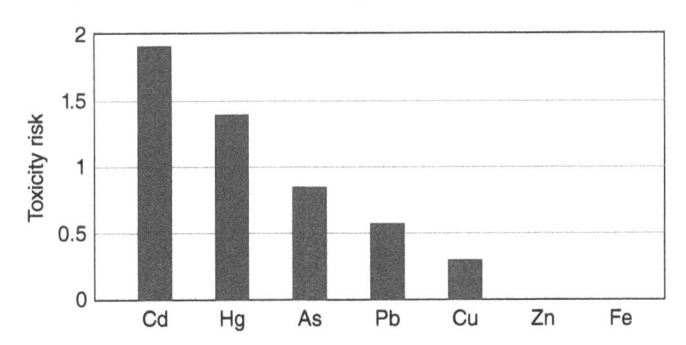

Figure 4.18 Toxicity risk potential of several common metals.

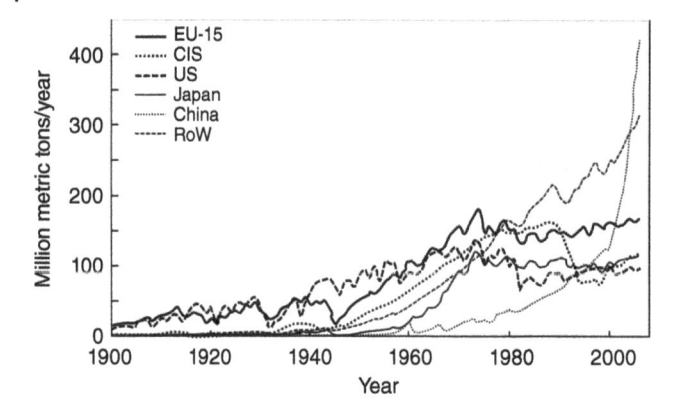

Figure 4.19 Steel production in different countries.

4.3.3 Overshooting

Occasionally, resource depletion (or at least a significant nonsustainable period of consumption) can happen without intention or notice. This concept is known as overshooting. This can be the result of growth coupled with a rapid change that results in overuse. Unfortunately, there can be a delay in the perception of the overuse and the responses that would strive to keep the system within its limits.

When overshooting occurs, resource stocks collapse, and waste and pollution can rapidly accumulate. As a result, economic and human investment is necessarily diverted from other functions to deal with the waste (e.g. wastewater treatment, disposal). Although some positive attributes can occur, such as the invention and adoption of technologies that lessen the need for a given resource, waste accumulation can overcome natural absorption processes, and the additional energy and resources expended to develop the resource can add to the waste generation.

Overshooting can result from unfavorable consumption patterns. With respect to consumption, three general categories have been developed (Princen 2001). *Background consumption* is the first category, and it is defined as the consumption needed by an organism for survival and reproduction. *Overconsumption*, the second category, is defined as the level of consumption that undermines a species' own life support system and which individuals and collective groups can make choices. *Misconsumption*, the third category, undermines an individual's own well-being even if there are no aggregate effects on the population or the species.

Consumption can also result in high rates of change with respect to resource use. For instance, iron and steel use/production in China has experienced a dramatic change. In 2002, China added more steel production capacity than that which existed in the United States (Figure 4.19). This behavior can be observed in other lesser-developed countries (LDCs). It is a common occurrence where LDCs are responsible for extraction of resources for products consumed by others (Figure 4.20).

4.3.4 Urban Metabolism

Consumption behavior may be modeled similar to a metabolic process. This concept, referred to as "urban metabolism," may be defined in terms of stocks and flows, with urban activities contributing according to stocks and flows. The urban stocks that occur via urban metabolism are of importance due to the quantity associated with these stocks. The nature of the stocks consists of significant amounts of recyclable materials, which can serve as a valuable resource in

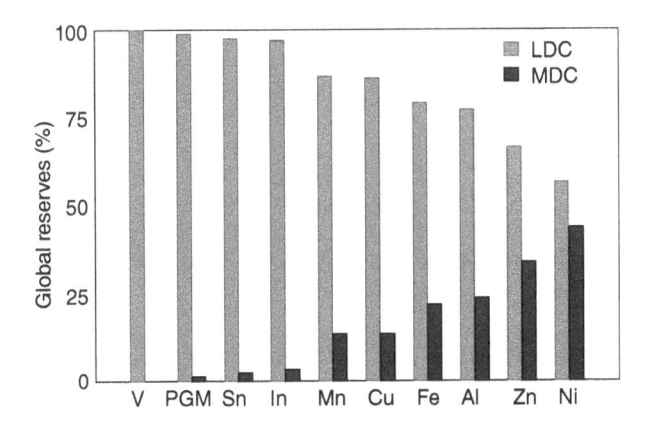

Figure 4.20 Global Reserves in LDCs and Moderately Developed Country of selected materials. Source: US Geological Survey (2008).

themselves. Additionally, the quantity of waste flows associated with the metabolism is important to design adequate waste and wastewater handling facilities. The location of these stocks and flows are also important considerations in devising systems and processes to utilize stocks and handle flows. Most urban areas continue to show per capita increases in the use of water, energy, and materials, as well as the generation of wastes and emissions. Further, urban growth can affect the microclimate and regional ecosystems due to resource use and urban heat island effects.

To date, urban metabolism studies are few and highly specialized and specific. Comparisons of cities and urbanized areas are not yet possible. To facilitate comparisons that could affect changes or mitigation activities, consistent metrics are necessary to compare performance and monitor progress.

4.4 Summary

Resources may be broadly classified into two types: renewable or nonrenewable. The greater attention is on nonrenewable resources, as overutilization of these resources can result in their depletion, which conflicts with the principles of sustainability.

Whether used for energy (fossil fuels), raw materials for manufacturing (petroleum, metals), or other essential processes (fresh water), if not preserved and protected, or at least used at sustainable levels, these resources may become overexploited, which may lessen or eliminate their ability to be utilized by future generations. Additionally, use and liberation of other key elements, such as carbon, sulfur, nitrogen, and phosphorus, can have detrimental effects to the environment when overexploited and redeposited in the environment at rates or volumes greater than what can be handled by the natural absorption capacity of the environment.

Overutilization of these resources can result in different complications. In the case of a general resource depletion, resources are used at a rate that exceeds the rate of generation. If this continues unabated, the reserves of the resource will become depleted. In some cases, "overshooting" may occur; this phenomenon can result in a series of events like a rapid collapse of resource availability.

In order to achieve sustainability, the long-term protection and preservation of natural resources is critical. In particular, given their finite nature, it is of utmost importance to preserve the nonrenewable resources currently available.

4.5 Questions

4.1 Name a few nonrenewable and renewable sources of energy.

4.2 Discuss the concept of "peak coal." How can estimates of "peak coal" vary by the metric used for estimation?

4.3 Explain how crude oil reserves and oil shale/tar sand exploitation is expected to have a more detrimental effect when considering sustainability over time.

4.4 What is the most common nuclear fuel used in nuclear energy generating facilities.

4.5 Discuss benefits and drawbacks associated with radioactive fuel use.

4.6 Explain why extraction of "virgin" copper reserves would be expected to become more expensive with time.

4.7 Select a consumer product and discuss its water cycle with respect to manufacturing.

4.8 Discuss adverse side effects associated with the mining and refining of heavy metals.

4.9 List specific examples how agricultural practices could have an effect on the environment with respect to sustainability.

4.10 What are the major causes for resource depletion?

4.11 What are the effects of resource depletion? Explain briefly.

4.12 What are the issues related to resource depletion from sustainability perspective? How would you assess resource availability?

4.13 Explain how embedded energy, local supply, and environmental impacts associated with recovery may be considered in resource material acquisition.

4.14 Explain how alternate material substitutions can impact resources and environment.

4.15 What is overshooting? Explain briefly.

4.16 Describe how overshooting could have a sudden and detrimental effect on human societal units.

4.17 Discuss the concept of urban metabolism.

4.18 How could cities improve metrics related to urban metabolism, and why would cities have a metabolic advantage over less densely populated areas, such as rural communities?

References

Bilgili, M., Ozbek, A., Sahin, B., and Kahraman, A. (2015). An overview of renewable electric power capacity and progress in new technologies in the world. *Renewable and Sustainable Energy Reviews* 49: 323–334.

Boryczko, B., Hołda, A., and Kolenda, Z. (2014). Depletion of the non-renewable natural resource reserves in copper, zinc, lead and aluminum production. *Journal of Cleaner Production* 84: 313–321.

BP (2015). Statistical Review of World Energy. https://www.bp.com/content/dam/bp-country/es_es/spain/documents/downloads/PDF/bp-statistical-review-of-world-energy-2015-full-report.pdf. (accessed 20 September 2018).

Bravo, G. (2014). The human sustainable development index: new calculations and a first critical analysis. *Ecological Indicators* 37: 145–150.

Brimblecombe, P., Hammer, C., Rodhe, H. et al. (1989). Human influence of sulfur cycle. In: *Evolution of the Global Biochemical Sulfur Cycle* (ed. P. Brimblecombe and A.Y. Lein). Wiley.

Chapman, I. (2014). The end of peak oil? Why this topic is still relevant despite recent denials. *Energ Policy* 64: 93–101.

Cronshaw, I. (2015). World energy outlook 2014 projections to 2040: natural gas and coal trade, and the role of China. *The Australian Journal of Agricultural and Resource Economics* 59 (4): 571–585.

Dill, H.G. (2010). The "chessboard" classification scheme of mineral deposits: mineralogy and geology from aluminum to zirconium. *Earth-Science Reviews* 100 (1–4): 1–420. https://doi.org/10.1016/j.earscirev.2009.10.011.

DOE (1994). *Handbook of Methods for the Analysis of the Various Parameters of the Carbon Dioxide System in Sea Water; Version 2* (ed. A.G. Dickson and C. Goyet). ORNL/CDIAC-74 https://www.nodc.noaa.gov/ocads/oceans/DOE_94.pdf.

Erisman, J.W., Galloway, J.N., Seitzinger, S. et al. (2013). Consequences of human modification of the global nitrogen cycle. *Philosophical Transactions of the Royal Society B* 368 (1621): 20130116.

US Geological Survey (2008). Mineral Commodity Summaries, 2009. Government Printing Office.

Gleick, P.H. (2003). Global freshwater resources: soft-path solutions for the 21st century. *Science* 302 (5650): 1524–1528.

Gordon, R.B., Koopmans, T.C., Nordhaus, W.D., and Skinner, B.J. (1987). *Toward a New Iron Age?: Quantitative Modeling of Resource Exhaustion*. Harvard University Press.

Haddeland, I., Heinke, J., Biemans, H. et al. (2014). Global water resources affected by human interventions and climate change. *Proceedings of the National Academy of Sciences* 111 (9): 3251–3256.

Lacy, P. and Rutqvist, J. (2015). On borrowed time. In: *Waste to Wealth*, 3–18. UK: Palgrave Macmillan.

Meadows, D.H., Meadows, D.H., Randers, J., and Behrens, W.W. III, (1972). *The Limits to Growth: A Report to the Club of Rome (1972)*. New York: Universe Books.

Meinert, L.D., Robinson, G.R., and Nassar, N.T. (2016). Mineral resources: reserves, peak production and the future. *Resources* 5 (1): 14.

Mikesell, R.F. (2013). *The World Copper Industry: Structure and Economic Analysis*. Routledge.

Milici, R.C., Flores, R.M., and Stricker, G.D. (2013). Coal resources, reserves and peak coal production in the United States. *International Journal of Coal Geology* 113: 109–115.

Miller, R.G. and Sorrell, S.R. (2014). The future of oil supply. *Philosophical Transactions. Series A, Mathematical, Physical, and Engineering Sciences* 372 (2006): https://doi.org/10.1098/rsta.2013.0179.

Mudd, G.M. (2007). The sustainability of mining in Australia: key production trends and their environmental implications. Department of Civil Engineering, Monash University and Mineral Policy Institute, Melbourne.

NEI (2017). Nuclear statistics. Nuclear Energy Institute. https://www.nei.org (accessed 11 April 2017).

Oil and Gas Journal (2012). Worldwide look at reserves and production. *Oil & Gas Journal* 110 (12).

Oki, T. and Kanae, S. (2006). Global hydrological cycles and world water resources. *Science* 313 (5790): 1068–1072.

Penuelas, J., Poulter, B., Sardans, J. et al. (2013). Human-induced nitrogen–phosphorus imbalances alter natural and managed ecosystems across the globe. *Nature Communications* 4.

Princen, T. (2001). Consumption and its externalities: where economy meets ecology. *Global Environmental Politics* 1 (3): 11–30.

Riazi, M.R. and Gupta, R. (Eds.). (2015). *Coal Production and Processing Technology*. CRC Press.

Schuckert, M., Beddies, H., Florin, H. et al. (1997). Quality requirements for LCA of total automobiles and its effects on inventory analysis. In: *Proceedings of the Third International Conference on Ecomaterials*, 325–329. Tokyo: Society of Non-Traditional Technology.

Stevenson, F.J. and Cole, M.A. (1999). *Cycles of Soils: Carbon, Nitrogen, Phosphorus, Sulfur, Micronutrients*. Wiley.

Sverdrup, H.U., Ragnarsdottir, K.V., and Koca, D. (2014). On modelling the global copper mining rates, market supply, copper price and the end of copper reserves. *Resources, Conservation and Recycling* 87: 158–174.

Trabalka, J.R. and Reichle, D.E. (1986). The changing carbon cycle. In: *A Global Analysis*. New York: Springer-Verlag. ISBN: 978-1-4757-1917-8.

Wikipedia (2018). Elemental abundances.svg (2018, June 5). Wikimedia Commons, the free media repository. Retrieved 12:02, September 7, 2018 from https://commons.wikimedia.org/w/index.php?title=File:Elemental_abundances.svg&oldid=304744160.

WNA (2016). Uranium mining overview. World Nuclear Association. http://www.world-nuclear.org (accessed 11 April 2017).

5

Disaster Resiliency

5.1 Introduction

Resiliency has become a buzzword lately in different fields such as ecology, climate science, community, cyberspace, energy, and infrastructure. Correspondingly, terms such as ecological resiliency, climate resiliency, cyber resiliency, energy resiliency, and infrastructure resiliency have become popular, depending on the field of consideration. In general, resiliency means the ability to recover from a shock or disturbance and adapt to such changes that may occur in future. Therefore, ecological resiliency refers to the capacity of an ecosystem to recover from perturbations; climate resiliency means the ability of systems to recover from climate change; cyber resiliency means the ability of a computer network to maintain its service at times such as when there is a breach in the information security; energy resiliency means the ability to generate and supply energy even in adverse conditions; and infrastructure resiliency means engineered infrastructure that can withstand, respond and/or adapt to a vast range of disruptive events. The concept of resilience during and after extreme and/or unexpected events was originally introduced by Holling (1973) and Timmerman (1981) for ecological systems and human communities.

Disaster resiliency is another term commonly used with reference to built or engineered infrastructure (especially civil infrastructure) to sustain impacts of extreme events associated with climate change. In this case, resiliency means the capability of an engineered system to anticipate, survive, and recover from climate impacts such as sea level rise, extreme flooding, extreme drought, wild fires, and other stressors. Unfortunately, the frequency and magnitude of these climate impacts are not accurately predicted or forecasted; hence, engineers are challenged to design infrastructure systems for nonstationary design criteria developed from observed impacts. Nevertheless, the ideal goal of designing resilient engineered systems is to have them undergo less damage during the extreme events and then recover to full function quickly.

Engineers have a moral obligation to consider resiliency in engineering designs to prevent loss of property and life while also protecting the environment. The expectation for these designs is that they will be capable of functioning to some degree following exposure to a threat, even if the threat creates conditions well beyond their design limits.

This chapter presents stressors associated with climate change and extreme events, the concept of resiliency, a discussion of a range of initiatives developed by different entities to address resiliency, resilient design frameworks, a general methodology for resilient design, examples of resilient designs, and finally, challenges to resilient engineering designs.

Sustainable Engineering: Drivers, Metrics, Tools, and Applications, First Edition.
Krishna R. Reddy, Claudio Cameselle, and Jeffrey A. Adams.
© 2019 John Wiley & Sons, Inc. Published 2019 by John Wiley & Sons, Inc.

5.2 Climate Change and Extreme Events

In Chapter 2, greenhouse gases, global warming, and climate change were discussed. Scientific data from the Intergovernmental Panel on Climate Change (IPCC), National Oceanic Atmospheric Administration (NOAA), United States Environmental Protection Agency (USEPA), United States Global Change Research Program (USGCRP), and Climate Impacts Group (CIG) Scientific Studies indicate that certain climate trends are occurring on global, national, and local scales. In May 2018, for the first time in the recorded history, the average monthly level of CO_2 in the atmosphere exceeded 411 parts per million (ppm), according to researchers from Scripps Institute of Oceanography at the University of California San Diego and NOAA. Research indicates that if this remains unchecked, increased CO_2 levels could cause pollution-related deaths to increase by tens of thousands. Further, since the CO_2 levels tend to be higher in indoor environments within urbanized areas, numerous urban populations are at risk from physiological effects of CO_2, including decreasing cognition skills. Of course, CO_2 also contributes to warming that causes extreme events such as sea level rise, heat waves, flooding, drought, wildfires, and hurricanes. Recently, increases in CO_2 levels have been accelerating when compared to historic trends. Some experts think that CO_2 levels could reach 550 ppm by the end of the twenty-first century, which could cause average global temperatures to rise as high as 6 °C.

Climatic effects have been estimated based on assumptions of varying degrees of continuing CO_2 emissions. Under a low greenhouse gas emissions scenario, thermal expansion of the oceans, and melting of small mountain glaciers is projected to result in approximately one foot of absolute sea level rise by 2100 on a global scale. Even with a drastic reduction in greenhouse gas emissions, sea level rise will continue through the end of the twenty-first century because of the latency of response of the oceans to temperature conditions at the Earth's surface. When assuming a high gas emissions scenario, these factors are combined with glacial and ice sheet melting resulting in an estimated 4 ft of sea level rise on a global scale by 2100 (NOAA 2017). When combining rising sea levels with high tide, storm surges and subsidence, a range of adverse physical effects may occur, including (i) an increase in the elevation, depth, or extent of inundation along the marine and coastal shorelines; (ii) an amplification of the inland reach of high tides, resulting in increased flooding further inland of the coastline, especially when compounded by severe storm events; (iii) movement of the saltwater intrusion further upstream in tidally influenced rivers; (iv) saltwater intrusion into groundwater; and (v) increased risks of landslide and rates of erosion along coastal bluffs.

By the 2080s, extreme precipitation events (24-hour rain events) are projected to increase in intensity by 22% on average, with a corresponding increase in the frequency (compared to the 1980s). This would be an increase to about 8 days per year, on average, as opposed to just 2 days per year in the past. The increasing frequency and intensity of atmospheric rivers – those moist airflows that extend from the tropical Pacific to the west coast of North America during winter – are expected to carry more moisture in the future, resulting in more frequent, more intense rainfall events (CIG 2018; OCCRI 2017).

With increased temperatures, more winter precipitation will fall as rain instead of snow, and snowmelt is expected to begin earlier in the spring. These increased temperature-induced changes are expected to result in a shorter snow season on average and earlier peak streamflow in rivers with a significant snowmelt component. These changing precipitation patterns contribute to flow changes in major snowmelt-influenced rivers with higher flows in winter and lower flows in summer, more frequent and severe river flooding, increased landslide risk due to saturation of soil, and increased erosion and riverine sediment transport in fall, winter, and spring (NOAA 2017; CIG 2018).

Average annual air temperatures are projected to increase. This means that the average temperatures by the 2050s will be higher than the warmest temperatures in the previous 100 years. Significant air temperature increases are projected after the mid-century. These changing temperature patterns contribute to more severe drought and potentially lower groundwater tables, more frequent and intense heat waves in summer, less frequent and intense cold events in winter, reduced amounts of snowpack and associated altered runoff patterns described above, and increased frequency and intensity of summer wildfires.

The Global Change Research Act of 1990 requires the USGCRP to develop national climate assessment every four years or less. These assessments are intended to evaluate the state of the science and the broad range of impacts of climate change in the United States. The draft Fourth National Climate Assessment (NCA4) report reviewed a wide range of topics of high importance to the United States and society more broadly, extending from human health and community well-being to the built environment, to businesses and economies, and to ecosystems and natural resources. The climate trends are expected to accelerate in the decades ahead, contributing to the increased extreme events occurring individually or in some cases in combination. The notable extreme events to be considered include sea level rise, coastal flooding, increased flooding due to tidal influences, increased frequency and intensity of precipitation or drought, snowpack melting, earlier peak spring stream flow, shrinking glaciers, increased average annual air temperature, increased frequency of nighttime heat waves, longer frost-free seasons, increased wildfires, and ocean acidification. The IPCC projections indicate that climate change will notably impact the production of crops, fisheries, and several animal and plant species are projected to be endangered. Moreover, infrastructure damage and associated consequences such as loss of life and property, can have enormous impact on the environment, economy, and society.

5.3 Impacts of Extreme Events

It is essential that climate change and related impacts be dealt with in a comprehensive manner. To mitigate the effects associated with the climate change, it is desirable, and likely more efficient, to address the root causes of climate change. IPCC links anthropogenic greenhouse gas emissions to the key drivers that include population size, economic activity, lifestyle, energy use, land use patterns, technology, and climate policy. The 2015 Paris Agreement under the UN Framework Convention on Climate Change (UNFCCC) targets voluntary mitigation efforts by signatory countries; however, the implementation falls well short of criteria that can have a material impact on climate change. Since the occurrence of extreme events and associated impacts on people and infrastructure have been quite evident, it is necessary to invest in adaptation to climate change. Designing built infrastructure that is adaptable to changing climate and capable of withstanding extreme impacts is necessary. Several notable recent natural disasters that overwhelmed existing infrastructure are described below.

5.3.1 The 2012 Hurricane Sandy in New York City

Hurricane Sandy (also known as Superstorm Sandy) made landfall in October 2012 and became the deadliest and most destructive hurricane to hit the metropolitan New York region. With a storm surge of more than 3 m, it flooded streets, tunnels and subway lines, and cut power in and around the city, inflicting nearly $70 billion in damage. The New York Stock Exchange was closed for two consecutive days. Numerous homes and businesses were destroyed by fire. Large parts of the city and surrounding areas lost electricity for several days. Several thousand people in midtown Manhattan were evacuated for six days due to a crane collapse. Hospitals were closed and evacuated. Flooding disrupted voice and data communication in lower Manhattan. At least 53 people died in New York as a result of the storm. Thousands of homes and an estimated 250 000 vehicles were destroyed during the storm. It is estimated that about $32 billion was required for restoration across the state. Overall, Hurricane Sandy revealed that extreme events could occur in an unpredictable manner, and current infrastructure is inadequate to prepare or mitigate the impacts.

5.3.2 The 2016 Chile's Wildfires by Drought and Record Heat

Wildfires occurred in Chile's central provinces during November 2016 to January 2017, with 119 active fires and forest fires, affecting an area of 1.35 million acres. At least 100 vineyards were damaged or destroyed. A total of 11 deaths were confirmed. Years of drought, combined with an extended period of extreme heat and dry weather, helped to fuel the historic wildfires. Over several preceding years, annual precipitation deficits ranged from 30% to 70%, resulting in acute drought and leading to long-lasting and widespread water shortage.

Multiple studies have identified human-caused climate change as a primary adverse factor on recent precipitation trends in Chile. Recent trends highlight increasing temperatures in central Chile, especially for the upper elevation stations, where precipitation amounts have also been decreasing. Higher temperatures lead to greater water loss from snow-covered areas (sublimation), crops and natural vegetation (evapotranspiration), and lakes and reservoirs (evaporation), which contributes to water deficits and increases wildfire risk.

5.3.3 The 2017 Worst South Asian Monsoon Floods

The worst monsoon rains in South Asia occurred during August 2017 (around the same time Hurricane Harvey inundated Houston). More than one-third of Nepal had reported flooding or mudslides from heavy rainfall, decimating crops across the country. In Bangladesh, record rainfall pushed rivers to some of the highest levels ever and submerged one-third of the country, damaging enormous areas of farmland. On India's west coast, Mumbai received extreme rainfall up to 322 mm (12.7 in) in 24 hours, causing widespread flooding. Many other cities in the region saw extreme 24-hour rainfall rates that caused widespread and deadly flooding.

5.4 What Is Resiliency?

The term *resiliency* is derived from the Latin word *resilio*, which means to jump back. Resiliency is defined differently in different fields such as ecology, social science, economy, and engineering, with relevant meanings and implications:

- **Ecological resilience** is defined as a measure of the persistence of systems and of their ability to absorb change and disturbance and still maintain the same relationships between populations or state variables. Stability represents the ability of a system to return to an equilibrium state after a temporary disturbance; the more rapidly it returns to equilibrium and the less it fluctuates, the more stable it would be (Holling 1973).
- **Social resilience** is defined as the ability of groups or societies to cope with external stresses and disturbances because of social, political, and environmental changes (Adger 2000).
- **Economic resilience** is defined as the inherent ability and adaptive response that enables individual business firms and entire regions to avoid maximum potential losses (Rose and Liao 2005).

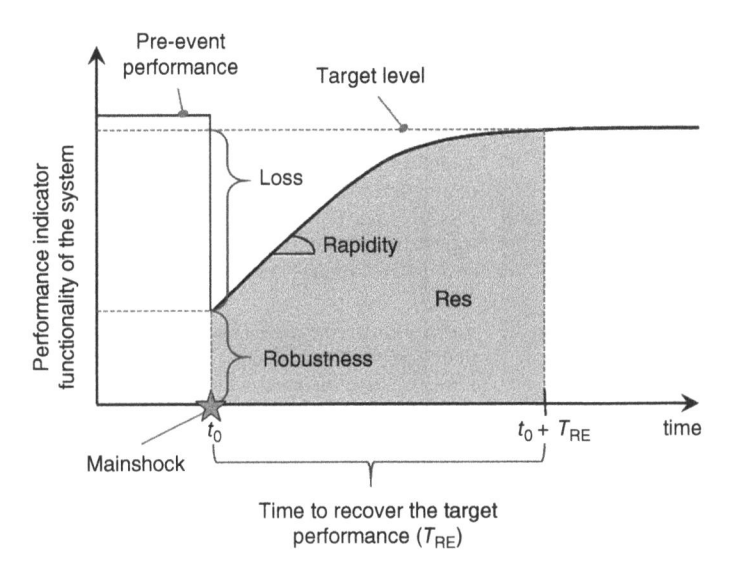

Figure 5.1 Concept of resilience. Source: Cimellaro et al. (2016).

- **Engineering resilience** is defined as the capability of a system to maintain its functionality and to degrade gracefully in the face of internal and external changes. In particular, engineered infrastructure systems (e.g. water supply, electricity, and transportation) is recognized as a key element for communities' resilience in the context of disasters.

The general concept of resilience is depicted in Figure 5.1, and some common elements of resiliency include the system design that performs its function safely under normal conditions, the vulnerability potential for an extreme event to occur, system absorbing/withstanding the effect of the event to maintain the function at some desired level, and recovering to achieve target level of functionality.

In general, disaster resilience is specifically used to refer to minimizing any reduction in quality of life (e.g. loss of life, injuries, disruption of services, and economic losses) due to disaster. Specifically, disaster resilience is characterized by (i) reduced failure probabilities – the reduced likelihood of damage and failures to critical infrastructure, systems, and components; (ii) reduced consequences from failures – in terms of injuries, lives lost, damage, and negative economic and social impacts; and (iii) reduced time to recovery – the time required to restore a specific system or set of systems to normal or predisaster level of functionality. Based on these characteristics, resilience can be enhanced by reducing the likelihood of failure of critical infrastructure (thereby, reducing their impacts) and speeding up the time it takes to make a full recovery.

Sometimes resilience is considered as one of the indicators of sustainability. However, the correlation between these two is more complicated. As stated in Chapter 1, the definition of sustainability given by the Brundtland Commission, formally known as the World Commission on Environment and Development (Brundtland 1987), is "the development that meet the needs of the present without compromising the ability of future generations to meet their own needs". ASCE defines sustainability as "a set of economic, environmental and social conditions (aka 'The Triple Bottom Line') in which all of society has the capacity and opportunity to maintain and improve its quality of life indefinitely without degrading the quantity, quality or the availability of economic, environmental and social resources" (ASCE 2018a). From a practical point of view, the triple bottom line impacts are optimized to design sustainable systems

Table 5.1 Similarities and differences among sustainability and resilience.

Categories	Sustainability	Resilience	Matching
Common definition	*Sustainable development* is development that meets the needs of the present without compromising the ability of future generations to meet their own needs (Brundtland 1987)	*Resilience* is the ability of human communities to withstand external shocks or perturbations to their infrastructure and to recover from such perturbations (Timmerman 1981)	o
Keywords often used for description in papers	Holistic, green, life cycle, life-cycle assessment, life-cycle costing, social costing, sustainable development, indicators, rating	Recovery, extreme events, disaster management, functionality, infrastructure, lifelines, networks, communities	+
Dimensions/pillars	Dimensions: economic, ecological, social Strategies: efficiency, sufficiency, consistency	Dimensions: technical, organizational, social, economic Means: redundancy, resourcefulness	++
Target	Reduction of impacts and resource consumption in the three dimensions, inter- and intra-generational fairness	Goals: robustness, rapidity results: higher reliability, fast recovery, lower risk	+
Measuring labels/medals	Labels: Platinum, gold, silver, certified, quantitative measures are constantly developed	A medal-based system is used in practice, but quantitative measures are becoming more popular among researchers	+
Quantification	Mostly based on indices summarizing different quantitative and qualitative indicators; result is a score	Often quantified by the index in Eq. (5.1) (Cimellaro et al. 2010a; Bocchini and Frangopol 2011)	o
Spatial scale	Most advanced on building level, activities on network level seldom included	Mainly community and network level (but also individual structure level)	o/+
Important instruments/calculation methods in science	LCA, LCC, external costs, multicriteria decision-making, energy modeling, building information modeling	LCC, user cost analysis, multicriteria decision making, extreme events simulation	++
Important instruments/calculation methods in practice	Decision supporting assessment schemes, mostly based on checklists compared with a reference structure	Classification of critical infrastructure in three levels Prioritization of interventions based on previous experience	+
Current relevance in science for infrastructure assessment	Interest growing since 2005, extremely forthcoming because of several research projects started in the last two years	Interest growing since 2000, very significant momentum in the last three years; very topical these days	++
Current relevance in practice for infrastructure assessment	Medium relevance for newly built structures; several new systems recognizable that will increase the relevance within the next five years. Work for the implementation in national and international standards underway	Approach very different from science; some associations are promoting the use of resilience criteria in the building codes, but not even a common vocabulary has been established	++

Note: ++ = perfect matching; + = good matching; o = no significant matching.

$$R = \frac{\int_{t_0}^{t_0+t_h} Q(t) dt}{t_h} \tag{5.1}$$

where R is the resilience index, t_h is the time horizon investigated, and Q is the percentage "functionality" (or "quality" or "serviceability" of the system), and t is time.
Source: From Bocchini et al. (2014).

(see, for example, Chapters 20–22). ASCE defines resilience as "the capability to mitigate against significant all-hazards risks and incidents and to expeditiously recover and reconstitute critical services with minimum damage to public safety and health, the economy, and national security" (ASCE 2018b). It is possible to have a sustainable design that can reduce resource and energy consumption, optimize waste management, and be economically efficient, but not necessarily maintain operating conditions in case of extreme events so that they are not resilient. However, such systems are not truly sustainable. Resilience is essential to be sustainable and should be taken into account simultaneously over the life cycle of projects. Table 5.1 summarizes the similarities and differences between sustainability and resilience.

Although both the sustainability and resilience are essential for future infrastructure, they might work against each other in some cases. Sustainable systems may not be resilient. Despite the holistic aspirations of both sustainability and resilience, it can be argued that systems that are resilient may not be sustainable, systems that are not resilient may not be sustainable, and systems that are not sustainable might be resilient. For example, dense and compact cities (e.g. featuring tall buildings, public transportation) are considered sustainable as they can reduce the energy consumption, but such dense urban systems can make cities more vulnerable to extreme events (nonresilient). So defining a limit for population density in a city might be the solution to have cities both sustainable and resilient. To the contrary, resilient systems may not be sustainable. For example, after Superstorm Sandy hit New York, sea gates were proposed at the narrow section of the New York Harbor entrance to avoid surge flooding and achieve resiliency. However, such a gate system would require long-time maintenance and could also adversely affect the surrounding ecology; hence, this system is not considered sustainable even though it is considered resilient.

5.5 Initiatives and Policies on Resiliency

Many organizations and authorities have recognized the value of resiliency and developed proposals, policies, standards, or guidelines to support resiliency in general and for resilient infrastructure design in particular. Some of the selected efforts are briefly presented below.

US Presidential Policy Directives (PPDs): The directives (PPD-8 and PPD-21) have defined resilience for communities to withstand and adapt to changing conditions (PPD 2011, 2013). Other organizations, such as The Infrastructure Security Partnership (TISP) and the Department of Homeland Security (DHS), have also suggested definitions for resilience for functionality of built systems. According to PPD21 (Critical Infrastructure Security and Resilience), the term resilience refers to the ability to prepare for and adapt to changing conditions and withstand and recover rapidly from disruptions. Resilience includes the ability to withstand and recover from deliberate attacks, accidents, or naturally occurring threats or incidents. The PPD-8 calls for the DHS to "coordinate a comprehensive campaign to build and sustain national preparedness, including public outreach and community-based and private-sector programs to enhance national resilience…" The efforts by the US government to enhance the resilience of the national infrastructure was one of the stimuli that led to the identification of principles that could be used by designers to create infrastructure systems that would enable these systems to avoid, withstand, and recover from a broad range of threats.

Rockefeller Foundation's 100 Resilient Cities Initiative: The Rockefeller Foundation has worked since December 2013 in helping cities around the world to become more resilient to physical, social, and economic challenges. The 100 Resilient Cities project aims to help individual cities

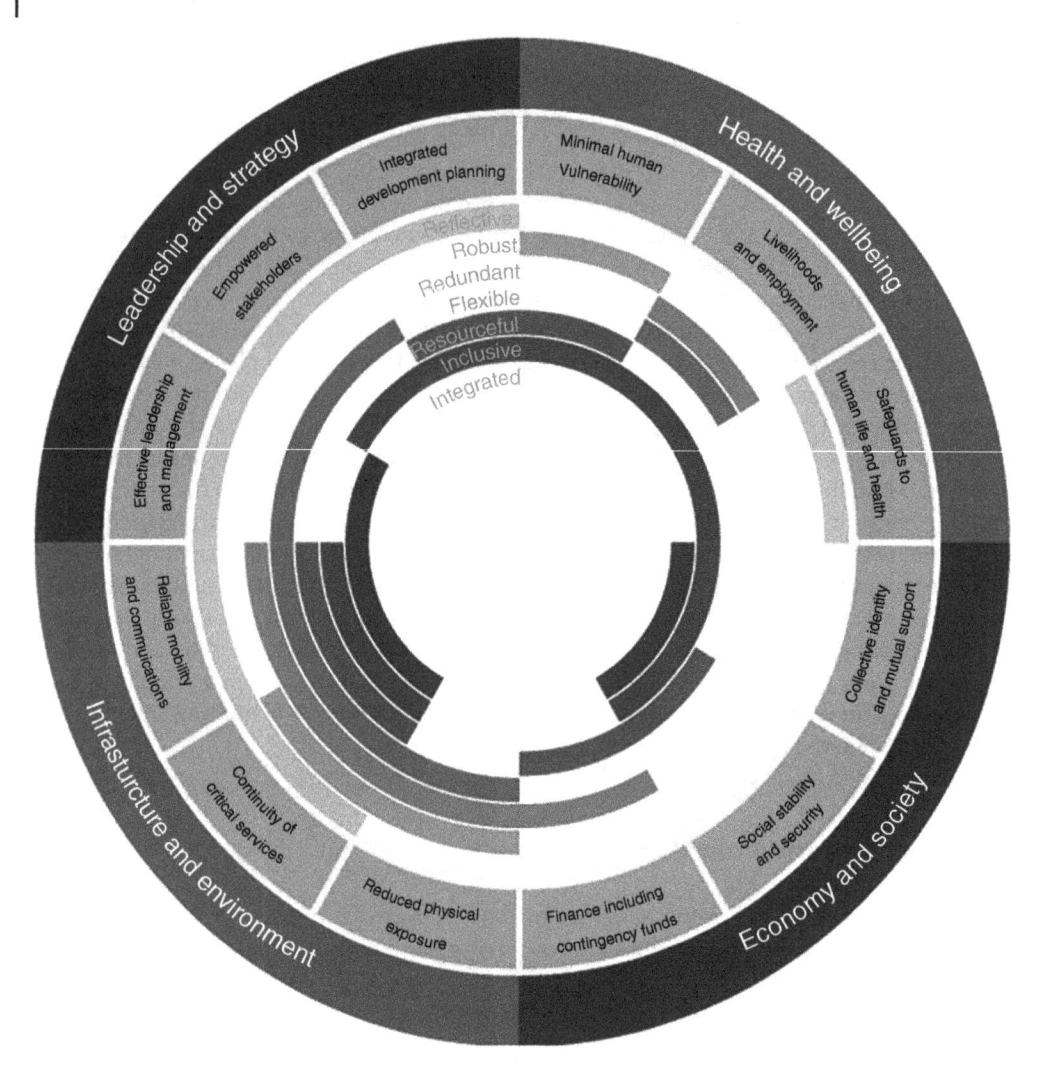

Figure 5.2 The framework of resilience design. Source: Rockefeller Foundation City-Resilience-Framework, http://www.100resilientcities.org, Nov 2015.

to become more resilient and simultaneously to facilitate the building of a global practice of resilience among governments, NGOs, the private sector, and individuals. It defined urban resilience as the capacity of individuals, communities, institutions, businesses, and systems within the city to survive, adapt, and grow with no matter of chronic stresses and acute shocks they experience. The Rockefeller framework is built on four essential dimensions of urban resilience: health and well-being, economy and society, infrastructure and environment, and leadership and strategy. Each of the dimensions is divided into three more detailed drivers as shown in Figure 5.2.

The United Nations (UN): The UN defines resilience as follows: "… ability of a system, community or society exposed to hazards to resist, absorb, accommodate to and recover from the effects of a hazard in a timely and efficient manner, including through the preservation and restoration of its essential basic structures and functions." The UN Office for Disaster Risk Reduction (UNISDR) began in March 2015 to implement a new ISO standard for resilient and sustainable cities (ISO 371207). The United Nations Office for Disaster Risk Reduction at the request of the UN General Assembly adopted The Sendai Framework for Disaster Risk Reduction 2015–2030 at the Third World Conference in Sendai, Japan on 18 March 2015. The Sendai Framework is the successor instrument to the Hyogo Framework for Action (HFA) 2005–2015: Building the Resilience of Nations and Communities to Disasters. The Sendai Framework also articulates the following: the need for improved understanding of disaster risk in all its dimensions of exposure, vulnerability and hazard characteristics; the strengthening of disaster risk governance, including national platforms; accountability for disaster risk management; preparedness to "Build Back Better"; recognition of stakeholders and their roles; mobilization of risk-sensitive investment to avoid the creation of new risk; resilience of health infrastructure, cultural heritage and work-places; strengthening of international cooperation and global partnership, and risk-informed donor policies and programs, including financial support and loans from international financial institutions (UNISDR 2015).

The Sendai Framework aims to achieve substantial reduction of disaster risk and losses in lives, livelihoods, and health and in the economic, physical, social, cultural, and environmental assets of persons, businesses, communities, and countries. To pursue and achieve this goal, it is essential to enhance the implementation capacity and capability of developing countries, least developed countries, small island developing states, landlocked developing countries, and African countries, as well as middle-income countries facing specific challenges, including the mobilization of support through international cooperation for the provision of means of implementation in accordance with their national priorities.

In order to support the assessment of progress toward the above-mentioned goals and expected outcomes, seven global targets have been agreed under the Sendai Framework listed as follows:

1. Substantially reduce global disaster mortality by 2030, aiming to lower the average per 100 000 global mortality rates in the decade 2020–2030 compared to the period 2005–2015;
2. Substantially reduce the number of affected people globally by 2030, aiming to lower the average global figure per 100 000 in the decade 2020–2030 compared to the period 2005–2015;
3. Reduce direct disaster economic loss in relation to global gross domestic product (GDP) by 2030;
4. Substantially reduce disaster damage to critical infrastructure and disruption of basic services, among them health and educational facilities, including through developing their resilience by 2030;
5. Substantially increase the number of countries with national and local disaster risk reduction strategies by 2020;
6. Substantially enhance international cooperation to developing countries through adequate and sustainable support to complement their national actions for implementation of the present Framework by 2030;
7. Substantially increase the availability of and access to multihazard early warning systems and disaster risk information and assessments to people by 2030.

Intergovernmental Panel on Climate Change (IPCC): The Fifth Assessment Report of the IPCC (IPCC 2015) defines resilience as "the capacity of social, economic, and environmental systems to cope with a hazardous event or trend or disturbance, responding or reorganizing in ways that maintain their essential function, identity, and structure, while also maintaining the capacity for adaptation, learning, and transformation." The report recommends creating a climate-resilient pathway as a measure to respond to climate change risks that may be concentrated in urban areas. This recommendation implies that, in terms of sustainable

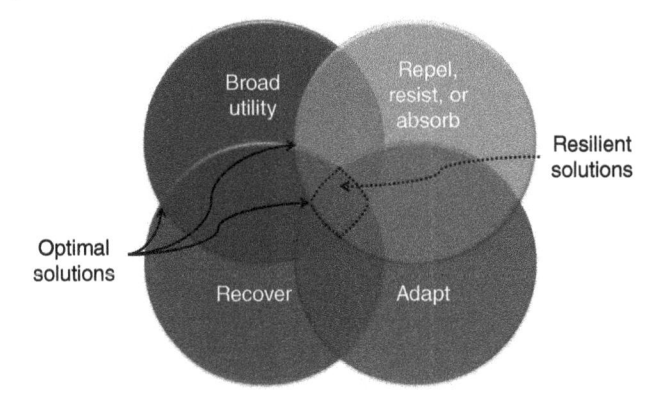

Figure 5.3 Four key properties of resilient DoD systems.

development, we must change our decision-making and behavior pertaining to the economy, society, technology, and politics.

US Department of Defense (DOD): The DOD considers four key properties for a resilient system: Repel/Resist/Absorb; Recover; Adapt; and Broad Utility as depicted in Figure 5.3. In this figure, the circles represent greater ability at the edges than in the center for each resilience property; the intersection of two or more circles can be viewed as providing a reasonably effective resilience solution. The intersection (i.e. nexus) of all four properties defines the space of resilient solutions with each solution in the space being proximal to an optimal solution for a particular property.

American Society of Civil Engineers (ASCE): ASCE Policy Statement 493 supports additional basic and applied research efforts and the development of national standards in support of design and construction initiatives that increase the reliability, safety, security, and survivability of the nation's vast infrastructure (e.g. water, energy, utilities, buildings, and transportation) against natural and man-made disasters. Another ASCE policy statement 500 supports initiatives that increase resilience of infrastructure against man-made and natural hazards through education, research, planning, design, construction, operation, and maintenance. ASCE recognizes the importance of development of performance criteria and uniform national standards that address interdependencies and establish minimum performance goals for infrastructure. An all-hazard, comprehensive risk assessment is recommended to be used for planning and design of infrastructure that considers event likelihood and consequence, mitigation strategies, monitoring, and recovery/return service. Designers and planners are encouraged to incorporate system resiliency (the ability of a project or system to withstand and recover to full functionality from extreme events quickly and efficiently) into the decision-making process. The increasing frequency and intensity of natural disasters as well as deliberate destructive events, combined with continually increasing population densities, reliance on technology, and interdependencies, are considered to be vulnerabilities in the nation's infrastructure. An important component of resilience is understanding the impact of the loss of infrastructure and the timeline and cost to restore its function following an extreme event.

5.6 Resiliency Framework

It should be noted that resilience is the ability to provide required capability in the face of adversity, which is different from risk. While risk pertains to the loss of value due to uncertain

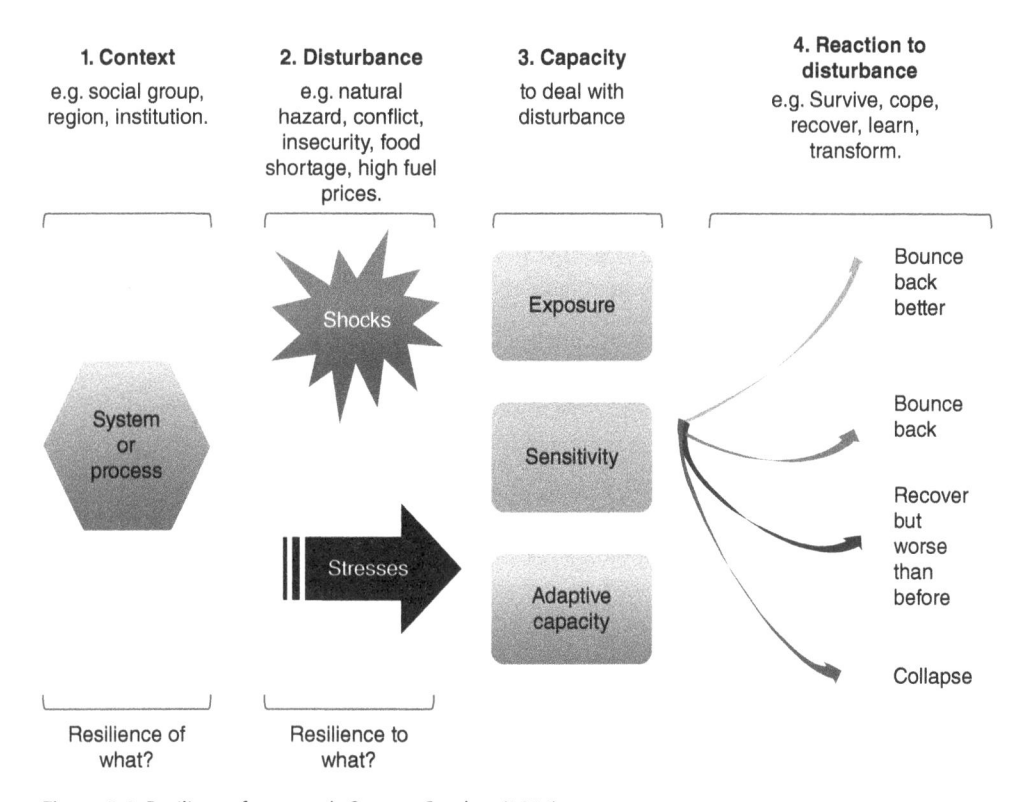

Figure 5.4 Resilience framework. Source: Combaz (2014).

future events, resilience has to do with designing a system to maintain a predesignated level of capability following a disturbance. Therefore, design for resiliency will require the engineer to take the existing design frameworks a step further and evaluate the robustness of the engineered systems during and after an extreme event. A resilient system should be able to recover from damage rapidly and regain functionality if lost, but at the same time needs to adapt and be prepared for the next extreme event that may be just as uncertain and disastrous, or more. Ideally, bouncing back or simply recovering is not considered sufficient to characterize a system as resilient. The resilient system can take advantage of the newly acquired knowledge and improve its response to the next extreme event. An ideal resilience framework is similar to community resilience frame consisting of four elements as shown in Figure 5.4.

In order to increase resilience, the concepts of monitor, respond, learn, and anticipate are considered. Monitoring is recommended that supports preparedness. The monitoring parameters should be carefully selected, and they should be able to identify positive or negative performance of the system, actual performance of the system in case of an event and help to infer what could happen if another event occurs and the response, depending on the monitoring data, to address impacts of the event. Knowing what happened after an impact, lessons could be learned and monitoring/operations could be adjusted or the system could be modified/redesigned. The major complexity with resilient design is the unpredictable nature of extreme events; therefore, one should anticipate occurrence of potential extreme event and preparing for it in light of project-specific demands and constraints and changing operating conditions.

The effectiveness of a resilient system depends upon its ability to anticipate, absorb, adapt to, and/or rapidly recover from a potentially disruptive event. Resiliency performance may be measured by metrics based on the data collected from previous shock situations

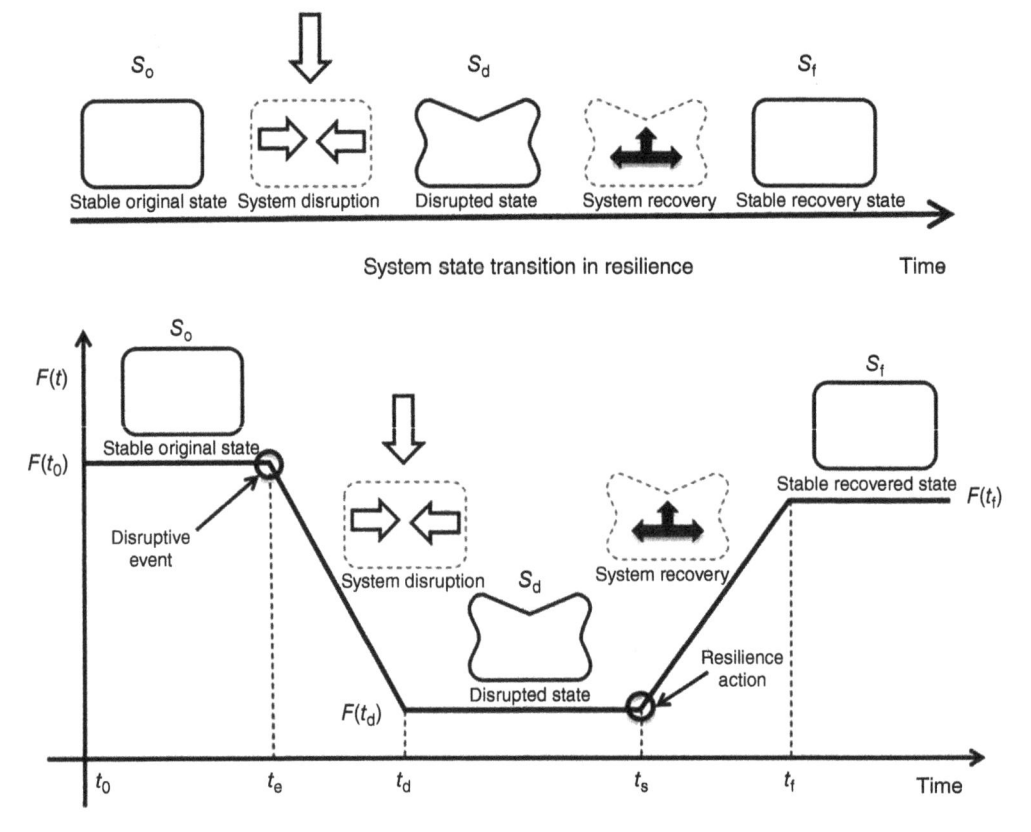

Figure 5.5 Delivery function transition in resilience. Source: Henry and Ramirez-Marquez (2012). Reproduced with permission of Elsevier.

(i.e. an extreme event). The data on the speed and time of recovery of the system (S_0, S_d, S_f, t_e, t_d, t_s, and t_f) as shown in Figure 5.5 may be determined and used as resilience metrics. A system is resilient if it can adjust its functioning prior to, during, or following extreme events, and thereby sustain desired system function under both normal conditions and unexpected extreme event conditions.

Resilience of infrastructure has received greater attention after Hurricanes Katrina, Irene, and Sandy crippled domestic infrastructure, including bridges, which stranded many people without basic infrastructure services or access to emergency facilities, and delayed the evacuation and recovery processes. Over the past two decades, several studies (e.g. Rinaldi et al. 2001; O'Rourke 2007) have investigated resilience determinants in infrastructure systems to provide a theoretical framework for identifying components and characteristics of resilient infrastructure systems. Some researchers have conceptualized resilience using four properties (known as 4Rs of resilience): robustness, redundancy, resourcefulness, and rapidity (Bruneau et al. 2003).

Bocchini et al. (2014) emphasize the need to consider both resiliency and sustainability in an integrated manner and proposed a unified approach to address resilient and sustainable civil infrastructure (e.g. bridges, transportation) simultaneously and quantitatively over infrastructure life cycle. The approach is based on a well-established framework of risk assessment (probability of occurrence and risk). The impact of infrastructure and its service states on the society in normal operational conditions are assessed by sustainability analysis based on three dimensions (triple bottom line) – ecology, economy, and society. The impacts to infrastructure after extreme events are assessed by resilient analysis based on ability to deliver certain service level after the event and then recover to the desired functionality quickly. All these aspects are weighted by the associated probabilities of occurrence and combined in a global impact assessment.

5.7 Resilient Infrastructure

Disasters such as Superstorm Sandy in New York have revealed that civil infrastructure (e.g. buildings, water, and transportation) is highly vulnerable to extreme events. Therefore, infrastructure adapting to climate change impacts is critical for the economic security, the health and safety of people, and the health of our environment and abundant natural resources. As stated before, climate impacts could include sea level rise and coastal inundation; hurricane surge; riverine flooding and extreme rain events; landslide and erosion; wildfire; and drought.

Although resilience is such an attractive concept, incorporating resilience in engineering practice for infrastructure design can be challenging. For the resilient infrastructure design, first and foremost a vulnerability assessment should be conducted in order to (i) identify climate change impacts that have the greatest potential to adversely affect the infrastructure; (ii) understand the scope of infrastructure vulnerability to these impacts, by learning what types of vulnerable infrastructure there are, and where they are located, and determining which specific types of infrastructure have high potential to be affected by climate change impacts, and what those vulnerabilities may be; (iii) develop a process for the project engineers to conduct a more detailed and site-specific vulnerability assessment; and (iv) inform the development of an adaptation design and operational strategy to increase resilience of infrastructure.

The prediction of climate change and associated impacts, including extreme events, has been a daunting task. Therefore, vulnerability assessment is generally based on the analysis of historical observations based on the assumption that the underlying environmental processes are stationary. However, the recent observed data shows that the environmental processes are not stationary and the past variability does not reflect future variability. This means that historical data may not be a good predictor of future variability, especially decades into the future when infrastructure designed today may well still be expected to be operable and in service.

Infrastructure along coastal areas or areas further inland may be vulnerable to high flooding hazards due to a number of regional and site-specific factors, such as proximity to tidally influenced freshwater rivers and tributaries affected by sea level rise; amount of developed areas with nonpermeable surfaces; amount of low-lying land; type and adequacy of flood controls; and locations within 100- or 500-year floodplains. For projects identified as vulnerable to climatic impacts, further evaluation of the risk should be done based on project-specific information.

Different standards are being used to address specific hazards (e.g. BRRT, FORTIFIED, and REDi) and many tools or assessment methods are being developed to measure project-scale resilience (e.g. Envision, BRLA, PEER, SITES, LEED Pilot Credits, RELi, ANCR, and ICCR). Financial models are used to calculate full benefit–cost ratios and make the business case for incorporating resiliency in design.

Many entities have initiated efforts to develop guidelines to incorporate the effects of extreme events such as sea level rise and flooding into infrastructure design. For example, the city of San Francisco (2015) has developed guidance for incorporating sea level rise into capital planning (http://onesanfrancisco.org/sea-level-rise-guidance). Similarly, the State of Washington (2017) developed guidance on incorporating climate impacts on contaminated site remediation. USEPA has developed guidance also on how to address climate impacts in selecting a remedial strategy or technology based on site-specific conditions.

Many analytical or decision tools are being developed for evaluating climate change vulnerabilities and impacts. For example, the following tools focus on the vulnerability of water resources (including stormwater, wastewater, and drinking water), sea level rise, and flooding:

- Sea Level Change Curve Calculator Tool (USACE 2014) is available to calculate three local sea level change scenarios (low, intermediate, and high curves) based on historic rates of sea level change, the modified National Research Council's projections, Intergovernmental Panel on Climate Change projections, and/or vertical land movement.
- National Stormwater Calculator with Climate Assessment Tool (USEPA 2017) was developed by USEPA to evaluate future climate vulnerabilities applicable to stormwater, based on national estimates of annual rainfall and frequency of runoff. Local soil conditions, land cover, and historic rainfall are also considered.
- Climate Resilience Evaluation and Awareness Tool (CREAT) 3.0 (USEPA 2016) is a valuable tool to assist site managers in understanding and assessing potential climate change impacts and risks to drinking water, wastewater, and stormwater systems.
- NOAA Sea Level Rise Viewer developed by the NOAA Office for Coastal Management (2018a) is a web mapping tool to visualize community-level impacts from coastal flooding or sea level rise (up to 10 ft above average high tides).
- NOAA Surging Seas is developed by the NOAA Office for Coastal Management (2018b). This web-based tool allows users to see areas potentially affected by sea level rise and storm surge, down to the neighborhood scale, and with risk timelines. The tool also provides population statistics, summaries of homes and land affected by sea level rise and storm surge.

Ideally, our target should be not only to design resilient infrastructure but also sustainable infrastructure. Hence, resilient assessments and sustainability assessments should be integrated and optimized resilient sustainable designs or solutions be developed. Such optimized approach is depicted in Figure 5.6. However, much more work is needed to establish integrated resilient and sustainable assessment frameworks and assessment/design methodologies.

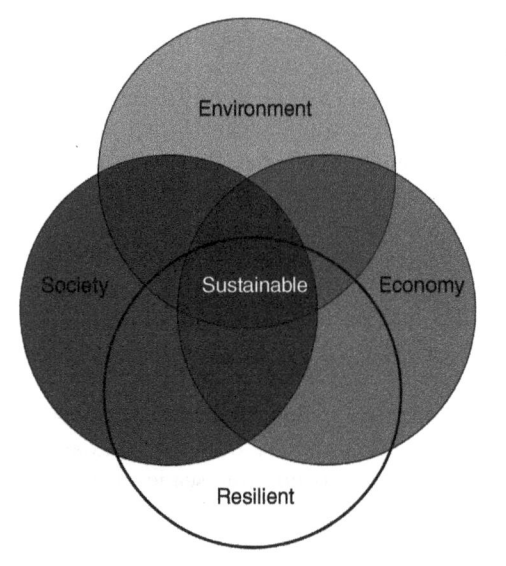

Figure 5.6 Integrated resilient and sustainable framework.

5.8 Resilient Infrastructure Examples

Several examples are presented in this section to understand how the efforts are being currently made addressing climate impacts in designing resilient engineering systems.

5.8.1 San Francisco Firehouse Resilient Design

San Francisco Public Works (SFPW) investigated floating pier (float) and fixed pier options and chose a float, in part to accommodate sea level rise. Swinerton Builders/Power Engineering Construction Joint Venture was awarded the project by SFPW. Liftech Consultants Inc. is designing the pier, access ramp, float, and piling; Shah Kawasaki Architects is designing the building.

Fire Station 35 is an Essential Facility with 50-year design life requiring resilience to 2070. Design sea level rise has been estimated at 2.8 ft by 2070, which is consistent with the "Guidance for Incorporating Sea Level Rise into Capital Planning in San Francisco: Assessing Vulnerability and Risk to Support Adaptation" dated 14 December 2015.

Figure 5.7a shows the project details. Facility has resilience for estimated sea level rise to year 2100, including a 1% storm (base flood elevation, or (BFE)) + wave runup and setup + sea level rise. Local conditions as shown in Figure 5.7b are considered in the final design. The project design components included float guide pile height; access ramp and gangway with hinges at the landside connection and wheel supports at the float; and utilities with flexible joints at moving interfaces. The final design of the floating pier resilient to sea level rise is shown in Figure 5.7c.

Some items considered while designing for sea level rise are as follows:

- Long-term regional plan for addressing sea level rise.
- The vicinity of site and how it will be affected by sea level rise; if only focused on the specific site, then solutions may be less effective (e.g. building a barrier around the site that rising water will eventually flow around).
- How all the elements of the project need to adapt (e.g. float is moving, so structures and utilities connected to float from landside need to accommodate that movement).
- Certain adaptations may be more favorable than others (e.g. float is more aesthetically appealing than a fixed pier several feet above surroundings).
- Required resiliency of the project given its exposure and sensitivity (e.g. fire station on the water is an essential facility with high exposure and sensitivity to sea level rise, so should have high resiliency).
- Preferred solution may be more expensive given above or other considerations, but cost estimates should be made on a long-term basis, which could include future renovations if initial design does not accommodate long-term uncertainty.

5.8.2 San Francisco Resilient CSD Design

San Francisco is implementing climate adaptation strategies in their Sewer System Improvement Program (SSIP). The SSIP is a programmatic approach to strategize, plan, and implement a portfolio of capital improvements in San Francisco's combined sewer system over a 20-year timeframe. In order to make this infrastructure resilient to seal level rise, the following design criteria is used:

- **New infrastructure** must accommodate expected sea level rise within the service life of the asset (i.e. 6 in by 2030, 11 in by 2050, 36 in by 2100), and be consistent with the City's Guidance for Incorporating Sea Level Rise into Capital Planning.

(a)

(b)

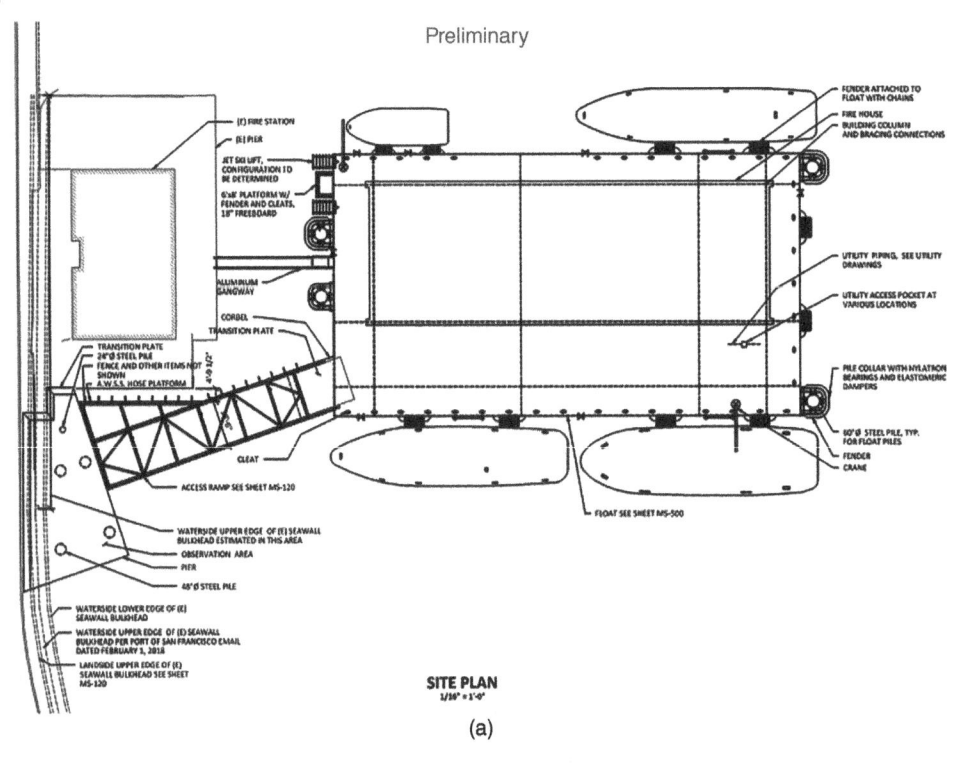

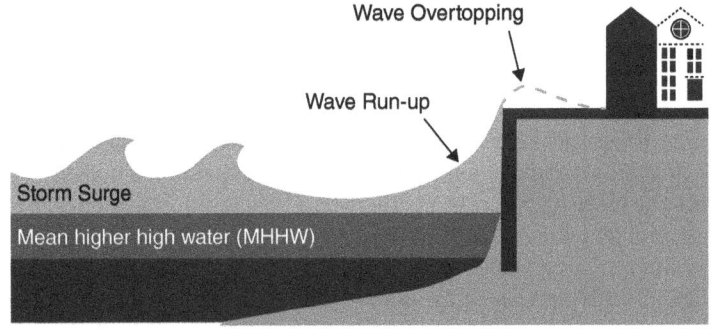

Factors that Influence Loacal Water Level Conditions in Addition to Sea Level Rise

Factors Affecting Water Level	Typical Range CCSF Pacific Shoreline (a)	Typical Range CCSF Bay Shoreline (a)	Period of Influence	Frequency
Tides	5 to 7 ft	5 to 7 ft	Hours	Twice daily
Storm Surge	0.5 to 3 ft	0.5 to 3 ft	Days	Several times a year
Storm Waves	10 to 30 ft	1 to 4 ft	Hours	Several times a year
El Niños (within the ENSO cycle)	0.5 to 3 ft	0.5 to 3 ft	Months to Years	Every 2 to 7 years

Sources:

(a) Typical ranges for tides, storm surge, and storm waves for the CCSF Pacific Coast: BakerAECOM 2012. Intermediate Data Submittal #1. Scoping and Data Review. San Francisco Country, California. California Coastal Analysis and Mapping Project / Open Pacific Coast Study. Submitted to FEMA Region IX. February 2012.

(b) Typical ranges for tides, storm surge, and storm waves for the CCSF Bay shoreline: DHI. 2010. Regional Coastal Hazard Modeling Study for North and Central Bay. Submitted to FEMA region IX. October 2011.

Figure 5.7 (a) Project details. (b). Water levels considered for design. (c). Floating fire house designed to be resilient to sea level rise. Source: City of San Francisco (2015).

(c)

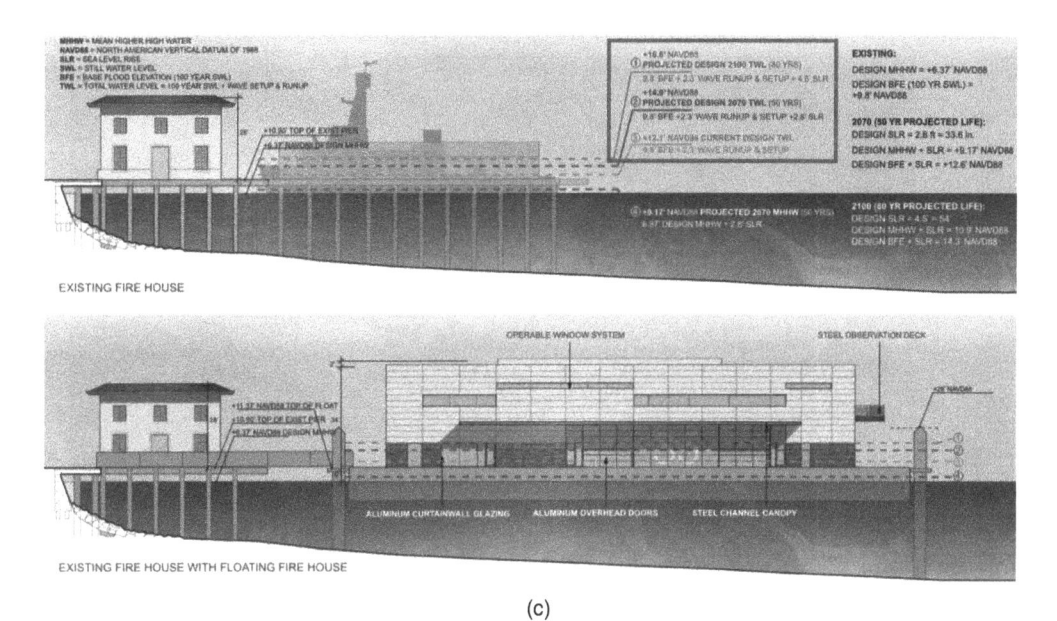

Figure 5.7 (*Continued*)

- **Existing infrastructure** that is impacted by sea level rise, within the service life of the asset, will be modified based on sea level rise projections (e.g. pump stations, combined sewer discharge (CSD) structures).

There are 29 CSD structures on the Bayside as shown in Figure 5.8a. The conceptual design of CSD is presented in Figure 5.8b, and how it can function during wet weather, dry weather, and a combination dry weather and sea level rise is depicted in Figure 5.8b–d, respectively. Tides have a diurnal pattern with short durations of high tides (20–120 minutes). Tidal inflow calculation tool was developed to estimate inflows. Modeled projected tide levels (2-year storm surge + sea level rise) are used to estimate inflow volume at each CSD location. In addition to projecting when CSDs may need backflow prevention, timing of condition improvements was also estimated using annual inspection data. The capital strategy aims to complete both types of work in one project for implementation efficiency.

5.8.3 Resilient Environmental Remediation

Washington State Department of Ecology summarized cleanup sites that have incorporated resilience into the cleanup design or adaptive management plans. They discuss incorporating sea level rise into the remedial design at landfill cleanup sites located along the marine waterfront, at contaminated soil and groundwater sites located along the marine waterfront, and at sediment cleanup sites in a marine embayment. They have also implemented adaptive management during post-cleanup construction at a sediment site to address damage from multiple severe storm events within three days of each other. In addition, they have incorporated protection from severe storm events into the remedial design for a sediment and upland shoreline site. One of these projects is described briefly below.

The March Point Landfill site (also referred to as Whitmarsh Landfill) is located along the shoreline of Padilla Bay in Anacortes. It is situated along a lagoon connected to Padilla Bay by a 100-foot wide channel under the Burlington Northern Santa Fe (BNSF) railroad embankment.

120 | *5 Disaster Resiliency*

(a)

(b)

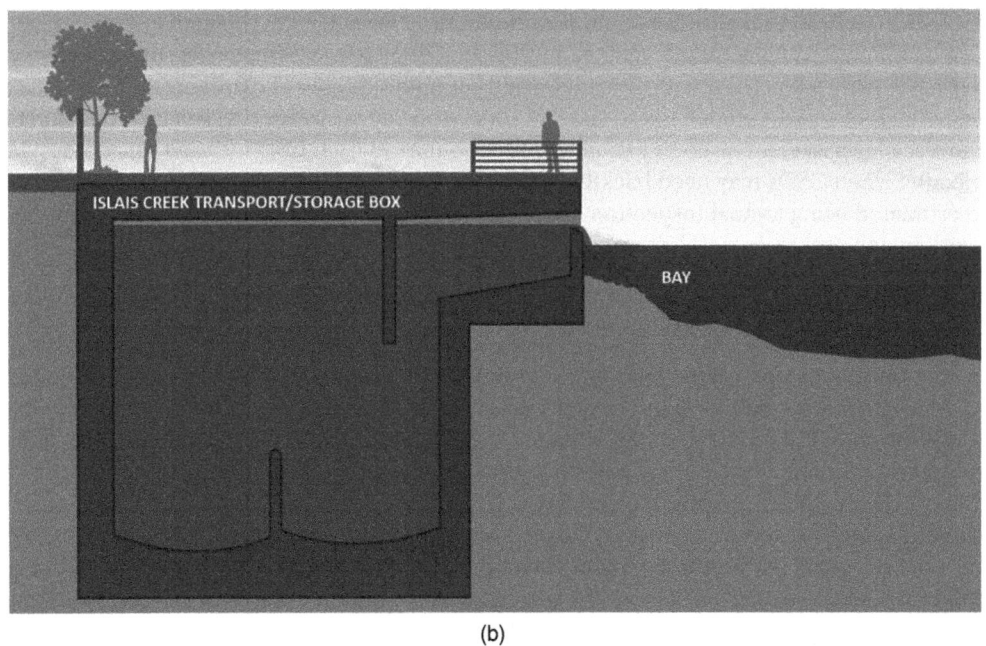

Figure 5.8 (a) Locations of combined sewer discharge structures on the bayside. (b) Conceptual rendering of CSD structure in wet weather. (c) Conceptual rendering of CSD structure in dry weather. (d) Conceptual rendering of CSD structure in dry weather-impacted by sea level rise. Source: City of San Francisco (2015).

(c)

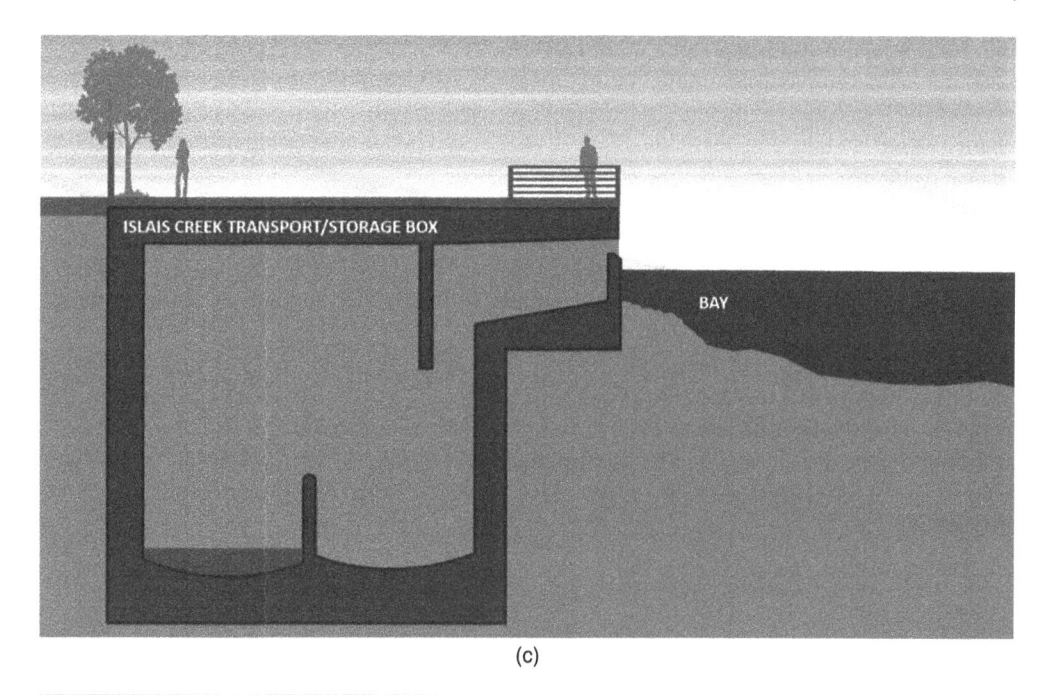

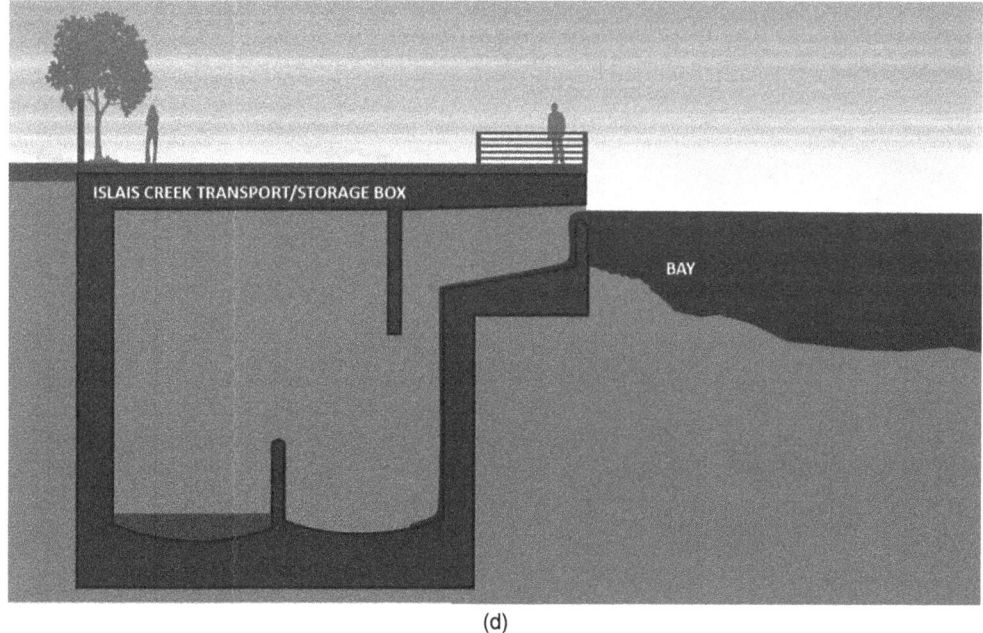

(d)

Figure 5.8 (Continued)

The landfill operated for more than 20 years as an unregulated public dump, then around 1973 served as a county disposal area for household, commercial, and industrial wastes. From the late 1980s to 2011, a sawmill was operated at the site, which resulted in accumulations of wood waste up to 10-ft thick over large portions of the landfill (Figure 5.9a). The contaminants of concern are numerous:

5 Disaster Resiliency

- *Soil*: solid waste, wood waste, and landfill gas (methane); metals, total petroleum hydrocarbons, benzene, semivolatile organic compounds (SVOCs), and PCBs.
- *Groundwater:* SVOCs, benzene, pesticides, PCBs, and metals.
- *Surface water:* metals, one pesticide, PCBs, SVOCs, and benzene seeping from the landfill to surface water.

The cleanup remedy consists of

- A design to address impacts from 100-year storm events and sea level rise of 7.6 ft above mean higher high water (MHHW) due to tsunamis (short-term impact) or climate change (long-term impact) (Figure 5.9b–d).
- Moving solid waste from the edges of the landfill inward.
- Grading the waste to a mound.
- Installing a passive landfill gas collection system to vent gases to the atmosphere.
- Installing an engineered cap over the landfill with standard geosynthetic clay laminated liner (GCLL) above 16 ft mean lower low water (MLLW) to minimize or eliminate infiltration into the landfill.

(a)

Figure 5.9 (a) March point landfill before cleanup showing the lagoon. (b) March point landfill plan view of extent of GCL liner and waste. (c) March point landfill details of enhanced GCL liner. (d) March point landfill details of enhanced GCL liner. Source: State of Washington (2017).

(b)

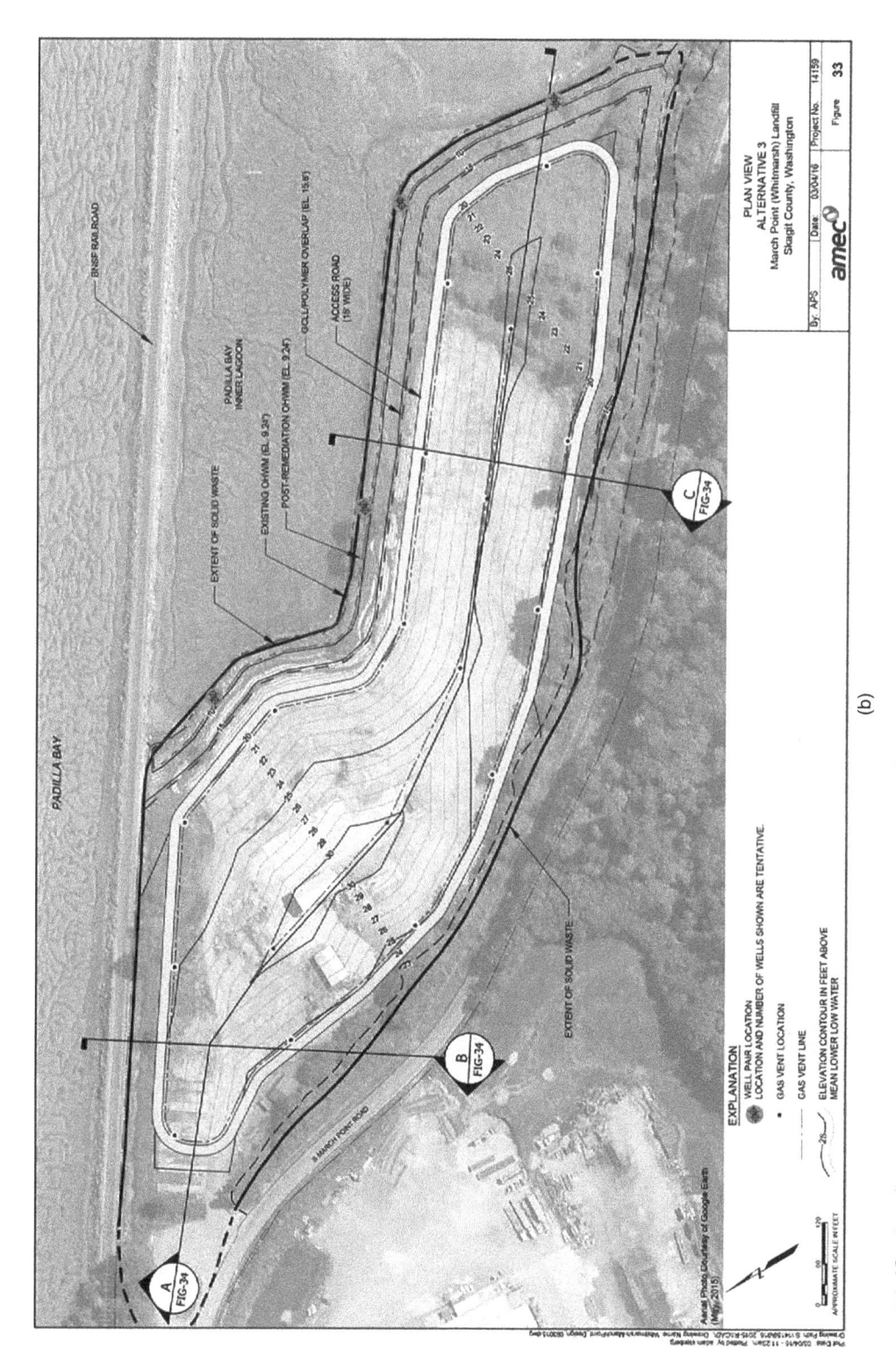

Figure 5.9 (Continued)

Figure 5.9 (Continued)

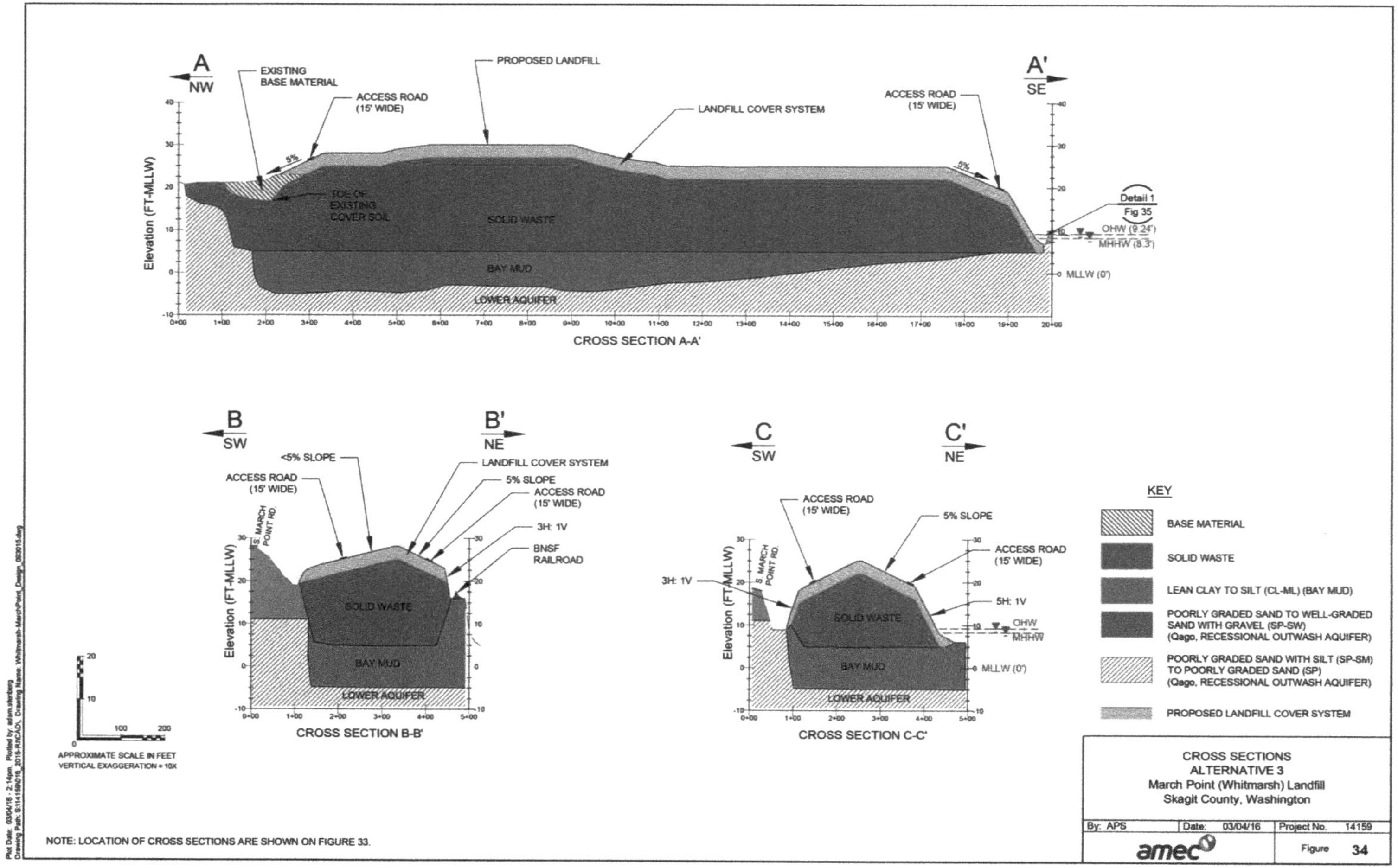

(d)

Figure 5.9 (Continued)

- Placement of an enhanced geosynthetic clay liner over the landfill along the shoreline up to 16 ft MLLW to minimize or eliminate discharge of groundwater to surface water (Figure 5.9b–d).
- Constructing a perimeter access road around the landfill.
- Requiring 30 years of operation and maintenance.
- A tidal study, and geotechnical and hydrogeological evaluations, showed that (a) the shallow nature of Padilla Bay, (b) the BNSF rail embankment, (c) nearby hillside, and (d) Highway 20 protected the lagoon and shoreline from increased wave heights that might result from sea level rise, as well as wave and current actions during 100-year storm events.

5.9 Challenges

It is projected that the United States and many other countries worldwide sustain greater economic hardship from climate impacts such as hurricanes, rising sea levels, extreme flooding, and high temperatures. Many entities are planning to address resiliency in infrastructure projects. Many challenges exist to achieve improved resilience, including those listed below:

- Uncertainty in prediction of magnitude and frequency of extreme events. Thus, there is inevitable need for adaptability and acceptance of some level of uncertainty.
- Climate change is a global problem and mitigation/adaptation strategies and development of standards require coordination of multiple nations and multiple jurisdictions.
- There is no acceptable way to quantify resilience; hence, it is not possible to assess alternatives to increase robustness to extreme events.
- There is a lack of census-based standard strategy to assess vulnerability to extreme events and resilient design.
- There are no mandates or incentives to design for resiliency, resulting lack of investments. Life cycle costs and benefits should be used to justify pursuing resilient designs.
- There is a lack of educational programs on resiliency, especially in areas of the world where resilient systems are utmost needed. Lack of transfer of technology and knowledge may be impediments to developing or implementing innovative resilient systems.
- Resilient thinking should be highlighted as a moral obligation of an engineer to ensure the safety of public.
- Lack of communication among the stakeholders (design engineers, policymakers, regulators, citizens, and investors) to build consensus for the need for design for resiliency and additional investments.

5.10 Summary

Resilience is a term common to many fields of study. Resilience helps systems to withstand, respond, and/or adapt to a vast range of disruptive events by preserving and even enhancing critical functionality. Greenhouse gases and climate change are found to be responsible for the growing concern of extreme events such as sea level rise, storm surge, extreme precipitation and flooding, extreme heat. Recently, Superstorm Sandy in New York caused extensive damage to the civil infrastructure, impacting quality of life, economy, and environment. This and such other extreme events have highlighted the need for resilient civil infrastructure. Many entities developed policies or guidance on pursuing resilient systems. The framework for incorporating resiliency in infrastructure design is not well established, but a vulnerability assessment is needed and adaptive resilient systems need to be designed. The effectiveness of a resilient infrastructure design depends upon its ability to anticipate, absorb, adapt to, and/or rapidly recover from a potential extreme event. Despite the growing importance of resilient infrastructure, there is a lack of guidance for engineers regarding how to specify, design, monitor, and assess the resilient civil infrastructure. In addition, educational and technology transfer programs as well as incentives to pursue resilience are needed.

5.11 Questions

5.1 What is meant by resiliency? What are the different fields that deal with resiliency and in what ways? Explain briefly.

5.2 Mention some of the disastrous climatic impacts.

5.3 Describe how changes in precipitation patterns and snowfall could affect runoff and flooding patterns.

5.4 What is disaster resilience and what are its characteristics?

5.5 Discuss in detail an infrastructure project of which you are familiar and identify ways to enhance its design to improve its resiliency.

5.6 List in detail some important indicators or inputs of a resiliency assessment tool that interests you.

5.7 What are the different policies and initiatives that were taken in regard to resilience? Explain briefly.

5.8 Mention some of the tools developed for evaluating climate change vulnerabilities and impacts.

5.9 What are the important considerations while designing for resilience against sea level rise?

5.10 What are the major challenges in adaptation and implementation of resiliency in engineering and infrastructure design?

References

Adger, W.N. (2000). Social and ecological resilience: are they related? *Progress in Human Geography* 24 (3): 347–364. https://doi.org/10.1191/030913200701540465.

American Society of Civil Engineers (ASCE) (2018a). Policy statement 418 – the role of the civil engineer in sustainable development. https://www.asce.org/issues-and-advocacy/public-policy/policy-statement-418---the-role-of-the-civil-engineer-in-sustainable-development/ (accessed 20 September 2018).

American Society of Civil Engineers (ASCE) (2018b). Working definitions. http://ciasce.asce.org/working-definitions (accessed 20 September 2018).

Bocchini, P. and Frangopol, D.M. (2011). Resilience-driven disaster management of civil infrastructure. In: *Computational Methods in Structural Dynamics and Earthquake Engineering* (ed. M. Papadrakakis, M. Fragiadakis and V. Plevris), 1–11. Keynote paper, Institute of Structural Analysis and Antiseismic Research, School of Civil Engineering, National Technical University of Athens (NTUA).

Bocchini, P., Frangopol, D.M., Ummenhofer, T., and Zinke, T. (2014). Resilience and sustainability of civil infrastructure: toward a unified approach. *ASCE Journal of Infrastructure Systems* 20 (2): 04014004.

Brundtland, G.H. (1987). *Our Common Future: Brundtland-Report*. Oxford: Oxford University Press.

Bruneau, M., Chang, S.E., Eguchi, R.T. et al. (2003). A framework to quantitatively assess and enhance the seismic resilience of communities. *Earthquake Spectra* 19 (4): 733–752.

Cimellaro, G.P., Reinhorn, A.M., and Bruneau, M. (2010a). Framework for analytical quantification of disaster resilience. *Engineering Structures* 32 (11): 3639–3649.

Cimellaro, G.P., Renschler, C., Reinhorn, A.M., and Arendt, L. (2016). PEOPLES: a framework for evaluating resilience. *Journal of Structural Engineering* 142 (10): 04016063.

City of San Francisco (2015). Guidance for incorporating sea level rise into capital planning in San Francisco: assessing vulnerability and risk to support adaptation. http://onesanfrancisco.org/sites/default/files/inline-files/Guidance-for-Incorporating-Sea-Level-Rise-into-Capital-Planning1.pdf (accessed 20 September 2018).

Climate Impacts Group (CIG) (2018). Climate change. Available at https://cig.uw.edu/learn/climate-change/ (accessed 20 September 2018).

Combaz, E. (2014). *Disaster Resilience: Topic guide*. Birmingham, UK: Governance and Social Development Research Centre, University of Birmingham.

Henry, D. and Ramirez-Marquez, J.E. (2012). Generic metrics and quantitative approaches for system resilience as a function of time. *Reliability Engineering and System Safety* 99: 114–122.

Holling, C.S. (1973). Resilience and stability of ecological systems. *Annual Review of Ecology and Systematics* 4: 1–23.

IPCC (2015). Climate change 2014: synthesis report. Intergovernmental Panel on Climate Change. http://ar5-syr.ipcc.ch/ipcc/ipcc/resources/pdf/IPCC_SynthesisReport.pdf (accessed April 2016).

NOAA (2017). National Centers for Environmental Information, State of the Climate: Global Climate Report for Annual 2016. https://www.ncdc.noaa.gov/sotc/global/201613 (retrieved 20 September 2018).

NOAA Office for Coastal Management (2018a). Sea level rise viewer. https://coast.noaa.gov/digitalcoast/tools/slr.html (accessed 20 September 2018).

NOAA Office for Coastal Management (2018b). Surging seas. https://coast.noaa.gov/digitalcoast/tools/surging-seas.html (accessed 20 September 2018).

O, Rourke, T.D. (2007). Critical infrastructure, interdependencies, and resilience. *The Bridge* 37: 22–29.

Oregon Climate Change Research Institute (OCCRI) (2017). Third oregon climate assessment report. Available at: http://www.occri.net/publications-and-reports/third-oregon-climate-assessment-report-2017/ (accessed September 2018).

Presidential Policy Directive (PPD) (2011). PPD-8: national preparedness. https://www.dhs.gov/presidential-policy-directive-8-national-preparedness (accessed 20 September 2018).

Presidential Policy Directive (PPD) (2013). PPD-21: critical infrastructure security and resilience. https://obamawhitehouse.archives.gov/the-press-office/2013/02/12/presidential-policy-directive-critical-infrastructure-security-and-resil (accessed 20 September 2018).

Rinaldi, S.M., Peerenboom, J.P., and Kelly, T.K. (2001). Identifying, understanding, and analyzing critical infrastructure interdependencies. *IEEE Control Systems Magazine* 21 (6): 11–25.

Rose, A. and Liao, S.-Y. (2005). Modeling regional economic resilience to disasters: a computable general equilibrium analysis of water service disruptions. *Journal of Regional Science* 45 (1): 75–112.

State of Washington (2017). Adaptation strategies for resilient cleanup remedies: a guide for cleanup project managers to increase the resilience of toxic cleanup sites to the impacts from climate change. https://fortress.wa.gov/ecy/publications/SummaryPages/1709052.html (accessed 20 September 2018).

Timmerman, P. (1981). Vulnerability. Resilience and the collapse of society: a review of models and possible climatic applications. In: *Environmental Monograph*, 1. Canada: Institute for Environmental Studies, University of Toronto.

UNISDR (2015). Sendai framework for disaster risk reduction 2015–2030. 32 p. https://www.unisdr.org/we/inform/publications/43291.

United States Army Corps of Engineers (USACE) (2014). Sea-level change curve calculator using the flood risk reduction standard for sandy rebuilding projects. https://www.usace.army.mil/corpsclimate/Climate_Preparedness_and_Resilience/App_Flood_Risk_Reduct_Sandy_Rebuild/SL_change_curve_calc/ (accessed 20 September 2018).

United States Environmental Protection Agency (USEPA) (2016). Climate resilience evaluation and awareness tool (CREAT). https://www.epa.gov/crwu/creat-risk-assessment-application-water-utilities (accessed 20 September 2018).

United States Environmental Protection Agency (USEPA) (2017). National stormwater calculator with climate assessment tool. https://toolkit.climate.gov/tool/national-stormwater-calculator%E2%80%94climate-assessment-tool (accessed 20 September 2018).

Section II

Sustainability Metrics and Assessment Tools

6

Sustainability Indicators, Metrics, and Assessment Tools

6.1 Introduction

In the first section of this book, we discussed the influence of human activities on the environment. We described their impacts on ecosystems, biodiversity and species survival, soil contamination, water and air quality, climate change, resource depletion, and others. These impacts on the environment, in turn, affect human health and their well-being, as well as the economy. Furthermore, current human activities and their impacts on the environment could be compromising to the development of future societies and individuals. This aspect signifies the importance of considering sustainability, which is defined as development that meets the needs of the present without compromising the ability of future generations.

Sustainability analysis is performed to identify and decide the most sustainable alternative among several, and often competitive, alternatives or solutions. As discussed in previous chapters, the triple bottom line of sustainability encompasses environmental, social, and economic metrics with the goal of maximizing the overall positive contribution from these three dimensions. To measure the degree of sustainability in engineering projects, it is essential to define and apply quantitative sustainability metrics. This section II presents common indicators, metrics, and tools that are used to quantify and assess sustainability of engineering projects.

6.2 Sustainability Indicators

Sustainability indicators are measurable aspects of environmental, economic, or social dimensions associated with potential engineering alternatives for a project (Bell and Morse 2012). Since they are measurable, these indicators can be estimated beforehand or monitored on a real-time or periodic basis. These are used to determine how a project, or a project alternative and its sustainability characteristics may positively (or negatively) contribute to human health and the environment. A sustainability indicator should have the following attributes defined by the "SMART" acronym:

- *Specific* – The indicator should target a specific area of interest for the analysis. It identifies the "what," "when," or "how" associated with the project. As an example, an indicator for a project alternative may be the minimization of emissions of particulate matter less than 10 μm in diameter (PM_{10}). PM_{10} emissions can have a deleterious effect on the health of people with respiratory ailments, such as asthma or chronic lung disease, and can also lead to other long-term adverse health effects. A specific indicator may be using emissions controls on heavy equipment at a project site.

Sustainable Engineering: Drivers, Metrics, Tools, and Applications, First Edition.
Krishna R. Reddy, Claudio Cameselle, and Jeffrey A. Adams.
© 2019 John Wiley & Sons, Inc. Published 2019 by John Wiley & Sons, Inc.

- *Measurable* – The indicator should be capable of being counted, compiled, analyzed, or tested so that a data set can be collected and assessed to determine the degree of success. In our example of PM_{10} emissions, filters or other measurement devices can be deployed at or near a project to measure PM_{10} emissions. Baseline or ambient conditions may also be established to determine the degree to which the project may contribute to the measured indicator.
- *Actionable/achievable* – The indicator should have a clear performance target that is easily understandable and may be realistically achieved at a project. For instance, if heavy diesel-powered equipment is to be used at a project, it may be unrealistic to expect 0 PM_{10} emissions; however, another performance standard, such as a 50% reduction compared to previous projects where emissions controls were not used may be appropriate and achievable.
- *Relevant* – The indicator should be selected such that it has a meaningful contribution to the overall goal or strategy associated with the project. Many indicators can be selected for a given project; however, they should be critically assessed for their overall meaningful contribution to the environmental, economic, or social dimensions of sustainability for a project. In the PM_{10} example, it is relatively easy to demonstrate that reduced PM_{10} emissions have a direct benefit to the project's environmental conditions as well as meaningful contributions to economy and society by protecting public health, quality-of-life, and associated economic benefits.
- *Timely* – The indicator should be achieved within an appropriate timeframe and/or subjected to the time constraints of the project. For this example, the 50% PM_{10} reduction may be assigned to the life of the project or over a specific subset of time, such as a period when equipment operations will be the greatest and PM_{10} reductions are most necessary.

The key indicators as discussed above may be objective or subjective. Sustainability can be viewed at global, continental, national, city, company (corporation), or specific project scale. Indicators are often defined and tracked to help assess progress towards sustainability or sustainable development. But, these indicators can vary widely depending on the scale at which sustainability is being considered. For example, the United Nations Department of Economic and Social Affairs (UNDESA) proposed 134 indicators to reflect sustainability at global level (UNDESA 2007). Figure 6.1 shows some of these UN measurable objective indicators for sustainable development.

Following the UN initiative, European Union (EU) has proposed over 100 sustainability indicators (EC 2009). Many countries around the world have also developed national-level sustainability indicators- for example the UK developed a national sustainable development strategy and proposed 68 indicators (DEFRA 2005). Many local authorities have also initiatives to address local sustainability and proposed sustainable development indicators- for instance the city of London (UK) proposed a set of 61 indicators to complement the UK Sustainable Development Strategy (DEFRA 2005). Even businesses and corporations are under growing pressure from stake holders to align with the principles of sustainable development, and they have developed eco-efficiency or sustainability indicators and tools to assess them. Global Reporting Initiative (GRI) is a framework for reporting corporate sustainability performance initiated by the US-based Coalition of Environmentally Responsible Economies (CERES) in partnership with the United Nations Environmental Program (UNEP) and proposed 79 indicators for sustainability assessment (GRI 2010). Numerous companies have used this GRI and in revised versions to cover environmental, economic, and social aspects. For example, BASF, one of the largest chemical companies have developed a tool called "Socio-Eco-Efficiency tool-SEEBalance®" that encompasses life cycle assessment (LCA), economic life cycle costing, and social indicators (BASF 2019). In considering engineering projects with respect to sustainability (focus of this book), key indicators are essential to the evaluation of a project, whether

Poverty
- Unemployment rate
- Poverty index
- Population living below poverty line

Population Stability
- Population growth rate trend
- Population density

Human Health
- Average life expectancy
- Access to safe drinking water
- Access to basic Sanitation
- Infant mortality rate

Living Conditions
- Urban population growth rate
- Floor area per capita
- Housing cost

Coastal Protection
- Population growth
- Fisheries yield
- Algae index

Agricultural Conditions
- Pesticide use rate
- Fertilizer use rate
- Arable land per capita
- Irrigation % of arable land

Ecosystem Stability
- Threatened species
- Annual rainfall

Atmospheric Impacts
- Greenhouse gas emissions
- Sulfur oxide emissions
- Nitrogen oxides emissions
- Ozone depleting emissions

Generation
- Municipal waste
- Hazardous waste
- Radioactive waste
- Land occupied by waste

Consumption
- Forest area change
- Annual energy consumption
- Mineral reserves
- Fossil fuel reserves
- Material intensity
- Groundwater reserves

Economic Growth
- GNP
- National debt/GNP
- Average income
- Capital imports
- Foreign investment

Accessibility
- Telephone lines per capita
- Information access

Figure 6.1 UN sustainability indicators. Source: United Nations, *Indicators of Sustainable Development*; World Bank, *World Development Indicators*.

they considered objective or subjective indicators. All aspects of an engineering project may be considered on a life-cycle basis (considering raw material acquisition, material manufacturing, construction, etc.) or on an individual, discrete basis, whether this constitutes the design phase, the construction phase, the operational phase, or the final disposal phase. Additionally, any combination of these project phases may be considered when assessing sustainability.

Further, when considering the sustainable aspects of an engineering project, it is essential to consider indicators representative of all the three dimensions that constitute the triple bottom line. Environmental indicators may include the following:

- Greenhouse gases and other air emissions
- Contributions to climate change
- Use of fresh water resources
- Impacts to soil
- Utilization of raw natural resources
- Impacts to surface water or groundwater
- Use of recycled/repurposed materials
- Overall waste generation
- Diversion of waste materials from or to landfill facilities.

Economic sustainability indicators that may be considered for the engineering projects may include the following:

- Direct and indirect job creation within the community
- Direct and indirect investment within the community
- Facilitation of government grants for the project and community as a whole

- Long-term tax and revenue generation within the enhanced community
- Degree of highest and best use (HBU) achieved by the engineered project.

When compared to environmental and economic dimensions, social sustainability indicators have not been incorporated as extensively, nor have they been as developed or refined. In general, social sustainability is focused on the impacts of engineering activity on society as a whole including the dimensions related to quality-of-life, diversity, cultural awareness, social cohesion, and harmony. Some key indicators of social sustainability include the following:

- Enhancement of community aesthetics
- Enhancement of quality-of-life features (e.g. improved transportation opportunities or recreational facilities)
- Public participation in decision-making
- Educational and job training opportunities
- Interaction between community groups
- Emotional ownership of the community in an engineering project
- Improved physical and mental health and well-being of members of the community
- Enhanced social opportunities for members of the community
- Strengthening or enhancement of existing community institutions (e.g. recreational organizations, charitable foundations, and houses of worship).

6.3 Sustainability Metrics

The indicators presented above provide key variables that may be assessed when evaluating the degree of sustainability for a specific engineering project or a product. The indicators presented earlier may not be easily measurable. However, numerical values or quantitative characteristics may be integrated with these indicators so that they may be objectively and accurately assessed. As a result, metrics can now be connected to the indicators. Sustainability metrics are numerical values that may be used to assess specific indicators related to sustainability, and they are vitally important for performing objective analysis with respect to engineering project sustainability (Sikdar 2003). These metrics should be incorporated into a range of sustainability measurement tools, which are discussed in greater detail in subsequent chapters in this book.

The metrics that may be used to assess sustainability of engineering projects are, in many cases, fairly straightforward. As a result, their ability to be accurately measured in many cases is well established. This is especially the case for economic and environmental sustainability. However, social sustainability indicators and metrics have not been as extensively defined or developed. Many of the social metrics can be evaluated only qualitatively or semi-quantitatively, which makes the determination of social impacts difficult.

Before some common metrics are presented, it is important to note that there is no standard established framework regarding an appropriate set of parameters to be used for sustainability evaluation of engineering projects, nor is there a consensus on what constitutes sustainable engineering. Additionally, there is a wide range of opinion regarding the degree to which individual metrics contribute to or affect sustainability. This is further reflected in the relatively wide range of scope inherent in several sustainability assessment tools that are presented in later chapters of this book. Further, there is no commonly accepted set of metrics used by engineering practitioners to evaluate whether an engineering project is sustainable.

Physical metrics may be used to assess the physical inputs and outputs of an engineering project alternative, including those focused or tailored for positive or negative contributions with respect to sustainability. Such examples include the following:

- Energy consumption (kWh or BTU)
- Renewable energy consumption (kWh or BTU or as a percentage of total energy consumption)
- Fresh water or recycled/reclaimed water consumption (gallons or liters)
- Air emissions (tons or kg)
- Greenhouse gas emissions (tons or kg)
- Carbon emission offset (tons or kg or a percentage of greenhouse gas emissions)
- Solid waste generation (tons or kg)
- Use of recycled solid materials (tons or kg).

Several of these indicators including energy (nonrenewable or renewable), water (fresh or reclaimed), or air emissions may be combined on a per unit basis (e.g. per treated unit mass/volume of soil or water). Of course, the indicators may further be coupled with time or monetary unit to determine these metrics based on a unit time or a unit cost. Further, other actions may be quantitatively assessed, including credits/offsets of ecological restoration, increased real estate value, and preservation or restoration of natural resources.

Since the potential list of sustainability metrics for engineering projects is enormous, and because there is a lack of consensus or standard regarding key indicators and related metrics, there has been a growing dialogue between a number of sustainability-focused organizations and regulatory agencies regarding potential efforts for standardization. One example is the USEPA's Tool for the Reduction and Assessment of Chemical and Other Environmental Impacts (TRACI). This environment impact assessment tool, utilized by several life cycle assessment (LCA) software programs, includes the following nine environmental impact categories (USEPA 2016a; Bare 2011):

- *Global climate change impact category* – Reported as carbon dioxide (CO_2) equivalents
- *Acidification impact category* – Reported as sulfur dioxide (SO_2) equivalents
- *Eutrophication impact category* – Reported as nitrogen (N) equivalents
- *Ozone depletion impact category* – Reported as trichlorofluoromethane (CFC-11) equivalents
- *Photochemical smog formation impact category* – Reported as ozone (O_3) equivalents
- *Human health particulate matter (PM) impact category* – Reported as fine particulate matter ($PM_{2.5}$) equivalents
- *Human health cancer impact category* – Reported as comparative toxicity unit cancer (CTU-cancer) equivalents
- *Human health noncancer impact category* – reported as comparative toxicity unit noncancer (CTUnoncancer) equivalents
- *Eco-toxicity impact category* – Reported as comparative toxicity unit eco-toxicity (CTUeco) equivalents.

Other impact categories, such as those associated with renewable energy and nonrenewable energy use may also be incorporated into an assessment when permissible by the LCA tool that is being used for an analysis.

6.4 Sustainability Assessment Tools

Several general and specific assessment tools have been developed to account for a wide range of metrics in engineering projects (Ness et al. 2007). Although many tools focus primarily on environmental related metrics, tools are being developed or revised to incorporate economic and social metrics as well. These assessment tools take on many forms and range from simple to complex. Some have been developed to focus with a local context, while others have been developed to address a more global context. These tools also range in precision ranging from generic to specific frameworks or metrics; some measure qualitative metrics, while others are focused on quantitative measurement. Many tools exist in the public domain, while others have been proprietarily developed for in-house corporate work and for its use by consultants.

In the subsequent chapters of this Section II, the following general tools used for sustainability assessment are presented:

- Although it is not an assessment tool for sustainability, the ***material flow and budget analysis*** can be performed to determine the reserves of resources, consumption and flows among extraction, consumption, and use of the materials. In general, the material flows are evaluated for a country or a region, and therefore, the import/export flows has to be considered. The analysis of the materials in use and the life expectancy of the products that contains these materials is an indicator of necessity of a material to a specific society and the amount of materials for recycling at the end of the useful life for respective products.
- The emission of greenhouse gases has a direct impact on the global warming and climate change. Thus, the ***GHG inventory*** and the so-called carbon footprint is an indicator of the impact of each activity or organization toward global warming and climate change. The reduction of the carbon footprint can be used as an indicator of environmental sustainability when deciding among different project alternatives or modifications of processes.
- One of the most powerful tools to assess the environmental impacts of a product or a process is the life cycle assessment (***LCA***). This tool can be used to perform a comprehensive analysis of all the environmental impacts of a product or a process across all the life cycle stages, from the extraction of raw materials, transformation, manufacturing, use, and disposal, including the transportation of materials or products between and within the life cycle stages. LCA can provide a detailed analysis of all the potential impacts and is an efficient tool to determine the critical stages or operations that result in the major environmental impacts. Unfortunately, this analysis is complex and require a significant amount of data, so it is very important to define the boundaries of our system and the scope of our study to accurately identify the data required for the analysis.
- ***Streamlined life-cycle assessment (SLCA)*** is another sustainability assessment tool that was developed to overcome the difficulties and limitations of LCA. The ***SLCA*** is based on the identification of the most significant environmental impacts across the life-cycle stages of a product. A matrix consisting of potential environmental impacts as the column headings and the different life-cycle stages of the project/product as the row headings is created. The matrix is populated with values that range from 0 (for very high impact) to 4 (for no impact) relating the magnitude of impact of a particular life-cycle stage to a specific environmental impact category. The sum of scores under an impact category in the matrix will give the global score for one alternative. Comparing these global scores for different alternatives quantifies the relative sustainability of the two products/alternatives. The graphical representation of the matrix either as a heat map or as a target plot will be useful to determine the most critical impacts in the life cycle stage of a product/project.
- ***Economic input–output (EIO) analysis*** is a powerful tool developed by Nobel Laureate Dr. Wassily Leontief. Many nations create input–output models of their economy that are useful in predicting the evolution of their economy and the definition of their national budgets. The EIO models can be applied to evaluate the environmental impacts and thereby perform an environmental sustainability analysis by defining the environment as an additional sector and quantifying the economic fluxes associated to the environmental impacts of various sectors. Alternatively, the EIO models can be defined in terms of materials fluxes instead of economic

fluxes. This method of analysis using material fluxes is called physical input–output (PIO) analysis. PIO models are more detailed than the aggregated economic models, and hence more difficult to construct, but the benefits of greater detail facilitate a more comprehensive analysis of environmental impacts.

- *Quantitative environmental health risk assessment* is a tool developed by USEPA to determine the risks from the exposure of chemicals to potential receptors (human health and ecosystem). The risk assessment method will allow for the evaluation of the risks with lower degree of subjectivity and lower uncertainty than the individual perception. Risks to humans and the environment are real and should be carefully evaluated, taking human perception into account.

In addition to the above general assessment tools, many specific tools have been reported for specific engineering applications. For example, Leadership in Energy and Environmental Design (LEED) is a sustainability rating tool developed for assessing the sustainable practices adopted in green buildings and further certifies those buildings to acknowledge the innovative and efficient use of resources while causing minimal harm to the human health and the environment (USGBC 2018). Envision™, like LEED, is another sustainability rating tool developed for assessing and acknowledging the sustainable initiatives used in constructing any civil infrastructure (ISI 2018). Likewise, Spreadsheets for Environmental Footprint Analysis (SEFA) (USEPA 2016b) and SiteWise™ (SURF 2018) are quantitative sustainability assessment tools developed for encouraging the incorporation of the principles of green and sustainable remediation in the remediation of contaminated sites. These frameworks are presented in greater detail throughout this book.

6.5 Summary

Sustainability is a functionally related to the concept of the triple bottom line. Therefore, to define or quantify sustainability, environmental, economic, and social indicators that can be expressed by metrics are needed. Sustainability indicators should be carefully selected and should be those that are meaningful, reliable, and quantifiable. We have addressed several key environmental metrics, including consumption of natural resources, climate change and greenhouse gas emissions, air pollution, water pollution, land use/degradation, energy consumption, and waste generation, among others. We have also addressed several key economic metrics and social metrics related to the anticipated returns and positive and negative side effects associated with economic activity. The key question is how to find a means to measure and properly account for the numerous metrics in a reasonable manner such that we can critically assess economic activities and/or find means to reduce impacts to identify more sustainable alternatives. Several general sustainability assessment tools are identified, and many tools for very specific applications are being developed by the engineering and practicing professionals.

6.6 Questions

6.1 What is the most general definition of sustainability?

6.2 Describe the aspects of a project or product that may be considered on a life-cycle basis.

6.3 What are the most desirable characteristics of a sustainability indicator? Explain briefly.

6.4 Mention five indicators for environmental, economic, and social sustainability.

6.5 What are the impact categories assessed in TRACI?

6.6 Select a project or a product and discuss actions that could be taken to optimize the effects of three environmental key indicators.

6.7 Select a project or a product and discuss actions that could be taken to optimize the effects of three economic key indicators.

6.8 Select a project or a product and discuss actions that could be taken to optimize the effects of three social key indicators.

6.9 Describe a scenario where trying to optimize a physical sustainability metric (i.e. recycling water use instead of fresh water) would NOT have a desirable effect on a project or a product.

6.10 Describe factors or activities you believe would be important to include in a greenhouse gas inventory assessment.

6.11 When might a Streamlined LCA be appropriate in place of a traditional LCA?

6.12 Discuss factors you believe would be important to include in an environmental health risk assessment.

6.13 Name some of the sustainability assessment tools.

References

Bare, J. (2011). TRACI 2.0: the tool for the reduction and assessment of chemical and other environmental impacts 2.0. *Clean Technologies and Environmental Policy* 13 (5): 687–696.

BASF (2019) SEEbalance®. Measuring sustainable development on a product level. https://www.basf.com/global/en/who-we-are/sustainability/management-and-instruments/quantifying-sustainability/seebalance.html (accessed January 22, 2019).

Bell, S. and Morse, S. (2012). *Sustainability Indicators: Measuring the Immeasurable?* Routledge.

Department of Environment, Food and Rural Affairs (DEFRA) (2005). *Securing the Future- the UK Government Sustainable Development Strategy*. London. https://www.gov.uk/government/publications/securing-the-future-delivering-uk-sustainable-development-strategy (accessed January 22, 2019).

European Communities (EC) (2009). 2009 monitoring report of the EU sustainable development strategy. https://ec.europa.eu/eurostat/web/products-statistical-books/-/KS-78-09-865 (accessed 22 January 2019).

Global Reporting Initiative (GRI) (2010). GRI Standards. https://www.globalreporting.org/standards/ (accessed 22 January 2019).

ISI (2018). Envision: Driving Success in Sustainable Infrastructure Projects. Institute for Sustainable Infrastructure https://sustainableinfrastructure.org/envision (accessed 25 July 2018).

Ness, B., Urbel-Piirsalu, E., Anderberg, S., and Olsson, L. (2007). Categorising tools for sustainability assessment. *Ecological Economics* 60 (3): 498–508.

Sikdar, S.K. (2003). Sustainable development and sustainability metrics. *AIChE Journal* 49 (8): 1928–1932.

SURF (2018). SiteWise version 3.1. Sustainable Remediation Forum. http://www.sustainableremediation.org/library/guidance-tools-and-other-resources/sitewise-version-31 (accessed 25 July 2018).

United Nations Department of Economic and Social Affairs (UNDESA) (2007). *Indicators for Sustainable Development: Guidelines and Methodologies*, 3rd Edition. United Nations, New York. https://sustainabledevelopment.un.org/index.php?page=view&type=400&nr=107&menu=1515 (accessed 22 January 2019).

USEPA (2016a). *Tool for Reduction and Assessment of Chemicals and Other Environmental Impacts (TRACI)*. US Environmental Protection Agency https://www.epa.gov/chemical-research/tool-reduction-and-assessment-chemicals-and-other-environmental-impacts-traci (accessed 25 July 2018).

USEPA (2016b). *Methodology & Spreadsheets for Environmental Footprint Analysis (SEFA)*. US Environmental Protection Agency https://clu-in.org/greenremediation/methodology/#SEFA (accessed 25 July 2018).

USGBC (2018). *Leadership in Energy and Environmental Design*. U.S. Green Building Council https://new.usgbc.org/leed (accessed 25 July 2018).

7

Material Flow Analysis and Material Budget

7.1 Introduction

The concept of sustainability is challenged by society's rapidly increasing use of nearly every resource on the planet. As an example, the increasing consumption of materials in the United States is presented in Figure 7.1. It emphasizes the trends of increased nonfuel resource use, with construction materials, such as aggregate, rock, and sand, constituting the majority of materials use. The consumption scenario in the United States is also prevalent among many other developed countries.

Figure 7.2 shows the ratio of the sum of anthropogenic flows to natural flows for the elements in the periodic table. The anthropogenic flows include mining and fossil fuel combustion, and the natural flows include erosion and air-blown dust. The flow of elements on the left in the figure is dominated by anthropogenic use, whereas natural processes dominate the element flows on the right. The four metals that constitute over 98% of the mass of all metal utilized by society are highlighted in Figure 7.2. Copper is anthropogenic flow-dominated, aluminum is natural flow-dominated, and iron and zinc are about evenly distributed among anthropogenic and natural flows. Figure 7.2 shows the fact that humans dominate the global cycles of about a third of the elements of the periodic table.

The objective of this chapter is to analyze how human activities dominate the flows of materials and how the present consumption rate of materials will lead to the depletion of natural resources, which will affect the capacity of future generations to meet their own needs. The analysis of natural material resources and their flows will help in the design of more sustainable processes.

7.2 Budget of Natural Resources

Before analyzing the flows and consumption of natural resources in material flow analysis (MFA), we need to define certain fundamental variables and parameters (Graedel and Allenby 2010). A budget is defined as the accounting of the receipts, disbursements, and reserves of the valued resource. The resource is stored in a container called reservoir. The actual amount of the resource is called the pool size. The resource enters and exits the reservoir at a rate called flux. The inflow and outflow streams are called sources and sinks.

As an example, we can model the components of a budget as a water trough. If the total source and sink flows are not equal, the pool size (i.e. the quantity of the material of interest in the reservoir) will change over time as indicated in Eq. (7.1). For any reservoir, the change in

Sustainable Engineering: Drivers, Metrics, Tools, and Applications, First Edition.
Krishna R. Reddy, Claudio Cameselle, and Jeffrey A. Adams.
© 2019 John Wiley & Sons, Inc. Published 2019 by John Wiley & Sons, Inc.

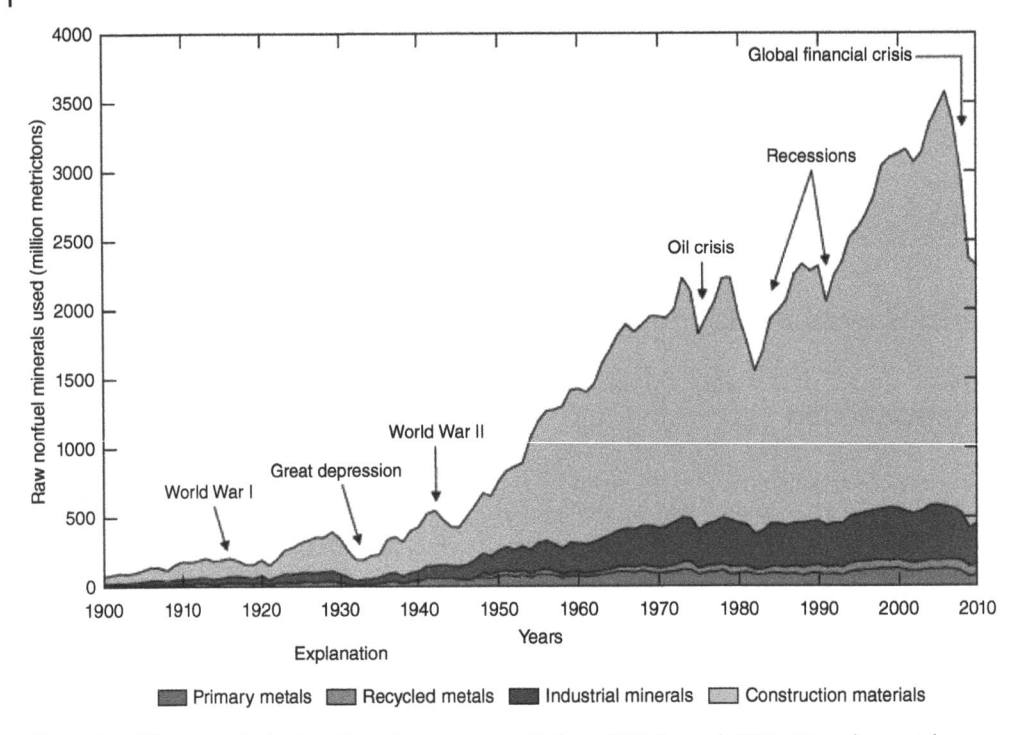

Figure 7.1 US raw nonfuel minerals put into use annually from 1900 through 2010. Mineral materials embedded in imported goods are not included. Source: Matos (2012).

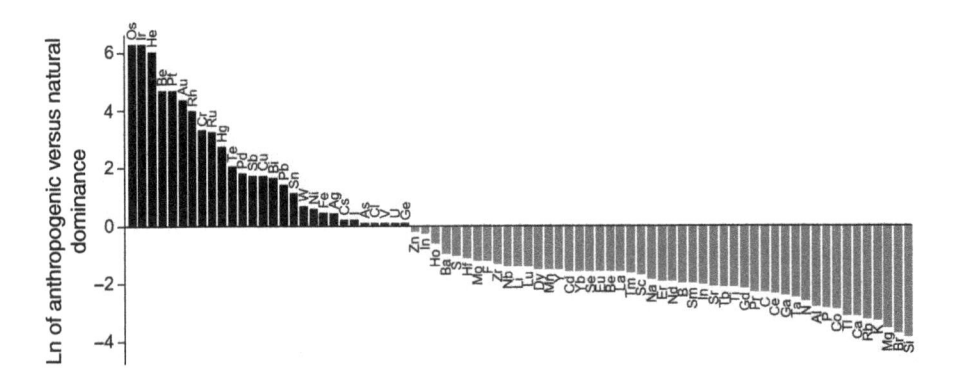

Figure 7.2 Ratio of anthropogenic/natural flows for elements. Source: Klee and Graedel (2004).

contents can be readily calculated once all flows are known. Conversely, measuring the change in contents can aid in determining an unknown flow.

$$\Delta_r = \int_{t_1}^{t_2} \left(\sum_i F_i - \sum_o F_o \right) dt \qquad (7.1)$$

A cycle is defined as a system of connected reservoirs that transfer and conserve a specific resource. In most material flow analyses, there are several reservoirs, each with several linkages to other reservoirs. A complete analysis quantifies all flows and the contents of each reservoir. Figure 7.3 shows a generic cycle for the human-driven flows of a metal with several reservoirs (ore-environment, processing, fabrication, etc.) and the sources and sinks that interconnect the

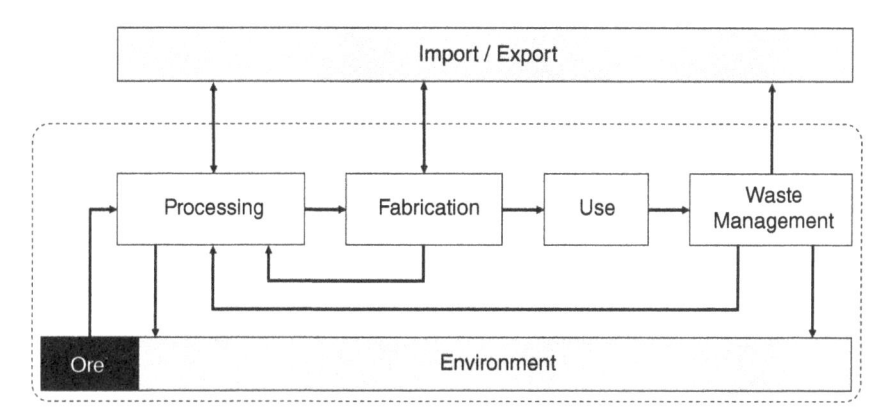

Figure 7.3 Generic cycle for the human-driven flows of a metal.

reservoirs. The actual cycles for metals are much more complex than what the simple example in Figure 7.3 suggests. Multiple reservoirs can be identified in production, fabrication, manufacturing, use, and waste facilities. The flows to other regions or countries must be considered since the analysis is done for a specific country or region (Reck et al. 2008). The inflows and outflows to the same reservoir are not necessarily the same, so the nickel in the reservoir increases or decreases, which may be accounted with Eq. (7.1). The average residence time of a metal (or any other resource) in a reservoir can be determined with Eq. (7.2), where Ψ indicates the fraction of the constituent having a residence time between τ and $\tau + d\tau$.

$$\tau_r = \int_{t_1}^{t_2} \tau \Psi(\tau) dt \qquad (7.2)$$

As an example, steel, which is widely used for a vast array of products, exhibits a significant range in residence time on a product-by-product basis, from a few years for kitchen gadgets to more than half a century for steel beams in buildings.

7.3 Constructing a Budget

The construction of a budget requires several tasks. The first step is the identification of the substance or substances to be evaluated, followed by reservoirs with the corresponding sinks and sources. Then, the limits or boundaries of the system may be defined. The second step requires the identification of the reservoir contents and the fluxes. This can be done with available data or with specific measurements wherever possible, and estimations of resource content and fluxes can be used if no other source of objective data is available. This has to be done for all sinks and sources, or at least for the most relevant contributions. Small fluxes can be disregarded if their contribution is minimal. The analysis of material flows can be tracked at different organizational levels: for a single facility, groups of facilities, at a regional or national level, or globally. In principle, a MFA can be conducted at any spatial or organizational level where a reservoir and the associated sources and sinks can be identified and measured.

7.4 Material Flow Analysis

Figure 7.4a depicts the entire chromium cycle for Japan in 2000, including multiple reservoir analysis. Japan mined no chromium, but imported large amounts in various forms, and fabricated, exported, used, and recycled it extensively. The fluxes of chromium among the different

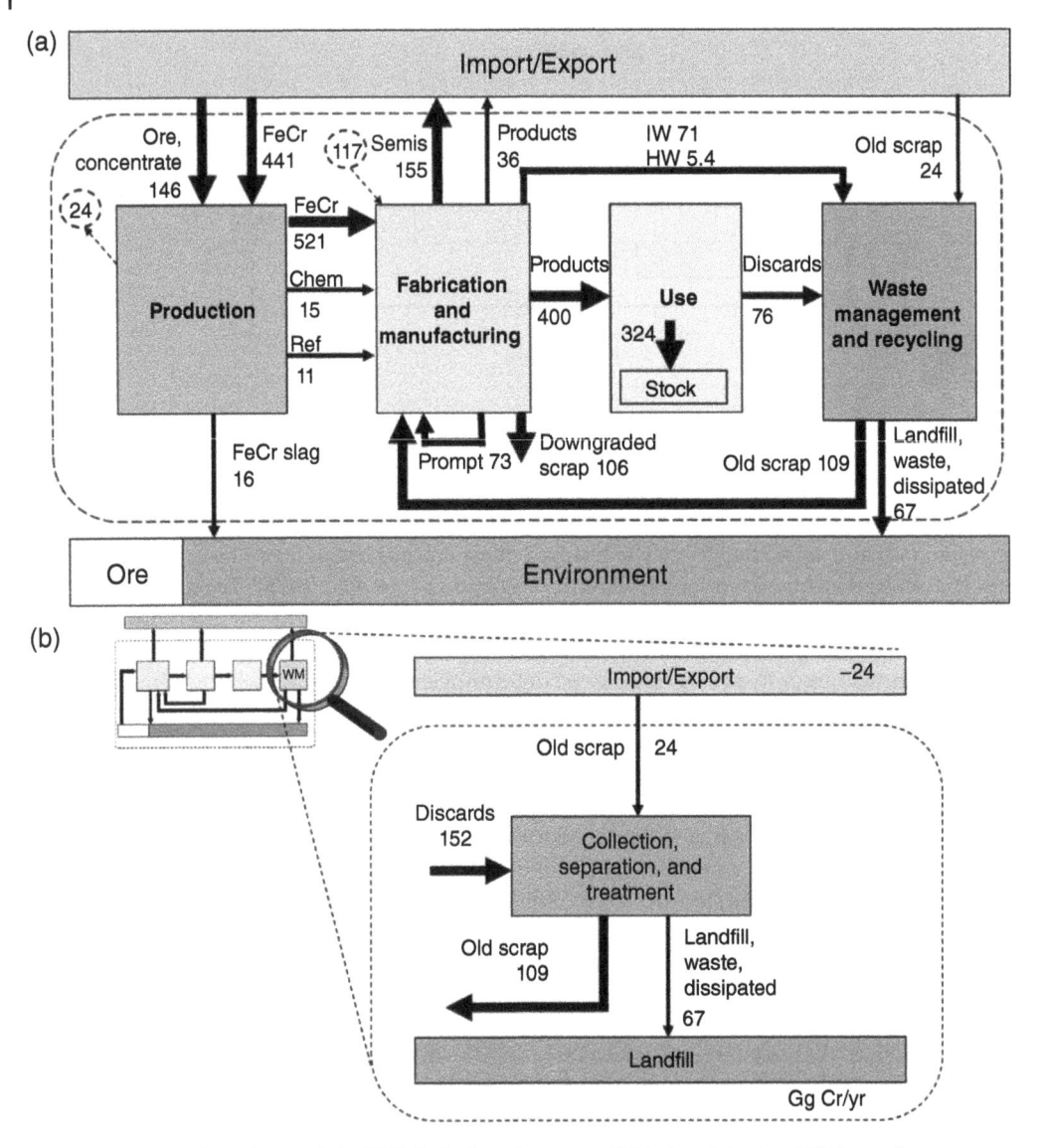

Figure 7.4 Japan chromium cycle in 2000. FeCr: ferrochromium; IW: industrial waste; HW: hazardous waste; Ref: refractories. (Gg Cr/yr). (STAF Project, Yale University).

reservoirs can be seen in the Figure 7.4a, where the fluxes are expressed in Gg (gigagrams, or thousand metric tons) of chromium per year. An example of a subset budget is presented in Figure 7.4b, which illustrates chromium inputs and outputs at the waste management stage of the life cycle in Japan in 2000. Of the ~150 units discarded, about two-thirds were recycled within Japan, along with a small amount of scrap that was imported. A little more than one-third was lost to landfilling. Similar material cycles can be prepared for different materials at country level or at a project level.

Many materials are mined or produced in a small number of countries and are then transported and sold in many other countries. This is very common for many materials traded relatively freely around the world. Thus, a multilevel cycle can be constructed by characterizing

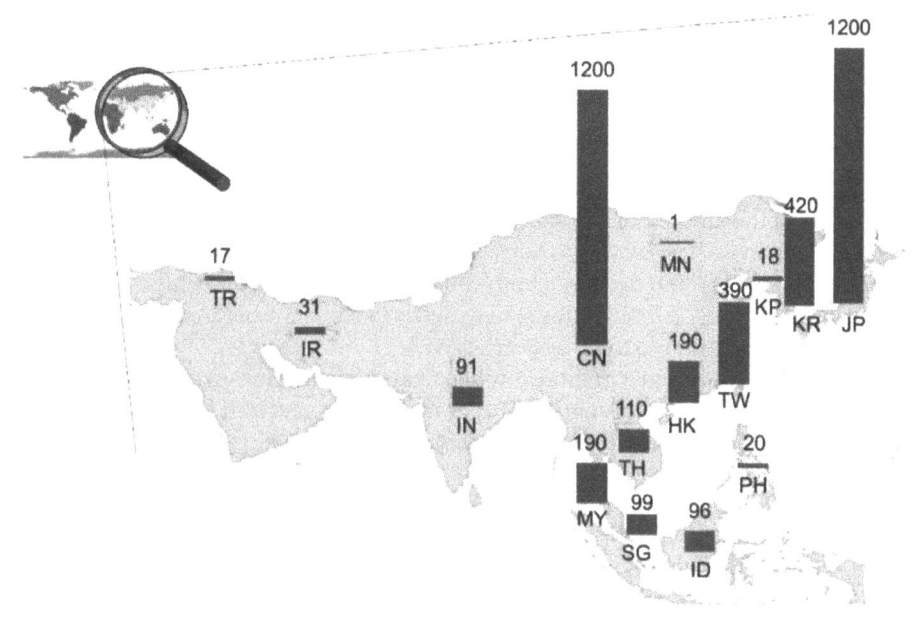

Figure 7.5 Country-level copper usage (Gg Cu/year).

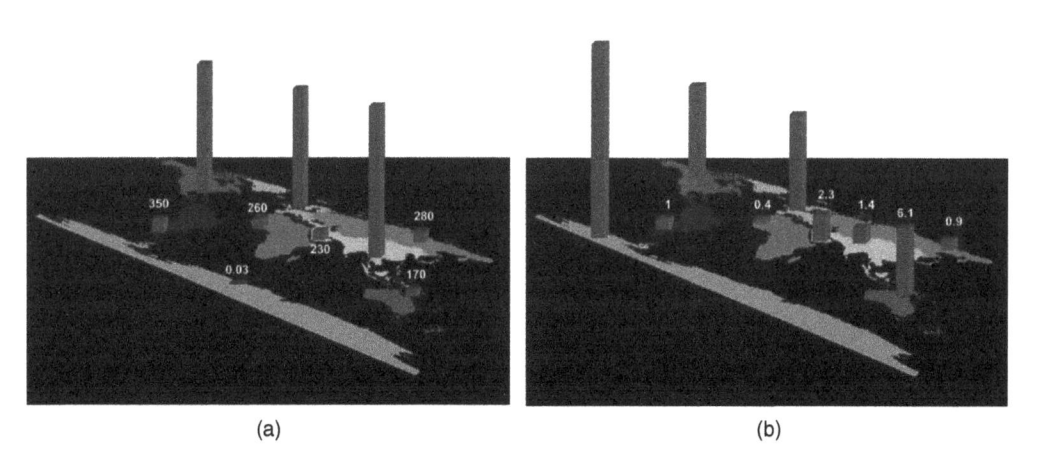

Figure 7.6 Copper use on a regional basis: (a) total use, (b) use per capita. Source: Reck et al. (2008). Reproduced with permission of American Chemical Society.

the country-level cycles and then aggregating them to the regional and global level. If MFAs for the same reservoir in several different industrial ecosystems are constructed, flows can be directly compared, as in a study of copper use in the mid-1990s (Figure 7.5), where countries are represented as ecosystems. A similar comparison for copper use in the mid-1990s at regional level may be performed (see Figure 7.6). North America, Europe, and Asia show a similar level of copper usage, whereas other regions (including Antarctica) use much less. When the analysis is made in terms of the amount of copper per capita, the results are very different. Antarctica now has a clear lead, reflecting the large quantity of resources needed to maintain a high-technology lifestyle in an extremely cold environment.

7.5 Material Flow Analysis: Wastes

An interesting application of MFA is the detailed analysis of waste management. As an example, Figure 7.7 depicts, on a global basis, the estimates of waste generation flow in seven categories during the mid-1990s. Industrial waste represents about 75% of the total, followed by municipal solid waste (12%) and construction and demolition debris (8%). A more detailed analysis permits an evaluation of the amount of metals in the waste flows. Copper and zinc are valuable metals that can be found in wastes. In most wastes, the concentrations are very low, but copper is quite prevalent in electronics, and the use of both copper and zinc is relatively high in vehicles (Figure 7.8). This suggests that the flows of these two metals in some wastes are high enough to have a relatively high contribution to their respective MFAs.

In calculating the MFAs, the product of total flow by metal concentration gives the total metal flow in the several discard streams (wastes). Copper is mainly concentrated in electronics and vehicles, whereas zinc is more evenly distributed among several waste fractions (Figure 7.9).

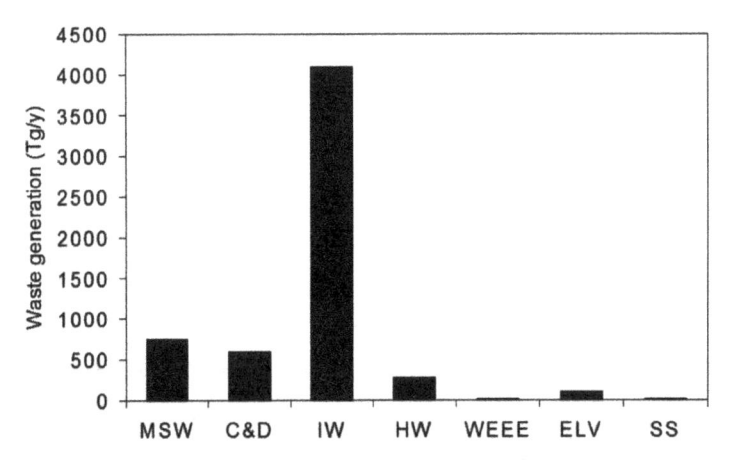

Figure 7.7 World waste generation flows in 1990s. MSW: municipal solid waste, C&D: construction and demolition debris, IW: industrial waste, HW: hazardous waste, WEEE: waste of electrical and electronic equipment, ELV: end of life vehicles, SS: sewage sludge.

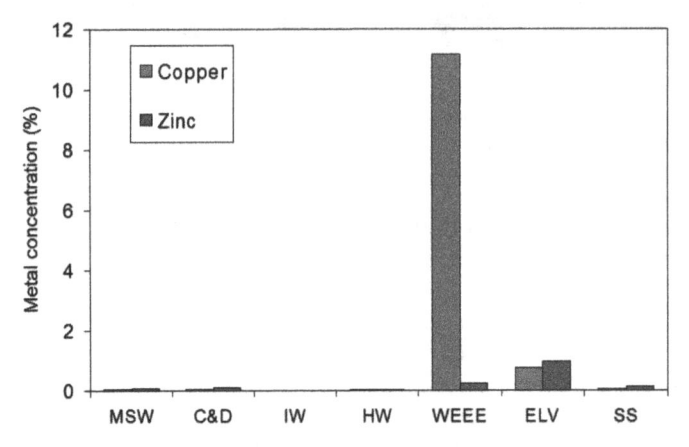

Figure 7.8 Copper and zinc concentration in wastes. MSW: municipal solid waste, C&D: construction and demolition debris, IW: industrial waste, HW: hazardous waste, WEEE: waste of electrical and electronic equipment, ELV: end of life vehicles, SS: sewage sludge.

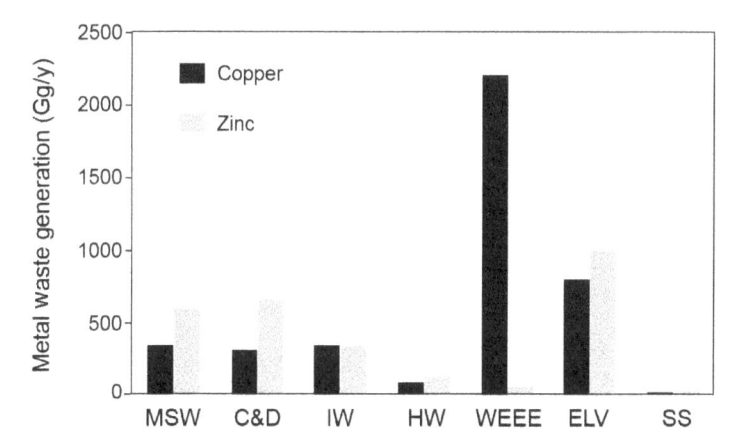

Figure 7.9 Global flows of copper and zinc in discarded materials (wastes). MSW: municipal solid waste, C&D: construction and demolition debris, IW: industrial waste, HW: hazardous waste, WEEE: waste of electrical and electronic equipment, ELV: end of life vehicles, SS: sewage sludge.

Thus, the results from the MFAs for copper and zinc suggest which materials may be more effectively recycled or repurposed compared to existing practices.

Material recycling is the primary alternative to extracting and processing new raw materials. The MFA for copper and zinc identifies the respective amounts of these metals in the different reservoirs, such as products in use and in wastes. However, it is important to note that these metals are incorporated into products with specific lifetimes; therefore, the metals will not be available for recycling until the useful life of these products has elapsed. Recycling streams incorporate several sequential processes, including development, use, discard, and reuse, with products having different lifetimes overlapping in the recycling stream.

An important question is to determine how much material is available in products in use that can serve as reusable resource in the future when the products are discarded; and a second question is to determine when the materials will become available for resource extraction and reuse. In other words, we need to determine the resources in use, their age, their probable lifetime until entering a waste stream, and where they can be reprocessed and reused. Equation (7.3) accounts for resources or stocks in use that corresponds with the amount of resource in the reservoir "use" in Figure 7.3.

$$S(t) = \sum_{i=1}^{n} N_i(t) \cdot \frac{M_i(t)}{N_i(t)} \tag{7.3}$$

where

S is the in-use stock,

N is the number of units of a particular product (automobiles, etc.) that use the resource to provide services of some sort, and

M_i/N_i is the material intensity per unit of product, all summed over the i types of products addressed by the analyst.

As an example, the calculation of the amount of copper in use requires the amount or concentration of copper in respective products as well as the number of those products in use. The first step is the identification of the products with the major amounts of copper. A typical car contains 21 kg of copper, and a typical house contains 200 kg of copper. The main product categories include buildings, equipment, transportation, and infrastructure. Figure 7.10 depicts copper stocks per capita in products in use at three different levels: city, state, and country. Differences

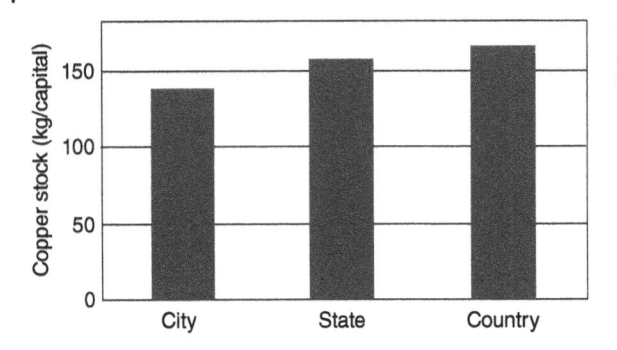

Figure 7.10 Copper in-use stock results at different spatial levels. City: New Haven, State: Connecticut, Country: USA. Source: Drakonakis et al. (2007).

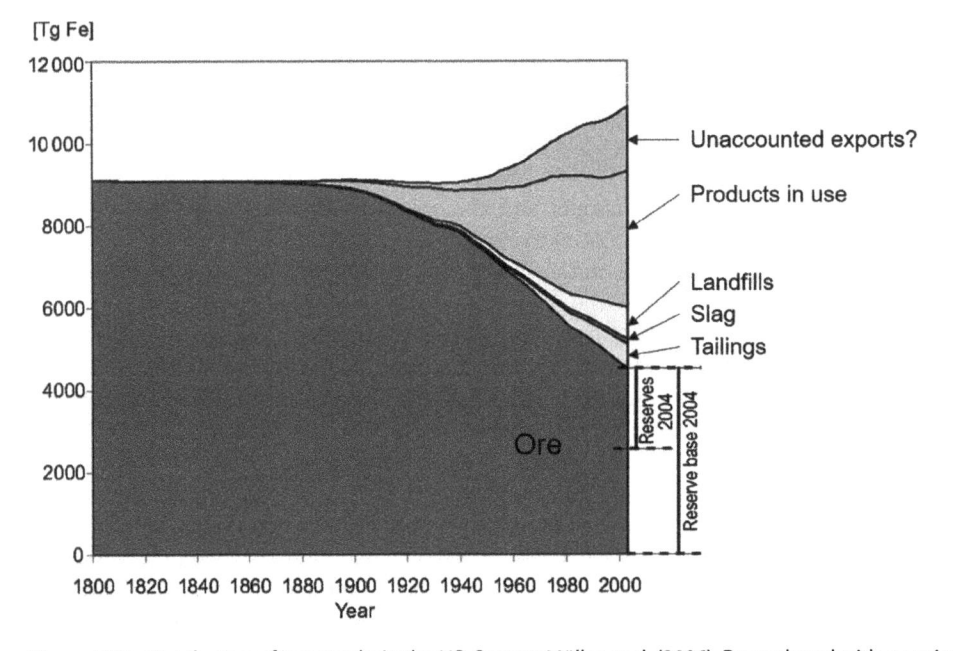

Figure 7.11 Distribution of iron stocks in the US. Source: Müller et al. (2006). Reproduced with permission of National Academy of Sciences, USA.

among the three levels exist because some reservoirs (off-shore oil platforms, military and naval equipment, etc.) are not presented at each level.

One of the most commonly used metals is iron. In a two-century study of iron stocks, Müller et al. (2006) demonstrate that about half of the mineable iron that was below ground in the United States in 1800 has been extracted, and most of that material is still in use. The total (above and below ground) stock of iron has increased in the last half-century as a consequence of extensive imports of iron in products (Figure 7.11). The iron stocks in use in the United States have continued to grow throughout the twentieth century; the largest fraction is contained in building and infrastructure construction. On a per capita basis, the stock reached a saturation around 1980, suggesting that by that point the United States had built about as many buildings, bridges, and other infrastructure products as were needed on a per person basis, and new construction was performed at a steady rate with population growth or was merely replacing old infrastructure.

7.6 National Material Account

The basic model for the national material account (NMA) was established by Matthews et al. (2000) and is presented in Figure 7.12. The NMAs attempt to count all flows crossing jurisdictional borders and those crossing the boundary between the ecosphere and the anthroposphere (except water and agricultural tillage). Minor flows are excluded even if they have significant environmental impacts. Material inputs from the environment are defined as intentional extraction or movement of natural materials by human or human-controlled means. Material outputs to the environment are defined as material flows released from the economic system into the environment. Table 7.1 listed the categories of material inputs and outputs.

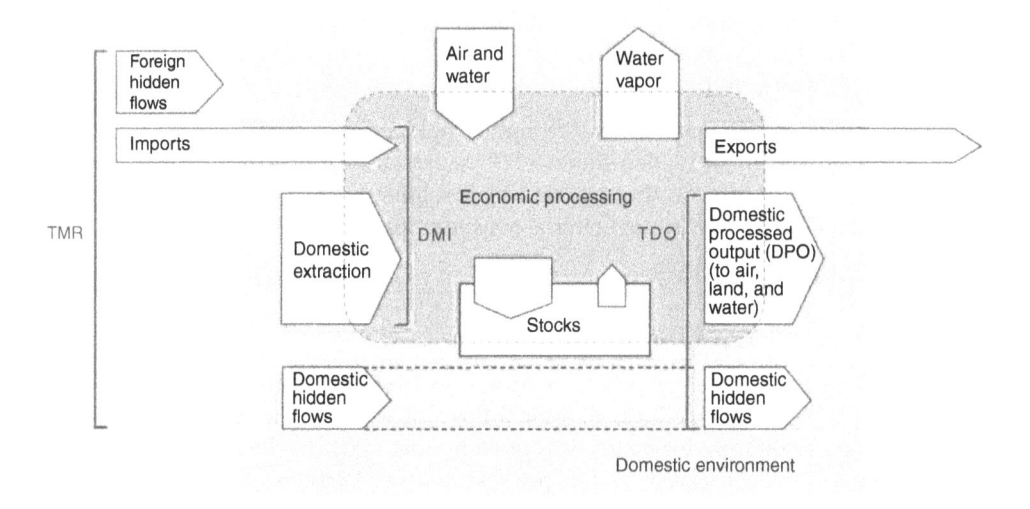

TMR (total material requirement)=DMI+domestic hidden flows+foreign hidden flows
DMI (direct material input) = domestic extraction + imports **TDO** (total domestic output)=DPO+domestic hidden flows
NAS (net additions to stock)=DMI−DPO−exports DPO (domestic processed output)=DMI−net additions to stock−exports

Figure 7.12 Basic model for a national material account (Matthews et al. 2000; World Resources Institute, USA). TMR: total material requirements, DMI: domestic material inputs, FHF: foreign hidden flows, DHF: domestic hidden flows, DE: domestic extraction, DPO: domestic processed output, TDO: total domestic output.

Table 7.1 Input and output material categories.

Material input categories	Material output categories
Domestic extraction (amount used)	Domestic processed output released to nature (e.g. carbon dioxide, herbicides)
Unused domestic extraction (domestic "hidden flows")	Disposal of unused domestic extraction (domestic "hidden flows")
Imports	Exports
Indirect flows associated with imports (foreign "hidden flows")	Indirect flows associated with exports

Hidden flows are associated with those materials associated with extraction (mining) or processing that are discarded and hence not accounted. For example, if a firm in Japan buys one kilogram of refined zinc, that zinc is mined as 50 kg of ore plus rock (2% zinc "ore grade"). It was then milled to remove the rock and leave 18 kg of zinc ore, smelted to produce 2 kg of impure zinc, and then refined to the purchased finished product (1 kg of zinc). Of the original 50 kg that was extracted, 49 kg constitutes a hidden flow invisible to the purchaser of the refined zinc, but which nonetheless had a physical effect (requiring handling as mine tailings or refining waste), mostly in the country within which the mining and refining took place.

The NMA is a mass balance of the flows in Figure 7.12. If we define:

- DE = domestic extraction
- Direct material input: DMI = DE + imports
- Direct material consumption: DMC = DMI − exports
- DPO = domestic processed output
- TDO = total domestic output.

DMI and DMC measure material use. Domestic hidden flows (DHF) account for unused domestic extraction. Foreign hidden flows (FHF) represent the upstream primary resource requirements of the imports. So the total materials requirements (TMR) of an economy are determined by Eq. (7.4), and the total material consumption (TMC) is determined by Eq. (7.5).

$$DMI + DHF + FHF = TMR \quad (7.4)$$

$$TMR - exports - export\ HF = TMC \quad (7.5)$$

Figure 7.13 shows an example of NMA for Austria in 1996. In Figure 7.13, in addition to the NMA flows explained in Figure 7.12, additional flows of water and air are bonded or mixed into that amount, accounting for nearly 40% of additional mass. In the Austrian NMA, DMI amounts to about 22 metric tons per capita per year, another 13 tons of air and water enter the production process. About half of the direct material input, 11 tons, is added to the existing

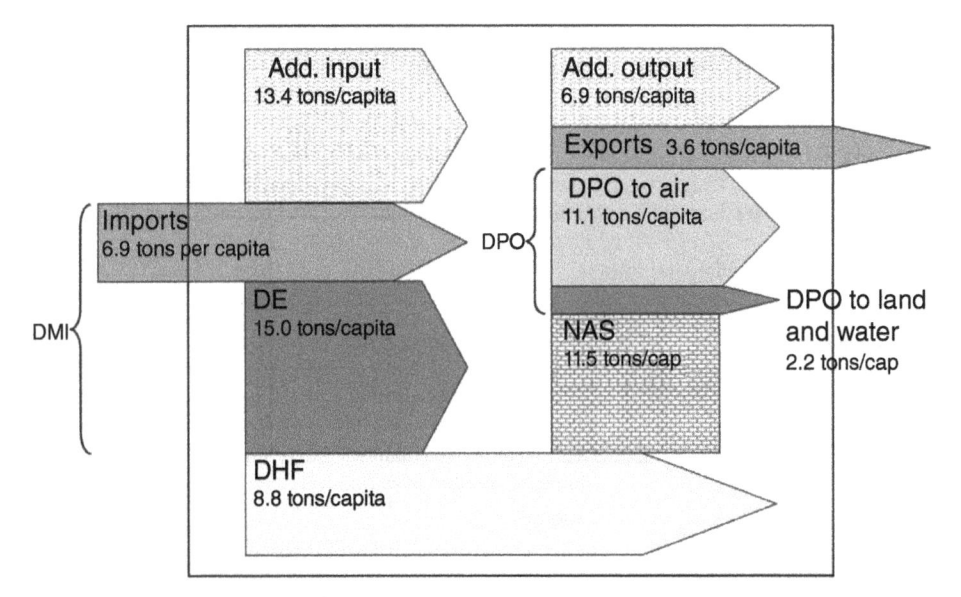

Figure 7.13 National material account (NMA) for Austria in 1996. Source: Fischer-Kowalski (2000). Reproduced with permission of John Wiley & Sons.

material stocks. On the output side, there are three components: exports to other countries, additional outputs of air and water that result from economic processing, and direct processed output (DPO), consisting of wastes, emissions, and dissipatedly used materials. This amounts to 11 tons per capita per year. Since stocks are growing, all outputs taken together amount to only about two-thirds of all input. DPO makes up about half of this output. This result is similar to other studied countries – it is smaller than DMI – in the Austrian case, 13 tons per capita per year.

NMA is also a useful source of information to assess the sustainability of the practices and activities in a country. It is often noted that large flows tend to have modest environmental impacts per unit weight, while small flows sometimes have large impacts. This is often referred to as the "elephants and scorpions" analogy.

Equation (7.6) defines how NMAs may be used to estimate the environmental impact associated with the use and consumption of materials. The total impact consists of the product of the weight of a material used and its potential impact. This approach has been used occasionally, especially in the Netherlands, to evaluate progress in environmental improvement and to identify potential areas of improvement to reduce the impact on the environment.

$$T_{m,e} = W_m * E_{m,e} \qquad (7.6)$$

where

$T_{m,e}$ = total impact of material m of type e;
W_m = annual weight of material m used;
$E_{m,e}$ = material m impact of type e.

NMA flows and results can be compared for different regions or countries, and this may help in understanding intrinsic characteristics of each country/society (Figures 7.14–7.17). Figure 7.14 shows total mass requirements per capita in various countries. As depicted in the figure, there are enormous variations among countries; one explanation can be found in the respective standard of living of each country or society. Figure 7.15 demonstrates that industrial societies typically exhibit direct material consumption rates at least threefold greater than less developed, agrarian-focused societies. Most of the difference is accounted for by mineral usage, and in some countries in fossil fuel usage. Figure 7.16 shows the use of different types of materials in various countries. For example, Germany was the largest per capita user

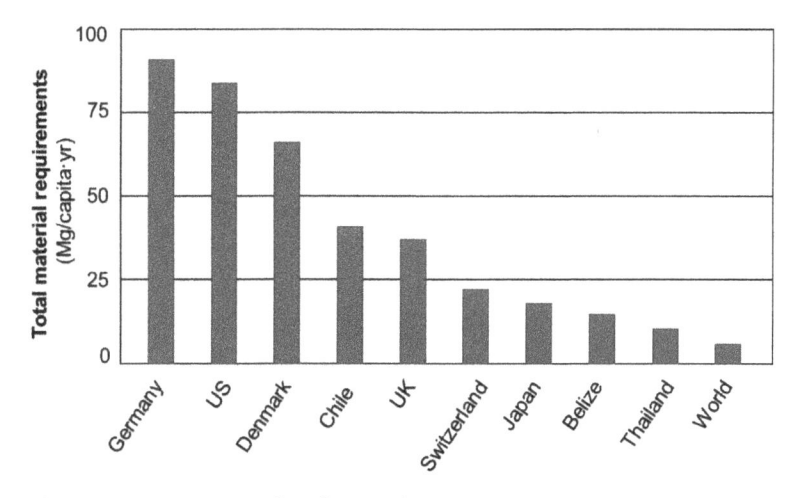

Figure 7.14 Comparison of total material requirements (TMR) across countries.

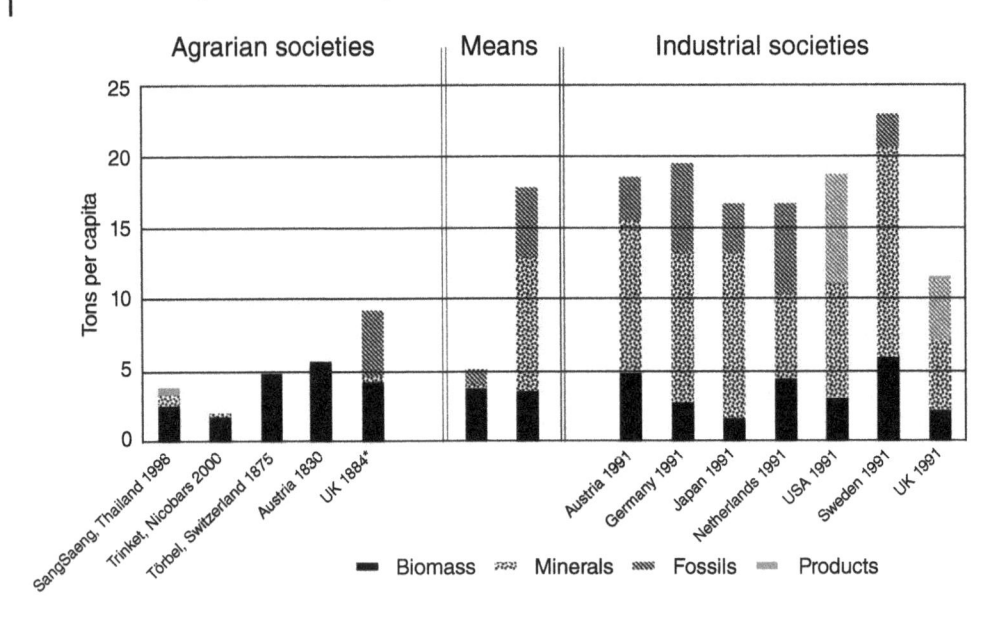

154 | *7 Material Flow Analysis and Material Budget*

Figure 7.15 Metabolic profiles by socioecological regimes (DMC/capita). Source: Fischer-Kowalski (2000). Reproduced with permission of John Wiley & Sons.

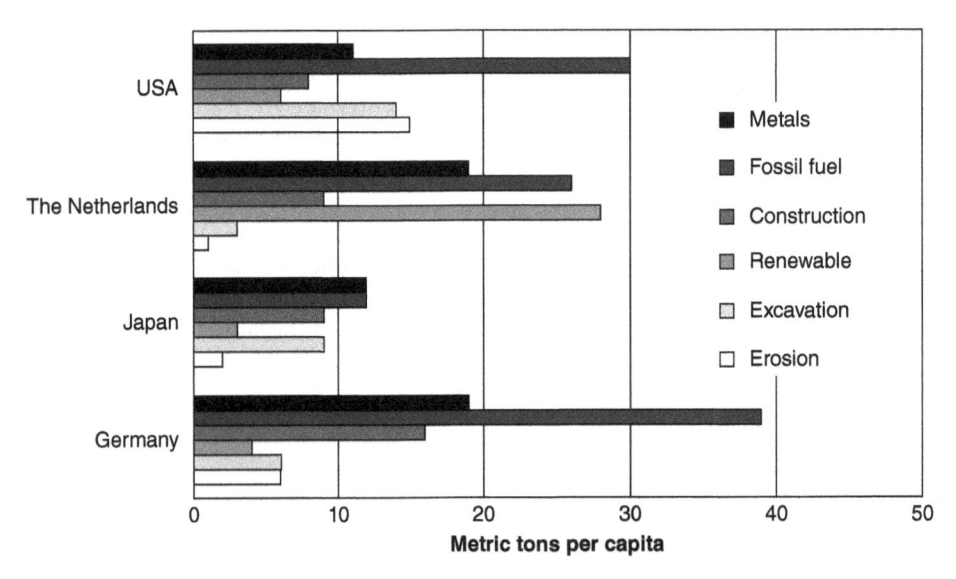

Figure 7.16 Sector contributions to TMR. Source: Matthews et al. (2000).

of fossil fuels in 1994, and the Netherlands led by far in renewable applications. Figure 7.17 compares the fraction of total material requirements that comes from imports rather than domestic supplies. The differences among countries are striking. The United States import fraction was very small, whereas Netherlands shows just the opposite situation. Japan shows a balanced point between import and export. Finally, it is important to note that most NMAs do not include the flows of energy, water, greenhouse gases, or pollutants. The addition of these flows is important to assess environmental impacts and sustainability of the related regions

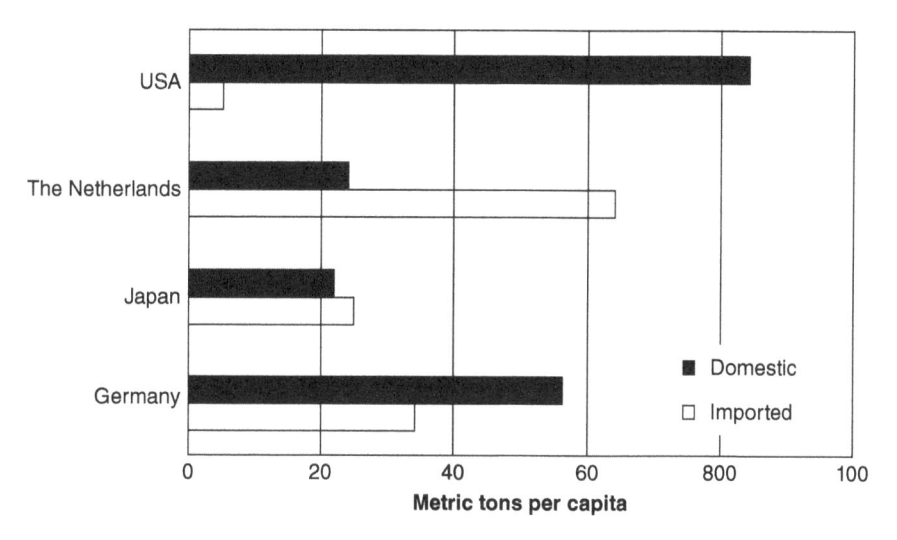

Figure 7.17 Domestic and foreign components of TMR. Source: Drakonakis et al. (2007).

or countries. While making a comparison, it is important to make sure that the assessments include equivalent types of flows.

Material accounts of smaller geographical units are obviously embedded in and are part of larger units. For example, the city of London is embedded in the UK, and for the purpose of this analysis, the UK is assumed to be included in the European Union. As it can be anticipated for a city, virtually all of London's materials come from outside its boundaries. Both the UK and the European Union import about 20% of DMI. The most interesting thing about this comparison is that the average Londoner has a DMI about half that of the average EU person. Does this suggest that cities really are more efficient from a materials standpoint? The likely answer to that question is no. The material analysis in a country includes industrial facilities and other major activities that account for large materials consumption and are often not accounted in major cities because they are located out of the areas of greatest population.

7.7 Summary

Few anthropogenic element cycles have been adequately characterized; however, humans appear to dominate most of the element cycles that have been studied. Approximately one-third of the elements in the periodic table seem to be dominated by anthropogenic processes instead of natural processes. The analysis of the flows of materials are useful in understanding the extraction, transformation, uses, and trade of the most commonly used materials: metals and construction materials. Material cycle analysis reveals the potential for the recovery and reuse of materials, but different materials require different strategies. Characterizing anthropogenic material cycles provides a wealth of useful information relating to resource availability, environmental impacts, and policy options.

Industrial economies are becoming more efficient in their use of materials, but consumption and waste generation continue to increase. Material cycles reveal quite different behavior at different spatial levels. Imports associated with resource-intensive extraction and processing are increasingly shifted to less-developed countries. The extraction and use of fossil energy resources dominate output flows in all industrial countries. Finally, over time, material flows of all kinds increase as populations become more affluent.

7.8 Questions

7.1 Discuss how metals may be considered "anthro-dominated" or "natural-dominated"?

7.2 What are examples of nonmetallic compounds that may be considered "anthro-dominated" or "natural-dominated"?

7.3 Discuss how a reservoir behaves if sources and sinks are in balance. Discuss if one is dominant compared to the other.

7.4 List reservoirs that may be considered for the manufacture of a product.

7.5 Consider crude oil extraction and refining. List reservoirs associated with these activities and discuss relative residence times within these reservoirs.

7.6 Describe the different tasks involved in the construction of a material budget.

7.7 What is material flow analysis? Explain briefly.

7.8 Define the terms budget, reservoir, pool size, flux in the context of material flow analysis.

7.9 Discuss how net consumers or net producers of a product would have a different material flow analysis.

7.10 How is the total material requirements and total materials consumption of an economy determined?

7.11 Describe how a material flow analysis may be used to devise effective or more optimal waste reclamation strategies.

7.12 Explain the application of material flow analysis in the context of waste management.

7.13 What is national material account? Explain briefly.

7.14 Describe how a national material account may be used to assess sustainability practices.

7.15 Discuss how subset material accounts (i.e. cities) may be used to assess national material accounts.

References

Drakonakis, K., Rostkowski, K., Rauch, J. et al. (2007). Metal capital sustaining a North American city: Iron and copper in New Haven, CT. *Resources, Conservation and Recycling* 49 (4): 406–420.

Fischer-Kowalski, M. (2000). The end of the pipe – material outflows from industrial economies. Presentation at the Year 2000 Gordon Conference on Industrial Ecology: Engineering Global systems, New London, USA (11–16 June 2000).

Graedel, T.E. and Allenby, B.R. (2010). *Industrial Ecology and Sustainable Engineering*. Prentice Hall, Upper Saddle River, NJ.

Klee, R.J. and Graedel, T.E. (2004). Elemental cycles: a status report on human or natural dominance. *Annual Review of Environment and Resources* 29: 69–107.

Matos, G.R. (2012). Use of raw materials in the United States from 1900 through 2010. U.S. Geological Survey. U.S. Department of the Interior. Fact Sheet 2012–3140.

Matthews, E., Amann, C., Bringezu, S. et al. (2000). *The Weight of Nations. Material Outflows from Industrial Economies*. Washington, DC: World Resources Institute.

Müller, D.B., Wang, T., Duval, B., and Graedel, T.E. (2006). Exploring the engine of anthropogenic iron cycles. *Proceedings of the National Academy of Sciences* 103 (44): 16111–16116.

Reck, B.K., Müller, D.B., Rostkowski, K., and Graedel, T.E. (2008). Anthropogenic nickel cycle: insights into use, trade, and recycling. *Environmental Science & Technology* 42 (9): 3394–3400.

8

Carbon Footprint Analysis

8.1 Introduction

Most human activities require the consumption of energy and materials resulting in the generation and emission of gases into the atmosphere that contribute to the greenhouse effect. Greenhouse gases (GHGs) naturally occur in the Earth's atmosphere and make life possible on Earth by trapping the Sun's energy and heating the surface of the planet to desirable temperatures. GHGs are responsible for maintaining the average temperature on Earth at about 15 °C (58 °F). In the absence of the GHGs in Earth's atmosphere, the Earth's surface temperature would have been around −18 °C (0 °F). In atmosphere, the key natural GHGs are water vapor (H_2O) and carbon dioxide (CO_2). However, changes in human activities in the last 150 years, such as rapid industrialization, massive use of fossil fuels, and deforestation, have resulted in increased levels of GHGs entering the Earth's atmosphere. The main GHGs of anthropogenic origin are listed in Table 8.1, including CO_2 and methane (CH_4). These gases are responsible for the increasing greenhouse effect in the atmosphere, and therefore, exacerbating global warming and climate change. These are two of the most challenging impacts to the environment resulting from human activities. Considering the importance of climate change and its impact on the future generations, the inventory of GHGs has been proposed as a metric to measure the sustainability of the human activities.

In this chapter, the inventory of GHGs is discussed as a metric to measure the sustainability of a product, project, or organization. The inventory of GHGs is usually referred as carbon footprint. The total emissions of GHGs into the atmosphere are an important environmental issue as it associates with the global warming and climate change. The changes in temperature of the planet, the erratic changes in climate, and the corresponding sea level rise will have an enormous impact on future generations.

8.2 Global Warming Potential and Carbon Footprint

The emissions of anthropogenic GHGs are a result of various activities (IPCC 2007). These GHG sources include agriculture, fossil fuel combustion, and industrial processes. Human-related sources are responsible for 38% of total GHG emissions (IPCC 2014). The emissions of CO_2 are mostly related to the combustion of fossil fuels and deforestation, which reduces the amount of CO_2 trapped in the biomass (that transforms to fossils). CH_4 emissions have been associated with livestock enteric fermentation (i.e. generated by cows), manure management, paddy rice farming, land use, and wetlands changes, as well as landfill emissions and losses in petroleum and natural gas exploitations and pipelines. Nitrous oxide (N_2O) has been used for its anesthetic and analgesic effects in the medical industry, as an oxidizer in rocketry, and in automobile

Sustainable Engineering: Drivers, Metrics, Tools, and Applications, First Edition.
Krishna R. Reddy, Claudio Cameselle, and Jeffrey A. Adams.
© 2019 John Wiley & Sons, Inc. Published 2019 by John Wiley & Sons, Inc.

Table 8.1 Greenhouse gases and global warming potential (IPCC 2007, 2014).

Greenhouse gas	Chemical formula	Life time (yrs)	Global warming potential (time horizon) 20 yrs	Global warming potential (time horizon) 100 yrs
Carbon dioxide	CO_2	Variable	1	1
Methane	CH_4	12.4	84	28
Nitrous oxide	N_2O	121	264	265
Hydrofluorocarbons (HFC)[a]	$C_xH_yF_z$	1.4–270	437–12 000	124–14 800
Perfluorocarbons (PFC)[a]	C_xF_z	1 000–50 000	5 210–8 630	7 390–12 200
Sulfur hexafluoride (SF6)	SF_6	3 200	16 300	23 500

a) See Table 2.1 for specific greenhouse gases.

Table 8.2 Changes in global warming potentials in different assessments (IPCC 1995, 2007, 2014).

Industrial designation or common name	Chemical formula	GWP values for 100-yr time horizon Second assessment report (SAR)	Fourth assessment report (AR4)	Fifth assessment report (AR5)
Carbon dioxide	CO_2	1	1	1
Methane	CH_4	21	25	28
Nitrous oxide	N_2O	310	298	265

racing to increase the power output of engines. Hydrofluorocarbons (HFCs), perfluorocarbons (PFCs), and sulfur hexafluoride (SF6) are manufactured gases commonly used in different industrial applications and products. SF6 is used as a gaseous dielectric medium for high voltage (35 kilovolts (kV) or above) in several electric equipment applications. PFCs have been used in refrigeration units as replacements for CFCs (chlorofluorocarbons), which were banned to protect the stratospheric ozone layer. HFCs, PFCs, and SF6 have been identified in the Kyoto protocol as gases of concern from the standpoint of the global warming.

Global-warming potential (GWP) is a relative measure of how much heat a GHG traps in the atmosphere. GWP compares the amount of heat trapped by a certain mass of the gas to the amount of heat trapped, normalized by the same mass of carbon dioxide. Each gas has a different GWP, which is a measure of its "strength" in contributing to global warming as compared to carbon dioxide. For example, methane has a GWP of 28, which means it is 28 times more potent as GHG than carbon dioxide in a time horizon of 100 years. Thus, 1 g of methane contributes equivalently to the global warming as 28 g of CO_2.

Total GHG emissions are calculated and reported as carbon dioxide equivalents (eCO_2 or CO_2e) by accounting for the GWP of each gas as reported in Table 8.1. The total GHG emission calculation expressed in CO_2e is known as the carbon footprint. In calculating the carbon footprint, it is important to use the present values of the GWP that have been updated for the major GHG gases (methane and nitrous oxide) presented in recent editions of the IPCC reports (Table 8.2).

The GHG inventory is the accounting of GHG emissions associated with the operations of an entity, a process, or a product. When calculating the total GHG emissions or the carbon footprint, it is essential to be aware of current emissions associated with the specific entity, process, or product for which the sustainability is being analyzed. The carbon footprint calculation is very useful in helping to predict the future emissions and identifying major emission sources. The data may then be used to plan, design, implement, and monitor mitigation strategies to reduce respective carbon footprints.

8.3 Measuring Carbon Footprint

The determination of the carbon footprint (Weidema et al. 2008) involves several steps as described below.

8.3.1 Define the Scope of Your Inventory

The first step is to establish the objective and purpose of measuring the carbon footprint (Bastianoni et al. 2004; Matthews et al. 2008). This facilitates the definition of the limits of the system to be evaluated. Once the limits of the system are defined, all the sources of GHGs have to be listed and accounted. The limits of the system must also include the unit and the time period to be used for accounting the GHG emissions. The unit could be a specific product, such as the manufacturing of a pair of shoes or a car, or it could be the generation of electricity in a power plant; in this case, the carbon footprint could be based on a specific unit of energy (kWh). The unit can also be a manufacturing process, such as the manufacturing of canned food, or an entity, such as an apartment building, residential subdivision, or a university, for a specific period of time.

8.3.2 Measure Emissions and Establish a Baseline

Once the limits of the system are defined, the next step is to identify the sources of GHGs and account for the emissions for each source. In this process, three groups of sources can be identified; these sources are usually referred to as Scope 1 through 3:

- Scope 1, or direct emissions, includes the GHGs generated in the limits of our system due to the generation of energy. For example, direct emissions may include the gases generated in the combustion of fossil fuels.
- Scope 2, or indirect emissions, are GHG emissions generated to produce energy that is purchased for use in our system; e.g. emissions for purchased electricity.
- Scope 3, or indirect emissions, are an optional category that includes GHGs generated for other external sources but associated with our system; e.g. employee business travel.

To calculate the carbon footprint, the GHG emissions must be accounted for all the sources considered. This always must include Scope 1 and 2, whereas Scope 3 is optional. The GHG emissions are determined using appropriate emission factors; for example, the amount of CO_2, CH_4, and/or N_2O for the combustion of fossil fuels (Table 8.3). Vehicular GHG emissions can be accounted for using the emission factors in Table 8.4, although the emissions may also be determined for a type of vehicle and number of passengers. The emission factor for electricity use has to be determined locally since the total gas emissions largely depends on the type of energy used in its production (power plant, nuclear, hydroelectric, photovoltaic, etc.). Finally, all GHG emissions must be accounted in terms of CO_2e using the GWP in Table 8.2 to obtain the total carbon footprint expressed in CO_2e.

Table 8.3 GHG emission factor for stationary combustion (USEPA 2015).

Fuel type	Heating value (mmBtu per short ton)	CO$_2$ factor (kg CO$_2$ per mmBtu)	CH$_4$ factor (g CH$_4$ per mmBtu)	N$_2$O factor (g N$_2$O per mmBtu)	CO$_2$ factor (kg CO$_2$ per short ton)	CH$_4$ factor (g CH$_4$ per short ton)	N$_2$O factor (g N$_2$O per short ton)
Coal and coke							
Anthracite coal	25.09	103.69	11	1.6	2602	276	40
Bituminous coal	24.93	93.28	11	1.6	2325	274	40
Sub-bituminous coal	17.25	97.17	11	1.6	1676	190	28
Lignite coal	14.21	97.72	11	1.6	1389	156	23
Fossil fuel-derived fuels (solid)							
Municipal solid waste	9.95	90.70	32	4.2	902	318	42
Plastics	38.00	75.00	32	4.2	2850	1216	160
Tires	28.00	85.97	32	4.2	2407	896	118
Biomass fuels (solid)							
Agricultural byproducts	8.25	118.17	32	4.2	975	264	35
Solid byproducts	10.39	105.51	32	4.2	1096	332	44
Wood and wood residuals	17.48	93.80	7.2	3.6	1640	126	63
	(mmBtu per scf)	(kg CO$_2$ per mmBtu)	(g CH$_4$ per mmBtu)	(g N$_2$O per mmBtu)	(kg CO$_2$ per scf)	(g CH$_4$ per scf)	(g N$_2$O per scf)
Natural gas							
Natural gas	0.001 026	53.06	1.0	0.10	0.054 44	0.001 03	0.000 10
Biomass fuels (gaseous)							
Landfill gas	0.000 485	52.07	3.2	0.63	0.025 254	0.001 552	0.000 306
Other biomass gases	0.000 655	52.07	3.2	0.63	0.034 106	0.002 096	0.000 413
	(mmBtu per gallon)	(kg CO$_2$ per mmBtu)	(g CH$_4$ per mmBtu)	(g N$_2$O per mmBtu)	(kg CO$_2$ per gallon)	(g CH$_4$ per gallon)	(g N$_2$O per gallon)

(*Continued*)

Table 8.3 (Continued)

Fuel type	Heating value (mmBtu per short ton)	CO$_2$ factor (kg CO$_2$ per mmBtu)	CH$_4$ factor (g CH$_4$ per mmBtu)	N$_2$O factor (g N$_2$O per mmBtu)	CO$_2$ factor (kg CO$_2$ per short ton)	CH$_4$ factor (g CH$_4$ per short ton)	N$_2$O factor (g N$_2$O per short ton)
Petroleum products							
Aviation gasoline	0.120	69.25	3.0	0.60	8.31	0.36	0.07
Butane	0.103	64.77	3.0	0.60	6.67	0.31	0.06
Ethane	0.068	59.60	3.0	0.60	4.05	0.20	0.04
Motor gasoline	0.125	70.22	3.0	0.60	8.78	0.38	0.08
Naphtha (<401 °F)	0.125	68.02	3.0	0.60	8.50	0.38	0.08
Residual fuel oil no. 5	0.140	72.93	3.0	0.60	10.21	0.42	0.08
Special naphtha	0.125	72.34	3.0	0.60	9.04	0.38	0.08
Still gas	0.143	66.72	3.0	0.60	9.54	0.43	0.09
Biomass fuels (liquid)							
Biodiesel (100%)	0.128	73.84	1.1	0.11	9.45	0.14	0.01
Ethanol (100%)	0.084	68.44	1.1	0.11	5.75	0.09	0.01
Rendered animal fat	0.125	71.06	1.1	0.11	8.88	0.14	0.01
Vegetable oil	0.120	81.55	1.1	0.11	9.79	0.13	0.01

Table 8.4 GHG emission factors for mobile combustion (USEPA 2015).

Fuel type	CO_2 (kg/gallon)
Aviation gasoline	8.31
Biodiesel (100%)	9.45
Diesel fuel	10.21
Ethanol (100%)	5.75
Kerosene-type jet fuel	9.75
Liquefied natural gas (LNG)	4.46
Liquefied petroleum gases (LPG)	5.68
Motor gasoline	8.78
Residual fuel oil	11.27

8.3.3 Develop Targets and Strategies to Reduce Emissions

The determination of the carbon footprint not only allows the calculation of the total GHG emission, but it also assists in identifying where these emissions are produced by analyzing the data of each GHG source. An action plan may be designed using this data to manage the emissions. This plan is often known as a Climate Change Action Plan. Carbon footprint reduction targets can either be absolute (e.g. A 10% reduction relative to the chosen baseline year) or intensity-based (e.g. a percentage reduction per unit produced or per full-time employee (FTE) of a services organization).

8.3.4 Off-set Unavoidable Emissions

Some sources of GHG are difficult to reduce; these are usually associated with the production of energy. To compensate for these emissions, organizations can then purchase carbon credits (or "offsets") to compensate for any unavoidable emissions. Offsets are derived from projects that are specifically implemented to avoid emissions. The carbon footprint reduction due to those projects can be used to compensate for unavoidable emissions.

8.3.5 Independent Verification

The final step in the determination of carbon footprint reduction efforts is the verification of the process. The verification or certification is typically completed by an external organization. The external certification increases the credibility of the process, enhances stakeholder trust, and increases senior management confidence. This external verification benefits the public corporate image of the company or the organization and may be used in the marketing of the company's products or services to customers.

8.4 Standards for Calculating the Carbon Footprint

Various standard methods have been developed for carbon footprint analysis. However, it is important to use recognized standards and approaches when calculating carbon footprint so that the results can be compared with other entities, process, or products. If different standards are used, the results may vary and an accurate, direct comparison may not be possible.

One of the first developed standards is the GHG Protocol (ghgprotocol.org), a corporate accounting and reporting standard issued by the World Business Council for Sustainable Development and the World Resources Institute. This was the most widely used international accounting tool for government and businesses to quantify and manage GHG emissions. The first edition of this standard method was published in 2001. Later, in 2006, the International Organization for Standardization (ISO) developed ISO 14064-I: Specification with Guidance at the Organization Level for Quantification and Reporting of GHG Emissions and Removals. Currently, the ISO 14064 is widely accepted in the determination of the carbon footprint (Weng and Boehmer 2006).

The ISO 14064 is organized into three parts or sections. Part 1 includes specifications for guidance at the organizational level for quantification and reporting of GHG emissions and reductions. Part 2 includes specifications for guidance at the project level for quantification, monitoring, and reporting of GHG reductions or removal enhancements. Finally, Part 3 includes specifications and guidance for validation, and verification of assertions.

The International Auditing and Assurance Standards Board (IAASB) is an independent standard-setting body that has developed an alternative standard for the evaluation of GHG emissions and carbon footprint. The standard, called the *International Standard on Assurance Engagements*, was designed for all assurance engagements other than audits or reviews of historical financial information.

Other methods and standards for GHG inventory analysis and carbon footprint analysis include the *Attest Engagements on GHG Information* developed by the American Institute of CPAs, the UE ETS (Emissions Trading System) in Europe, and the DEFRA guidelines in the UK.

The existence of several standards resulted from multiple organizations or countries moving to develop or adopt a standard to normalize the results of the GHG in the absence of a widely recognized standard. Currently, the ISO 14064 is considered to be the most appropriate standard, although other standards are still preferred in specific countries or sectors. The existence of several standards can create difficultly when side-by-side analyses or comparisons are desired; however, a level of assurance is provided when the GHG inventory is performed using several standards.

8.5   GHG Inventory: Developments in the United States

The United States issued Executive Order (EO) 13514, "Federal Leadership in Environmental, Energy, and Economic Performance," which was signed by President Obama on 5 October 2009. The EO 13514 specifically outlined how the United States Federal government was planning to approach the problem of GHG emissions. The GHG inventory in EO 13514 was organized in three scopes. Scope 1 includes GHG emissions from sources owned or controlled by a Federal agency. Scope 2 includes GHG emissions resulting from the generation of electricity, heat, or steam purchased by a Federal agency. Scope 3 includes GHG emissions from sources not owned or directly controlled by a Federal agency but related to agency's activities, such as vendor supply chains, delivery services, and employee travel and commuting.

The first step in the application of the EO 13514 was the development of goals for GHG emissions reduction by the agency. Each agency was directed to prepare a comprehensive GHG inventory, and based on the results, was asked to define percentage reduction targets for Scope 1, 2, and 3 emissions with respect to 2008 emissions. EPA developed the tool Development of

Agency Reduction Targets (DARTs) for agencies to use for the calculation of Scope 1 and 2 GHG emissions. The Federal Energy Management Program (FEMP) developed the Annual GHG and Sustainability Data Report, a tool for agencies to use for developing their comprehensive GHG inventories.

8.6 USEPA: Greenhouse Gas Reporting Program

The USEPA issued the Mandatory Reporting of GHGs Rule (74 FR 56260) in response to the FY2008 Consolidated Appropriations Act (H.R. 2764; Public Law 110-161). This rule entitled the large sources and generators of GHG in the United States to report the GHG data and other relevant information to USEPA. The purpose of the rule is to have an accurate inventory and timely data of GHG emissions to undertake future policy decisions. The rule is referred to as 40 CFR Part 98: Greenhouse Gas Reporting Program (GHGRP). This rule establishes a mandatory reporting of GHGs for those suppliers or facilities that meet at least one of the following conditions: suppliers of certain products that would result in GHG emissions if released, combusted, or oxidized; facilities that inject CO_2 underground for sequestration or otherwise; and facilities that emit greater than 25 000 metric tons per year of GHGs. They are required to report the GHG emissions annually.

The emissions for the year 2010 were reported to USEPA via the electronic GHG reporting tool (e-GGRT) in September 2011. USEPA made the GHGRP data for the first year of the reporting period to be publicly available through their data publication tool in January 2012 and will continue to update the data. This data is available at ghgdata.epa.gov/ghgp.

8.7 Tools for GHG Inventory

There are several tools and software programs available for the quantitative determination or estimates of GHGs/carbon footprint. Life cycle assessment (LCA) includes the inventory of GHG in its procedure, but there are other specific software tools developed for the calculation of carbon footprint as well as web calculators and simple carbon footprint spreadsheets as listed below.

- *Climate leadership in parks tool (CLIP)* – A tool that can be used to estimate GHG emission from sources in the parks such as vehicle combustion and wastewater treatment. It aids in promoting ways and creating awareness among the users to reduce the emissions from parks.
- *Comparable emissions database (CEDB)* – A tool that calculates potential emissions due to moving of passengers and freight from a wide range of vehicles across all transportation modes.
- *Employee commute emissions calculator* – This tool helps users measure emissions from commuting of employees via different transportation modes.
- *EPA's electronic greenhouse gas reporting tool (e-GGRT)* – Is a tool used for reporting of the GHG emissions under the Greenhouse Gas Reporting Program (GHGRP).
- *Durable goods calculator (DGC)* – This tool calculates emissions and benefits from energy saving from different disposal methods for disposal of 14 typical durable goods.
- *Waste reduction model (WARM)* – Is a tool to estimate the GHG emissions impacts of the waste management practices (e.g. landfilling, recycling, incineration, and composting).
- *Events emissions calculator* – Is a tool that can be used to estimate the GHG emissions from events based on the length, the commute of the people attending the event, modes of transportation, etc.

- *Fleet emissions calculator* – Is a tool used to measure the emissions from the fleet using the information on the type of vehicle and the distance traveled.
- *GHG protocol calculation tools and guidance* – A tool to estimate indirect emissions from the service sector (e.g. banks and hospitals) and office-based organizations.
- *GHG inventory guidance, calculator, and inventory management plan for small and medium-sized organizations and service sector businesses* – A tool used to estimate emission from low-emitting businesses (e.g. service-industry company with 1000 branch offices, a company with one small manufacturing plant, and an owner of a single office building).
- *MOtor Vehicle Emission Simulator (MOVES)* – A tool to estimate emissions from on-road and nonroad mobile sources.
- *National recycling coalition's (NRC) environmental benefits calculator* – A tool that helps in determining the GHG and energy benefits from current waste disposal practices.
- *Northeast recycling council's environmental benefits calculator* – A tool used to calculate the environmental benefits based on waste management practices.
- *Recycled Content (ReCon) tool* – A tool used to estimate life cycle energy and GHG impacts of purchasing or manufacturing some specific materials. The tool also calculates GHG and energy benefits based on recycled content in those materials.
- *Shipping emissions calculator* – A tool used to measure emissions from shipment of items based on distance of shipping, mode of shipping and weight of item.
- *Office footprint calculator*™ – A tool used to calculate emissions from an individual's office-related activity (e.g. energy use and waste stream) and the resulting CO_2 emissions from it.
- *Optimization model for reducing emissions of greenhouse gases from automobiles (OMEGA)* – A tool used to estimate the cost of technology for automobile manufacturers to achieve variable fleet-wide levels of vehicle GHG emissions.
- *Paper use emissions calculator* – A tool used to estimate emissions from paper use.
- *Personal GHG emissions calculator* – A tool used to estimate a person's or a family's GHG emissions from their daily activities.
- *GREET model: The GHGs, regulated emissions, and energy use in transportation model* – A life-cycle modeling tool developed by Argonne National Laboratory to evaluate fuel combustions of various vehicle on a fuel-cycle/vehicle-cycle basis. The tool can also be used to evaluate total fuel consumption, energy use, GHG emissions, and emission of criteria pollutants.

8.8 UIC Carbon Footprint Case Study

The University of Illinois at Chicago (UIC) is an organization that produces more than 25 000 metric tons of GHGs per year, so it is subjected to the Greenhouse Gas Reporting Program (GHGRP, 40 CFR Part 98). Thus, UIC determines its GHG inventory and measures its carbon footprint (UIC 2011).

The carbon footprint can be determined using the Campus Carbon Calculator™ (http://www.cleanair-coolplanet.org/toolkit/inv-calculator.php). This tool incorporates the carbon footprint into three scopes: (i) direct emissions owned or controlled by entity; (ii) purchased electricity, steam or heat; and (iii) other indirect sources owned or controlled by another entity

The resulting emissions for each GHG can be calculated with Eq. (8.1):

$$A \times F_g = E_g \tag{8.1}$$

where

E_g = resulting emissions for GHG g
F_g = emission factor of GHG g
A = quantification of an activity in units.

As an example, the value of A could be the amount of natural gas consumed (expressed in MMBtu), the gallons of heating oil consumed, the consumption of electricity (kWh), or the miles traveled by personnel (faculty and staff) and students.

To calculate the actual GHG emission for CO_2, CH_4, and N_2O, Eqs. (8.2–8.4) can be used

CO_2 emissions (metric tons/yr)
$$= (1\,\text{MT}/1000\,\text{kg}) \times \text{annual fuel consumption (mmscf/yr)}$$
$$\times \text{average annual higher heating value (HHV) (MMBtu/mmscf)}$$
$$\times CO_2 \text{ emission factor (kg } CO_2/\text{MMBtu)} \tag{8.2}$$

CH_4 emissions (metric tons/yr)
$$= (1\,\text{MT}/1000\,\text{kg}) \times \text{annual fuel consumption (mmscf/yr)}$$
$$\times \text{average annual HHV (MMBtu/mmscf)}$$
$$\times CH_4 \text{ emission factor (kg } CO_2/\text{MMBtu)} \tag{8.3}$$

N_2O emissions (metric tons/yr)
$$= (1\,\text{MT}/1000\,\text{kg}) \times \text{annual fuel consumption (mmscf/yr)}$$
$$\times \text{average annual HHV (MMBtu/mmscf)}$$
$$\times N_2O \text{ emission factor (kg } CO_2/\text{MMBtu)} \tag{8.4}$$

where:

Annual Fuel Consumption is the mass or volume of fuel combusted during the year from company records or directly measured by a fuel flow meter (express volume in standard cubic feet for gaseous fuel and volume in gallons for liquid fuel);

Annual Average HHV is the annual average higher heating value of the fuel from all valid measurements for the reporting year (MMBtu per mass or volume).

The emission factors must be selected for each GHG and each activity or source. For example, the emission factors for natural gas are the following:

- CO_2 emission factor for natural gas: 53.02 kg CO_2/MMBtu
- CH_4 emission factor for natural gas: 0.001 kg CH_4/MMBtu
- N_2O emission factor for natural gas: 0.0001 kg N_2O/MMBtu.

The total emissions of each gas (carbon dioxide, methane, nitrous oxide, sulfur hexafluoride, HFCs, and PFCs) multiplied by its global warming potential, GWP (Table 8.1), is the emissions expressed in CO_2e. The total sum of these emissions will be the total carbon footprint of UIC determined for a one-year period. The total GHG emissions or carbon footprint can be calculated with Eq. (8.5):

$$\text{Total GHG emissions } (CO_2e) = \sum_g E_g \cdot GWP_g \tag{8.5}$$

Year 2004

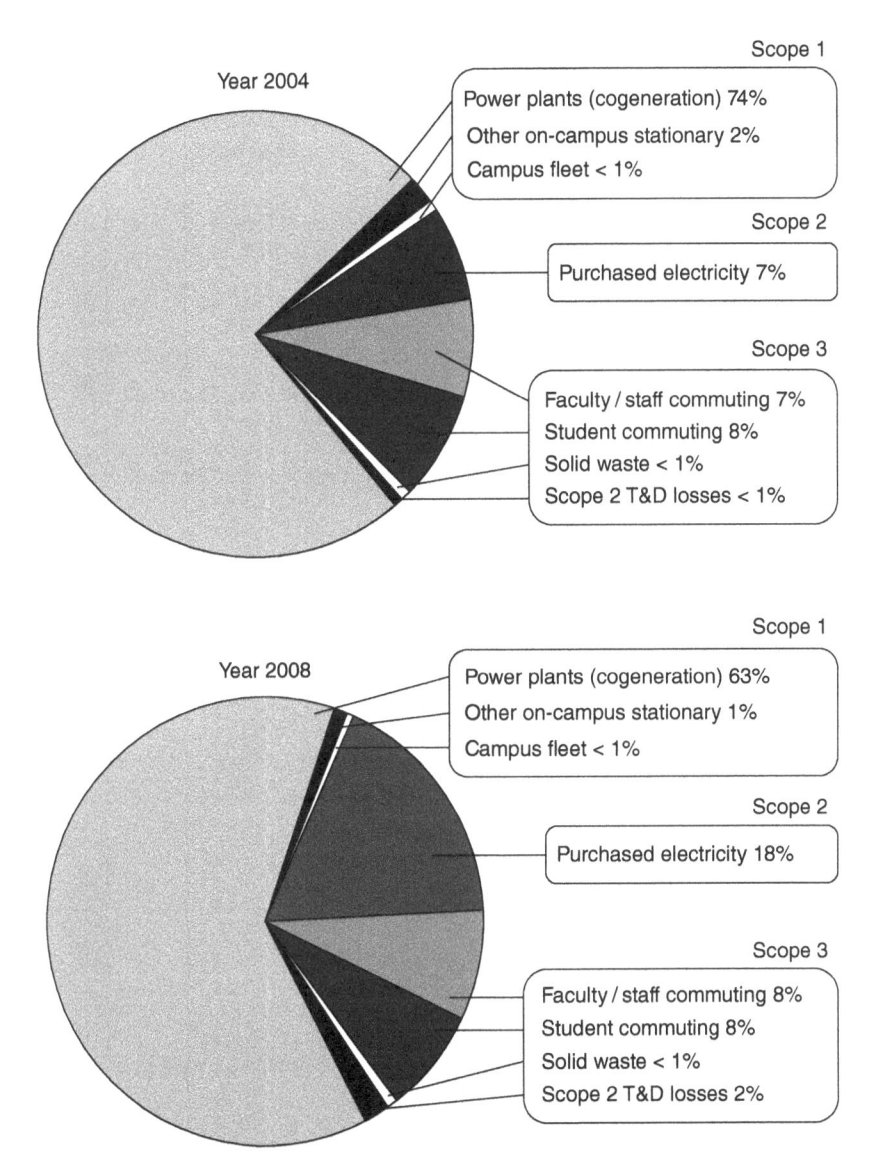

Scope 1
- Power plants (cogeneration) 74%
- Other on-campus stationary 2%
- Campus fleet < 1%

Scope 2
- Purchased electricity 7%

Scope 3
- Faculty / staff commuting 7%
- Student commuting 8%
- Solid waste < 1%
- Scope 2 T&D losses < 1%

Year 2008

Scope 1
- Power plants (cogeneration) 63%
- Other on-campus stationary 1%
- Campus fleet < 1%

Scope 2
- Purchased electricity 18%

Scope 3
- Faculty / staff commuting 8%
- Student commuting 8%
- Solid waste < 1%
- Scope 2 T&D losses 2%

Figure 8.1 Carbon footprint for the UIC in the years 2004 and 2008.

Figures 8.1 and 8.2 show the carbon footprint for UIC in the years 2004, 2008, and 2011. The carbon footprint is organized in the plots by Scope 1 – direct emissions; Scope 2 – indirect emissions associated to purchased energy; and Scope 3 – other indirect emissions. Figure 8.3 shows the contribution of buildings, transportation, and wastes to the carbon footprint in 2011. It is clear that the buildings are the major contributor to the GHG emissions; this is associated to the heating/cooling systems and electricity consumption. These data can be used to develop a specific climate change action plan with specific mitigation strategies to reduce GHG emissions. Figure 8.4 shows the projection of the carbon footprint of UIC under different scenarios; the

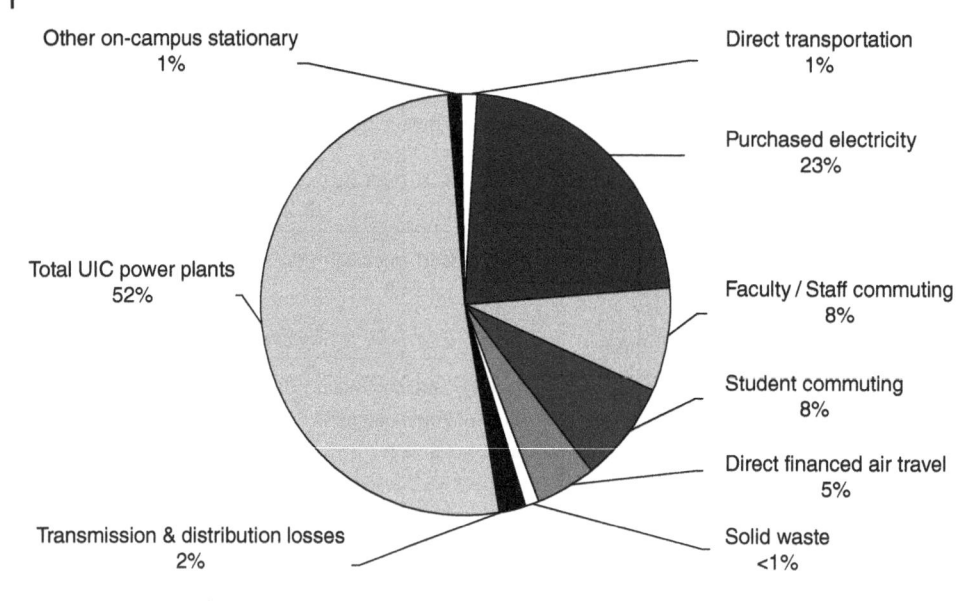

Figure 8.2 Carbon footprint for the UIC in the year 2011.

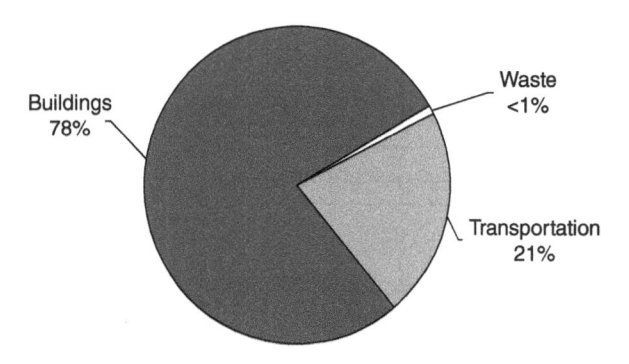

Figure 8.3 Contribution of buildings, transportation, and waste to the carbon footprint of the UIC in 2011.

use of biogas as an alternative source of energy, purchase of electricity from renewable sources, energy efficiency, and conservation show a major influence in the reduction of the future carbon footprint. The UIC Climate Action Plan includes the following guidelines:

- *Energy efficiency and conservation* – Reduce energy consumption increase the efficiency of equipment and reduce losses.
- *Clean and renewable energy sources* – Change from conventional energy sources (fossil fuels) to other clean and renewable energy sources with fewer emissions.
- *Improved transportation options* – Reduce the carbon footprint of transportation.
- Improved Grounds Operations.
- *Recycling and reduced waste streams* – Reduce the generation of wastes and favor recycling.
- Employment strategies.
- *Education, research and public engagement* – Favor individual actions.

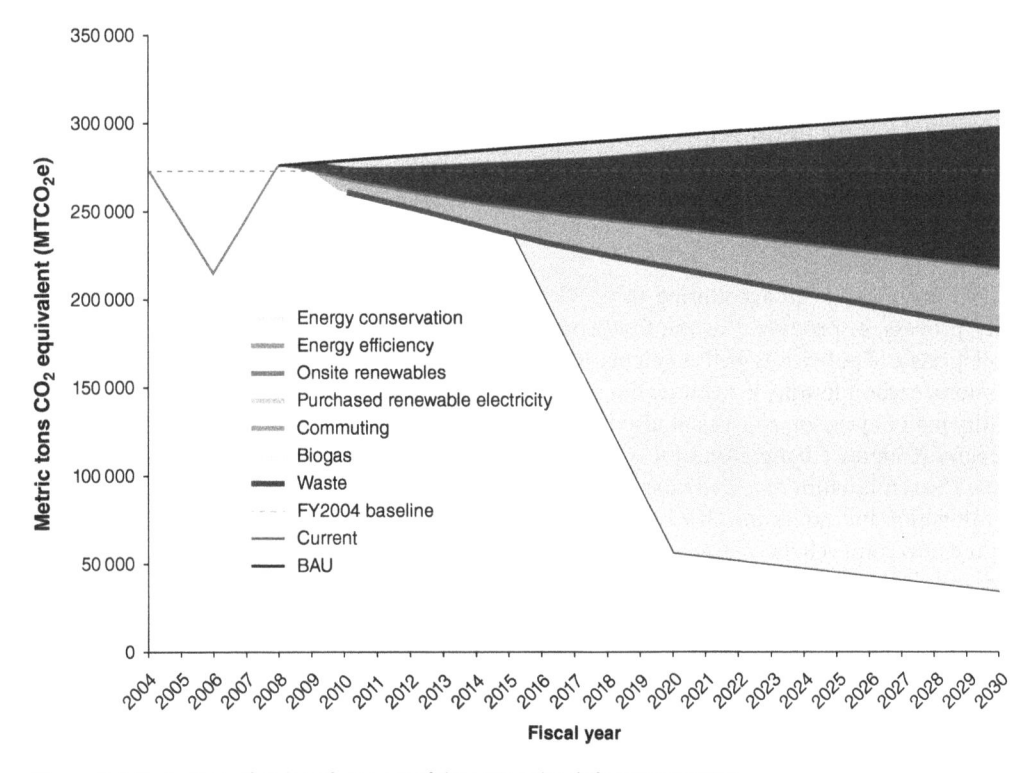

Figure 8.4 Projection of carbon footprint of the UIC under different scenarios.

8.9 Programs to Mitigate GHG Emissions

Organizations have developed several programs to mitigate GHG emissions and their effects on global warming and climate change (Dixit et al. 2016). The production of energy is one of the main sources of GHG emissions. Any action in energy saving and efficiency is going to reduce GHG emissions, but the main objective to achieve a significant reduction in the GHG is to change from a fossil fuel energy-based system to renewable energy with minimum GHG generation (Pierie et al. 2016). Other potential actions for reducing GHG emissions include better practices in livestock management and agriculture activities. Cities are also a major source for GHG emissions mainly due to transportation and heating systems and electricity consumption, so many cities try to implement action plans to mitigate the GHG generation locally (Larsen and Hertwich 2009). The following are some selected programs aimed at mitigating GHG emissions:

1. AgSTAR Program
2. California Climate Action Registry
3. Climate Change Technology Program (CCTP)
4. Climate Friendly Parks
5. Energy Star
6. Global Climate Change Monitoring
7. Intergovernmental Panel on Climate Change
8. NCAnet

9. Pew Center on Global Climate Change
10. SF6 Emission Reduction Partnership for Electric Power Systems
11. US Global Change Research Program (USGCRP)
12. World Resource Institute (WRI).

8.10 Summary

A GHG inventory is an accounting of GHG emissions associated with the operations of the entity, process, or product. Carbon footprint is the total GHGs expressed in CO_2 equivalents. Regulations and policies as well as calculation tools are evolving. It is essential to calculate GHG emissions/carbon footprint to know the current emissions, to predict future emissions, and to identify major emission sources. It also helps plan, design, implement, and monitor mitigation strategies (Climate Change Action Plans). Numerous GHG inventory calculators are available for us. The semiquantitative and quantitative outputs of these calculators may be used for identify mitigation initiatives for GHG emission reductions. Several of these initiatives have been adopted into comprehensive frameworks that have been promulgated by a wide variety of local, state, and Federal agencies.

8.11 Questions

8.1 What do you mean by carbon footprint?

8.2 What are the steps involved in the determination of carbon footprint? Explain briefly.

8.3 What are the three groups of sources of greenhouse gas emissions? Explain briefly.

8.4 What is the USEPA's greenhouse gas reporting program? What does it entail?

8.5 List any 5 tools used for determining the carbon footprint for different applications. Explain any two of the tools in detail.

8.6 What is a climate change action plan?

8.7 Name some of the programs aimed at mitigating the greenhouse gas emissions. Explain any two programs in detail.

References

Bastianoni, S., Pulselli, F.M., and Tiezzi, E. (2004). The problem of assigning responsibility for greenhouse gas emissions. *Ecological Economics* 49 (3): 253–257.

Dixit, M.K., Dixit, M.K., Culp, C.H. et al. (2016). Reducing carbon footprint of facilities using a facility management approach. *Facilities* 34 (3/4): 247–259.

IPCC (1995). IPCC second assessment. In: *Climate Change 1995. A Report of the Intergovernmental Panel on Climate Change*. IPCC.

IPCC (2007). Climate change 2007: synthesis report. In: *Contribution of Working Groups I, II and III to the Fourth Assessment Report of the Intergovernmental Panel on Climate Change* [Core Writing Team, (ed. R.K. Pachauri and A. Reisinger), 104. Geneva, Switzerland: IPCC.

IPCC (2014). Climate change 2014: synthesis report. In: *Contribution of Working Groups I, II and III to the Fifth Assessment Report of the Intergovernmental Panel on Climate Change*. Geneva, Switzerland: IPCC.

Larsen, H.N. and Hertwich, E.G. (2009). The case for consumption-based accounting of greenhouse gas emissions to promote local climate action. *Environmental Science and Policy* 12 (7): 791–798.

Matthews, H.S., Hendrickson, C.T., and Weber, C.L. (2008). The importance of carbon footprint estimation boundaries. *Environmental Science & Technology* 42 (16): 5839–5842. https://doi.org/10.1021/es703112w.

Pierie, F., Bekkering, J., Benders, R.M.J. et al. (2016). A new approach for measuring the environmental sustainability of renewable energy production systems: Focused on the modelling of green gas production pathways. *Applied Energy* 162: 131–138.

UIC (2011). *UIC carbon footprint. Office of Sustainability.* Chicago, IL: University of Illinois (UIC).

USEPA (2015). Emission factors for greenhouse gas inventories. EPA Center for Corporate Climate Leadership. https://www.epa.gov/sites/production/files/2015-12/documents/emission-factors_nov_2015.pdf.

Weidema, B.P., Thrane, M., Christensen, P. et al. (2008). Carbon footprint. *Journal of Industrial Ecology* 12 (1): 3–6.

Weng, C.K. and Boehmer, K. (2006). Launching of ISO 14064 for greenhouse gas accounting and verification. *ISO Management System* 15.

9

Life Cycle Assessment

9.1 Introduction

Life cycle assessment, or life cycle analysis (LCA), is a methodology developed to measure the impact of products, processes, or services on the environment, allowing for a measurement of the environmental sustainability of such products, processes or services. LCA is now increasingly being used to assess engineering designs and/or processes and to facilitate selection of more environmentally sustainable options. Before we delve into the specifics of LCA, let us discuss an example of sustainability measurement as it relates to alternative products. Consider a coffee cup – specifically, three product alternatives – a polyfoam cup, a coated paper cup, and a ceramic mug. Which type of cup do you think is more sustainable? At a first glance, most people would argue that the ceramic cup is the most sustainable choice. Intuitively, one may think that the ceramic cup can be reused, while the polyfoam cup and the coated paper cup are most often disposed after one use. Further, many would argue that the coated paper cup is more sustainable than the polyfoam cup – it is generated from renewable sources (trees and paper), whereas the polyfoam cup is generated from nonrenewable petroleum products.

However, the purpose of a comprehensive sustainability assessment such as LCA is to look past initial intuition. Let us take a deeper look at several parameters associated with the life cycle of these three products. When comparing processing water effluent, the manufacturing of a typical polyfoam cup generates approximately half of the effluent as compared to a ceramic cup; however, fabrication of a polyfoam cup results in twice as many air emissions of those for a ceramic cup. A ceramic cup requires significantly more energy input than paper or polyfoam cups. Further, polyfoam cups are relatively easier to recycle as compared to coated paper cups, and it is nearly impossible to recycle a ceramic cup. Of course, a ceramic cup may be used hundreds, if not thousands of times and may be repurposed, as opposed to the typical single-use polyfoam and paper cups. Reuse does require washing, which in turn consumes energy (hot water generation) and effluent (wash water, detergents, etc.).

As you can see, once these preuse, "operation," and postuse factors are incorporated, the answer as to the relative degree of sustainability considering these three options becomes unclear. This is the motivating factor behind conducting an LCA, with the goal to perform a structured and comprehensive environmental assessment of a product, process, or activity during its life cycle. A life cycle refers to the major activities in the course of the product's lifespan from its manufacture, use, and maintenance to its final disposal, including the raw material acquisition required for manufacturing of the product (USEPA 2006).

This chapter provides (i) various definitions of LCA, (ii) a brief history of LCA, (iii) and a systematic standardized framework for performing LCA. In addition, challenges involved in different phases of LCA are also highlighted. Several software packages are available to perform

LCA, but SimaPro software developed by PRé Consultants in the Netherlands is the most commonly used software to perform LCA. This chapter presents the main features of SimaPro, followed by a simple example highlighting the LCA methodology and software application.

9.2 Life Cycle Assessment

9.2.1 Definition and Objective

There are several definitions for LCA. According to the Society of Environmental Toxicology and Chemistry (SETAC), LCA is defined as follows:

> *"a process to evaluate the environmental burdens associated with a product, process, or activity by identifying and quantifying energy and materials used and wastes released to the environment; to assess the impact of those energy and materials used and releases on the environment; and to evaluate and implement opportunities to effect environmental improvements. The assessment includes the entire life cycle of the product, process, or activity, encompassing, extracting, and processing raw materials; manufacturing, transportation, and distribution; use, re-use, maintenance, recycling, and final disposal"* (Consoli 1993).

Specifically, the performance and completion of LCAs attempt to satisfy several specific goals with respect to the environment:

- The assessments serve to ultimately reduce or minimize the generation and magnitude of pollution as well as the conservation of nonrenewable resources.
- LCAs may be used to differentiate the impacts between two comparable products, as well as to assess design options for the same product. In doing so, there is a focus on attempting to maximize recycling of materials and a reduction of waste materials, which in turn can help preserve and conserve ecological systems.
- Further, by identifying processes or alternative products, LCAs can help drive the development or utilization of cleaner technologies.
- Finally, when unavoidable side effects are identified and quantified, measures can be implemented to abate or prevent environmental pollution.

LCA studies focus on the potential impacts that occur throughout a particular activity's life cycle – beginning with the acquisition of raw materials that form the activity's inputs all the way through final disposal. This approach, referred to as "cradle to grave," considers all the environmental impacts that occur, including use of natural resources, human health, and ecological consequences. Specifically, the side effects of production, use, and disposal, including related depletion or exhaustion of input resources, emissions, effluents, and wastes that could have a negative impact on the environment has to be properly considered. It is also useful to identify where specifically in a product or in an activity's life cycle that actions may be applied to target reductions in the environmental impact. Figure 9.1 presents a schematic that depicts the interaction of inputs and outputs with respect to these human activities.

9.2.2 Procedure

Life cycle assessment determines the environmental impacts of products, processes, or services, through production, usage, and disposal (Hauschild et al. 2018). Several key steps with respect

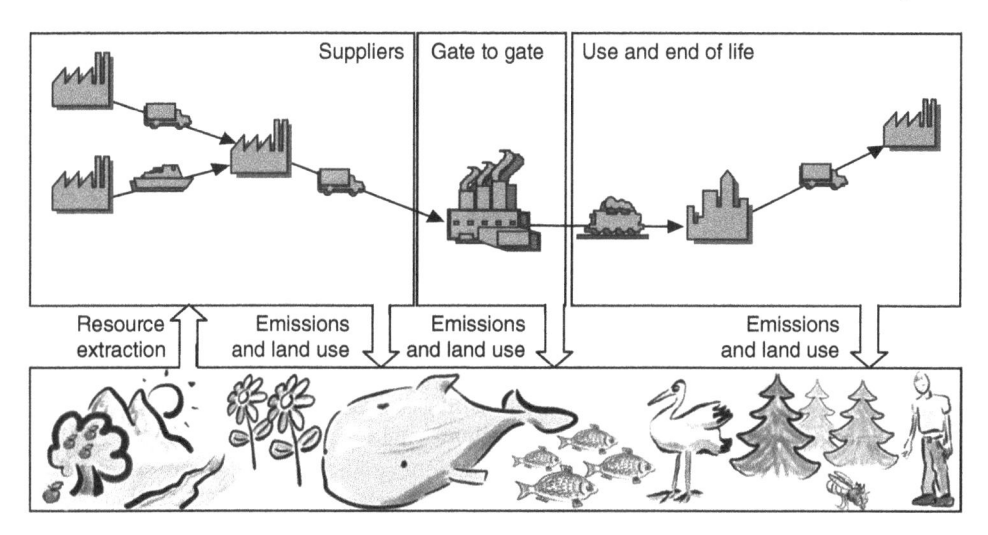

Figure 9.1 Environmental impacts of a product throughout a product life cycle. Source: With permission from Raul Carlsson, Swedish Research Institutes (RISE), Sweden 2018.

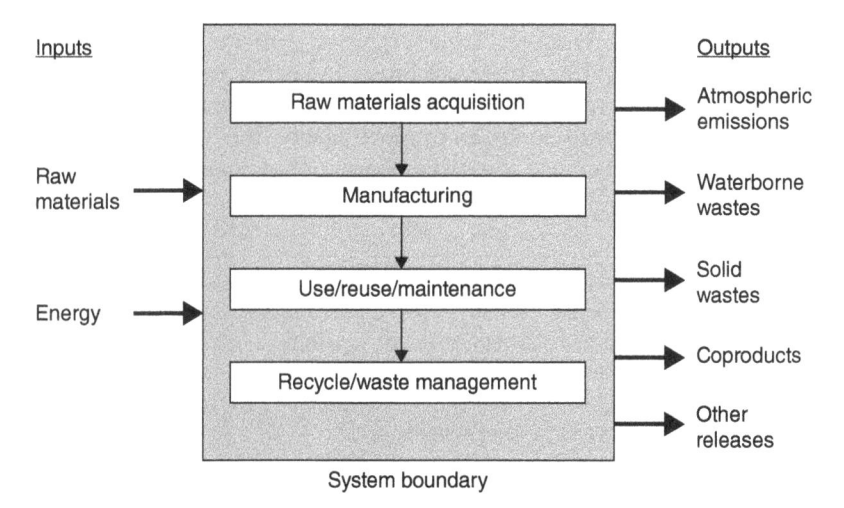

Figure 9.2 Inputs and outputs and stages in the life cycle of a product. Source: USEPA (2006).

to a product are included in an LCA analysis. Figure 9.2 depicts the flow of these key LCA activities as well as the interaction between them and the inputs and outputs to be considered in the LCA process. These include the following:

- *Raw materials extraction and harvesting* – Includes emissions and waste generated during the harvesting process.
- *Materials processing* – Refinement of the materials into some stable, usable form.
- *Manufacturing of the product* – This includes waste generated by the manufacturing process as well as wastes generated by the operation of equipment to generate the product, include energy inputs.
- *Use of the product* – An appropriate usage timeline/service life should be considered.
- *Disposal and or recovery* – Includes emissions related to disposal, incineration, as well as other potential environmental impacts related to these tasks.

- *Remanufacturing/recycling* – If a product is to be reformulated or repurposed, this includes all energy inputs and effluent/emissions outputs to get the product into a usable form.
- *Transportation* – All transportation activities, along with energy inputs and emissions between the above activities are also accounted.

As shown, inputs such as materials, energy, water, and air are considered. The activities that utilize these inputs include materials acquisition; formulation, processing, and manufacturing; product distribution; product use; recycling of products, components, and materials; and waste management. Outputs that are considered include principal products, coproducts, water effluents, airborne emissions, solid waste, and other environmental interactions.

Assessing the environmental burdens of a product, process, or service can be daunting. The conventional approach to completing an LCA is through the use of a process-based LCA method. In a process-based LCA, the inputs and outputs into a specific product or an activity are itemized. While performing a comprehensive LCA, it is important to perform several key steps, including a compilation of relevant inputs and outputs, evaluating potential environmental impacts, and interpreting the results in relation to the objectives of the particular study.

9.2.3 History

The origin of LCA can be dated in the early 1960s (Table 9.1), when concerns about the rapid depletion of fossil fuels and other natural raw materials arose. Industries began to account for energy requirements of processes and products, with the ultimate objective to reduce energy consumption wherever possible with no loss of product quality. The LCA idea rapidly developed, and in 1963, the first LCA report about the energy requirements in chemical

Table 9.1 History of life cycle assessment.

Early 1960s	Concerns about the rapid depletion of fossil fuels, which sparked interest in finding ways to account energy use and to project future resource supplies and use
1963	Harold Smith published one of the first LCA report on energy requirements for the production of chemical intermediates at World Energy Conference
Late 1960s	The studies in *The Limits to Growth* and *A Blueprint for Survival* initiated the trend on predicting finite resources in the world
1969	The Coca-Cola Company conducted an LCA study on different beverage containers
1970–1975	USEPA refined The Coca-Cola LCA methodology and created the *Resource and Environmental Profile Analysis*
Late 1970s–1980s	Environmental concerns shifted to issues of hazardous waste management and solid waste. The life cycle logic was incorporated into emerging method of risk assessment and analyzing the environmental problems
Late 1980's	A broad base of consultants and researchers across the globe has been further refining and expanding the LCA methodology
1991	11 State Attorneys General in the USA denounced the use of LCA results to promote products until uniform methodology and consensus were reached on how environmental comparison can be advertised nondeceptively
1992	ISO 1400 family grew out of ISO's commitment to support the objective of sustainable development discussed at the United Nations Conference on Environment and Development, in Rio de Janeiro
1993	ISO launched the new technical committee, ISO/TC207, Environmental Management
1997–2010	ISO developed the LCA standards

Table 9.2 ISO's LCA and other related standards.

ISO 14040:2006[a]	Environmental management – Life cycle assessment – Principles and framework
ISO 14044:2006[a]	Environmental management – Life cycle assessment – Requirements and guidelines
ISO 14045:2012	Environmental management – Eco-efficiency assessment of product systems – Principles, requirements and guidelines
ISO/TR 14047:2012	Environmental management – Life cycle assessment – Illustrative examples on how to apply ISO 14044 to impact assessment situations
ISO/TS 14048:2002	Environmental management – Life cycle assessment – Data documentation format
ISO/TR 14049:2012	Environmental management – Life cycle assessment – Illustrative examples on how to apply ISO 14044 to goal and scope definition and inventory analysis
ISO 14046:2014	Environmental management – Water footprint – Principles, requirements and guidelines
ISO/TS 14071:2014	Environmental management – Life cycle assessment – Critical review processes and reviewer competencies: Additional requirements and guidelines to ISO 14044:2006
ISO/TS 14072:2014	Environmental management – Life cycle assessment – Requirements and guidelines for organizational life cycle assessment
ISO/TR 14073:2016	Environmental management – Water footprint – Illustrative examples on how to apply ISO 14046

a) Replaced ISO Standards 14040, 14041, and 14042.

intermediates production was published. Then, in 1969, the Coca-Cola Company conducted an LCA study to determine the impact of the different beverage containers with the idea to adopt the most beneficial. The Coca-Cola LCA procedure was later refined by the USEPA to create the first attempt of LCA standardization: the so-called "Resource and environmental profile analysis."

During the 1980s, numerous attempts were undertaken around the world to refine and expand the LCA methodology to any field or activity. In 1992, as a result of the Rio summit, the ISO 1400 standards evolved to establish a standard protocol for sustainable development. After some modifications (Table 9.2), the present versions of the ISO standards for LCA were issued in 2010.

9.3 LCA Methodology

The United Nations Environment Program (UNEP) and the International Organization for Standardization (ISO), a worldwide federation of national standards bodies, have standardized the LCA framework and guidance within the following standards:

- *ISO 14040:2006, Environmental Management – Life Cycle Assessment – Principles and Framework*: Provides a clear overview of the practice, applications, and limitations of LCA.
- *ISO 14044:2006, Environmental Management – Life Cycle Assessment – Requirements and Guidelines*: Provides guidance on life cycle inventory (LCI) analysis, impact assessment, the interpretation phases of LCA, and data collection and quality issues.

These standard procedures replaced an earlier series of standard procedures on LCA, and the two standards were reviewed and reaffirmed in 2010.

As defined by the ISO standards, the LCA methodology framework consists of the following phases:

- *Goal and scope definition* – Identifying the product, process, or service to be assessed, a functional basis for comparison of alternative products or services being assessed, the required level of detail, and the boundaries which are being adopted.
- *Inventory analysis* – Identifying the inputs, usually energy and materials, involved in the different stages of the life cycle as well as the outputs, namely the products, liquid, solid, and gaseous emissions/wastes and land uses required to support the activity. The LCI is typically depicted in a flow chart.
- *Impact assessment* – Identifying qualitatively and, where possible, quantitatively, the effects of resource use and the emissions generated. These are usually grouped into impact categories that can then be weighted based on significance or importance.
- *Interpretation* – Representing the result in the most informative manner with descriptions of the opportunities to reduce the impacts of the product, process, or service.

Figure 9.3a,b illustrates an LCA methodology framework that depicts the interaction between these key actions. More details on these various phases are provided in the following sections.

9.3.1 Goal and Scope Definition

The first step of an LCA is to define the scope and goal(s) of the study. LCA studies are commonly used to improve the design of a product or the components of an activity to be more protective of human health and the environment. The studies may also be used to select among a range of products or activities that are meant to achieve a need but where the choice incorporates environmentally protective features (Curran 2016). More specifically, some common goals include the following:

- *Assessment of environmental impacts of current products and activities* – These studies assess the overall burden to environmental health and the environment resulting from all inputs (e.g. materials and energy) and outputs (e.g. emissions and waste generation) associated with the manufacturing, use, and disposal of a product, including those associated with materials, transport, and packaging of the product. These studies may also be applied to a specific activity or a process. For the case of activities, baseline assessments may also be undertaken to determine the net effect of changes to the activity in question. In both cases, the information from the LCA can be used to guide the development of completely new products or activities.

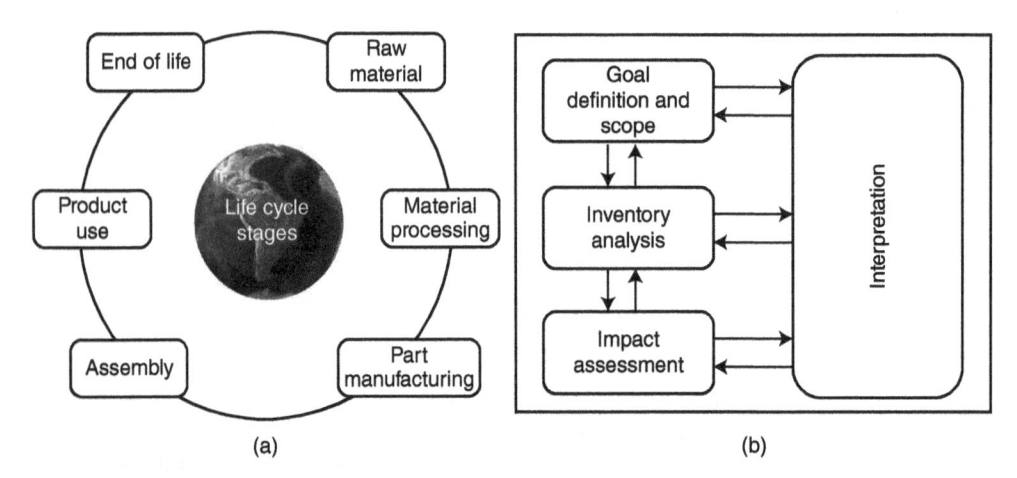

Figure 9.3 Life cycle assessment framework. a) life cycle stages, b) methodology.

- *Assessment of environmental impacts when selecting between several product and activity options* – These studies assess the relative burdens between a range of product or activity options including the "do-nothing" alternative. The benefits and drawbacks among these options are quantitatively assessed such that the overall environmental burdens compared between options may also be combined with social and economic considerations to make an appropriate decision.
- *Support of public policy initiatives and legislation* – Public policies and legislative actions often require the consensus of a range of stakeholders. However, it is often difficult to reach a consensus on meaningful pieces of legislation, as the interests of the stakeholders may be varied and often at odds with each other. LCA studies can often provide conclusive data analysis that may be desired or even necessary to convince groups of the merits of such initiatives. The studies can also be useful to guide the decisions and enforcement of such initiatives.
- *Data gap analysis* – Perhaps ironically or paradoxically, LCA studies can identify data gaps inherent in such studies; identifying areas where little knowledge may exist, or areas where such additional knowledge can prove to be informative with respect to the product, the activity, or the LCA study itself.

When determining the scope and the goal of the LCA, it is also important to assess who the target audience for the LCA may be. If the LCA is meant to support confidential or proprietary development decisions within an organization, proper precautions to be protective of sensitive information such as trade secrets should be taken. For public entity audiences, similar precautions should be considered with respect to sensitive aspects of the study, although it is likely the expectation that such assessments may become public. Further, for suppliers or producers that occupy a specific location or range on an activity or product value chain, it is important to assess upstream or downstream perspectives.

Additionally, it is important to set a range of study expectations when developing the LCA scope and goal. For instance, it should determine if all potential uses for a specific product/activity are to be considered, or if a specific range or single use should be considered. Expectations regarding data accuracy, the expected statistical significance, and potential effects of data variance or inaccuracy should be determined at this stage. Procedures for handling data gaps, establishing compatible units of measurement, and other aspects regarding data quality and measurement should also be determined. Finally, feedback mechanisms regarding data quality assurance/control, the impact of assumptions, and "real-time" determinations if the assessed data satisfies the originally stated goals and scope can be implemented at this time.

9.3.2 Life Cycle Inventory (LCI)

In order to perform an LCA, it is critical to properly account for the inputs and outputs associated with a system that is under study (Curran 2012). This task is known as the LCI. The LCI incorporates a significant array of potential data into the study. Ultimately, to facilitate the LCA studies, several key tasks should be incorporated, including the development of a flow diagram that incorporates all the key processes that are under study, the development of a data collection plan, the collection of data, and the evaluation of the results of the study, as well as the provision to report the results.

When assessing a system using an LCA, a *system boundary* must be defined first. The boundary serves to illustrate a separation of the system from its surroundings. The boundary encompasses the system, which is defined as a collection of materially and energetically connected operations that perform a defined function. The greater the environment lies to the outside of the system boundary; it acts as the source of all inputs into the system and acts as a sink for all outputs of the system. A schematic depiction of a system, its boundary, and the greater environment is depicted in Figure 9.4.

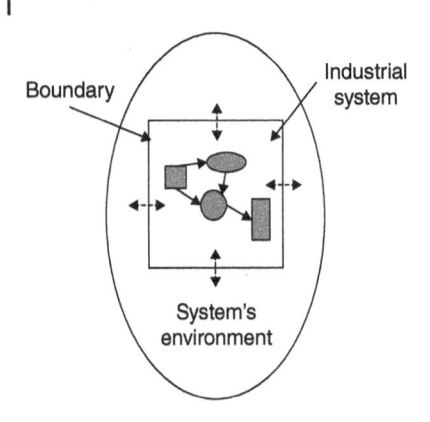

Figure 9.4 Industrial system, its boundaries and the inputs–outputs with the environment.

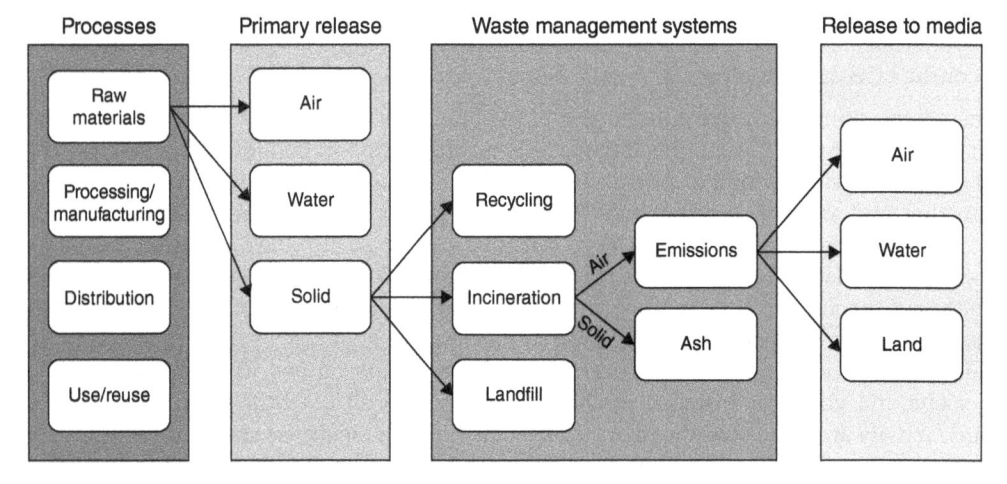

Figure 9.5 Outputs from raw materials.

Inputs (i.e. raw materials, energy) and outputs (i.e. atmospheric emissions, waterborne wastes, solid wastes, coproducts, and other releases) must be properly identified during the inventory analysis. When considering the inputs and outputs, it is important to specifically determine the following:

- Which materials are used by the system?
- What are the quantities of the materials that are consumed?
- How much energy is consumed at the various stages?
- What feedstock is used to generate energy?
- What wastes or emissions are produced?
- Are any of the wastes or emissions toxic?

Outputs occur from each stage of the life cycle; this is depicted in Figure 9.5. As shown in the figure, one of the processes or stage of the life cycle is selected; in this case it is raw materials acquisition. As indicated, the extraction of the raw materials may result in primary releases to air, water, or soil. Solids, in turn, may be recycled, incinerated, or landfilled. If materials are incinerated, ash (solid) and emissions (air) will be generated. The air emissions may result in air pollution, or subsequent deposition from airborne pollutants may be deposited into soil or surface waters, leading to the impacts on these media.

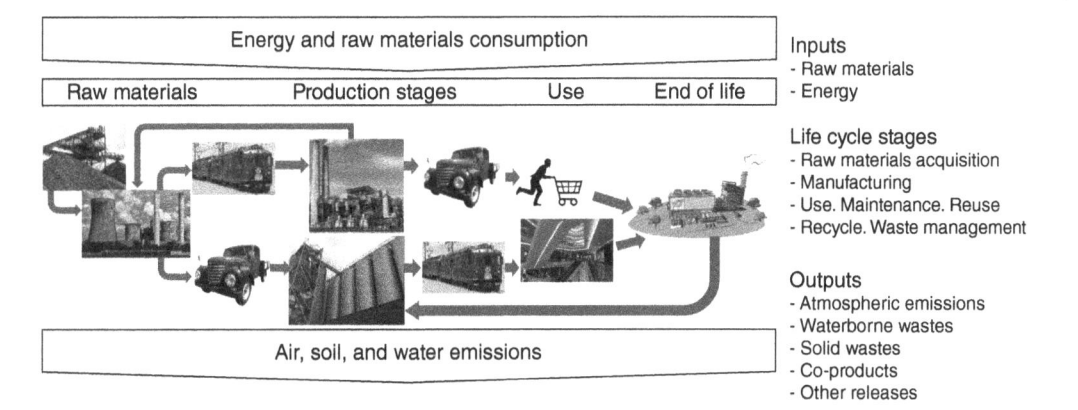

Figure 9.6 Stages of the entire life of a product.

An LCI incorporates inputs and outputs to track *material flows within systems*. For example, Manufacturer A produces intermediate materials from raw materials, such as plastic pellets from a petroleum feedstock. Manufacturer B produces components from intermediate materials, such as cases for laptops using materials derived from Manufacturer A. Manufacturer C produces components or assembles final products using intermediate materials produced in part by Manufacturer B, such as laptop computers.

Another concept to be considered in LCI analyses is the *feedback loop*; that is, the reversion of material inputs and associated outputs to generate other inputs. For instance, coal and iron are used to generate steel, which may then be fabricated into a shovel. This shovel, in turn, is used to extract more iron ore and coal to make more shovels. These "reverse" flows should also be accounted for in the LCI.

In addition to inputs and outputs, the *specific stages for the life cycle* are also defined. Some of these include raw materials acquisition and processing, manufacturing, use, maintenance, and reuse, recycling, and waste management. Each of these specific activities requires specific inputs and result in outputs that may lead to environmental impacts. Figure 9.6 depicts an entire life cycle that incorporates successive steps of the raw materials producer, the manufacturer, and the consumer, including final product disposal.

When performing an LCI, it is important to determine an equivalent *unit of function*. For instance, a 12-oz aluminum can, a 20-oz plastic bottle, and a 2-l plastic bottle are all single containers that are used to package soft drinks; however, it is clear to see that the raw materials and resulting waste generations from usage of these containers are different. Therefore, a functional unit needs to be defined. This allows for all data to be normalized to a baseline unit. Once selected, other similar products may be related to the functional unit taking into consideration size, time, and intensity scales associated with the inputs and outputs as well as their relative effect to the environment.

The *treatment of data collection* is as important as including inputs, outputs, and system boundaries. It is very important to determine the goals for data quality, which can then be used to determine cost/benefit for accurate data collection, as well as data quality indicators, which serve as a benchmark for determining if the collected data is useful and of a sufficient quality for the study. The types and sources of data should be determined as well, including published research data, government statistics, laboratory results, "real-time" equipment sources, and a wide range of publicly available references. The data may be directly sampled, modeled, measured, based on regional or spatial assumptions or estimates, or general assumptions, especially where more refined data measurement or modeling may be difficult, time-consuming, or cost-prohibitive. Further, consideration should be given to a sensitivity analysis such that determinations can be made regarding the overall consequences to the study when data is erroneous, missing, or misrepresented. Finally, these decisions can also be used to determine when minor, secondary, or "background" data may be neglected or eliminated because it is of minor or no consequence to the quality of the overall study.

Finally, the results of the LCI must be analyzed, interpreted, and reported in a *format* that will be useful for the audience or stakeholders to understand its significance. The results may be used to determine direct scores between assessment alternatives or to identify trends, either due to the whole, product or activity or associated with a specific input or an aspect associated with the product or activity. The resulting data may also be used to determine if spatial/geographical or temporal trends may exist with respect to the findings of the study. The context of the result with respect to the overall effects to human health or the environment may also be assessed. Clear representation of the data is also critical to demonstrate the findings of the study, whether the data and findings are presented in graphical or text form. At this time, informed decisions regarding product/activity design or selections among several alternatives may be undertaken.

Perhaps intuitively, the results of the LCI may also answer questions regarding the model. If there is a sense that the data and the results gleaned from the study appear to be counterintuitive, misleading, or even inaccurate, the model assumptions, including model boundary, inputs, outputs, and processes may be reconsidered to assess if the model and the underlying assumptions, data, and data collection are accurate. If omissions or errors regarding data or the model itself are identified, changes can be made to these assumptions, and a reassessment may be undertaken.

LCI analyses can be very beneficial and powerful; however, there are *several unresolved problems* that should be recognized. First, the definition of a functional unit may be difficult where a product has multiple functions. Next, the selection of the system boundary must be done carefully; if too broad, the LCI process may be unnecessarily complex or lack sensitivity; if too narrow, important inputs and outputs may be erroneously neglected. Additionally, there is discretion of when to stop adding selection criteria to the inventory analysis, or put differently, what may be neglected from the model and its overall effects need to be considered. Finally, there has to be a discretion and judgment on how to best allocate environmental burdens.

9.3.3 Life Cycle Impact Assessment (LCIA)

The LCI stage is very important when a comprehensive LCA is performed. However, it is equally important to determine the context of the measured parameters, namely what their relative effects are to human health and the overall environment. These effects are measured and accounted in a life cycle impact assessment (LCIA). The objective of an LCIA is to evaluate the impacts to human health and the environment resulting from emissions, resource extraction, and other interventions from human activities and technical systems. To emphasize, an LCIA should address ecological effects and human health effects as well as resource depletion. Figure 9.7 depicts the elements of an LCIA.

An LCIA evaluation aims to assess subjective descriptions of process to a human activity in objective physical terms. The assessment is meant to determine and assess the relative differences in potential ecological and human health effects for each option associated with the activity. Some of the environmental impacts that are considered include the use of renewable or nonrenewable resources, the generation of toxic substances, the release of greenhouse gas emissions or potentially ozone-depleting gases, the generation of liquid and solid wastes, the release of excess nutrients to the environment, and the detriment of land resources through mineral extraction or deforestation.

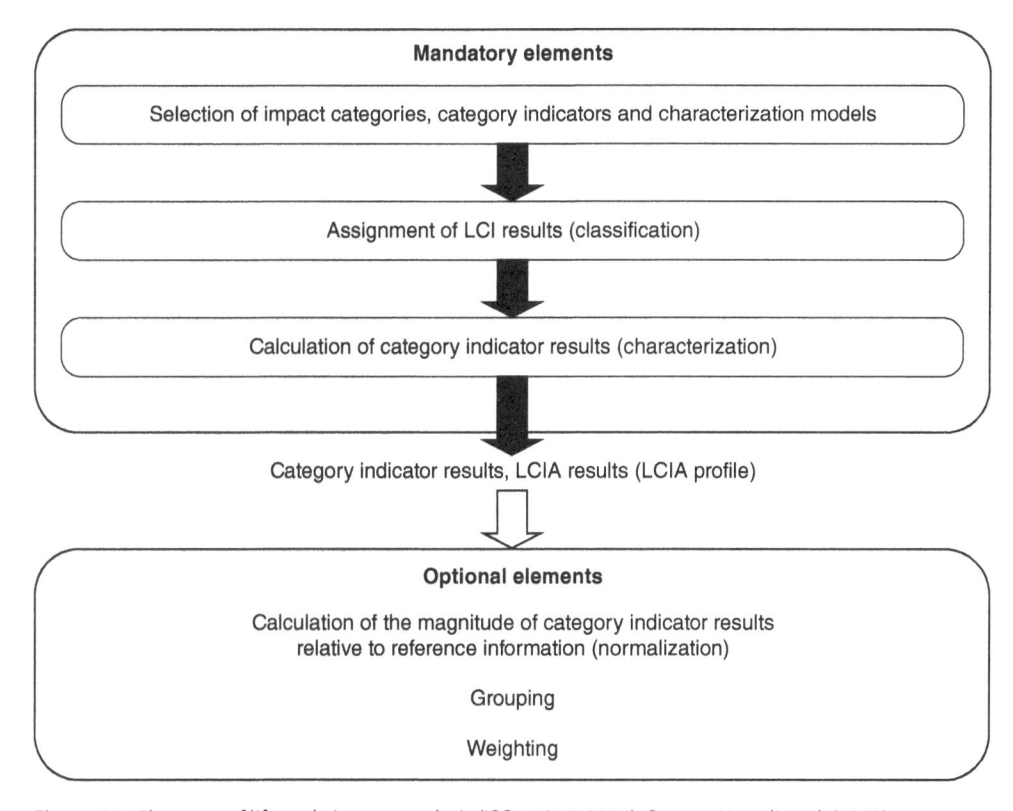

```
┌─────────────────────────────────────────────────────────────────┐
│                      Mandatory elements                          │
│  ┌───────────────────────────────────────────────────────────┐  │
│  │ Selection of impact categories, category indicators and   │  │
│  │            characterization models                         │  │
│  └───────────────────────────────────────────────────────────┘  │
│                              ▼                                   │
│  ┌───────────────────────────────────────────────────────────┐  │
│  │       Assignment of LCI results (classification)           │  │
│  └───────────────────────────────────────────────────────────┘  │
│                              ▼                                   │
│  ┌───────────────────────────────────────────────────────────┐  │
│  │  Calculation of category indicator results (characterization) │
│  └───────────────────────────────────────────────────────────┘  │
└─────────────────────────────────────────────────────────────────┘
                               ▼
        Category indicator results, LCIA results (LCIA profile)
                               ⇩
┌─────────────────────────────────────────────────────────────────┐
│                       Optional elements                          │
│       Calculation of the magnitude of category indicator results │
│           relative to reference information (normalization)      │
│                                                                  │
│                            Grouping                              │
│                                                                  │
│                            Weighting                             │
└─────────────────────────────────────────────────────────────────┘
```

Figure 9.7 Elements of life cycle impact analysis (ISO 14040, 2006). Source: Hemdi et al. (2009).

An LCIA evaluation differs from other evaluation techniques. For example, a *risk assessment* is used to assess rules and regulations about handling chemicals. An *environmental impact assessment* is used to assess the application of rules and regulations pertaining to improvements and enhancements of the built environment. A *cost–benefit analysis* is used to evaluate the meaningfulness of measures (with respect to financial considerations) to decrease environmental impacts. As discussed above, numerous environmental effects may be incorporated into an LCIA. These may be further categorized as local, regional, or global effects:

- On a local basis, some cause-and-effect examples include the generation, combustion, and disposal of chemicals and related chronic and acute health effects; industrial and household processes and the release of related chemicals to the environment; and chemicals in an industrial working environment and their effects with respect to human health and system functions.
- On a regional basis, some examples include the release of by-product gaseous emissions and the formation of photochemical or particulate smog; the release of chemical effluents and its effects to large bodies or water, including acidification or eutrophication, and the release of chemical compounds to the subsurface with the resulting fouling of surface and groundwater resources.
- On a global level, some relationships include the overexploitation of mineral wealth and the reduction of the potential utilization by future generations; the release of greenhouse gas emissions and their resulting effect on climate change; and the release of ozone-depleting chemicals and their potential effects to the protective ozone layer, resulting in greater surface levels of ultraviolet radiation.

One example of the cause and effect of these relationships is the effects of CFCs/Halons. When these emissions occur, chemical reactions release chlorine and bromine ions. When the *midpoint effect* is considered, these ions act to destroy ozone layer; and a midpoint analysis assesses the resulting depletion of the ozone layer. Taken further, an *endpoint analysis* may be performed with a broader perspective of the environmental impacts. In such analysis, the resulting effects of skin cancer incidents, crop damage, immune system suppression, damage to materials, damage to marine life, and human optical maladies, such as cataracts, are measured.

The EPA Science Advisory Committee has attempted to *classify environmental effects* based on several criteria. First, greater weight is placed on larger-scale impacts as compared to smaller-scale impacts. The severity of a hazard is ranked by the relative danger posed; for instance, more toxic substances are of a greater concern than less toxic substances. The potential for exposure is also considered – well-sequestered substances pose less of a threat than easily mobilized substances. Finally, a greater weight or emphasis is applied to the "penalty for being wrong" – incidents that result in greater remediation times are of greater concern than those requiring less remediation times.

Applying actual effects to these categories, the EPA has classified the problems into high, medium, or low risk as shown by the following examples:

- Relatively high-risk problems include habitat alteration and destruction, species extinction and loss of biological diversity, stratospheric ozone depletion, and global climate change.
- Relatively medium-risk issues include herbicide and pesticides in the environment, toxics, nutrients, biochemical oxygen demand, and turbidity in surface waters, acid deposition, and airborne toxics.
- Relatively low-risk issues include oil spills, groundwater pollution, the presence of radionuclides in the environment, acid runoff to surface waters, and thermal pollution.

The LCIA has four steps, as depicted in Figure 9.7: classification, characterization, normalization, and valuation (weighing).

Step 1 – Classification: Resource usage and emissions/releases are assigned to respective environmental impacts. It is common for multiple intertwined relationships to exist (e.g. sulfur dioxide emissions resulting in damage to surface water bodies and the built environment due to acid rain), and it is also common for many intertwined cause-and-effect ("stressors" and "impacts") chains.

Step 2 – Characterization: This is accomplished by the quantitative determination of the impacts resulting from the stress indicated by the LCI values by using the characterization equation, which is defined as the following:

$$S_j = \Sigma\, C_{i,j} \cdot E_i \tag{9.1}$$

where S_j = Category stress indicator for category j; $C_{i,j}$ = "Characterization factor" for species i and category j (i.e. what level of environmental stress of category j is caused by a unit mass of species i); and E_i = Mass flow identified for species i in the inventory assessment. When performing the characterization, it is critical to apply appropriate weighting to properly capture the impacts associated with inventory data. To accomplish this, it is important that the impact be defined and that methods exist to quantify the contribution of the impact. As presented in Chapter 8, greenhouse gas contributions may be expressed in terms of a global warming potential (GWP). As a base unit, carbon dioxide (and its associated impact is defined as 1; or alternatively, GWP is expressed as CO_2 equivalents. Methane, for instance, has a 28-fold greater effect as a greenhouse gas as compared to carbon dioxide on a molecular basis. Therefore, methane has a GWP of 28.

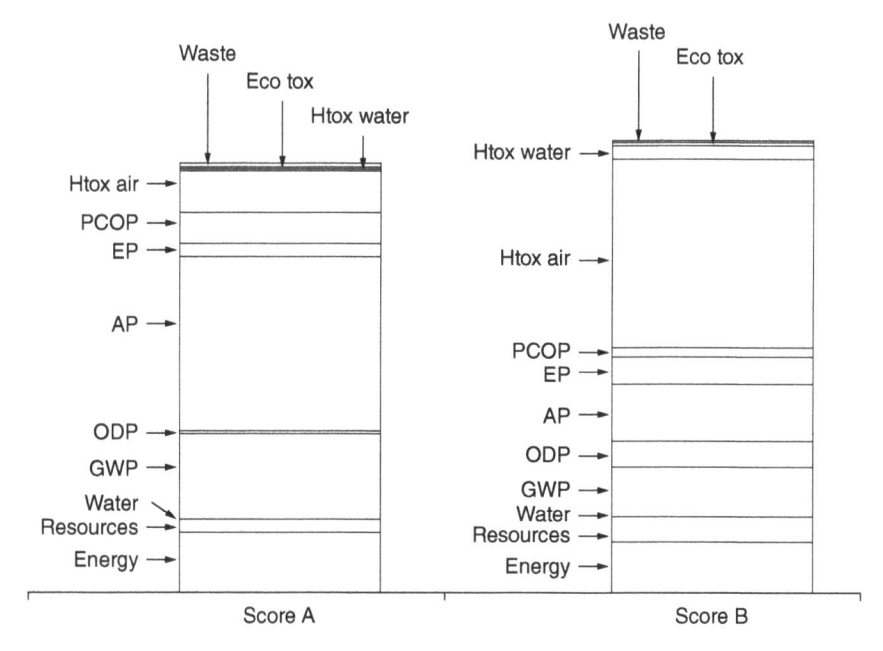

Figure 9.8 Valuation with normalization of two optional product designs.

Step 3 – Normalization: During the normalization process, the impact values resulting from the characterization stage are related to a common set of reference values. During the normalization process, the normalization equation is used, which is defined as the following:

$$N_j = S_j/R_j \tag{9.2}$$

where N_j = Normalized indicator; S_j = Category stress indicator for category j; and R_j = Reference value. Figure 9.8 depicts a graphical representation of the normalization process for two optional product designs for a variety of stress indicators. As shown, the normalized indicators associated with each category stress indicator may be summed to determine the cumulative effect of each option to determine the relative effects associated with each option.

Step 4 – Valuation process: During the valuation process, weighting factors are assigned to different impact categories based on their perceived relative importance. The valuation equation is applied in this step and is defined as follows:

$$W_j = \Omega_j/N_j \tag{9.3}$$

where W_j = Weighted impact indicator; Ω_j = Weighting factor; and N_j = Normalized indicator. Once the weighted impact indicators have been computed, the overall LCA score may be calculated as follows:

$$I = \Sigma_j W_j \tag{9.4}$$

where I = LCA score; and W_j = Weighted impact indicator.

Figure 9.9 summarizes the flow of the LCIA process. As shown in the figure, following the completion of an inventory assessment, associated resource usage and emissions/releases are assigned to respective environmental impacts during the classification stage, and the quantitative impacts resulting from the stress indicated by the LCI values are determined during the

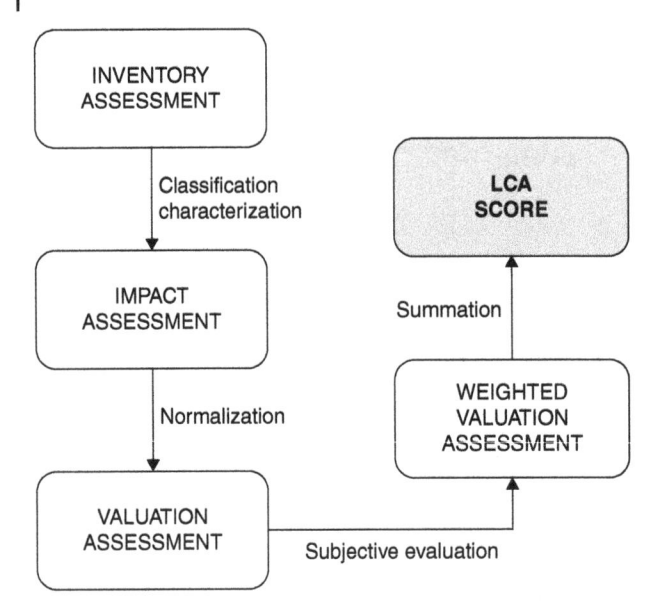

Figure 9.9 The LCIA sequence.

characterization stage. The assessed impacts are then normalized, where a valuation assessment is performed. Weighting is applied on a subjective basis, the resulting weighted factors are summed, and an LCA score is determined.

As with LCI studies, the results of the LCIA may be assessed to determine if the results appear to be counterintuitive, misleading, or erroneous. If so, the parameters may be reconsidered, including several of the somewhat subjective factors used for the quantitative assessments explained above. If the data appears to be accurate and logical, and appropriate, accurate conclusions may be drawn, recommendations may be derived and presented. However, it is important to acknowledge the assumptions, subjectivity, and potential limitations of the data due to these factors.

Several issues still remain with respect to the use of LCIAs:

- LCIA are subject to the heuristics inherent in predefined software methodologies. There could be differences in the results using two different software programs.
- There are effects associated with spatial variation. For example, impacts of manufacturing 1 kg of copper could be different in USA, China, and Finland.
- There is the issue of capturing local environmental uniqueness, as well as the dynamics of the overall environment. For example, local air pollution emission could affect the nearby receptors nonuniformly depending on the wind direction. Such differences are not accounted; instead average values are used.
- The weighting process can be quite subjective, resulting in appreciable variability between assessments. Individuals perceive the risk from hazards differently, thereby deriving acceptable set of weightings will be difficult.

9.3.4 Interpretation

Following the computational steps associated with the LCIA process, the results are reported and interpreted in a systematic way so that the need and opportunities to reduce negative

9.4 LCA Tools and Applications

impacts of the product or process or service on the environment may be identified and mitigated. Sensitivity and uncertainty analyses results could help make better interpretation of the results leading to better decision-making.

9.4 LCA Tools and Applications

Several software packages are developed to perform LCA as summarized in Table 9.3. Among these, SimaPro developed by PRé Consultants in the Netherlands is commonly used. SimaPro is software specifically designed for LCA of an organization, a project, or a product. This software makes sustainability decisions measurable, thus the LCA results can be used to implement changes at various levels, so the organization or product evolves to more sustainable conditions.

The software requires the inventory of relevant energy and material inputs to the process and the outputs to the environment. The software is designed to assess the potential impacts associated with identified inputs and releases. Several methodologies and databases can be used for the assessment: TRACI, BEES, or ReCiPe are commonly used.

TRACI, "Tool for Reduction and Assessment of Chemicals and Other Environmental Impacts", is an environmental impact assessment tool developed by the USEPA. This tool allows the evaluation of factors for LCIA, industrial ecology, and sustainability metrics. Those factors

Table 9.3 Software packages and solutions for LCA.

BEES	Building for Environmental and Economic Sustainability NIST, USA	nist.gov/bees/
Eco-Bat	Eco Balance Assessment Tool HEIG-VD, Switzerland	eco-bat.ch
EcoScan	Environmental impact and cost of products TNO, The Netherlands	tno.nl
EIME	Environmental Improvement Made Easy Bureau-Veritas	bureauveritas.com
EPD Tools Suit	Enthought Python Distribution Enthought Scientific Computing Solutions	enthought.com
GaBi	GaBI product sustainability software Thinkstep Global Headquarters, Germany	gabi-software.com
JEMAI-LCA Pro	Japan Environmental Management Association for Industry	jemai-lca-pro.software.informer.com
openLCA	Open source Life Cycle Assessment software	openlca.org
LTE/OGIP	ENERGIE UND STOFFFLÜSSE Switzerland	rb-i.ch/de/produkte/OGIP
SimaPro	Life Cycle Assessment Software PRé Consultants, The Netherlands	Simapro.com
TEAM™	Life Cycle Assessment Software Ecobilan, France	ecobilan.pwc.fr
The Boustead Mode	Boustead Consulting UK	boustead-consulting.co.uk
US Life Cycle Inventory Database	National Renewable Energy Laboratory	nrel.gov/lci/

quantify the potential impacts of the project inputs and outputs in specific categories, including ozone depletion, climate change, acidification, eutrophication, smog formation, human health impacts, and ecotoxicity. The factors are normalized against the TRACI database so the importance of impact in each category can be compared. The other aforementioned tools include BEES, the Building for Environmental and Economic Sustainability from NIST, and ReCiPe, established by the Dutch National Institute for Public Health and the Environment (RIVM).

The last step in the LCA is the interpretation of the program results by the user, thus allowing for informed decisions for the project toward more sustainable conditions. SimaPro incorporates several features that help in the interpretation of the LCA results for a specific project and the comparison among several alternate solutions for the project.

The data for input and output flows in the project can be processed in SimaPro using formulas designed by the user. The data and plots can be exported to a spreadsheet, so the user can adapt the results into presentations with specific formats or purposes. However, the more useful features are related with the presentation/comparison of results and the sensitivity analysis of data. The so-called pedigree matrix quantifies the uncertainty of the amount of a specific input or output, allowing for easier detection of errors or sensitive points. The Parameters test tab permits the user to vary the values of different parameters, which is especially useful in the performance sensitivity analysis.

SimaPro presents results in several standard graphs. As an example, the "Network" is a graphical representation of a model that permits identification of the main sources of impacts in a graphical way using thin or thick arrows to represent the flow of materials and impacts through all the stages of the project.

LCAs can be applied to make decisions regarding environmental sustainability of alternate processes or designs as well as optimize the components of a process or design for improved sustainability. Several detailed examples of the applications of LCA in engineering projects are provided in Chapters 20–22.

9.5 Summary

LCA is a useful means to measure the impact of products, processes, or services in the environment, allowing for a measurement of the sustainability of such products, processes, or services. LCA studies are commonly used to improve the design of a product or the components of an activity to be more protective of human health and the environment.

The first step of an LCA is to define the goal(s) and scope of the study. As demonstrated in this chapter, selecting the system and system boundary for an LCA is of prime importance. However, there is an inclination to draw a broad boundary with the goal of developing a comprehensive model. This can lead to problematic side effects with respect to the lack of data, complexity, or expense associated with performing such studies. As a result, the models can become unworkable. The next step is to prepare LCI of all materials and processes involved in each life cycle stage. LCI results could be used to assess the environmental impacts, but LCIA is a systematic process to evaluate the impacts to human health and the environment resulting from emissions, resource extraction, and other interventions from human activities and technical systems. Finally, LCIA results are interpreted to address ecological effects and human health effects as well as resource depletion.

The mere use or attempt for use of LCA can be beneficial to raise awareness with respect to measures that may be used to mitigate environmental impacts associated with activities or products. Therefore, regardless of the problems that may arise for LCA complexity, it is still beneficial to pursue more efficient and useful LCA for projects.

9.6 Questions

9.1 What is the purpose of a LCA?

9.2 What are the major phases involved in LCA methodology established by ISO?

9.3 Why is LCA important in assessing the sustainability of a product, a process, or an organization? Explain your answer with an example.

9.4 Name some of the major life cycle stages involved in a typical life cycle of a product.

9.5 Select a common household product and two alternatives. Using a qualitative framework, discuss the degree of sustainability for manufacture, operation, maintenance, and disposition of the product and the alternatives.

9.6 What is a functional unit? Why is it an essential part of life cycle analysis?

9.7 What do you mean by a system boundary in the context of LCA? What is the significance of the system boundary in LCA?

9.8 What are the effects of a "small" system boundary? What are the effects of a "large" system boundary?

9.9 How can a data gap effectively be used to assess an LCA and the product/project that is being assessed?

9.10 Discuss how a feedback loop is handled in an LCI.

9.11 Describe the importance of assessment format in preparing an LCI.

9.12 What are the four key steps of an LCIA? Discuss in detail the importance of each step.

9.13 Name three life cycle impact assessment methods mentioned in this chapter.

9.14 Provide detailed examples of local, regional, and global effects that may be identified during an LCIA.

References

Consoli, F. (1993). Guidelines for life-cycle assessment: a "code of practice". In: *Proceedings of the Fifth Technical Workshop on LCA Sponsored by SETAC at Sesimbra, Portugal 31 March – 3 April 1993*. Pensacola, FL: Society of Environmental Toxicology and Chemistry (SETAC).

Curran, M.A. (ed.) (2012). *Life Cycle Assessment Handbook: A Guide for Environmentally Sustainable Products*. Wiley.

Curran, M.A. (ed.) (2016). *Goal and Scope Definition in Life Cycle Assessment*. Springer.

Hauschild, M., Rosenbaum, R.K., and Olsen, S.I. (2018). *Life Cycle Assessment – Theory and Practice*. Cham: Springer International Publishing.

Hemdi, A.R., Saman, M.Z.M., and Sharif, S. (2009). An indicator for measuring sustainable product design: a review and future research. Conference on Scientific & Social Research CSSR08'09. 14–15 March 2009. Paper number: 8695303.

USEPA (2006). Life cycle assessment: principles and practice. U.S. Environmental Protection Agency. Report No: EPA/600/R-06/060.

10

Streamlined Life Cycle Assessment

10.1 Introduction

The usefulness of a full-scale life cycle assessment (LCA) to determine the degree of sustainability of an organization, project, or product was discussed in Chapter 9. However, such a full-scale study can be significantly time-consuming and expensive. In fact, an LCA typically utilizes significantly expensive software and databases, and the output of the assessment may be questioned since the LCA results often show some variation depending on the software/database used. Moreover, an LCA requires itemization of inputs (materials and energy resources) and outputs (emissions and wastes to the environment) for a given step in producing a product. Even for a very simple product, this process-based LCA method can quickly spiral into an overwhelming number of inputs and outputs to be considered.

As seen in Chapter 9, even for a simple case of disposable paper cup, the LCA might list the paper and glue for the materials, the electricity or natural gas for operating the machinery to form the cup for the inputs, scrap paper material, waste glue, and low-quality cups that become waste for the outputs. For a more complex product such as an automobile, with over 20 000 individual parts, or a process such as electricity generation, the complexity is much higher. When considering the broad life cycle perspective, this same task must be done across the entire life cycle of the materials for the manufacturing and use of the product.

Again, in the case of the drinking paper cup, one needs to identify the inputs, such as pulp, water, and dyes to make the paper, the trees and machinery to make the pulp, and the forestry practices to grow and harvest the trees. Similarly, one needs to include inputs and outputs for packaging the cup for shipment to the store, the trip to the store to purchase the cups, and that result from throwing the cup in the trash and eventually being landfilled or incinerated.

Overall, the LCA method can be so complex and expensive, it may become unworkable or cost-prohibitive. Furthermore, it can be quite inefficient and even contentious among stakeholders. Even though LCAs can be significantly beneficial for both their output as well as their ability to increase awareness of environmental impacts, they are not commonly used as a routine assessment tool.

How can the benefits of the LCA process be realized without the drawbacks related to cost, time, and complexity? The idea is to develop a method that retains the quintessential aspects of the LCA methodology while eliminating some of the complexity associated with it to dramatically improve the efficiency. This alternative method is called streamlined life cycle assessment (SLCA) (Graedel and Allenby 2010; Weitz et al. 1999). The goal of SLCA is to retain the breadth and approach and awareness-raising aspects of a full-scale LCA (as detailed in Chapter 9), but simplifying the process such that the overall efficiency of the process is greatly improved. This chapter provides SLCA methodology, simple applications, and highlights other applications.

Sustainable Engineering: Drivers, Metrics, Tools, and Applications, First Edition.
Krishna R. Reddy, Claudio Cameselle, and Jeffrey A. Adams.
© 2019 John Wiley & Sons, Inc. Published 2019 by John Wiley & Sons, Inc.

10.2 Streamlined LCA (SLCA)

To conduct a product-related SLCA, it is important to identify the key life cycle stages in SLCA. It should be noted here that "product" could also be "engineered system." As depicted in Figure 10.1, the SLCA includes the following stages: premanufacture, product manufacture, product packaging, product use, and product recycling/disposal. Additionally, it is important to identify the key environmental concerns to be included in the SLCA, as depicted in Figure 10.2. These concerns include the following: materials choice, energy use, solid residues, liquid residues, and gaseous residues. Both the life cycle stages and the environmental concerns listed here are similar to those considered in the LCA method discussed in Chapter 9.

Once the key life cycle stages and environmental concerns have been identified, a matrix may be established to account for the resulting environmental concerns with each respective life cycle stage. This matrix, called the environmentally responsible product matrix (ERP matrix), is shown in Figure 10.3. The columns are populated by the respective environmental concerns (depicted along the horizontal axis), and the life cycle product stages are shown on the vertical axis (each with a corresponding row). Each resulting square in the matrix represents the specific environmental concern associated with a specific stage of the life cycle. For instance, the middle square or cell of the matrix (3, 3) corresponds to the solid residue that results from product packaging.

The matrix is populated with numerical values based on the relative environmental impact associated with an activity of the product's life cycle. The goal is to efficiently determine an entry for each element in the matrix as an "informed estimate" of the overall life cycle impact. One possible format, which seems to be very effective in characterization of the environmental impacts, is to assign only integers between 0 and 4; descriptions of these values are presented in Table 10.1. The top and bottom evaluations, (i.e. 4 – No environmental impact; 0 – very high environmental impact) tend to be obvious in their assignment. If those numbers are not assigned for a specific impact and stage of life cycle, the choices are restricted to 3, 2, and 1 – better, equal to, or poorer than expected.

For example, while considering material choice for packaging (row 3; column 1), the cell would receive a value of 4 if no packaging were used for the specific product. A score of 0 would be assigned if the package used three or more materials that were not recyclable. Scores of 3, 2, or 1 would be assigned by the relative performance of a number of factors, including the

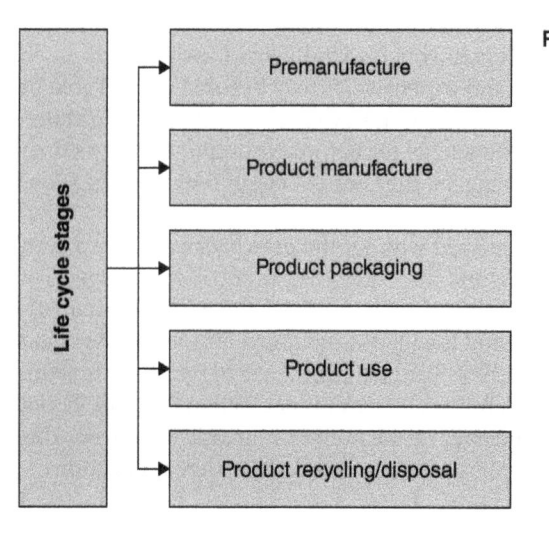

Figure 10.1 Life cycle stages for product LCA.

Figure 10.2 Environmental concerns for product life cycle assessment.

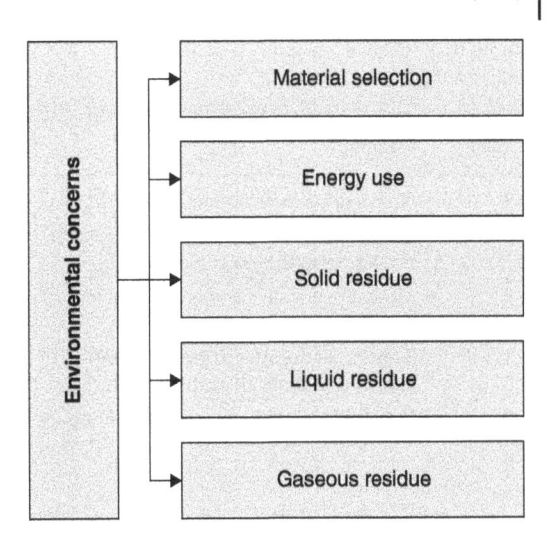

Figure 10.3 Matrix for accounting the environmental impacts of each life cycle stage.

use of recycling packaging materials, minimization of material diversity, existence of recycling infrastructure, and the relative involvement of the packaging engineer with the product design team to identify other efficiencies.

Once each of the matrix cells have been populated, the overall ERP rating can be computed by summing all of the matrix cell values as shown in Eq. (10.1):

$$R_{erp} = \Sigma M_{i,j} \tag{10.1}$$

where

R_{erp} = product rating;
$M_{i,j}$ = value corresponding to cell with row i, column j.

When all the cells within the matrix are populated, the total score of this 5× 5 matrix is the result of cumulatively added values of all cell, resulting in a value between 0 and 100 for the matrix, with a similar meaning as shown in Table 10.1. If the total score is 100, the product shows minimal environmental impact. For total score of 0, the product shows a very high impact.

Table 10.1 Relative environmental impact of each activity of a product life cycle.

Value	Meaning
4	No negative environmental impact
3	Minimal negative environmental impact (i.e. less than the expected average)
2	Moderate negative environmental impact (i.e. about the expected average)
1	Substantial negative environmental impact (i.e. above the expected average)
0	Very high negative environmental impact

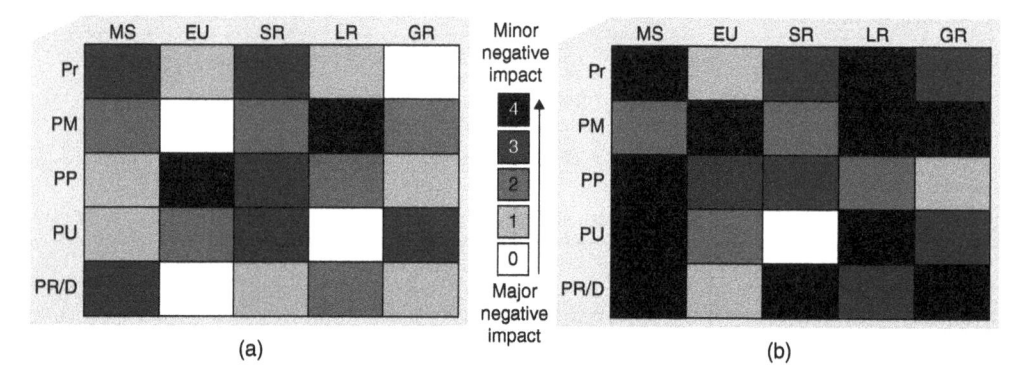

Figure 10.4 Graphical representation of the SLCA matrix using coded-colors instead of integer values, for (a) an irresponsible product, and (b) a more sustainable product.

Any intermediate value defines the degree of sustainability of the product and the room for improvement to get a better product with less impact.

Figure 10.4 depicts the graphical representation of the SLCA matrices for two different products, using a color-coded system instead of the 0–4 range of numeric values for the scoring. Figure 10.4a shows a completed matrix for an environmentally undesirable product, considering the low scoring values in most of the matrix cells. The overall score is 45 out of 100. Figure 10.4b shows a matrix for a relatively more ERP, earning a score of 73 out of 100. Note that even a product with a high score retains areas where improvement is possible; perhaps because the original design was forced to use a toxic material, or an alternative end-of-life approach was too costly. Additionally, Figure 10.5 shows an alternative graphical representation – this "rose" or "target" plot shows the product environmental concerns and life cycle categories in radial form. The target plot is a circle divided into five shaded sectors that correspond with the five stages in the product life cycle: premanufacture, manufacture, packaging, product use, and recycling. The scoring for each environmental impact and each stage of the product life is represented as a black dot over five concentric circles. These circles represent the scoring from 0 (the outer circle) to 4 (the inner circle). Points are closer to the center, or "target" for life cycle categories with less relative environmental impact. A target plot is a graphical alternate means for displaying the matrix results, with high scores near the center and low scores near the edge (as with a rifle-range target). In this plot, it is easier to identify groupings and extreme values.

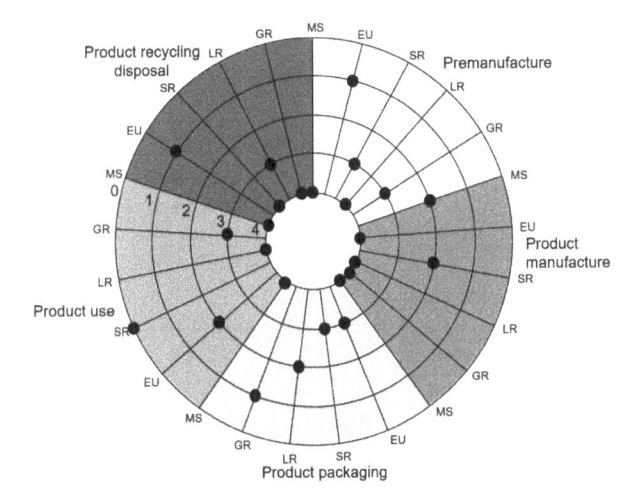

Figure 10.5 Target plot for the graphical representation of the SLCA matrix scoring. It is the same product as Figure 10.4b.

10.3 Expanded SLCA

The SLCA matrix considers the environmental impacts of the product across its life cycle, but other categories can be added to the matrix (Figure 10.3) to expand the life cycle stages and the possible impacts of the product and make the calculation of its environmental sustainability more precise. Health and safety components may be included in the matrix and calculations. We can identify specific health/safety issues in the five life cycle stages described before (premanufacture, manufacture, product delivery, product use, and disposal/recycling), and a new life cycle stage is added now to the matrix: the field maintenance or service.

The health/safety issues include physical hazards, chemical hazards, noise hazards, shock hazards, and ergonomic hazards (Figure 10.6). The health and safety issues can be organized into a matrix (Figure 10.7) where scores are again assigned to cells using a scale between 0 and 4. This is a 6×5 matrix, the maximum scoring is 120, and the final scoring is normalized by dividing it by 120 and expressing as percentage to obtain a final score ranging from 0 to 100.

Figure 10.6 Product health and safety assessment.

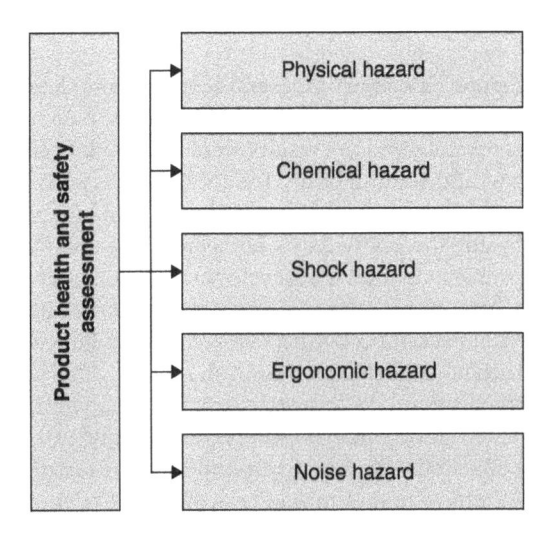

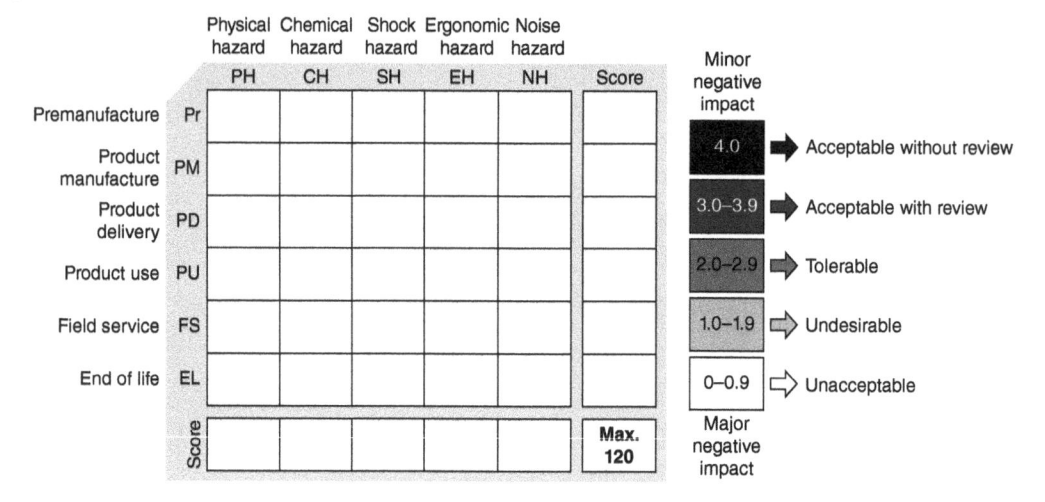

Figure 10.7 SLCA health and safety matrix.

Figure 10.8 Health and safety matrix for a transportation product.

Figure 10.8 shows an example of the completed matrix for a transportation product. The reason for the fractional scores is that the four major subsystems in the product were evaluated separately, and the results were averaged. In this case, the values in each cell are summed on a row and column basis. The total score is 94.8 out of 120, which corresponds to a final score of 79% after normalization. It is clear that chemical hazard is the most problematic of all of the health and safety issues (as seen by the column scores), and product manufacture is the most problematic of the life cycle stages (as seen by the row scores).

Next, the health and safety matrix can be integrated in the product SLCA matrix. Thus, the values assigned to each cell are summed up on a row basis (Figure 10.9), and these values are then normalized to result in an average value between 0 and 4. The values are then inserted into the product SLCA matrix (originally shown in Figure 10.3) as an additional column (Safety). The resulting matrix is depicted in Figure 10.10. Again, all the cells in the matrix have to be populated with the corresponding value for all the environmental, health and safety issues, and they may be summed and normalized to determine a percentage score.

10.3 Expanded SLCA | 199

Safety matrix

		Physical hazard PH	Chemical hazard CH	Shock hazard SH	Ergonomic hazard EH	Noise hazard NH	Average
Premanufacture	Pr	2	3	1	4	1	2.2
Product manufacture	PM	2	2	3	2	3	2.4
Product delivery	PD	3	3	2	2	4	2.8
Product use	PU	3	4	2	3	1	2.6
Field service	FS	1	1	4	0	2	1.6
End of life	EL	2	0	1	3	3	1.8

Procduct matrix

		Material selection MS	Energy use EU	Solid residue SR	Liquid residue LR	Gaseous residue GR	Safety	Score
Premanufacture	Pr						2.2	
Product manufacture	PM						2.4	
Product delivery	PD						2.8	
Product use	PU						2.6	
Field service	FS						1.6	
End of life	EL						1.8	
Score							13.4	

Figure 10.9 Integration of the safety matrix in the into the product SLCA matrix.

		MS	EU	SR	LR	GR	S	Score
Premanufacture	Pr							
Product manufacture	PM							
Product delivery	PD							
Product use	PU							
Field service	FS							
End of life	EL							
Score								Max. 144

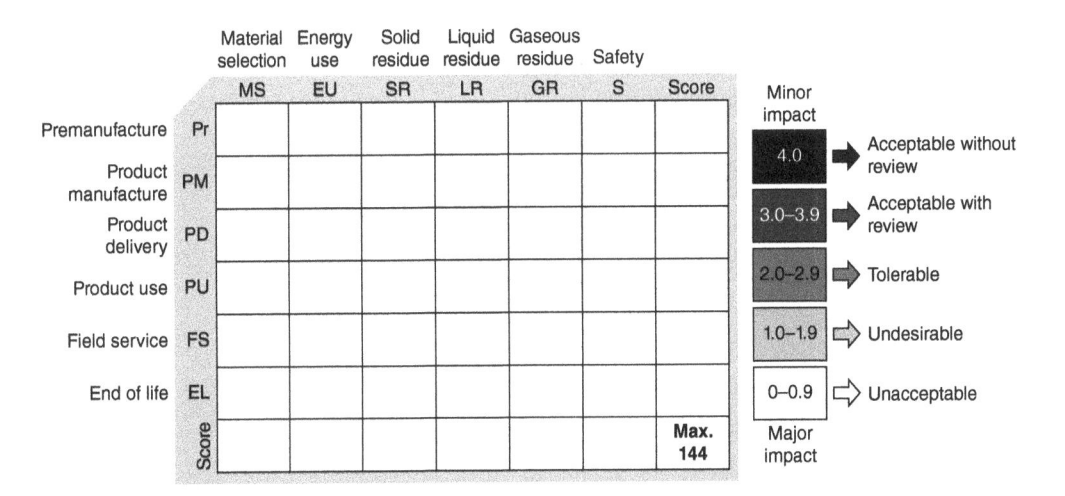

Minor impact
- 4.0 ⇒ Acceptable without review
- 3.0–3.9 ⇒ Acceptable with review
- 2.0–2.9 ⇒ Tolerable
- 1.0–1.9 ⇒ Undesirable
- 0–0.9 ⇒ Unacceptable

Major impact

Figure 10.10 Expanded product matrix (6 × 6). The total score is now the result of summing all the matrix element values and dividing by 144 to get the percentage score.

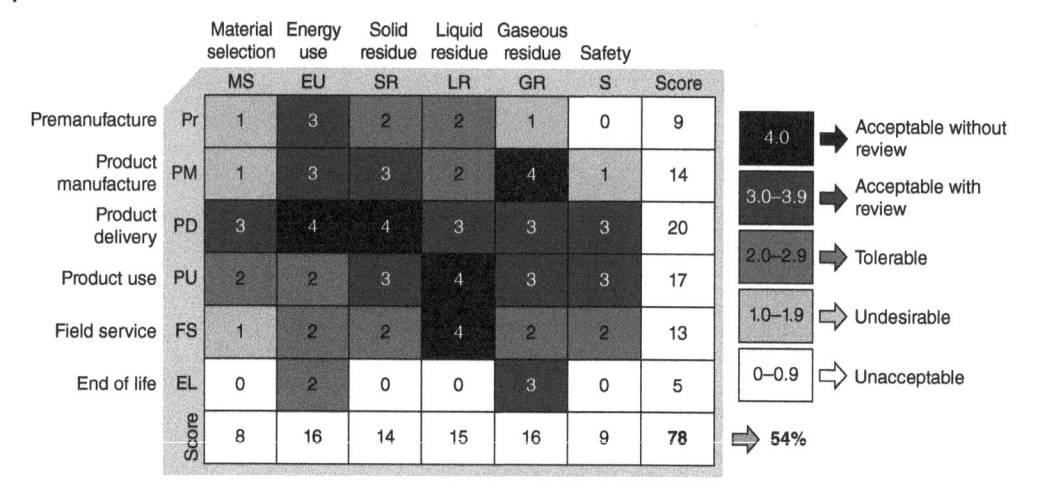

		Material selection	Energy use	Solid residue	Liquid residue	Gaseous residue	Safety	Score
		MS	EU	SR	LR	GR	S	Score
Premanufacture	Pr	1	3	2	2	1	0	9
Product manufacture	PM	1	3	3	2	4	1	14
Product delivery	PD	3	4	4	3	3	3	20
Product use	PU	2	2	3	4	3	3	17
Field service	FS	1	2	2	4	2	2	13
End of life	EL	0	2	0	0	3	0	5
Score		8	16	14	15	16	9	78 ⇨ 54%

- 4.0 ⇨ Acceptable without review
- 3.0–3.9 ⇨ Acceptable with review
- 2.0–2.9 ⇨ Tolerable
- 1.0–1.9 ⇨ Undesirable
- 0–0.9 ⇨ Unacceptable

Figure 10.11 Completed example of the expanded SLCA matrix for a specific product.

An example of an expanded SLCA matrix for a specific product is shown in Figure 10.11. The total score after normalization is 54%, indicating clear opportunities for improvement in the sustainability of this product. As shown in Figure 10.11, the premanufacture and end of life stages need the most improvement, and material selection and safety are of more concern than other environmental, health, and safety issues.

10.4 Simple Example of SLCA

An example of the application of SLCA is presented in Table 10.2 and Figures 10.12–10.14. This example depicts the SLCA of two typical automobiles manufactured in the 1950s and the 2000s (Graedel and Allenby 2010). Table 10.2 depicts several characteristics assigned to both automobile models with regard to the materials, fuel efficiency, use of catalytic devices for exhaust gases, and the use of ozone-depleting gases for air conditioning. The values in Table 10.2 suggest important changes in the manufacturing of automobiles from the 1950s to the 2000s toward more sustainable products, but the SLCA can give a better picture of the

Table 10.2 Materials and fuel efficiency for automobile models (Graedel and Allenby 2010).

	1950s	2000s
Materials		
Plastic	0	101
Aluminum	0	68
Steels	1290	793
Fluids	96	81
Other	515	391
Total (kg)	1901	1434
Fuel efficiency (mi/gal)	15	27
Exhaust catalyst	No	Yes
Air conditioning	CCF-12	HFC-134a

Figure 10.12 Typical 1950s and 2000s automobile.

1950s automobile

	MS	EU	SR	LR	GR
Pr	2	2	3	3	2
PM	0	1	2	2	1
PP	3	2	3	4	2
PU	1	0	1	1	0
PR/D	3	2	2	3	1

Score: 46

2000s automobile

	MS	EU	SR	LR	GR
Pr	3	3	3	3	3
PM	3	2	3	3	3
PP	3	3	3	4	3
PU	1	2	2	3	2
PR/D	3	2	3	3	2

Score: 68

Figure 10.13 SLCA matrix for two automobiles (scoring as per Table 10.1).

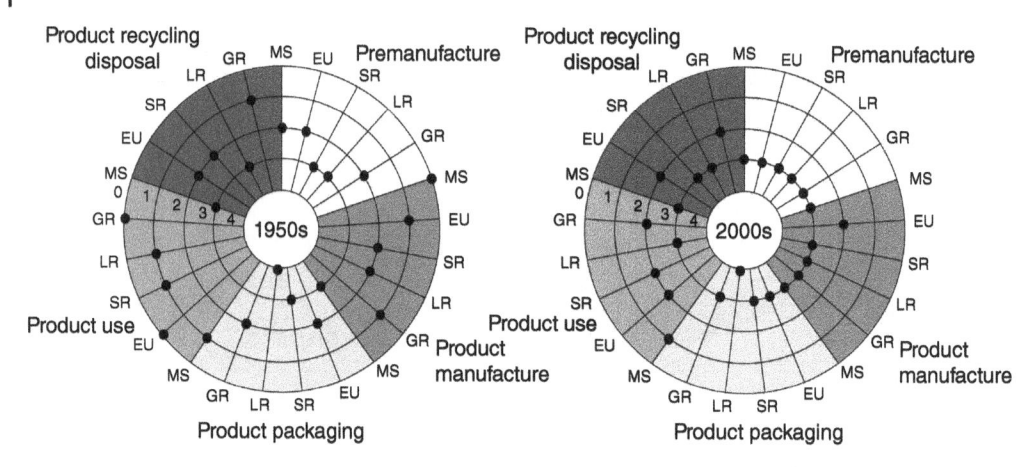

Figure 10.14 Target plot for the SLCA for two automobiles. Source: Graedel et al. (1995).

real changes. Figure 10.13 shows the relative scoring for the two models in the SCLA matrix, whereas Figure 10.14 shows the scoring depicted on a target plot. As shown in these figures, there is a significant difference in the relative environmental impacts associated with these two car models. The 1950s automobile earned a global scoring of 46. As an example, some individual scoring in the matrix is notable. Matrix element 2,1 is 0 because of the use of toxics in manufacture. Element 1,3 is 3 resulting from the generation of waste rock in metal mining. Element 5,3 is 2 because the steel in the automobile is highly recyclable, even if the design for recycling was not performed. Element 3,4 is 4 because product delivery involves no liquid waste. Element 4,4 is 0 because no exhaust emissions equipment was present.

The matrix for the 2000s automobile shows a total score of 68 – much better than the 1950s car but still offering much opportunity for improvement. The individual impacts of the two cars can be seen in Figure 10.14. The target plot for 1950 shows that one could improve the score by addressing almost any stage of the life cycle. The 2000s target plot clearly indicates that the use life cycle stage and, to a lesser extent, the recycling life cycle stage are where improvements may be realized.

10.5 Applications of SLCA

SLCAs are not only useful for products, but they may also be very useful to assess processes and facilities. As an example, consider several common life stages associated with a generic process – resource provisioning, process implementation, primary process operation, complementary process operation, and process termination (Figure 10.15). Once again, similar environmental concerns may be applied as in our product example – material selection, energy use, solid residues, liquid residues, and gaseous residues. Figure 10.16 shows a matrix constructed with these respective life cycle stages and environmental concerns.

Table 10.3 and Figures 10.17 and 10.18 show an example of an SLCA comparison – once again for 1950s and 2000s automobiles – but this time focused on the manufacturing processes. The improvement in the manufacturing process of automobiles is evident in Figure 10.18. The scoring points are all over the target plot for the manufacturing process of the 1950s, whereas the points tend to concentrate in the center for the 2000s manufacturing process, indicating a higher score and lower environmental impacts.

SLCAs can be applied to a variety of facets or positions on product/service value chains. As another example, consider the life cycle stages of an industrial facility (Figure 10.19). Some important stages include site development/implementation, facility operations, products,

Figure 10.15 Life stages of an industrial process.

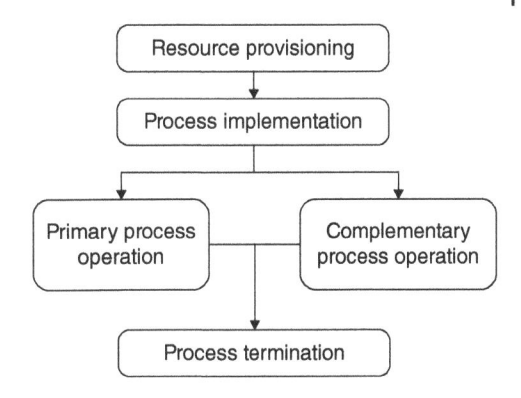

```
Resource provisioning
        ↓
Process implementation
        ↓
Primary process ─── Complementary
operation           process operation
        ↓
Process termination
```

		Material selection	Energy use	Solid residue	Liquid residue	Gaseous residue
		MS	EU	SR	LR	GR
Resource provisioning	RP					
Process implementation	PI					
Primary process operation	PPO					
Complementary process operation	CPO					
Recycle/disposal	R/D					

Figure 10.16 Environmentally responsible process matrix.

Table 10.3 Characteristics of generic automobile manufacturing process.

	1950s	2000s
Material source	Virgin	Some recycled
Energy source	Very large	Large
Metal trimming	Common	Common
Metal cleaning	CFCs	Detergents
Metal plating	Ubiquitous	Common
Consumables recycling	No	Yes
VOC emissions	Large	Modest

processes, and site closure. Again, these stages may be combined with related environmental concerns – biodiversity, energy use, solid residues, liquid residues, and gaseous residues for SLCA. A matrix for the SLCA of the facility is constructed, similar to those for a product or a process SLCA. The matrix is completed with the individual scores for each life cycle and environmental concern to obtain the absolute assessment or a comparative assessment as shown in previous examples. In the case of SLCA of industrial facilities, the information from the SLCA for the products and processes of such facility can be integrated to obtain the assessment of the facility, as shown in Figure 10.20.

204 | *10 Streamlined Life Cycle Assessment*

(a)

		Material selection MS	Energy use EU	Solid residue SR	Liquid residue LR	Gaseous residue GR
Resource provisioning	RP	4	0	1	4	1
Process implementation	PI	0	0	2	2	2
Primary process operation	PPO	0	1	1	1	1
Complementary process operation	CPO	0	2	2	0	1
Recycle/disposal	R/D	3	3	1	2	4

(b)

		Material selection MS	Energy use EU	Solid residue SR	Liquid residue LR	Gaseous residue GR
Resource provisioning	RP	4	2	2	4	3
Process implementation	PI	3	2	2	4	3
Primary process operation	PPO	2	2	1	3	3
Complementary process operation	CPO	3	3	3	2	3
Recycle/disposal	R/D	4	1	3	4	4

Figure 10.17 SLCA matrix for the manufacturing process of (a) 1950s and (b) 2000s generic automobiles.

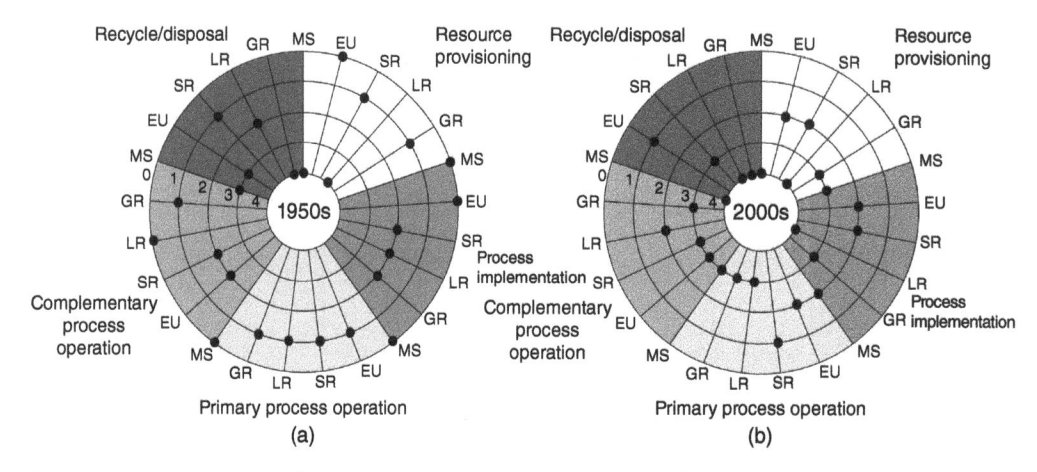

Figure 10.18 Target plot for the manufacturing process of (a) 1950s and (b) 2000s generic automobiles. Source: Adapted from Graedel and Allenby (2010).

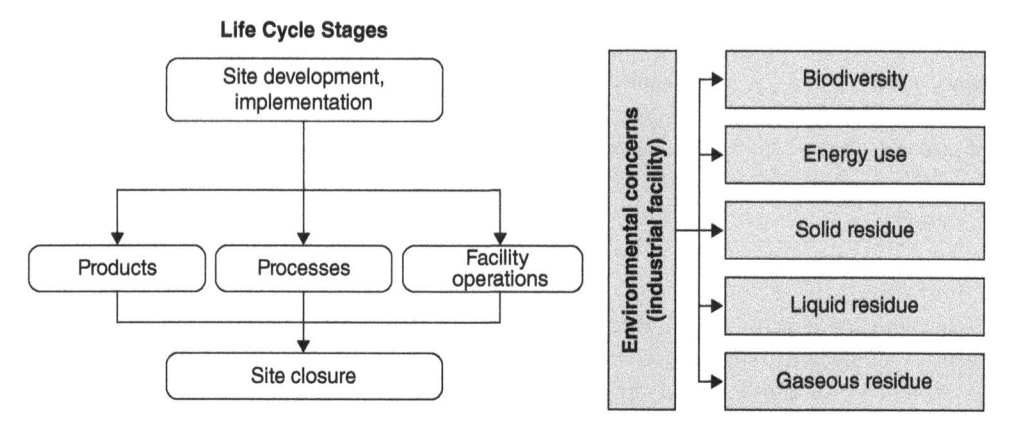

Figure 10.19 Life cycle stages and environmental concerns for an industrial facility.

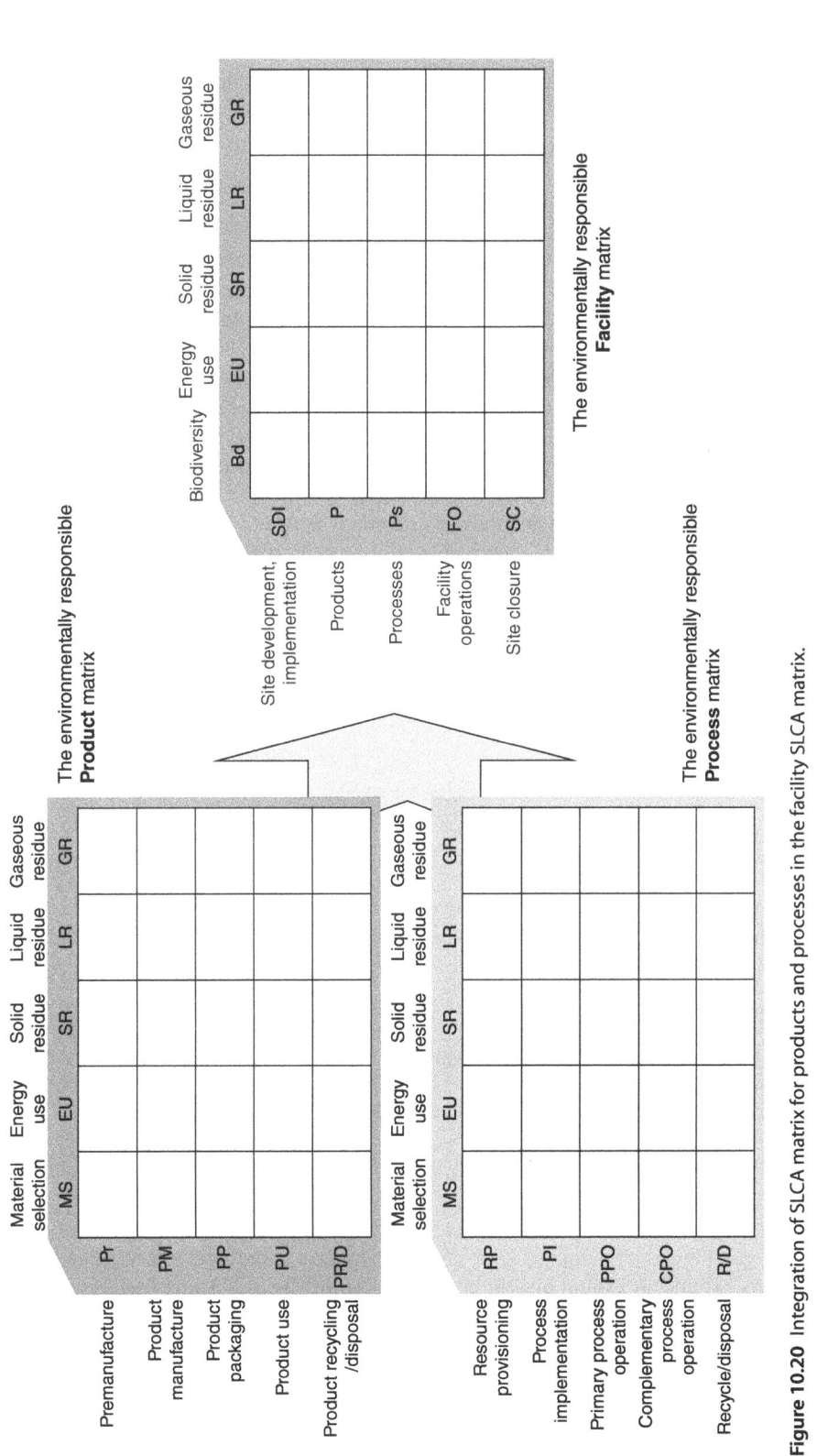

Figure 10.20 Integration of SLCA matrix for products and processes in the facility SLCA matrix.

SLCA can also be easily adapted to assess social aspects of products or designs as well. Chapters 20–22 provide example application of SLCA in different ways for alternate engineering designs.

10.6 Summary

SLCA can be done efficiently and can reveal most or all the concerns identified by traditional LCA. Matrix plots or target plots facilitate identification of opportunities for improvement. The concept may be extended in a straightforward manner to address the health and safety considerations of processes and facilities.

As demonstrated, SLCAs are relatively straightforward and efficient, and they can be used to assess or reveal most, or all of the concerns raised during the course of a more comprehensive "traditional" LCA analysis. Graphical representations of these matrices, whether in the form of heat-mapped matrices or as target plots, can identify opportunities for improvement with respect to processes, activities, or products. The concepts discussed with respect to SLCAs can be extended to a range of processes, facilities, or other aspects of the product/service value chain. However, the results of SLCA can vary based on individual's subjective judgment of potential and importance of environmental concerns at each life cycle stage.

10.7 Questions

10.1 What is SLCA?

10.2 Discuss factors that may lead you to choose an SLCA over a traditional LCA.

10.3 What are the key life cycle stages in SLCA?

10.4 Assuming a product category such as environmental impacts of manufacturing and a scale from 0 to 4. List for each score from 0 to 4 characteristics that would be assigned to each numerical value.

10.5 Select a common household product. Using the example shown in Figure 10.3, prepare a SLCA matrix for the product. Calculate the corresponding score for the product.

10.6 Prepare a rose diagram of the SLCA matrix results.

10.7 Discuss specific health and safety metrics that may be included in an SLCA matrix.

10.8 Consider an SLCA that identified a clear advantage of product A over product B. Assuming product A is more expensive to produce, prepare a memo that describes the SLCA and why product A should be selected. Discuss how the SLCA was performed and why aspects other than cost should be considered.

References

Graedel, T.E. and Allenby, B.R. (2010). *Industrial Ecology and Sustainable Engineering*. Upper Saddle River, NJ: Prentice Hall.

Graedel, T.E., Allenby, B.R., and Comrie, V.R. (1995). Matrix approaches to abridged life cycle assessment. *Environmental Science & Technology* 29 (3): 134–139. https://doi.org/10.1021/es00003a751.

Weitz, K., Sharma, A., Vigon, B. et al. (1999). Streamlined life-cycle assessment: a final report from the SETAC North America streamlined LCA workgroup. Society of Environmental Toxicology and Chemistry (SETAC) and SETAC Foundation for Environmental Education.

11

Economic Input–Output Life Cycle Assessment

11.1 Introduction

As presented in Chapter 9, a full-scale life cycle assessment (LCA) is the best option for analyzing a project or a product because it can give a complete picture of the environmental impacts associated with a product, an activity, or an organization. However, the full-scale LCA can get very complex and prohibitive. Even for the environmental assessment of a very simple product, the full LCA may incorporate an overwhelming number of inputs that makes the process extremely complicated. The streamlined life cycle assessment (SLCA) concept was presented in Chapter 10 as an efficient alternative to full-scale LCAs, which can be a simplified approach toward achieving the same general goals of LCA.

An alternative to LCA and SLCA, an economic input–output (EIO) model with LCA, often known as EIO-LCA, can be used. An EIO model is used to represent the monetary transactions between industry sectors in a mathematical form. It is a means to assess what outputs (goods or services) of an industry are used as inputs or consumed by other industries. This EIO-LCA model was designed to assess the environmental impacts of a product or a project while overcoming the complexity of the full-scale LCA as well as the issues associated with boundary definition and circularity.

11.2 EIO Model

EIO models are useful to analyze and predict the behavior of the economy in a country (Rose 1995). Many nations create EIO models of their economy at various degrees of specificity or frequency. For example, the U.S. EIO models are created every five years and represent the transactions among and between approximately 400 industry sectors. National budgets are based on National Financial Accounts (NFAs), which are obtained from EIO models. Accuracy in NFAs is critical; even small errors can result in budget estimates that are erroneous by wide margins. These errors, if they occur, can have notable consequences because these budgets are used to allocate funding and other similar expenditures.

When determining NFAs and gross domestic products (GDPs), it is critical to have an understanding of the inputs and outputs related to various sectors of the economy. A method to do this, the Input–Output Analysis, was invented by Russian-American economist and Nobel laureate Dr. Wassily Leontief. Born in Germany in 1905, Dr. Leontief developed this method to analyze the intricacies of NFAs, and this information can be used to make predictions about the economic effects of a change in actions in different economic sectors (Leontief 1986). Specifically, his research focused on how an economic change in a given sector could affect

---

Sustainable Engineering: Drivers, Metrics, Tools, and Applications, First Edition.
Krishna R. Reddy, Claudio Cameselle, and Jeffrey A. Adams.
© 2019 John Wiley & Sons, Inc. Published 2019 by John Wiley & Sons, Inc.

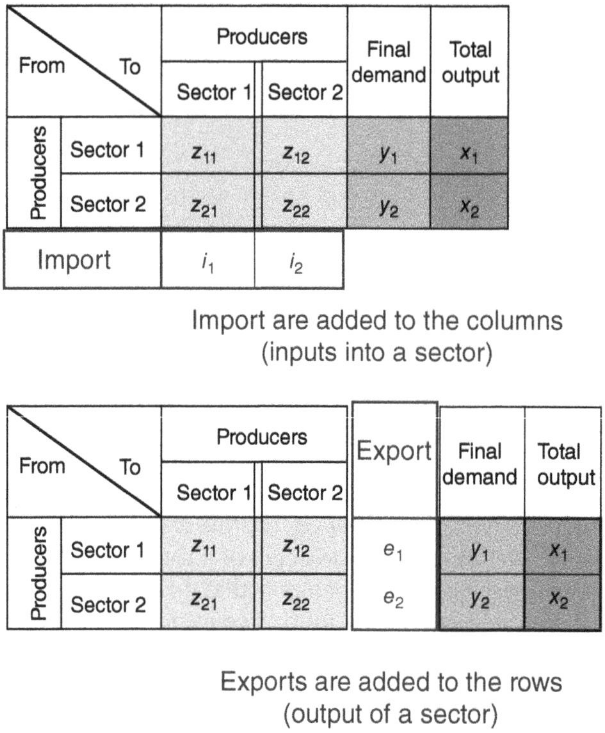

11 Economic Input–Output Life Cycle Assessment

From \ To	Producers Sector 1	Producers Sector 2	Final demand	Total output
Producers Sector 1	z_{11}	z_{12}	y_1	x_1
Producers Sector 2	z_{21}	z_{22}	y_2	x_2
Import	i_1	i_2		

Import are added to the columns
(inputs into a sector)

From \ To	Producers Sector 1	Producers Sector 2	Export	Final demand	Total output
Producers Sector 1	z_{11}	z_{12}	e_1	y_1	x_1
Producers Sector 2	z_{21}	z_{22}	e_2	y_2	x_2

Exports are added to the rows
(output of a sector)

Figure 11.1 Input–output transaction table for a two sector economy showing the imports and exports for each sector.

other sectors. The importance of this cannot be understated, as the heart of the EIO model is the "input–output transaction table." An assumption in this model is that each sector produces a homogeneous output from a product composition and price standpoint as explained below.

Figure 11.1 shows an example of a transaction table for two sectors (Graedel and Allenby 2010). As mentioned above, for this description, it is assumed that each sector produces a homogeneous product (i.e. a homogeneous composition and price). Sectors appear in both rows and columns. In Figure 11.1, z_{ij} define intra- and inter-industry flows; y_i indicates flows entering use, or final demand; and x_i indicates total production, or output, of a sector. Columns represent the input, and rows represent the output of a sector, respectively. Imports are entered to the columns (represented as inputs) to a sector, while exports are entered to rows as outputs of a sector.

In constructing a basic EIO model for a two-sector economy, the total production for Sector 1 is presented in Eq. (11.1):

$$x_i = a_{ij}x_i + y_i \tag{11.1}$$

where

a_{ij} = flow from sector i to sector j,
x_i = production of sector i,
y_i = final demand of sector i.

The same equation presented in matrix notation (Eq. (11.2)) can be used to define the total production of the two sectors:

$$\mathbf{x} = \mathbf{A}\mathbf{x} + \mathbf{y} \tag{11.2}$$

where

A is the 2 × 2 matrix populated by a_{ij} values,
x is the vector populated by x_i values,
y is the vector populated by y_i values.

Constant technical coefficients z_{ij} are equal to $a_{ij} \times x_j$. Mass balance markets x_i are equal to the addition across rows of z_{ij} and y_i: $x_1 = z_{11} + z_{12} + y_1$; $x_2 = z_{21} + z_{22} + y_2$. This is equivalently represented as **x** = **Ax** + **y** (Eq. (11.2)), where **x** = the production vector; **y** = the final demand vector, and **A** = the technical coefficient matrix. Now, to solve, **x** = **Ax** + **y**, first, **A** is subtracted from the Identity Matrix **I**, resulting in (**I** − **A**)**x** = **y**. Premultiplying with (**I** − **A**)$^{-1}$, results in **x** = (**I** − **A**)$^{-1}$ × **y**, where (**I** − **A**)$^{-1}$ is the Leontief inverse.

EIO models are typically presented in a large-scale matrix form. Each economy or industrial sector is represented in the matrix as a row and a column. The intersection of the row and the column for a specific sector is the output from this sector that is used as an input in the same sector. This format provides two helpful characteristics. First, it can be determined if the output of a given industry sector is used as an input for the same sector (mathematically, this is represented as a nonzero value along the diagonal of a matrix). Second, using linear algebra computations, the total effects of changes (direct and indirect) can be calculated for a specific economy.

EIO models have several important uses and applications when studying an economy. First, they can be used to study the changes in supply, demand, and other relevant changes of an economy. Additionally, they can identify when economic shifts have occurred or when outputs from specific industry sectors have increased or decreased. This is quite a common scenario over a given time period, such as the instance when output increases from service-based sectors within advanced economies, coupled with a noticeable shift from low-tech manufacturing outputs. Further, EIO models can be used to aid decision-makers to determine cause-and-effect relationships within an economy, especially when connections or linkages have not been previously understood or observed between seemingly unrelated sectors. Pronounced or drastic changes in economies can also be mapped in such a manner.

EIO modeling does have some limitations. First, as an economic-based model, it is not particularly useful in tracking physical transactions, such as materials usage or quality of recycled materials. The solutions provided by the models can be limited in nature – an example of an EIO model outcome would be to determine the impact of producing 1 kg of metal instead of 1$ worth of the same metal. Further, there are questions on how materials and energy flows can be incorporated into related models.

11.3 EIO-LCA

The limitations of EIO models may be overcome through the coupling with life cycle assessments. For instance, an additional row and column can be added to incorporate the output and input of the "environment" sector. In this way, the utilization of natural resources as well as the effects of emissions and waste generation can be added to the EIO model, as well as the influences on and resulting from interactions with other sectors. Such an approach also eliminates two major difficulties of conventional LCA studies – boundary definition and circularity effects. First, the boundaries of an LCA study are, by definition, very broad and inclusive because all inputs and outputs among all industry sectors are included. Second, the circularity effects are considered in the assessment because self-consumption in the sectors is included in the matrix.

Formally speaking, the combination of an EIO and a life cycle assessment (EIO-LCA) is a mathematical method to estimate the amount of energy and materials consumed, and the emissions generated from all economic activities (Matthews and Small 2000; Nakamura and Kondo 2009). The method uses industry transactions, namely, purchases of materials among industries, and the direct environmental emissions from each industry, to estimate the total amount of emissions throughout the life cycle of a product, service, or sector. This mathematical procedure is able to determine the effect of changing the output of a single sector in the whole system just using economic and environmental data. The method can be applied to any economy or economic system if the information of the transactions between sectors is available. Each application of the method defines an EIO-LCA model.

With sufficient computing power readily available such that the requisite large-scale matrix manipulations could be performed in a short span of time, researchers at the Green Design Institute of Carnegie Mellon University developed a computing application of the Leontief's method in the mid-1990s. The researchers subsequently developed an EIO-LCA computational tool capable of quickly evaluating a commodity, service, or representative supply chain. This tool is available online at www.eiolca.net. The online tool provides information about the impacts of products, materials, services, or industries with respect to resource consumption and emissions throughout a supply chain. As an example, this tool permits to evaluate effects of producing an automobile, including not only the impacts in the final assembly facility but also those associated with natural resources extraction (e.g. mining and materials extraction), fabrication of parts, etc., that constitute an entire supply chain.

EIO-LCA models may be applied to state or national economies at various levels of detail. Each national EIO model is developed using a base of publicly available resource use and emissions data. Since the mid-1990s, the method has been applied to develop economic models of the United States (on a federal basis and individual state basis), Canada, and several member countries of the European Union. Table 11.1 provides a listing of several available EIO-LCA models.

Table 11.1 Available EIO-LCA models at eiolca.net.

United States models	2002 Benchmark producer price
	2002 Benchmark purchaser price
	1997 Industry Benchmark producer price
	1997 Industry Benchmark purchaser price
	1992 Industry Benchmark
	1992 Industry Mini
	1997 Pennsylvania (PA) state
	1997 West Virginia (WV) state
	1997 PA + WV combined state
International models	2002 Canada Industry Account
	1995 Germany Industry Account
	2002 Spain Industry Account
Customized models	Build-a-product option based on the US 1997 model
	Build-a-sector option based on the US 1997 model
	MATLAB models (may require license)

11.4 EIO-LCA Model Results

11.4.1 Interpretation of Results

When interpreting the results of EIO-LCA models, it is recommended to focus first on interpreting the economic results and then focus on a single economic metric of choice. The environmental information about resource consumption and emission metrics must be interpreted in a similar way. Additionally, the environmental metrics can be combined and interpreted in conjunction with the economic results to obtain more detailed information about the direct and indirect impact of each product or sector; therefore, it is important to understand both the sets of results.

When the interpretations are made, several assumptions made in these models should be kept in mind. First, these models are linear. The model results represent the impacts through the production phase of the sector, but the use phase and the end-of-life phases are not directly included in the results. However, additional analyses with the EIO-LCA method could be performed to account for these life cycle stages. Further, many assumptions are incorporated in the creation of the impact vectors, i.e. the values for material consumption, emissions, and environmental effects. The input–output (IO) models used in the development of EIO-LCA models are defined for the economy of a single nation. However, imports and exports represent a major contribution to any economy's transactions. In the IO models, the imports are implicitly assumed to have the same production characteristics as similar products made in the country of study.

11.4.2 Uncertainty

When discussing the use and applicability of these EIO-LCA models, it is important to discuss sources of uncertainty. First, the models are typically constructed using available data, which in the best case is one or two years old and that data is going to be used to develop a model for the following year. The economic data may vary widely in relatively short periods of time. In a similar fashion, environmental data can vary significantly over time due to operational changes in industrial processes and process efficiency, the regulatory framework with respect to emissions or waste generation, or production levels. Further, there is uncertainty not just with respect to changes in data over time; there is also uncertainty with respect to the original or baseline data collected for the model. The data used in EIO-LCA models is typically obtained from surveys and forms elaborated by governments for statistical purposes and submitted to the industries. The uncertainty in sampling, the degree of response from participants, biases inherent in these responses, missing or incomplete data, estimations or guessing incorporated into the specific responses, etc., can serve as a source of uncertainty in the modeling. Further, the baseline data may also be incomplete or have an unknown factor affecting some degree of the data. Often, the unknown factor or factors may later become known and incorporated, which can skew results due to the deviation with respect to the baseline data. A good example is toxicological or toxic release data.

Not only may there be errors, unknowns, or omissions in the baseline data, it can often be difficult to aggregate and/or characterize the data. Often, important collected data may be difficult to be categorized considering the EIO sectors of the IO matrix. For example, electricity consumption in commercial buildings may be aggregated based on industry sector (e.g. accounting, medical, engineering, and legal) or can be organized by the type of activity, such as office buildings, retail, etc. Therefore, the aggregation and allocation of data have been done considering the most appropriate industry sector and the objective of the study. Further, aggregation of sectors may introduce uncertainty. One of the most interesting characteristics of the EIO-LCA method is their capacity to represent or estimate the impacts from a change in the demand of a specific industrial sector. An industry sector includes a collection of various industry types and this is defined in the selected model. Thus, the IO model and the aggregation of industrial activities in sectors may lead to uncertainty due to how well a specific industry is modeled based on the aggregation into sectors.

11.4.3 Other Issues and Considerations

Several other issues are of importance when an EIO-LCA model is being constructed. For instance, each EIO-LCA model uses the currency of the country of origin, and the monetary values are based on when the model was constructed. Rates on inflation within a country as well as currency exchange rates between different countries over time may have a significant effect on the accuracy or applicability of the model. Another example is the correct use of producer versus purchaser prices. For many goods, producer prices can be significantly less than purchaser prices. As an example, producer prices associated with leather goods are typically 35% of final purchaser prices in the United States. Further, EIO-LCA models are incomplete in analyzing environmental effects. There is a very limited number of usable metrics to be included into EIO-LCA models. Further, there is no data in a format to be incorporated in the models for environmental impacts such as habitat destruction, nonhazardous waste generation, and/or nontoxic water pollution. Finally, the EIO-LCA method estimates the resource consumption and emissions or associated with the life cycle of an industry sector. So the EIO model and results are representative of the inventory stage of the LCA, but the actual environmental or human health impacts associated to the emissions and consumption are not included.

11.5 Example of EIO-LCA Model

The input–output transactions table represents the monetary flow between sectors. These transactions tables are constructed from "make" and "use" data, which represents the sales and purchases between two particular sectors. Transactions include the purchases between two sectors and the total demand that is the sum of the purchases of one sector with all the others. Further, the sum of value-added (noninterindustry purchases) and final demand is GDP. From the input–output transactions table, a technical requirements matrix is created by dividing each column by total sector input matrix. Entries represent direct inter-industry purchases per dollar of output. An example of a transactions table is shown in Table 11.2.

Table 11.3 presents an example of a two-sector EIO model. As shown in the table, reading across row 1, Sector 1 sells $150 of output to Sector 1 itself, $500 of output to Sector 2, and $350 of output to consumers. In the first column, Sector 1 purchases $150 of output from Sector 1 itself, $200 of output from Sector 2, and adds $650 of value to produce its output. Similar readings may be made with respect to Sector 2. The transaction flows are presented in the Transactions Matrix in Table 11.4.

A requirements matrix is presented in Table 11.5. The matrix entries are calculated as $A_{ij} = X_{ij}/X_j$, where X equals output, i equals intra-sector or inter-sector output for Sector i, and j equals total sector output for Sector j. As shown in the flowchart in Figure 11.2, for production of Good 1 in our two-sector model, to produce $1 of output from Sector 1 requires $0.15 of goods from Sector 1, and $0.20 of goods from Sector 2. For the production of Good 2,

Table 11.2 Example of transactions table.

Output from sectors	Input to sectors				Intermediate output O	Final demand F	Total output X
	1	2	3	n			
1	x_{11}	x_{12}	x_{13}	x_{1n}	O_1	F_1	X_1
2	x_{21}	x_{22}	x_{23}	x_{2n}	O_2	F_2	X_2
3	x_{31}	x_{32}	x_{33}	x_{3n}	O_3	F_3	X_3
n	x_{n1}	x_{n2}	x_{n3}	x_{nn}	O_n	F_n	X_n
Intermediate input I	I_1	I_2	I_3	I_n			
Value added V	V_1	V_2	V_3	V_n		GDP	
Total input X	X_1	X_2	X_3	X_n			

Table 11.3 EIO Example for two sector.

	1	2	Final demand
1	150	500	350
2	200	100	1700
Value added	650	1400	2050

Table 11.4 Transactions Matrix for a two-sector EIO example.

	Sector 1	Sector 2	Final demand	Total output
Sector 1	150	500	350	1000
Sector 2	200	100	1700	2000
Value added	650	1400	2050	
Total input	1000	2000		

Table 11.5 Requirements matrix for a two-sector EIO example.

	Sector 1	Sector 2
Sector 1	150/1000 = 0.15	500/2000 = 0.25
Sector 2	200/1000 = 0.2	100/2000 = 0.05

as shown in Figure 11.2, to produce $1, $0.05 of goods are required from Sector 2, and $0.25 are required from Sector 1. The Leontief Inverse matrix is then calculated using $[I - A]^{-1}$.

Final demand is next added. For example, to determine the effects of an additional $100 of demand from Sector 1, using the equation $X = [I - A]^{-1} y$, the resulting total outputs of Sector 1 and Sector 2 are $125.40 and $26.40, respectively, or $151.80 total. The direct intermediate inputs of $15 for Sector 1 and $20 for Sector 2 for output of $100, or $135 total.

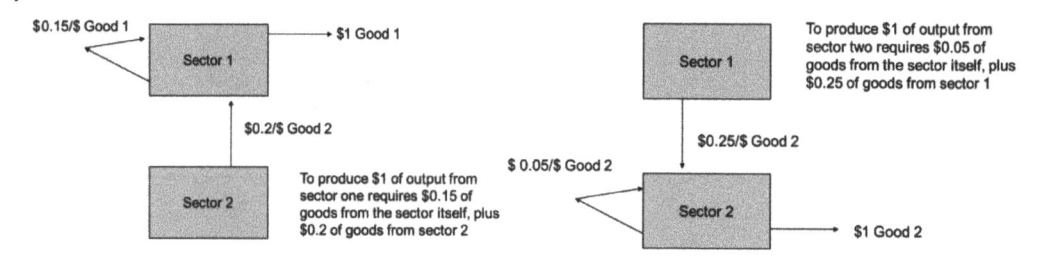

Figure 11.2 Production of two goods (Good 1 and Good 2) in a two-sector EIO model.

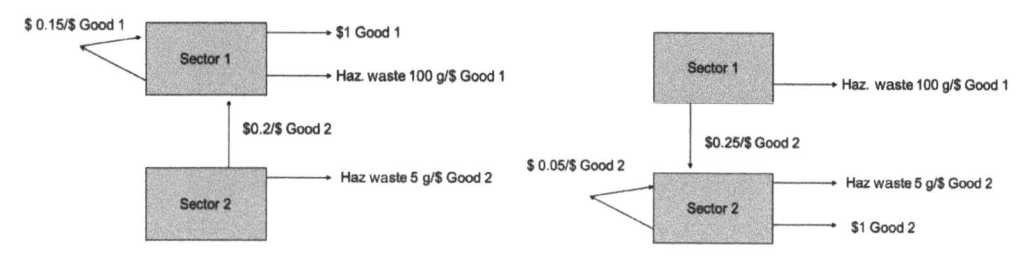

Figure 11.3 Production of Good 1 and Good 2 in a two-sector EIO model, including the costs of the generated waste.

Further, environmental effects may be added such as the sector-level environmental impact matrices R may be constructed, which represents the effect per monetary output for a particular sector. As an example, assume 100 g of hazardous waste generation per dollar in Sector 1, and 5 g of hazardous waste generation per dollar in Sector 2. Introducing these values into our two-sector model, the waste generation is accounted for as outflows from each sector as shown in our revised flowcharts in Figure 11.3.

Finally, based on the matrix calculations and calculating waste production B as the product of matrices B and X, 12 540 g of hazardous waste is generated by Sector 1 and 132 g of hazardous waste are generated by Sector 2, for a total hazardous waste generation of 12 672 g.

11.6 Conventional LCA versus EIO-LCA

The following illustrates characteristics, strengths, and drawbacks for both conventional LCA and EIO-LCA:

Conventional LCA

- Provides detailed and specific analysis of processes
- Provides product comparisons
- Allows for process improvements or weak point analyses
- Suitable for future product development assessments
- Subjective establishment of the system boundary
- Tend to be time-sensitive and costly
- Difficult design of new processes
- Uses proprietary data

- Modeling cannot be replicated if confidential data is used
- There is uncertainty in the data.

EIO-LCA

- Provides economy-wide, comprehensive assessment in which all direct and indirect environmental effects may be included
- Allows for sensitivity analyses and scenario planning
- Use publicly available data, allowing for independently reproducible results
- Suitable for future product development assessments
- Incorporates information on every commodity in the economy
- Some product assessments contain aggregate data
- There is a difficulty in linking dollar values to physical units
- Economic and environmental data may reflect past practices
- Process assessments can be difficult
- Imports are treated as US products
- Can be difficult to apply to an open economy with substantial noncomparable imports
- Non-US data availability can be a problem
- There is uncertainty in the data.

With respect to boundary definition, there is always a question of how to set the model boundaries for an LCA. For instance, "conventional" LCAs include all processes; but when time and financial limitations may constrain the extent and detail of the model, it is important to draw a boundary such that the most important processes are captured. For an EIO-LCA, the boundary, by definition, is the entire economy along with the interrelationships between industrial sectors. Additionally, in an EIO-LCA, the products described by a sector are representing an "idealized" or "average" product – not a real specific product. Further, when implementing an EIO-LCA model on a large scale, a matrix (incorporating hundreds of rows and columns representing the sectors of the US economy) may be incorporated. These can be augmented with sector-level environmental impact coefficient matrices that represent the amount of impact per dollar of output. Assuming this matrix $\mathbf{R}$ incorporates these impact values on its diagonal, the environmental impact $\mathbf{E}$ may be calculated as follows:

$$\mathbf{E} = \mathbf{R}\,[\mathbf{I} - \mathbf{A}]^{-1}\,\mathbf{y}$$

where

$\mathbf{I}$ = identity matrix;
$\mathbf{y}$ = vector representing final desired demand;
$\mathbf{A}$ = requirements matrix.

As a brief example, consider the switch from paper to electronic versions (e.g. a CD) for a set of conference proceedings. In calculating costs, and assuming a $3 unit cost of a CD (also assuming 500 copies, or $500), this compares to the production of 200 pages for 300 people, or 60 000 pages at 5 cents per page, or $3000. Table 11.6 shows the associated environmental costs that may be factored, and Table 11.7 shows a listing of several physical environmental effects.

In further demonstrating how complex the input–output vectors can be, Figure 11.4 represents the supply chain for the manufacture of an automobile as well as a component engine. The vectors are presented in terms of actual cost (in dollars) as well as a relative percentage of the total cost. As shown, there may be a significant number of entries.

Table 11.6 Economic effects of using paper or CDs for conference proceedings.

Paper		CDs	
Sector	$	Sector	$
Pulp mills	3120	Magnetic media	1500
Logging	500	Misc. plastics	250
Industrial chemicals	400	Wholesale trade	170
Wholesale trade	260	Plastics	110
Sawmills	230	Industrial chemicals	80
Forestry products	150	Tracking services	70
Crude petroleum	140	Electric components	60
Tracking services	130	Paper mills	50
Electric utilities	130	Electric utilities	50

Table 11.7 Environmental effects of using paper or CDs for conference proceedings.

Emission	Paper (lb)	CDs (lb)
SO_2	35	2
Particulates	10	0.1
GWP	3000	600
RCRA	175	60
TRI chemicals	20	2

I–O and supply chains

Example: Requirements for car and engine

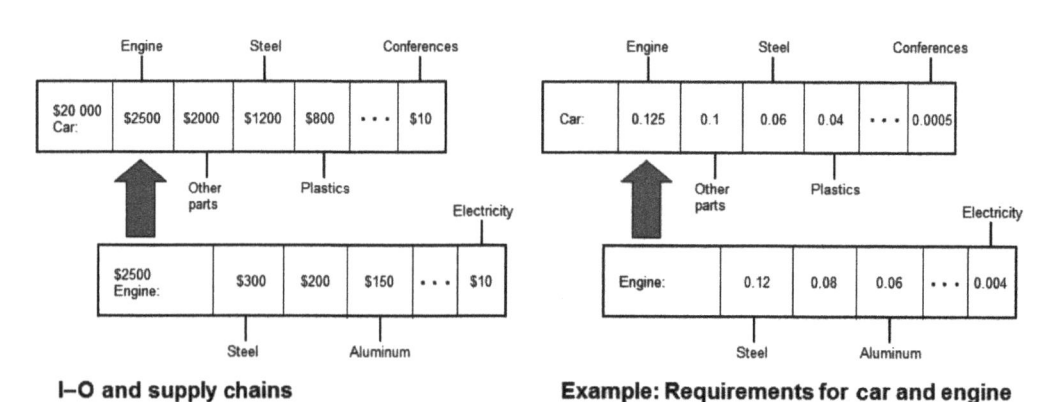

Figure 11.4 Input–output in supply chains for automobile and engines.

11.7 EIO versus Physical Input–Output (PIO) Analysis

In addition to EIO analyses, similar analyses may be performed using physical units as opposed to monetary inputs. Physical input–output (PIO) analyses are often quite detailed and require a significant amount of effort as compared to the typically highly aggregated nature of EIO analyses.

11.7 EIO versus Physical Input–Output (PIO) Analysis

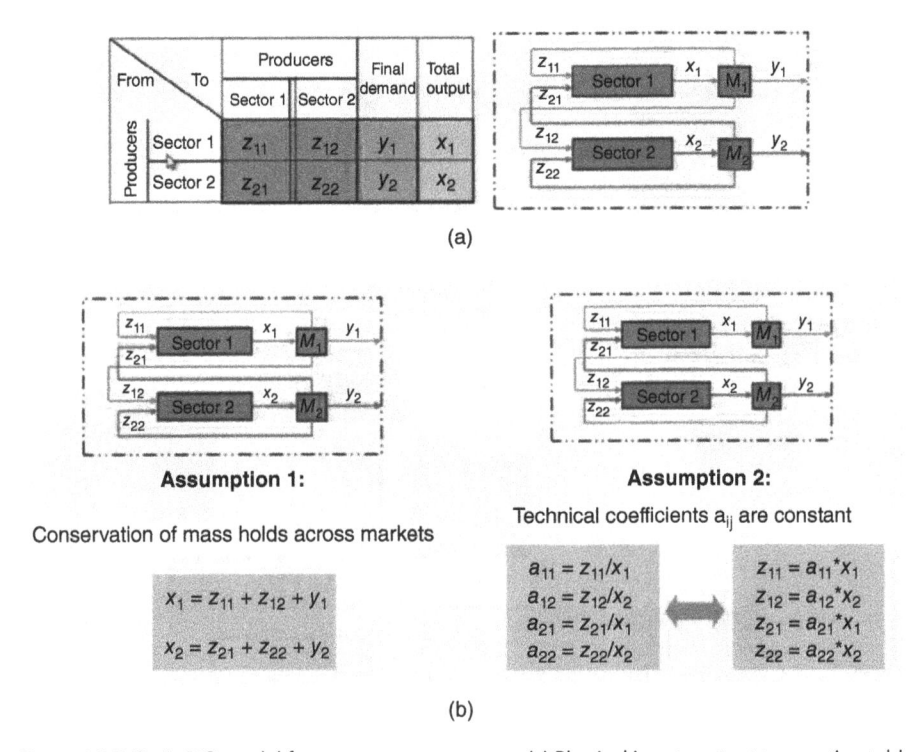

Figure 11.5 Basic PIO model for a two sector economy. (a) Physical input–output transaction table. (b) Assumption 1 and 2.

Figure 11.5a shows a typical PIO transaction table and basic PIO model for a two-sector economy. When creating the model, two assumptions are commonly made. First, it is assumed that the conservation of mass holds across all markets. Second, it is assumed that technical coefficients a_{ij} are constant (Figure 11.5b).

As an example, suppose a decision is being considered that would replace leaded solder (60% tin; 40% lead) with nonlead solder (95.5% tin, 3.9% silver, 0.6% copper). To illustrate this use and its effect, Figure 11.6 depicts a flow of metals among production processes. In considering the example further, Figure 11.5a also depicts a PIO for Japan's integrated metals production. As shown, columns represent the inputs of a metals industry sector, while the rows depict the outputs of a metals industry sector. Following the computations of the PIO, the results can be summarized in Figure 11.7. Figure 11.7 depicts the implications of nonlead solder – first, the percent change in needed supplies of the constituent metals; second, the demand for silver when considering both types of solder. Figure 11.8 further depicts the final destination of silver for different industry uses considering both types of solder.

In building PIO models and analyses, there are several challenges that should be considered with respect to data. First, it is often difficult to match economic value to physical flows – even if prices were known, it is difficult to accurately decouple economic data because it is often aggregated into a different format than physical units. As a result, physical data needs to be used, but these are often not directly available for use in modeling. Additionally, there are many aspects of the economy that cannot be measured directly in physical data terms. Services, for example, cannot accurately be assigned physical units. As a result, a mixed, or a hybrid, system can be used where physical units are important for accurate modeling.

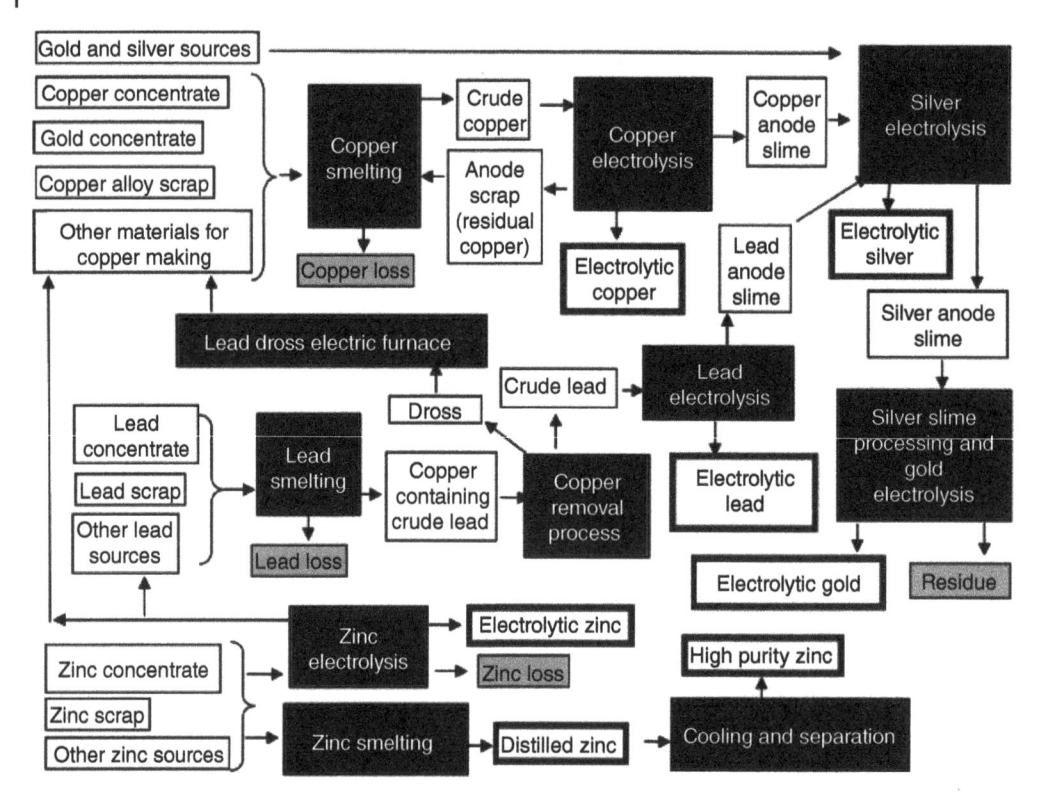

220 | *11 Economic Input–Output Life Cycle Assessment*

Figure 11.6 Flow of metals among production process. Source: Nakamura et al. (2008). Reproduced with permission of American Chemical Society.

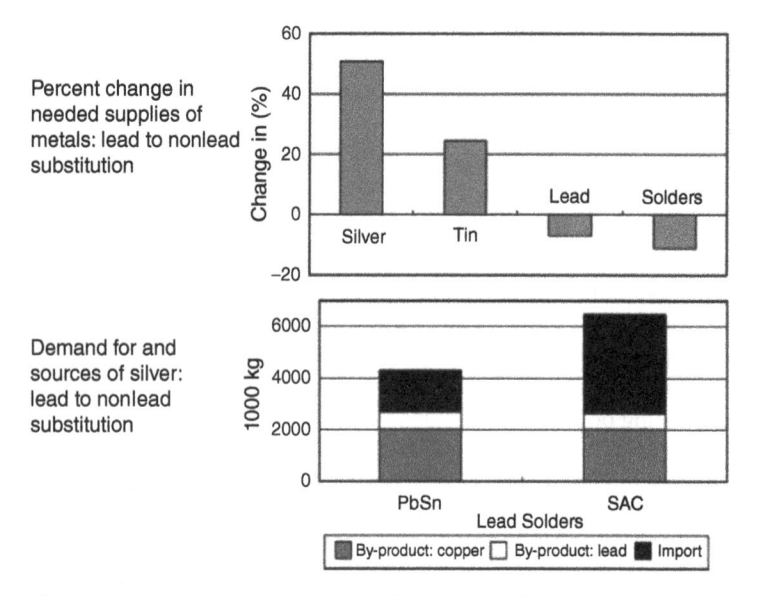

Figure 11.7 Implications of nonlead solder. Source: Nakamura et al. (2008). Reproduced with permission of American Chemical Society.

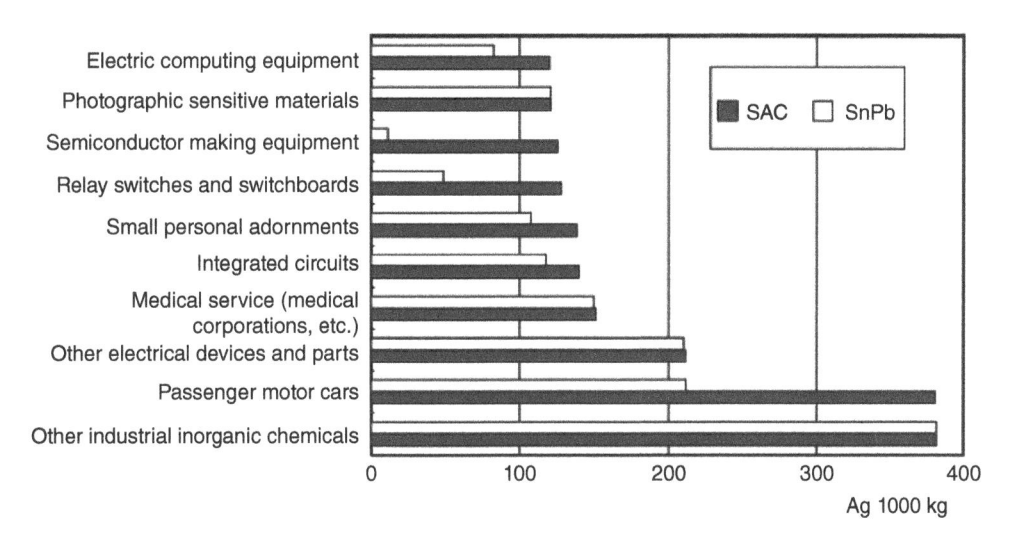

Figure 11.8 Final destination of silver under alternative solder approaches. Source: Nakamura et al. (2008). Reproduced with permission of American Chemical Society.

11.8 Summary

National material accounts describe the inputs and outputs of entire societies, and they also indicate both indirect and direct flows. EIOs further model these activities by measuring and modeling domestic monetary exchanges among industrial sectors. EIO-LCA is a mathematical method that allows to determine the effect of changing the output of a single sector using existing economic and environmental data. The use of EIO-LCA models does in fact offer several advantages over conventional LCA modeling. Finally, PIOs describe the domestic physical material exchanges that occur among industrial sectors.

11.9 Questions

11.1 What is an economic input–output model? Why was this developed?

11.2 Who developed the method of input–output analysis?

11.3 What are the major advantages of EIO-LCA over conventional LCA?

11.4 List some limitations associated with EIO modeling.

11.5 What are some of the limitations associated with an EIO-LCA model?

11.6 Discuss in detail some sources of uncertainty that may occur during the use of EIO-LCA models.

11.7 What are the major assumptions commonly made in PIO models?

References

Graedel, T.E. and Allenby, B.R. (2010). *Industrial Ecology and Sustainable Engineering*. Upper Saddle River, NJ: Prentice Hall.

Leontief, W.W. (1986). *Input-Output Economics*. Oxford University Press.

Matthews, H.S. and Small, M.J. (2000). Extending the boundaries of life-cycle assessment through environmental economic input-output models. *Journal of Industrial Ecology* 4 (3): 7–10.

Nakamura, S. and Kondo, Y. (2009). *Waste Input-Output Analysis: Concepts and Application to Industrial ecology*, vol. 26. Springer Science and Business Media.

Nakamura, S., Murakami, S., Nakajima, K., and Nagasaka, T. (2008). Hybrid input-output approach to metal production and its application to the introduction of lead-free solders. *Environmental Science and Technology* 42 (10): 3843–3848. https://doi.org/10.1021/es702647b.

Rose, A. (1995). Input-output economics and computable general equilibrium models. *Structural change and economic dynamics* 6 (3): 295–304.

12

Environmental Health Risk Assessment

12.1 Introduction

As seen in Chapter 10, when performing a life cycle impact assessment, the final objective is to evaluate ecological and human health effects as well as resource depletion resulting from emissions, resource extraction, and other material fluxes across the life cycle of a product. The results of LCIA for different product alternatives are useful for relative comparison of potential damage on human health and the environment of the different products. However, this method does not provide a quantitative indicator of the absolute risk or actual damage that may have been caused by a product. Moreover, LCIA involves many stressors, locations, and impact categories, which makes it impossible to perform an accurate risk assessment. In this context, it is necessary to develop a method or a tool to perform a quantitative environmental risk assessment so that the risks can be accurately determined.

The objective of this chapter is to analyze how the risks can be evaluated as a necessary step for developing methods of risk management. It is essential to develop methods or tools to determine human health risks resulting from exposure to the environment and workplace to be able to minimize the negative mid-term or long-term effects on health. Considering the interdisciplinary field of this topic, synergic working groups should be encouraged where scientists, engineering, policymakers, and regulatory agencies work together in developing the methods and rules to minimize the risk to human health and prevent future damages at the environmental, economic, and societal levels.

12.2 Emergence of the Risk Era

As increasing amounts of chemicals exhibiting harmful or toxic characteristics were utilized by industry and in consumer products, a greater need arose to evaluate the risks of human exposure to these compounds. The US Congress issued a directive to study the institutional means for risk assessment. In 1981, the Committee of Institutional Means for Assessment of Risks to Public Health was formed. The conclusions of its work were reported in the eponymous *RED BOOK*: *Risk Assessment in the Federal Government: Managing the Process*, published in 1983 (National Research Council 1983). Experts from the administration, industry, and academia formed the committee, which reviewed available literature on risk assessment and analyzed the regulations and operations of the federal agencies in relation with the use of harmful substances. The conclusions and recommendations in the *Red Book* have been a valuable contribution for risk assessment and environmental health.

During the 1970s, public interest increased with respect to the effects of technology to the environment and to public health. By that time, evidence had been accumulated regarding the carcinogenic effects of some commonly used substances and materials, such as asbestos. The concerns of the public regarding health and environmental risk and safety motivated politicians and regulatory agencies to act in developing objective and scientific-based methods to estimate the risks to human health posed by these products and activities. The study conducted by the Committee of Institutional Means for Assessment of Risks to Public Health was focused primarily, although not exclusively, on increased risk of cancer due to exposure to chemicals in the environment. One key finding of the study was the necessity of separating risk assessment methods from risk management. Further, the study identified the necessity of performing risk analyses based on uniform guidelines for all regulatory agencies, and a uniform framework of risk assessment should include the following four steps: hazard identification, dose–response assessment, exposure assessment, and risk characterization. The commission also suggested the necessity of a single organization to be designated to perform risk assessment for all regulatory agencies.

12.3 Risk Assessment and Management

As discussed above, risk assessment should be differentiated from risk management. Risk assessment can be defined as the process to identify risk and facilitate risk reduction for individuals implementing control systems and regulations for those hazards identified in the environment. The management of risk is the process of balancing available policy alternatives to select the most appropriate regulatory action. This is done by considering the data from the risk assessment and environmental, economic, and social concerns to reach a decision that permits the minimization of effects resulting from exposure to hazardous materials or activities. Figure 12.1 establishes the differentiation between risk assessment and management and includes the four steps of risk assessment, which can be defined as follows:

- Hazard identification is the identification of those chemicals with a causal link to particular diseases or health effects.
- Dose–response assessment is the relation between the exposure to specific concentrations of hazards and the probability of development a specific disease or health problem.
- Exposure assessment is the determination of the degree of exposure to hazards before and after the application of the risk management actions.

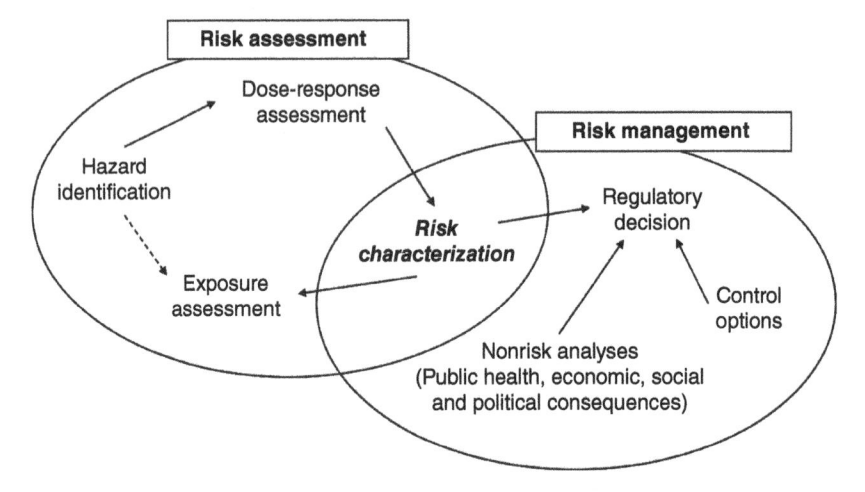

Figure 12.1 Risk assessment and management.

- Risk characterization is the definition of the nature and intensity of the risks to human health and environment, including the degree of uncertainty of the effects after the exposure to particular hazards.

12.3.1 Hazard Identification

A hazard is anything that can produce the conditions to generate a damage in health or safety. In general, we identify hazards as chemicals that can produce an accident (e.g. gasoline may explode or burn at a gas station) or affect our health through exposure (e.g. organic solvents inhaled in a workplace, smoke from a cigarette). Other elements or situations can be identified as hazards if they can threaten our safety or health; for instance, the storage of harmful chemicals close to a community.

The first step in risk assessment is the identification of those substances or chemicals that exhibits any specific toxic effect; in other words, the first step involves identifying if a causal link exists to particular health effects at environmentally relevant concentrations. The problem with the identification of the toxic capacity of chemicals is the huge number of chemicals that are available. More than 50 000 chemicals exist, and it is reported that each year about new 1000 chemicals are synthetized. As a result, the identification of the risks for humans and environment associated to the exposure to these chemicals is daunting due to the limitation of available data to assess the possible effect of chemicals in different scenarios. Additionally, the presence of multiple compounds may have an additive or multiplicative relationship with respect to risk. Further, some chemicals at very low concentrations may exert a significant risk over a mid- or long-term period (Neale et al. 2015).

Substances that are suspected to show specific toxic effects for living organisms are listed as chemicals of potential concerns (COPCs). With COPCs, precautions are encouraged when managing these substances until further studies definitively conclude if a particular substance causes health problem or is linked with any specific disease. It is also important to identify if the exposure to a specific substance results in critical health effects, such as the development of cancer or serious noncancer maladies. For these compounds, additional study is needed to establish if scientific evidence exists of a specific critical health effect or effects.

The identification of a hazard is not a problem in itself. The important thing is the probability of a dangerous incident, an accident, or an exposure that may occur due to the hazards. This defines risk as the probability of an incident of exposure associated with a hazard that may occur.

Once a chemical has been identified as being capable of producing harmful or toxic effects, the next step is the determination of the concentration in the environment and the possible means of exposure that can produce those harmful effects. The exposure to chemicals may take place by contact with skin, eyes, ingestion, or inhalation. The concentration of chemicals or contaminants in air, water, soil, vegetables or food, etc., is indicative of the level of risk for humans and ecosystems.

12.3.2 Dose–Response Assessment

The health response to specific doses of chemicals helps to define the toxicity of the studied compound. Thus, it is possible to determine the relationship between the magnitude of the administered, applied, or absorbed dose and the probability of occurrence and magnitude of health effect(s). Figure 12.2 shows the possible route through which a chemical may enter the body and the possible organs where related toxic effects may appear, as well as the ways the chemical can be excreted from the body.

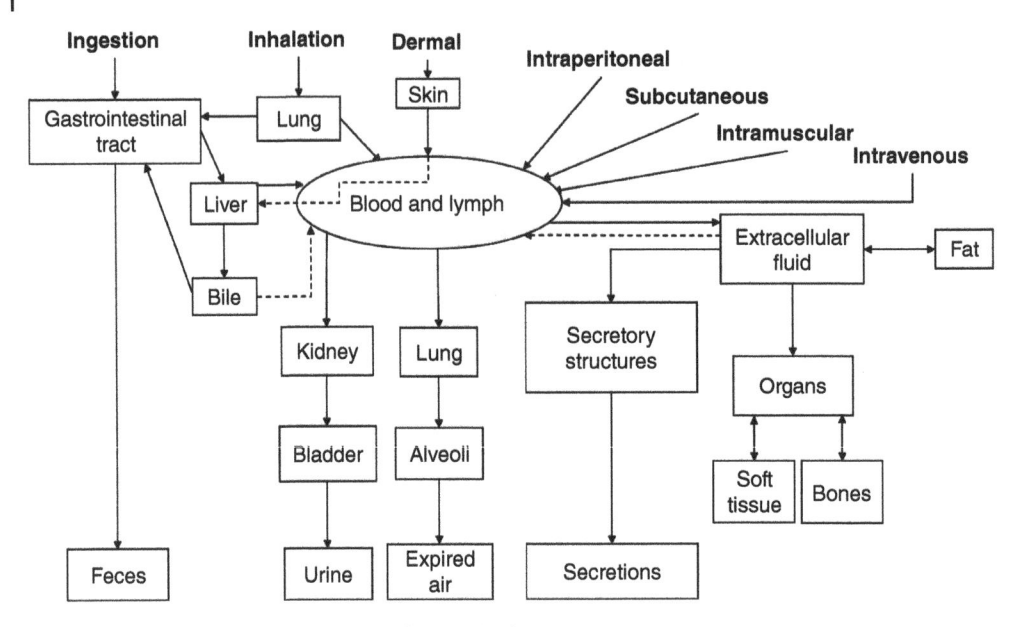

Figure 12.2 Exposition to hazards and effects on body.

There are several methods of determining how an organism responds to a specific dose of a contaminant/chemical. The dose of the contaminant is the amount of contaminant received by the individual, and it is commonly expressed as milligrams (mg) of contaminant per kilogram (kg) of body mass. The most common and objective method of knowing the safe dosage is through laboratory testing of animals, such as rodents or other small mammals. The exposition of rodents to specific doses of a chemical or increasing doses may determine the risk of exposure to that chemical. The results of the toxicity test may be reported as follows:

- *Effective concentration (EC)* – The received concentration by the subject necessary to induce a specific reaction or effect, e.g. leg paralysis.
- *Lethal concentration (LC)* – The necessary concentration to produce the death of the subject in a specific period of time. Usually, it is reported as LC_{50}, which is the dose to produce the death of half of the individuals in the test.
- *Lowest observable effect concentration (LOEC)* – The lowest dose resulting in a detectable specific health effect in the subject.
- *No observed adverse effects level (NOAEL)* – The maximum dose with no measurable negative effect in the subject.
- *Maximum allowable toxicant concentration (MATC)* – The maximum dose that does not provoke any significant negative impact in the subject.

Then, the laboratory results can be extrapolated to humans and ecosystem to determine the maximum limits of exposure to prevent negative impacts to human or environmental health. Alternative methods include medical examinations or checkups for people commonly exposed to a chemical or a specific environment (e.g. factory workers who may be exposed to organic solvents in the air). Thus, the analysis of the impact of diseases to specific group of people who may be exposed to a chemical or a specific environment may help determine if a specific compound or group of compounds may induce the development of specific diseases.

Figure 12.3 shows a typical dose–response curve for a specific substance used to identify the risk of carcinogenic effects compared with the frequency of tumors when there is exposure to a noncarcinogen. The comparison of the two plots may be used to determine if the tested

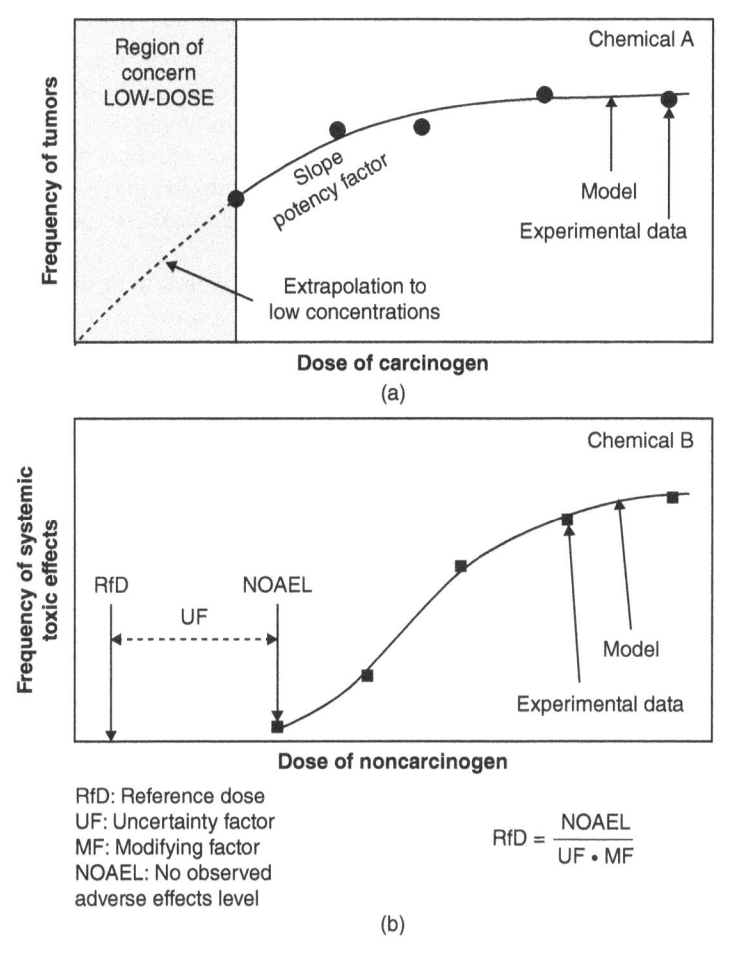

Figure 12.3 Determination of reference dose. Source: USEPA (1993).

substance is carcinogenic or if it is not carcinogenic as well as to identify the dose where carcinogenic effects begin to appear. For carcinogens (Figure 12.3), slope factor (SF) or cancer potency factor (CPF) is determined. For noncarcinogens, the reference dose (RfD) is determined. The RfD, as defined by USEPA (1993), is an estimate, with uncertainty spanning approximately an order of magnitude, of a daily oral exposure to the human population (including sensitive subgroups) that is likely to be without an appreciable risk of deleterious effects during a lifetime. RfD is commonly reported in units of milligrams per kilogram per day (mg/kg d). In the determination of the RfD, the NOAEL is considered, and the uncertainty factor (UF) is incorporated, which accounts for the different effects of toxicity between humans and animals used in laboratory toxicity tests and variation among several human populations. USEPA uses the RfD as the limit of exposure to chemicals to assure little or no adverse health effects during a lifetime duration.

12.3.3 Exposure Assessment

When a substance is determined to have a negative health effect, and there is a causality between the exposure to a substance and the development of specific disease(s), it is important to assess the exposure on a qualitative or quantitative basis. The exposure to a harmful chemical may occur through the medium or media where the chemical is present: air, water, or soil. It is necessary to determine the concentration in the medium to determine the risk. The receptor can be any organism that comes in contact with the medium. A contaminant in the air can be inhaled by humans or by animals; if it is present in the water, the aquatic organisms would be directly affected, but any other human or animal may be exposed to the chemical via ingestion, dermal absorption, or inhalation of vapors emanating from impacted water. Soil impacts may directly affect plants, but humans and animals may also be exposed when these plants are introduced into the food chain.

In the exposure assessment, the concentration of the chemical is critical to determine the risk in the exposure, but it has been demonstrated that chronic, continuous exposure to even low concentrations may have negative effects that could be worse than short-term, acute exposure to high concentrations. There is a cumulative exposure effect that needs to be considered when evaluating risk. Furthermore, some classes of contaminants tend to be accumulated within organisms (bioaccumulation). For instance, some metals or organics tend to accumulate within fatty tissues, and the toxic effects appear when the concentration within the organism reaches a specific level, even if the concentration in the environment is much lower. This is defined as the accumulation factor, which is the ratio of the contaminant in an organism to the concentration in the ambient, steady-state environment, where the organism can intake the contaminant via the exposure pathways described above.

The quantification of the exposure has to include the concentrations in the medium, the possible intake, and the time period of exposure. As an example, the exposure of air pollutants associated with automobile exhaust is going to be more significant in policemen or bridge toll collectors who may be exposed for several hours on a repetitive basis, as compared to other people who are infrequently exposed to exhaust and for short periods.

Equation (12.1) is the general formula for estimating chemical exposures by inhalation.

$$I = \frac{C \cdot CR \cdot ET \cdot EF \cdot ED}{BW \cdot AT} \tag{12.1}$$

where

I is the chemical intake (mg/kg d)
C is the chemical concentration (mg/m^3)
CR is the contact rate (m^3/h)
ET is the exposure time (h/day)
EF is the exposure frequency (day/year)
ED is the exposure duration (years)
BW is the body weight (kg)
AT is the averaging time (days)

12.3.4 Risk Characterization

Risk characterization is a complex process where all the data from the risk assessment has to be considered. Thus, the data from hazard identification and dose–response assessment, as well as the exposure assessment may help in characterization of risk, keeping in mind that there will be in general a degree of uncertainty. The risk characterization should include the list of identified hazards, their effects, modes of exposure, transportation, and methods of prevention. This information will be used to develop the risk assessment.

The risk characterization can be defined as the likelihood of adverse effects for human health due to the exposure to a hazard and defines the maximum exposure level to avoid such negative effects. Thus, the cancer risk equation (Eq. (12.2)) permits the determination of the likelihood of developing a tumor due to exposure to a substance with carcinogenic potential. The cancer risk calculated in Eq. (12.2) is defined as the probability of an individual developing cancer:

$$\text{Cancer risk (CR)} = I \times \text{CPF} \tag{12.2}$$

where

I = chronic daily intake averaged over 70 years (mg/kg yr); CPF or SF = cancer slope (or potency) factor obtained from the toxicity assessment (mg/kg/day) as shown in Figure 12.3a.

The likelihood of noncancerous health effects is presented in Eq. (12.3). The Hazard Quotient (HQ) is defined as the ratio of the potential exposure to the substance and the level at which no adverse effects are expected.

$$\text{HQ} = I/\text{RfD} \tag{12.3}$$

where

I = exposure level (or chemical intake) obtained from exposure assessment; RfD = reference dose obtained from the toxicity assessment for the same exposure pathway as I.

When the HQ is less than or equal to one (HQ ≤ 1) adverse noncancer effects are not likely to occur from the exposure pathway and to that particular chemical assuming the exposure conditions factored into Eq. (12.3). An HQ greater than 1 indicates that a risk has exceeded the assumed threshold for a noncancerous health effect to occur given, how much an exposure concentration exceeds the RfD. An HQ greater than 1 means that the exposure to a specific chemical is higher that RfD, and that increasing values of HQ means increasing potential for adverse effects.

If an individual is exposed to several substances with health risk potential for the same organ, the HQ for each chemical can be summed to evaluate the total noncarcinogenic effects. The sum of the HQ for each chemical is called Hazard Index (HI). Similarly, exposures below 1.0 for HI likely will not result in adverse health effects over a lifetime of exposure and would commonly be considered acceptable. It should be noted that the HI is only an approximation of the aggregate effect of several chemicals exposure on the body because different substances might cause their negative effects by different (i.e. nonadditive) mechanisms.

While the concept of risk assessment can readily be appreciated, in practice it is extremely challenging. Figure 12.4 is from a study relating land use policy to exposure to air toxics.

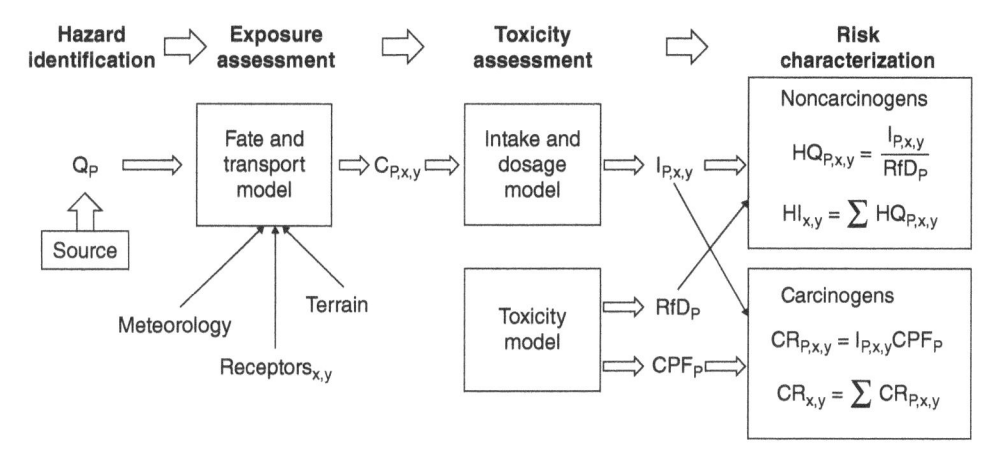

Figure 12.4 Risk assessment sequence. Source: Willis and Keller (2007). Reproduced with permission of Elsevier.

A detailed computer model, diagrammed here, was required for the analysis. The variables on the diagram are as follows: Q = volatilization rate per unit area; C = contaminant concentration, I = chemical intake; HQ = hazard quotient; HI = hazard index; RfD = reference dose; CR = carcinogenic risk; SF = cancer slope (or potency) factor. Each of these variables must be quantified in order to carry out the computation.

12.4 Ecological Risk Assessment

The ecological risk assessment can be defined as the process to determine how likely it is that the environment would be impacted due to the exposure to one or more stressors, such as contaminant chemicals, climate change, diseases, changes in land use, or other factors (USEPA 2017). The process requires several steps and phases of gathering information, analyzing, and proposing conclusions as depicted in Figure 12.5.

The first step is planning, which can be considered as an initial step preceding the actual ecological risk assessment. The main objective and activity of planning is to define the scope of the study and to gather basic information as described below:

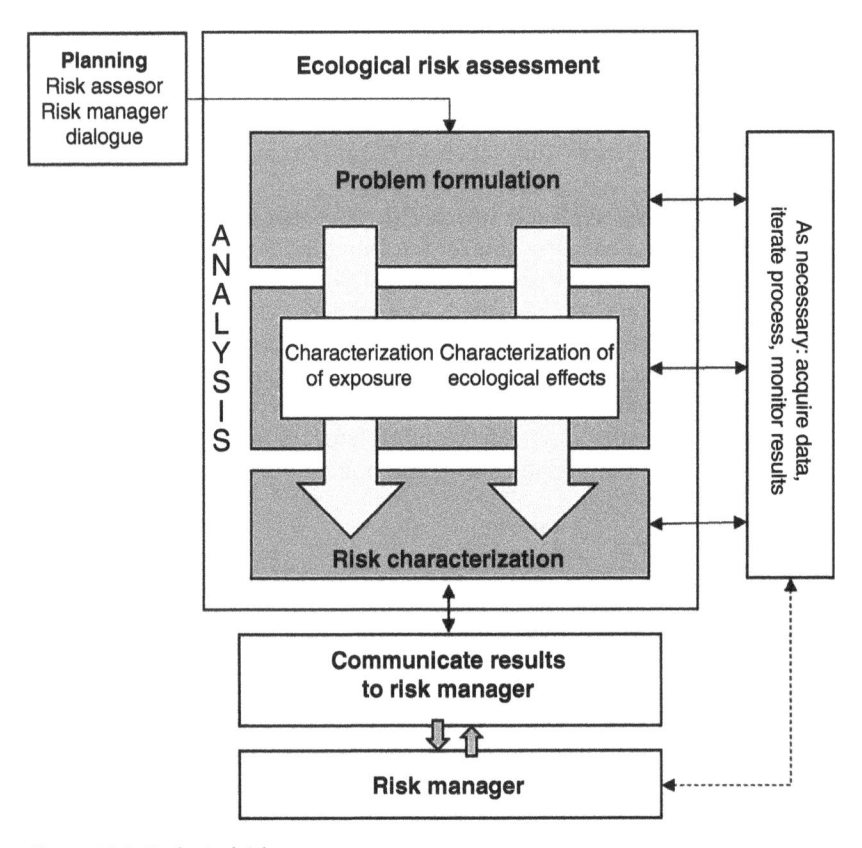

Figure 12.5 Ecological risk assessment.

- Define the individuals, species, communities, population, or ecosystems affected that are going to be the target of the risk assessment;
- Define the stressors to be analyzed: contaminant chemicals, microbiological (diseases) or biological (invasive species) elements, radiation, physical factors (changes in habitats, flood, etc.), and others;
- Define the source of the stressors: point sources (discharge of an industrial effluent) or dispersed sources (acid rain);
- Determine the exposure pathways (air, water, soil, waste, and food) and routes (inhalation, skin contact, ingestion, etc.);
- Determine the mechanisms of accumulation/assimilation/metabolization of the contaminants in the living organisms;
- Define the ecological impacts of the stressors: diseases and health disorders, tumors, mortality, reproduction effects, etc.;
- Define the duration of the impact: chronic, periodic or intermittent, immediate, and acute.

The ecological risk assessment, once the planning and scope are defined, is comprised of three sequential steps (Figure 12.5): Problem formulation, analysis, and risk characterization.

The problem formulation includes the identification of the ecological entity to be analyzed and protected. The ecological entity could be an individual, species, group of species, ecosystem, or a habitat. It is then necessary to identify the property or characteristic of the ecological entity to be protected and how the stressors affect those characteristics.

The analysis phase is designed to describe and predict the response of the ecological entity to the stressors under predefined exposure conditions. In the analysis phase, various parameters have to be calculated to quantitatively determine the exposure. These parameters include the HQ, ingestion rate, bioaccumulation rate, bioaccumulation factor, bioavailability, and others.

The objective of the risk characterization is to estimate the risk posed to ecological entities using the data from the analysis phase. The conclusions of this phase have to include the severity of the effect and their duration (acute or chronic), the frequency of the effects, and the individuals or population affected. The conclusions must also include evidence for risk estimates, the overall degree of confidence, identification of uncertainties, and the ecological consequences of the effects.

12.5 Summary

A risk assessment can provide quantitative information to evaluate the risk due to the exposure to the stressors. This information will be very valuable to analyze the risk under different environmental conditions and exposure scenarios. The perception of risk can be very different depending on the familiarity of the individual with the risks. The exposure to a well-known situation can result in different valuations of risk by different individuals. Furthermore, it can be difficult for individuals to assess risk from exposure to unknown stressors or conditions. The risk assessment method permits an evaluation of the risk with a lower degree of subjectivity, which lowers the uncertainty and variation among the individual perception on multiple individuals. Risks to humans and the environment are real, and should be carefully evaluated, taking human perception into account.

12.6 Questions

12.1 What is risk assessment and what are the four steps involved in risk assessment?

12.2 What is hazard identification and what does it encompass?

12.3 Discuss the terms of Eq. (12.1), and how you may select "baseline" representative values for modeling.

12.4 Discuss the differences between acute and chronic health or ecological effects.

12.5 Explain briefly how is a dose–response assessment performed?

12.6 What are the different metrics used for reporting the toxicity of a chemical?

12.7 What are the different exposure routes for contaminant exposure to human and ecological entities?

12.8 What are some compounds or exposure scenarios where bioaccumulation effects are prominent?

12.9 Define the terms: Cancer Risk, Hazard Quotient, and RfD.

12.10 When is the Hazard Quotient approach used?

12.11 Discuss how the presence of multiple chemical could affect a Cancer risk or Hazard Quotient calculation.

12.12 What are the steps involved in ecological risk assessment? Explain briefly.

References

National research Council (1983). *Risk Assessment in the Federal Government: Managing the Process*. National Academies Press.

Neale, P.A., Ait-Aissa, S., Brack, W. et al. (2015). Linking in vitro effects and detected organic micropollutants in surface water using mixture-toxicity modeling. *Environmental Science & Technology* 49 (24): 14614–14624.

USEPA (1993). Reference dose (RfD): description and use in health risk assessments. US Environmental Protection Agency. https://www.epa.gov/iris/reference-dose-rfd-description-and-use-health-risk-assessments#1.1 (accessed May 2017).

USEPA (2017). Risk assessment. US Environmental Protection Agency. https://www.epa.gov/risk (accessed May 2017).

Willis, M.R. and Keller, A.A. (2007). A framework for assessing the impact of land use policy on community exposure to air toxics. *Journal of Environmental Management* 83 (2): 213–227.

13

Other Emerging Assessment Tools

13.1 Introduction

As discussed in Chapters 6–12, several initiatives and approaches have been developed toward the assessment and quantification of the degree of sustainability associated with a product/project. The qualitative tools that we have discussed earlier do not thoroughly assess environmental, economic, and social impacts. However, several actions taken in the form of best management practices could aid in incorporating some degree of sustainable practice. Semiquantitative tools rule out the ambiguity and uncertainty in sustainability assessment by incorporating ratings or scores to identify relative impacts among different alternatives. However, they may be of limited utility due to subjectivity of the person who is performing the assessment. Quantitative tools like life cycle assessment (LCA) are perhaps the most standardized and adequate way of assessing the environmental sustainability. However, they do have limitations such as the large amount and detailed data required for inventory analysis and reliability of the data among others. In lieu of this, many new and innovative sustainability assessment tools with specific applicability are being developed. This chapter identifies and summarizes some of these emerging tools that try to address the issues and challenges that are being faced with the existing tools for sustainability assessment and highlights the major improvements in these tools.

13.2 Environmental Assessment Tools/Indicators

The importance of environmental protection and awareness toward resource depletion has gained significant importance over the last two decades. Further, the grand challenges of global climate change and population growth have made sustainable development and practices an imperative in building and civil infrastructure. The same principles are generally being applied to engineering and technology in general to realize sustainable development. This, in fact, has spurred a great interest in development of tools to assess the environmental impacts of a product/project and the associated sustainability. Over the years, several qualitative tools such as simple yes/no tools or spreadsheets, semiquantitative tools such as relative rating/scoring tools, and quantitative tools such as LCA have been developed for environmental impact assessment. Several new environmental assessment tools (qualitative, semiquantitative, and quantitative) have been developed specifically for the impact assessment of building and urban development. Some of the qualitative and semiquantitative tools are BREEAM Communities, CASBEE-UD, SBTool^{PT}-UP, and Pearl Community Rating System (PCRS). LEED is specifically developed to ascertain green buildings, while Envision is developed to assess the sustainability of any other civil infrastructure. A detailed explanation regarding the LEED and its application to green

Sustainable Engineering: Drivers, Metrics, Tools, and Applications, First Edition.
Krishna R. Reddy, Claudio Cameselle, and Jeffrey A. Adams.
© 2019 John Wiley & Sons, Inc. Published 2019 by John Wiley & Sons, Inc.

buildings design is presented in Chapter 16. Likewise, the Envision rating system is described in Chapter 17.

BREEAM tool was launched in 1990 in the UK by the Building Research Establishment (BRE). It was developed for the assessment of green buildings at first and was later extended for the assessment of urban community; hence, the name BREEAM Community. Likewise, CASBEE-UD launched in 2007 by Japan Sustainable Building Consortium and the Japanese Green Building Council is a tool for assessment of the environmental performance of buildings. SBTool^{PT}-UP was established by the International Initiative for a Sustainable Built Environment in 2005. SBTool^{PT} specifically assesses the building sector in Portugal. Later, the scope of the tool was also expanded to the development of sustainable urban planning and design. PCRS developed by Abu Dhabi Urban Planning Council in 2010 is the first sustainability assessment tool in the Middle East. PCRS also included "culture" as a fourth dimension of sustainability (apart from environmental, economic, and social) to give spatial effect to the assessment process. PCRS has been developed based on the differences and similarities in BREEAM and LEED and by also accounting for the spatial and cultural dimensions of UAE (Ameen et al. 2015).

Carbon footprint is one of the major environmental indicators for the assessment of environmental sustainability. Since the building sector is highly material and energy intensive, it induces a much larger carbon footprint than many other sectors. However, the embodied GHG emissions in buildings and other features of the built environment are often unaccounted. This is particularly important when trying to select a low-carbon footprint material for building construction. Realizing this difference, Chen and Ng (2015) established a comprehensive model that accounts for the building's embodied GHG emissions, which was integrated into one of the most widely accepted building environmental assessment scheme, LEED.

The existing sustainability assessment methods focus more on environmental aspects with limited or less attention toward economic and social aspects of sustainability. A multicriteria decision analysis method that incorporates a wide range of issues in the analysis is essential. Over the years, many researchers have focused on developing computer-based sustainability assessment tools that involve a life-cycle approach in assessing environmental impacts. In a similar manner, Kumanayake and Luo (2017) proposed a conceptual framework with an associated model for developing an automated sustainability assessment tool. The tool assists in building material-related decision-making in early design stages within a computerized environment based on embodied carbon emissions and life-cycle cost with an LCA approach. Numerous studies have shown that embodied energy can account for about 40–60% of total energy use of low-energy buildings (Kumanayake and Luo 2017). The type of materials used in the construction of a structure have a significantly high impact on its embodied energy content and carbon emissions due to large quantities of materials used. This concept can also be extended to environmental impact assessment in other sectors/domains.

The framework developed by Kumanayake and Luo (2017) includes an input module, assessment module, and output module, as shown in Figure 13.1. The goal of the proposed framework is to perform sustainability assessment of a building. The criterion for the environmental performance is taken as the embodied carbon. The life-cycle stages include raw material extraction, material manufacturing, transportation, construction, maintenance, and demolition. All of these stages contribute to embodied carbon emission of buildings in many some way. The life-cycle cost is selected as the criterion for economic performance and from the various cost components associated with the building life cycle, initial investment cost, maintenance cost, and deconstruction cost, which are directly relevant to material related decisions are considered. The input module is comprised of building modeling software (Autodesk), which is further integrated with a building material database developed using structured query language (SQL).

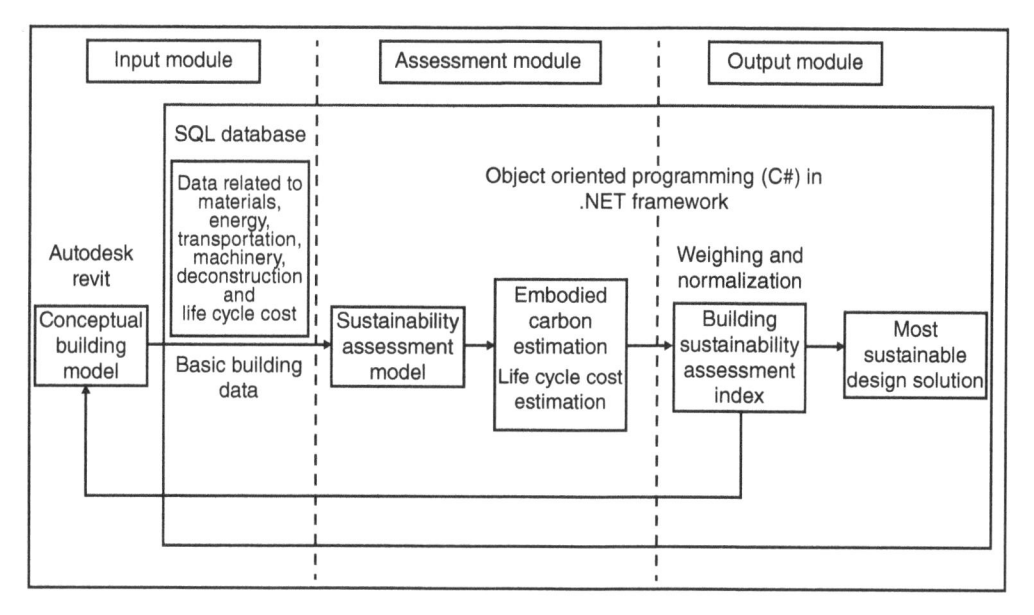

Figure 13.1 Schematic of the framework for integrated sustainability assessment model developed by Kumanayake and Luo (2017) for assessing building sustainability.

This provides the necessary inventory required for the construction of the entire building as conceptualized in Autodesk.

The assessment module as shown in Figure 13.1 identifies the scope and impact assessment method to be used for LCA. The quantitative data regarding the environmental performance is obtained for the different life-cycle stages, including material extraction, manufacturing, and transportation. Likewise, the direct costs incurred in each of the life-cycle stages is estimated (specific to the country) to evaluate the economic performance. Finally, in the output module, the environmental and economic performances are quantified, and the integrated performance score of the building (regarding both environmental and economic aspects) is computed. Suitable weights for each criterion were assigned, and the results are aggregated into a single index to finally compute the "Building Sustainability Assessment Index." It should be noted that the proposed framework does not include the social aspects of sustainability. However, the entire framework provides a good and a reasonably accurate assessment of the environmental and economic sustainability of the building.

13.3 Economic Assessment Tools

Development of innovative and practical sustainability assessment tools has been increasing in recent years. In addition, the use of a life-cycle approach in the environmental, economic, and social sustainability evaluation is given due importance. However, the appropriate integration and coherence in a combined sustainability assessment involving LCA, life-cycle costing (LCC), and cost–benefit analysis (CBA) is lacking. The interdependency of environmental and economic concerns and their related costs and benefits needs to be accurately considered to achieve an accurate and meaningful life-cycle sustainability assessment. In a similar manner, Hoogmartens et al. (2014) developed a framework to integrate LCA, LCC, and CBA, which is an element of LCC. A schematic of the framework developed by Hoogmartens et al. (2014) illustrating the interactions between different individual assessment tools is shown in Figure 13.2.

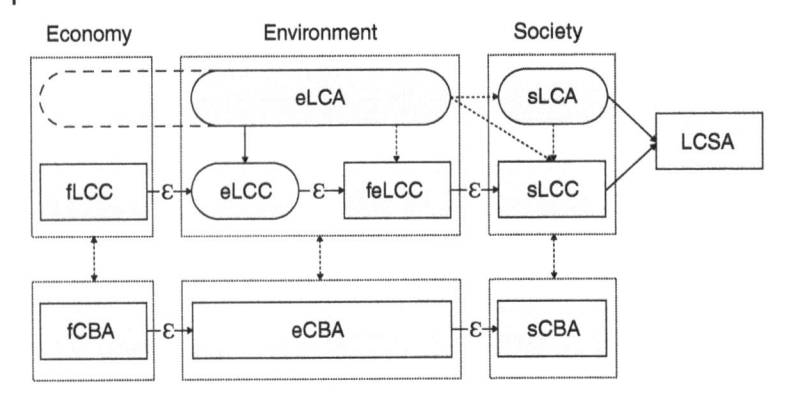

Figure 13.2 Schematic of the integrated framework for life-cycle sustainability assessment. Source: Hoogmartens et al. (2014). Reproduced with permission of Elsevier.

This figure shows that within widely recognized tools, such as LCA, LCC, and CBA, many other methodologies may be distinguished that focus on different aspects of sustainability namely, environmental LCA (eLCA), social LCA (sLCA), financial LCC (fLCC), environmental LCC (eLCC), full environmental LCC (feLCC), societal LCC (sLCC), financial CBA (fCBA), environmental CBA (eCBA), and social CBA (sCBA) (Hoogmartens et al. 2014).

13.3.1 Life-Cycle Costing

Life-cycle cost is similar to life-cycle analysis except that it involves evaluating the impacts of different processes and materials in the life-cycle stages in terms of a monetary value. The costs within the life cycle of a product or a project would include the direct costs such as the cost of materials, labor, equipment, and transportation. Indirect costs, which are often linked with environmental and societal aspects, can also be evaluated from assessing the monetary impacts on the submethodologies that were mentioned earlier.

Specifically, to deal with environmental, social, and financial concerns, four LCC types have been introduced: namely, fLCC, eLCC, feLCC, and sLCC. The definitions and a brief explanation about each of these LCC types as presented in Hoogmartens et al. (2014) is given below.

Conventional LCC assessments that focus on private investments from a firm or a consumer are called as fLCC. Since the economic lifetime is a major factor, the fLCC does not account for the complete life cycle. An example for fLCC includes investment costs, research and development (R&D) costs, and sales revenues (all of which are negative costs).

An eLCC builds upon data of fLCC and extends it to life-cycle costs borne by other firms. The focus in eLCC is on the real internalized cash flows with no conversion to monetary measures from environmental emissions (e.g. implementation of CO_2 taxes, waste disposal costs, and global warming adaptation costs).

feLCC is an extension of the eLCC with environmental costs that can be identified by methods such as eLCA. But it is important to note that the conversion of environmental impacts into monetary values is not straightforward.

Lastly, in the sLCC, the assessment is performed on the costs incurred by anybody in the society, be it today or in the future. In addition, the social impacts associated with all the life cycle stages of a product are considered. Public health impacts must be quantified and converted into monetary values, which is often a challenge in carrying out this type of assessment.

13.3.2 Cost–Benefit Analysis

In simple terms, CBA is a method to evaluate the net monetary value at a discount rate associated with a product or project. If the net value is positive, the product/project achieves profitability. The CBA can be influenced by financial, environmental, and social concerns, and accordingly, there are three types of CBA: namely, fCBA, eCBA, and sCBA, respectively. There can always be a combination of the three CBA types mentioned for an overall CBA analysis.

The fCBA is a tool for estimating the profitability of individual firm based on direct costs and cash flows in a project. This CBA; however, does not reflect the environmental and social aspects in the CBA that are essential for a more comprehensive sustainability assessment. The eCBA is the tool for analyzing the indirect costs and profitability caused by environmental impacts. This is obtained by expressing the damage caused by environmental impacts in terms of their equivalent monetary value. Some of the examples of costs pertaining to eCBA include pollution costs (e.g. GHG emissions), damage to ecosystems, groundwater pollution, and surface water pollution. In addition, the eCBA integrates financial aspects as well. The sCBA assesses a product/project by considering the societal benefits and well-being. However, the measure of welfare of the society is in terms of the monetary value to be able to compare this aspect across different product/project alternatives. Some of the examples for social benefits include development of commercial and recreational centers for communities, reduction in noise pollution, reduction in congestion, increase in employment opportunities, and increase safety in the community. Similar to eCBA, the sCBA integrates the information of fCBA and eCBA, since it comprises of both financial and environmental aspects.

13.4 Ecosystem Services Valuation Tools

Ecosystem services are the benefits that ecosystems provide to people (Millennium Ecosystem Assessment (MA), 2005). In recent years, there has been an increasing importance given for the integration of ecology, economics, and geography to support the decision-making on ecosystem services. However, the ecosystem service assessments need to be quantifiable, replicable, credible, flexible, and affordable (Bagstad et al. 2013). A variety of decision support tools (DSTs) have been developed in support of a more systematic assessment of ecosystem services. Numerous approaches for integrating ecosystem services into both public- and private-sector decision-making process are being developed. Some of the tools developed are generalizable while many others are site-specific. Further, these tools may differ in the way they handle economic valuation, spatial and temporal representation of services and incorporation of existing biophysical models (Bagstad et al. 2013). The tools used for assessment of ecosystem services constitute the evaluation of resource conservation, land-use planning, and hydrologic and ecological modeling that are in fact the important aspects of sustainable development. A study by Bagstad et al. (2013) presented an extensive critical review of 17 ecosystem services assessment tools that assess, quantify, model, value, and/or map ecosystem services by applying these tools for a common site. The 17 tools reviewed are listed as follows:

- Ecosystem Services Review (ESR)
- Integrated Valuation of Ecosystem Services and Tradeoffs (InVEST)
- Artificial Intelligence for Ecosystem Services (ARIES)
- LUCI
- Multiscale Integrated Models of Ecosystem Services (MIMES)

- EcoServ
- Co$ting Nature
- Social Values for Ecosystem Services (SolVES)
- Envision
- Ecosystem Portfolio Model (EPM)
- InFOREST
- EcoAIM
- ESValue
- EcoMetrix
- Natural Assets Information System (NAIS)
- Ecosystem Valuation Toolkit
- Benefit Transfer and Use Estimating Model Toolkit.

The tools were assessed against eight evaluation criteria that are important tool-characteristics desirable of an analytical ecosystem services tool. The eight criteria are listed as follows: (i) quantification and uncertainty, (ii) time requirements, (iii) capacity for independent application, (iv) level of development and documentation, (v) scalability, (vi) generalizability, (vii) nonmonetary and cultural perspectives, and (viii) affordability, insights, and integration with existing environmental assessment. In general, the authors concluded that the tools evaluated differed greatly in their performance against the eight evaluation criteria used.

13.5 Environmental Justice Tools

Environmental justice as defined by USEPA is "the fair treatment and meaningful involvement of all people regardless of race, color, national origin, or income with respect to the development, implementation, and enforcement of environmental laws, regulations, and policies." Environmental justice forms one of the prime aspects of social sustainability, as it deals with ensuring the facilities and benefits to all people in the near community or the society in general. There have been some legal tools developed as a part of the formulation of the idea of environmental justice by the USEPA (Ewall 2012). However, several other tools, such as the Environmental Justice Strategic Enforcement Assessment Tool (EJSEAT) developed for the EPA office of enforcement and compliance assurance. This tool helps identify areas with potentially high and adverse environmental and public health problems. The EJSEAT uses a simple method to identify such areas by using 18 federally recognized and managed databases (EJSEAT 2018).

Environmental Justice Screening and Mapping Tool (EJSCREEN) is another peer-reviewed tool, with a web-based geographic information system and data, developed by EPA for nationally consistent environmental justice (EJ) screening and mapping. This tool builds upon the national environmental justice advisory committee report on EJ screening and prior EPA experiences. EJSCREEN is a "pre-decisional screening tool" that highlights the places with potential EJ concerns. The EJ index is a combination of environmental and demographic information. There are 11 EJ Indexes in EJSCREEN reflecting the 11 environmental indicators. These indicators indirectly aid in evaluating the environmental impacts and thereby the environmental sustainability at a region based on actual historically collected data. The 11 EJ Index names are as follows:

- National Scale Air Toxics Assessment Air Toxics Cancer Risk
- National Scale Air Toxics Assessment Respiratory Hazard Index

- National Scale Air Toxics Assessment Diesel PM (DPM)
- Particulate Matter (PM2.5)
- Ozone
- Lead Paint Indicator
- Traffic Proximity and Volume
- Proximity to Risk Management Plan Sites
- Proximity to Treatment Storage and Disposal Facilities
- Proximity to National Priorities List Sites
- Proximity to Major Direct Water Dischargers.

EJSCREEN helps users identify areas with problems such as minority and/or low-income populations, potential environmental quality issues. However, many environmental concerns are not yet included in comprehensive, nationwide databases such as accurate data on drinking water quality and indoor air quality data are not available. EJSCREEN involves substantial uncertainty in the estimates for small areas. Even for indicators that directly estimate risks or hazards, the estimates have substantial uncertainty.

13.6 Integrated Sustainability Assessment Tools

Development of sustainability assessment models has evolved over time from less accurate qualitative tools toward more reliable quantitative tools. However, there is still a great scope for integrating the three aspects (environmental, economic, and social) of sustainability for a complete and adequate assessment of sustainability. An integrated sustainability assessment model or a tool is the one that integrates life-cycle environmental impacts, life-cycle cost (including the indirect costs and benefits) and life-cycle social impacts. A few frameworks exist that integrate the three aspects of sustainability. Fuzzy Evaluation for Life Cycle Integrated Sustainability Assessment (FELICITA), a novel model developed by Kouloumpis and Azapagic (2018), integrates LCA, LCC, and sLCA into a fuzzy inference framework. The use of fuzzy logic in sustainability assessments is most desirable due to the imprecise and uncertain data that is dealt with in the assessment. A good collection of articles on the use of fuzzy logic for LCA to determine the environmental impacts with uncertain data is presented in Kouloumpis and Azapagic (2018).

Fuzzy logic is a method used to describe the degree of truth when the values of variables cannot be determined exactly, which is mostly the case in sustainability assessment. The nature of the variables is often descriptive rather than numerical. The input data for the FELICITA model are the values for the sustainability indicators estimated through LCA, LCC, and sLCA. After the application of the fuzzy logic and the subsequent inference of the data, the individual sustainability indicators are aggregated into composite LCA, LCC, and sLCA indices. This process is repeated, using the composite indicators as the inputs to obtain an overall life-cycle sustainability indicator for each alternative considered and finally determine the most sustainable option. A schematic representation of the FELICITA model is shown in Figure 13.3. A case study describing the functionality and applicability of the FELICITA model is presented in Kouloumpis and Azapagic (2018).

The FELICITA model aids in resolving the issues with uncertain data with regard to sustainability indicators by using fuzzy inference techniques. The model can address a differing number of indicators within the three sustainability aspects without affecting the overall results. The model can also avoid the need for elicitation of preferences for different indicators and aspects of sustainability, reducing the subjectivity in the decision-making process. The model

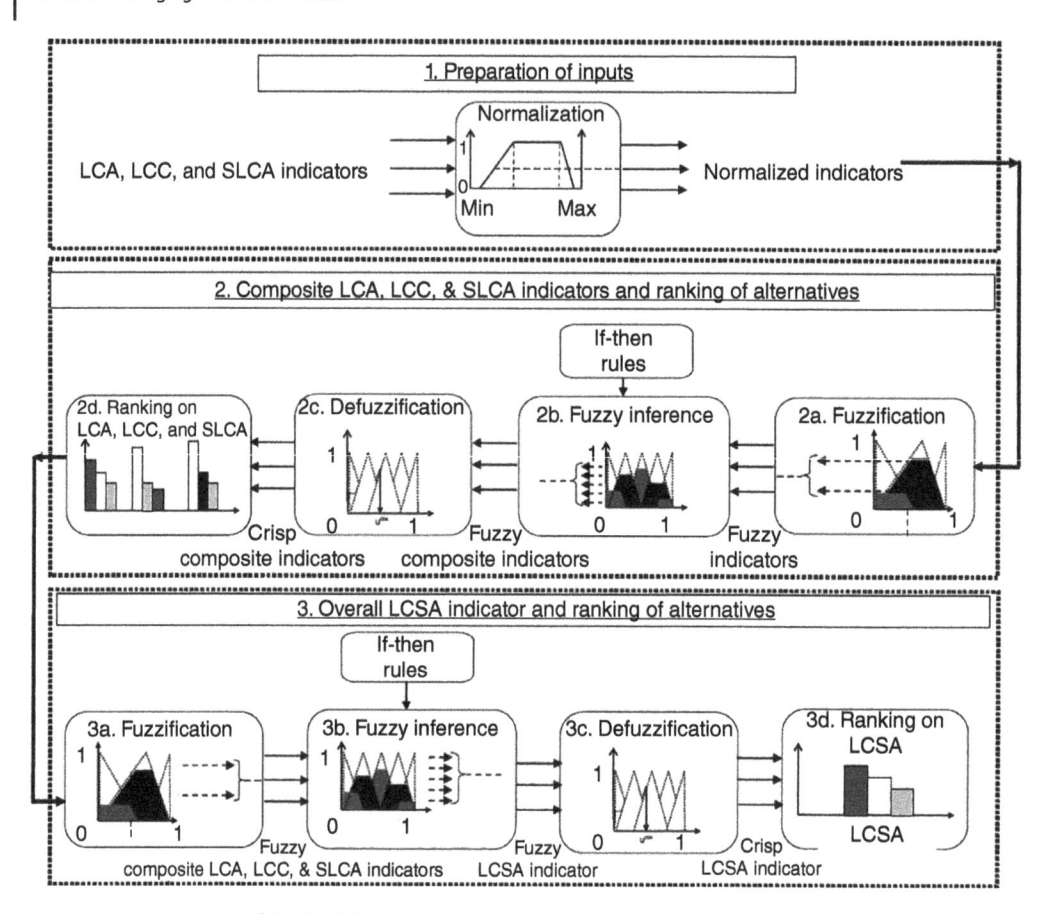

Figure 13.3 Schematic of the FLECITA model. Source: Kouloumpis and Azapagic (2018). Reproduced with permission of Elsevier.

is capable of accommodating different stakeholder preferences through the choice of different rule bases (neutral, optimistic, or pessimistic). Its further advantage is that it can be easily replicated and adjusted to reflect various changes in the assessment process. The FELICITA model can serve as a useful tool for integrated life cycle sustainability assessments. It can be used by companies for integrated sustainability assessment of the products and services as well as by policy-makers to examine the consequences of policies on the overall sustainability.

The Quantitative Assessment of Life Cycle Sustainability (QUALICS) is another integrated sustainability assessment framework/tool that quantifies the triple bottom line aspects (environmental, economic, and social) with a life-cycle perspective. The QUALICS framework is based out of combining two multi-criteria evaluation methods namely the Integrated Value Model for Sustainable Assessment (MIVES) and analytic hierarchy process (AHP). Multi-criteria decision analysis (MCDA) is a vital component of sustainability assessment tools as it allows for assessing the uncertainty associated with the data used for sustainability assessment and also identify the relevance and/or importance of each criteria used in sustainability evaluation. Specifically, the combination of two multicriteria methods MIVES and AHP has proved to be an effective tool to support decision-making in the selection of sustainable options. MIVES is a methodology that combines the concepts of MCDA and value engineering. It uses a generic value function, that standardizes each indicator, and the AHP method to determine the weights for the different criteria and indicators which essentially make up the requirements of the sustainability assessment. A detailed description of the QUALICS framework and its application to sustainability assessment of remedial options for contaminated site remediation is presented in Trentin et al. (2019).

13.7 Summary

Numerous tools have been developed to estimate/quantify sustainability of a product/a project. Some of the tools are specific in their applicability, while others can be extended to evaluate sustainability in several other domains. Several new qualitative and quantitative tools have been developed in recent years. Most of the recent efforts are focused on incorporating LCA approach into an integrated sustainability assessment framework. Although environmental sustainability assessment is well established, there is still a major need for adequate and appropriate inclusion of economic and social aspects. Several qualitative tools developed for environmental assessment of buildings have been developed. Further, many tools are also being developed for systematic assessment of ecosystem services, which are an essential part of sustainable development. Some of the tools that have been developed rigorously account for economic sustainability assessment by including CBA with LCC. This way the tools account for broader economic impacts that are not immediately apparent. In order to realize social sustainability, it is essential to involve all the stakeholders in the decision-making process and make all the voices heard. In this regard, several federally recognized tools such as EJSEAT and EJSCREEN have been developed to uphold environmental justice.

13.8 Questions

13.1 Name some of the tools developed for the environmental sustainability assessment of building and urban planning and development. Explain these tools briefly.

13.2 Name any 10 ecosystem services assessment tools.

13.3 List and discuss the importance of criteria used to evaluate ecosystem services assessment tools.

13.4 What is life-cycle cost analysis? How is it different from cost–benefit analysis?

13.5 What are the different components/aspects of life-cycle cost analysis? Explain briefly.

13.6 What are the different components/aspects of cost–benefit analysis? Explain briefly.

13.7 What do you mean by environmental justice? How is it related to sustainability assessment or sustainable development of a community?

13.8 Name some of the tools used by USEPA to evaluate environmental justice?

13.9 List the indicator categories into which the EJSAT data is divided into. Explain these categories briefly.

13.10 List the 11 environmental justice indices used by EJSCREEN tool developed by USEPA.

References

Ameen, R.F.M., Mourshed, M., and Li, H. (2015). A critical review of environmental assessment tools for sustainable urban design. *Environmental Impact Assessment Review* 55: 110–125.

Bagstad, K.J., Semmens, D.J., Waage, S., and Winthrop, R. (2013). A comparative assessment of decision-support tools for ecosystem services quantification and valuation. *Ecosystem Services* 5: 27–39.

Chen, Y. and Ng, S.T. (2015). Integrate an embodied GHG emissions assessment model into building environmental assessment tools. *Procedia Engineering* 118: 318–325.

Environmental Justice Strategic Enforcement Assessment Tool (EJSEAT) (2018). http://www.nam.org/Issues/Energy-and-Environment/Affordable-Energy/Domestic-Energy/Environmental-Justice-Strategic-Enforcement-Assessment-Tool/ (accessed 23 October 2018).

Ewall, M. (2012). Legal tools for environmental equity vs. environmental justice. *Sustainable Development Law and Policy* 13 (1): 4–13. 55–56.

Hoogmartens, R., Van Passel, S., Van Acker, K., and Dubois, M. (2014). Bridging the gap between LCA, LCC and CBA as sustainability assessment tools. *Environmental Impact Assessment Review* 48: 27–33.

Kouloumpis, V. and Azapagic, A. (2018). Integrated life cycle sustainability assessment using fuzzy inference: a novel FELICITA model. *Sustainable Production and Consumption* 15: 25–34.

Kumanayake, R. and Luo, H. (2017). Development of an automated tool for buildings' sustainability assessment in early design stage. *Procedia Engineering* 196: 903–910.

Millennium Ecosystem Assessment (2005). https://www.millenniumassessment.org/en/Index-2.html (accessed 20 September 2018).

Trentin, A.D.S., Reddy, K.R., Kumar, G. et al. (2019). Quantitative Assessment of Life Cycle Sustainability (QUALICS): framework and its application to assess electrokinetic remediation. *Chemosphere*. (In Review).

Section III

Sustainable Engineering Practices

14

Sustainable Energy Engineering

14.1 Introduction

A diverse range of energy sources is critical for any healthy nation and economy. Over the past couple of centuries, much of the developed world's energy has been derived from fossil fuels – coal, oil, and natural gas. While these energy sources have proven to be useful, as described throughout this text, numerous detrimental side effects are produced by their utilization as energy sources. Such detrimental effects are related to the emission of contaminants and greenhouse gases. Moreover, many fossil fuel reserves are concentrated in a few countries, leading to significant financial and environmental impacts related to materials transport to manage supply and demand. As a result, the economies of many developed and developing countries are directly affected by supplies and prices of fossil fuels in the open market. Because of the economic and environmental stresses associated with fossil fuel use, an increased focus has been placed on renewable energy sources.

Figures 14.1–14.3 depict several key statistics pertaining to energy sources and consumption in the United States. Figure 14.1 shows the percentages of respective energy sources in the United States in 2010. As shown, an overwhelming majority of energy sources are constituted by fossil fuels – over 85%. An additional 8.8% of energy is derived from nuclear fuels; virtually all of this is used for electric power generation. Despite the significant attention that sustainable energy sources have received over the past decade, they still only constitute less than 6% of energy sources.

Figure 14.2 highlights these usage trends when considering electricity generation. Electricity generation constitutes roughly 39% of the total energy consumption in the United States. Approximately 72% of electricity generation is derived from fossil fuel energy sources. Less than 9% of electricity generation in the United States is derived from sustainable sources – including hydroelectricity, biomass, wind, geothermal, and solar. Figure 14.3 depicts the resulting energy consumption with respect to energy source categories. As shown, fossil fuels have steadily remained the primary energy source as compared to renewable sources or even nuclear energy sources (EIA 2017a).

This chapter discusses environmental impacts of energy generation using fossil fuels, discusses nuclear energy, strategies for clean energy, and finally presents various renewable energy sources for sustainability.

Sustainable Engineering: Drivers, Metrics, Tools, and Applications, First Edition.
Krishna R. Reddy, Claudio Cameselle, and Jeffrey A. Adams.
© 2019 John Wiley & Sons, Inc. Published 2019 by John Wiley & Sons, Inc.

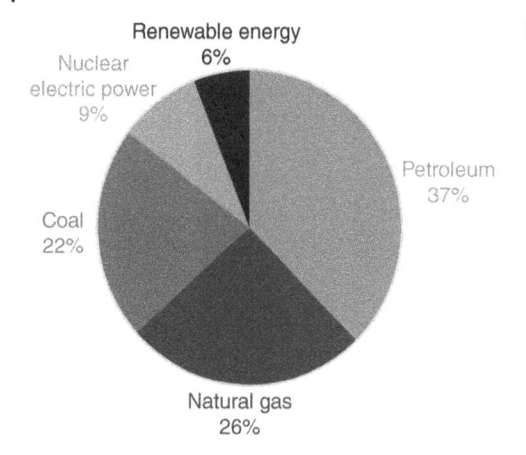

Figure 14.1 Energy sources in US in 2010.

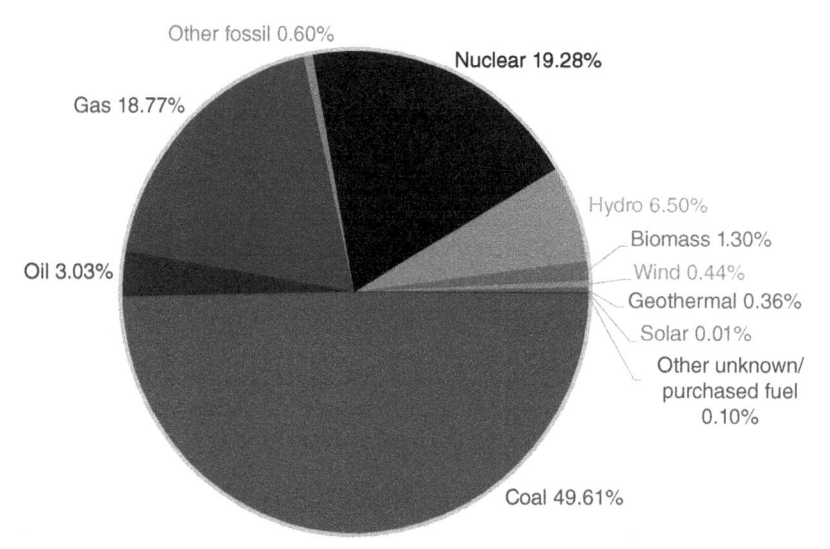

Figure 14.2 Fuel mix for electricity generation in the United States in 2010.

14.2 Environmental Impacts of Energy Generation

Energy is essential for civilization, especially for urban civilizations that consume significant amounts of energy in homes, commercial centers, heating systems, and transportation, as well as in the industrial production of goods. However, the generation, transportation, and utilization of energy sources results in several key environmental impacts. Many of these are direct effects and are quite intuitive, while others are more indirect and not commonly considered. A few of the key effects – air emissions, water resource use, water discharges, solid waste generation, and land resource use – are discussed here in.

14.2.1 Air Emissions

The generation and emission of airborne pollutants are a primary side effect of energy-related activities. Over 85% of anthropogenic greenhouse gas emissions result from energy-related activities; most of these are in the form of carbon dioxide resulting from the consumption of fossil fuels (IPCC 2014). More than one-half of these emissions result from large-scale,

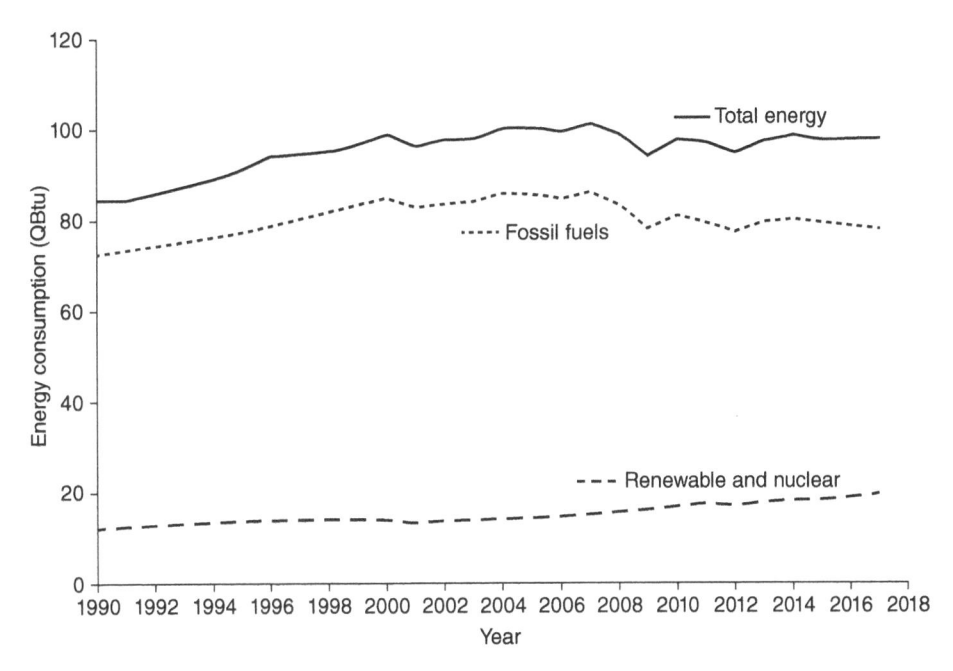

Figure 14.3 Energy consumption (quadrillion BTU) in the US in the period 1990–2017. Source: EIA (2017a) and data from USEPA.

stationary sources, while approximately one-third are generated from transportation-related sources. When considering these sources further, electrical power plants result in the largest-percentage of these emissions. Fossil fuel-fired power plants are responsible for the generation of 67% of the United States' sulfur dioxide emissions, 23% of the nitrogen oxide emissions, and 40% of anthropogenic carbon dioxide emissions (De Gouw et al. 2014). These emissions can lead to several detrimental effects on the environment, including smog, acid rain, and other aesthetic and public-health threatening consequences. Additionally, the emission of greenhouse gases can contribute to climate change. To further demonstrate the emission of greenhouse gases, Figure 14.4 presents a summary of these emissions by energy sector. As shown, fossil fuel combustion constitutes the highest percentage of these emissions. Figures 14.5–14.8 further describe the sectors that utilize fossil fuels and result in the generation of emissions. As depicted in these figures, quite intuitively, electrical generation and transportation activities consume the most fossil fuels and result in the highest percentages of emissions.

Coal has long been considered the "dirtiest" of the fossil fuels, or more clearly, results in the greatest generation of deleterious emissions. When coal is burned, carbon dioxide, sulfur dioxide, nitrogen oxides, and mercury compounds are released as emissions into the atmosphere (Miller et al. 2004; Pacyna et al. 2006). Because of these deleterious compounds, coal-fired boilers are required to have control devices in place to reduce the amount of emissions that are released into the atmosphere. For a megawatt-hour (MWh) of electrical generation, 2249 pounds of carbon dioxide, 13 pounds of sulfur dioxide, and 6 pounds of nitrogen oxides are generated, respectively. The mining, processing, and transportation of coal from a mine to a power plant generates additional emissions. Further, because of its common presence with coal, methane is often vented to the atmosphere during the mining and processing of coal. As previously discussed, methane is a more potent greenhouse gas when compared to carbon dioxide.

Though less commonly used, oil is considered a "cleaner" alternative for use in electricity generation. The combustion of oil results in the emission of nitrogen oxides, sulfur dioxide,

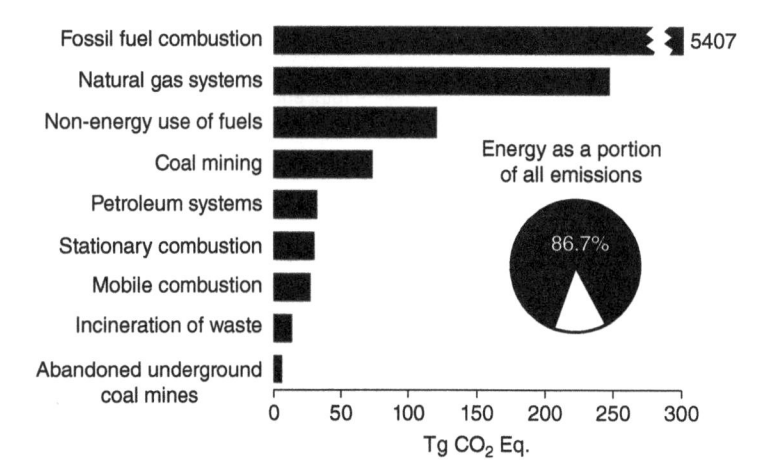

Figure 14.4 Emission of greenhouse gases expressed as CO_2 equivalents by energy production sector.

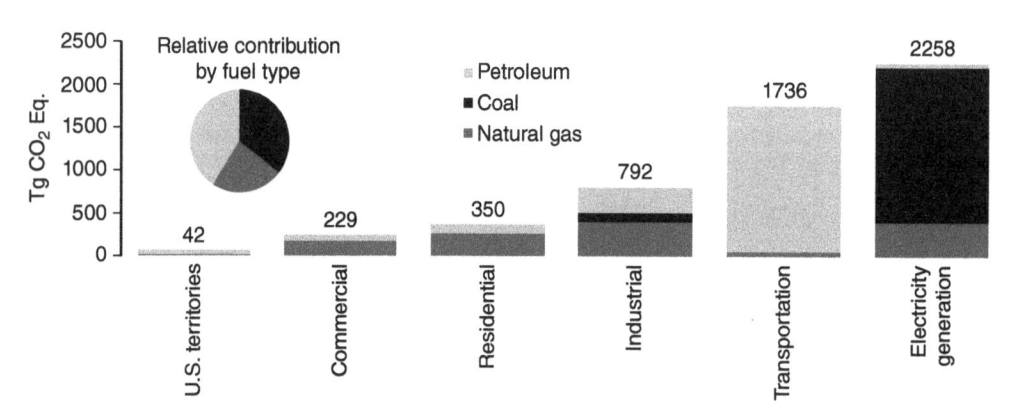

Figure 14.5 Emission of greenhouse gases (as CO_2 equivalent) in 2010 from fossil fuel combustion by sector and fuel type.

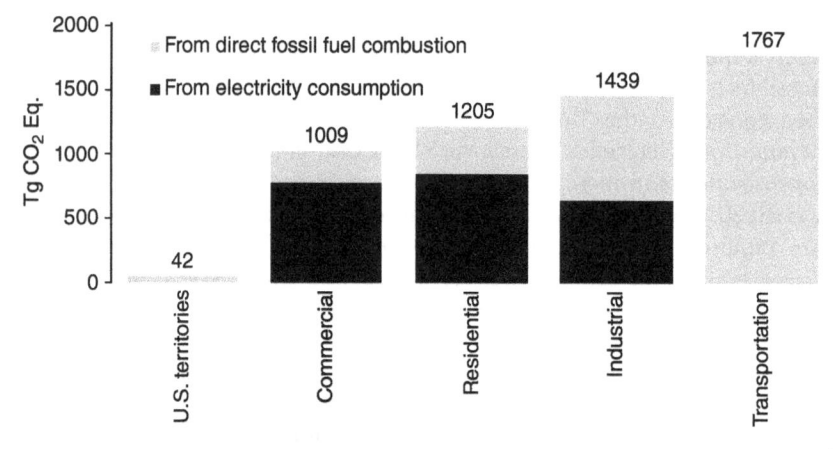

Figure 14.6 End-use sector emissions of greenhouse gases (as CO_2 equivalent) in 2010 from fossil fuel combustion.

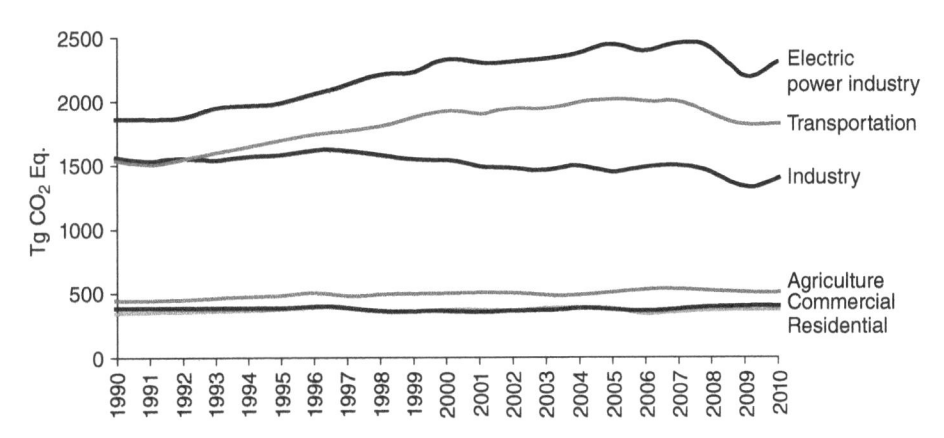

Figure 14.7 Emission of greenhouse gases (as CO$_2$ equivalent) allocated to economic sectors.

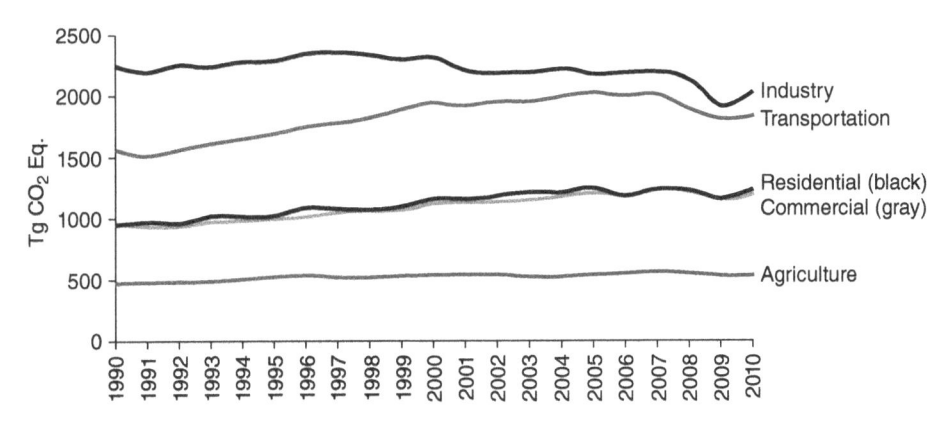

Figure 14.8 Emission of greenhouse gases (as CO$_2$ equivalent) from electricity distributed to economic sectors.

carbon dioxide, methane, and mercury compounds; although the amounts of sulfur dioxide and mercury compounds can vary widely depending on the relative content of sulfur and mercury present in the oil being consumed. Emissions are far more favorable for oil as compared to coal because a megawatt-hour of electrical generation by oil generates 1672 pounds of carbon dioxide, 12 pounds of sulfur dioxide, and 4 pounds of nitrogen oxides, respectively (EIA 2016). However, oil wells and oil collection equipment also result in the emission of methane. Further, oil extraction, transportation, and refining are highly energy-intensive processes that result in the consumption of vast quantities of natural gas or fuels, resulting in additional emissions.

Because of these concerns and regulations related to emissions, natural gas, considered to be the "cleanest" of all the fossil fuels, is being increasingly used for electricity generation. Nevertheless, natural gas consumption does result in the generation of nitrogen oxides and carbon dioxide. Further, when not completely burned or when leaking from pipelines or other similar facilities, methane can be emitted directly to the atmosphere. Again, emissions are far more favorable as compared to coal and oil; for a megawatt-hour of electrical generation, 1135 pounds of carbon dioxide, 0.1 pounds of sulfur dioxide, and 1.7 pounds of nitrogen oxides are generated, respectively (EIA 2016). As shown, natural gas combustion results in half of the emission of carbon dioxide and significantly less sulfur dioxide and nitrogen oxides when compared to coal combustion. Although the extraction and transportation of natural gas results in emissions, these are also lower as compared to oil and coal (Entrekin et al. 2011; Brandt et al. 2014).

14.2.2 Solid Waste Generation

The processing and consumption of fossil fuels (particularly coal and oil) result in the generation of solid waste materials in addition to gaseous emissions. During mining of coal, significant quantities of soil and rock overburden are generated. These materials, referred to as tailings, require active management and disposal. The burning of coal also results in ash. This waste, typically consisting of metal oxides and alkalis, can constitute up to 10% of the combusted coal. Further, solid wastes are generated during processing as well as during emissions control (through the use of "scrubbers") when pollutants are removed from stack gas. Much of this waste needs to be disposed in landfills or in abandoned mines, although some of this is being recycled and diverted into useful products, including cement and building materials (Deonarine et al. 2013; Yao et al. 2015). Oil refining also creates solid waste and is commonly in the form of wastewater sludge and other solids that can contain high concentrations of heavy metals and other toxic compounds (Hu et al. 2013). Further, solid residues can be generated during the combustion process during electricity generation.

14.2.3 Water Resource Use

Fuel processing and electricity generation requires the use of significant quantities of water in various stages of the process: extraction, transportation, and energy production process. Water is used primarily as a heat transfer substrate – it is commonly converted from liquid form into steam to drive turbines and electrical generator equipment, and is also used as a coolant. However, such use can result in detrimental effects to the resource itself or the greater environment (Miller et al. 2004).

When coal is processed at the mine, significant quantities of water are used to remove impurities. Stormwater can also pick up impurities when it comes in contact with uncovered coal stockpiles. When discharged, the impurity-laden wash water can harm the environment, including land and water bodies, due to the presence of heavy metals. Commonly, this process is closely monitored and requires permitting and/or regulatory oversight. Water is also used for steam generation and cooling purposes at the power plant (Spang et al. 2014). Extraction of the required water (e.g. river, lake, or groundwater) can affect aquatic ecosystems. Additionally, once consumed in the power generation or cooling cycle, pollutants can accumulate in the water. When released into surface waters, these pollutants can also have a negative effect on ecosystem health. Further, thermal pollution (e.g. the releases of warm or hot water into a surface water body) can have a detrimental effect on ecosystem health.

Electricity generation-related oil and natural gas combustion results in similar detrimental effects that occur with coal combustion, namely pollutant loading/release and thermal pollution during the release of wastewater. Additionally, oil drilling/extraction utilizes significant quantities of water. As with coal, this water use and release can result in the contamination of surface water bodies or groundwater, typically through the release of hydrocarbons. Further, transportation of oil is a relatively risky process; water-borne transportation is often required, and spills can have catastrophic effects to wildlife and the aquatic environment.

14.2.4 Land Resource Use

The extraction, generation, and disposal of wastes associated with electricity also require the use of applicable land resources. First, the energy sources themselves must be extracted from the Earth, either in the form of mines or extraction wells. In addition to consuming large quantities of land, these extraction activities are extremely invasive and result in unavoidable alteration or destruction of natural habitat areas, including deforestation, soil erosion, and impacts to surface water and groundwater. These materials also need space for processing (e.g. refineries, processing plants) to yield usable materials. Electricity generation plants also occupy large tracts of land, both for the structures needed for generation, control, and transmission as well as for the storage of fuel materials (Giampietro et al. 2014). Finally, the wastes generated need to be properly stored or disposed, commonly in landfills, abandoned mines, or specialized storage facilities.

14.3 Nuclear Energy

Before we discuss the advantages and drawbacks of several renewable energy sources, it is important to discuss nuclear energy. Nuclear energy has been a popular energy source worldwide; however, because of real or perceived dangerous side effects, it has been arguably the most controversial energy source on the planet.

The theory behind nuclear energy is relatively straightforward – when atoms are split, significant energy is liberated, and this energy may be captured to generate electricity. This process is known as nuclear fission. In a nuclear power plant, nuclear fission occurs within a nuclear reactor. Because of its relative ease of use in the fission process, uranium is commonly used for nuclear fuel. The uranium atom is relatively large and relatively easy to split, securing it as a popular choice for fission. Because uranium is a mined resource (derived from mineral ores), it is not considered renewable; however, a relatively minute amount of uranium is needed to derive vast amounts of energy as compared to other sources.

Figure 14.9 provides a more detailed schematic of how energy is derived from a nuclear source. In a nuclear reactor, fuel rods (consisting of uranium pellets) are placed into a water vessel. A reaction is induced in which the uranium atoms are split, releasing energy in the form of heat. The prodigious amount of heat that is liberated generates steam, which in turn is directed to a turbine. The turbine powers an electrical generator, creating electricity. The steam cools and is transformed back into liquid water. It is circulated back into the system, where the cycle repeats. In many cases, excess heat is released into the atmosphere via cooling towers.

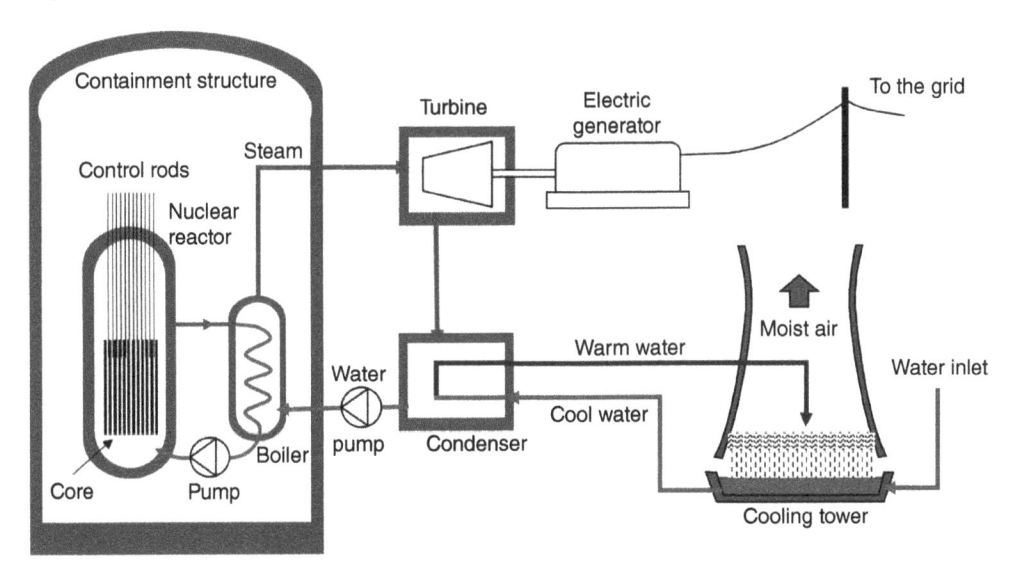

Figure 14.9 Sketch of a nuclear power plant.

Nuclear energy has several attractive benefits. As a result, nuclear reactors are used in many developed countries to generate electricity (NEI 2017). The United States has 104 nuclear reactors, accounting for approximately one-fifth of the electrical generation in the United States. The use of nuclear fuel does not result in the burning of fuels; therefore, zero greenhouse gas emissions are generated (EIA 2016). Further, nuclear fuel is extremely efficient – one ton of uranium can produce 40 million kilowatt-hours of electricity, which is equal to approximately 16 000 tons of coal or 80 000 barrels of oil. Not only are the emissions associated with fossil fuel use eliminated, the fuel used to transport a ton on uranium is obviously significantly less than the energy required to transport equivalent energy productive quantities of coal or oil (Corner et al. 2011).

Nevertheless, nuclear energy also has several unique drawbacks. The mining and refining of uranium ore requires significant amounts of energy input, but the main drawback is related to the spent nuclear fuel; once the uranium has been used up in a nuclear reactor, highly radioactive waste remains. These wastes will remain dangerously radioactive for centuries, and the proper, safe disposal of these materials are still the subject of much contentious debate (Krauskopf 2013). Further, nuclear fuels can be refined into weapons-grade materials; therefore, the proliferation of nuclear energy for peaceful purposes can be used to develop weapons-grade material for nuclear weapon arsenals.

Nuclear energy electricity generation, like the use of fossil fuels, can result in thermal pollution when wastewater is discharged to the environment. Fortunately, steam and wastewater from the cooling process are not radioactive, as they do not contact the fuel rods. Nevertheless, stormwater runoff at mines/processing facilities can contain heavy metals and/or trace amounts of radioactive materials. In addition, major risks for the environment are associated with the handling of solid waste from the nuclear reactor. Every 18 to 24 months, nuclear power plants must shut down so that the spent uranium fuel rods can be moved from the reactor. Although no longer useful as a fuel material, this material is still highly radioactive. Currently, rods are stored at the facility where they were used, typically in steel and concrete-lined vaults. Approximately 2000 metric tons of these materials are produced annually in the United States. Additionally, much of the electrical generation equipment also becomes radioactively contaminated. All of these materials will remain dangerously radioactive for centuries. Although the packaging, transport, storage, and disposal of these wastes are strictly regulated by the US Department of Transportation and the US Nuclear Regulatory Commission, a permanent storage solution has not yet been determined (Ewing and Von Hippel 2009).

14.4 Strategies for Clean Energy

The consumption of fossil fuel for energy production, especially for transportation and electric energy generation, has significant negative side effects for the environment and public health. Thus, several strategies have been developed to counteract these negative effects. Some of these include the following:

- *Energy efficiency and reduced consumption* – Implementation of best management practices and technologies to reduce energy consumption.
- *Pollution controls* – Minimization of emissions through the adoption and implementation of new technologies and controls.
- *Carbon capture and sequestration* – Interim or long-term diversion, capture, and sequestration of carbon-laden emissions, either into water bodies or underground.
- *Renewable energy* – Increased use of energy from renewable sources coupled with a reduction of fossil fuel consumption.

While these strategies are broad-based, several specific initiatives can be implemented to pursue these goals. Some of these initiatives are as follows:

- *Energy efficient buildings* – Buildings account for 30% of all greenhouse gas emissions in the United States. These emissions can be reduced through the adoption of a range of technology-focused strategies, including the use of more efficient heating, air conditioning, and lighting systems, as well as the adoption of architectural principles and materials selection that can reduce the need for these systems. Figure 14.10 shows the use of electric energy in the American home; almost 50% is associated to heating and cooling systems and appliances (EIA 2017b).
- *Green vehicles* – Cars, trucks, buses, airplanes, trains, and other vehicles account for almost one-third of energy consumption in the United States, and they result in approximately one-third of greenhouse gas emissions. Vehicles are increasingly being developed that can use less energy. Cars are becoming more fuel efficient due to the incorporation of lighter materials and more efficient engines, resulting in less fuel consumption and related emissions per mile traveled. Alternative fuel vehicles, including hybrid vehicles or those that operate on electricity (Figure 14.11), standard battery cells, or hydrogen fuel cells are eliminating the need for petroleum fuels. Flexible fuel vehicles run on gasoline but can accommodate up to 85% ethanol mixtures, which are considered a renewable fuel.
- *Carbon capture and underground storage* – While there is a growing desire to move toward renewable or zero emission sources for electricity generation, it will be decades before these technologies will replace fossil fuel in these applications on a grand scale. In the meantime, there is an increased focus on the development of ways to capture carbon dioxide from power plants and factories and safely store it underground to prevent their release into the atmosphere. This can be accomplished through carbon capture and sequestration. The carbon dioxide-containing emissions can be diverted deep underground into rock formations where it can be stored safely on a permanent basis. These storage reservoirs are typically placed more than a half-mile below the ground surface. Monitors are used to assure that the captured carbon dioxide does not escape toward the surface. More details are provided in Chapter 19.

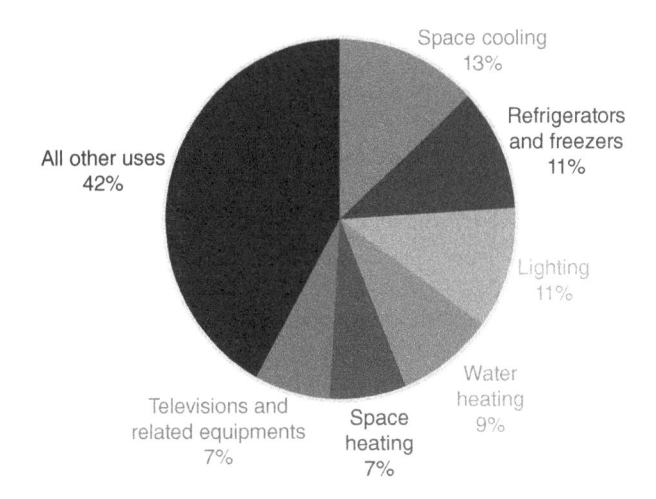

Figure 14.10 Residential major uses of electricity in the United States, 2015. Source: EIA (2017b).

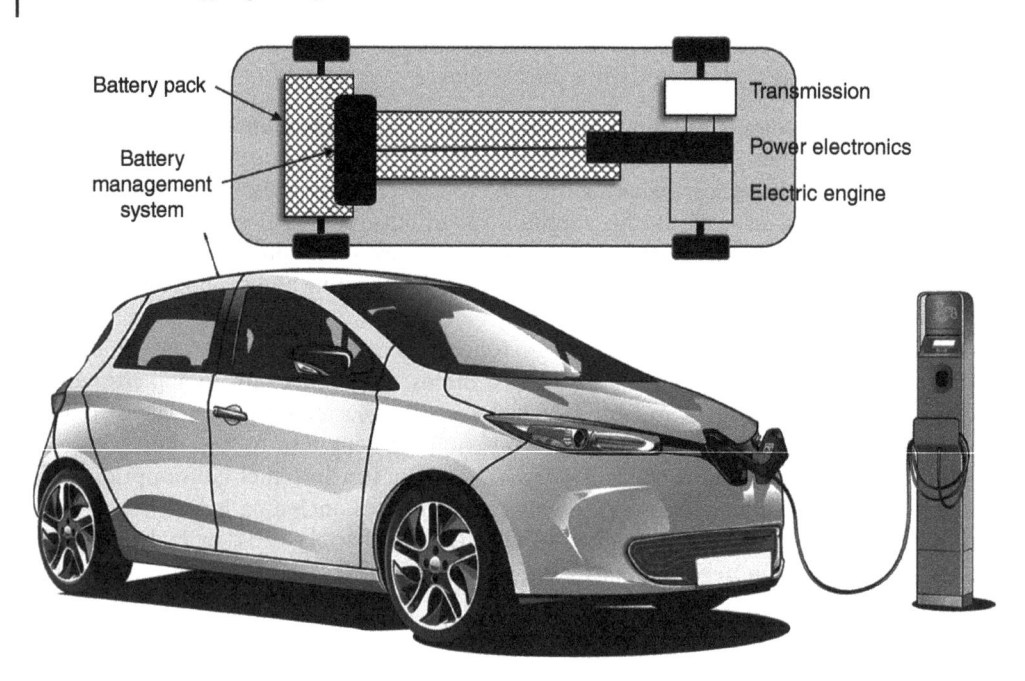

Figure 14.11 Electric car.

14.5 Renewable Energy

The adverse effects caused by the massive use of nonrenewable fossil fuels have resulted in an increasing interest in the adoption of energy derived from low or zero emission renewable energy sources (Dincer 2000). The goal of the use of such energy sources is to reduce the overall carbon footprint associated with energy consumption. Common renewable energy sources include solar, wind, water, geothermal, biomass, or landfill gas/biogas. Energy derived from renewable sources can result in reduced emissions and air pollution, lower energy bills, enhanced economic development and technological advancement, improved reliability and security of the nation's energy system, diversification of energy sources, and less reliance on imported energy sources, which often consist of fossil fuels (Twidell and Weir 2015).

However, the consideration of renewable energy sources should be approached with the same caution when considering fossil fuel sources – no single technology exists that meets all needs or eliminates all problems. When considering such systems, it is important to assess the availability of such resources, consider the costs of those systems, examine direct and indirect costs, consider permitting requirements associated with facility development and siting, the concerns of project stakeholders, and available financing or economic incentives (Evans et al. 2009).

14.5.1 Solar Energy

Solar energy is energy derived from the light of the sun (EIA 2017c). It is a renewable and practically infinite source of energy. This energy can be harnessed in one of three ways:

Photovoltaic cells – These cells, typically contained in familiar solar panels, absorb sunlight and directly convert it into electricity.

Solar thermal technology – Sunlight is gathered and focused (typically through the use of mirrors) to generate steam, which in turn power turbines and electrical generators. Once the

Figure 14.12 Solar energy for electricity production, (a) photovoltaic, (b) Solar thermal, (c) Photovoltaic plant.

water cools, it can be recirculated and converted to steam again. Often, another material is used (such as molten salt) to transfer heat to water. The molten salt can retain heat for a long period of time, allowing these systems to operate during nighttime hours.

Passive solar heating – Architectural finishes, including properly sized/sited windows, awnings and window covers can utilize solar heating in the wintertime and deflect sunlight in the summertime, lessening the requirement for heating and cooling derived from other sources.

Figure 14.12 shows the available technologies for electricity production from solar energy, and Figure 14.13 depict areas suitable to the use of solar energy resources for electricity generation.

Solar reserves are located everywhere in the United States, although some areas receive more light than others, depending on the climate and latitude. In general, the greatest reserves are located in the southwestern portion of the United States. It has been reported that enough sunlight falls on a 100-mile by 100-mile area to meet the electrical needs of the entire country. Solar systems do not result in air emissions, and water discharges are negligible. Water may be used in closed loop form within a solar thermal system. Additionally, negligible solid wastes are generated; mostly associated with equipment repair and de-commissioning as well as fabrication of components. With respect to land resources, solar systems do require area. However, photovoltaic systems can be placed on existing structures. Solar thermal systems do require larger areas, depending on the array of mirrors required. Although the systems do not create waste, the installation of these systems can affect wildlife habitat due to their presence in previously undeveloped or wild land areas.

14.5.2 Wind Energy

Wind energy has been used for thousands of years. Uneven heating of the Earth plus the Earth's rotation results in the generation of wind due to the movement of warmer and cooler air masses. From antique windmills to modern wind turbines, harnessing wind power has been beneficial for civilization. Wind turbines capture the passing wind; the spinning blades in turn spin an electrical generator and convert it to electricity. Windmills and wind farms are best suited in areas of consistent winds, including hilltops/ridges, open plains, and shorelines. Large wind farms are commonly connected to the electricity grid, allowing for transmission

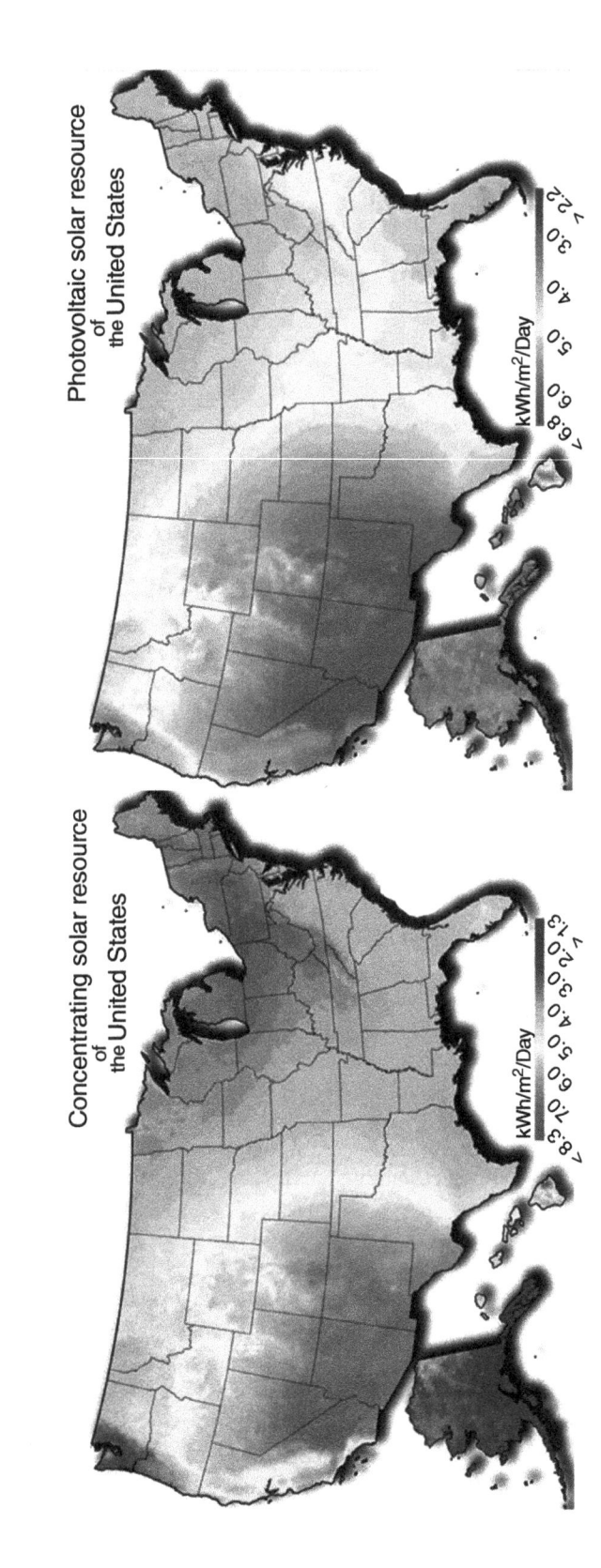

Figure 14.13 Suitable zones in the United States for the use of photovoltaic energy or concentrating solar energy for electricity production. Source: EIA, https://www.eia.gov/energyexplained.

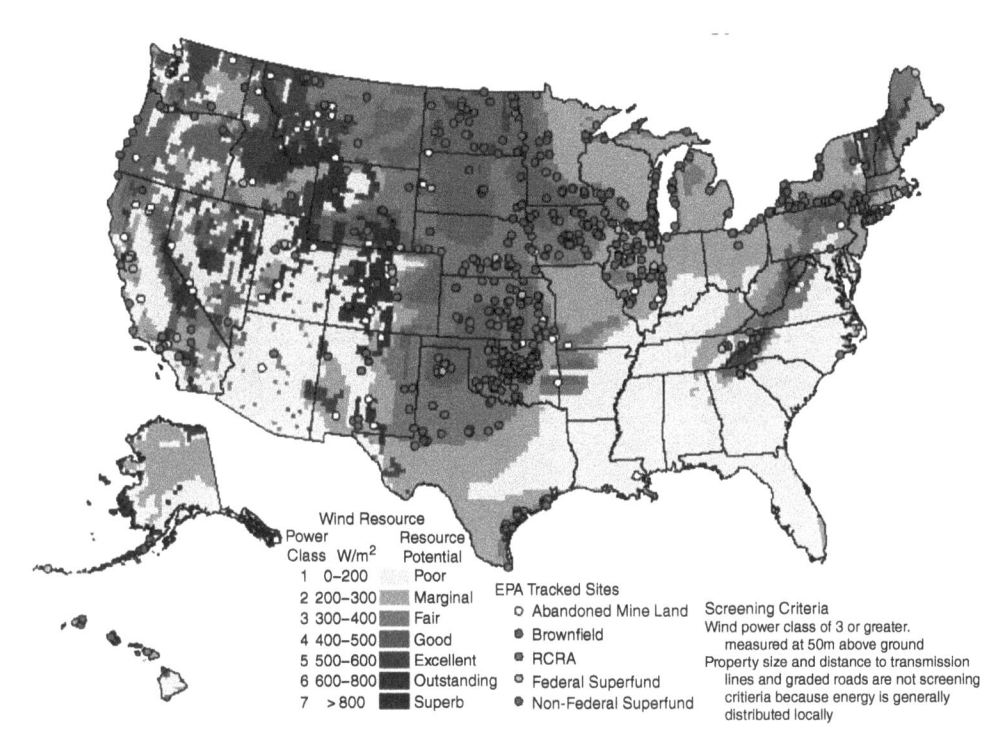

Figure 14.14 Potential areas for nongrid connected wind energy generation. Source: EIA, https://www.eia.gov/energyexplained.

of the generated electricity. Figures 14.14 and 14.15 show areas of wind energy potential on a private, community, and utility-scale level, respectively.

Wind systems can be used in various parts of the country, although the best areas include the following states: North Dakota, Texas, Kansas, South Dakota, Montana, Wyoming, Nebraska, Oklahoma, Minnesota, Iowa, Colorado, New Mexico, California, Wisconsin, and Oregon. Wind turbines are best deployed in plains and mountain passes within these areas. For instance, winds are consistent and strong enough in the Rocky Mountain and Great Plains states to meet between 10% and 25% of energy requirements in these states (Bradbury et al. 2014). Wind turbines require little to no water, other than for cleaning of blades. They do not generate air emissions, water discharges, or any appreciable solid wastes. Wind farms do require large areas of land or offshore space, and they may be considered aesthetically displeasing from a noise and visual standpoint. They can; however, be incorporated into large, open spaces (e.g. cattle grazing areas) with little effect on land use. However, they have been known to increase bird and bat mortality, and if installed improperly, they can result in soil erosion.

14.5.3 Water Energy

Flowing water also contains energy that can be harnessed for useful purposes. As with the case of wind power, this technology has been harnessed for centuries, from riverside mills to large-scale hydroelectric dams. Water energy may be harnessed for electrical generation in several ways, but all rely on the flow of rushing water to spin an electrical generator. Hydroelectric dams capture energy from flowing river water (Figures 14.16 and 14.17). Dam operators can control the flow of this water by restricting the water flow rate, which will often result in the

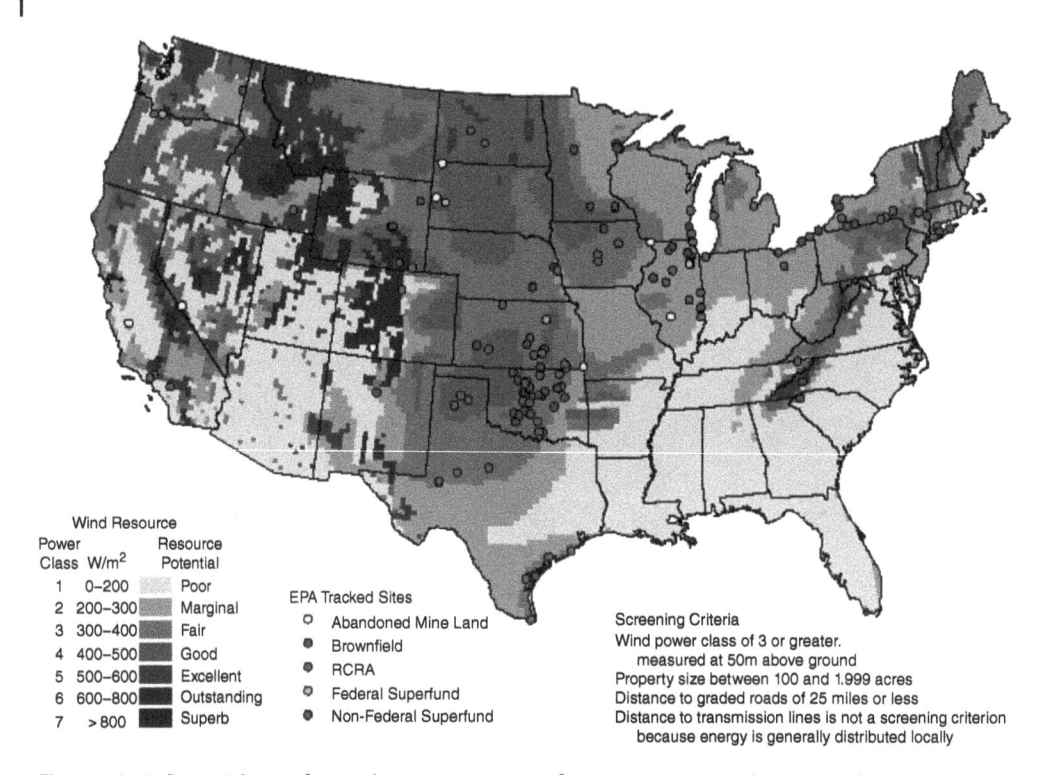

Figure 14.15 Potential areas for wind energy generation for community uses. Source: EIA, https://www.eia.gov/energyexplained.

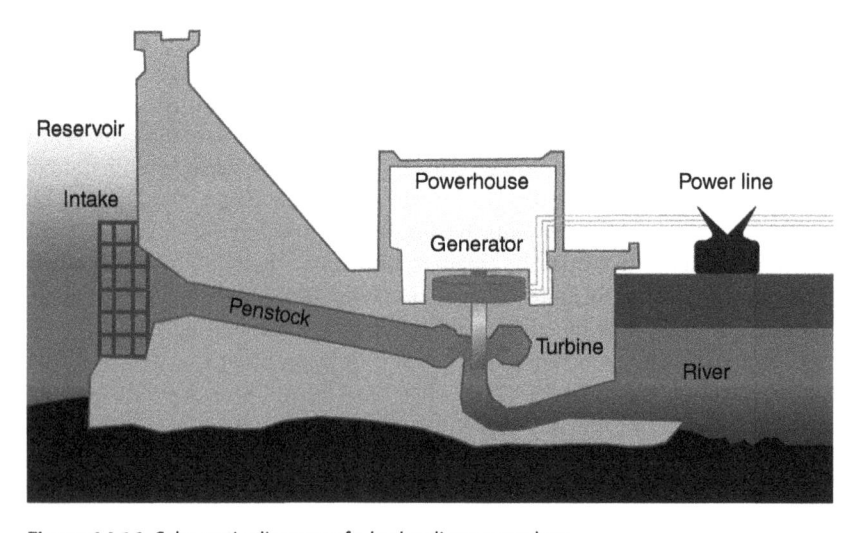

Figure 14.16 Schematic diagram of a hydraulic power plant.

creation of lakes/reservoirs behind the dams. These water bodies also can be used as valuable resource for domestic water, recreation, and wildlife habitat (Kao et al. 2015). Wave power can be captured along an ocean coastline by special buoys or floating devices. Tidal power can also be used: ebb or flood tides flowing out or in toward a coastline can also be captured for electricity generation (Uihlein and Magagna 2016).

Million megawatthours

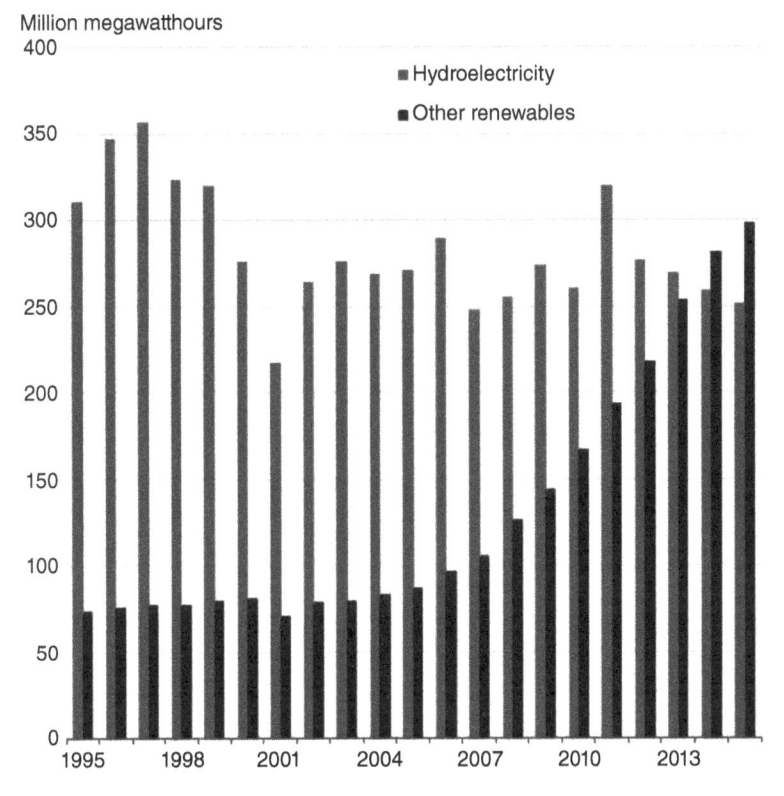

Figure 14.17 Production of hydropower and other renewable energies in the United States in the period of 1995–2005. Source: US Energy Information Administration, Monthly Energy Review, April 2016.

Water-derived power does not result in air emissions or solid waste generation; however, water-derived energy does come with several pitfalls. First, altered water flow can result in excessive vegetation, which, when decaying, can result in methane production. The use of dams can also greatly affect the flow and environment of rivers, leading to catastrophic effects on migrating fish populations. Additionally, thermal vertical gradients can result, affecting oxygen levels throughout the water column and adversely affecting wildlife and aquatic populations. Further, large-scale discharges from dams, either controlled during high flow periods or uncontrolled breaches or failures, can have adverse consequences on downstream areas. Water discharges do occur, and thermal pollution may result, but no contaminant loading occurs.

14.5.4 Geothermal Energy

Because temperatures generally increase with depth below the ground surface, in many cases, energy (geothermal energy) can be captured and used. Geothermal energy is continuously produced beneath the Earth's surface due to the extreme heat contained in molten rock (magma) within the Earth's core. In some cases, this heat is exposed to water, resulting in hot water or steam generation. This can be piped to the surface for electrical generation through turbine operation. Alternatively, geothermal energy can be tapped by piping and exposing water to hot, dry rocks, where the heated water can be re-circulated toward the surface (Figure 14.18).

Geothermal energy can be utilized under two scenarios. The first one is the capture of the Earth's heat in geothermal power plants to generate steam, and in turn, generate electricity.

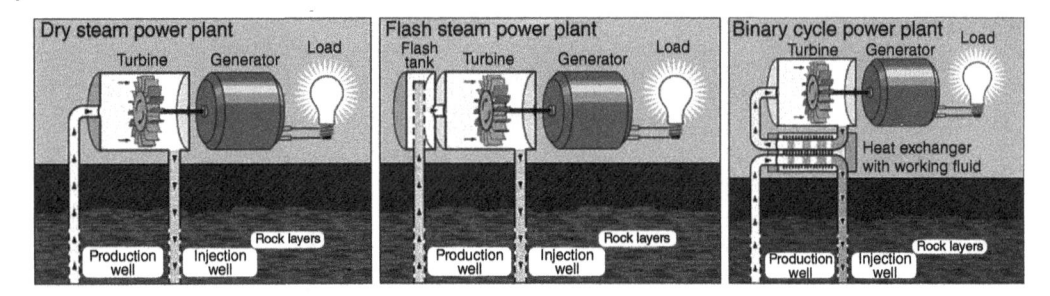

Figure 14.18 Geothermal energy power plants. Source: US EIA.

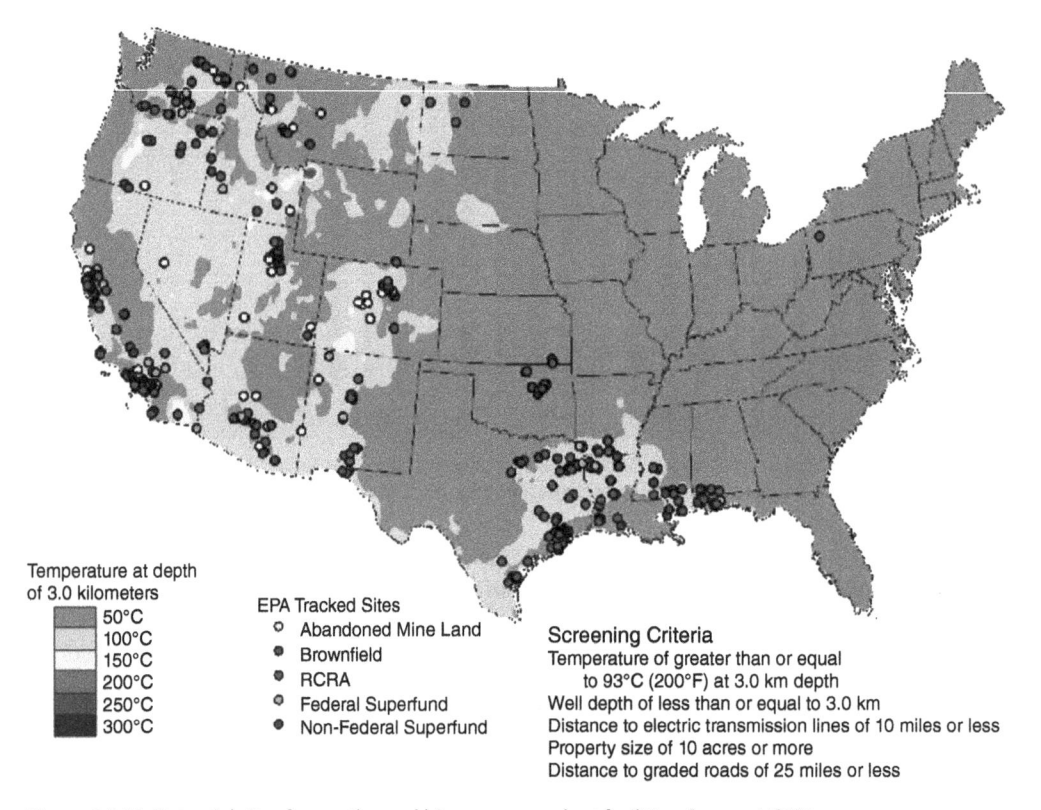

Figure 14.19 Potential sites for geothermal binary power plant facilities. Source: USEPA.

Wells are commonly drilled one to two miles below the Earth's surface to pump steam or hot water toward the surface. This is best applied in areas with hot springs, geysers, or volcanic activity. The second scenario is through the use of heat pumps. Heat that is present close to the surface is tapped and used to heat water or provide heat to buildings. These temperatures need not be hot; this technology is feasible in many locations where the near-surface temperature is commonly between 50 and 60 °F. The systems, used for heating in winter can be reversed to provide cooling in the summer. Figures 14.19–14.21 show geothermal energy potential considering these scenarios.

Geothermal reserves exist everywhere; although practically speaking, areas with known hot springs, geysers, and volcanic activity may be most easily and cost effectively utilized, including portions of California, Oregon, Nevada, Arizona, Washington, Idaho, Montana, Alaska, Utah,

14.5 Renewable Energy | 261

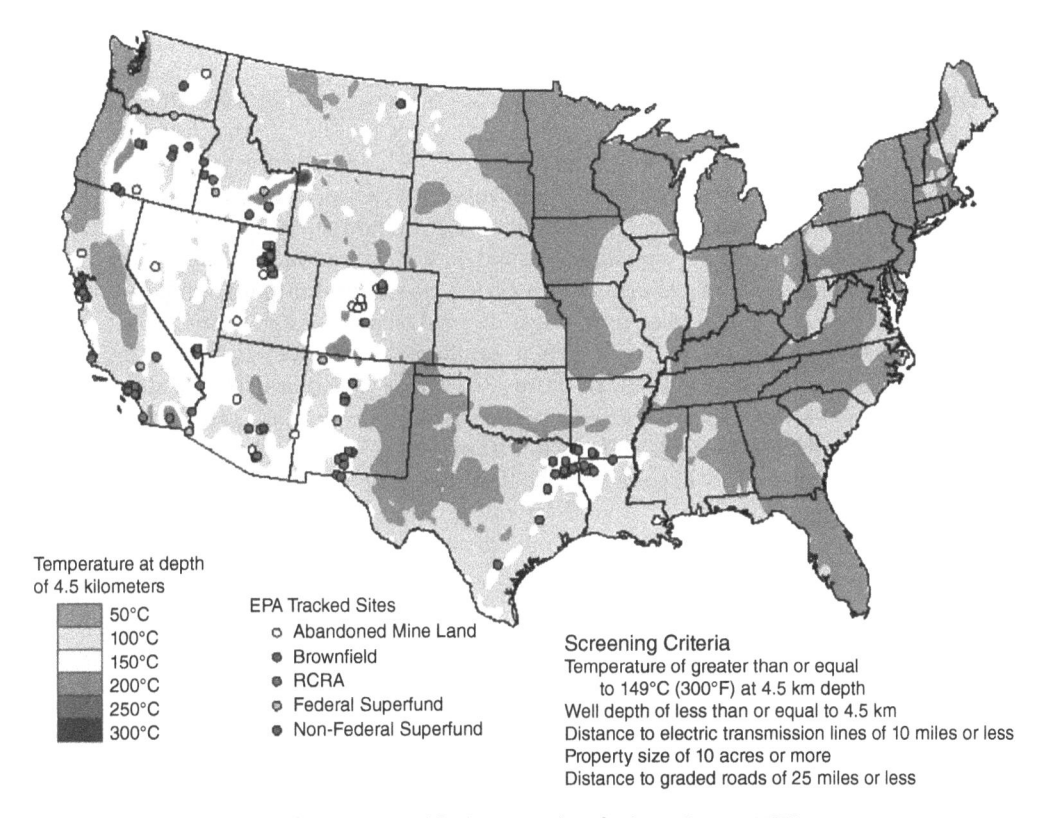

Temperature at depth of 4.5 kilometers
- 50°C
- 100°C
- 150°C
- 200°C
- 250°C
- 300°C

EPA Tracked Sites
- ○ Abandoned Mine Land
- ● Brownfield
- ● RCRA
- ○ Federal Superfund
- ● Non-Federal Superfund

Screening Criteria
Temperature of greater than or equal to 149°C (300°F) at 4.5 km depth
Well depth of less than or equal to 4.5 km
Distance to electric transmission lines of 10 miles or less
Property size of 10 acres or more
Distance to graded roads of 25 miles or less

Figure 14.20 Potential sites for geothermal flash power plant facilities. Source: USEPA.

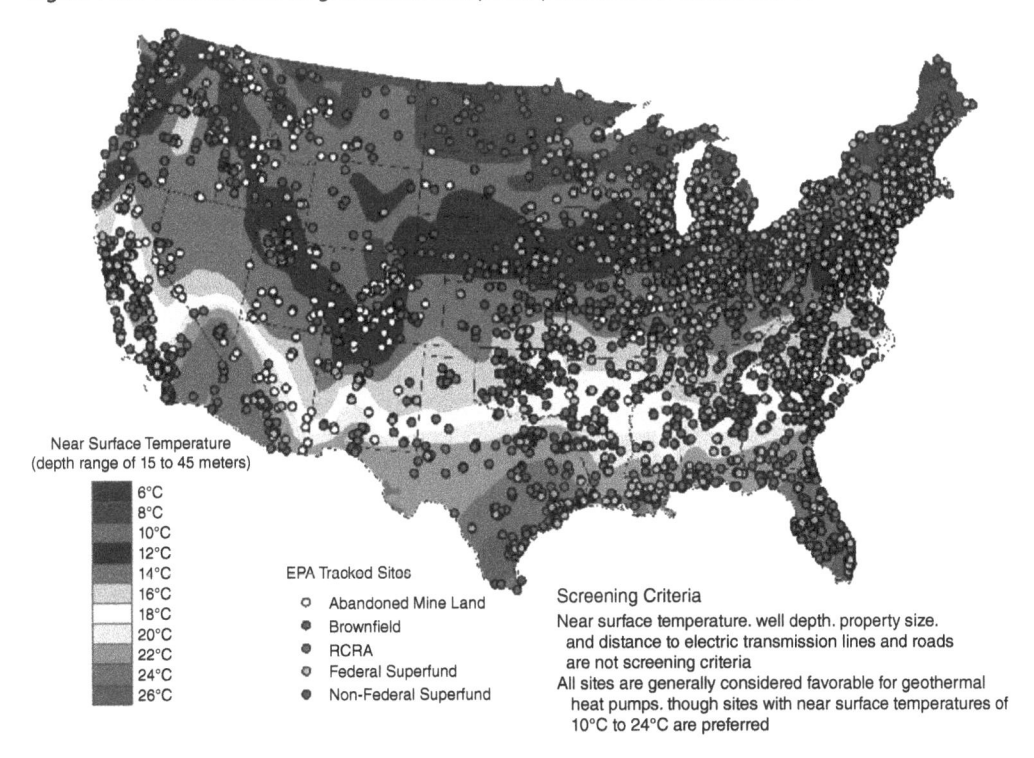

Near Surface Temperature (depth range of 15 to 45 meters)
- 6°C
- 8°C
- 10°C
- 12°C
- 14°C
- 16°C
- 18°C
- 20°C
- 22°C
- 24°C
- 26°C

EPA Tracked Sites
- ○ Abandoned Mine Land
- ● Brownfield
- ● RCRA
- ○ Federal Superfund
- ● Non-Federal Superfund

Screening Criteria
Near surface temperature, well depth, property size, and distance to electric transmission lines and roads are not screening criteria
All sites are generally considered favorable for geothermal heat pumps, though sites with near surface temperatures of 10°C to 24°C are preferred

Figure 14.21 Suitable areas for geothermal heat pump systems. Source: USEPA.

and Hawaii. In 2003, a total of 2300 MW was generated from geothermal sources. It is believed that at least 20 000 MW of sources may be effectively used.

Inherently, geothermal energy uses water resources; however, the hot water and steam are often injected back in a quasi-closed loop process. If water is used to extract heat from dry rocks, water will need to be used from an outside source. Additionally, the water is commonly injected into the subsurface and is free of contaminants, though well drilling may result in localized contamination to the subsurface, depending on the method of drilling used. Geothermal energy does not result in any air emissions or solid waste generation. In general, there is no need for a significant use of land resources; although if extracted groundwater is not re-injected, land subsidence may occur. However, in large geothermal facilities a significant impact can be observed in the subsoil and groundwater. When significant amounts of energy are dissipated or removed from the subsoil, especially in those sites with no geothermal activity, important changes in the temperature of groundwater can be measured. Florea et al. (2017) reported temperature changes in the subsoil and groundwater from 5 to 17 °C depending on the size of the geothermal system and the flow of groundwater. These temperature changes may have an enormous impact on the geochemistry of groundwater with premature carbonate precipitation and pore soil clogging. Further research in this topic is necessary to evaluate the environmental impact of geothermal systems.

14.5.5 Biomass Energy

Biomass energy is derived from materials based from plants and animals. Often these materials may be burned directly or transformed into other substances that can be burned to obtain energy. Wood or peat are common examples as materials that can be directly burned to obtain energy (IEA 2007). Biomass contains stored energy derived from sunlight converted during photosynthesis. While the burning of these materials release carbon dioxide and cannot be considered a zero or low emissions energy source, these plants absorb and sequester carbon dioxide, fixating it as the plant grows leaves, flowers, branches, and stems. Many different types of biomass may be used for energy, including wood chips, corn, and garbage materials. Some types can also be converted to liquid fuels (biofuels) for use in transportation activities, including plant and animal fat-based biodiesel and corn, sugar cane, and rice straw-derived ethanol. Although these materials may come from the primary crop yield (such as corn kernels), in many cases, biomass sources consist of plant or animal waste products, including peanut shells, sludge, manure, corn husks, and plant clippings. Because of the carbon sequestration, these may be considered carbon neutral. Additionally, because of the commonly rapid growing cycles, these materials are considered renewable (McKendry 2002).

In many cases, biomass cannot be utilized exclusively as a fuel; it may need additional fuels to burn efficiently. This process, called co-firing, often utilizes coal as a co-fuel (Dai et al. 2008). The coal use allows for higher temperature combustion, and the use of biomass greatly reduces net emissions (Koppejan and Van Loo 2012). Figures 14.22 and 14.23 show biomass potential considering the power and refining basis.

While the biomass which consists of crop yield material or "fresh" waste may be burned directly, methane, derived from decaying waste products, may also be burned to derive energy. Trash that ends up in landfills will produce methane as it decomposes. Methane can also be generated and captured from farm digesters – large tanks that generate methane from decaying manure. The methane is often purified or combined with other combustible gases and is burned to generate heat or electricity. Figure 14.24 depicts the pipe system for the collection and recovery of methane from landfill. Figure 14.25 shows the potential sites for landfill gas energy recovery as well as instances of current activity. Municipal solid waste (MSW) may also

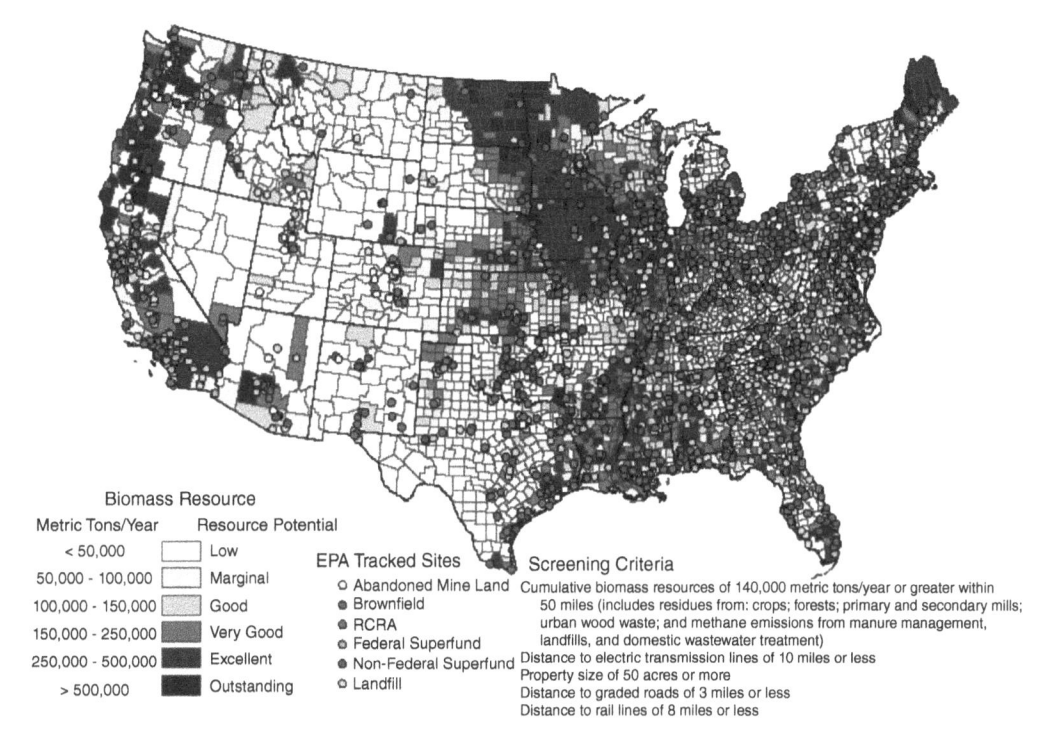

Figure 14.22 Potential areas for biopower facility siting. Source: USEPA.

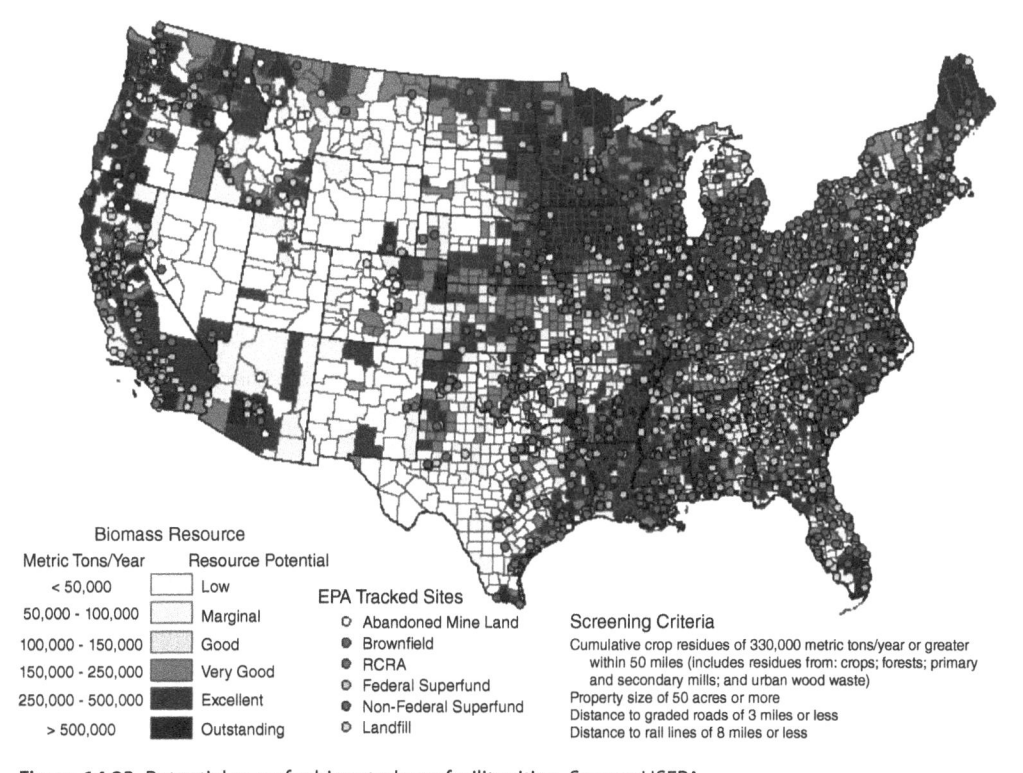

Figure 14.23 Potential areas for biopetroleum facility siting. Source: USEPA.

264 | *14 Sustainable Energy Engineering*

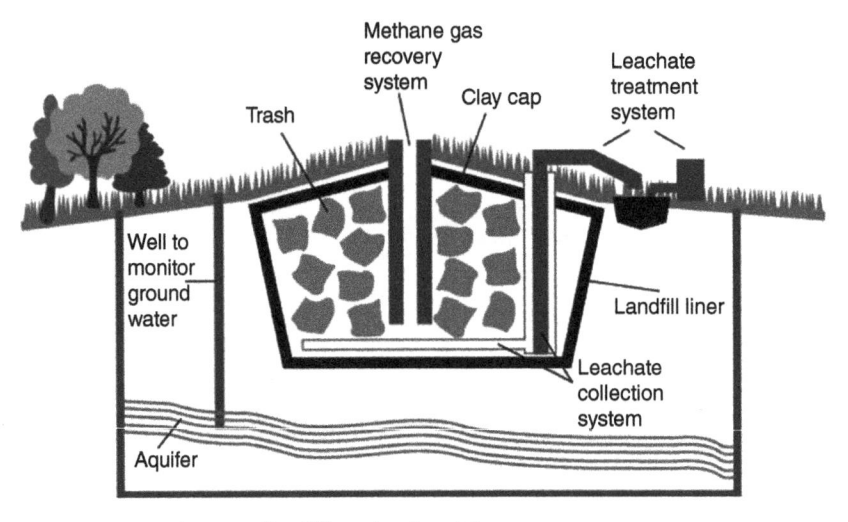

Figure 14.24 Collection of landfill gas (methane) for energy recovery.

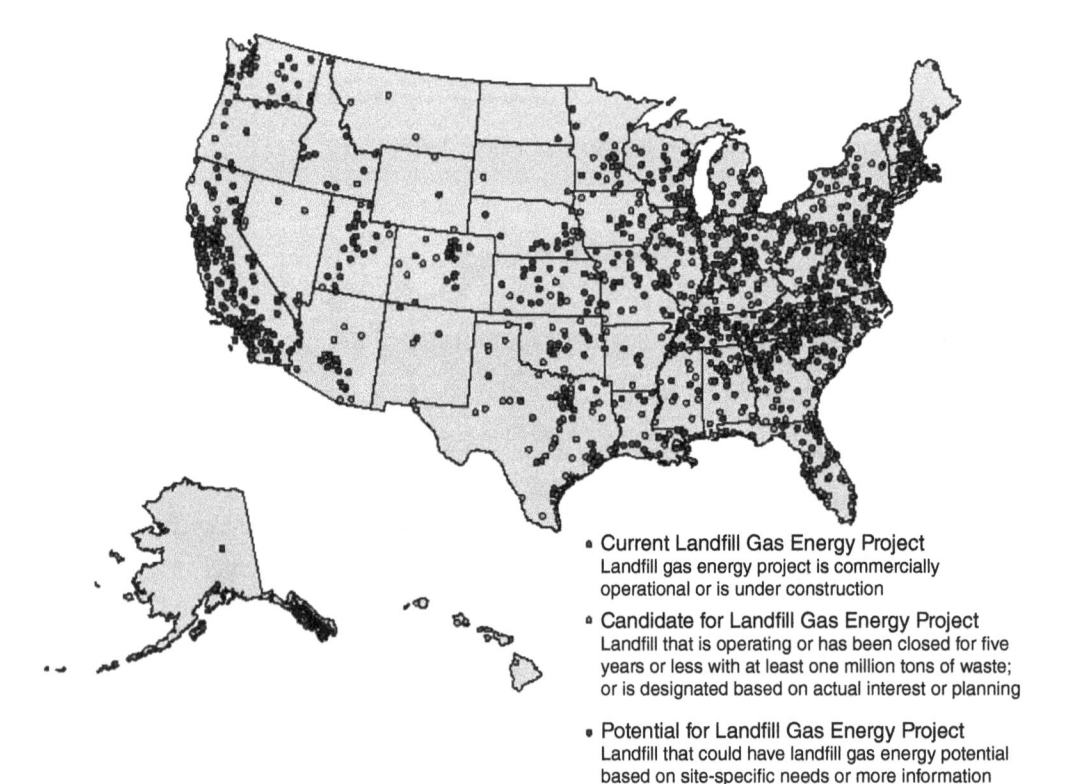

- Current Landfill Gas Energy Project
  Landfill gas energy project is commercially operational or is under construction
- Candidate for Landfill Gas Energy Project
  Landfill that is operating or has been closed for five years or less with at least one million tons of waste; or is designated based on actual interest or planning
- Potential for Landfill Gas Energy Project
  Landfill that could have landfill gas energy potential based on site-specific needs or more information

Figure 14.25 Potential sites for landfill gas energy recovery. Source: USEPA.

be incinerated to generate heat or electricity; however, doing so results in the generation of carbon dioxide, sulfur dioxide, and nitrogen oxides comparable to emissions generated during combustion of coal.

With respect to biomass, only 14 million dry tons of a currently estimated 590 million tons are available for utilization. Biomass sources are available throughout the United States. The burning of biomass results in the generation of air emissions, including nitrogen oxides and sulfur dioxide; however, the relative amounts are heavily dependent on the types of biomass burned. These emissions tend to be less than those associated with coal and therefore, as previously described, biomass can be attractive to co-fire with coal. Biomass burning has similar water re-use requirements as fossil fuel burning; namely, water needs to be extracted for steam generation and cooling. Cooling water also needs to be discharged into the surface water systems after its intended use. As with fossil fuels, water extraction and discharge can affect aquatic habitats by disrupting water levels and introducing thermal gradients. Discharged water may also result in pollutant loading of receiving waters. Solid waste (ash) is also generated during burning; however, the concentrations and mass of deleterious compounds (e.g. heavy metals) is much lower as compared to fossil fuels (IRENA 2012). From a land perspective, biomass requires productive agricultural area for cultivation as well as area associated with power generation. The cultivation of biomass-related products can incorporate sustainable procedures, including reduced use of agricultural chemicals and farming methods that reduce the dependence on water use and is focused on soil conservation.

Landfill gases, similar to biomass, results in the generation of air emissions, including nitrogen oxides and sulfur dioxide; however, the relative amounts are heavily dependent on the quality of the landfill gas burned. The emissions are often similar in constituency and magnitude to those associated with natural gas. Again, since landfill gas is the result of decaying organic matter, it is considered part of the natural carbon cycle, and the emission of carbon dioxide is considered neutral from a carbon standpoint. Water utilization as well as water and solid waste generation are similarly low in magnitude as compared to natural gas. Since landfill gases evolve from land use already devoted to waste-related activities, landfill gas combustion does not impact land resources.

14.6 Summary

While a diverse range of energy sources is necessary for a nation's economic health and quality of life, an overreliance on fossil fuels – coal, oil, and natural gas – has resulted in numerous detrimental side effects to the environment, including solid wastes, liquid wastes, and the generation of air emissions, including greenhouse gases. The overutilization of fossil fuels has taxed already depleting reserves, leaving the economies of developed and developing countries susceptible to market and pricing pressures.

Although there has been an increased focus on renewable energy sources, and interest in nuclear power sources, an overwhelming majority of energy sources utilized in the United States are still constituted by fossil fuels. Nevertheless, sustainable energy sources – including hydroelectricity, biomass, wind, geothermal, and solar, can be utilized to greatly reduce wastes, emissions, and other stressors to the environment. Further, the appropriate use of these technologies based on the geographic/special availability of natural resources/phenomena from which they can be derived can lead to energy independence derived from a diverse mix of energy sources, providing a greater range of energy sources in the energy market. This can lead to improved macroeconomic health as well as localized economic stimulus through lower energy prices and job creation as these energy sources are developed and utilized.

14.7 Questions

14.1 What are the environmental impacts of energy generation?

14.2 Name some of the major sectors that utilize large amounts of energy.

14.3 Rank the order of the three major fossil fuel energy sources based on their carbon emissions.

14.4 Describe the contributions to air emissions from different fossil fuel energy sources.

14.5 What are the three major nonrenewable fossil fuel sources of energy?

14.6 Name a few sources of renewable energy.

14.7 What are the advantages and challenges of renewable energy sources?

14.8 Explain the principles behind the generation of nuclear energy and geothermal energy.

14.9 Discuss the benefits and drawbacks of nuclear energy use.

14.10 What are the few strategies developed to curb the negative impacts of energy generation using nonrenewable sources?

14.11 Compare and contrast strategies used to reduce energy demand.

14.12 Discuss some potential reuse options of solid waste generated from fossil fuel energy source use.

14.13 Discuss some of the environmental side effects that may arise from renewable energy source use.

14.14 Discuss potential sources and uses of biomass as an energy source.

References

Bradbury, K., Pratson, L., and Patiño-Echeverri, D. (2014). Economic viability of energy storage systems based on price arbitrage potential in real-time US electricity markets. *Applied Energy* 114: 512–519.

Brandt, A.R., Heath, G.A., Kort, E.A. et al. (2014). Methane leaks from North American natural gas systems. *Science* 343 (6172): 733–735.

Corner, A., Venables, D., Spence, A. et al. (2011). Nuclear power, climate change and energy security: exploring British public attitudes. *Energy Policy* 39 (9): 4823–4833.

Dai, J., Sokhansanj, S., Grace, J.R. et al. (2008). Overview and some issues related to co-firing biomass and coal. *The Canadian Journal of Chemical Engineering* 86 (3): 367–386.

De Gouw, J.A., Parrish, D.D., Frost, G.J., and Trainer, M. (2014). Reduced emissions of CO_2, NO_x, and SO_2 from US power plants owing to switch from coal to natural gas with combined cycle technology. *Earth's Future* 2 (2): 75–82.

Deonarine, A., Bartov, G., Johnson, T.M. et al. (2013). Environmental impacts of the Tennessee Valley Authority Kingston coal ash spill. 2. Effect of coal ash on methylmercury in historically contaminated river sediments. *Environmental Science & Technology* 47 (4): 2100–2108.

Dincer, I. (2000). Renewable energy and sustainable development: a crucial review. *Renewable and Sustainable Energy Reviews* 4 (2): 157–175.

EIA (2016). Carbon dioxide emissions coefficients. U.S. Energy Information Administration. https://www.eia.gov/environment/emissions/co2_vol_mass.cfm (accessed 12 April 2017).

EIA (2017a). Total energy. US Energy Information and Administration. Independent Statistics and Analysis. https://www.eia.gov/totalenergy (accessed April 2017).

EIA (2017b). Electricity explained – use of electricity. US Energy Information and Administration. Independent statistics and analysis. https://www.eia.gov/energyexplained/ (accesses April 2017).

EIA (2017c). Energy from the Sun. US Energy Information and Administration. Independent Statistics and analysis. https://www.eia.gov/energyexplained (accessed 12 April 2017).

Entrekin, S., Evans-White, M., Johnson, B., and Hagenbuch, E. (2011). Rapid expansion of natural gas development poses a threat to surface waters. *Frontiers in Ecology and the Environment* 9 (9): 503–511.

Evans, A., Strezov, V., and Evans, T.J. (2009). Assessment of sustainability indicators for renewable energy technologies. *Renewable and Sustainable Energy Reviews* 13 (5): 1082–1088.

Ewing, R.C. and Von Hippel, F.N. (2009). Nuclear waste management in the united states – starting over. *Science* 325 (5937): 151–152. https://doi.org/10.1126/science.1174594.

Florea, L.J., Hart, D., Tinjum, J., and Choi, C. (2017). Potential impacts to groundwater from ground-coupled geothermal heat pumps in district scale. *Groundwater* 55 (1): 8–9.

Giampietro, N., Aspinall, R.J., Ramos-Martin, J., and Bukkens, S.G.F. (2014). *Resource Accounting for Sustainability Assessment. The Nexus between Energy, Food, Water and Land Use*. New York, NY: Routledge.

Hu, G., Li, J., and Zeng, G. (2013). Recent development in the treatment of oily sludge from petroleum industry: a review. *Journal of Hazardous Materials* 261: 470–490.

IEA (2007). *Bioenergy Project Development and Biomass Supply*. Paris, France: International Energy Agency.

IPCC (2014). Climate Change 2014: Synthesis Report. Contribution of Working Groups I, II and III to the Fifth Assessment Report of the Intergovernmental Panel on Climate Change. Intergovernmental Panel on Climate Change, Geneva, Switzerland, 151 pp.

IRENA (2012). Biomass for power generation. IRENA working paper. Renewable energy technologies: cost analysis series. International Renewable Energy Agency. Abu Dhabi,

Kao, S.C., Sale, M.J., Ashfaq, M. et al. (2015). Projecting changes in annual hydropower generation using regional runoff data: an assessment of the United States federal hydropower plants. *Energy* 80: 239–250.

Koppejan, J. and Van Loo, S. (eds.) (2012). *The Handbook of Biomass Combustion and Co-firing*. Routledge.

Krauskopf, K. (2013). *Radioactive waste disposal and geology*, vol. 1. Springer Science & Business Media.

McKendry, P. (2002). Energy production from biomass (part 1): overview of biomass. *Bioresource Technology* 83 (1): 37–46.

Miller, P.J., Atten, C.V., and Bradley, M.J. (2004). *North American Power Plant Air Emissions. Commission for Environmental Cooperation of North America*. Quebec, Canada: Montreal.

NEI (2017). World statistics. Nuclear Energy Around the World. Nuclear Energy Institute. https://www.nei.org/Knowledge-Center/Nuclear-Statistics/World-Statistics (accessed 12 April 2017).

Pacyna, E.G., Pacyna, J.M., Steenhuisen, F., and Wilson, S. (2006). Global anthropogenic mercury emission inventory for 2000. *Atmospheric Environment* 40 (22): 4048–4063.

Spang, E.S., Moomaw, W.R., Gallagher, K.S. et al. (2014). The water consumption of energy production: an international comparison. *Environmental Research Letters* 9 (10): 105002.

Twidell, J. and Weir, T. (2015). *Renewable Energy Resources*. Routledge.

Uihlein, A. and Magagna, D. (2016). Wave and tidal current energy–a review of the current state of research beyond technology. *Renewable and Sustainable Energy Reviews* 58: 1070–1081.

Yao, Z.T., Ji, X.S., Sarker, P.K. et al. (2015). A comprehensive review on the applications of coal fly ash. *Earth-Science Reviews* 141: 105–121.

15

Sustainable Waste Management

15.1 Introduction

Waste is the inevitable by-product of all human activities. Virtually every activity generates some type of physical by-product. When the materials that result from any activity are no longer useful or have degraded such that they no longer fulfill their original or otherwise useful purpose, they are classified as a waste material. These wastes are commonly classified based on their respective generative sources. Some of these activities include those associated with residences, commercial and retail businesses, institutions, construction and demolition activities, municipal services, water/wastewater treatment, and incineration facilities. Several industrial processes generate waste materials, including industrial construction and demolition, fabrication, manufacturing, chemical refining and synthesis, and nuclear power and defense purposes. Additionally, these waste materials may be classified depending on their inherent characteristics and physical properties (USEPA 2015). The storage, handling, transport, and disposal of these materials may be handled in a variety of ways, and new methods are under consideration to make these processes more sustainable.

Waste management is an area of increasing importance. Waste generation, both hazardous and nonhazardous wastes, continues to increase at an appreciable rate (USEPA 2017a). For instance, industries generate and dispose over 7.6 billion tons of industrial solid wastes each year. Over 40 million tons of this material is considered hazardous. Additionally, both nuclear and medical wastes are increasing in quantity on an annual basis.

Proper waste management is critical to achieve sustainability. This chapter presents waste types and their impacts, waste management approach (hierarchy), and evolving integrated/sustainable waste management strategies and approaches. Finally, the concept of circular economy is presented.

15.2 Types of Waste

The Resource Conservation and Recovery Act (RCRA) has statutorily defined solid wastes as any garbage or refuse, sludge from a wastewater treatment plant, residues from an air pollution control facility, and other discarded material, including solid, semisolid, liquid, or contained gaseous material resulting from commercial, industrial, mining, and/or agricultural operations and community activities. Solid wastes may be further categorized as nonhazardous or hazardous wastes (USEPA 2015). Regardless of this categorization, the generation of solid waste represents pollution and, in many cases, an unnecessary waste of resources. In particular, hazardous waste generation may lead to environmental degradation, degradation of natural resources, health problems, and premature death through chronic or acute exposures.

Sustainable Engineering: Drivers, Metrics, Tools, and Applications, First Edition.
Krishna R. Reddy, Claudio Cameselle, and Jeffrey A. Adams.
© 2019 John Wiley & Sons, Inc. Published 2019 by John Wiley & Sons, Inc.

15.2.1 Nonhazardous Waste

Nonhazardous waste typically is composed of trash or garbage generated from residential households, commercial offices, and other similar sources. These materials are generally classified as municipal solid waste (MSW). Industrial solid wastes such as mining wastes and coal combustion residuals are nonhazardous material resulting from the manufacturing and production of products by different manufacturing industries.

Regulations pertaining to the handling and disposal of nonhazardous waste are presented in 40 CFR Parts 239–259. These are commonly referred to as RCRA Subtitle D regulations (USEPA 2017b). These regulations prohibit the open dumping of solid waste, establish criteria for MSW landfills and other solid waste disposal facilities, and mandate the development of comprehensive plans for the management of MSW as well as nonhazardous industrial solid waste. Further, because these wastes are not considered hazardous, they present a potential for recycling and reuse of these materials as potentially valuable resources. The programs enacted by industries and government agencies are increasingly facilitating these actions.

15.2.2 Hazardous Waste

As opposed to nonhazardous solid waste, hazardous solid wastes consist of materials that are dangerous or potentially harmful to human health and the environment. Hazardous waste is also addressed by RCRA regulations; the legal framework in place for hazardous waste and its classification, identification, management, and disposal is described in 40 CFR Parts 260 through 279. This framework is also commonly referred to as RCRA Subtitle C regulations (USEPA 2017b). These regulations control hazardous waste from the time they are generated until their final disposal; a timeframe commonly referred to as "cradle to grave."

15.3 Effects and Impacts of Waste

Waste generation and management can have many adverse effects on public health and the environment. A wide range of chemicals is present within waste materials, many of which pose a threat to public health and the environment. Air pollution and greenhouse gas emissions could be a major problem. Leachate may be generated from these materials, and if not controlled, can enter the soil and groundwater, leading to contamination of these resources. When released into the environment, contaminants can alter the chemistry of receiving waters, leading to the destruction of aquatic life and aquatic ecosystems. Complex species on the food chain are also impacted, as species they prey upon are reduced or eliminated. Additionally, contaminants that enter the food chain through plants or microbiological organisms bioaccumulate and are passed on to other organisms following the food chain (Parfitt et al. 2010).

While RCRA was developed to handle both hazardous and nonhazardous wastes, many sites have been adversely impacted through previous land use or disposal practices (USEPA 2018). For instance, more than 36 000 environmentally impacted sites have been identified for potential consideration and addition to the Comprehensive Environmental Response, Compensation, and Liability Act (CERCLA) National Priority List (NPL). A total of over 1400 sites have been placed on this list, necessitating cleanup due to contamination that poses an acute, imminent threat to public health and the environment. The USEPA has also identified approximately 2500 additional contaminated sites that will require remediation. The US Department of Defense maintains approximately 19 000 sites, many of which have been extensively environmentally impacted. Further, approximately 400 000 underground storage tanks have been confirmed or are suspected to be leaking contents, affecting both soil and groundwater (USEPA 2018). Remediation of contaminated sites is a daunting task and more about this will be discussed in Chapter 18.

15.4 Waste Management

As mentioned, wastes are being generated at an increasing rate. Population growth and urbanization, coupled with increased industrial, commercial, and institutional activity, contribute to the increased waste production, as well as rapid economic growth and industrialization throughout the developing world. These socioeconomic dynamics have led to increased consumption of raw materials, processed goods, and services. Although increased economic activity has, without question, improved the quality of life for billions of people, drastic environmental consequences have also resulted. It is necessary to properly manage a range of wastes in order to protect public health and the environment while also ensuring steady economic growth.

Figure 15.1 depicts a schematic for the hierarchy of waste management. As shown, the pyramid shape indicates less-favored options toward the bottom (e.g. disposal, energy recovery), with more preferred alternatives higher up the pyramid, including waste minimization and prevention.

At the top of the pyramid is the most desirable alternative, waste prevention. Because prevention reduces the challenges of waste management, it is a fundamental goal of all waste management strategies. Several technologies can be employed throughout the manufacturing, use, or postuse portions of product life cycles to eliminate waste, and in turn, reduce or prevent pollution. Some strategies include the use of manufacturing methods that incorporate less hazardous or harmful materials, the use of modern leakage detection systems for material storage, innovative chemical neutralization technologies to reduce reactivity, or water saving technologies that reduce the need for fresh water inputs. Pollution prevention can also be accomplished by the practice of green chemistry. All of these waste management options are discussed in detail in the following sections.

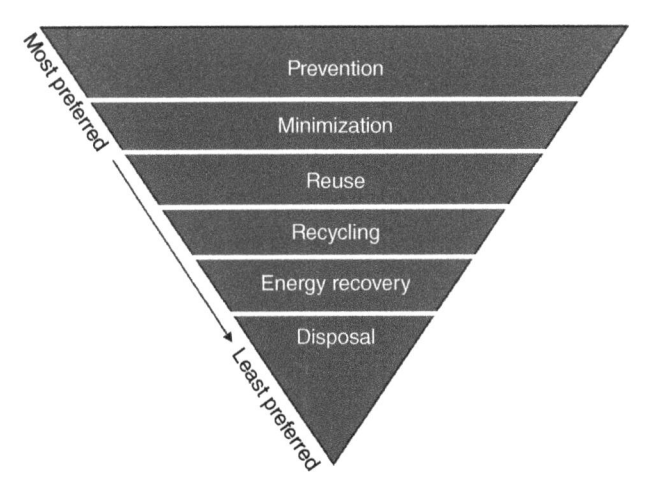

Figure 15.1 Hierarchy diagram on waste management.

15.4.1 Pollution Prevention

One ideal approach for pollution prevention (P2) consists of actions that contribute to reducing the generation of hazardous substances or dissipation of natural resources (Bishop 2002). There are two goals associated with a P2 strategy: (i) every atom that enters a manufacturing process should leave the process as a saleable product (Trost 1991); and (ii) every erg of energy used in manufacturing should produce a desired material transformation. In pursuing these dual goals, it is important to assess current or future processes associated with waste generation. Some tasks that may be performed include constructing a process flow diagram, evaluating environmental impacts and issues, identifying pollution prevention opportunities, analyzing alternatives, and documenting results. In performing these assessments, P2 analyses are targeted on single facilities and on processes, not products. Further, P2 is generally focused on the reduction of discarded or dissipated materials, energy, and water as opposed to finding suitable alternatives for their use. When processes have been assessed, P2 actions may result in process modification, technology modification, better housekeeping practices, substitution of inputs, and on-site and off-site reuse. Further, these actions may be applied to multiple scales. On a micro-scale level, chemicals and materials may be modified. On the meso-scale, the existing process may be modified. Finally, on a macro-scale, beneficial reuses of waste products may be identified.

A P2 strategy provides several benefits with respect to waste-generating processes and subsequent waste management. Significant financial savings can be achieved with respect to compliance costs, energy and materials purchases, waste treatment, and disposal costs. With less waste and emissions generation, work places are often made healthier. Because of less waste handling and disposal, legal liabilities are often reduced. Additionally, using an active process to reduce environmental impacts often leads to a positive association with a proactive commitment to protecting the environment.

An active P2 strategy differs in several ways from mandated environmental compliance. First, when used properly, P2 can be used as a proactive core organizing principle, leading to efficiencies throughout a process value chain, as opposed to often inefficient reactive or "retrofitting" measures to achieve compliance. The use of P2 encourages a systematic evaluation of a production process, which also can lead to other positive unintended side effect or discovery of other efficiencies. Further, because it results in a holistic view of a particular process, a proactive P2 strategy can often enhance profitability through reduced costs and efficiency gains (USEPA 2017c).

15.4.2 Green Chemistry

Green chemistry is the vital aspect of pollution prevention, and it is the practice of chemical science and manufacturing that is sustainable, safe, and nonpolluting, consuming minimum amounts of energy and material resources while producing virtually no wastes (Sheldon et al. 2007). Figure 15.2 demonstrates a schematic of the process, including the use of renewable feedstocks (including those from sources such as biomass), proper controls, internal recycling, and the minimization on nondegradable wastes when they cannot be avoided.

With respect to feedstock, biomass sources are especially attractive to replace petroleum. Not only is petroleum a depleting resource, partially oxidized biomass materials can actually be more easily applied to processes when compared to partially oxidized petroleum materials. Ideally, feedstocks are renewable, pose no toxicity issues, may be converted to a desired end-product in a few steps, resulting in 100% yield, or 100% atom economy. The percentage of atom economy (Trost 1991) is defined as the total mass of product divided by the total mass of reactant (Figure 15.3).

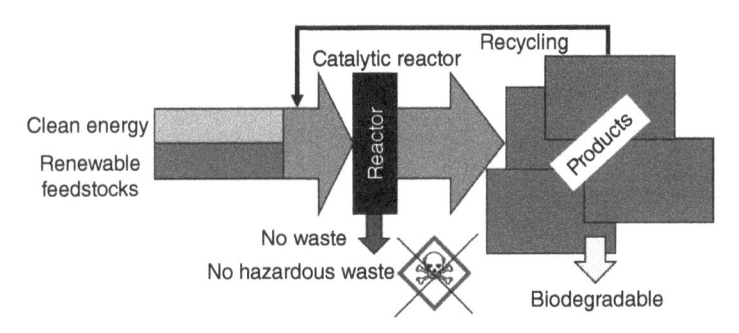

Figure 15.2 Production process based on green chemistry transformation of renewable feedstock.

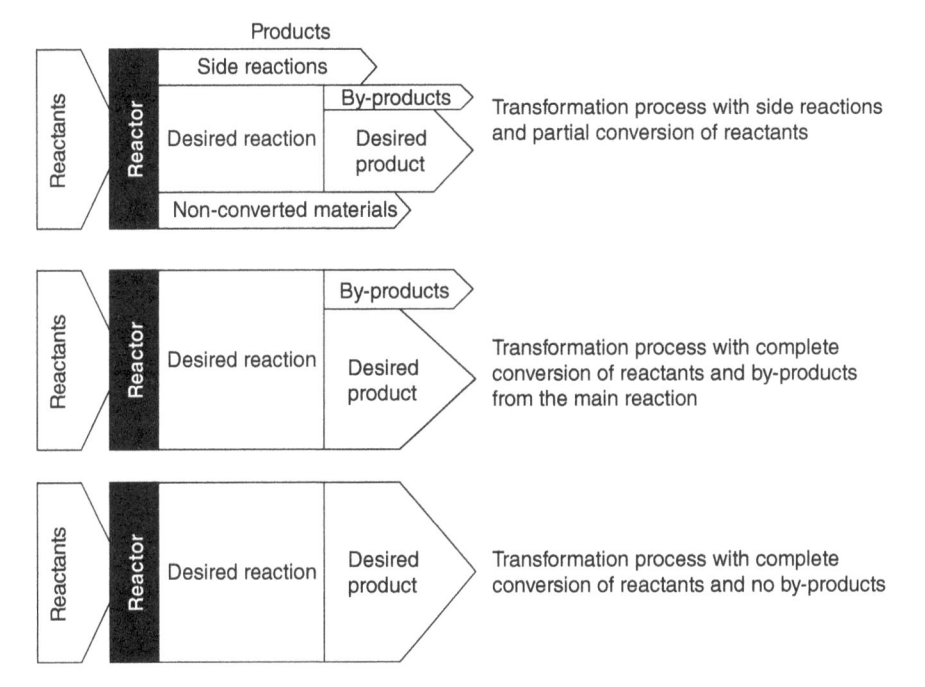

Figure 15.3 Objectives of the atom economy: transform every atom of reactant in saleable product increasing conversion and selectivity of the transformation process and decreasing side reactions and waste generation.

When considering chemical processes, green chemistry focuses on improvements in several areas. First, it involves the elimination of toxic materials or processes. Second, consideration is given to the use of existing chemical synthesis processes but using safer, less polluting processes as well as identifying means to make reagents greener, safer, or less likely to result in adverse side effects that can lead to pollution (Bozell 2008). In pursuing these goals, green synthetic chemistry focuses on finding ways to make new chemicals and new ways to make existing chemicals (Figure 15.4). This involves either using existing feedstocks in processes that are more benign or using alternative feedstocks in similar processes.

Ultimately, green chemistry is driven by several key principles (Anastas and Warner 1998). It focuses on the prevention of waste as opposed to its treatment. Energy use, the use of toxic reagents and products, and the use of solvents/extractants are all minimized. Atom economy is practiced, and an emphasis is placed on the use of renewable raw materials. Finally, there is

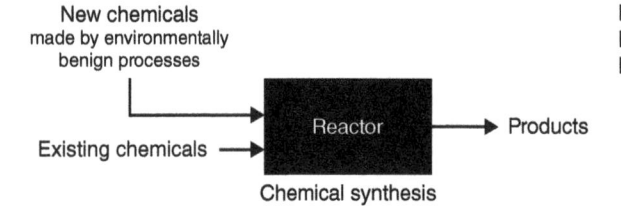

Figure 15.4 Substitution of existing chemical by new chemicals made by environmentally benign processes.

a greater emphasis of catalysis than stoichiometry. Several of these key principles are inherent to the goals of green engineering (Anastas and Zimmerman 2003, Tang et al. 2008).

15.4.3 Waste Minimization

While waste prevention or green chemistry techniques are viewed as superior alternatives with respect to waste management, waste minimization also offers an attractive alternative. Several strategies may be used to minimize and/or reduce waste generation (Clark 2012). First, manufacturing processes and products may be redesigned to use less material and energy. Moreover, increasing the efficiency of the transformation process results in less side products and less waste production, which in turn, improves the economy of the global process. The modifications in the manufacturing process may reduce the utilization or production of less harmful chemicals, so waste management will incur relatively minor risks, and waste treatment may be less expensive with more options available for recycling of nonhazardous waste. The manufacturing processes may also be redesigned such that less waste and/or pollution is produced. The implementation of a quality system to minimize production errors and out-of-specification products will result in less waste generation. Products may also be designed such that they are easier to repair, reuse, remanufacture, compost, or recycle. Additionally, packaging systems may be redesigned to eliminate unnecessary materials. Finally, end-of-use strategies may be implemented to facilitate a greater focus on recycling or repurposing – either through incentive-based disposal options (i.e. fee per bag) or by establishing cradle-to-grave responsibility for manufacturers with respect to wastes generated during manufacturing (Fiksel and Fiksel 1996).

15.4.4 Reuse/Recycling

The use of recycled materials shows many advantages not only for the environment and the conservation of natural resources, but for the global economy of the production processes. When usable materials have already been created, it is also beneficial to focus on means to reduce the fabrication of identical materials when existing versions can be captured and reused. Reusing items decreases the use of material and energy resources as well as emissions and environmental degradation associated with manufacturing. Recycling captures the materials for refabrication, albeit with a reduction or elimination of the need for material inputs. With reuse, products are captured, cleaned, or processed, and re-entered into the market for an identical use. Figure 15.5 shows the energy consumption associated with different types of 350 ml containers, and Table 15.1 shows the energy savings in obtaining materials from wastes compared to raw materials (USEPA 2016).

Recycling may consist of primary, closed-loop recycling, or secondary recycling. Wastes may be classified either into preconsumer (internal to manufacturing processes) or postconsumer (captured after use in the marketplace) streams. With respect to household wastes, the materials

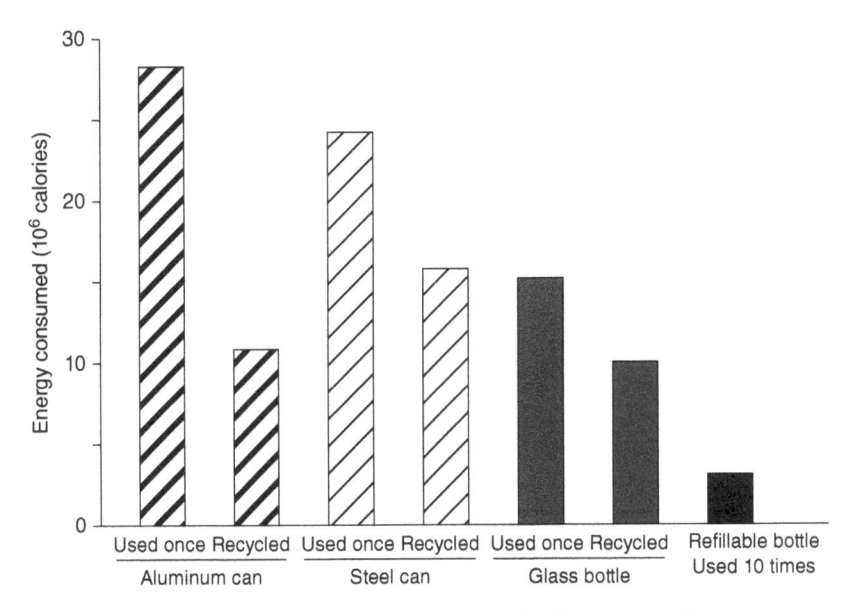

Figure 15.5 Energy consumption for the production of different types of 350 ml containers.

Table 15.1 Energy saving for recycled materials compared to the original raw material.

Material	Energy consumption (kcal) in the production of 1 kg		Energy saving (%)
	Raw materials	Recycled materials	
Glass	1 200	800	35
Iron	10 300	5 100	50
Paper	3 700	1 100	70
Polyethylene	4 500	500	89
Aluminum	47 000	1 400	97

may be sorted or not sorted prior to diversion to a material recovery facility (MRF). Further, materials recycling fees may be charged either on a set action basis ("pay as you throw") or on a unit basis ("fee-per-bag"). Some selected materials may be recycled as follows:

- *Biodegradable wastes* – Includes green wastes and other wastes that can easily breakdown in the natural environment – these materials may be composted on an individual or municipal basis.
- *Paper* – Recycling of paper is easy, effective, and substantially reduces energy and water inputs as well as pollution. Additionally, increased focus is being placed on the reduction of chlorine-based bleaching agents with hydrogen peroxide or oxygen products.
- *Metals* – The recycling of metals, such as aluminum or steel, reduces the need for mining and use of ore materials. Inputs and waste products may be greatly reduced.
- *Plastics* – Only approximately 4% of plastics are recycled. Further, recycling of plastics is relatively difficult, and it is often cheaper to fabricate new plastic products.

Nevertheless, recycling and reuse can be encouraged. Governments can facilitate these practices by providing subsidies or other financial incentives, as well as attach fees or taxes to the use of virgin materials. Additionally, public interest groups as well as private groups continue to encourage these practices. Many industrial wastes are found to be suitable to use as construction materials, consuming large quantities of wastes for beneficial purpose and thus avoiding the need for new landfills (Sharma and Reddy 2004).

15.4.5 Energy Recovery

Waste materials may also be utilized for the generation of energy. Commonly, this is done with waste-to-energy incinerators. Approximately 600 of these incinerators exist worldwide, although most are located in Great Britain. Figure 15.6 illustrates a typical waste-to-energy incinerator as a part of comprehensive MSW management.

Waste-to-energy incineration offers several advantages (Astrup et al. 2015). First, the combustion greatly reduces the volume of waste materials, thereby reducing the need for landfill capacity. The pollution of water can be sufficiently minimized and hazardous waste residues are consolidated into a small volume by incineration. The resulting energy may be sold, reducing the overall cost associated with incineration. Additionally, some facilities are equipped to recover and sell metallic materials. Finally, modern facilities and controls greatly reduce resulting air emissions. However, incineration has some disadvantages. First, facilities are relatively expensive to construct and permit. They can often be expensive to operate as compared to local landfills with short transportation routes. These facilities often face significant public opposition, especially because older facilities do in fact contribute significant amounts of air emissions. Even new facilities are responsible for emitting carbon dioxide and other compounds (Lombardi et al. 2015). Further, there can be "competition" with recycling facilities, as incinerators divert combustible materials that could otherwise be recycled. Finally, to maintain operations, incinerators need input materials; therefore, they provide little to no incentive to reduce waste generation.

15.4.6 Landfilling

Solid waste of all types is quite commonly placed into landfill facilities (Zaman 2010; Alam and Alam 2014). While open dumps in the United States have essentially been eliminated, they still exist elsewhere. These facilities have been replaced in the United States with engineered landfills. Figure 15.7 shows a typical landfill facility, including protective and monitoring systems. Landfills are engineered structures meeting the regulatory requirements and consist of liner and cover systems, and leachate and gas collection and removal systems (Sharma and Reddy 2004). Engineered landfills eliminate the need to burn garbage materials, and when constructed properly, minimize odor and potential soil and groundwater pollution. Engineered landfills can be constructed quickly, can handle large quantities of waste, and modular design allows for efficient increases in capacity. They can be sited on lands that offer marginal benefits for other purposes, are relatively low cost to operate, and can be reclaimed for other purposes when landfill space is filled. However, landfills use does have several drawbacks. They result in the generation of noise, traffic, dust, and other nuisance conditions. Air emissions, both from equipment and from decaying waste, can result in contributions to greenhouse gases. Decomposition occurs slowly, limiting potential reuses on reclaimed land following filling. Finally, as with incineration, there is no incentive to reduce waste generation.

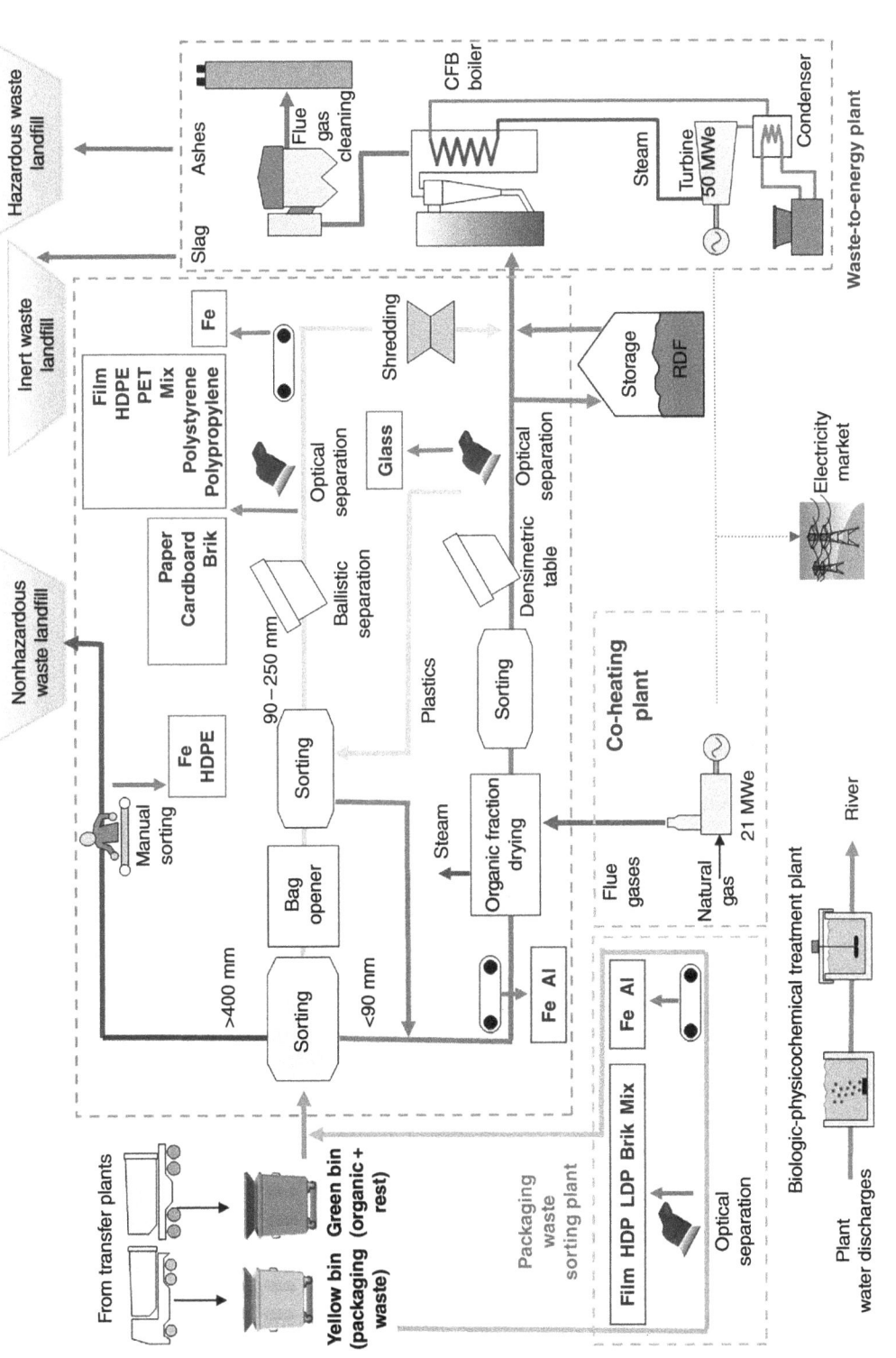

Figure 15.6 MSW management plant (SOGAMA, Galicia, Spain).

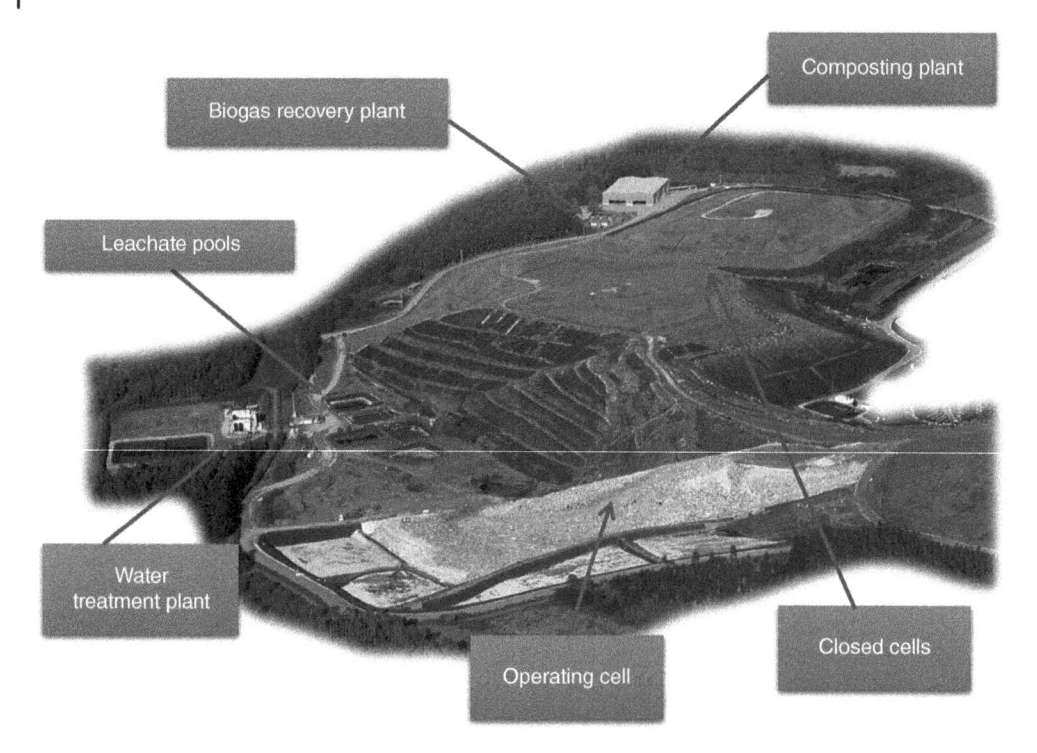

Figure 15.7 Landfill for nonhazardous waste (SOGAMA, Galicia, Spain).

15.5 Integrated Waste Management

The waste management pyramid, or "hierarchy," presented in Figure 15.1 intuitively offers an optimal system to minimize the negative effects of waste generation at all levels; however, the system has its own limitations. First, it has no measurable scientific basis. Second, it does not address combinations of technologies. Finally, it does not address cost considerations. Because so many potential management techniques (each with several notable advantages) exist for dealing with waste materials, it is often appropriate to implement an integrated management approach (McDougall et al. 2008). In this approach, a combination of strategies is selected and used as appropriate. Figure 15.8 depicts how such a strategy may be applied for waste management for a combination of waste stream types. When considering hazardous wastes, it is also advantageous to convert materials into less hazardous forms or implement means to assure safe long-term storage.

When managing nonhazardous wastes, the approach techniques may be prioritized as shown in Figure 15.9. The first priority is the primary pollution/waste prevention. It encompasses making change in the industrial processes to eliminate use of harmful chemicals and use of less harmful materials. Reduce the use of material in packaging and products. To make products that are durable, recyclable, reusable, or are easier to repair.

The second priority is secondary pollution/waste prevention. It consists of reuse, repair, recycling/composting, and use of recycled products. The last priority is the waste management which consists of treating waste to reduce toxicity. Incineration of waste to reduce the volume of the waste generated. The final option is the landfilling of waste.

When managing hazardous wastes, the approach techniques may also be prioritized as follows: The first priority is to produce less waste by changing industrial processes to eliminate

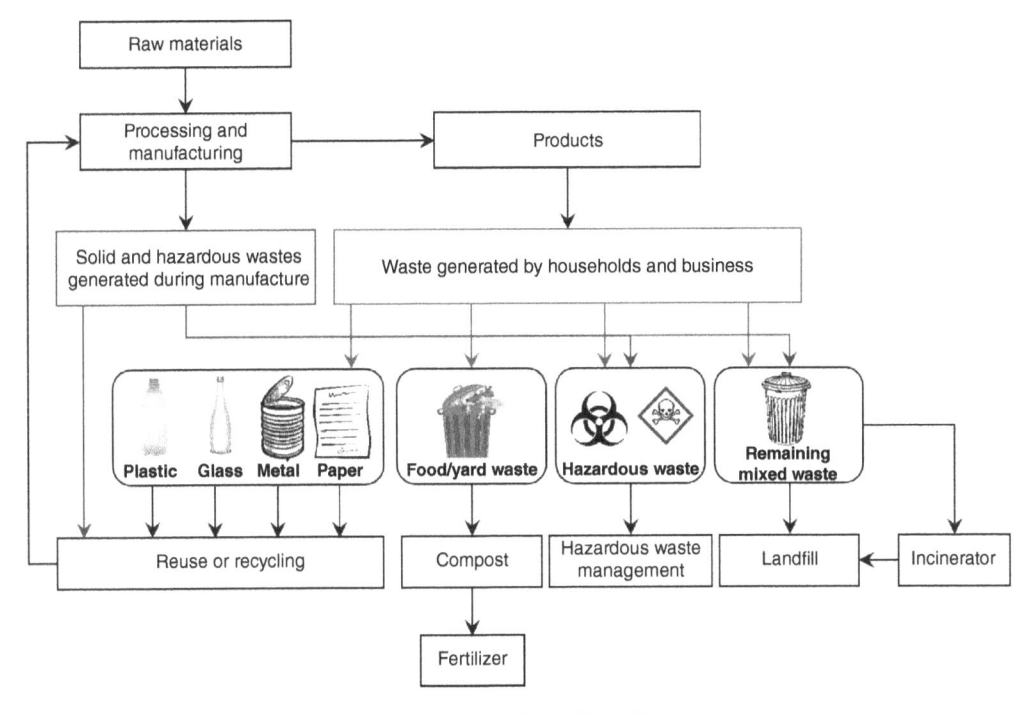

Figure 15.8 Integrated approach for minimization and recycling of wastes.

- **Primary pollution and waste prevention**
  - Remove harmful chemicals in industrial processes
  - Reduce packaging materials
  - Increase lifespan of products, easy to repair
  - Products easily recyclable or reusable

First priority

- **Primary pollution and waste prevention**
  - Reuse
  - Repair
  - Recycle
  - Compost

Second priority

- **Waste management**
  - Waste treatment for toxicity reduction
  - Waste incineration
  - Bury waste in landfills
  - Waste release into environment for dilution or dispersal

Last priority

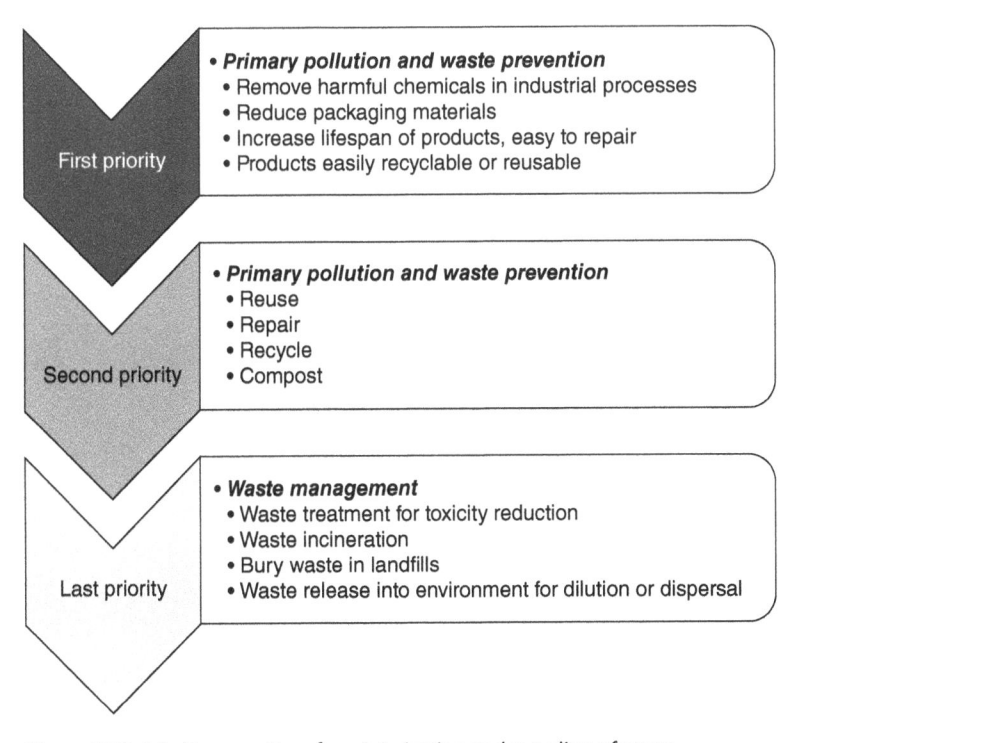

Figure 15.9 Priority operations for minimization and recycling of waste.

hazardous waste production and recycle/reuse of hazardous waste. The second priority is to convert it to less hazardous material. This can entail natural decomposition, incineration, thermal treatment, chemical, physical, or biological treatment, and dilution in air and water. The last priority is to put waste in perpetual storage, such as underground injection wells, surface impoundments, landfilling of waste, and underground salt formations.

Hazardous waste materials can be processed and converted into less harmful materials. They may be detoxified using a variety of methods, including physical methods, chemical methods, bioremediation, phytoremediation, the use of nanomagnets, incineration, or plasma arc torch methods (Figure 15.10). Figure 15.11 illustrates the use of phytoremediation to transform hazardous waste materials to less hazardous materials. Phytoremediation is easy to establish, inexpensive, uses low energy inputs, can reduce materials dumped into landfills, and produces relatively little air pollution (Batty and Dolan 2013). On the other hand, it is slow, effective only

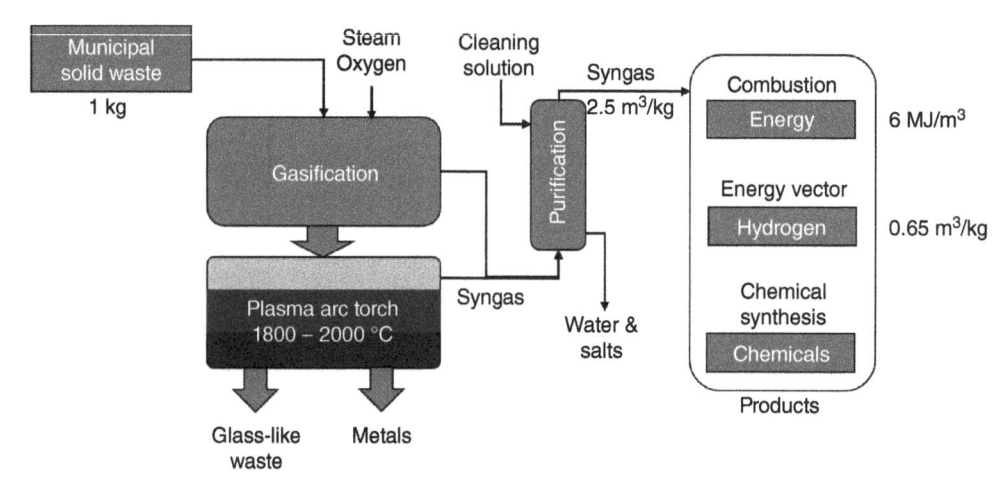

Figure 15.10 Waste gasification by plasma arc torch technology.

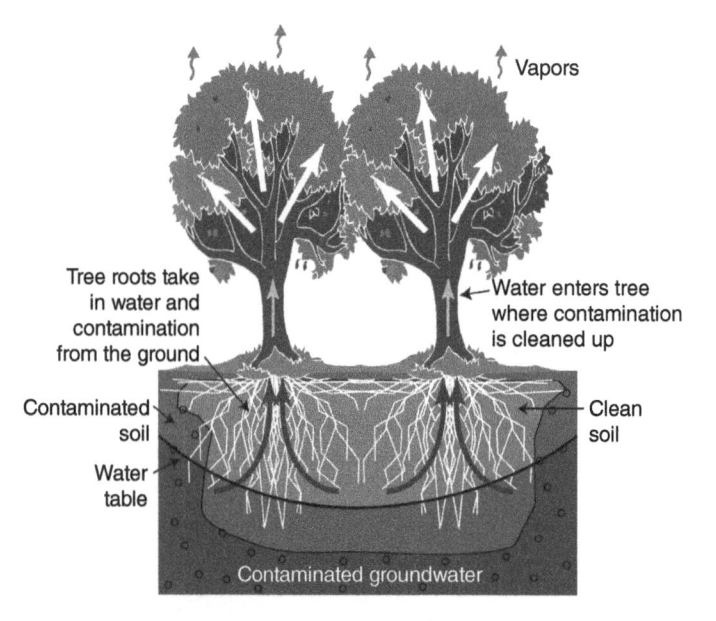

Figure 15.11 Phytoremediation technology for the cleanup of contaminated land. Source: USEPA (2012).

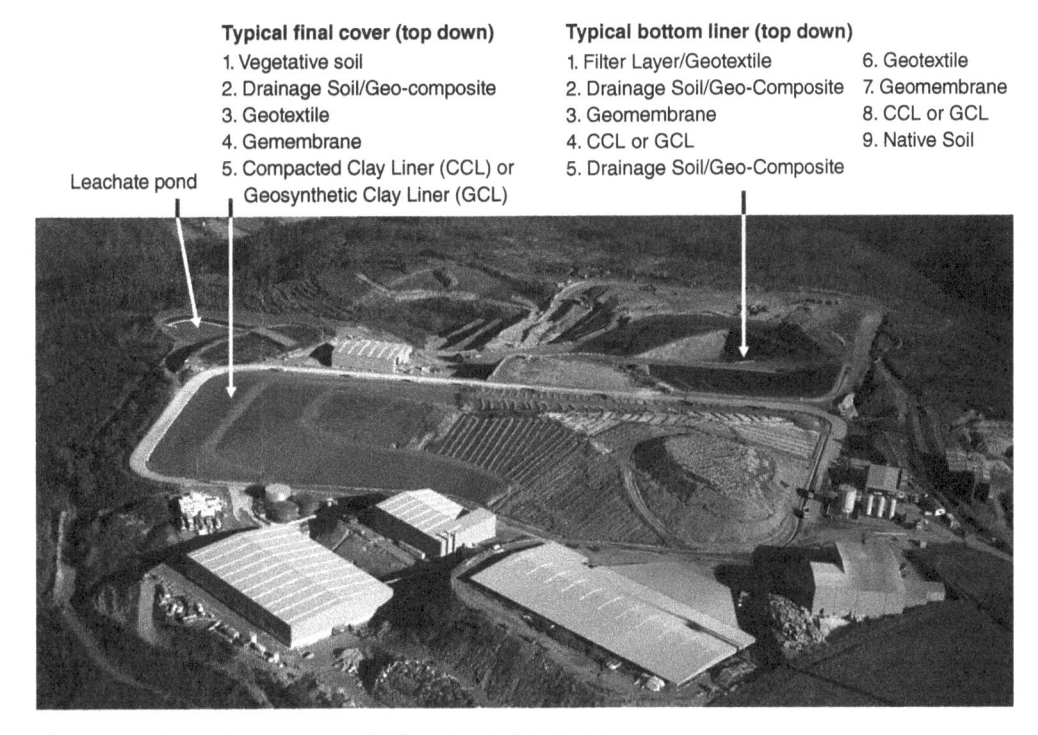

Typical final cover (top down)
1. Vegetative soil
2. Drainage Soil/Geo-composite
3. Geotextile
4. Gemembrane
5. Compacted Clay Liner (CCL) or Geosynthetic Clay Liner (GCL)

Typical bottom liner (top down)
1. Filter Layer/Geotextile
2. Drainage Soil/Geo-Composite
3. Geomembrane
4. CCL or GCL
5. Drainage Soil/Geo-Composite
6. Geotextile
7. Geomembrane
8. CCL or GCL
9. Native Soil

Leachate pond

Figure 15.12 Hazardous waste landfill (Sogarisa, Galicia, Spain).

to a depth that can be reached by roots, allows for emission of some toxic materials through plant respiration, and can render the plant materials themselves toxic.

Hazardous wastes can also be stored within the subsurface. Some methods include deep-well disposal, surface impoundments, and hazardous waste landfills (Blackman Jr 2016). Deep wells can be safe at appropriate sites. They are easy to implement, relatively low cost, and wastes can often be retrieved if necessary. However, they may lead to leaks or surface spills, failure of casings, and fractures within rock formations can lead to unanticipated migration. Surface impoundments are also low cost and easy to implement. The disposed waste can also be retrieved if necessary. These facilities can be constructed quickly and can be secured when implementing appropriate liner technology. However, groundwater can be impacted if liners fail, and volatile materials may escape and enter the atmosphere. Finally, landfills may be specially designed for hazardous waste. Many protective measures should be incorporated to minimize the potential for releases into the environment (Sharma and Reddy 2004). Figure 15.12 depicts a typical hazardous waste landfill facility with clay and flexible membrane liners, leachate collector systems, and landfill covers to contain and control the waste.

15.6 Sustainable Waste Management

Lately, there is a greater emphasis on sustainable waste management (Ai and Leigh 2017). Further, there is a growing interest in environmental justice awareness, and advocacy for economically depressed areas where a disproportionate share of hazardous waste disposal facilities and other related facilities are located. As a result, sustainable waste management aims to be environmentally effective, economically affordable, and a socially acceptable strategy for waste management. The integrated waste management strategies presented earlier provides an

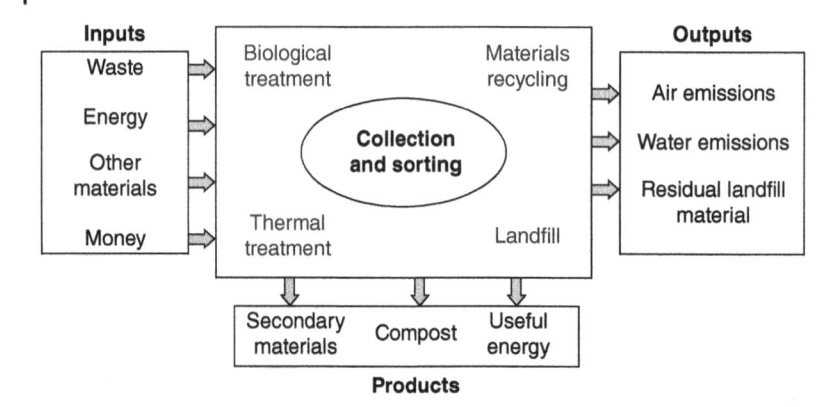

Figure 15.13 Life Cycle Assessment of an integrated waste management.

acceptable means to achieve sustainable waste management. Alternatively, the use of life cycle assessments (LCAs) can also be coupled with integrated waste management strategies, thereby leading to sustainable waste management (Laurent et al. 2014a,b).

As described in earlier chapters, waste management typically occupies the final step in a product's life cycle. Therefore, an LCA boundary can be drawn around waste management steps. An LCA model for solid waste calculates total energy consumption, emissions to air or water, final solid waste, and the overall economic costs associated with these effects. Figure 15.13 presents a schematic of an LCA for integrated waste management. Inputs consist of waste, energy, other materials, and financial costs. The integrated waste management framework consists of activities such as biological treatment, thermal treatment, landfilling, recycling, collection, and sorting. Products include secondary materials, compost, and useful energy capture. Finally, outputs include air emissions, water emissions, and residual landfill material. The LCA captures key metrics, such as net energy consumption, air emissions, water emissions, residual landfill volume, recovered materials, and secondary materials (such as compost). The goal of the LCA is to devise more sustainable strategies, including more useful products, fewer emissions, less final inert waste, and less energy consumption. Additionally, economic sustainability is a goal as well. Further, using an existing waste management strategy as a baseline, the LCA can be used to make comparisons among potential strategic alternatives, allowing for the selection of the optimal waste management strategy based on the needs of the local environment, economy, and the society. Ultimately, the selection may be made based on a single criterion (e.g. available landfill space); multiple criteria where more than one issue are important (e.g. energy consumption and air emissions); "less is better," where one option is superior in all categories, or an impact analysis, where categories are combined to assess an overall effect, such as global warming.

15.7 Circular Economy

Construction and product consumption are often defined in a linear framework. Projects are manufactured or constructed, which includes refinement of raw materials, development of intermediate products, and manufacturing or erection of a useable product or project. The project has a design life during which it is operated, and occasionally repaired or renovated to extend the useful life. At some point, it is removed from useful service, at which time it is disposed and reclaimed to some extent. This make/use/dispose pattern is referred to as the linear economy.

As an alternative, there is growing interest in the pursuit of a "circular economy." A circular economy is an industrial system that is restorative or regenerative by intention and design in that it replaces the linear "make/use/dispose" model with one that incorporates new raw materials on a limited or "as-needed" basis. The circular economy focuses on restoration, a greater emphasis on the use of renewable energy, minimizes or eliminates the use of toxic chemicals, and looks to minimize waste generation through optimized designs of materials, products, systems, and business models (Ellen MacArthur Foundation 2013).

The ultimate goal of the circular economy is to eliminate waste through better product or project design. Products and projects are designed and optimized for a cycle of disassembly and reuse, which also look to achieve savings in labor and energy inputs. Second, the circular economy approach looks to differentiate between consumable goods and durable goods/projects. Because of their rapid use cycle, consumable goods should be designed with the goal of safe and useful components and end-products so they may be returned safely to the environment. Durable goods/projects that may contain materials that are less friendly to the environment should be designed to maximize reuse alternatives. Finally, steps should be taken to incorporate renewable energy sources into the economic cycle in order to decrease resource dependence and increase economic resilience (Ellen MacArthur Foundation 2013).

15.8 Summary

Waste materials are generated by virtually every human activity. These wastes may be classified into a variety of hazardous and nonhazardous waste categories. Regardless of the classification, they may be managed using a variety of approaches. These approaches may be thought of as a "pyramid," with less-favored options toward the bottom (e.g. disposal, energy recovery), and more preferred alternatives higher up the pyramid, including minimization and prevention.

Since several potential management techniques (each with several notable advantages) exist for managing the waste materials, it is often appropriate to implement an integrated management approach that utilizes a combination of different strategies for waste management.

The integrated waste management concept, combined with an LCA approach, can improve the degree of sustainability of waste management. The incorporation of LCA results in a more informed, scientific-based approach to strategic development than other arbitrary approaches. As with other LCA applications, better data and more comprehensive data can lead to more accurate LCA modeling, but it will also result in more complicated modeling. Using these methods as well as others, integrated waste management strategies may be developed that can better meet the needs of the local population and the greater environment. Consideration of the pursuit of a circular economy should be of the highest priority to prevent waste generation and economic resilience.

15.9 Questions

15.1 What are the major waste management options in waste management hierarchy?

15.2 What constitutes a municipal solid waste?

15.3 What is the major focus of pollution prevention strategy for effective waste management?

15.4 Discuss how wastes are classified using the RCRA regulatory framework.

15.5 What are characteristics that are used to determine if waste is hazardous under the RCRA framework?

15.6 Why was there a need to develop the CERCLA regulatory framework?

15.7 What are some of the site classifications covered by CERCLA?

15.8 Name three advantages of pursuing pollution prevention (P2) strategies.

15.9 Discuss the benefits of using renewable feedstocks from a green chemistry standpoint.

15.10 Describe in detail advantages and drawbacks of recycling programs with respect to specific waste stream materials.

15.11 Discuss specific strategies to render hazardous waste less harmful or nonhazardous.

15.12 From an integrated waste management approach, what are some strategies for waste treatment/management? What are benefits and drawbacks of these strategies?

15.13 What is sustainable waste management and how can one achieve it?

References

Ai, N. and Leigh, N.G. (2017). *Planning for Sustainable Material and Waste Management*. PAS Report 587, American Planning Association, Chicago, IL.

Alam, P. and Alam, M. (2014). Bioreactor landfills: new trends in landfill design. *Applied Sciences and Engineering Research* 3 (1): 187–193.

Anastas, P.T. and Warner, J.C. (1998). *Green Chemistry: Theory and Practice*. New York, NY: Oxford University Press.

Anastas, P.T. and Zimmerman, J.B. (2003). The twelve principles of green engineering. *Environmental Science & Technology* 38: 94A–101A.

Astrup, T.F., Tonini, D., Turconi, R., and Boldrin, A. (2015). Life cycle assessment of thermal waste-to-energy technologies: review and recommendations. *Waste Management* 37: 104–115.

Batty, L.C. and Dolan, C. (2013). The potential use of phytoremediation for sites with mixed organic and inorganic contamination. *Critical Reviews in Environmental Science and Technology* 43 (3): 217–259.

Bishop, P.L. (2002). *Pollution Prevention: Fundamentals and Practice*. New York: McGraw-Hill.

Blackman, W.C. Jr., (2016). *Basic Hazardous Waste Management*. CRC Press.

Bozell, J.J. (2008). Feedstocks for the future–biorefinery production of chemicals from renewable carbon. *CLEAN–Soil, Air, Water* 36 (8): 641–647.

Clark, J.H. (2012). *Chemistry of Waste Minimization*. Springer Science & Business Media.

Ellen MacArthur Foundation (2013). *Towards the Circular Economy 1*. Ellen MacArthur Foundation.

Fiksel, J. and Fiksel, J.R. (1996). *Design for Environment: Creating Eco-Efficient Products and Processes*. McGraw-Hill Professional Publishing.

Laurent, A., Bakas, I., Clavreul, J. et al. (2014a). Review of LCA studies of solid waste management systems–Part I: Lessons learned and perspectives. *Waste Management* 34 (3): 573–588.

Laurent, A., Clavreul, J., Bernstad, A. et al. (2014b). Review of LCA studies of solid waste management systems–Part II: Methodological guidance for a better practice. *Waste Management* 34 (3): 589–606.

Lombardi, L., Carnevale, E., and Corti, A. (2015). A review of technologies and performances of thermal treatment systems for energy recovery from waste. *Waste Management* 37: 26–44.

McDougall, F.R., White, P.R., Franke, M., and Hindle, P. (2008). *Integrated Solid Waste Management: A Life Cycle Inventory*. Wiley.

Parfitt, J., Barthel, M., and Macnaughton, S. (2010). Food waste within food supply chains: quantification and potential for change to 2050. *Philosophical Transactions of the Royal Society B: Biological Sciences* 365 (1554): 3065–3081.

Sheldon, R.A., Arends, I., and Hanefeld, U. (2007). *Green Chemistry and Catalysis*. Wiley.

Sharma, H.D. and Reddy, K.R. (2004). *Geoenvironmental Engineering: Site Remediation, Waste Containment, and Emerging Waste Management Technologies*. Wiley.

Tang, S.Y., Bourne, R.A., Smith, R.L., and Poliakoff, M. (2008). The 24 principles of green engineering and green chemistry: "improvements productively". *Green Chemistry* 10 (3): 268–269.

Trost, B.M. (1991). The atom economy – a search for synthetic efficiency. *Science* 254: 1471–1477.

USEPA (2012). Citizen's Guide to Phytoremediation. EPA 542-F-12-016, Office of Solid Waste & Emergency Response, Washington, DC.

USEPA (2015). *Waste Classification*. US Environmental Protection Agency. ISBN: 978-1-84095-601-6.

USEPA (2016). *Advancing Sustainable Materials Management: 2014 Fact Sheet*. US Environmental Protection Agency.

USEPA (2017a). *Advancing Sustainable Materials Management: Facts and Figures*. US Environmental Protection Agency https://www.epa.gov/smm/advancing-sustainable-materials-management-facts-and-figures (accessed 12 April 2017).

USEPA (2017b). *Resource Conservation and Recovery Act (RCRA) Laws and Regulations*. US Environmental Protection Agency https://www.epa.gov/rcra (accessed 12 April 2017).

USEPA (2017c). *Pollution Prevention (P2)*. US Environmental Protection Agency https://www.epa.gov/p2 (accessed 12 April 2017).

USEPA (2018). *Superfund: National Priorities List (NPL)*. U.S. Environmental Protection Agency https://www.epa.gov/superfund/superfund-national-priorities-list-npl (accessed 25 July 2018).

Zaman, A.U. (2010). Comparative study of municipal solid waste treatment technologies using life cycle assessment method. *International Journal of Environmental Science & Technology* 7 (2): 225–234.

16

Green and Sustainable Buildings

16.1 Introduction

Structures are an essential element for the existence and survival of human civilization. On a most basic level, they provide shelter for humans from the often harsh and dangerous effects of weather. With increased complexity, they provide a means of function for virtually every human activity and institution. The design and operation of structures can also have profound influence, both positive and negative, on human health and the environment. As a greater awareness of the environmental function and impacts of buildings evolves, "green building" approaches that utilize more resource-efficient design, implementation, and decommissioning of buildings and infrastructure have continued to grow in popularity.

The USEPA defines green building, or sustainable and/or high-performance building, as the practice of creating structures and using processes that are environmentally responsible and resource-efficient throughout a building's life cycle, from siting to design, construction, renovation, operation, maintenance, and demolition (USEPA 2013). These design principles are incorporated into the traditional design approach concerning economy, durability, utility, aesthetics, and comfort. Considering all life-cycle stages from conceptual design to eventual decommissioning or demolition, this design approach seeks to optimize resource inputs and/or the resulting environmental impacts from the use of the built environment. These inputs and impacts include energy, water, materials, natural resources, waste, air pollution, water pollution, indoor pollution, heat islands, stormwater runoff, noise, harm to human health, and environmental degradation/contamination (Retzlaff 2009).

Green buildings are designed to reduce the overall impact of the built environment on human health and the natural environment by utilizing efficient resource use, such as water, energy, and construction materials, protecting occupant health and enhancing productivity, and reducing waste, pollution, and other deleterious environmental effects. Some examples of these practices include the use of renewable, recycled, and repurposed materials in construction, incorporation of healthier indoor environments, such as through the use of low-emissions materials and natural lighting, and green landscaping principles, including use of native plant palettes, drought-tolerant plants, and the use of recycled/reclaimed water for irrigation. This chapter presents the history and importance of green buildings, green building concepts/components, green building rating systems (e.g. LEED), and general principles of green buildings/infrastructure design and operation.

Sustainable Engineering: Drivers, Metrics, Tools, and Applications, First Edition.
Krishna R. Reddy, Claudio Cameselle, and Jeffrey A. Adams.
© 2019 John Wiley & Sons, Inc. Published 2019 by John Wiley & Sons, Inc.

16.2 Green Building History

Several design approaches considered to be green have been used for millennia. For instance, architectural designs addressing local climate patterns and using local, renewable, and indigenous materials and building practices have been used throughout the history of human civilization. Yet, a greater contemporary focus on these principles arose out of increased environmental awareness spurred by incidents and conditions of the 1960s and 1970s. Further, economic shocks arising from oil crises of the 1970s inspired significant research and activity to improve energy efficiency and to utilize renewable energy resources. Green building principles continued to evolve and became more formalized in the 1990s, as architects, engineers, planners, and government agencies attempted to formalize procedures, guidelines, and codes. Several milestones that occurred during this period include the following:

- In 1989, the American Institute of Architects (AIA) formed the Committee on the Environment;
- In 1992, the USEPA funded the publishing of the Environmental Resource Guide, which was published by the AIA;
- In 1992, the USEPA and the United States Department of Energy launched the ENERGY STAR program;
- In 1992, the city of Austin, Texas, launched the first local green building program;
- In 1993, the United States Green Building Council (USGBC) was founded;
- In 1998, the USGBC launched the pilot "1.0" program of Leadership in Energy and Environmental Design (LEED).

16.3 Why Build Green?

Because buildings consume such a significant percentage of natural resources for construction and ongoing operation, green building design and operation principles can facilitate significant reductions in resource consumption and waste generation. For instance, in the United States, buildings account for 39% of total energy use, 12% of water consumption, 71% of total electricity consumption, and 39% of carbon dioxide emissions (Figure 16.1). Green building practices and technologies are estimated to reduce energy use by 30–50%; reduce carbon emissions by 35%; reduce water use by 40%; and reduce solid waste by 70% (Figure 16.2). Thoughtful incorporation

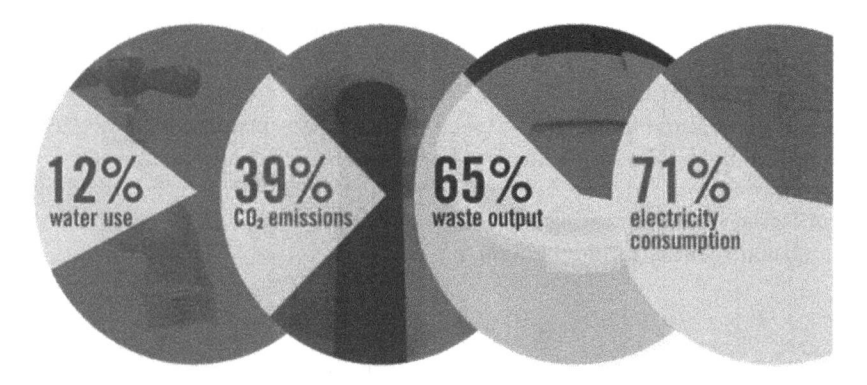

Figure 16.1 Impact of buildings in water and electricity consumption, waste generation, and carbon emissions.

Figure 16.2 Average savings on energy, water, solid waste, and carbon emissions of green buildings.

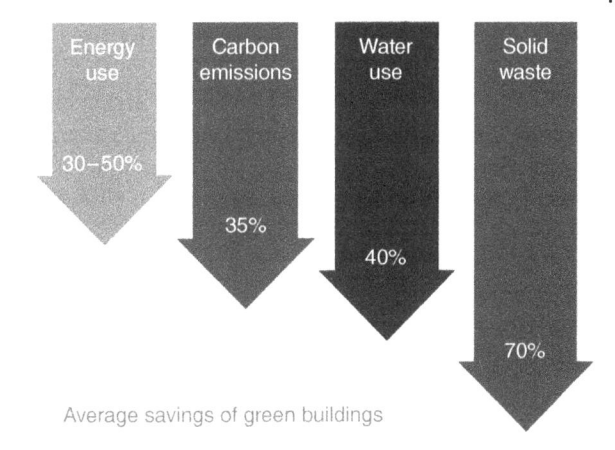

Energy use 30–50%
Carbon emissions 35%
Water use 40%
Solid waste 70%

Average savings of green buildings

of green building strategies can result in energy savings, reductions in materials use and waste generation, and operational cost savings, while creating attractive and appealing occupational conditions.

Green building design and operation can provide several other environmental benefits. First, these principles can be used to enhance adjacent or nearby ecosystems and protect biodiversity. Indoor and outdoor air and water quality can be improved by appropriate material selection and installation of control systems. Waste streams can be reduced during construction and operation, which can lead to materials savings and resource use. From an economic standpoint, green building can reduce operating costs, create, expand, and shape markets for green products and services, improve occupant productivity, and optimize life-cycle economic performance. From a social standpoint, green building practices can enhance occupant comfort and health, heighten aesthetic qualities, minimize strain on local infrastructure, and improve overall quality of life (Willard 2012). Buildings may be called sustainable buildings when environmental, economic, and social impacts are considered.

16.4 Green Building Concepts

The technologies, methods, and design of green buildings incorporate a range of concepts. Some of these include energy efficiency, the incorporation of energy from renewable resources, water efficiency, environmentally protective and beneficial building materials and specifications, waste reduction, toxics reduction, indoor air quality, and smart growth and sustainable development principles. These strategies may be applied to a range of structures, including single-family and multi-family residential structures, retail and commercial buildings, schools, government facilities, industrial buildings, health care facilities, and government/institutional facilities.

Because of the materials-intensive nature of building construction, a key component of green building is the careful consideration of how construction and demolition materials are generated and managed (Kibert 2016). Some of the principal building materials warranting consideration include concrete, aggregate, cement, asphalt, steel, zinc, copper, and aluminum. Many of these materials are recyclable and may be reused in subsequent construction applications. Because of increased global demand for these materials, design principles that lead to reduced materials usage can achieve significant cost savings. Figure 7.2 shows how materials demand has precipitously increased over the past century (Matos 2012).

16.5 Components of Green Building

Figures 16.3–16.6 depict green building concepts, including conceptual drawings of stormwater harvesting, solar power, hot water generation and circulation, and geothermal systems, respectively. Through increased efficiency and incorporation of renewable, zero emissions energy sources, these systems can lessen or eliminate use of fossil fuels or fresh/potable water

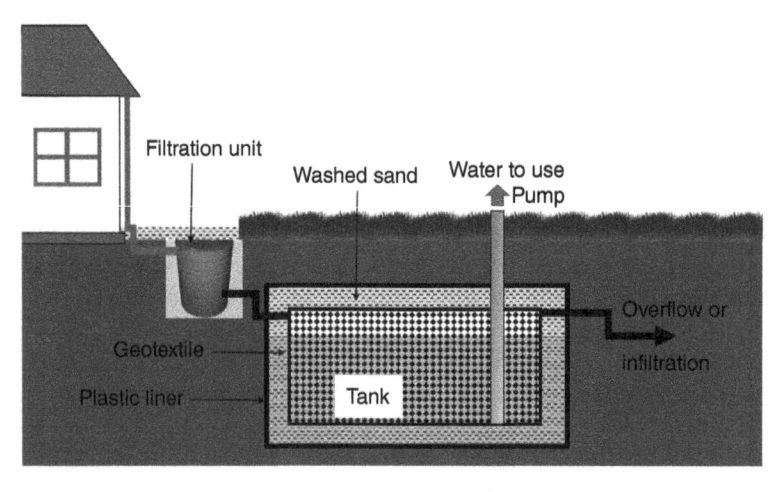

Figure 16.3 Features of green buildings: stormwater harvesting.

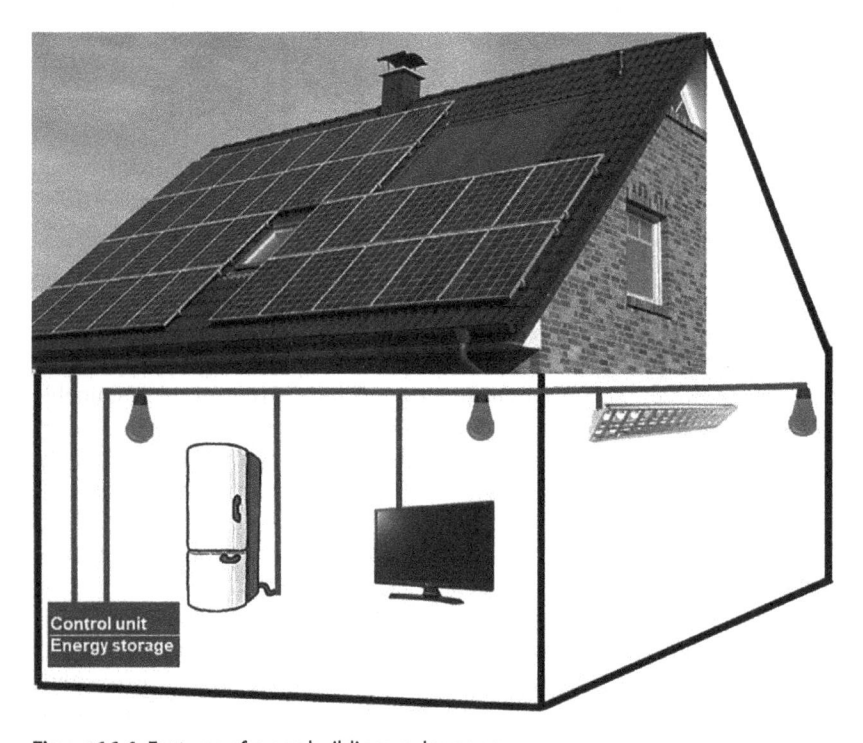

Figure 16.4 Features of green buildings: solar power.

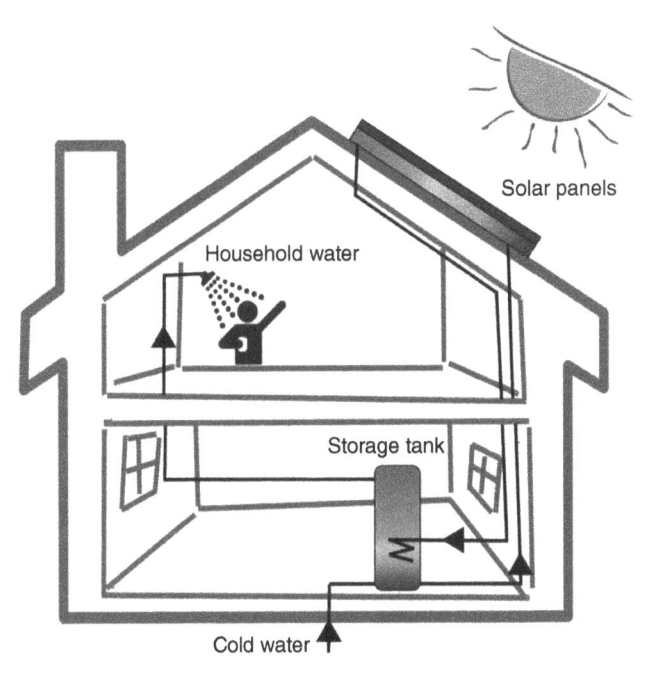

Figure 16.5 Features of green buildings: hot water generation with solar energy and circulation system.

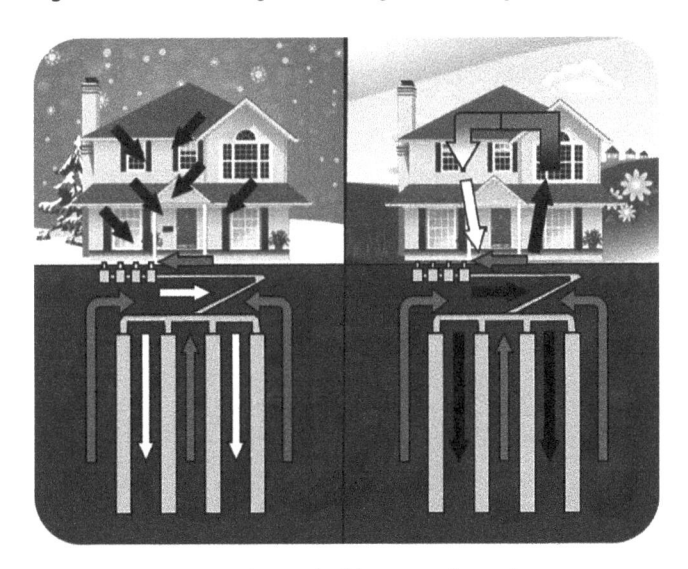

Figure 16.6 Features of green buildings: geothermal energy system.

resources. Figure 16.7 depicts several green building concepts that can be incorporated into structures. These include the following:

- Home orientation to optimize solar exposure;
- High-efficiency Low E glass systems reduce losses of warm or cool air;
- Rain gardens to collect, control, and utilize stormwater;
- Native landscaping that reduces the need for irrigation or agricultural chemicals;

Generation of **photovoltaic** energy

Exposure to the sun
Reduction of energy consumption in heating and lighting
Protection form excessive solar heating and UV light

Heating/cooling
Use renewable energies: solar, geothermal
Use passive systems in solar heating

Low energy consumption
Use daylight-controlled lighting systems
Use occupancy sensors for lighting

Efficient electrical appliances
Helps to reduce the electricity bill

Rain gardens
Favor rain infiltration in soil and reduce storm water run off

Gardens with native plants
Require less maintenance and irrigation

Insulation
Walls and windows insulation to reduce energy use
Use air sealing windows
Minimizing thermal bridging

Sustainable building materials
Use of recycled materials in floors and walls
Help to reduce the use of raw materials/natural resources

Insulated basement floors
Eliminate dampness and reduce utility costs

Rain water harvesting system
The water can be used in toilets, to wash cars or water plants

Water saving
Use duo-flushing toilets and water saving faucets
Use rain sensors to reduce lawn sprinkling

Figure 16.7 Features of green buildings.

- Interior water conservation measures, such as low-flow plumbing fixtures and dual-flush or low-flow toilets;
- High efficiency mechanical and heat ventilation and air-conditioning (HVAC) systems that reduce reliance on energy derived from nonrenewable sources;
- Energy-efficient appliances;
- Insulated floors and walls that reduce loss of warm air or cool air;
- Recycled building materials, such as wood for framing or decking, plastics and metals for finished surfaces, and concrete/asphalt for paving materials;
- Other considerations, such as low VOC paints, "whole-house" attic ventilation fans, green flooring materials, energy-efficient lighting systems, and smart control systems.

16.6 Green Building Rating – LEED

As the interest in green/sustainable buildings has evolved, objective assessment tools that may be used to determine the degree of green and sustainable design have been developed (Gowri 2004). Several rating systems have been developed, some of which are listed as follows:

- Green Star – Australia
- Protocollo Itaca – Italy
- Green Mark – Singapore
- Building Research Establishment Environmental Assessment Method (BREEAM) – United Kingdom
- Leadership in Energy and Environmental Design (LEED) – United States.

LEED was developed by the USGBC as a consensus-based system to provide an independent, third-party verification that structures or communities were designed and constructed with the goal of achieving high performance in key areas of human and environmental health. Some of these areas include sustainable site development, water savings, energy efficiency, materials selection, and indoor environmental quality (USGBC 2018). Several of these key parameters are described as follows.

Integrative process – This aspect of rating system goes beyond any checklist and is an imperative to any green building design. LEED encourages integration right from the beginning and early stages of building design while clarifying owner's aspirations, building performance goals, and project needs. An integrative process identifies the interactions among all the systems of the building and the site. The credit is given to a comprehensive approach toward integrating project team members of different systems and components to achieve mutual advantages and thereby leading to achieving high levels of building performance, human comfort, and environmental benefits. LEED promotes collaboration to enhance the efficiency and effectiveness of every system involved in the building design.

Locations and transportation – Buildings have an unavoidable effect and connection to their surrounding community. LEED encourages construction on previously developed or infill sites and away from environmentally sensitive areas. An emphasis is placed on building construction near existing infrastructure, community resources, and transit systems as well as in locations that promote walking, physical activity, and time spent outdoors.

Sustainable sites – LEED discourages development on previously undeveloped land, seeks to minimize a building's impact on ecosystems and waterways, encourages regionally appropriate landscaping, including the use of native species, rewards smart transportation choices, and promotes control and reduction of stormwater runoff, erosion, light pollution, heat island effects, and construction-related pollution.

Water efficiency – Since a wide range of building systems use potable water, LEED encourages efficient and optimized use of water through conservation, use of reclaimed/recycled water, use of water-saving fixtures and appliances, and water-saving vegetation and irrigation systems that reduce water consumption.

Energy and atmosphere – LEED encourages energy-saving strategies, including energy-efficient design and construction, monitoring, energy-efficient appliances, enhanced refrigerant management systems, and lighting, renewable/zero emission energy resource use, and other measures, such as use of materials that reduces energy or thermal losses.

Materials and resources – LEED encourages the selection of sustainably grown, harvested, produced, and transported products and materials. LEED also promotes waste reduction/diversion, reuse, and recycling and places a specific emphasis on waste reduction at a product's source. LEED encourages demonstration of reduced environmental impacts by reusing existing building resources or demonstrating a reduction in materials use through life-cycle assessment. From a long-term impacts perspective, LEED also promotes to reduce the release of persistent, bioaccumulative and toxic chemicals associated with the life cycle of building materials.

Indoor environmental quality – The USEPA estimates that people spend approximately 90% of their day indoors. Therefore, indoor air quality has a major impact on overall health. LEED promotes strategies that promote healthy indoor environments, including improved indoor air quality, naturally ventilated spaces, increased use of natural light, reduced exposure to tobacco smoke indoors, enhanced views, enhanced acoustics, and improved thermal comfort.

Innovation – LEED encourages innovative design by providing credit for the use of strategies and technologies that improves a building's performance well beyond a "baseline" performance if such strategies were not used. Rewards are also given to projects that utilize a LEED Accredited Professional on the project team to ensure a holistic, integrated approach to design and construction.

Regional priority – USGBC's regional councils, chapters, and affiliates have identified the most important local environmental concerns, and six LEED credits addressing these local priorities have been selected for each region of the country. Projects can earn credit by achieving one or more of these regional priorities.

LEED may be applied to all life-cycle aspects of a project or may be applied to a range of projects and project types, including new construction, existing building operations and maintenance, commercial interiors, core and shell, schools, retail, health care, homes, and neighborhood development (Figure 16.8). Points are tallied and based on performance in the categories described above (also shown in Figure 16.9); the projects may earn one of the following four designations (Figure 16.10):

- *Certified* – 40–49 points
- *Silver* – 50–59 points
- *Gold* – 60–79 points
- *Platinum* – 80–110 points.

To pursue certification, a project is registered, progress is documented, and applied for certification. Some examples of building types eligible for the certification include offices, retail and service establishments, libraries, schools, museums, religious institutions, hotels, and residential buildings of four or more habitable stories. LEED points are awarded on a 100-point scale, and credits are weighted to reflect their potential environmental impacts. A total of 10 bonus points are available, including those associated with regional-specific parameters. A project must satisfy all prerequisites and earn a minimum number of points to be certified.

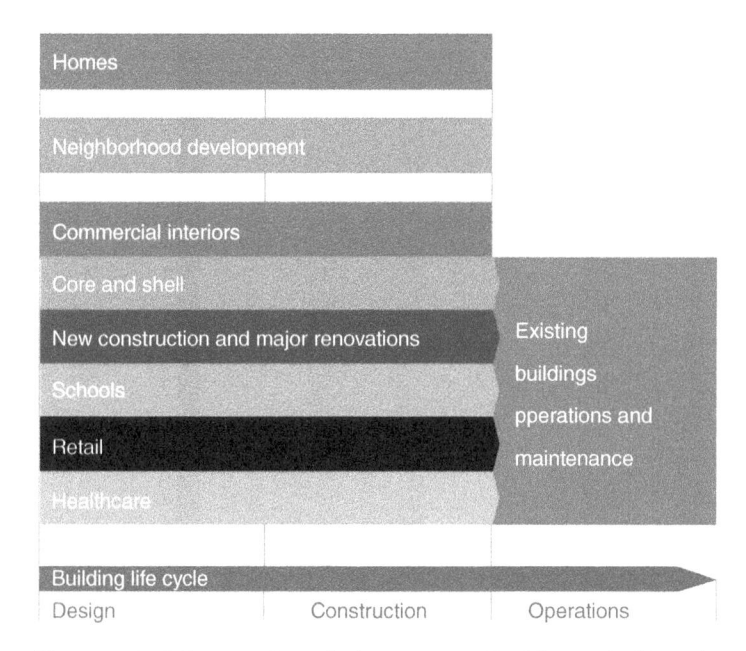

Figure 16.8 LEED system is applied to any type of building and address the complete lifetime of a building, from design and construction to use and decommissioned.

The Green Building Certification Institute (GBCI) administers LEED certification for all commercial and institutional projects registered under any LEED Rating System. By participating in LEED design professionals, operations professionals, and project proponents have the ability to objectively assess the degree of green and sustainable design and operation their project may achieve. While the environmental benefits that may be achieved are explicitly understood, LEED certification can achieve enhanced financial performance. LEED-certified buildings are commonly designed to achieve lower operating costs through more efficient energy and water use, reduced waste generation and related costs, and reduced occupation health impacts. Additionally, LEED-certified structures are often eligible for tax incentives, enhanced or favorable zoning regulations, and other local and national incentives. Potential tenants often consider LEED certification an attractive benefit, which can lead to a premium for rental income. Finally, because of reduced operating costs and greater income potential, the underlying asset is more valuable in the marketplace for the project owner.

LEED credentials are sought by a range of technical and market professionals, including structural and landscape architects, interior designers, engineers, facility managers, real estate professionals, contractors, lenders, and government officials. A range of available LEED credentials include the following:

- LEED Green Associate
- LEED AP Building Design and Construction
- LEED AP Operations and Maintenance
- LEED AP Interior Design and Construction
- LEED AP Homes
- LEED AP Neighborhood Development
- LEED AP Without Specialty
- LEED Fellow.

LEED v4 for BD+C: New Construction and Major Renovation
Project Checklist

Project Name:
Date:

Y	?	N			
			Credit	Integrative Process	1

Location and Transportation — 16

0	0	0			
			Credit	LEED for Neighborhood Development Location	16
			Credit	Sensitive Land Protection	1
			Credit	High Priority Site	2
			Credit	Surrounding Density and Diverse Uses	5
			Credit	Access to Quality Transit	5
			Credit	Bicycle Facilities	1
			Credit	Reduced Parking Footprint	1
			Credit	Green Vehicles	1

Sustainable Sites — 10

0	0	0			
		Y	Prereq	Construction Activity Pollution Prevention	Required
			Credit	Site Assessment	1
			Credit	Site Development - Protect or Restore Habitat	2
			Credit	Open Space	1
			Credit	Rainwater Management	3
			Credit	Heat Island Reduction	2
			Credit	Light Pollution Reduction	1

Water Efficiency — 11

0	0	0			
		Y	Prereq	Outdoor Water Use Reduction	Required
		Y	Prereq	Indoor Water Use Reduction	Required
		Y	Prereq	Building-Level Water Metering	Required
			Credit	Outdoor Water Use Reduction	2
			Credit	Indoor Water Use Reduction	6
			Credit	Cooling Tower Water Use	2
			Credit	Water Metering	1

Energy and Atmosphere — 33

0	0	0			
		Y	Prereq	Fundamental Commissioning and Verification	Required
		Y	Prereq	Minimum Energy Performance	Required
		Y	Prereq	Building-Level Energy Metering	Required
		Y	Prereq	Fundamental Refrigerant Management	Required
			Credit	Enhanced Commissioning	6
			Credit	Optimize Energy Performance	18
			Credit	Advanced Energy Metering	1
			Credit	Demand Response	2
			Credit	Renewable Energy Production	3
			Credit	Enhanced Refrigerant Management	1
			Credit	Green Power and Carbon Offsets	2

Materials and Resources — 13

0	0	0			
		Y	Prereq	Storage and Collection of Recyclables	Required
		Y	Prereq	Construction and Demolition Waste Management Planning	Required
			Credit	Building Life-Cycle Impact Reduction	5
			Credit	Building Product Disclosure and Optimization - Environmental Product Declarations	2
			Credit	Building Product Disclosure and Optimization - Sourcing of Raw Materials	2
			Credit	Building Product Disclosure and Optimization - Material Ingredients	2
			Credit	Construction and Demolition Waste Management	2

Indoor Environmental Quality — 16

0	0	0			
		Y	Prereq	Minimum Indoor Air Quality Performance	Required
		Y	Prereq	Environmental Tobacco Smoke Control	Required
			Credit	Enhanced Indoor Air Quality Strategies	2
			Credit	Low-Emitting Materials	3
			Credit	Construction Indoor Air Quality Management Plan	1
			Credit	Indoor Air Quality Assessment	2
			Credit	Thermal Comfort	1
			Credit	Interior Lighting	2
			Credit	Daylight	3
			Credit	Quality Views	1
			Credit	Acoustic Performance	1

Innovation — 6

0	0	0			
			Credit	Innovation	5
			Credit	LEED Accredited Professional	1

Regional Priority — 4

0	0	0			
			Credit	Regional Priority: Specific Credit	1
			Credit	Regional Priority: Specific Credit	1
			Credit	Regional Priority: Specific Credit	1
			Credit	Regional Priority: Specific Credit	1

0	0	0	TOTALS	Possible Points:	110

Certified: 40 to 49 points, Silver: 50 to 59 points, Gold: 60 to 79 points, Platinum: 80 to 110

Figure 16.9 LEED scoring system for new constructions.

Figure 16.10 LEED certification level based on the scoring system.

State and local governments are increasingly adopting LEED for publicly owned and funded buildings, including governmental, educational, and other institutional uses. Various federal agencies, including the Department of Defense, Department of Agriculture, Department of Energy, and Department of State have also developed LEED initiatives. Further, LEED projects are in design or operation around the world.

16.7 Summary

Buildings can often last decades, even centuries; therefore, good or bad design will live on in use for many years. In recent years, a greater environmental awareness has led to an increased focus on green and sustainable building design and operation. Green and sustainable design and operation, whether applied to buildings as described in this chapter or to infrastructure as described in Chapter 17, should follow several guiding principles. First, development sites and rights-of-way should be chosen to minimize disturbance of the natural environment. To the extent practicable, material and energy inputs should be derived from renewable resources as well as recycled and repurposed materials. Designs and operations should be pursued that select nonhazardous materials to a practicable degree, incorporate processes and systems that maximize mass, energy, and space efficiency, and minimize waste generation in all aspects of the project.

Repetitive or universal, "one-size-fits-all" project designs should be avoided; instead, designs that incorporate the unique attributes of a given development site, including terrain, climate, and directional exposure should be strongly considered and incorporated into design. Designs should be pursued that optimize maintenance, refurbishment, and provide flexibility should be incorporated into designs that allow for future renovation or expansion. Finally, systems should be designed to enable and encourage recycling at the end of the design/operational life.

16.8 Questions

16.1 What is the definition of a green building as per USEPA?

16.2 What are the advantages of green buildings over conventional buildings?

16.3 Mention some of the important components of a green building.

16.4 Describe some key green building concepts and how they can be incorporated into buildings or regional development.

16.5 Name some of the rating systems developed for certification of green buildings.

16.6 What are the key parameters involved in LEED certification of green buildings?

16.7 Describe in detail how green building principles can be incorporated into ongoing building operations.

16.8 Discuss the LEED credentials that are available to practicing professionals.

16.9 What are the four major designations earned by green buildings through LEED certification?

References

Gowri, K. (2004). Green building rating systems: An overview. *ASHRAE Journal* 46 (11): 56.

Kibert, C.J. (2016). *Sustainable Construction: Green Building Design and Delivery*. Wiley.

Matos, G.R. (2012). *Use of Raw Materials in the United States from 1900 through 2010*. Reston, VA: U.S. Geological Survey.

Retzlaff, R.C. (2009). Green buildings and building assessment systems: a new area of interest for planners. *CPL Bibliography* 24 (1): 3–21.

USEPA (2013). Sustainable design and green building toolkit for local governments. US Environmental Protection Agency. Report No. EPA-904-B-10-001.

USGBC (2018). LEED v4 rating system – green building rating system. www.usgbc.org/leed (accessed 1 September 2018).

Willard, B. (2012). *The New Sustainability Advantage: Seven Business Case Benefits of a Triple Bottom Line*. New Society Publishers.

17

Sustainable Civil Infrastructure

17.1 Introduction

Infrastructure consists of the basic physical and organizational structures needed for the operation of a society or an enterprise, or the services and facilities necessary for an economy to function (Hayes 2005). Infrastructure can be further classified as "hard" or "soft." Soft infrastructure refers to institutions required for economic, public health, cultural, and social standards of developed nations, such as those that are associated with the financial systems, education systems, health-care systems, government systems, and law enforcement systems, as well as institutions that provide emergency services.

Hard infrastructure refers to the large physical networks necessary for the functioning of a nation. Transportation infrastructure consists of road and highway networks, bridges, tunnels, culverts, railroads, canals, seaports, and airports. Energy infrastructure includes the electrical transmission grid and related generation stations, natural gas pipelines, and petroleum pipelines. Water and environment infrastructure consists of the drinking water supply, wastewater treatment facilities, stormwater systems, irrigation systems, flood control systems, coastal structures and facilities, municipal solid waste collection and sorting facilities, landfills, incinerators, and recycling facilities. Telecommunications infrastructure includes telephone networks, television and radio transmission facilities, data transmission and Internet facilities, and communications satellites/relay systems.

While civil engineering traditionally has been focused on delivering technical solutions to meet society's infrastructure needs, there is a growing interest to seek sustainable infrastructure solutions given increased recognition of the finite nature of natural resources and a recognized growth in demand for these resources due to ever-growing populations (Fenner et al. 2014). Further, sustainable solutions are sought given increased demands from the regulatory environment, a recognition to address aging infrastructure cost effectively, and the need to address these factors with limitations in government funding to support critical and necessary infrastructure programs (Frangopol and Liu 2007). Since infrastructure plays a critical role in providing a physical linkage between the social and economic aspects of a functioning society, and sustainability is focused on deriving benefit with respect to environmental, social, and economic conditions, there is a natural inspiration toward sustainable infrastructure approaches.

As seen earlier, Chapter 16 dealt with the design and operational considerations for green buildings. While this chapter aims to discuss civil infrastructure in general, the need and principles of sustainable infrastructure and the Envision tool that is used to rate the sustainability of civil infrastructure are also discussed. As an example, sustainable green infrastructure to address water-related issues, specifically stormwater, is presented.

17.2 Principles of Sustainable Infrastructure

In his book *Strategy for Sustainability: A Business Manifesto* (Werbach 2009), environmentalist and sustainability expert Adam Werbach defined a successful sustainability strategy as "different and much bigger that just green." Sustainable strategies should focus not only on effects and interaction with the natural environment but also on interactions with social, economic, and cultural dimensions. Because of the inherent large scale of major infrastructure projects, virtually any design or operational aspect of these projects will have a profound impact toward these dimensions. Further, with increased anticipated impacts of climate changes, infrastructure designs need to be adaptive to account for variances and extremes for key design variables that may be affect during the design life of infrastructure projects.

As outlined in the previous chapter, with respect to the design of sustainable infrastructure, regardless of the type, several guiding principles are useful and should be followed (Dasgupta and Tam 2005; Sarté 2010), including siting that protects the natural environment, materials selection that emphasizes renewable resources and recycled materials, waste minimization, and flexible, adaptive designs. As infrastructure is constructed or rehabilitated, sustainable designs should be sought that strive to be more durable, resilient, and increase the probability of full operation under a wider range of extreme conditions. To facilitate this approach, new rating systems have emerged to assess the degree of sustainability for these projects. These rating systems aim to assess performance in extreme conditions as well as superior overall performance and expected durability within and beyond a conventional service life. To fully realize these assessment goals, a new holistic framework and sustainability assessment system for infrastructure projects is needed (Sahely et al. 2005). The optimization of various infrastructure elements will often have a community-wide impact. Finally, an ideal sustainability rating system for infrastructure should encourage conservation and restoration of resources and natural systems.

17.3 Civil Infrastructure

As stated in the introductory section, infrastructure consists of facilities and networks necessary to allow society to function. These facilities may be highly visible or barely noticeable facilities in cities, towns, and open areas. Regardless of their degree of visibility or obscurity, objectively speaking, the condition of civil infrastructure in the United States desperately needs improvement. Civil infrastructure includes bridges, dams, roads, water supply/treatment, waste management, and other similar facilities. Figure 17.1 shows a civil infrastructure report card issued by the American Society of Civil Engineers (ASCE). In summary, ASCE concluded that United States infrastructure earned an overall "D+" grade, indicating a poor overall condition (Figure 17.2). Furthermore, it was estimated that approximately $2.2 trillion was needed for infrastructure investment over the next five years. It is important to note that this investment is needed just to recover the utility and performance of seriously deficient systems. Beyond this investment, new infrastructure and upgrades and maintenance of existing systems are needed to enhance quality of life, promote economic competitiveness, and support job creation.

17.3 Civil Infrastructure | 301

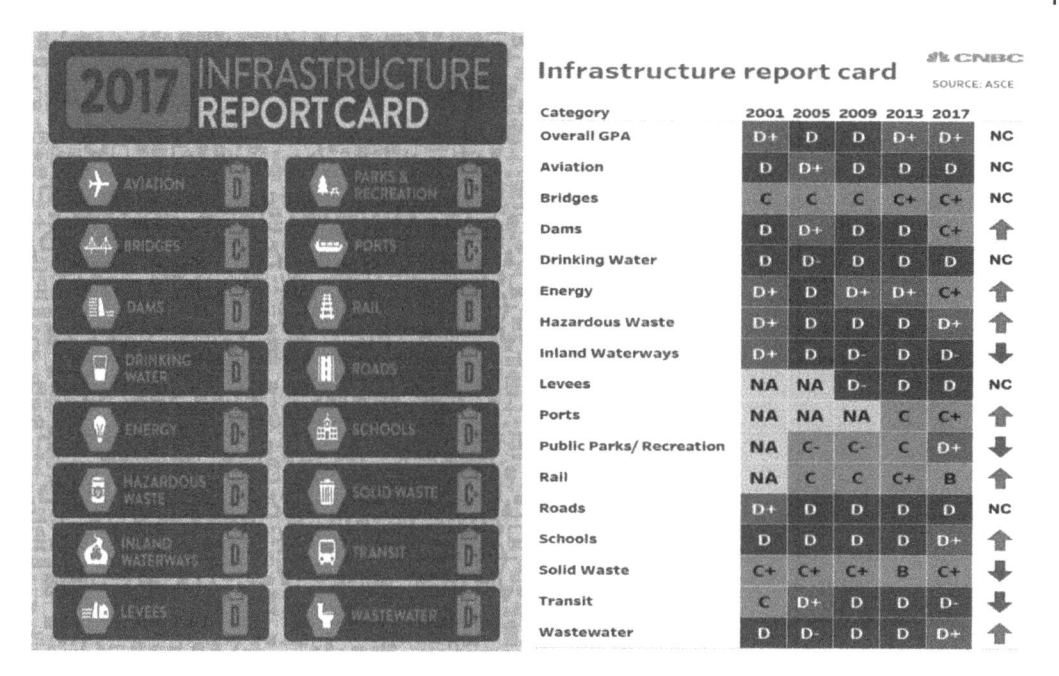

Category	2001	2005	2009	2013	2017	
Overall GPA	D+	D	D	D+	D+	NC
Aviation	D	D+	D	D	D	NC
Bridges	C	C	C	C+	C+	NC
Dams	D	D+	D	D	C+	↑
Drinking Water	D	D-	D	D	D	NC
Energy	D+	D	D+	D+	C+	↑
Hazardous Waste	D+	D	D	D	D+	↑
Inland Waterways	D+	D	D-	D	D-	↓
Levees	NA	NA	D-	D	D	NC
Ports	NA	NA	NA	C	C+	↑
Public Parks/ Recreation	NA	C-	C-	C	D+	↓
Rail	NA	C	C	C+	B	↑
Roads	D+	D	D	D	D	NC
Schools	D	D	D	D	D+	↑
Solid Waste	C+	C+	C+	B	C+	↓
Transit	C	D+	D	D	D-	↓
Wastewater	D	D-	D	D	D+	↑

Figure 17.1 Infrastructure report grading the state of main US infrastructure in 2017, and their evolution from 2001. Source: ASCE, www.asce.org/infrastructure.

Figure 17.2 Overall grade of the US main infrastructure in 2017. Source: ASCE, www.asce.org/infrastructure.

17.4 Envision™: Sustainability Rating of Civil Infrastructure

To meet the demand for an objective sustainability rating system for civil infrastructure, ASCE has been partnering with the American Council of Engineering Companies (ACEC), the American Public Works Association (APWA), and formed the Institute for Sustainable Infrastructure (ISI), which further collaborated with the Zofnass Program for Sustainable Infrastructure at Harvard University to develop Envision™. The Envision rating system, which assesses triple bottom line parameters, provides an objective framework for assessing the degree sustainability of civil infrastructure projects, while describing the elements of sustainable projects. Envision also identifies means to optimize projects as well as the benefits of sustainable project aspects to designers, regulators, and project owners (ASCE 2018).

The purpose of Envision's web-based platform is to enhance the sustainability of the nation's infrastructure, excluding occupied buildings. Envision assesses civil infrastructure using two parallel criteria. First, the "pathway contribution" is assessed. In considering the needs of the community to be served by the civil infrastructure project under consideration, the assessment can select an appropriate solution that aims to "do the right thing" for the given community. Dialogue between the project team and proponents and the community-at-large is highly encouraged to identify key factors important to the community. Second, the "performance contribution" is assessed, in which the assessment is focused on how the project design and delivery can achieve a desired degree of performance. With this criterion, the emphasis is on "doing things right" by identifying and incorporating best management practices (BMPs) and encouraging higher degrees of sustainability with respect to the project.

Envision may be used for projects that exhibit a wide range of size and complexity. Instead of providing a uniform set of methods for projects, Envision incorporates the specific needs and circumstances of a community to optimize the performance of a project for that particular community. Additionally, as a voluntary assessment framework, Envision may either be used for internal self-assessment or for audit and/or independent verification purposes. By recognizing progress and contributions to sustainable principles, the use of Envision is intended to pursue balanced approaches that maximize triple bottom line parameters, leading to conservation and restoration of natural resources and ecological systems as well as the strengthening of the social fabric of communities (Pedersen 2012; Behr 2014).

Envision provides a comparison of infrastructure alternatives across a range of life cycle categories. The results of an Envision assessment may be incorporated into a sensitivity analysis that assesses design variables of potential infrastructure alternatives with respect to the triple bottom line, allowing project stakeholders to optimize a project's outcome in these categories. Costs and benefits may be assessed over the life cycle, including environmental benefits and performance objectives, to achieve higher levels of sustainability. It is also intended to critically assess the entire life cycle of a project, including decommissioning activities.

Envision may be applied to numerous civil infrastructure categories, including roads, bridges, pipelines, railways, airports, power transmission systems, telecommunication towers, dams, levees, solid waste landfills, water supplies and conveyance systems, wastewater treatment plants, and public spaces. During the evaluation, Envision assesses the overall sustainability of a project not simply as a singular improvement but in terms of its overall contribution to the community that is served. Further, it is not intended to replace existing performance rating systems; rather, it is meant to provide an essential context for ratings results.

Envision incorporates four stages of assessment as follows:

- In Stage 1, a self-assessment checklist is applied, and a focus is placed on the education of the sustainability aspects of a particular infrastructure.

- In Stage 2, the third-party objective rating of the project is emphasized, allowing project proponents to submit a particular project for verification and public recognition. This stage includes guidance with respect to scoring. It is essential that sustainability achievements associated with a project are properly documented at this stage and that the independent third-party verification be performed by a qualified individual or entity.
- Additional stages (Stages 3 and 4) are focused on accounting for complex/multi-stage projects as well as optimization of a particular project during design, respectively.

The Stage 2 scorecard metrics are divided into five sections or categories: quality of life (QL), leadership (LD), resource allocation (RA), natural world (NW), and climate and resilience (CR) as shown in Table 17.1. A maximum of 64 credits can be earned that amounts to a maximum of 1000 points that may be achieved. Each credit is described in a two-page summary, as shown in table 17.1, that includes the intent and purpose of the particular metric, the basis and documentation of the evaluation, the achieved score, and a summary of how a higher score may be achieved. More details can be found in Envision Manual (ASCE 2018). Examples of Envision applications are given in Chapters 20–22.

Table 17.1 Categories for ENVISION rating system (ASCE 2018).

1			QL1.1 Improve Community Quality of Life
2			QL1.2 Enhance Public Health & Safety
3		WELLBEING	QL1.3 Improve Construction Safety
4			QL1.4 Minimize Noise & Vibration
5			QL1.5 Minimize Light Pollution
6	QUALITY OF LIFE		QL1.6 Minimize Construction Impacts
7			QL2.1 Improve Community Mobility
8		MOBILITY	QL2.2 Encourage Sustainable Transportation
9			QL2.3 Improve Access and Wayfinding
10			QL3.1 Advance Equity & Social Justice
11		COMMUNITY	QL3.2 Preserve Historic & Cultural Resources
12			QL3.3 Enhance Views & Local Character
13			QL3.4 Enhance Public Spaces & Amenities
14			LD1.1 Provide Effective Leadership & Commitment
15		COLLABORATION	LD1.2 Foster Collaboration & Teamwork
16			LD1.3 Provide for Stakeholder Involvement
17			LD1.4 Pursue Byproduct Synergies
18	LEADERSHIP		LD2.1 Establish a Sustainability Management Plan
19		PLANNING	LD2.2 Plan for Sustainable Communities
20			LD2.3 Plan for Long-Term Monitoring & Maintenance
21			LD2.4 Plan for End-of-Life
22			LD3.1 Stimulate Economic Prosperity & Development
23		ECONOMY	LD3.2 Develop Local Skills & Capabilities
24			LD3.3 Conduct a Life-Cycle Economic Evaluation

(Continued)

Table 17.1 (Continued)

#	Category	Subcategory	Credit
25	RESOURCE ALLOCATION	MATERIALS	RA1.1 Support Sustainable Procurement Practices
26			RA1.2 Use Recycled Materials
27			RA1.3 Reduce Operational Waste
28			RA1.4 Reduce Construction Waste
29			RA1.5 Balance Earthwork On Site
30		ENERGY	RA2.1 Reduce Operational Energy Consumption
31			RA2.2 Reduce Construction Energy Consumption
32			RA2.3 Use Renewable Energy
33			RA2.4 Commission & Monitor Energy Systems
34		WATER	RA3.1 Preserve Water Resources
35			RA3.2 Reduce Operational Water Consumption
36			RA3.3 Reduce Construction Water Consumption
37			RA3.4 Monitor Water Systems
38	NATURAL WORLD	SITING	NW1.1 Preserve Sites of High Ecological Value
39			NW1.2 Provide Wetland & Surface Water Buffers
40			NW1.3 Preserve Prime Farmland
41			NW1.4 Preserve Undeveloped Land
42		CONSERVATION	NW2.1 Reclaim Brownfields
43			NW2.2 Manage Stormwater
44			NW2.3 Reduce Pesticide & Fertilizer Impacts
45			NW2.4 Protect Surface & Groundwater Quality
46		ECOLOGY	NW3.1 Enhance Functional Habitats
47			NW3.2 Enhance Wetland & Surface Water Functions
48			NW3.3 Maintain Floodplain Functions
49			NW3.4 Control Invasive Species
50			NW3.5 Protect Soil Health
51	CLIMATE AND RESILIENCE	EMISSIONS	CR1.1 Reduce Net Embodied Carbon
52			CR1.2 Reduce Greenhouse Gas Emissions
53			CR1.3 Reduce Air Pollutant Emissions
54		RESILIENCE	CR2.1 Avoid Unsuitable Development
55			CR2.2 Assess Climate Change Vulnerability
56			CR2.3 Evaluate Risk and Resilience
57			CR2.4 Establish Resilience Goals and Strategies
58			CR2.5 Maximize Resilience
59			CR2.6 Improve Infrastructure Integration

17.5 Sustainable Infrastructure Practices: Example of Water Infrastructure

Civil infrastructure is multifaceted and can include bridges, dams, roads, water supply/treatment, stormwater management, waste management, and other similar facilities. To demonstrate how sustainability principles may be pursued across a comprehensive infrastructure system, water management/infrastructure is considered as an example herein. The infrastructure-related management and use of water may include domestic water treatment and conveyance, wastewater management, stewardship of groundwater and surface water, protection of coastal areas, watersheds, and wetlands, and stormwater treatment and conveyance.

- A public drinking water system, or public water system (PWS), is a system that provides water for human consumption through at least 15 service connections or regularly serves (at least 60 days per year) at least 25 individuals. PWSs are responsible for providing drinking water to 90% of the population of the United States.
- Groundwater serves as a source for drinking water, either for public supply systems or for private well systems. Groundwater is recharged as precipitation infiltrates through permeable ground surfaces and migrates through soil and rock into aquifers. Many communities derive water sources from underground aquifers. Water supply wells are constructed by drilling through soil until the aquifers are reached. Under most conditions, pump systems are used to extract these groundwater sources to the surface for treatment, storage, and conveyance.
- Stormwater runoff is generated when precipitation events result in surface flows over land or impervious surfaces, including paved streets, parking lots, and building rooftops. During overland flow, stormwater accumulates debris, chemicals, sediment, and other pollutants that could adversely affect the quality of receiving waters if the runoff is not treated prior to discharge. Large areas of connected impervious surfaces exist, and continued development and ongoing conversion toward impervious surfaces can dramatically increase the volume and rate of stormwater discharges, leading to increased pollutant delivery, flooding, and erosion as well as decreased groundwater recharge. As a result, the impacts of stormwater runoff within urbanized areas can be extensive (Lee and Bang 2000).
- Many stormwater discharges are considered point sources and require coverage under an NPDES (National Pollutant Discharge Elimination System) permit. Understanding the impacts of urban stormwater is important to preserve and protect aquatic ecosystems and human health. The goal of sustainable stormwater infrastructure is to manage the small, frequent storms that account for most pollutant loading (Barbosa et al. 2012). A range of systems collectively called BMPs are available to control and treat these flows as well as attempt to mimic to the extent practicable pre-urbanized stormwater runoff.

To assess the effectiveness of green infrastructure in protecting our water resources, researchers must link the performance of green infrastructure initiatives such as those listed above to water quality and ecological outcomes in receiving waters. Research indicates that green infrastructure and low impact development (LID) initiatives offer an effective approach to improving receiving water quality and flow. Further, a growing knowledge base is increasingly being shared by the regulatory and development communities with respect to the costs and benefits of sustainable stormwater control systems.

For example, sustainable stormwater infrastructure and its use to control flow volumes, rates, and pollutant loading provide many beneficial effects to the physical environment, including reduced pollutant loading and reduced alteration of flow rates, velocities, and volumes to receiving surface water bodies. Groundwater conditions are also improved; stormwater infrastructure can induce groundwater recharge, mimicking rates that would be expected in a natural setting prior to addition of impervious surfaces or other additions to the built environment. Additionally, bioretention facilities can mitigate pollutant loading to the subsurface. The use of vegetation in planter boxes, rain gardens, wetlands, bioswales, green roofs, and tree canopies can facilitate improved air quality and sequester carbon dioxide from the atmosphere. These techniques can also moderate ambient air temperatures in the vicinity of the built environment, leading to a reduced reliance on HVAC output and a reduction in related energy consumption. Natural habitats may also be enhanced – open space can be preserved or enhanced, providing urban wildlife habitat, increasing biodiversity, and propagating native species in a controlled manner. Finally, sustainable stormwater infrastructure enhances urban quality of life through aesthetic improvements and dedication of open space to enhance physical activity and public health. Nevertheless, pros and cons of any planned infrastructure under short-term or long-term conditions should be evaluated.

A brief description of these green stormwater infrastructure options is presented below.

17.5.1 Green Roofs

Green roofs incorporate actively growing vegetation into roof surfaces to reduce stormwater runoff volumes, regulate building temperatures, reduce urban heat island effects, and provide urban wildlife habitat (Figures 17.3 and 17.4). Active research is underway with respect to the selection of plant species, propagation and establishment methods, plant succession, carbon sequestration potential, and water and nutrient requirements (Oberndorfer et al. 2007).

17.5.2 Permeable Pavements

Permeable pavements may be used as an alternative to impermeable concrete and asphalt (Figures 17.5 and 17.6). These materials allow a portion of stormwater runoff to infiltrate into the ground surface instead of contributing to overland flow and runoff. Typically, these materials incorporate specially designed air-entrained concrete or the use of discrete permeable paving blocks. Permeable pavements are among the most extensively studied and well-understood infrastructure practices. Assuming geotechnical and geologic conditions will not be compromised, the use of these materials address infiltration capacity and pollutant removal efficiency, adding to the long-term beneficial effects on surface water quality. Additionally, when systems are properly designed, they can be protective of groundwater quality (Scholz and Grabowiecki 2007).

Figure 17.3 Green roof helps cool the building and minimize water run-off.

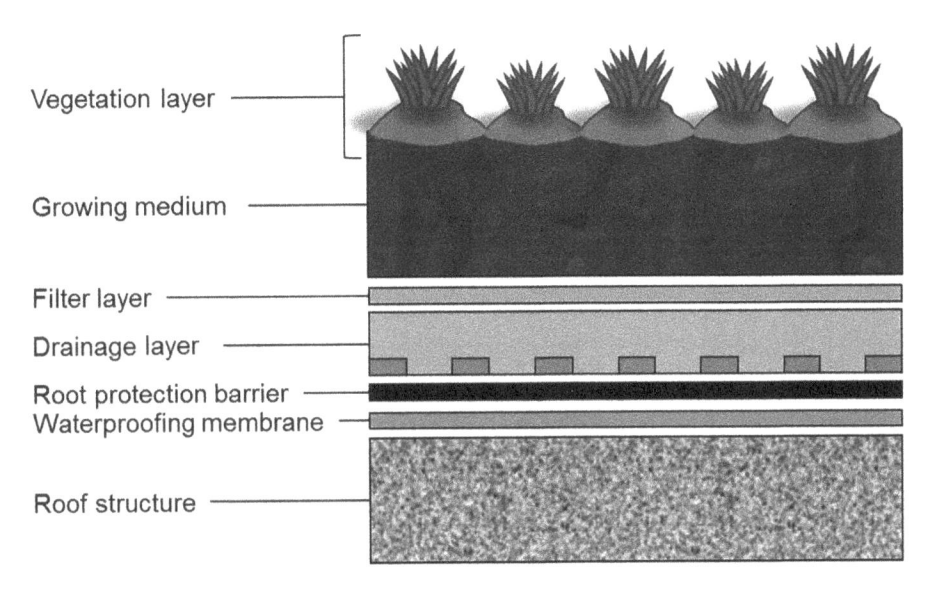

Figure 17.4 Green roof structure.

Figure 17.5 Permeable pavement. Source: USEPA.

17.5.3 Rainwater Harvesting

Recent droughts and potential impacts of climate change have generated interest in rainwater harvesting as an approach to water conservation as well as stormwater management (Figures 17.7 and 17.8). Research is being conducted on sizing methodologies, public perceptions, and the water quality of harvested rainwater. Ironically, in several generally arid western states, the legality of rainwater harvesting is questionable with respect to water rights. These longstanding, and in many cases, antiquated statutes are fortunately being revisited to address rainwater harvesting potential (Boers and Ben-Asher 1982).

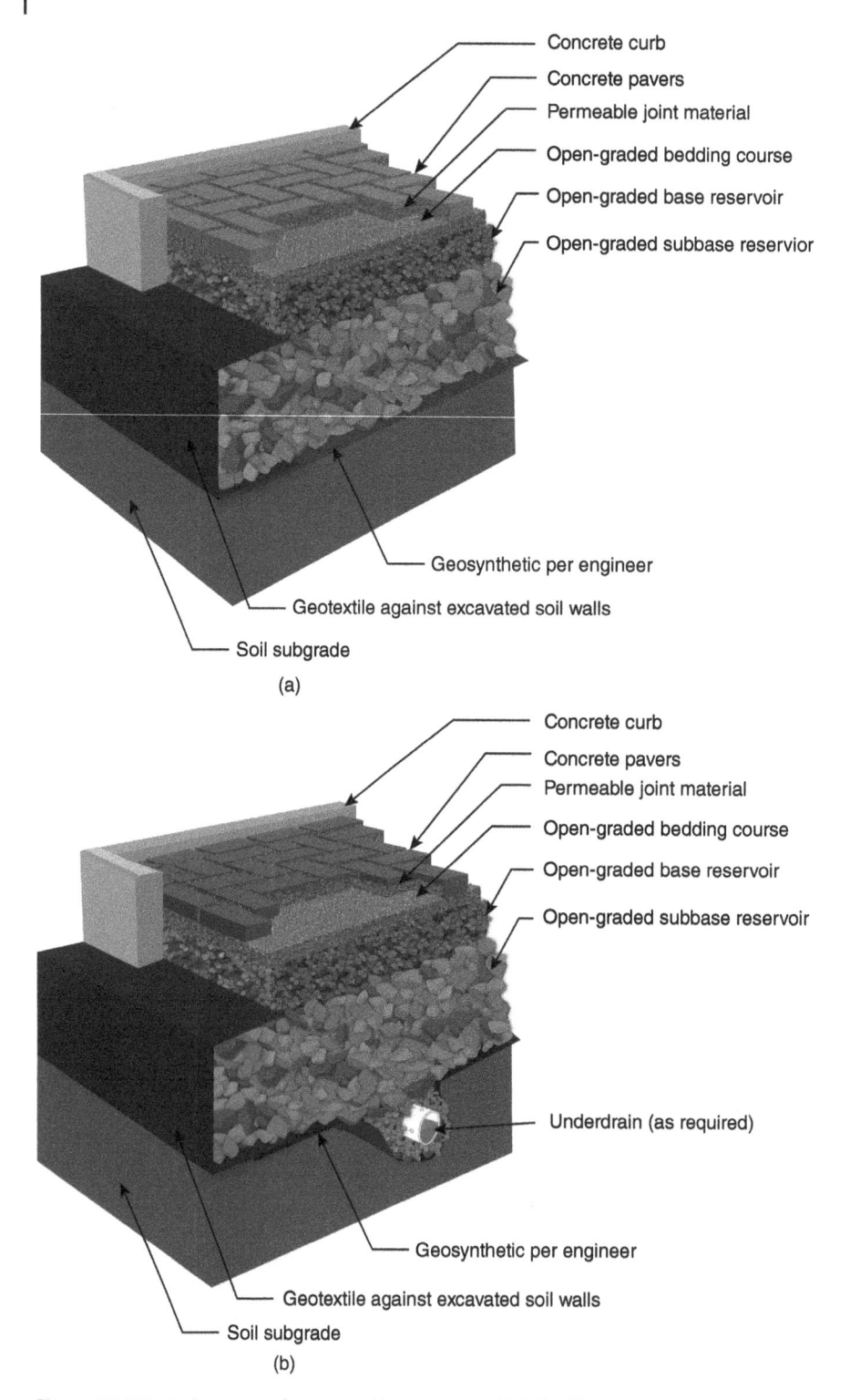

Figure 17.6 Typical structure for permeable pavement. (a) Full infiltration, (b) partial infiltration. Source: Courtesy of ICPI. Smith (2017).

Figure 17.7 Rainwater harvesting. Source: USEPA.

17.5.4 Rain Gardens and Planter Boxes

Rain gardens and planter boxes are both forms of bioretention (Figure 17.9). The concept of bioretention is simple – stormwater runoff is captured and infiltration rates are controlled to provide a sufficient residence time within a treatment zone, allowing microorganisms within soil material to degrade a range of pollutants. Bioretention areas provide a source for groundwater recharge, pollutant removal, and runoff detention as well as an effective means for stormwater management where open space is limited. Research is underway to assess the factors that are incorporated into bioretention design and operation (Davis et al. 2009).

17.5.5 Bioswales

Similar in many ways to rain gardens and planter boxes, bioswales are another form of bioretention. These swales are used to convey stormwater within a shallow, vegetation-covered channel. Bioswales can be effective in attenuating peak flows, reducing flow velocity, reducing pollutant loads, and enhancing biodiversity (Figures 17.10 and 17.11). Performance is sensitive to design parameters that are incorporated (Mazer et al. 2001).

17.5.6 Constructed Wetlands and Tree Canopies

Constructed wetlands are designed to facilitate stormwater collection and detention on a grander scale into a wetlands habitat. Constructed wetlands are complex systems with many

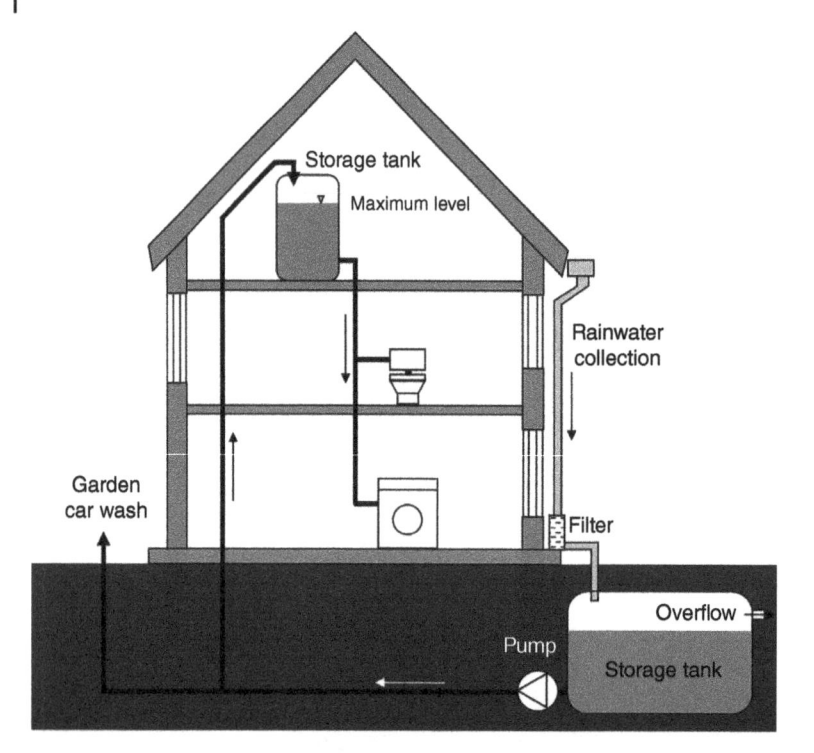

Figure 17.8 Functioning of a rainwater harvesting system.

Figure 17.9 Rainwater runoff collection in a rain garden.

design parameters; proper selection of species, topography, and other physical parameters are essential to create a successfully operating project (Figures 17.12 and 17.13). Previous studies have examined not only wetland performance and design but also wetland management and evaluation as well. Tree canopies in an urban environment can effectively mitigate air pollution, reduce runoff quantity, and reduce energy use while enhancing habitat and aesthetics (Figure 17.14). Additionally, research is being performed on the interaction between urban soils and urban trees (Wu et al. 2015).

17.5 Sustainable Infrastructure Practices: Example of Water Infrastructure | 311

Figure 17.10 Treatment of water runoff in a bioswale. Source: USEPA.

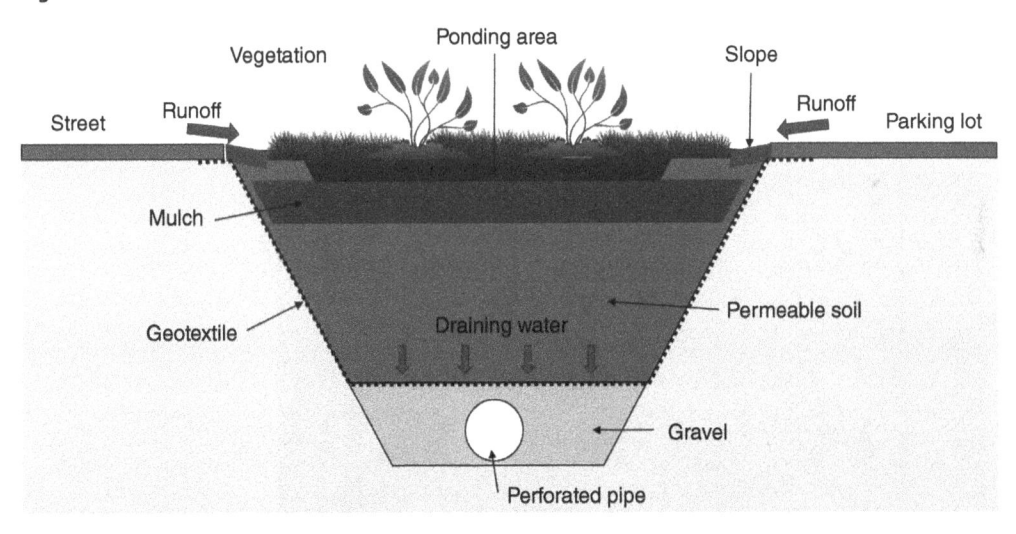

Figure 17.11 Bioswale structure.

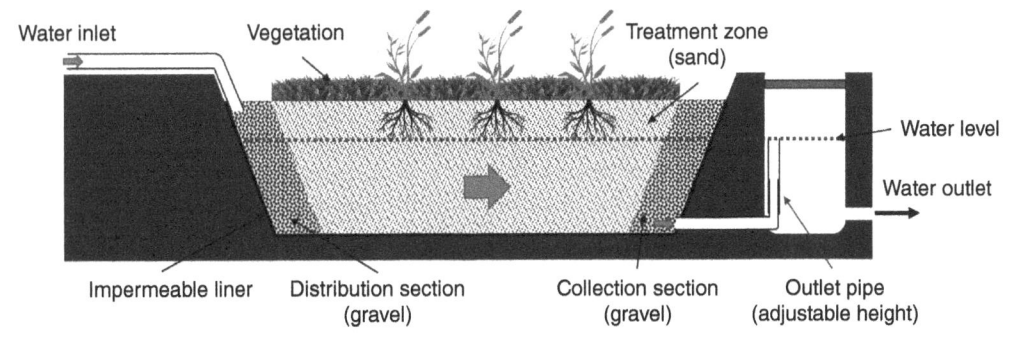

Figure 17.12 Structure of constructed wetlands for water treatment.

Figure 17.13 Wastewater treatment in a constructed wetland. Source: Ecocelta, Spain.

Figure 17.14 Tree canopies reduce storm water runoff and mitigate pollution.

17.6 Summary

Both "soft" and "hard" infrastructure is critically necessary for the function of society. While civil engineering has traditionally been focused on delivering technical solutions to meet society's needs and to protect the environment, only recently has increased awareness of nature and natural resources inspired a search for sustainable infrastructure. Several additional key factors have affected the design of civil infrastructure, including additional demands from the regulatory environment, a focus on cost-effective solutions to aging infrastructure, and the desire to accomplish this with limitations in government funding.

There is a well-established connection between sustainability concepts and environmentally focused engineering practices, and interest continues to grow with respect to the benefits of sustainability in other disciplines. As infrastructure provides a critical physical linkage between the social and economic aspects of a functioning society, and sustainability is focused on deriving benefit with respect to environmental, social, and economic conditions; there is a natural connection to devise sustainable infrastructure.

17.7 Questions

17.1 What are the two major classifications of infrastructure and what do they comprise of?

17.2 What is the principle behind sustainable infrastructure?

17.3 Discuss in detail long-term benefits associated with sustainable infrastructure approaches.

17.4 Why has existing infrastructure received poor ratings during infrastructure assessments?

17.5 Who developed the Envision™ rating tool?

17.6 What is the purpose of Envision tool?

17.7 Discuss the importance of the stages of Envision™ ratings.

17.8 What are the different methods to achieve sustainable water infrastructure?

17.9 What do you mean by green roofs and bioswales? Explain briefly.

17.10 What is the concept behind bioretention systems? Discuss the importance of ongoing maintenance activities for tree canopies and constructed wetlands.

17.11 Discuss the importance of ongoing maintenance activities for tree canopies and constructed wetlands.

17.12 Discuss technical challenges that may be encountered during the use of stormwater treatment systems (permeable pavers, bioswales, and bioretention) and discuss strategies to mitigate these challenges.

References

ASCE (2018). ENVISION. American Society of Civil Engineers. https://www.asce.org/envision/ (accessed September 2018).

Barbosa, A.E., Fernandes, J.N., and David, L.M. (2012). Key issues for sustainable urban stormwater management. *Water Research* 46 (20): 6787–6798.

Behr, C. (2014). A value-based rating system for envision. In: *ICSI 2014: Creating Infrastructure for a Sustainable World*, 744–754. Long Beach, CA: American Society of Civil Engineers.

Boers, T.M. and Ben-Asher, J. (1982). A review of rainwater harvesting. *Agricultural Water Management* 5 (2): 145–158. doi: 10.1016/0378-3774(82)90003-8.

Dasgupta, S. and Tam, E.K. (2005). Indicators and framework for assessing sustainable infrastructure. *Canadian Journal of Civil Engineering* 32 (1): 30–44.

Davis, A.P., Hunt, W.F., Traver, R.G., and Clar, M. (2009). Bioretention technology: overview of current practice and future needs. *Journal of Environmental Engineering* 135 (3): 109–117. https://doi.org/10.1061/(ASCE)0733-9372(2009)135:3(109).

Fenner, R.A., Cruickshank, H.J., and Ainger, C. (2014). Sustainability in civil engineering education: why, what, when, where and how. *Proceedings of the Institution of Civil Engineers: Engineering Sustainability* 167 (5): 228–237. https://doi.org/10.1680/ensu.14.00002.

Frangopol, D.M. and Liu, M. (2007). Maintenance and management of civil infrastructure based on condition, safety, optimization, and life-cycle cost*. *Structure and Infrastructure Engineering* 3 (1): 29–41.

Hayes, B. (2005). *Infrastructure: A Guide to the Industrial Landscape*. New York: W. W. Norton and Co. Ltd. http://www.gettextbooks.com/isbn/9780393349832/.

Lee, J.H. and Bang, K.W. (2000). Characterization of urban stormwater runoff. *Water Research* 34 (6): 1773–1780.

Mazer, G., Booth, D., and Ewing, K. (2001). Limitations to vegetation establishment and growth in biofiltration swales. *Ecological Engineering* 17 (4): 429–443. https://doi.org/10.1016/S0925-8574(00)00173-7.

Oberndorfer, E., Lundholm, J., Bass, B. et al. (2007). Green roofs as urban ecosystems: ecological structures, functions, and services. *BioScience* 57 (10): 823–833. https://doi.org/10.1641/B571005.

Pedersen, T.A. (2012). Institute for Sustainable Infrastructure Envision™ Rating System: Applicability to Water Infrastructure Projects. *Proceedings of the Water Environment Federation* 2012 (15): 1849–1856.

Sahely, H.R., Kennedy, C.A., and Adams, B.J. (2005). Developing sustainability criteria for urban infrastructure systems. *Canadian Journal of Civil Engineering* 32 (1): 72–85.

Sarté, S.B. (2010). *Sustainable Infrastructure: The Guide to Green Engineering and Design*. Wiley.

Scholz, M. and Grabowiecki, P. (2007). Review of permeable pavement systems. *Building and Environment* 42 (11): 3830–3836. https://doi.org/10.1016/j.buildenv.2006.11.016.

Smith, D.R. (2017). *Permeable Interlocking Concrete Pavements*, 5e. Interlocking Concrete Pavement Institute (ICPI).

Werbach, A. (2009). *Strategy for Sustainability: A Business Manifesto*. Harvard Business Press.

Wu, H., Zhang, J., Ngo, H.H. et al. (2015). A review on the sustainability of constructed wetlands for wastewater treatment: design and operation. *Bioresource Technology* 175: 594–601.

18

Sustainable Remediation of Contaminated Sites

18.1 Introduction

The United States Environmental Protection Agency (USEPA) has estimated that there are thousands of sites that have been contaminated in the United States, and over 294 000 of these sites require urgent remedial action (Figure 18.1). The contaminants encountered include organic compounds, heavy metals, and radionuclides. A variety of sources can cause the subsurface contamination, as depicted in Figure 18.2, and these sources of contamination include land disposal of solid and liquid wastes, accidental spills, waste dumps and landfills, leakage from underground storage tanks (USTs), and mining operations (leaching of the spoil material, milling wastes, etc.). The cost to cleanup these sites is estimated to exceed $209 billion (US dollars) (USEPA 2004).

Traditional, risk-based site remedial approaches have not always been sustainable because they often do not account for broader environmental impacts such as extraction and use of natural resources, wastes created, energy used, transportation of equipment and materials, and related greenhouse gas (GHG) emissions for on- and off-site operations. These approaches do not explicitly account for the net environmental benefit when all relevant environmental parameters are considered. To address this, principles of "green remediation" and "sustainable remediation" have emerged. "Sustainable remediation" is defined as a remedy or combination of remedies whose net benefit on human health and the environment is maximized through the judicious use of limited resources (Baker et al. 2009). On the other hand, "Green remediation" is defined as the practice of considering all environmental effects of remedy implementation and incorporating options that maximize the net environmental benefit of cleanup actions (USEPA 2008).

Green and sustainable remediation (GSR) has emerged as a remediation strategy (involving one or more remediation technologies) whose net benefit to human health and the environment is maximized through an optimized selection of remediation technologies and resource use that consider how the community, global society, and the environment would benefit, or be adversely affected by its implementation (ITRC 2011a,b).

This chapter presents an overview of sustainability decision frameworks, metrics, and assessment tools. This is followed by a discussion and analysis of case studies with respect to sustainability and the degree of success achieved with each of the respective studies. Finally, an outlook of the future evolution of this innovative approach to environmental remediation is presented. More detailed information can be found in Reddy and Adams (2015).

Sustainable Engineering: Drivers, Metrics, Tools, and Applications, First Edition.
Krishna R. Reddy, Claudio Cameselle, and Jeffrey A. Adams.
© 2019 John Wiley & Sons, Inc. Published 2019 by John Wiley & Sons, Inc.

18 Sustainable Remediation of Contaminated Sites

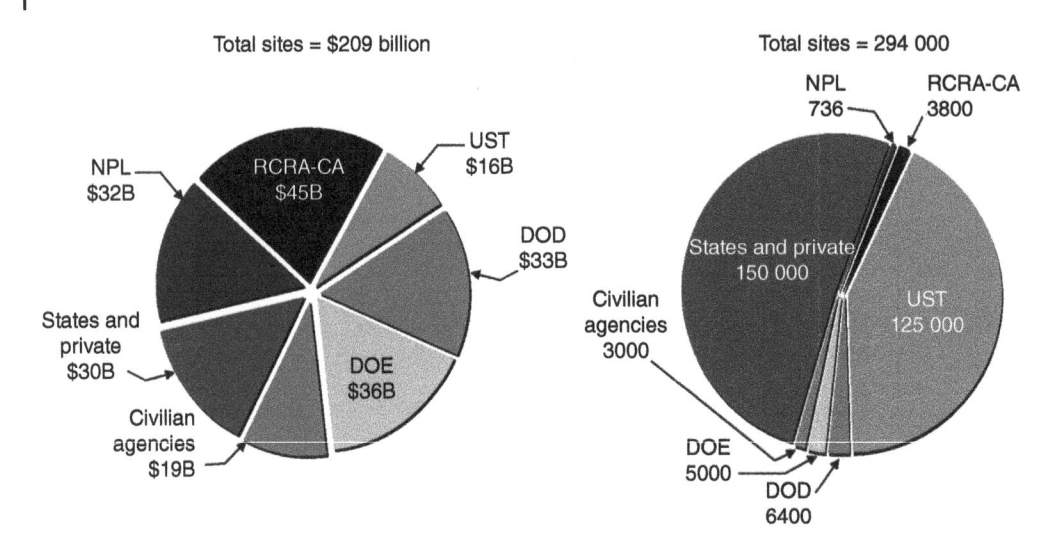

Figure 18.1 Estimated number of contaminated sites in the United States. (Cleanup horizon: 2004–2033). Source: USEPA (2004).

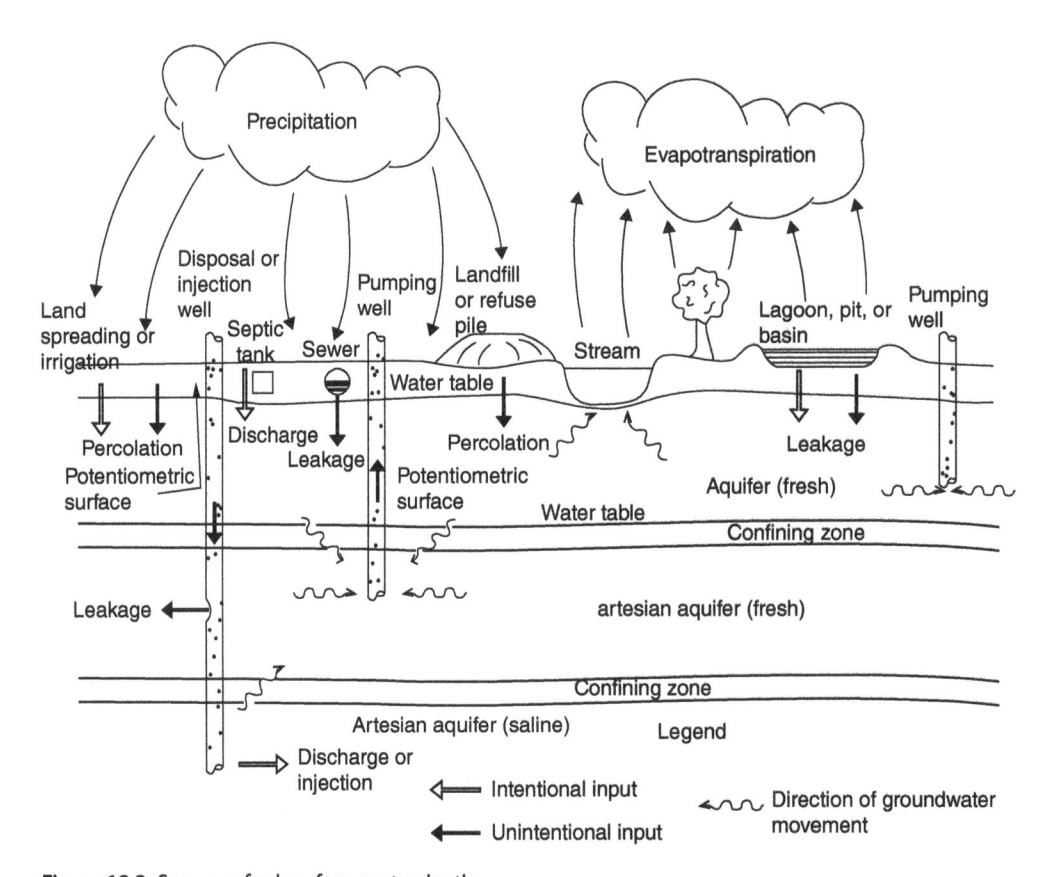

Figure 18.2 Sources of subsurface contamination.

18.2 Contaminated Site Remediation Approach

When considering financial and timing implications, a systematic approach is necessary for the characterization and remediation of contaminated sites. The most important tasks of such a systematic approach include: (i) site characterization, (ii) risk assessment, and (iii) the selection of an effective remedial action. Figure 18.3 outlines such a systematic approach. In addition, innovative integration of various tasks of remediation can often lead to a faster, cost-effective remedial program.

To develop a remediation strategy, site characterization is usually the first step. It consists of the collection and assessment of data representing contaminant type and distribution at a site under investigation. The results of a site characterization form the basis for risk assessment and decisions concerning the requirements of remedial goals and action. Additionally, the results serve as a guide for design, implementation, and monitoring of the remedial system. An effective site characterization includes the collection of data pertaining to (i) site geologic data, including site stratigraphy and important geologic formations, (ii) hydrogeological data, including major

Figure 18.3 Systematic approach for contaminated site remediation.

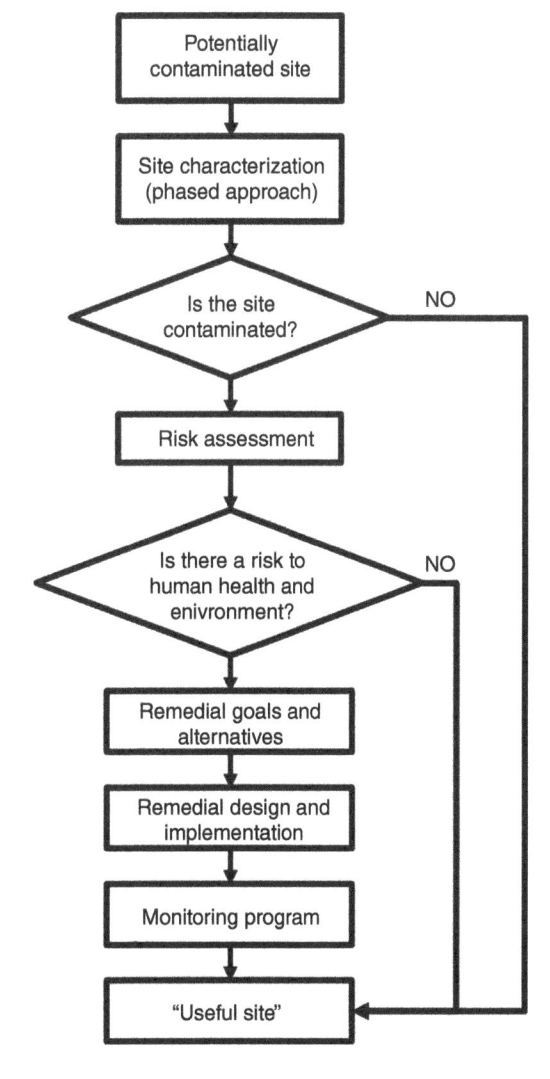

water-bearing formations and their hydraulic properties, such as groundwater depth, direction of flow, and gradient, and (iii) site contamination data, including type, concentration, phase, and distribution, and lateral and vertical extent of contamination. Additionally, surface conditions both at and around the site must be taken into consideration.

Once characterization activities have confirmed the presence of site contamination, a risk assessment may be performed. The USEPA has developed comprehensive risk assessment procedures. The USEPA procedure was originally developed by the United States Academy of Sciences in 1983. It was adopted with modifications by the USEPA for use in Superfund feasibility studies and RCRA corrective measure studies (USEPA 1989). This procedure provides a general, comprehensive approach for performing risk assessments at contaminated sites. It consists of four steps: (i) hazard identification, (ii) exposure assessment, (iii) toxicity assessment, and (iv) risk characterization. The detailed information on the four steps is provided in Chapter 12.

Remediation is required when risks posed by the contamination at the site are deemed unacceptable. Generally, remediation methods are divided into two categories: in-situ and ex-situ remediation methods. In-situ methods treat contaminated soils and/or groundwater in-place, eliminating the need to excavate the contaminated soils and/or extract groundwater. In-situ methods are advantageous because they are often less expensive, less invasive, often allowing site surface activities to continue, and they provide increased safety to both the on-site workers and the public within the vicinity of the remedial project. Successful implementation of in-situ methods requires a thorough understanding of subsurface conditions. In-situ containment, which may include the use of bottom barriers, vertical walls, and caps, may be a feasible strategy to minimize the risk posed by the contamination at some sites. Ex-situ methods are used to treat excavated soils and/or extracted groundwater. Surface treatment may be performed either on-site or off-site, depending on site-specific conditions. Ex-situ treatment methods are attractive because it may lessen the need extensive in-situ monitoring during remediation of subsurface conditions in such cases. Ex-situ treatment also is often faster than in-situ treatment and offers easier control and monitoring during remedial activity implementation.

18.3 Green and Sustainable Remediation Technologies

Remedial technologies are classified into two groups based on their scope of application: (i) vadose zone or soil remediation technologies, and (ii) saturated zone or groundwater remediation technologies. The reader should refer to Sharma and Reddy (2004) for more detailed information on these remediation technologies. The most common practice used to remediate vadose zone contamination is excavation. Impacted soil is typically characterized through testing, allowing for appropriate transportation and disposal measures, commonly involving landfill disposal. Following excavation activities, confirmation samples are often collected from the sidewalls and base of the resulting excavation to ensure remedial goals were achieved, and clean fill materials are used to backfill the resulting excavation. When the excavation of contaminated soils is not a feasible option, several conventional and innovative treatment methods may be utilized. These methods may either be in-situ or ex-situ methods (Figure 18.4). Common vadose zone soil remedial methods include the following:

- Soil vapor extraction (SVE) consists of three basic components, an extraction system, an airflow system, and an off-gas treatment system. By applying a vacuum to the subsurface within the contaminant zone, the extraction system induces the movement of volatile organics and facilitates their removal and collection. Collected vapors pass through the airflow system

18.3 Green and Sustainable Remediation Technologies

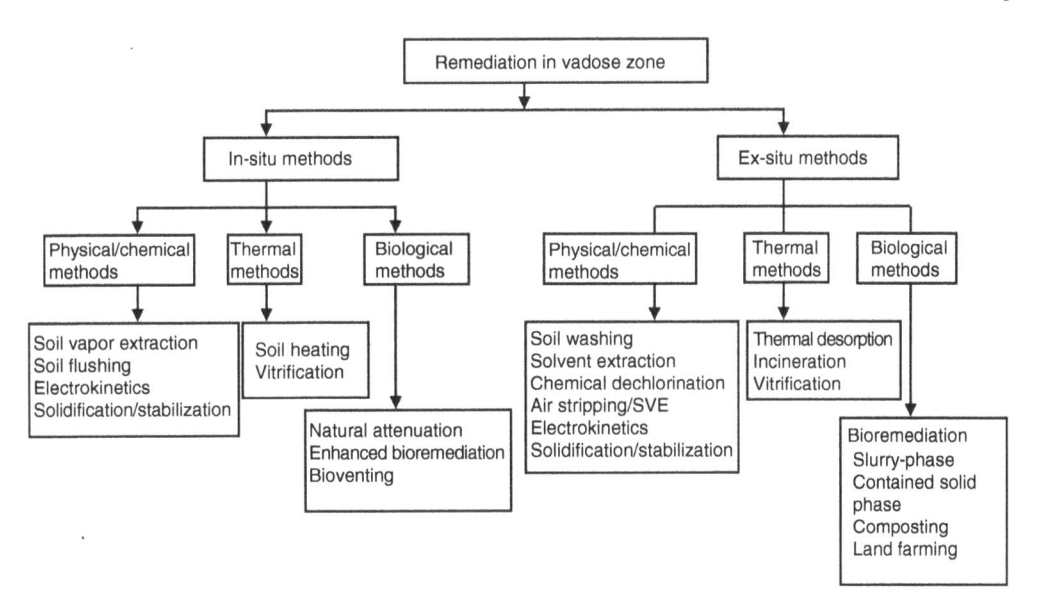

Figure 18.4 Vadose zone (soil) remediation technologies.

and are delivered to the off-gas treatment system, or, if regulatory limits permit, are emitted directly to the atmosphere.
- In-situ soil flushing involves the extraction of contaminants from the soil using water or other selected aqueous wash solutions. The flushing agent may be introduced into the subsurface in several ways, and once introduced, the flushing agent moves downward through the contaminated zone. Once the migrating agent/contaminant solution encounters the water table, it will mix with the groundwater and flow downgradient to a withdrawal point where it may be extracted, often via conventional extraction wells.
- Chemical oxidation technologies have also evolved as a preferred remedial alternative for in-situ or ex-situ remediation of soils and groundwater. With this technology, an oxidizing agent is introduced and mixed into the subsurface. Chemical oxidation typically involves reduction/oxidation (redox) reactions that chemically convert hazardous contaminants to nonhazardous or less toxic compounds that are more stable, less mobile, or inert.
- Soil stabilization and solidification involve applying additives or processes to contaminated soil to chemically bind and immobilize contaminants, preventing mobility.
- Electrokinetics involves the application of a low electric potential gradient across a contaminated soil zone to induce contaminant movement toward electrodes for subsequent removal.
- Bioremediation utilizes microorganisms to biologically degrade contaminants into harmless end-products.
- Vitrification (ISV) employs electrical power to heat and melt contaminated soil to destroy organic contaminants and stabilize inorganic contaminants. Alternatively, low-temperature soil heating is used to decontaminate soils through vaporization, steam distillation, and stripping.

If groundwater remediation is required, some of the aforementioned remedial technologies may be applied to saturated soils, including soil flushing, electrokinetics, and bioremediation. In addition, the following remedial methods may also be used:

- Groundwater may be extracted for treatment ("pump-and-treat") to remove any free-phase contaminants and/or contaminated groundwater with dissolved contaminants. While

pump-and-treat may be successful during the initial stages of implementation, performance drastically decreases at later times. As a result, significant amounts of residual contamination can remain, unaffected by continued treatment. Due to these limitations, the pump-and-treat method is now primarily used for free product recovery and to control contaminant plume migration.
- Air sparging involves injecting a gas, usually air, into the saturated soil zone below the lowest known level of contamination. Due to the effect of buoyancy, the injected air will rise toward the surface. As the air comes in contact with contamination, it will, through a variety of mass transfer and transport mechanisms, strip the contaminant away or assist in in-situ degradation.
- Dual-phase extraction, also known as vacuum-enhanced recovery, is a hybrid remediation technique that combines technology from pump-and-treat and SVE. During implementation, groundwater is extracted through the application of a vacuum, allowing for the removal of the dissolved contaminants within the extracted groundwater as well as the contaminant vapors due to the applied vacuum.
- Permeable reactive barriers (PRBs) incorporate a permeable, reactive media to adsorb, degrade, or destroy contamination within groundwater as it passes through the permeable barrier. Common reactants include zero-valent iron, zeolites, organobentonites, and hydroxyapatite.

In some cases, it may be impractical or undesirable to actively remediate contamination in soils and/or groundwater via in-situ methods. This can be due to the presence of surface obstructions, such as structures or utilities, or the lateral/vertical position of contamination such that it cannot be readily addressed. In such situations, containment systems may be considered (Figure 18.5). Surface capping involves the installation of a surface barrier that prevents or limits the ability of underlying contaminated subsurface media to be encountered. Vertical cutoff barriers, vertical cutoff walls, or barrier walls are embedded or keyed into a low permeability formation, preventing migration of contamination. Horizontal configuration can be circumferential, downgradient, or upgradient (Figure 18.5a).

In some cases, groundwater pumping well systems are used as active containment systems to manipulate and manage groundwater for removing, diverting, and containing a contaminated plume or for adjusting groundwater levels to prevent plume movement (Figure 18.5b). Subsurface drains are an alternative to pumping wells for the containment of contaminated groundwater (Figure 18.5c). They consist of drainpipe surrounded by filter and backfill to intercept a plume hydraulically downgradient and then divert for collection and discharge to a treatment plant.

Using just one technology may not be adequate to remediate some contaminated sites where different types of contaminants exist (e.g. heavy metals combined with volatile organic compounds (VOCs)) and/or when the contaminants are present within a complex geological environment, such as a heterogeneous soil profile consisting of lenses or layers of low permeability zones surrounded by high permeability soils. Under these situations, different remediation technologies can be used sequentially to achieve remedial goals. The use of such multiple remediation technologies is often referred to as "treatment trains." Typical treatment trains used at contaminated sites include soil flushing followed by bioremediation, SVE followed by soil flushing, SVE followed by stabilization/solidification, and thermal desorption followed by solidification/stabilization, which is then followed by soil flushing. Alternatively, different remediation technologies can be used concurrently, such as SVE and air sparging, electrokinetics and bioremediation, and soil flushing and bioremediation.

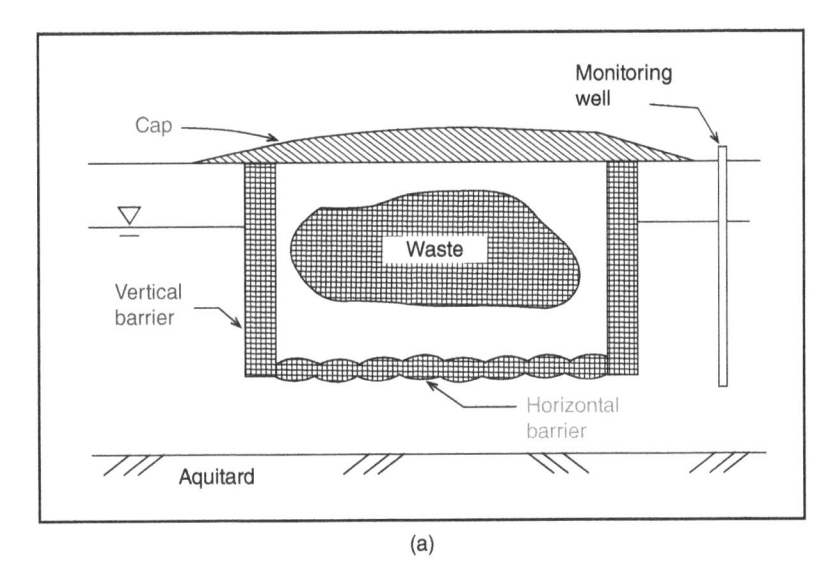

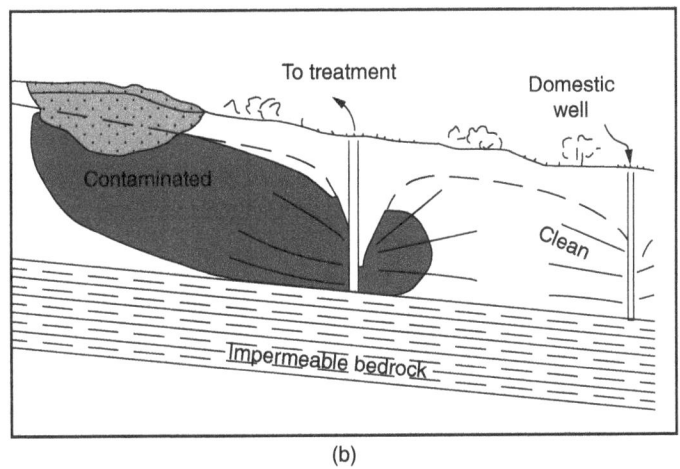

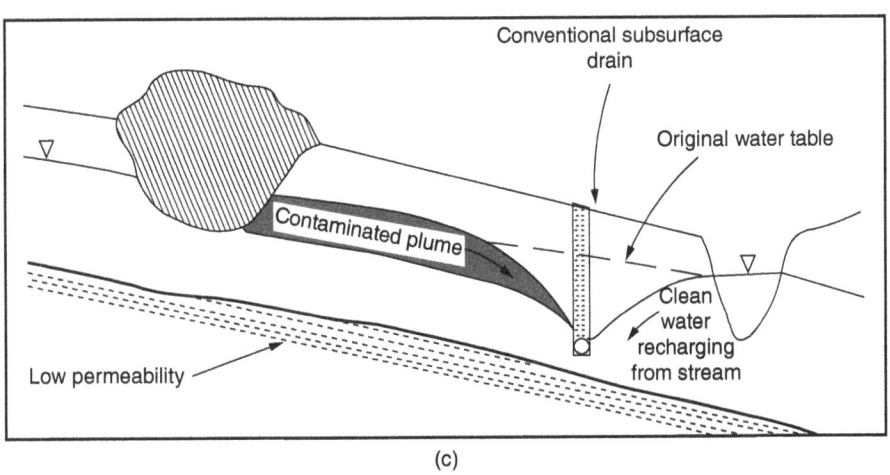

Figure 18.5 Containment technologies: (a) cap, vertical barrier and bottom barrier; (b) pumping well systems, and (c) subsurface drain system.

It can be challenging to incorporate sustainability parameters into the process of selecting remedial technologies. Some technologies, such as pump-and-treat operations and incineration, are known to be energy-intensive and may not meet GSR criteria. An ideal remediation technology (and all associated on-site or off-site actions) should aim to:

- Minimize the risk to public health and the environment in a cost-effective manner within a reasonable time;
- Minimize the potential for secondary waste and prevent uncontrolled contaminant mass transfer from one phase to another;
- Provide an effective, long-term solution;
- Minimize the impacts to land and ecosystem;
- Facilitate appropriate and beneficial land use;
- Minimize or eliminate energy input. If required, renewable energy sources (e.g. solar, wind, etc.) should be used;
- Minimize the emissions of air pollutants and GHGs;
- Eliminate freshwater usage while encouraging the use of recycled, reclaimed, and storm water. Further, the remedial action should minimize impact to natural hydraulics water bodies;
- Minimize material use while facilitating recycling and/or the use of recycled materials.

Technologies that encourage uncontrolled contaminant partitioning between media (i.e. from soil to liquid or from liquid to air) or those that generate significant secondary wastes/effluents are not sustainable. Rather, technologies that destroy contaminants, such as bioremediation or chemical oxidation/reduction, minimize energy input, and minimize air emissions and wastes, are preferred. In-situ systems are often attractive, as they typically minimize GHG emissions and limit disturbance to ground surface and the overlying soils.

A variety of remedial technologies satisfy core GSR criteria; however, the project life cycle for a specific technology should be considered to determine if it is appropriate for use at a given site. For example, ex-situ biological soil treatment may be considered an appropriate GSR technology; however, the impacts of transporting soil (if off-site treatment is required) should be evaluated. Similarly, enhanced in-situ bioremediation is also considered as an attractive GSR technology, but the cumulative impacts that occur during its characteristically long treatment duration should be compared to those of other active remediation technologies that require less time. In general, passive containment systems such as phytoremediation and PRBs utilize little mechanical equipment and minimize energy input while resulting in minimal waste/effluent.

A single remediation technology often cannot cost-effectively address the technical challenges posed by contamination at a particular site. Based on the site-specific conditions, multiple technologies may be sequentially or concurrently used for remediation. Further, technologies not typically considered sustainable may be combined with other technologies to develop multicomponent remedial programs that are sustainable.

Some popular technologies used to treat residual contaminant concentrations are not considered effective in treating source remediation. Groundwater plumes with moderate to high dissolved contaminant concentrations may require a brief implementation of active remediation technologies to expedite contaminant mass reduction. Alternatively, many technologies appropriate for source removal are often ineffective in treating residual or lower concentrations that result from reduced contaminant diffusion and dissolution. Under such conditions, GHG emissions and energy usage associated with aggressive technologies may outweigh further contaminant mass removal/destruction. Further, a technology with lower energy requirements and emissions may be used to treat residual contamination. Large, dilute groundwater plumes may be treated using lower-energy passive technologies; this may extend the duration of the remediation program, but it will reduce overall net impacts to the environment.

The duration of the remediation program can itself be a major governing factor in remediation system selection. Remediation technologies such as bioremediation may require lower energy input, but they require longer treatment time. Further, given the duration of the remediation, cumulative energy use can often be greater as compared to a shorter but energy-intensive remediation program. Other anticipated or unanticipated side effects, such as incomplete mineralization, can render these as ineffective alternatives. Further, even energy-intensive aggressive technologies, such as thermally enhanced remediation, may become attractive from a sustainability standpoint if renewable energy sources are used.

Opportunities exist for reducing energy and carbon footprints from existing remediation systems. In particular, energy efficiency can be maximized by optimizing existing treatment systems, by critically evaluating the remedial design, and upgrading the equipment. In addition, alternative sources of energy, including solar, wind, landfill gas, biomass, geothermal, tidal/wave, and cogeneration can be incorporated into existing systems. A growing number of existing projects have started to use solar or wind energy sources.

18.4 Sustainable Remediation Framework

A sustainable remediation framework is a systematic basis or means by which the sustainability of a remediation project may be assessed with respect to environmental, social, and economic factors. This assists in decision making to increase sustainability of a remediation project. Although a universally acceptable standardized framework has not yet been developed, several agencies and organizations in the United States and other countries have been active in developing frameworks for measuring and facilitating sustainability in remediation of contaminated sites. The frameworks are developed by USEPA, SURF, ITRC, and ASTM and are explained in detail in Reddy and Adams (2015). A brief explanation on these frameworks is provided:

- *USEPA framework* – The framework emphasizes green remediation concepts and techniques that take into consideration a range of environmental effects. In emphasizing green remediation, the goal of the framework is to evaluate and select remediation alternative and options that achieve maximum net environmental benefit during all phases of site characterization, remediation system implementation and operation, and postremediation monitoring. The green remediation framework has incorporated five core elements to minimize environmental impacts as shown in Figure 18.6.

Figure 18.6 Core elements of the USEPA green remediation framework. Source: USEPA (2008).

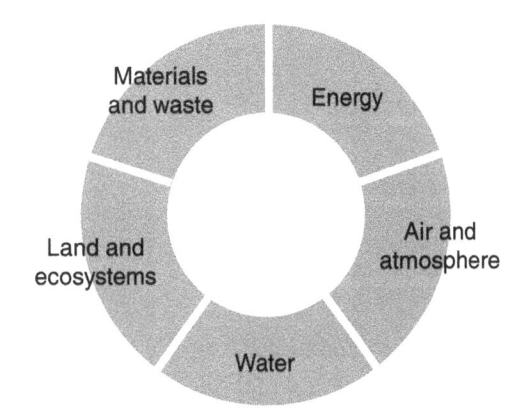

- *SURF (Sustainable Remediation Forum) framework* – The SURF framework consists of a tiered decision-making process that considers each phase of a remediation project: site characterization, remediation alternative analysis and selection, remediation system design and construction, operations and maintenance, postmonitoring, and closure. Tier 1 consists of a standardized, non-project-specific, qualitative evaluation that utilize checklists, lookup tables, guidelines, results from past project experience, rating systems, and matrices to identify best management practices (BMPs) that maximize positive sustainability impacts. Tier 2 consists of a semiquantitative approach using project-specific and non-project-specific information as well as greater stakeholder involvement. The project-specific information can be evaluated using various assessment tools such as emission calculations, exposure calculations, scoring and weighing systems, spreadsheet-based tools, and simple cost–benefit analyses. This tier 2 evaluation is best suited for sites that are moderately complex and requires greater involvement of stakeholders. Tier 3 is the most comprehensive, detailed, quantitative evaluation for sustainability based on project-specific information. The tier 3 requires a large quantity of project-specific data and utilizes sophisticated tools such as life cycle assessment for the sustainability evaluation of the project.
- *ITRC (Interstate Technology and Regulatory Council) framework* – Figure 18.7 depicts five generalized steps of this framework that can be performed to user-desired detail during each phase of the project. These steps include the following: (i) evaluation/update of a conceptual site model (CSM); (ii) establishment of GSR goals for the project; (iii) project stakeholder involvement; (iv) selection of GSR metrics, evaluation level, and boundaries; and (v) documentation of GSR efforts. A three-level approach is recommended for evaluating and selecting GSR metrics: Level 1 consists of "common-sense"-based BMPs. These are selected to promote resource conservation and process efficiency. The net impact on the environment, community, or economics is not evaluated with this approach. Level 2 consists of the selection and implementation of BMPs with some degree of qualitative and/or semiquantitative evaluation. Qualitative evaluations may reflect trade-offs associated with different remedial strategies or use value judgments for different GSR goals to determine the best way to proceed. Semiquantitative evaluations are those that can be completed by using simple

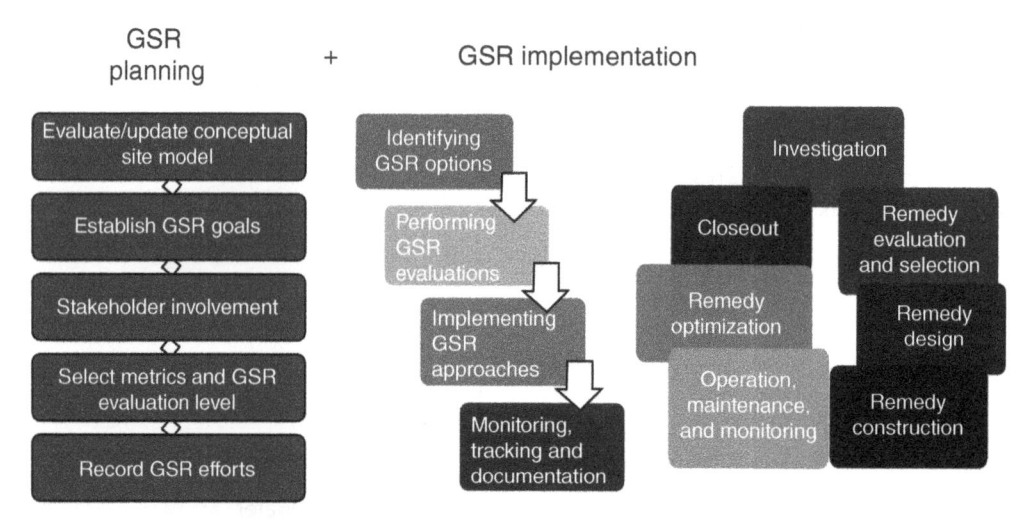

Figure 18.7 ITRC green and sustainable remediation framework. Source: ITRC (2011a,b).

mathematical calculations or intuitive tools (e.g. conversion factors, online calculators, and spreadsheet-based programs). Level 3 consists of selection and implementation of BMPs using a comprehensive quantitative evaluation. The evaluation often relies on life-cycle assessment (LCA) or footprint analysis approaches. This level requires more time, expense, and expertise.

- *ASTM (American Society of Testing and Materials) framework* – ASTM has developed a standard guide for sustainable remediation specifically focused on "greener" cleanups, in their ASTM E2893 standard. The standard describes a process for identifying, evaluating, and incorporating BMPs and, as appropriate, integrating a quantitative evaluation that facilitates an overall net reduction in environmental impact associated with remediation projects. This guide addresses the five core elements outlined in the USEPA green remediation framework as described above. The standard provides detailed guidance on planning and scoping a remediation project, implementation of appropriate BMPs, employing a quantitative evaluation when appropriate, and documentation and reporting of sustainability-related performance. ASTM has developed another standard for sustainable remediation projects (Figure 18.8), ASTM E2876, which provides a framework for integrating environmental, economic, and social aspects into remediation projects. BMPs implemented under this guide can incorporate all three aspects of sustainability (environmental, economic, and social) into remediation projects that are designed to address human health, public safety, and ecological risks (Figure 18.9).

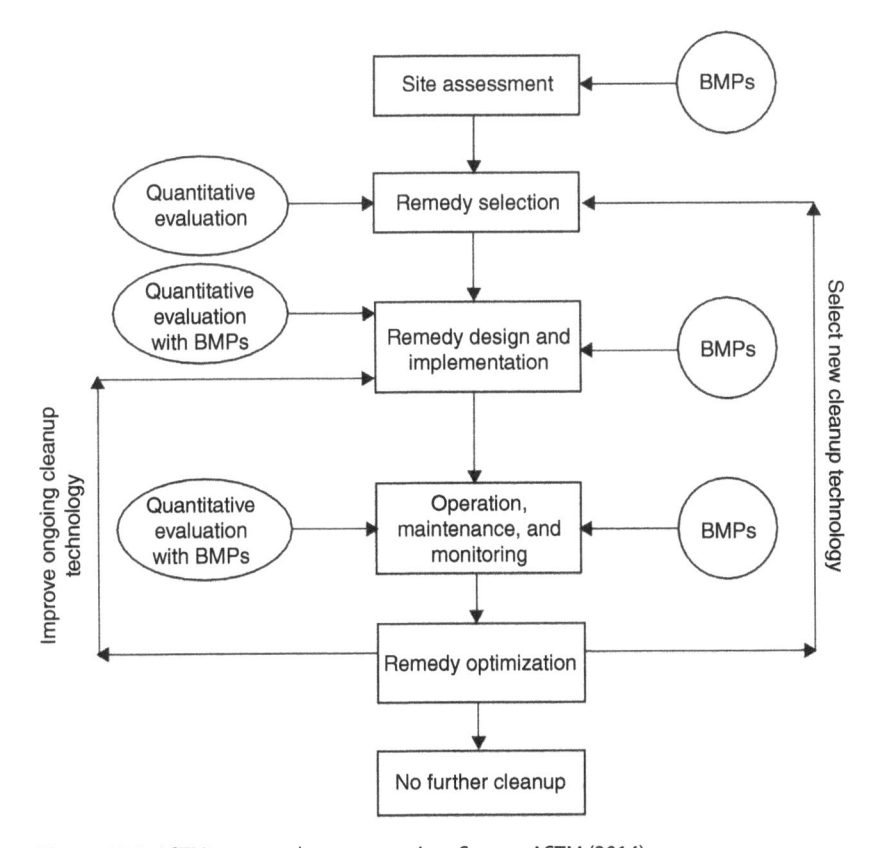

Figure 18.8 ASTM greener cleanup overview. Source: ASTM (2014).

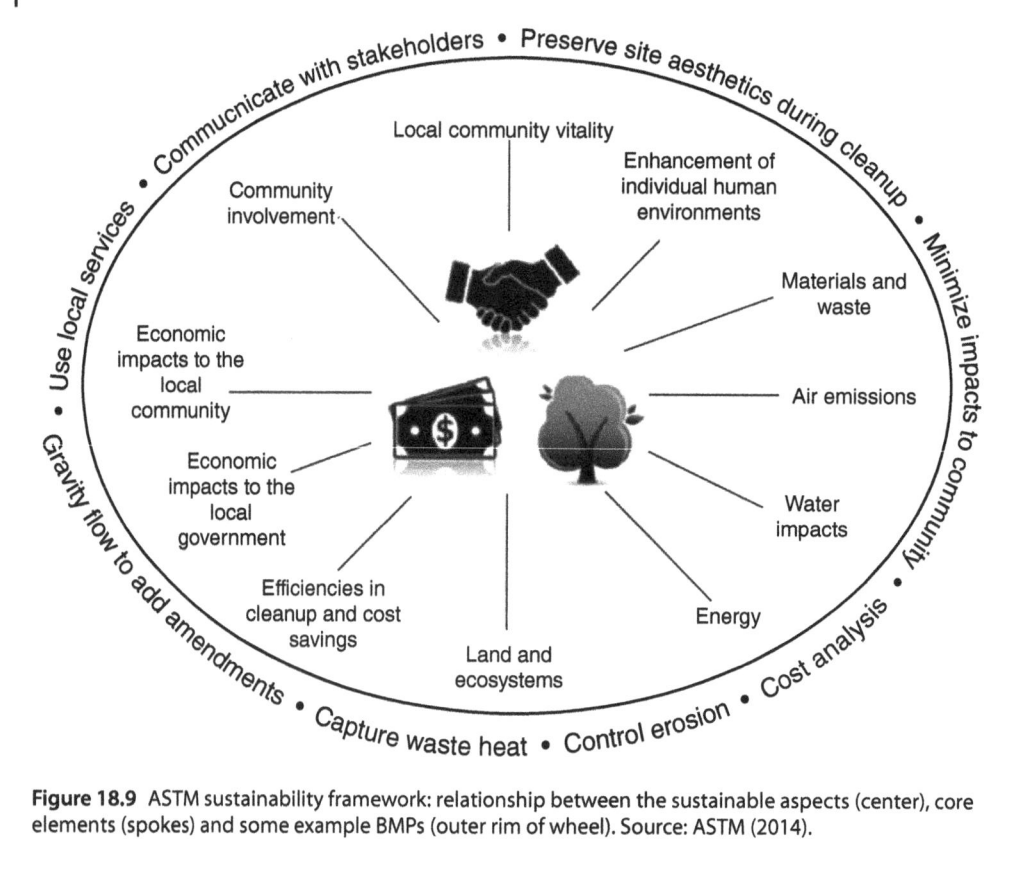

Figure 18.9 ASTM sustainability framework: relationship between the sustainable aspects (center), core elements (spokes) and some example BMPs (outer rim of wheel). Source: ASTM (2014).

18.5 Sustainable Remediation Indicators, Metrics, and Tools

As stated in Chapter 5, sustainability indicators must have attributes of "SMART" and should be conducive to express the environmental impact in quantitative terms. Environmental indicators may include the following:

- GHG and other air emissions;
- Contributions to climate change;
- Use of freshwater resources;
- Impacts to soil;
- Utilization of natural resources;
- Impacts on surface water or groundwater;
- Use of recycled/repurposed materials;
- Overall waste generation;
- Diversion of waste materials from or to landfill facilities.

Economic sustainability indicators that may be considered for the remediation project include the following:

- Direct and indirect job creation within the community;
- Direct and indirect investment within the community;
- Facilitation of government grants for the project and community as a whole;
- Long-term tax and revenue generation within the enhanced community;

- Degree of highest and best use (HBU) achieved by the remediated property;
- Potential to "upzone" the property and nearby properties due to remediation activity.

Some key indicators of social sustainability include:

- Enhancement of community aesthetics;
- Enhancement of quality-of-life features (e.g. improved transportation opportunities or recreational facilities);
- Public participation in decision making;
- Educational and job training opportunities;
- Interaction between community groups;
- Emotional ownership of the community in a remediation project;
- Improved physical and mental health and well-being of members of the community;
- Enhanced social opportunities for members of the community;
- Strengthening or enhancement of existing community institutions (e.g. recreational organizations, charitable foundations, houses of worship).

Once key indicators and related metrics have been devised for sustainability analyses, they may be formally evaluated using a sustainability assessment tool. These tools can range from simple decision trees or spreadsheets to full LCAs. Several qualitative, semiquantitative tools, and quantitative tools have been developed (Reddy and Adams 2015) and are briefly summarized as follows.

The purpose of qualitative assessment tools is to screen remediation technology and/or BMP alternatives based on anticipated impacts across the environmental, economic, and societal dimensions of sustainability. These commonly consist of guidance documents or advisory manuals that outline an appropriate selection process, including relevant criteria. Two examples of qualitative tools have been developed by public regulatory agencies, including the Illinois EPA Greener Cleanup Matrix and the Minnesota Pollution Control Agency Toolkit for Greener Practices.

While qualitative tools offer a screening tool of BMPs or other remediation-related factors, semiquantitative tools offer more rigor in the analysis of sustainability. Typically, these tools will offer a "scorecard"-like approach in which potential quantitative factors may be ranked and scored, resulting in a weighted average or cumulative score that allows for a direct numerical comparison among several potential remediation alternatives. These semiquantitative tools are typically straightforward and do not incorporate advanced numerical modeling; rather, they may be used for screening or feasibility assessment when considering remediation alternatives for a project as well as alternative applications or design of a particular remediation technology that may have been selected for a project. Examples of semiquantitative assessment tools are California Green Remediation Evaluation Matrix (GREM) and Social Sustainability Evaluation Matrix (SSEM) (Reddy and Adams, 2015).

For many projects, the use of a qualitative or semiquantitative analysis tool will prove to be useful for analyzing sustainability aspects of one or more remediation alternatives. This is especially the case when a project is relatively simple or straightforward, or when the tool is applied as a screening tool to assess feasibility for a remediation project. In many instances, however, the results of a qualitative or semiquantitative analysis may be too limited to be of much use for sustainability analysis. This is especially the case for more complex remediation projects where a wide range of parameters must be carefully and thoroughly assessed.

When warranted by the degree of complexity of a project, quantitative analysis tools should be incorporated for sustainability analysis. These advanced tools for sustainability evaluations typically offer a far more detailed and rigorous assessment of the environmental, social, and economic impacts of remediation. Some common tools developed and used in practice include SiteWise™, USEPA Environmental Footprint Analysis Tool (SEFA), and general ISO LCA. For details, see Reddy and Adams (2015).

No single tool can cover every type of project. Rather, it is important to assess several key aspects of a project, which can then be used to select the most appropriate tool for analysis.

18.6 Case Studies

Several examples have been reported where sustainability evaluation was performed to select the most sustainable remediation based on the site-specific conditions (Sadasivam and Reddy 2017; Reddy and Adams 2015). Three case studies that use different sustainability assessment tools are presented:

- *Case study #1* – An industrialized site contaminated with VOCs, polycyclic aromatic hydrocarbons (PAHs), pesticides, metals, total organic content, and abnormal pH levels. Several potential soil and groundwater contamination remediation technologies have been considered for the site based on applicability, cost range, limitations, and commercial availability. For soils, excavation/disposal, phytoremediation, in-situ chemical oxidation, and solidification/stabilization have been identified as potential remediation alternatives. For groundwater, pump-and-treat, in-situ flushing, PRB, and monitored natural attenuation (MNA) have been identified as potential remediation alternatives. A comparative assessment of potential remedial technologies was performed based on the BMPs as well as qualitative and quantitative assessments. The general BMPs for the selected technologies have been assessed based on the BMPs listed in the Greener Cleanup Matrix developed by the Illinois EPA and the Toolkit for Greener Practices developed by the Minnesota Pollution Control Agency (ITRC 2011a,b). In addition to BMPs, the GREM tool was used to perform a qualitative comparison of remediation technologies for sustainability and to evaluate adverse environmental impacts. A quantitative assessment was also performed based on the sustainability metrics. The sustainability metrics for the selected potential technologies were calculated using two tools: The Sustainable Remediation Tool (SRT) and SiteWise. A combination of remediation methods was identified as the best alternatives for the site. Solidification/stabilization (to be applied in areas of high contaminant concentrations) and phytoremediation (to be applied in other contaminated areas) have been recommended for remediation of soils with PAHs and heavy metals, while MNA and phytoremediation have been recommended for the treatment of impacted groundwater.
- *Case study #2* – A degraded wetland site known as Indian Ridge Marsh (IRM) was found to be contaminated by heavy metals, pesticides, VOCs, PAHs, pesticides, and one observed instance of an LNAPL plume containing petroleum hydrocarbons. Qualitative and quantitative analyses were conducted to evaluate potential environmental impacts associated with each remedial option using tools such as the GREM, SiteWise, and the SRT. Following a qualitative evaluation of sustainability metrics using GREM (i.e. noise; worker safety; aesthetics), a quantitative evaluation of energy/resource consumption was conducted using both SRT and SiteWise considering several project phases, including the remedial investigation, remedial action construction, operations and maintenance, and long-term monitoring. Additionally, the SSEM tool was applied to the IRM project to evaluate the social impacts of both remedial alternatives. The recommended strategy for remediation of IRM consists of the phyto-EB option. This alternative will act to stimulate existing soil microorganisms to

enhance degradation of organic contaminants at all identified AOCs. Native tree species with high growth and transpiration rates, deep rooting depths, and the ability to accumulate and/or sequester contaminants of concern will be employed.
- *Case study #3* – At a former zinc smelting site, soils were found to be contaminated with lead, arsenic, and copper. The SimaPro software was used to evaluate the life cycle impact of two common methods of treatment: landfilling (excavation and hauling), and in-situ treatment by solidification and stabilization. The results of this analysis showed that excavation and hauling result in greater impacts in every category except human health – cancer, as compared to solidification/stabilization. Ultimately, this analysis indicated that given the large costs and disturbances associated with excavation and hauling, the solidification/stabilization is the more attractive and sustainable remediation option.

18.7 Challenges and Opportunities

While there is growing interest in incorporating sustainability into environmental remediation, there are several challenges and opportunities that have to be addressed to promote sustainable remediation in practice. Some of these challenges include:

- *"Greenwashing"* – Greenwashing refers to situations where sustainable remediation options have not been evaluated and backup documentation is lacking but claims exist that say sustainable remediation approaches have been implemented. Greenwashing in a larger sense is commonly associated with a wide range of approaches, from consumer product marketing to legislative initiatives in which dubious "green" or "sustainable" claims are made. These claims may be misleading or outright false.
- *Lack of financial incentives of sustainable remediation* – A perception exists among many taxpaying citizens that only publicly funded projects are capable of absorbing the additional cost burden associated with incorporating sustainability measures, which does not motivate the engineering projects to inculcate sustainable engineering.
- *Lack of a regulatory mandate* – While many regulatory oversight agencies encourage sustainability-focused activities and efforts, no clear mechanism requiring the incorporation of such measures exists as applied to many remediation projects. Incorporation of BMPs and other activities that enhance the dimensions of sustainability in many cases are optional. With many agencies, the use of such measures may be encouraged, but in many cases, these activities are not required.
- *Lack of public awareness* – Despite the virtues of government-based incentives or mandates (the proverbial "carrot" or "stick") that would encourage the application of sustainability-focused activities or practices for site remediation, government will not act in such a manner if it is not the will of the public. In many ways, the general public is aware of environmental issues.
- *Lack of specialty training on LCA, carbon balance, and other assessment tools for professionals* – Many remediation professionals do have a desire to incorporate sustainable measures into remediation projects; however, in many cases, they lack the skill set or knowledge of assessment tools to demonstrate the related benefits.
- *Greater academic focus* – This academic/practitioner model has worked for countless technological advances and will continue to advance the applications of environmental remediation. With a greater emphasis of sustainability as a common interest between academia and practitioners, sustainability-related applications would also be expected to evolve at an accelerated pace.

- *Further refinement and development of assessment tools and frameworks* – A move toward standardization, at least among similar assessment tools and frameworks, would be beneficial to the environmental remediation practice. Even if differences among computational processes within different tools remained, increased standardization in terms of reported output units, indicators, and metrics considered, and greater agreement on the range of remediation activities that could be handled by different remediation tools would eliminate confusion.
- *Improved assessment of indirect consequences* – The difficultly in accounting for the indirect benefits exists for two primary reasons. First, system boundary selection will invariably affect the number of indirect consequences that are accounted for in an analysis. The system boundary expansion leads to a significant increase in the complexity and difficultly of an analysis, in terms of both time and cost associated with the analysis. The second reason that indirect consequences are not often properly accounted for by existing assessment tools, regardless of the choice of system boundary. Benefits such as reduced emissions, resource utilization, or construction jobs created by a remediation project would be considered typical and easily handled by a comprehensive analysis tool. However, benefits, such as increased neighborhood tax receipts, increased life expectancy for residents near a project site, or increased number of species present in a rehabilitated natural habitat can be difficult to quantify with existing tools. As is the case with enhanced assessment tools, practitioners and academia could successfully collaborate to identify key indicators and metrics of indirect benefits resulting from sustainability-focused remediation projects as well as ways to incorporate and quantify into existing and future assessment tools.
- *Improved metrics/tools to address social issues* – Assessment frameworks and tools evolved, the focus was primarily placed on environmental and economic dimensions because metrics associated with these dimensions were relatively easy to quantify and analyze. Further, whether associated on costs or physical units, economic and environmental metrics are relatively easy to objectively compare among remediation project alternatives. While there has been a general interest in the measurement of social-related sustainability impacts, tools and frameworks other than those cited in this book have been lagging the development of other more economically and environmentally focused tools.

18.8 Summary

Environmental pollution, including soil and groundwater contamination, is one of the major problems faced by developed countries across the globe. Thousands of contaminated sites necessitating urgent remedial action have been identified. Environmental policies have evolved with respect to the characterization and remediation of contaminated sites. Systematic frameworks and approaches to contaminated site characterization and remediation have also evolved, such as the well-established procedures developed by the USEPA to appropriately characterize, assess, and remediate contaminated sites.

Several in-situ and ex-situ remediation technologies have been developed, and new and innovative technologies continue to be developed to address various contaminants and contaminated media. However, the net overall (environmental, economic, and social) benefit associated with using one remediation technology over other potential remedial alternatives for remediation of a contaminated site requires ongoing study. If not carefully assessed, the choice of a remediation technology and an overall remediation strategy itself may cause more harm than good when considered on a holistic or a triple-bottom-line basis.

GSR is emerging as a new paradigm for remediation of contaminated sites. Several national and international organizations have developed qualitative, semiquantitative, and quantitative sustainability assessment tools/frameworks that aid in adopting sustainable practices and alternatives in remediation strategy while ensuring minimizing risk to human health and the environment. Despite these efforts, there is still a lack of appropriate indicators, metrics, and assessment tools that can comprehensively analyze the environmental, economic, and social impacts of the remediation activities/strategies. Several challenges exist, which pose as opportunities for academia and practicing professionals to developed innovative and reliable tools to incorporate sustainability principles into contaminated site remediation.

18.9 Questions

18.1 What is green remediation?

18.2 What is the definition of green and sustainable remediation as per ITRC?

18.3 What is the difference between green remediation and green and sustainable remediation?

18.4 Explain the systematic approach developed by USEPA for the remediation of contaminated sites.

18.5 What are the steps of the approach necessary for the characterization and remediation of contaminated sites?

18.6 What are the major tasks involved in the risk assessment procedure of contaminated sites developed by USEPA.

18.7 What are the two major categories of remediation technologies? Mention any two major differences between the two categories.

18.8 Discuss the advantages and disadvantages of both in-situ and ex-situ remediation.

18.9 List any five remediation technologies used for remediation of contaminated soil in vadose zone. Explain the technologies briefly.

18.10 List any five remediation technologies used for remediation of contaminated groundwater. Explain the technologies briefly.

18.11 Describe the trade-offs of cost, timing, and environmental side effects (both positive and negative) associated with three remediation technologies. Also, describe a situation(s) where some of these factors may be significant and another situation where they may be negligible.

18.12 What are the attributes of an ideal remediation technology?

18.13 Explain the sustainability assessment framework developed by SURF.

18.14 Explain the sustainability assessment framework developed by ITRC.

18.15 Mention any five environmental, economic, and social sustainability indicators for sustainable remediation.

18.16 What are the challenges in implementing sustainable remediation of contaminated sites?

References

ASTM (2014). *ASTM E2893-16e1, Standard Guide for Greener Cleanups*. West Conshohocken, PA: ASTM International.

Baker, C.B., Smith, L.M., and Woodward, D.S. (2009). *The Sustainable Remediation Forum*. Wiley Periodicals, Inc. Wiley Interscience.

ITRC (Interstate Technology & Regulatory Council) (2011a). *Green and Sustainable Remediation: State of the Science and Practice. GSR-1*. Washington, DC: Interstate Technology & Regulatory Council, Green and Sustainable Remediation Team www.itrcweb.org.

ITRC (Interstate Technology & Regulatory Council) (2011b). *Green and Sustainable Remediation: A Practical Framework, GSR-2*. Washington, DC: ITRC (Interstate Technology & Regulatory Council).

Reddy, K.R. and Adams, J.A. (2015). *Sustainable Remediation of Contaminated Sites*. NY: Momentum Press.

Sadasivam, B.Y. and Reddy, K.R. (2017). Approaches to selecting sustainable technologies for remediation of contaminated sites: case studies. In: *Sustainability Issues in Civil Engineering*, 271–306. Singapore: Springer.

Sharma, H.D. and Reddy, K.R. (2004). *Geoenvironmental Engineering: Site Remediation, Waste Containment, and Emerging Waste Management Technologies*. Hoboken, NJ: Wiley.

USEPA (1989). Risk Assessment Guidance for Superfund. Volume 1. Human Health Evaluation Manual (Part A). U.S. Environmental Protection Agency. Report number EPA/540/1-89/002)

USEPA (2004). Cleaning Up the Nation's Waste Sites: Markets and Technology Trends. 2004 Edition. US Environmental Protection Agency. Office of Solid Waste and Emergency Response. Report number: EPA 542-R-04-015.

USEPA (2008). *Green Remediation: Incorporating Sustainable Environmental Practices into Remediation of Contaminated Sites*. EPA 542-R-08-002. Washington, DC: Office of Solid Waste and Emergency Response.

19

Climate Geoengineering

19.1 Introduction

As discussed in previous chapters, increased levels of atmospheric carbon dioxide and other greenhouse gases (GHGs) from anthropogenic sources have contributed to an increase in global temperatures. This effect, in turn, is contributing to changes in global climate and to adverse side effects that result from climate change. As depicted in Figure 19.1, reflected solar radiation and radiation emitted from the earth's surface can be trapped by these increased GHGs thus maintaining a steady temperature on the Earth's surface. Figure 19.2 depicts a model of sources and sinks associated with the global carbon cycle. This figure demonstrates the flux that occurred during the preindustrial age as well as the increases that have occurred as a result of human industrial activity. Figures 19.3–19.5 further highlight the increases in carbon dioxide emissions and how the accumulation of carbon dioxide and other GHGs in the atmosphere may affect the measured global surface temperature change, which are captured by a range of climate models (IPCC 2000, 2007). Figure 19.3 shows the GHG emissions that constitute the driving factors in the IPCC scenario models. The shaded area indicates the range of the scenarios, and the solid and dashed lines indicate individual paths for selected examples from the scenario ranges shown in the figure at the right. Figure 19.4 depicts the average surface temperature resulting from human activity and GHG emissions. The figure depicts an average increase of 0.5 °C during the twentieth century, but the possible scenarios predict a range of 0.5–3.5 °C increase in global mean temperatures by 2100, much above 1990 levels. Figure 19.5 shows anticipated temperature changes for three models for the 2020s and 2090s. The extent of predicted warming differs with each scenario, but the increase is typically greatest in polar regions.

As climate change evolves, several public and private entities ranging from small, local groups to large, international organizations have proposed and sought means to mitigate climate change and its effects. Many of these entities have pointed to a general solution – a drastic reduction in GHG emissions to approximate levels that were present before the industrial revolution. A range of sustainable engineering approaches have been implemented, including renewable and zero emissions energy sources, the use of green and sustainable buildings and infrastructure, and other strategies (Gibbins and Chalmers 2008). However, all the efforts in reducing the GHG emissions have not been enough to reduce the emissions below a confidence level to avoid dangerous impacts on Earth's temperature and climate (Ehrlich and Ehrlich 2013). As more stringent emissions targets are proposed and adopted, it is hoped that greater actions for emissions reduction will occur, including the identification and implementation of emissions curbing strategies. However, there is still a risk that those actions will not be implemented in time to avoid irreversible changes and effects on climate (IPCC 2014).

Sustainable Engineering: Drivers, Metrics, Tools, and Applications, First Edition.
Krishna R. Reddy, Claudio Cameselle, and Jeffrey A. Adams.
© 2019 John Wiley & Sons, Inc. Published 2019 by John Wiley & Sons, Inc.

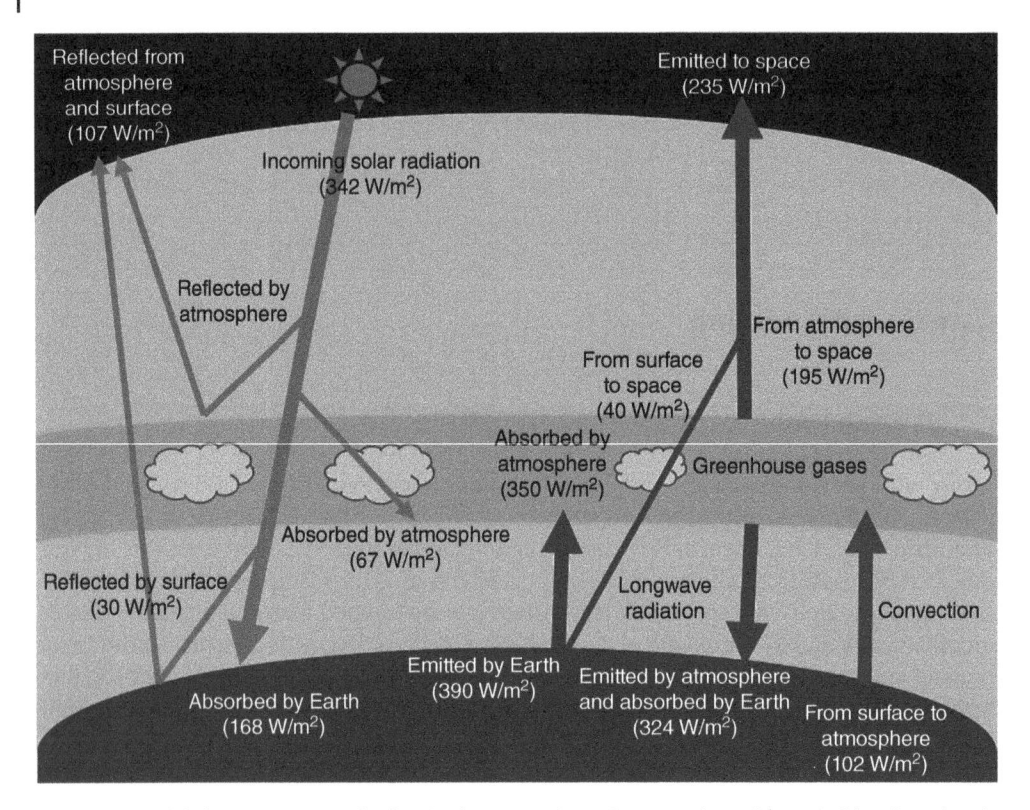

Figure 19.1 Global average energy budget in the atmosphere. Source: Adapted from Kiehl and Trenberth (1997). © American Meteorological Society.

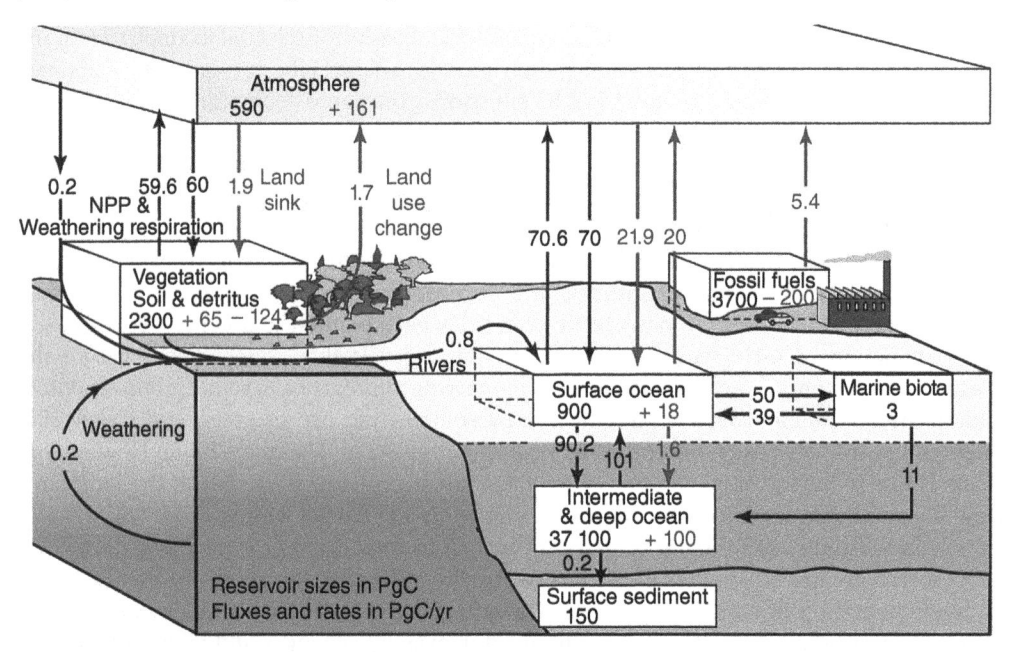

Figure 19.2 Global carbon cycle. Source: Sarmiento and Gruber (2002). Reproduced with permission of American Institute of Physics.

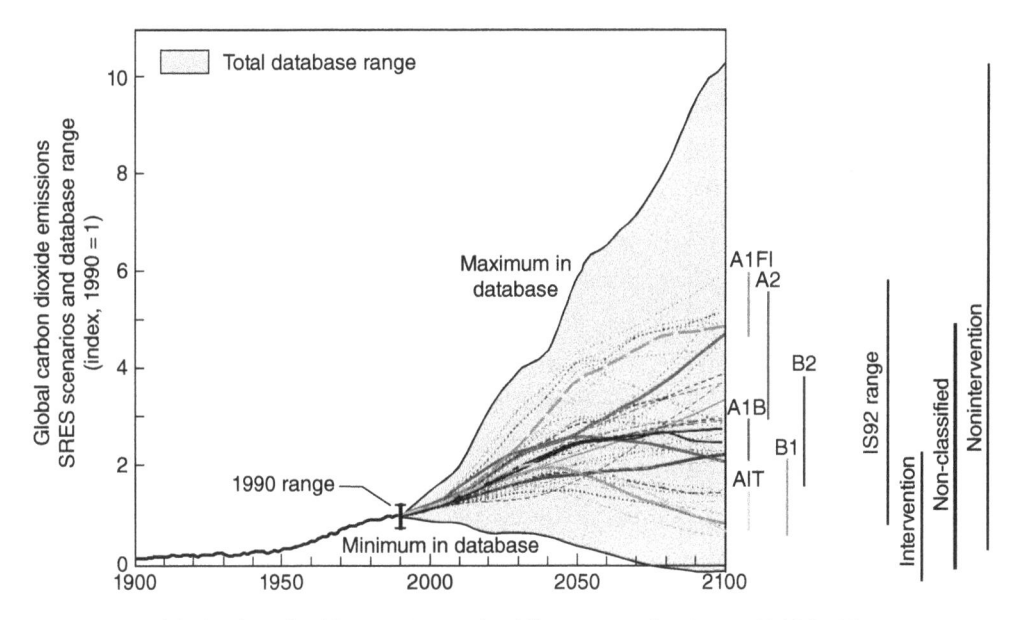

Figure 19.3 Global carbon dioxide emissions under different scenarios. Source: IPCC (2000).

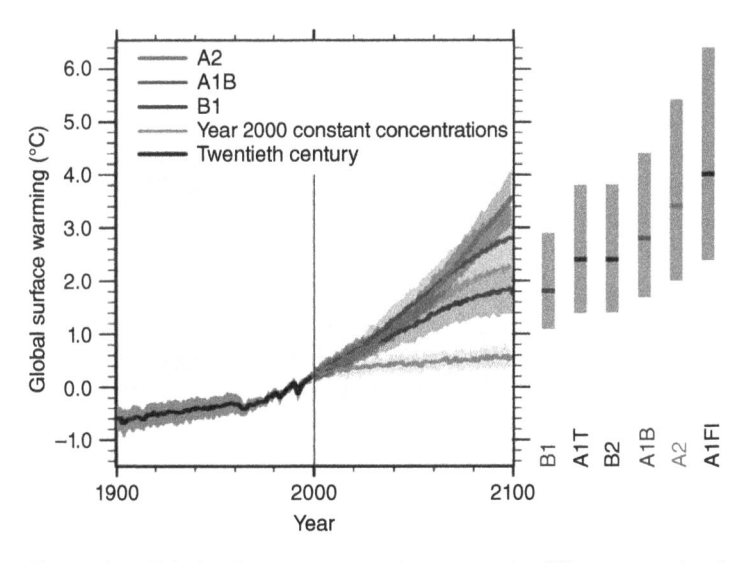

Figure 19.4 Global surface temperature increase under different scenarios. Source: IPCC (2007).

As presented in Figure 19.4, it is likely that surface temperatures will increase 2 °C or greater during this century, unless the global GHG emissions are reduced by at least 50% of 1990 levels by 2050 and even more thereafter (IPCC 2014). Further, there is no credible predictive models in which the global mean temperatures shows a peak by the year 2100 and then start to decline. Unless future reductions of GHG are much more successful than they have been so far, additional action will be necessary to cool the Earth this century. This chapter discusses such actions (also known as climate geoengineering) briefly.

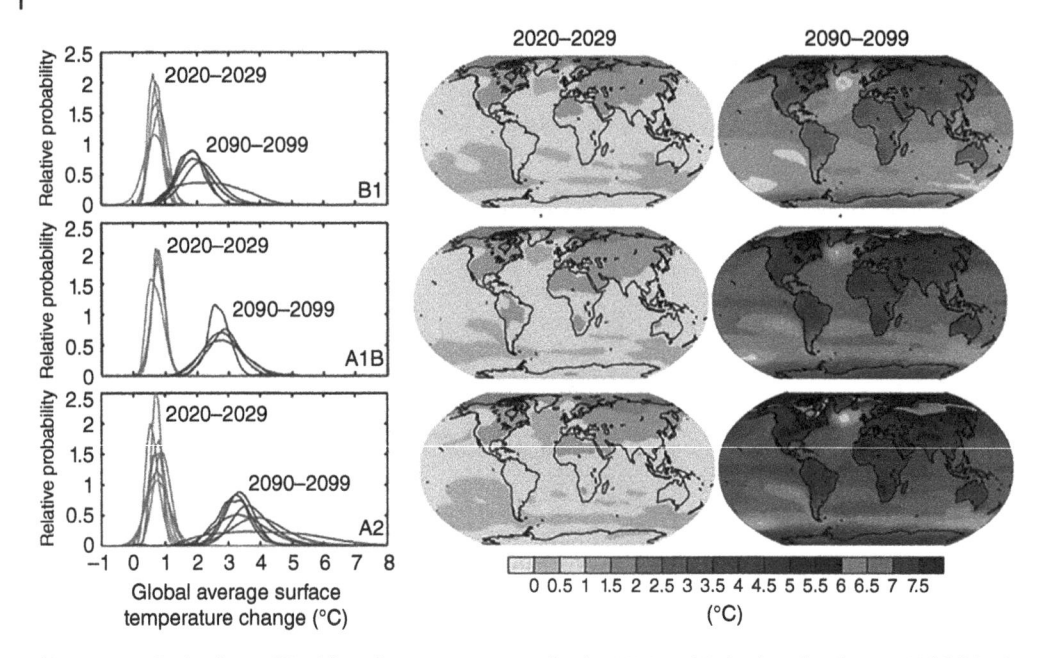

Figure 19.5 Projections of Earth's surface temperatures for the 2020 and 2090 decades. Source: IPCC (2007).

19.2 Climate Geoengineering

Climate geoengineering is an evolving field that seeks large-scale intervention strategies to mitigate climate change. Such strategies are deliberate interventions that need to be of a large enough scale to have a measurable effect on the Earth's climate (Shepherd 2012). These strategies, while evolving and highly innovative, generally fall into one of the two categories (Zhang et al. 2015) below:

- Carbon dioxide removal (CDR) methods address the original cause of climate change. CDR methods are designed to remove GHGs from the atmosphere, focusing in sequestration of carbon dioxide, which allows more outgoing long-wave radiation to escape back into space. These methods may remove GHGs directly or seek to influence natural processes to remove gases indirectly. These methods are comprehensive but may take years to fully work (Leung et al. 2014).
- Solar radiation management (SRM) methods attempt to offset the effects of increased GHG concentrations in the atmosphere reducing the absorption of solar radiation by the Earth. Various methods have been designed to deflect the incoming solar radiation back toward space (Ming et al. 2014).

19.3 Carbon Dioxide Removal (CDR) Methods

Four CDR methods include subsurface sequestration, surface sequestration, marine organism sequestration, and direct engineered capture.

19.3.1 Subsurface Sequestration

Subsurface sequestration involves the capture, liquification, and injection of carbon dioxide into underground geologic formations (Holloway 2005). Figure 19.6a shows underground geologic

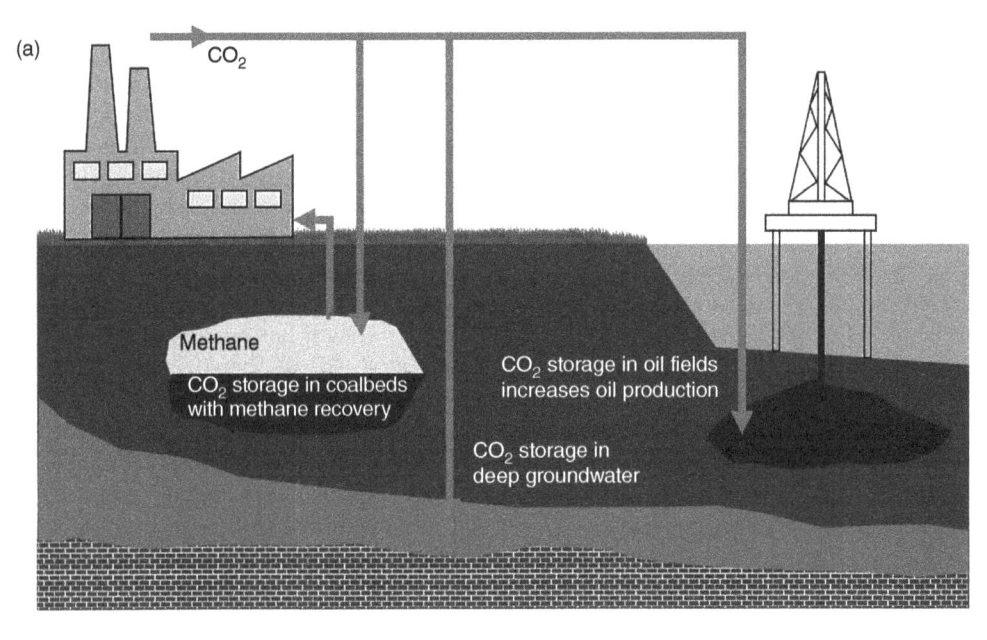

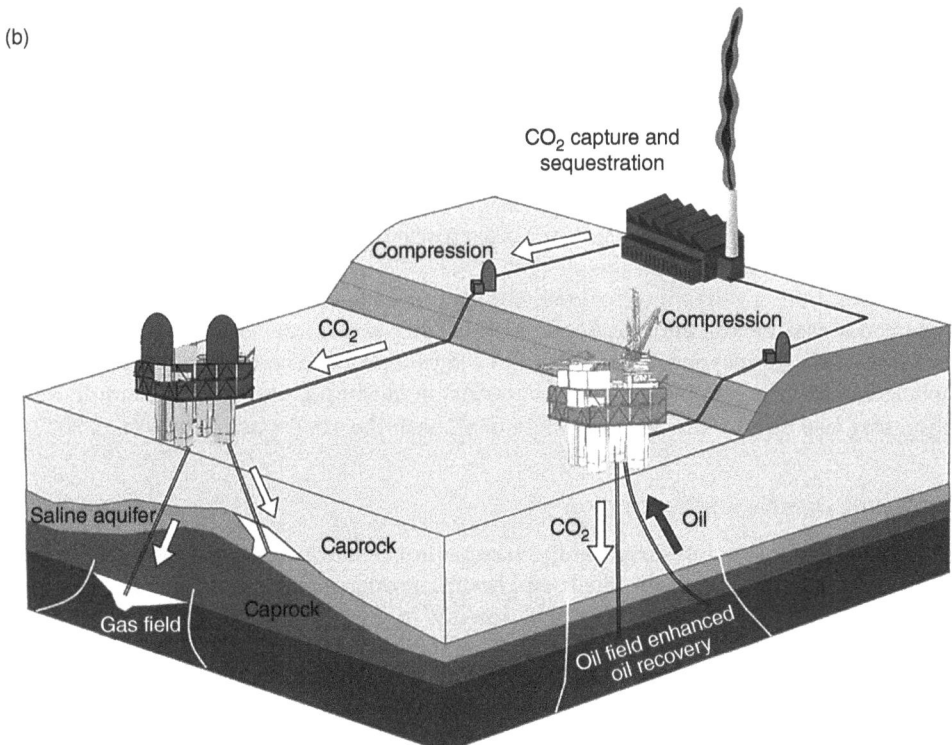

Figure 19.6 (a). Underground geologic carbon sequestration. (b). Carbon dioxide sequestration on a platform in the sea, similar to an oil drilling platform.

carbon sequestration. The geologic carbon sequestration primarily involves separation and capture of CO_2 at the point sources of emissions followed by its storage in deep underground geologic formations. This is sometimes also referred to as carbon capture and storage (CCS). The physical means of geologic sequestration involves capturing CO_2 and trapping it into some natural or man-made cavities (e.g. caverns, mines) or in the underground rock formations (e.g. depleted oil and gas reservoirs, aquifers). The injection of CO_2 and oil extraction efforts may be combined for simultaneous enhanced oil recovery and CO_2 sequestration (Pham and Halland 2017). The United States is a leading nation in enhanced oil recovery technology. The United States as of 3 January 2018, has injected about 16 million metric tons of CO_2 in the United States as a part of DOE's Clean Coal Research, Development, and Demonstration programs (DOE 2018). Figure 19.6b shows a depiction of a sequestration implementation on a platform, similar to an oil drilling platform, in the North Sea. This system has been tested to evaluate the long-term stability of CO_2 storage (Heinemann et al. 2013). The chemical means of carbon sequestration involves transforming the CO_2 into another stable substance via chemical reactions (e.g. dissolving CO_2 in underground water, forced/accelerated mineral carbonation forming stable precipitates, and adsorption). The subsurface sequestration methods have been estimated to cost between $500 billion and $1 trillion per year. Successful implementation is dependent on cooperation between corporate and governmental entities. From a technical standpoint, there is uncertainty regarding the stability of sequestered carbon, and suitable repositories must be present near the carbon source that needs to be sequestered. Some of the other major concerns regarding geological sequestration include the longevity of the captured and sequestered CO_2 to remain trapped, the hazardous risks of underground pressurized explosion and triggering of faults in rocks, the net energy use and carbon emission reduction from building and operating the carbon sequestration facilities, and cost-effectiveness of the sequestration technologies.

19.3.2 Surface Sequestration

Surface sequestration involves the planting of fast-growing trees and vegetation to incorporate carbon dioxide into cellulosic plant matter (Hui et al. 2017). As with subsurface sequestration, surface sequestration requires international cooperation and agreement between numerous corporate and governmental entities. Sequestration rates remain uncertain, and such strategies have substantial implications with respect to widespread forest stewardship. Further, there are implications for nitrogen budget and other ecosystem factors. The development of urban vegetation may also help to mitigate the CO_2 emissions to the atmosphere (Fares et al. 2017).

19.3.3 Marine Organism Sequestration

Marine organism sequestration involves the stimulation of marine organism growth, during which CO_2 is incorporated into the shells and bodies of organisms. One means that has garnered much attention (and at least an equal amount of controversy) is the use of iron fertilization (Chisholm 2000). The concept is fairly straightforward – an iron-based compound would be released into the ocean as a "seed" – it would act to stimulate plankton growth, which would result in the adsorption of CO_2 from the atmosphere due the increased metabolic processes (Figure 19.7). Upon death, the plankton settles out of the water column, resulting in a

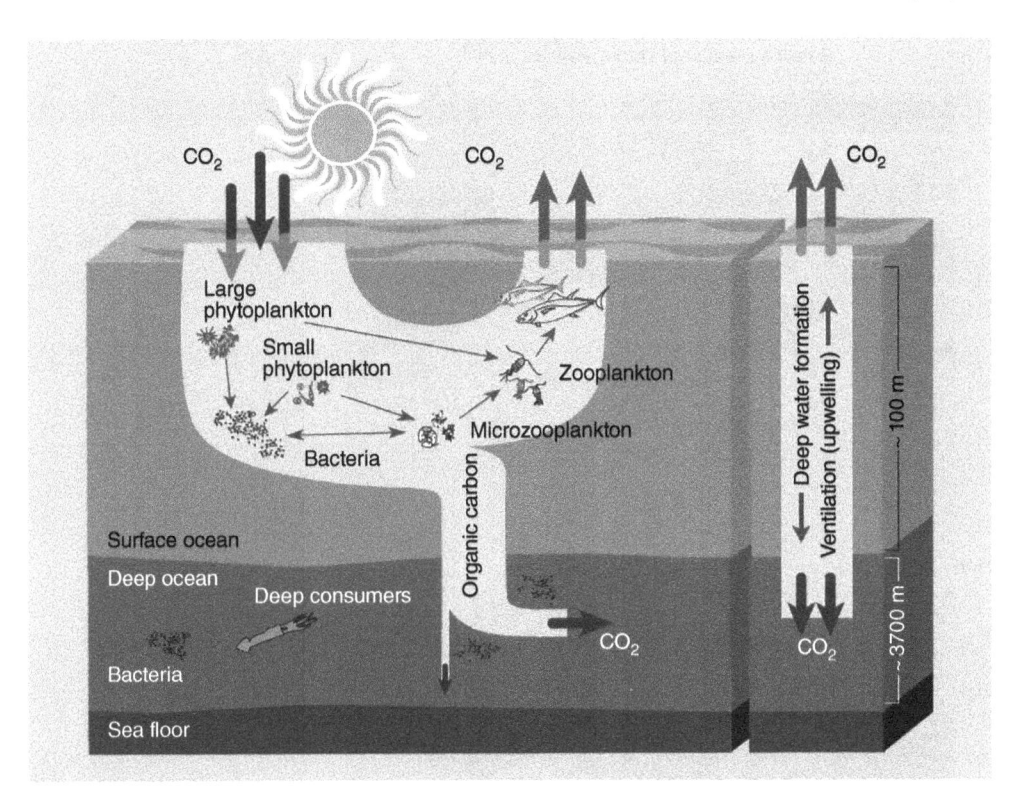

Figure 19.7 Iron fertilization of the oceans to CO_2 storage. Source: Chisholm (2000). Reproduced with permission of Springer Nature.

sequestration of the atmospheric carbon. Figures 19.8 and 19.9 show the results of an experiment performed in the southern hemisphere and reported by Chisholm (2000). As demonstrated, the seeding did result in an increase in algal growth and a measurable reduction in atmospheric CO_2. However, the full extent of effects to marine ecosystems is not understood, and this technique is highly controversial. If it were to be performed on a wide scale, it would be estimated to cost about $10–$20 billion, which represents an annual per capita cost of $10–$20 per person in the developed world. International agreements would be required, and even advocates acknowledge that the ocean would be acidified, affecting marine life and marine shell production. Further, such a large-scale plan would require a large fleet of ships and an enormous amount of iron. Finally, it is debatable as to how much of the ocean is in an iron-limited state, which would allow this technique to be successful (Hauck et al. 2016).

19.3.4 Direct Engineered Capture

Direct engineered capture is also a straightforward method for carbon sequestration. Sorbent materials are deployed, capturing and sequestering carbon dioxide flowing through the adsorbent medium (Figure 19.10). Common adsorption materials include calcium hydroxide or calcium oxide (Socolow et al. 2011).

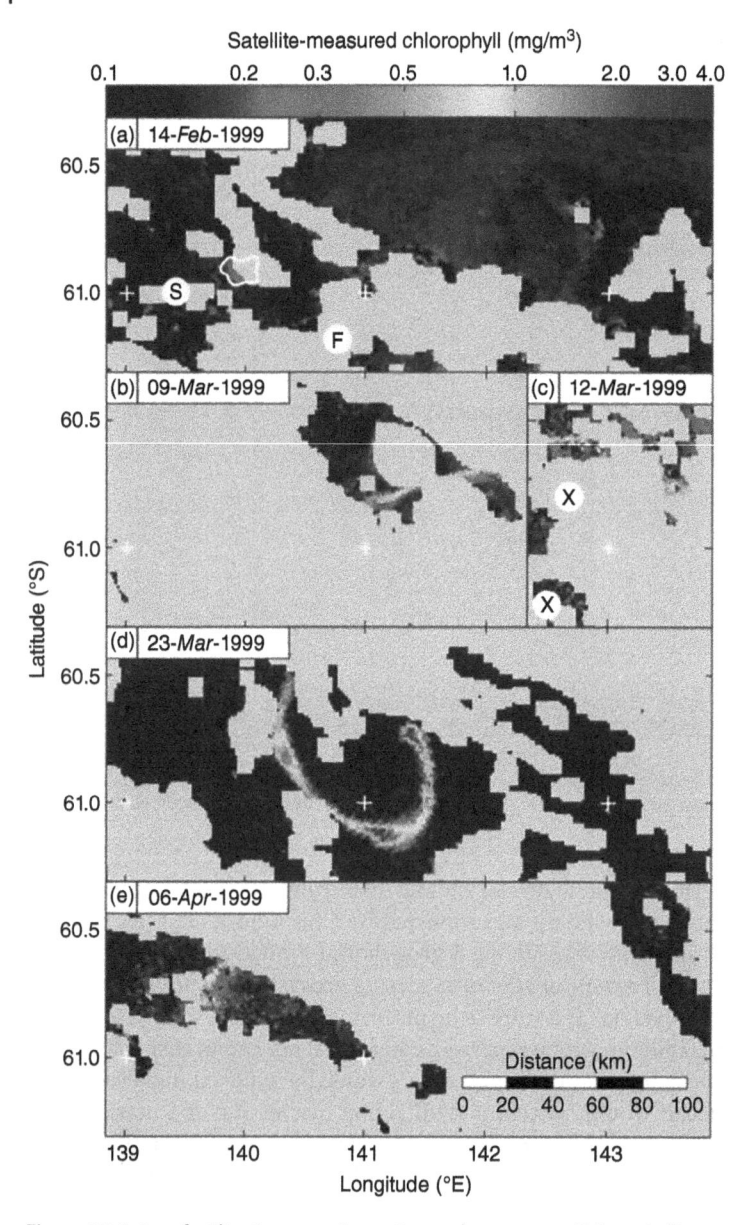

Figure 19.8 Iron fertilization experiment in southern ocean. Chlorophyll concentration. Source: Abraham et al. (2000). Reproduced with permission of Springer Nature.

19.4 Solar Radiation Management (SRM) Methods

SRM methods act to increase the amount of radiation reflected away from the Earth, thereby reducing the amount of solar radiation that reached the surface and is absorbed by the Earth. On a small scale, some thought has been given to increase the surface reflectivity, or "albedo," of the Earth through a range of methods, including white structural roofs ("cool roofs"), planting crops with a high relative reflectivity, or covering large areas of the desert with reflective material

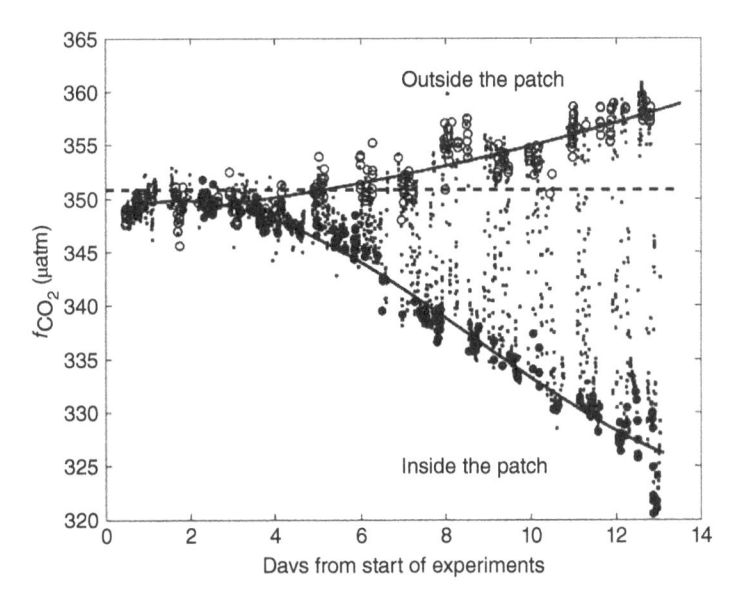

Figure 19.9 Iron fertilization experiment in southern ocean. Carbon dioxide reduction as a result of the enhanced plankton growth. Source: Watson et al. (2000). Reproduced with permission of Springer Nature.

Figure 19.10 Direct engineered capture of CO_2 using adsorbent materials (Climeworks, Switzerland).

(Ming et al. 2014). Methods have also been proposed to enhance the reflectivity of marine clouds. On a grander scale, methods have been developed to inject sulfate aerosols into the lower stratosphere, mimicking the effect of large-scale volcanic eruptions (Kalidindi et al. 2015). On an even larger scale, consideration has been given to deploy large-scale shields or reflectors in space to deflect a fraction of the solar energy that reaches the Earth.

SRM technologies are designed to directly modify the inflow and outflow of solar energy to the Earth. Large-scale deployment of these methods would likely have an effect on climate within a few years. These methods do not deal with the original cause of the climate change, the increasing concentration of GHG in the atmosphere. However, the quick response offered could be useful in a time of emergency, such as when a climate "tipping point" is approached.

19.4.1 Sulfur Injection

Sulfur injection involves the injection of sulfate aerosol particles into the stratosphere to reflect sunlight away from the Earth, reducing the amount of solar radiation reaching the Earth's surface, and acting to counteract climate change resulting from the greenhouse effect. Past evidence from large-scale volcanic eruptions, such as the eruption of Mt. Pinatubo in 1991, indicates that the addition of aerosols to the stratosphere can cool the planet (McCormick et al. 1995). Mt. Pinatubo's massive eruption resulted in a global scale cooling of 0.5 °C in its aftermath (Figure 19.11). Given the proper size and large-scale deployment of aerosol particles, it is theoretically possible to cool the Earth, as demonstrated in Figure 19.12, which shows modeling of reduced aerosol temperatures resulting from deployed aerosols (Rasch et al. 2008). While sounding somewhat farcical, such an approach has proponents. For instance, Paul Crutzen, a 1995 Nobel Laureate in chemistry, opined that research on the potential of increasing the Earth's albedo by large-scale injection of sulfur aerosols into the stratosphere should be performed (Crutzen 2006).

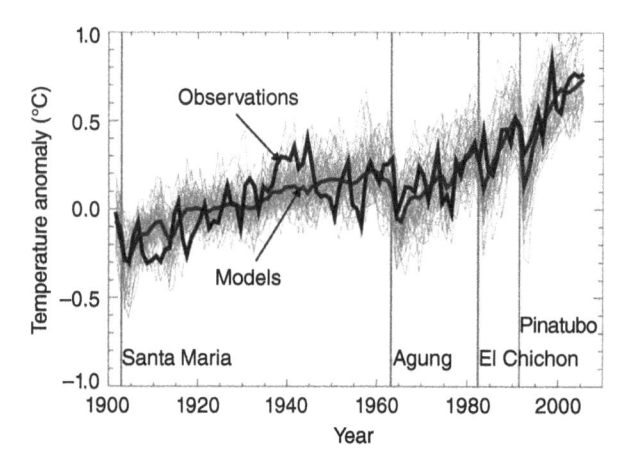

Figure 19.11 Effect of volcanic eruptions in the Earth's surface temperature due to the emission of sulfate aerosols (IPCC 2007).

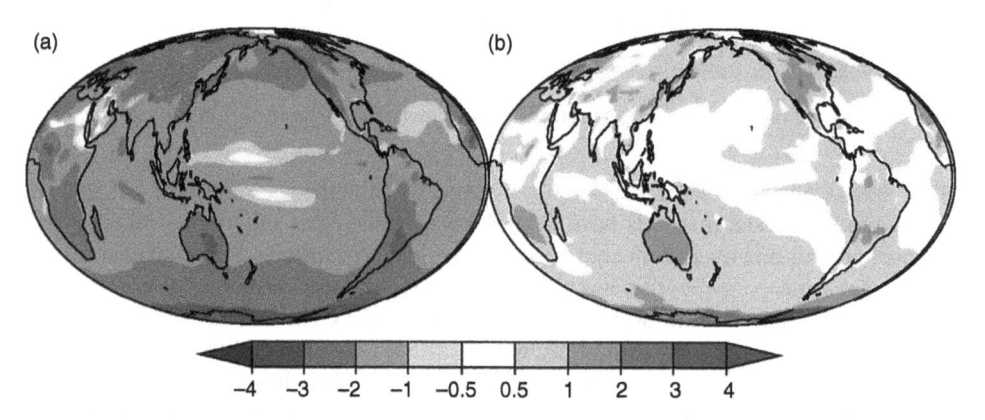

Figure 19.12 Surface temperature change from (a) present-day to (b) a doubling of current carbon dioxide levels scenario with the dispersion of 2 Tg/yr sulfur particles.

If such an approach was to be employed, in order to be successful, between 1 and 2 Tg of aerosols would need to be injected annually. The large-scale injection could be performed in one of three ways. In the first alternative, ballistic shells or missiles (likely on the order of several thousand per day) could be fired into the stratosphere. As a second alternative, several thousand daily injections could be made from large aircraft delivery in the stratosphere. As a third alternative, ground releases could be made in the tropics. Using one of the three alternatives would require an annual expenditure of between $25 and $50 billion dollars, which represents an annual per capita cost of $25–$50 per person in the developed world. The need for international agreements for this practice should be explored. While never tested on any appreciable scale, volcanic eruptions do provide evidence that the mechanics of this technique would have an effect on influencing climate. Additionally, since its effects are temporary, a long-term commitment would not be necessary.

19.4.2 Reflectors and Mirrors

Equally grand or revolutionary, if not unconventional, is the potential use of reflectors or mirrors in space. The goal of such an application would be to deploy mirrors in space to deflect incoming solar radiation (Govindasamy et al. 2003). Figure 19.13 demonstrates the deployment location at the Lagrangian Point, located at an intermediate point between the Earth and the Sun. The system could be opaque reflectors or mirrors that would totally divert some fraction of sunlight away, or they could consist of a system that would diffract sunlight to some lesser degree, allowing a fraction of impacted sunlight to get through to the earth (Figure 19.14). In either case, a space vehicle or satellite would be needed to deploy the reflection media to a

Figure 19.13 Mirror at the L1 Lagrangian point to partially block the solar radiation on Earth.

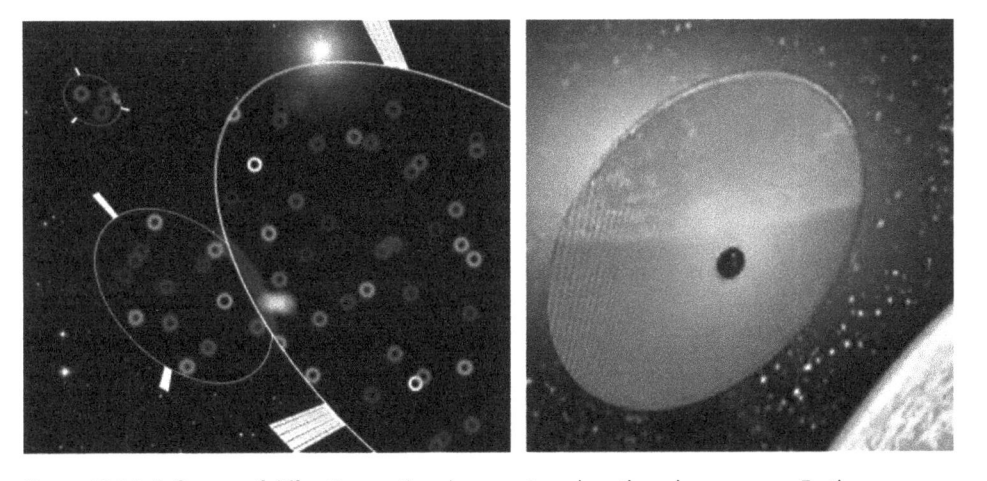

Figure 19.14 Reflectors of diffraction gratings in space to reduce the solar energy on Earth.

proper location. To maintain a proper distance from the Earth during its revolution around the Sun, an active positioning system, likely consisting of compressed gases, would need to be used on an ongoing basis. Because these gases would eventually be exhausted, it would be necessary to maintain or replace reflective instruments on a regular basis. The cost of deployment would likely be between $50 and $500 billion on an annual basis, which represents an annual per capita cost of $50–$500 per person in the developed world. International agreements would be required for deployment, and a long-term commitment would be required due to ongoing maintenance or replacement.

19.5 Applicability of CDR and SRM

The CDR and SRM climate geoengineering methods are very ambitious in reducing the effects of global warming a climate change. They could be considered for implementation but only as part of a wider set of options to reduce the GHG emissions, global warming and climate change (Blackstock and Long 2010). When comparing these two groups of methods, CDR methods are preferable to SRM methods because they mitigate the negative effects of GHG in the long term. CDR methods show various characteristics that make them preferable over other methods: they are safe, effective, sustainable, and affordable. Furthermore, they can be developed and applied alongside other conventional mitigation methods as soon as the technological applications are developed and the cost is affordable. SRM methods may be applied only when there is an urgent need to limit or reduce the average global temperatures. SRM methods may be an effective tool in case of necessity of rapid reduction of temperatures. However, there is limited information about the effects of SRM at medium and long term. Thus, these methods must only be considered for application for a short period, in conjunction with other GHG removal methods, CDR, or conventional CO_2 mitigation methods. Thus, the SRM methods can be scaled back or discontinued at an appropriate time. As depicted in Figure 19.15, techniques from these two classes could be applied together in an integrated application.

When considering both CDR and SRM technologies, their deployment cost and their effect on the environment are of a very large scale. Arguably, these would be considered drastic, "last resort" measures. However, some argue that climate change has already reached, or may soon reach, a "tipping point" or a "point of no return" in which climate effects and the effects imparted

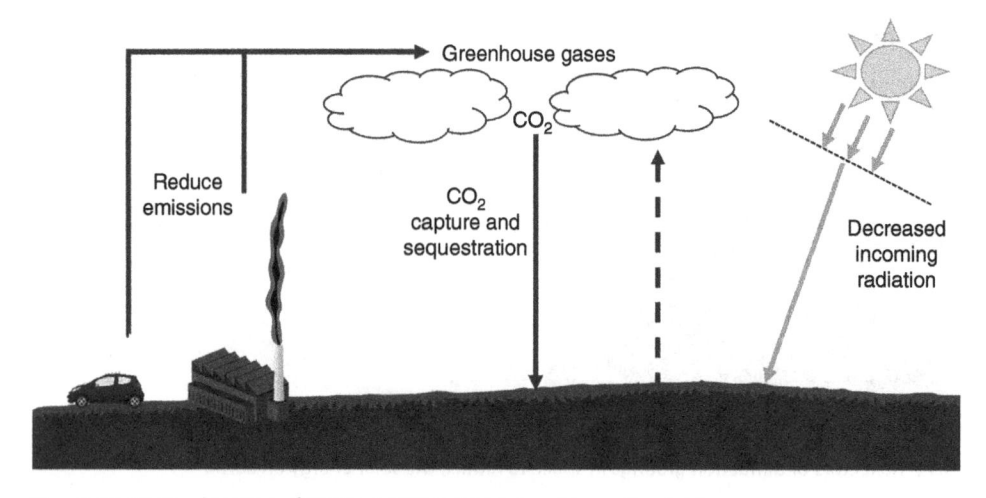

Figure 19.15 Combination of CDR and SRM methods to address climate change.

to the environment may be irreversibly damaging. There is a fear that these dynamics may result in a new stable state in which accelerated positive feedback effects occur. One example would be the collapse of Arctic ice, which could trigger the release of methane on a large scale from permafrost in Siberia. A "nightmare" scenario exists where a "domino effect" would occur, with successive parts of the climate system collapsing in succession, facilitating additional strain on subsequent parts and leading to a rapid, sudden climate change.

19.6 Climate Geoengineering – A Theoretical Framework

Because of the grand scale of deployment, and the exorbitant costs that would be associated with such use, as well as the international agreement and cooperation that would be necessary, it is critical to consider a framework in which such technologies would be analyzed, selected, and implemented with strong international consensus. First, there is general agreement that such technologies should only be considered when large-scale intervention is necessary, and then only to the extent that is needed. Further, a thoughtful investigation is needed to integrate economic and cultural dimensions, as well as a focus on systems as opposed to subsystems. Such an approach should also be inclusive and transparent with clear accountability. Metrics have to be developed and implemented that will allow for tracking of progress and the identification of problems or undesirable side effects. Further, there is a need to realize that any approach would involve high levels of uncertainty and should allow for mid-course intervention and correction. Systems should be designed to allow for incremental adjustment or reversibility as well as resiliency and redundancy.

19.7 Risks and Challenges

Such a large-scale approach with such widely ranging uncertainty inherently carries risks and has resulted in criticisms. First, the deployment costs would dwarf any large-scale engineering project contemplated to date in history. However, as demonstrated in Figures 19.16 and 19.17, once deployed, the incremental cost of mitigation with respect to a unit reduction of emissions mitigation declines drastically and may be attractive in the long run as compared to a comparable large-scale deployment of conventional approaches. Due to technical and implementation concerns, some have called for a moratorium on climate geoengineering research and consideration. Some of these reasons include concerns over effectiveness, control, predictability, unintended side effects, reliability, the potential of weaponization, effects on sunlight and cloud cover, moral objections, a lack of global control, and termination shock. Further, from an ethical standpoint, there is a concern that humans should not view the Earth as their property or artifact, and that to engineer systems that could strongly affect every global ecosystem is not considered ethical. Additionally, there is a concern that such an approach does not address the cause of the problem (increasing GHG emissions), but only mitigates the effects. Finally, there is a strong concern regarding unpredictability of these systems and that their implementation cannot be adequately or accurately modeled.

All of the proposed climate geoengineering systems require large-scale implementation to have any appreciable effect on climate change. Even the least expensive technologies would require expenditures in billions of dollars, and some of the more complex technologies could be significantly more. Further, for some of the more complex systems, a robust technological development process is necessary even before a prototype or pilot study could be performed, let alone operational deployment. At this time, no clear institutional framework is in place to

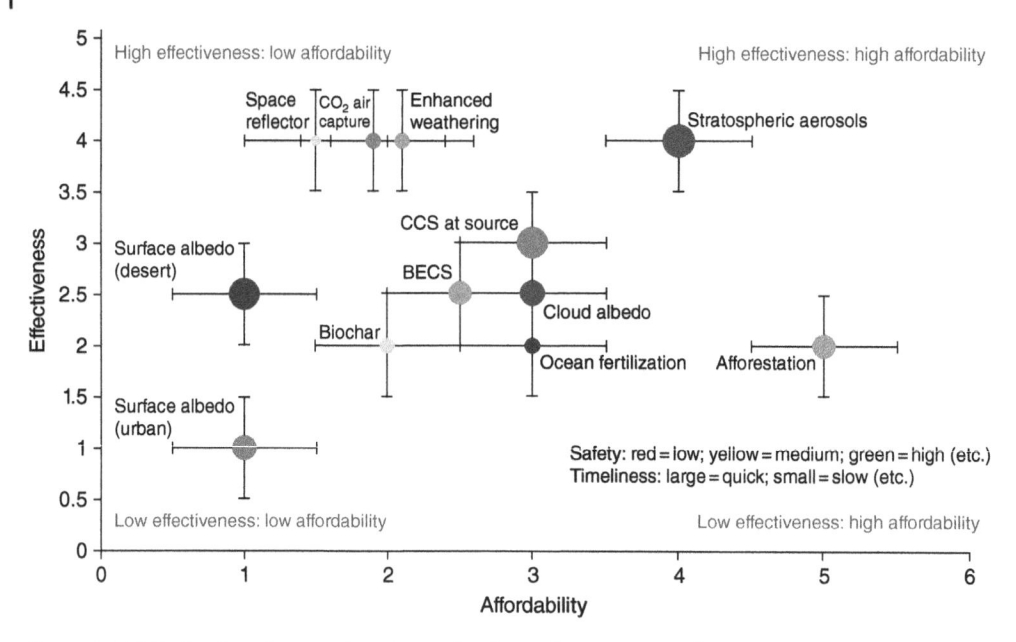

Figure 19.16 Evaluation of geoengineering technologies for mitigation of climate change. Source: Shepherd (2009).

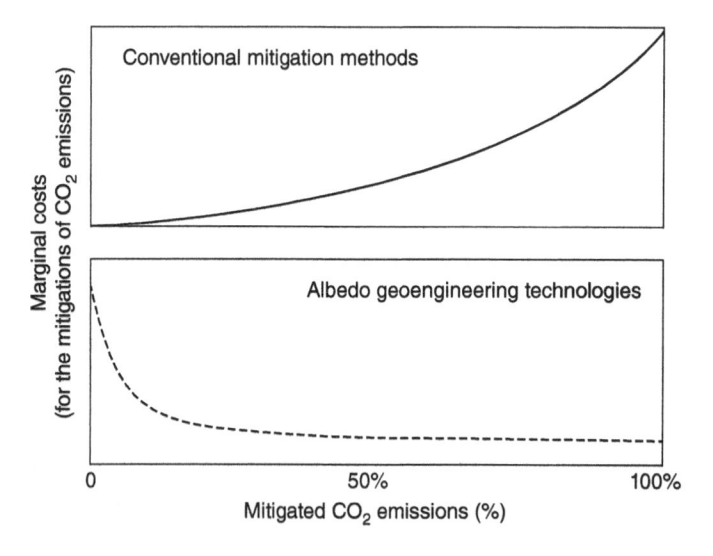

Figure 19.17 Comparative cost analysis of the conventional mitigation of climate change and geoengineering methods. Source: Keith (2000).

handle the related research and development program. As a result, many potentially feasible technologies do not have adequate development or experimental data to demonstrate their efficacy.

Even if a promising technology has been developed, it will be at least as difficult to implement. It is unclear who would have the authority or the ability to implement a global-scale technology. It is equally unclear how to allocate responsibility for implementation and operation as well as how to fund use. Further, use could result in significant ecosystem changes on a global scale, resulting in billions of people being adversely affected while billions of others benefit. Any deployment would surely result in effects that cross many, if not all international borders. The related legal issues and liability are even difficult to fathom. It is critical to develop these regulatory, legal, and economic frameworks before deployment could be contemplated.

While many challenges and concerns exist within the scientific and regulatory community, these concerns are dwarfed by the concerns that will likely develop within the general public. Public attitudes toward climate geoengineering as well as transparent public engagement in development will play a critical role for the potential implementation of climate geoengineering projects in the future. It is important to note that perceptions of risks involved, the level of trust in ongoing research, the transparency of all actions taken, and a clear delineation of vested interest will ultimately decide the political will and feasibility for future use. If there is to be even a marginal deployment of climate geoengineering methods in the future, an active international societal dialogue will be important to satisfy the demand of information for the possible environmental, social, and economic impacts and unintended consequences.

From a regulatory standpoint, no governing policies currently exist with respect to climate geoengineering. Parties to the United Nations Framework Convention on Climate Change (UNFCCC) should make increased efforts to adapt to and mitigate the effects of climate change, in particular by pursuing ongoing reductions of global GHG emissions, regardless of the potential use of climate geoengineering in the future. The UNFCCC has acknowledged that stabilization of emissions and the mitigation of climate change effects should be pursued.

Because of all these challenges with regard to climate change the ongoing research and development should continue to advance. Such work can better inform decisions about implementation and climate change effects can be more accurately modeled and predicted. Additionally, ongoing research and development may result in the development of lower risk technologies that could be more attractively deployed, either on a regional or a global scale. Further, a greater data set could be collected in the meantime with respect to climatic models, including the development of carefully planned and executed experiments.

19.8 Summary

Because of the costs involved and the likely dramatic effects that may result from implementation, climate geoengineering, while offering the promise of being a significant mitigation approach to climate change, remains highly controversial. Many even argue that to pursue with such an approach would be unethical, both from a professional and a moral standpoint. However, if faced with an imminent and dire climate crisis, it would be difficult to justify not acting with such an approach if a reasonable chance of success existed. Further, there are questions as to how to proceed if significant uncertainty of effectiveness and unintended consequences exist.

Several key organizations have at least implicitly agreed that climate geoengineering approaches should at least be considered and studied. These agencies include the Parliament of the United Kingdom, the Royal Society, the Institute of Mechanical Engineers, and NASA. On the other hand, some groups have been vocally opposed, including Greenpeace and Friends of the Earth. Controversy exists because it is generally believed that the most effective way to moderate the climate change is to reduce the GHG emissions before irreversible changes takes place. No climate geoengineering technology can provide an easy or readily acceptable alternative solution to mitigate climate change. However, all the methods described in the chapter could be useful in the future to mitigate, in part, the climate change, and therefore a more detailed research and analysis will be necessary before their practical implementation.

Geoengineering of the Earth's climate could very likely be technically feasible. However, the potential technical methodologies are in their beginning stages, and there are major uncertainties regarding effectiveness, cost, and environmental impact, both intended and unintended. Further international collaboration is critical in defining a research and development approach on promising technologies as well as a critical study of the feasibility, costs, benefits, risks, and opportunities presented by these technologies. Further, there is a pressing need to develop a governance and implementation framework to guide research, development, funding, and deployment, and engagement of all stakeholders though a transparent dialogue.

19.9 Questions

19.1 What is climate geoengineering? Explain the two major categories of climate geoengineering strategies.

19.2 Discuss climatic or environmental scenarios where geoengineering interventions could be necessary.

19.3 Name a few carbon dioxide removal methods.

19.4 Describe specific drawbacks associated with oceanic or marine carbon sequestration.

19.5 What are subsurface and surface carbon dioxide sequestration methods?

19.6 Name a few solar radiation management methods.

19.7 What are the disadvantages of sulfur injection method for solar radiation management?

19.8 Assess in detail the financial and political considerations of CDR and SRM approaches.

19.9 What are the challenges associated with implementing CDR and SRM methods for climate change mitigation? Explain briefly.

19.10 Develop and discuss a financial incentive program that could be used to encourage developed and developing countries to collaborate on CDR and SRM approaches.

References

Abraham, E.R., Law, C.S., Boyd, P.W. et al. (2000). Importance of stirring in the development of an iron-fertilized phytoplankton bloom. *Nature* 407 (6805): 727.

Blackstock, J.J. and Long, J.C. (2010). The politics of geoengineering. *Science* 327 (5965): 527–527.

Chisholm, S.W. (2000). Oceanography: stirring times in the Southern Ocean. *Nature* 407 (6805): 685–687.

Crutzen, P.J. (2006). Albedo enhancement by stratospheric sulfur injections: a contribution to resolve a policy dilemma? *Climate Change* 77 (3): 211–220.

DOE (2018). Carbon Storage Research. Office of Fossil Energy. https://www.energy.gov/fe/science-innovation/carbon-capture-and-storage-research (accessed September 2018).

Ehrlich, P.R. and Ehrlich, A.H. (2013). Can a collapse of global civilization be avoided? *Proceedings of the Royal Society B: Biological Sciences* 280 (1754): 20122845. https://doi.org/10.1098/rspb.2012.2845.

Fares, S., Paoletti, E., Calfapietra, C. et al. (2017). Carbon sequestration by urban trees. In: *The Urban Forest*, 31–39. Springer International Publishing.

Gibbins, J. and Chalmers, H. (2008). Carbon capture and storage. *Energy Policy* 36 (12): 4317–4322.

Govindasamy, B., Caldeira, K., and Duffy, P.B. (2003). Geoengineering Earth's radiation balance to mitigate climate change from a quadrupling of CO_2. *Global and Planetary Change* 37 (1): 157–168.

Hauck, J., Köhler, P., Wolf-Gladrow, D., and Völker, C. (2016). Iron fertilisation and century-scale effects of open ocean dissolution of olivine in a simulated CO_2 removal experiment. *Environmental Research Letters* 11 (2): 024007.

Heinemann, N., Wilkinson, M., Haszeldine, R.S. et al. (2013). CO_2 sequestration in a UK North Sea analogue for geological carbon storage. *Andean Geology* 41 (4): 411–414.

Holloway, S. (2005). Underground sequestration of carbon dioxide—a viable greenhouse gas mitigation option. *Energy* 30 (11): 2318–2333.

Hui, D., Deng, Q., Tian, H., and Luo, Y. (2017). Climate change and carbon sequestration in forest ecosystems. In: *Handbook of Climate Change Mitigation and Adaptation*, 2e (ed. W.-Y. Chen, M. Lackner and T. Suzuki), 555–594. New York: Springer. ISBN: 978-3-319-14408-5.

IPCC (2000). *Special Report on Emissions Scenarios*. Cambridge, UK: Cambridge University Press. ISBN: 0521804930.

IPCC (2007). Climate change 2007 – the physical science basis. In: *Contribution of Working Group I to the Fourth Assessment Report of the IPCC*. New York, NY: Cambridge University Press. ISBN: 978 0521 88009-1.

IPCC (2014). Climate change 2014: synthesis report. In: *Contribution of Working Groups I, II and III to the Fifth Assessment Report of the Intergovernmental Panel on Climate Change*. Geneva, Switzerland: IPCC.

Kalidindi, S., Bala, G., Modak, A., and Caldeira, K. (2015). Modeling of solar radiation management: a comparison of simulations using reduced solar constant and stratospheric sulphate aerosols. *Climate Dynamics* 44 (9-10): 2909–2925.

Keith, D.W. (2000). Geoengineering the climate: history and prospect. *Annual Review of Energy and the Environment* 25 (1): 245–284.

Kiehl, J.T. and Trenberth, K.E. (1997). Earth's annual global mean energy budget. *Bulletin of the American Meteorological Society* 78 (2): 197–208.

Leung, D.Y., Caramanna, G., and Maroto-Valer, M.M. (2014). An overview of current status of carbon dioxide capture and storage technologies. *Renewable and Sustainable Energy Reviews* 39: 426–443.

McCormick, M.P., Thomason, L.W., and Trepte, C.R. (1995). Atmospheric effects of the Mt Pinatubo eruption. *Nature* 373 (6513): 399.

Ming, T., Liu, W., and Caillol, S. (2014). Fighting global warming by climate engineering: Is the Earth radiation management and the solar radiation management any option for fighting climate change? *Renewable and Sustainable Energy Reviews* 31: 792–834.

Ming, T., de_Richter, R., Liu, W., and Caillol, S. (2014). Fighting global warming by climate engineering: is the Earth radiation management and the solar radiation management any option for fighting climate change? *Renewable and Sustainable Energy Reviews* 31: 792–834.

Pham, V. and Halland, E. (2017). Perspective of CO_2 for storage and enhanced oil recovery (EOR) in Norwegian North Sea. *Energy Procedia* 114: 7042–7046. https://doi.org/10.1016/j.egypro.2017.03.1845.

Rasch, P.J., Tilmes, S., Turco, R.P. et al. (2008). An overview of geoengineering of climate using stratospheric sulphate aerosols. *Philosophical Transactions of the Royal Society A: Mathematical, Physical and Engineering Sciences* 366 (1882): 4007–4037. https://doi.org/10.1098/rsta.2008.0131.

Sarmiento, J.L. and Gruber, N. (2002). Sinks for anthropogenic carbon. *Physics Today* 55 (8): 30–36.

Shepherd, J.G. (2009). *Geoengineering the Climate: Science, Governance and Uncertainty*. Royal Society.

Shepherd, J.G. (2012). Geoengineering the climate: an overview and update. *Philosophical Transactions of the Royal Society A: Mathematical, Physical and Engineering Sciences* 370 (1974): 4166–4175.

Socolow, R., Desmond, M., Aines, R. et al. (2011). *Direct Air Capture of CO_2 with Chemicals: A Technology Assessment for the APS Panel on Public Affairs (No. EPFL-BOOK-200555)*. American Physical Society.

Watson, A.J., Bakker, D.C.E., Ridgwell, A.J. et al. (2000). Effect of iron supply on Southern Ocean CO_2 uptake and implications for glacial atmospheric CO_2. *Nature* 407 (6805): 730.

Zhang, Z., Moore, J.C., Huisingh, D., and Zhao, Y. (2015). Review of geoengineering approaches to mitigating climate change. *Journal of Cleaner Production* 103: 898–907.

Section IV

Sustainable Engineering Applications

20

Environmental and Chemical Engineering Projects

20.1 Introduction

As demonstrated in previous chapters, comprehensive sustainability assessments incorporating environmental, economic, and social aspects can be applied to a range of projects and products. This chapter presents several examples of detailed sustainability assessments of environmental and chemical engineering projects. These examples show the methodology, tools, and general interpretation of the results for sustainability assessment. It should be noted that sustainability assessment results depend on site-specific conditions; therefore, the conclusions from each example cannot be generalized. Similar methodologies could be adopted to assess sustainability of any specific environmental and chemical engineering projects.

20.2 Food Scrap Landfilling Versus Composting

The USEPA has identified food waste recovery as an issue of concern and a key opportunity for reducing greenhouse gas emissions. The USEPA's Food Recovery Hierarchy "prioritizes actions that the organizations can take to prevent and divert wasted food." The options include source reduction, alternative uses of food waste as feed or industrial inputs, and processing options such as composting and digestion. Landfilling and incinerating food waste are the least preferable disposal scenarios, from the USEPA's standpoint.

USEPA has collected and reported the generation of municipal solid waste (MSW) from 1960. MSW has been growing from 88.1 million tons in 1960 to 251 million tons in 2012, with an average generation of 4.38 pounds per person per day. Total MSW recovery in 2012 was almost 87 million tons, but food waste accounts for only 2% (Figure 20.1) of it. After MSW recovery through recycling and composting, 164 million tons of MSW were discarded in 2012. Food waste is the largest component of the discards at 21%.

In this study, the last two options on the USEPA's hierarchy, composting and landfilling, are analyzed and compared for the environmental, economic, and social impacts with an objective of selecting the most sustainable option for food waste management.

20.2.1 Background

MSW facilities must meet certain environmental, design, monitoring, and safety standards, as outlined by the US Resource Conservation and Recovery Act (RCRA). As noted by the USEPA, "Modern landfills are well-engineered facilities that are located, designed, operated, and monitored to ensure compliance with the federal regulations." Even in landfill facilities that meet

Sustainable Engineering: Drivers, Metrics, Tools, and Applications, First Edition.
Krishna R. Reddy, Claudio Cameselle, and Jeffrey A. Adams.
© 2019 John Wiley & Sons, Inc. Published 2019 by John Wiley & Sons, Inc.

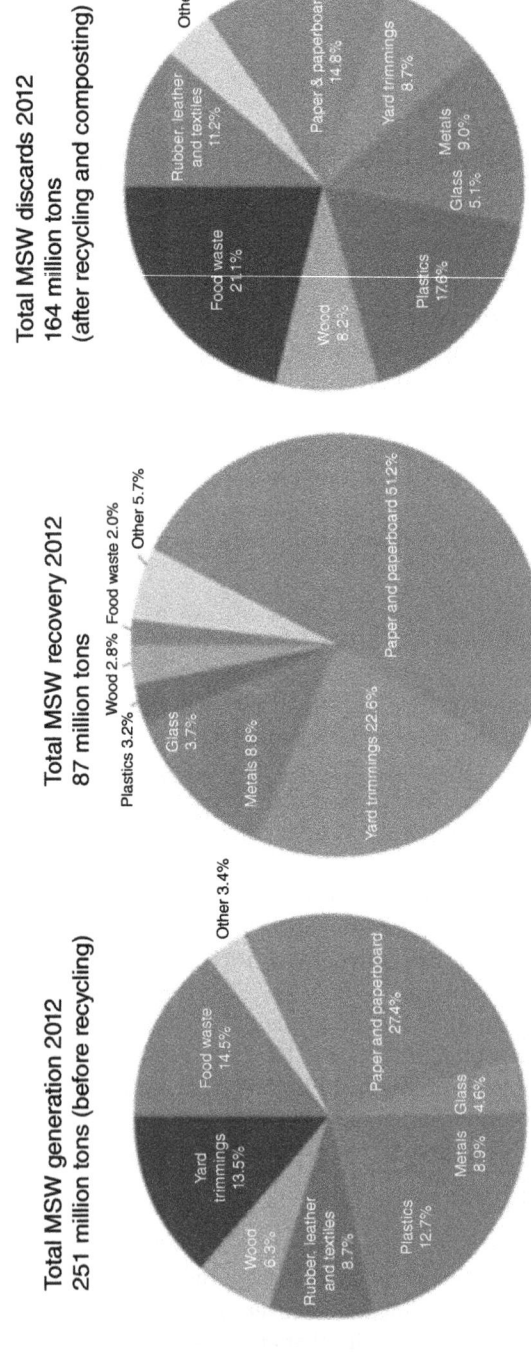

Figure 20.1 Municipal solid waste in the US in 2012 (USEPA 2017a).

the RCRA requirements, two primary limitations or concerns exist: greenhouse gas emissions and material efficiency.

Organic materials within the landfill cells, with little or no oxygen, are consumed by anaerobic bacteria and undergo biochemical reactions yielding biogas, a mixture of methane (CH_4) and carbon dioxide (CO_2). The USEPA estimates that the landfills account for roughly 18% of CH_4 emissions in the United States. CH_4 is a greenhouse gas with a global warming potential (GWP) of 24–28 times higher than CO_2. Landfills mitigate (but do not eliminate) methane emissions by installing and operating a biogas collection system, and the collected biogas can further be used for the production of electricity.

Once dumped and covered over with soil material (daily cover), landfilled material is difficult and costly to extract. In the cases where the landfilled material is readily recyclable or reusable in some way, this represents a loss of value to the local and/or national economy. The recycling industry has grown out of this concern, yet only 35% of the waste stream in the United States is currently captured for recycling. In Illinois, where the per-capita waste generation rate is 8.1 lbs. per person per day, only 19% of this material is diverted for recycling.

An alternative for the organic fraction of MSW is composting. This is a biological process where the inputs to the process are material feedstock, air, and water, and the outputs are compost, heat, water vapor, and carbon dioxide. Compost itself is a fine mixture of decomposed organic debris – typically brown and fluffy – which contains important nutrients in a form that can be readily accessed by plant root systems. Several materials, such as plant trimmings and yard debris, food scraps, fruit and vegetable waste, compostable paper, and often cooked food and meats can be used for composting. Composting is a way of recycling organic waste that can be used in soil fertilization.

20.2.2 Methodology

20.2.2.1 Goal and Scope

Composting and landfilling are two alternative processes for the management of the organic fraction of MSW. Both scenarios were analyzed using food scrap generation and waste management data available for Champaign County, IL, USA. The sustainability of the two scenarios has been analyzed considering environmental, economic, and social impacts. The analysis includes an environmental impact assessment, direct and indirect economic analysis, and social sustainability considerations.

20.2.2.2 Study Area

The location for this study is Champaign County in Illinois. This is a particularly fitting site for the analysis as it is between two active landfills (Clinton Landfill and Danville Landfill) and has an industrial composting facility within the County borders. Additionally, Champaign residents are likely to be interested in pursuing more sustainable ways of managing food waste. The County has many environmentally conscious university students and is home to the Illinois Green Business Association (IGBA). Nearly 20 businesses in Champaign County are IGBA-certified. Figure 20.2 shows the location of Champaign County, the two landfills, and the landscape recycling center (LRC).

20.2.2.3 Technical Design

System boundaries: The life cycle analysis requires the definition of a functional unit of food scraps and the boundaries of the system to be analyzed, from the generation of food waste to its final disposition. Thus, it is possible to determine the various assembly, transport, and processing steps through which the food scrap travels in its life cycle. The boundaries are important because food scraps are part of a much larger food system, which is complex and beyond the scope of this analysis.

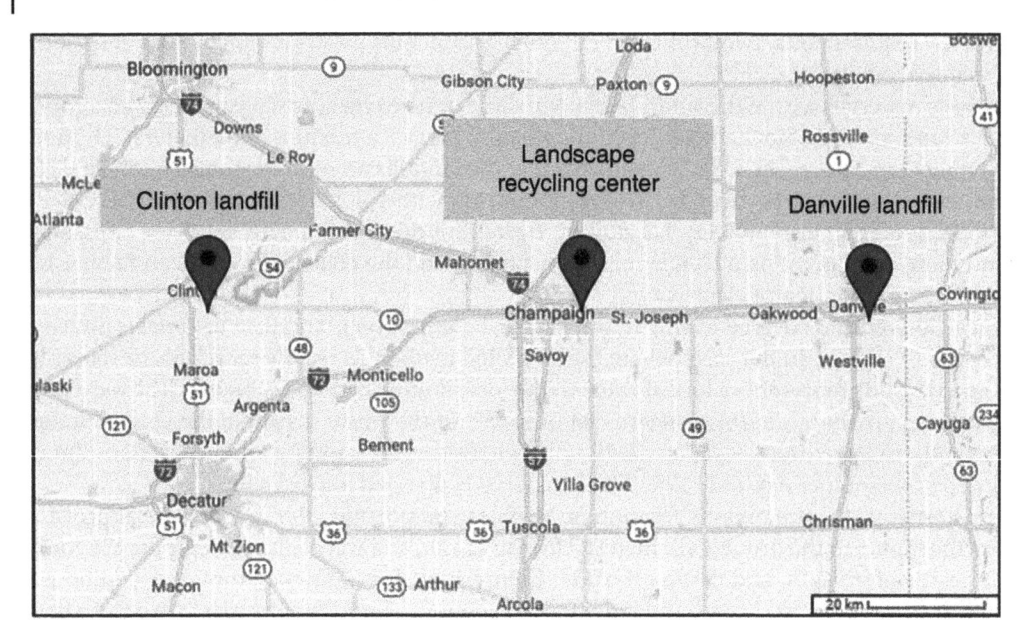

20 Environmental and Chemical Engineering Projects

Figure 20.2 Study area – Champaign County – with analyzed disposal sites.

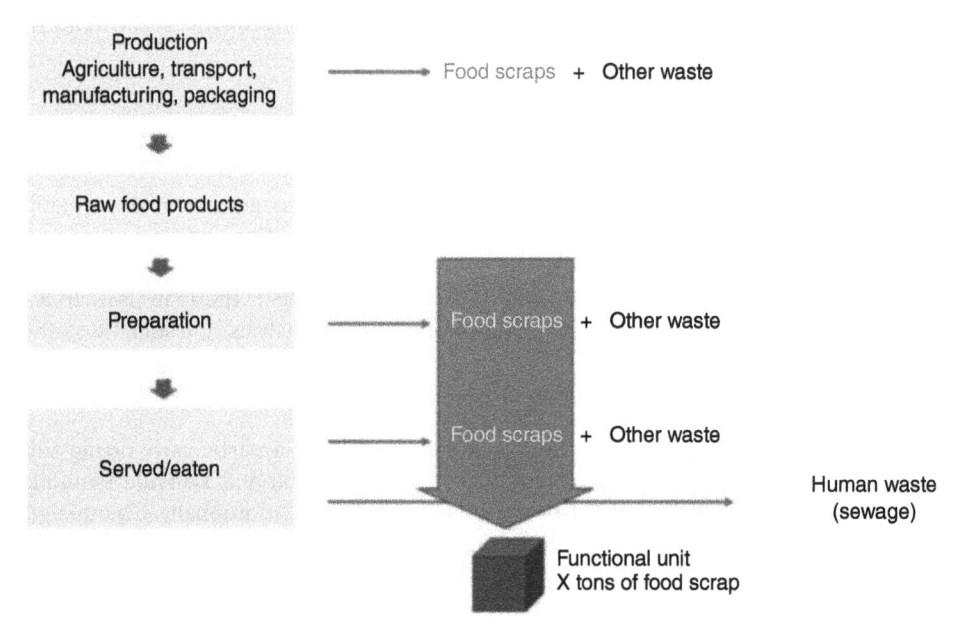

Figure 20.3 Life cycle stages of food (gray boxes) and system analysis boundary for this study (gray shaded arrow).

The basic life cycle of food from agriculture/production to its end-use as a consumption product is depicted in Figure 20.3, where the various waste outputs are shown. The selected life cycle boundaries are the food scrap (gray shaded arrow) waste generated from the preparation (restaurants and grocery stores) and consumption (households) of food. The functional unit in this study is the food scrap produced (expressed in tons). Figures 20.4 and 20.5 delineate the life cycle inputs and outputs for landfilling and composting referred

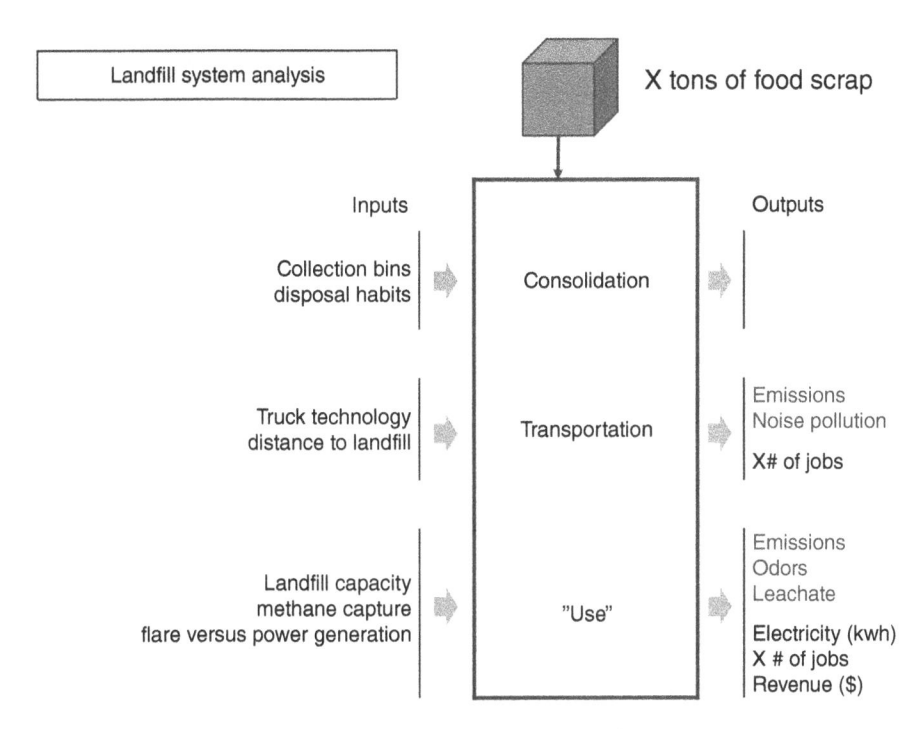

Figure 20.4 Life cycle inputs and outputs for landfill system analysis.

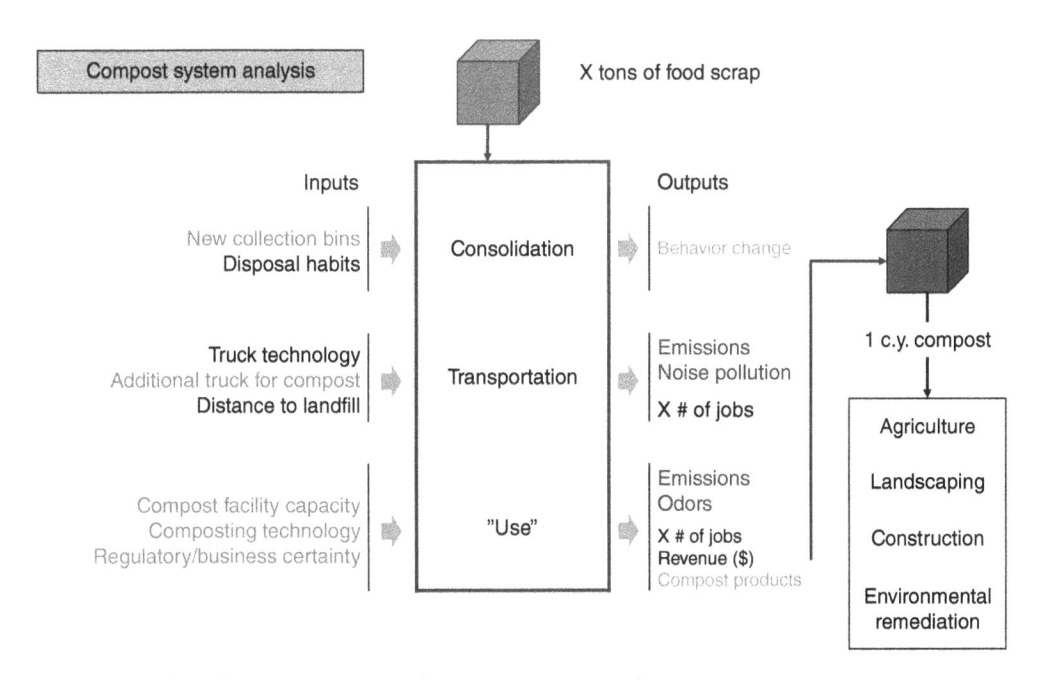

Figure 20.5 Life cycle inputs and outputs for compost system analysis.

Table 20.1 Evaluation of the transport of food waste in ton-miles.

Food scrap generator	Distance to facility			Tons/week captured	Ton-miles per week		
	Urbana LRC	Clinton Landfill	Danville Landfill		To compost	To Clinton	To Danville
Grocery	3.6	38.5	31.8	8.8	32.6	346.9	286.0
Restaurant	3.7	38.8	31.7	16.0	59.9	620.7	507.0
Households	1.5	40.6	31.2	75.0	107.8	2844.8	2180.5
						3812.4	2973.5
				Compost	200.4	Landfill average	3393.0

to the food scrap unit. Note that the Compost System Analysis includes a final stage of "use" of the compost product. This stage is factored into the environmental, economic, and social analyses.

Functional unit: The design of a food scrap waste management system requires a knowledge of generation rate (potentially compostable material generated), expected capture rate (percentage of material successfully separated from garbage and collected for composting), and processing capacity (existing infrastructure). The waste generation and characterization study for Illinois (Illinois Recycling Association 2009) estimated that the Champaign County generates 27 740 tons of food scrap annually. Assuming a scenario where 20% of the food scrap that currently goes into the garbage is separated for composting, the functional unit for this study was selected to be 100 tons/wk.

Transportation: To analyze the impact of the transportation, travel distances from grocery stores, restaurants, and households to the disposal facilities (landfill and recycling centers) are required. A total of 10 grocery stores and 30 restaurants near Champaign have been selected randomly. Thereafter, the respective addresses were geocoded using the software ArcGIS (ArcGIS 2017). Finally, the average distances from the grocery stores and restaurants to the disposal facilities were measured. As the households in Champaign County are evenly distributed across the County, the centroid of Champaign was used to calculate the distance from households to disposal facilities. The measured distances were then multiplied by the tons of food scrap captured per week to calculate the ton-miles of energy expended for transportation per week (Table 20.1).

20.2.3 Environmental Sustainability

EPA created the Waste Reduction Model (WARM) to help solid waste planners and organizations track and voluntarily report greenhouse gas emissions reductions from several different waste management practices. WARM calculates the total green house gases (GHGs) emissions of baseline and alternative waste management practices, source reduction, recycling, combustion, composting, and landfilling. In this study, landfilling the 100 tons/wk food scrap is defined as the baseline scenario. The alternative scenario of waste management is composting the food scrap. Two landfills, Clinton Landfill with no landfill gas (LFG) recovery and the Danville Landfill with LFG recovery for energy production, are considered for this study. Emissions that occur during transport of materials to the management facility are included in WARM model. Using the results from waste collecting model, there will be about 200.4 ton-miles of energy expended per week for the composting scrap management option, 3812.4 ton-miles of energy expended per week for Clinton Landfill, and 2973.5 ton-miles of energy expended per week for Danville Landfill due to transportation.

According to the results of WARM (Table 20.2), GHG emissions from the baseline waste management scenarios will be 84 MT CO_2e (metric tons of carbon dioxide equivalent) for Danville Landfill and 207 MT CO_2e for Clinton Landfill. These results indicate that a landfill with LFG recovery for energy is a more environmentally beneficial option for waste management. If 100 tons of food scrap per week are diverted from landfills to the composting facility, the alternative scenario will have 13 MT CO_2e reduction in GHG emissions. Thus, the net change of GHG emission resulting from diversion of food scrap disposal from Danville Landfill and Clinton Landfill to the compost facility will be 97 MT CO_2e and 220 MT CO_2e, respectively.

20.2.4 Life Cycle Assessment

The "food waste" product was analyzed through the landfill and compost waste scenarios in life cycle assessment (LCA) using SimaPro, yielding the environmental impact results shown in Figure 20.6. The majority of impacts were found to occur during agricultural and food production stages, indicating that source reduction should always be a priority when pursuing environmental impact reduction. However, the analysis focused on the operating impact of a landfill versus compost facility. To model such a scenario in SimaPro, a dummy "food waste" product was created, weighing 100 tons, but having no associated processes in manufacturing it. This dummy product was used in the SimaPro disposal scenarios: MSW Deposition, Landfill with Gas Utilization, and Treatment of Biowaste (Composting).

Using the IMPACT 2002+ database, the two scenarios yielded the normalized damage assessment scores shown in Figure 20.7. The network diagrams of the two scenarios are shown in Figures 20.8 and 20.9. Composting is clearly less environmentally impactful on the whole, with lower impact on Ecosystem Quality, Climate Change, and Resources. The higher impact on Human Health is mostly due to the particulate matter impacts of open windrow composting. The constant grinding and turning of material results in increased emissions of particulates into the air, compared with landfilling processes. It is important to note that the climate impact comparison validates the findings of the EPA WARM model.

20.2.5 Economic Sustainability

Operation and maintenance costs: Tables 20.3 and 20.4 show the costs of operation and maintenance (O&M) for the composting of food scrap and the landfilling with LFG recovery, respectively. Only the operation and maintenance cost of each process is considered. The results show higher operating and maintenance costs for the landfilling option compared to the composting option.

Cost of transportation: One of the main costs in both processes is the fuel cost of transporting the food scrap to waste management facilities. The Danville and Clinton Landfills are located far from the center of the Champaign County. The average energy expended in transportation from both landfills is 3393 ton-miles/wk. However, the energy for transportation is only 200.4 ton-miles/wk for the transportation to the composting facility, which is in the center of Champaign County. The assumed average fuel consumption of the collection trucks in the region of study is 3 miles/gal (Chandler et al. 2001) and the assumed price of diesel fuel is $3.50 per gallon, resulting in respective costs of $3958.50 for 100 tons food scraps transportation cost for landfilling option and $233.80 for 100 tons food scraps transportation cost for composting option. This significant difference in fuel consumption costs between two options is due to the long distances to the landfills.

Table 20.2 EPA'S WARM results for waste management (a) sending 100 tons food scrap to Danville landfill versus diverting the food scrap to composting facility (b) sending 100 tons food scraps to Clinton landfill versus diverting the food scrap to composting facility.

(a) Danville landfill scenario

GHG emissions from baseline waste management scenario (MTCO$_2$e)	84
GHG emissions from alternative waste management scenario (MTCO$_2$e)	−13
Total change in GHG emissions (MTCO$_2$e)	−97

	Baseline scenario							Alternative scenario					
Material	Tons recycled	Tons landfilled	Tons combusted	Tons composted	Total (MTCO$_2$e)	Tons source reduced	Tons recycled	Tons landfilled	Tons combusted	Tons composted	Total (MTCO$_2$e)	Change (Alt-Base) (MTCO$_2$e)
Food waste (non-meat)	N/A	100	0	0	84	0	N/A	0	0	100	−13	−97

(b) Clinton landfill scenario

GHG emissions from baseline waste management scenario (MTCO$_2$e)	207
GHG emissions from alternative waste management scenario (MTCO$_2$e)	−13
Total change in GHG emissions (MTCO$_2$e)	−220

| Food Waste (non-meat) | N/A | 100 | 0 | 0 | 207 | 0 | N/A | 0 | 0 | 100 | −13 | −220 |

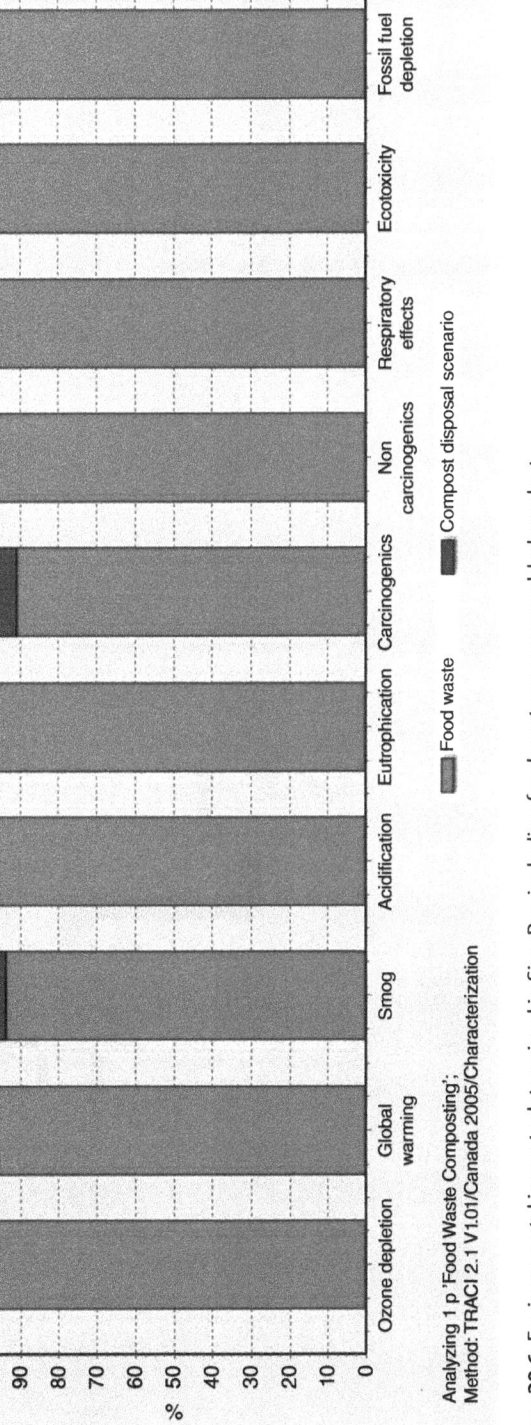

Figure 20.6 Environmental impacts determined in SimaPro, including food waste as an assembled product.

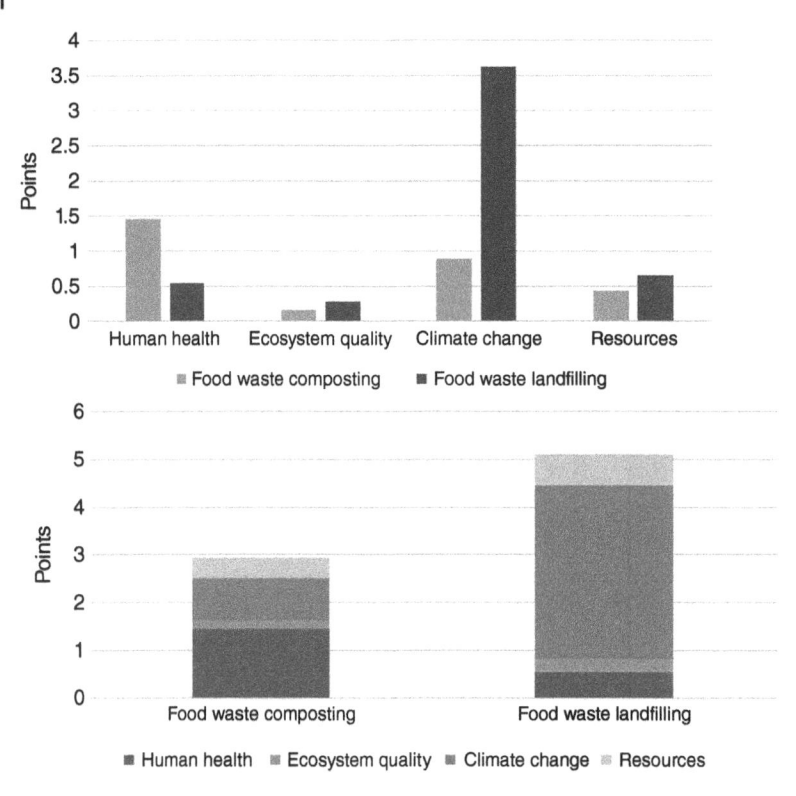

Figure 20.7 Environmental impacts determined in SimaPro: composting–landfilling comparison.

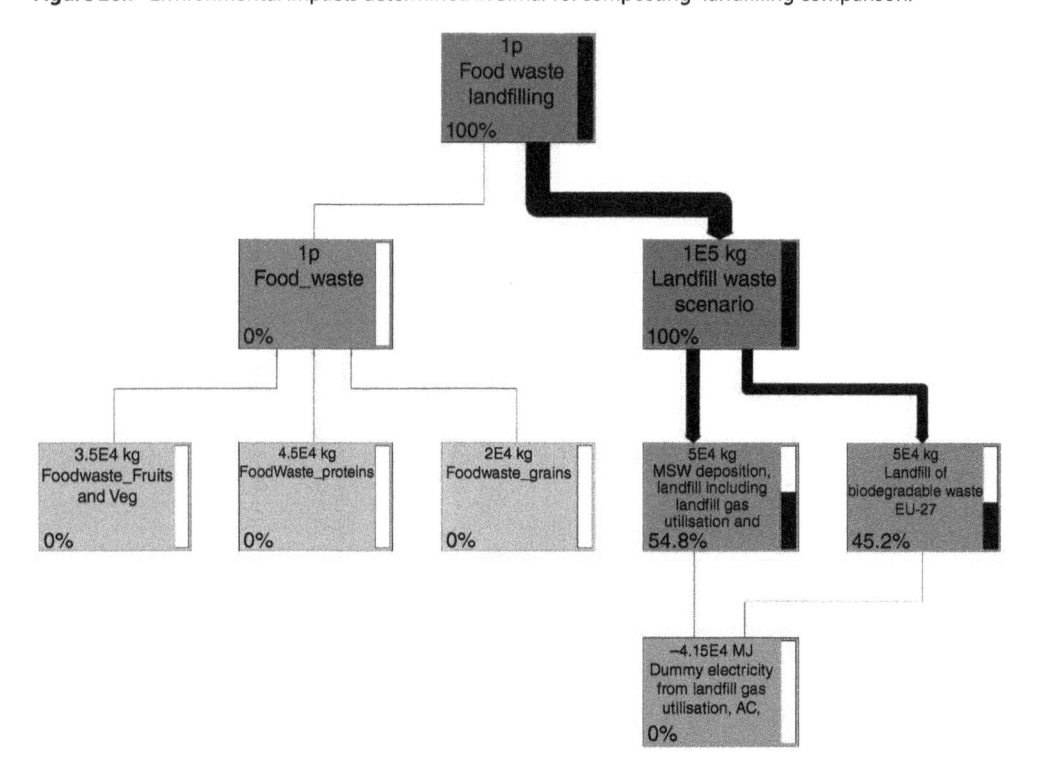

Figure 20.8 Network diagram of the environmental impacts of landfilling.

20.2 Food Scrap Landfilling Versus Composting | 363

Figure 20.9 Network diagram of the environmental impacts of composting.

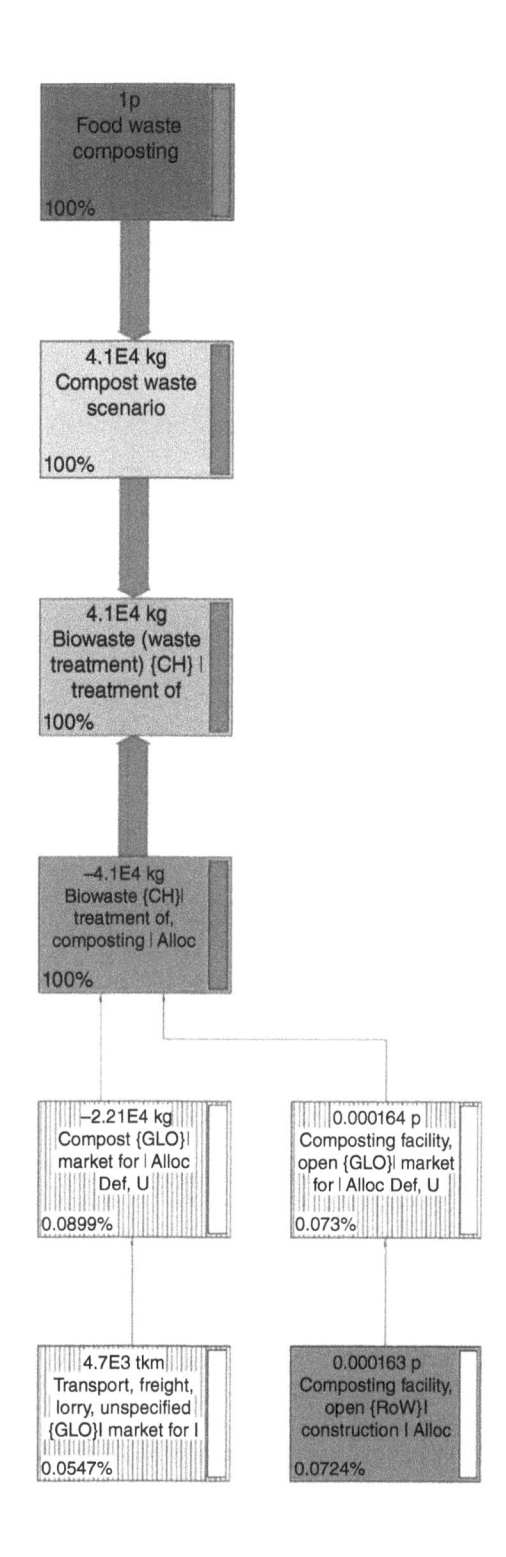

```
                            1p
                        Food waste
                        composting
                          100%
                            │
                            ▼
                        4.1E4 kg
                      Compost waste
                         scenario
                          100%
                            │
                            ▼
                        4.1E4 kg
                     Biowaste (waste
                     treatment) {CH} |
                      treatment of
                          100%
                            ▲
                        −4.1E4 kg
                      Biowaste {CH}|
                      treatment of,
                     composting | Alloc
                          100%
                         ╱     ╲
          ┌─────────────┘       └─────────────┐
     −2.21E4 kg                          0.000164 p
    Compost {GLO}|                    Composting facility,
    market for | Alloc                open {GLO}| market
       Def, U                          for | Alloc Def, U
      0.0899%                              0.073%
         ▲                                    ▲
         │                                    │
      4.7E3 tkm                          0.000163 p
   Transport, freight,                Composting facility,
   lorry, unspecified                    open {RoW}|
    {GLO}| market for |               construction | Alloc
      0.0547%                              0.0724%
```

Table 20.3 Operation and maintenance costs (O&M) of food scrap for the windrow composting method.

O&M breakdown	Cost (per ton)
Labor, data collection, analysis and reporting	$6.5
Rental fee of a tub grinder	$2.5
Turning rows two times per week	$4.5
Temperature monitoring and testing	$2.2
Screening compost	$3.8
Watering	$2.5
Windrowing	$3.5
Chipping	$1.5
Total	$27

Table 20.4 Operation and maintenance costs (O and M) of food scrap for the landfill with LFG recovery.

O and M breakdown	Cost (per ton)
Waste placement	$5.20
Electricity use for liquid injection and recirculation system	$0.74
Electricity use for gas collection and removal (anaerobic phase)	$0.1
Electricity use for air injection and collection (aerobic phase)	$5.10
Maintenance and replacement of flow meters	$1.49
Maintenance of liner and leachate pump	$0.99
Leachate, gas and solids sampling, monitoring, and testing	$12.38
Labor, data collection, analysis, and reporting	$23.81
Total	$49.81

The total cost of these processes can be the combination of operations and maintenance costs and fuel costs in each waste management options. The total costs of landfilling process and composting process are $8939.50 and $2933.80 per week for 100 tons food scraps, respectively. The total cost of landfilling is almost three times higher than the total cost of composting.

Indirect costs for landfilling versus composting: The monetary value of each pollutant is based either on the estimated real financial costs to society, in terms of environmental degradation and human health, or on the actual market value of the pollutant's emissions established through trading schemes such as the USEPA's sulfur dioxide emission permits under the Clean Air Act provisions for controlling acid rain. Table 20.5 lists the value of environmental impacts for each category emissions reductions per ton.

According to this table, a value of $36 per ton of carbon dioxide equivalent (CO_2e) is used based on GHG offset valuation used by Seattle City Light. The value of a ton of carbon dioxide varies in the North American (unregulated) marketplace, from $1 to $4 per ton with values exceeding $100 per ton in jurisdictions where carbon trading is regulated. The Stern Review (Nordhaus 2007) on the economics of climate change estimates the environmental cost of a metric ton of CO_2e emissions at $85. Therefore, a value of $36 can be considered as conservative.

Table 20.5 Value of environmental impact category emissions reductions per ton.

Climate change	Human health – particulates	Human health – toxics	Human health – carcinogens	Eutrophication	Acidification	Ecosystems toxicity
eCO_2	$ePM2.5$	eToluene	eBenzene	eN	eSO_2	E2,4-D
$36	$10 000	$118	$3030	$4	$661	$3280

Table 20.6 Economic benefits of composting and landfilling.

Composting (fertilizer)	
Total fertilizer value per ton of compost	$102.81
Price per ton delivered	$60
Revenue per ton	**$42.81**
Landfilling (electricity)	
Electricity generation	$0.03 per kW/h
Revenue per ton of food scraps	**$3.80**

Using the net change of GHG emission from EPA's WARM to monetize the indirect costs, it is seen that if the food scrap from Danville Landfill (with LFG recovery for energy) and Clinton Landfill (no LFG recovery for energy) are diverted to a compost facility, it can save $3852 per week and $8730 per week, respectively (Table 20.6).

20.2.6 Social Sustainability

Social sustainability indicators relevant to this project are first weighted to three classes (high, medium, and low) according to their importance. Then, a score between 1 and 3 is assigned based on their impact, where a higher score represents high positive impact. Results are shown in Table 20.7. After analyzing the social sustainability indicators and results in Table 20.7, it is found that the grand total for composting is 86 and for landfilling is 48, which clearly indicates that composting has a higher positive social impact.

20.2.7 ENVISION™

As explained in Chapter 17, Envision™ is a tool to rate sustainable infrastructure systems. It was developed by the collaboration of Zofnass Program for Sustainable Infrastructure at the Harvard University Graduate School of Design and the Institute for Sustainable Infrastructure. Envision provides a complete outline for evaluating and rating the community, environmental, and economic benefits of all types and sizes of infrastructure projects. It quantifies and gives recognition to infrastructure projects that use transformational, collaborative approaches to assess the sustainability indicators over the course of the project's life cycle. Furthermore, Envision has five major categories, which includes quality of life, leadership, resource allocation, natural world, and climate and risk. Different aspects of sustainability are assessed in these categories. The Envision checklist (Chapter 17) can be used to compare between two alternatives from sustainability point of view or can act as a preliminary assessment for applying Envision Sustainable Infrastructure Rating System (Figure 20.10). In this study, the Envision checklist is

Table 20.7 Social sustainability indicators.

Indicator	Compost	Landfill
Social sustainability indicators with high weight		
Achieve food security	3	1
Promote economic growth	2	1
Create jobs within planetary boundaries	3	2
Improve agriculture system	3	1
Access to affordable waste management	3	2
Total	14	7
Weight (total × 3)	42	21
Social sustainability indicators with medium weight		
Increase social equity	3	1
Achieve health and wellbeing	2	1
Improve air quality	2	1
Improve water quality	2	1
Curb human induced climate change	2	1
Biodiversity	2	1
Improve quality of life	2	1
Land use	2	1
Total	17	8
Weight (total × 2)	34	16
Social sustainability indicators with low weight		
End hunger	1	1
Youth transition into the labor market	2	1
Prevent and eliminate violence	1	1
Urban green space per capita	1	1
Ensure sustainable energy	1	1
Local noise effects	2	1
Local odor effects	1	3
Education	1	2
Total	10	11
Weight	10	11
Social sustainability	86	48

used to verify if the above triple bottom line assessment conclusions are valid or not regarding sustainability of composting versus landfilling (ASCE 2018).

As stated earlier, the goal of this project is to divert the food wastes from landfill to composting facility. The initial target is to capture 5548 tons of food waste per year and then divert those food wastes to composting facility. The impact of achieving that target is evaluated by using the Envision Sustainable Infrastructure Rating System. The results are shown in Figure 20.11.

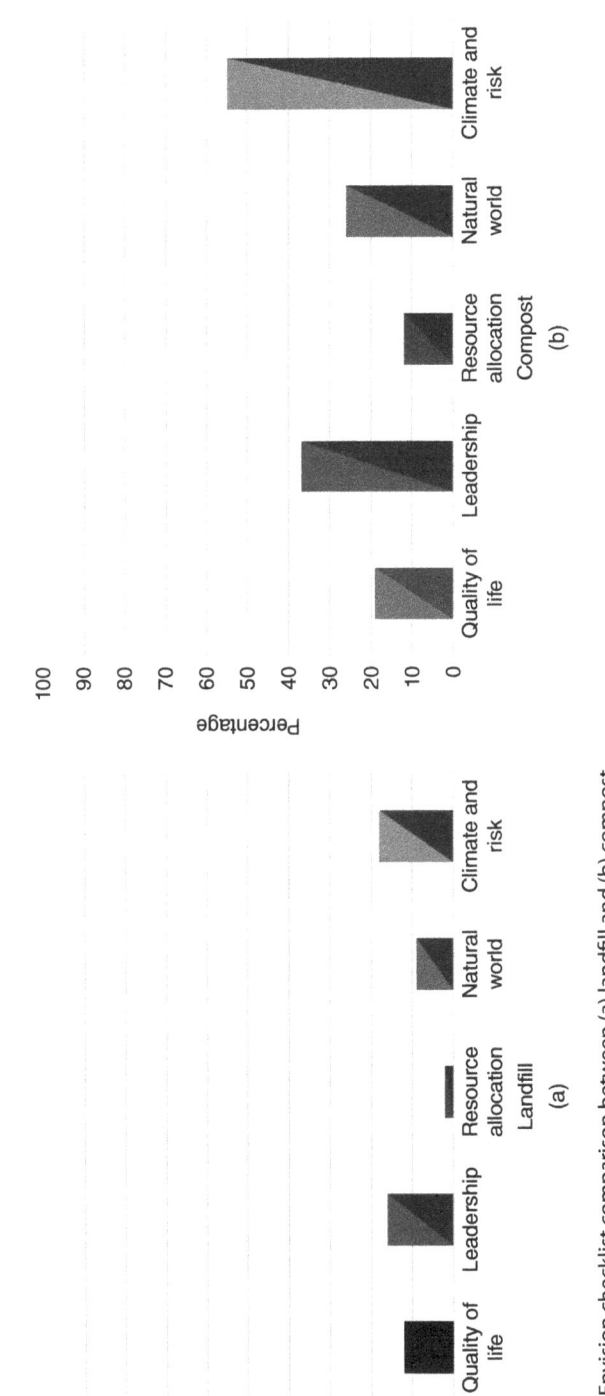

Figure 20.10 Envision checklist comparison between (a) landfill and (b) compost.

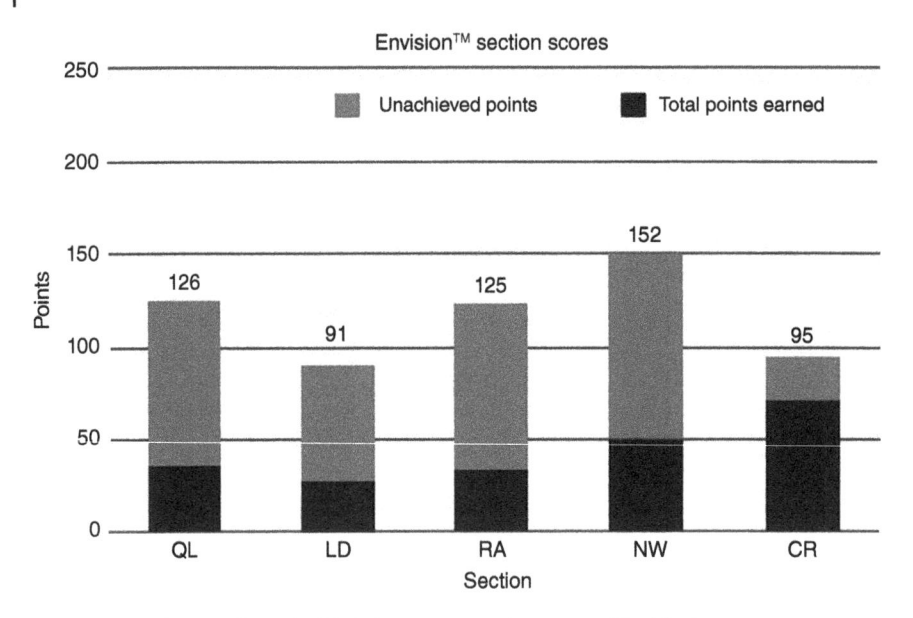

Figure 20.11 Scores of sustainable infrastructure rating system (Envision).

20.2.8 Conclusions

This study compared the sustainability performance of two food scrap management scenarios – Landfilling and Composting – using Champaign County as the region under study. In this region, a goal of diverting 20% of food scrap from landfill to the existing windrow composting facility, the Urbana LRC, was set. The functional unit – 100 tons/wk of food scrap – is diverted from both the Danville Landfill (equipped with LFG-to-energy conversion) and the Clinton Landfill (no LFG-to-energy). The analysis was performed based on LCA framework. Using real data from Illinois and Champaign County, along with standard measurement tools for sustainability assessment, it was concluded that composting is a more sustainable option compared to landfilling on all three aspects of the triple bottom line.

20.3 Adsorbent for the Removal of Arsenic from Groundwater

The presence of toxic pollutants in water is of major concern due to their harmful effects on human health and the environment. Arsenic is a toxic metalloid element, and its presence in water is associated with improper waste disposal practices in the past and accidental spills as well as natural geologic environment. In Tolono, IL, a number of wells are contaminated with arsenic at concentrations above the health-based screening levels. One way to remove arsenic from drinking water is the use of filtration technique using a selective adsorbent. In this study, two adsorbent materials for the retention of arsenic are analyzed for their sustainability using a triple bottom line assessment.

20.3.1 Background

In 2011, Illinois State Water Survey (ISWS) Public Service Laboratory analyzed a well water sample obtained from a home near Tolono. The total arsenic in the sample was 44 µg/l, which is nearly four times that of the maximum concentration limit (MCL) of 10 µg/l for drinking

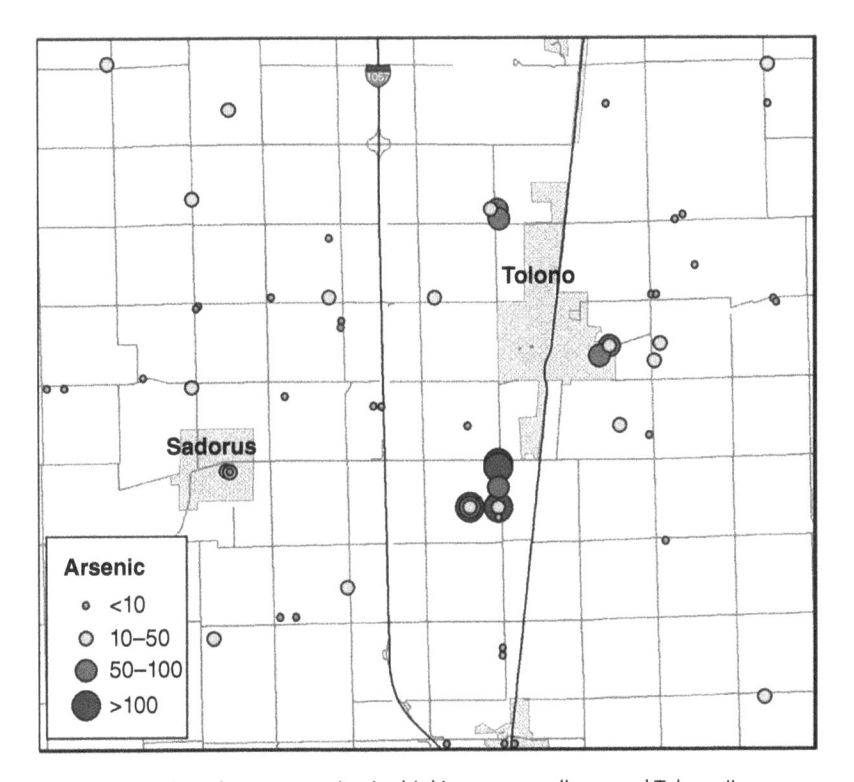

Figure 20.12 Arsenic concentration in drinking water wells around Tolono, IL.

water. ISWS conducted a study in the Tolono region to evaluate arsenic contamination in groundwater, which was supplemented with historical data from the ISWS Groundwater Quality Database (GWQDB).

Aquifers in Tolono area are present within glacial sand and gravel deposits interlayered with glacial tills. Arsenic concentrations in groundwater have been observed to be related to hydrogeological and geochemical conditions, but especially to redox conditions. Both iron and arsenic have been found to be associated in the solid phase. It has been largely hypothesized that dissolution of iron oxyhydroxides and oxidation of organic matter in sediments cause the release of arsenic into solution.

The study identified 24 sites with arsenic concentrations exceeding the MCL value. Ten sites exceeded the earlier arsenic MCL of 50 μg/l, including four sites with arsenic concentrations greater than 100 μg/l (Figure 20.12). The study observed considerable spatial heterogeneity in arsenic concentrations, but, in general, the wells with high arsenic concentrations extended to depths of 165–180 ft below the ground surface, though not all wells drilled to the deeper sand layer exhibited excessively high arsenic levels (Figure 20.13).

20.3.2 Methodology

20.3.2.1 Goal and Scope

The goal of this study is to identify the most sustainable adsorbent for arsenic removal from the groundwater based on a triple bottom line sustainability assessment. A process unit is to be designed and installed in Tolono to reduce arsenic concentrations in groundwater using a pump-and-treat method. Two different adsorbents that are considered to be effective for adsorption of arsenic are selected: activated alumina (AA), and bayoxide E33. The study

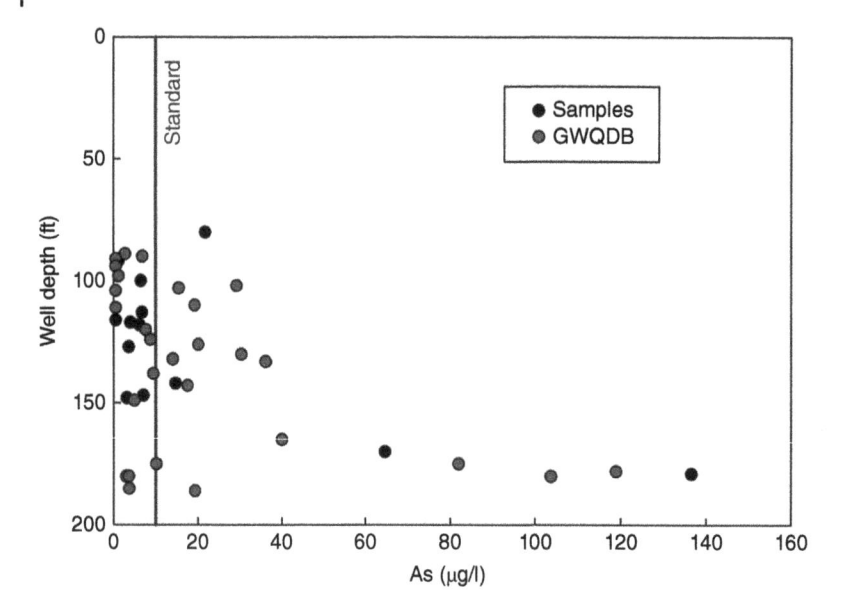

Figure 20.13 Arsenic concentrations as a function of well depth.

Table 20.8 Well water characteristics.

pH	7.4	As(III)	131 µg/l
SpC	696 µS/cm	Fe	1.14 mg/l
DO	0.28 mg/l	Mn	22.3 µg/l
DOC	7.36 mg/l	F$^-$	0.73 mg/l
Color	9 PCU	SiO$_2$	11.8 mg/l
Turbidity	4.1 NTU	TDS	394 mg/l
Hardness	190 mg/l	Alkalinity	374 mg/l

compares these two specific adsorbents based on their useful life, operation costs, recycle/disposal, direct environmental impacts, and related social impacts. The functional unit to compare both absorbents is the volume of water to be treated in one year of operation, which is estimated to be 157 680 kgal.

20.3.2.2 Site Location
The site of interest in this study is the village of Tolono, about 10 miles south of Champaign, IL. Based on the ISWS report (Figures 20.12 and 20.13), one of the wells showing high arsenic concentrations was selected for this study. The chemical properties of groundwater from this well are listed in Table 20.8.

20.3.2.3 Technical Design
The selection of the adsorptive media and pretreatment methods were based on a number of factors that affect the system performance, including arsenic concentration and speciation, pH, and the presence of competing anions, as well as media-specific characteristics such as costs, operating life, and empty-bed contact time requirements.

Properties	Activated alumina
Particle form	Spheres
Color	White
Particle size (mm)	0.2 ± 0.02
Surface area (m²/g)	365 ± 5
Pore volume (cm³/g)	0.42
Bulk density (g/cm³)	0.800

Figure 20.14 Properties of activated alumina adsorbent media.

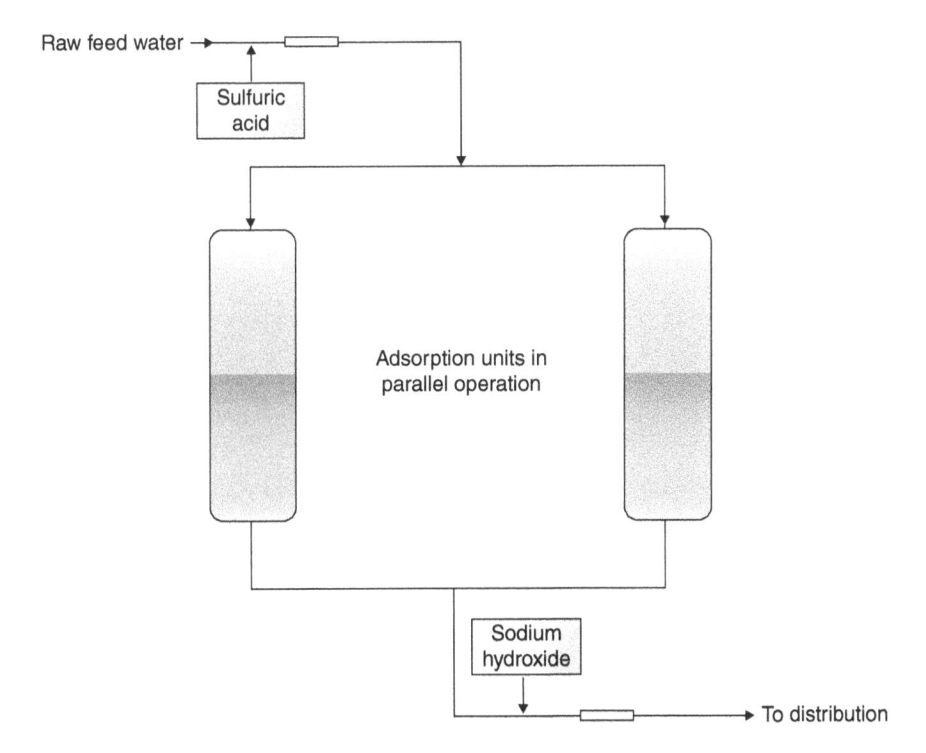

Figure 20.15 Process flow schematic for arsenic adsorption with activated alumina.

Activated Alumina (AA) is partially hydroxylated aluminum oxide (Figure 20.14) with a surface area in the range of 200–350 m²/g. This is a versatile adsorbent that has been used for the removal of As(V), PO_4^{3-}, and F^- from water. Alumina is particularly effective in the removal of As(V) in the form of $H_2AsO_4^-$. At the optimum pH for arsenic removal, fluoride, selenium, some organic molecules, and some trace heavy metal ions are adsorbed. Since these ions compete for the same adsorptive sites with arsenic, their presence might deplete the AA capacity for arsenic. Figure 20.15 depicts the schematic process flow for the treatment of water. The pH of the raw feed water is adjusted to 5.5 to effectively remove As(III) and As(V). Two tanks filled with AA in down-flow mode retain the arsenic that is sent to a storage tank before consumption. Periodically, the AA beds are regenerated backwashing with a 5% NaOH solution. Then, the treatment bed is neutralized with sulfuric acid to pH 5.5.

Table 20.9 Physical and chemical properties of bayoxide E33 adsorbent media.

Physical properties	
Parameter	Value
Physical form and appearance	Amber, dry granular solid
Matrix	Iron oxide composite
Bulk density (lb/ft^2)	28.1
BET area (m^2/g)	142
Attrition (%)	0.3
Moisture content (%)	<15 wt%
Base polymer	Macroporous polystyrene
Particle size distribution (U.S. Standard Mesh)	10 × 35
Crystal size (Å)	70
Crystal phase	α-FeOOH
Chemical analysis	
Constituents	Concentration (wt%)
FeOOH	90.1
CaO	0.27
MgO	1.00
MnO	0.11
SO$_3$	0.13
Na$_2$O	0.12
TiO$_2$	0.11
SiO$_2$	0.06
Al$_2$O$_3$	0.05
P$_2$O$_5$	0.02
Cl	0.01

BET, Brunauer–Emmett–Teller.

Bayoxide E33 (E33) media is an adsorbent in which the main component is goethite, iron oxide hydrate (Table 20.9). The E33 media was designed with a high adsorption capacity for both As(III) and As(V). The treatment process involves a prechlorination step to oxidize arsenic and iron to favor the adsorption efficiency of arsenic. The E33 bed requires routine backwash to remove sludge. At the end of its useful life, estimated to be four months for this application, the E33 cannot be regenerated and it is disposed of in a landfill. The spent E33 media passed the Toxicity Characteristic Leaching Procedure (TCLP) test, and it was classified as a nonhazardous waste.

20.3.3 Environmental Sustainability

The environmental sustainability analysis for the adsorption of arsenic comparing the two adsorbent materials was performed based on LCA using SimaPro software. The input data to SimaPro for the two materials and operation stage is presented in Table 20.10. The Tool for the Reduction and Assessment of Chemical and other environmental Impacts (TRACI) was used in this study. TRACI is a midpoint-oriented life cycle impact assessment (LCIA) methodology

Table 20.10 SimaPro input for arsenic adsorption with activated alumina and bayoxide E33.

Component	Source	Distance (km)	Quantity (ton)	t-km
Activated alumina (AA)				
Sulfuric acid	Willow Springs IL	229	200.1	45 823
Sodium hydroxide	Willow Springs IL	229	130.62	29 912
Activated alumina	Park Ridge IL	266	8.44	2 245
Bayoxide E33 (E33)				
AdEdge bayoxide E33 media	Buford GA	966	16.22	15 669
Sodium hypochlorite	Willow Springs IL	229	0.84	192

developed by the USEPA specifically for the life cycle inventory (LCI). The TRACI tool is a collection of LCIA models that provide relative comparison of the potential harm to human health and the environment. TRACI includes impact indicator models for the following 10 impact categories: acidification, ecotoxicity, eutrophication, global warming, human cancer, human noncancer, human health criteria, ozone depletion, smog formation, and fossil fuel use.

In the TRACI impact assessment method, normalization is used to check for inconsistencies of LCI and LCIA results, to provide information on the relative significance of the indicator results, and prepare for additional procedures, such as grouping, weighting, or life cycle interpretation. Normalization is defined as "calculating the magnitude of indicator results relative to reference information." Normalization of LCI and LCIA results adjust the characterization results to a common unit. For example, TRACI characterizes the potential ecological harm of a unit quantity of substance released into the environment in ecotoxicity potentials (ETPs) measured in 2,4-dichlorophenoxyacetic acid (2,4-D) equivalents for 161 substances released to air and water.

The results of the environmental analysis using the TRACI method (Figure 20.16) show that the impacts of the water treatment process using AA media are largely affected by the use of sodium hydroxide and sulfuric acid. Production and transport of both chemicals contribute significantly to the prevalence/emission of carcinogenic compounds in the atmosphere and increased ecotoxicity. Fuel used for transport of the materials and chemicals and electricity consumption contribute toward global warming, ozone depletion, fossil fuel depletion, and respiratory and smog effects.

For the water treatment process using E33 adsorption media, lower quantities of chemical usage significantly reduce the adverse impacts on the environment. Fuel use in transport of materials, particularly in the procurement of the E33 media, is a major stressor. Despite the longer life span and regenerability, higher negative impacts are resulted due to required large quantities of AA media, thus making AA media less environmentally sustainable as compared to E33.

20.3.4 Economic Sustainability

The capital cost and operation and maintenance cost are shown in Table 20.11 for the two alternative adsorbent materials. The capital cost includes the cost for equipment, site engineering, and installation. The operation and maintenance cost includes the cost for media replacement and disposal, electrical power consumption, and labor. The unit capital cost is calculated assuming that the system is operated 24 h/d, 7 days a week, at the design flowrate of 300 gpm to

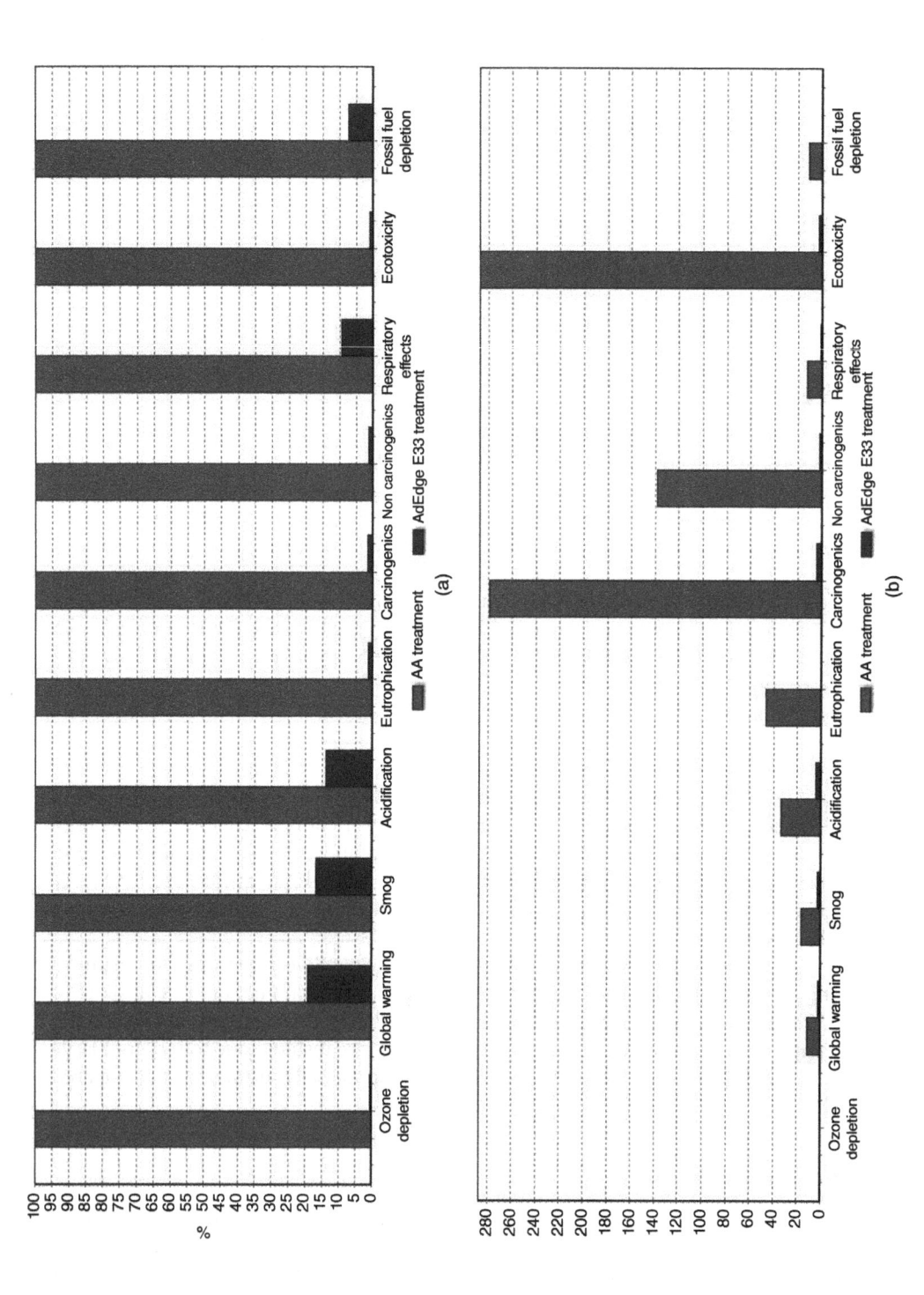

Figure 20.16 Comparison of the (a) environmental impacts of activated alumina and bayoxide E33 and (b) normalized impacts (TRACI 2008 method).

Table 20.11 Capital cost and operation and maintenance cost for water treatment with activated alumina and bayoxide E33.

Description	Activated alumina	Bayoxide E33
Equipment and materials	$214 990	$159 670
Engineering (labor and materials)	$58 015	$21 500
Installation (labor and materials)	$63 105	$19 950
Total capital investment	$336 110	$201 120
Unit capital cost per 1000 gal	$2.13	$1.28
Total O&M Cost	$268 208	$170 735
Total O&M Cost per 1000 gal	$1.70	$1.08

produce 157 680 kgal/yr. As reported in Table 20.11, the capital cost and the operation and maintenance costs are significantly higher for the AA adsorbent compared to the E33.

20.3.5 Social Sustainability

The perception of social impacts of any project may vary across the different strata of the society, thus, a standardized quantitative approach may not be feasible. In this study, social sustainability was determined with a semiquantitative method as shown in Table 20.12 (Reddy et al. 2014). This method considers multiple social impact indicators categorized into four dimensions namely social, socioinstitutional, socioeconomic, and socioenvironmental. Table 20.12 shows the matrix with the specific indicators that are related to this water treatment project. The rating system consists of scores ranging from -2 (unacceptable negative impact) to $+2$ (ideal, positive impact). Based on the scores awarded for each social impact indicator (Table 20.13), the treatment using bayoxide E33 media was found to be more socially sustainable (Figure 20.17).

20.3.6 Streamline Life Cycle Assessment (SLCA)

In addition to the above triple bottom line assessment, a streamlined life cycle assessment (SLCA) was performed to evaluate health and safety issues of the water treatment with the two adsorbents. The SLCA matrix addresses the health and safety issues in terms of choice of materials, energy usage, chemical residue, and operator safety. The life cycle stages included are chemical transport, chemical storage, treatment process, backwash, regeneration (only for AA media), and waste disposal. Each element of the matrix received a rating ranging from 0 to 4 along with an assigned color shade (Figure 20.18). A higher rating implied higher the safety of the design. The transport, storage, and use of large quantities of sodium hydroxide and sulfuric acid make the treatment with AA media a significant threat to human health and safety. The transport of the E33 media over 400 miles to the project location, small particle size of the media and possible respiratory effect, and the release of chlorine gas from sodium hypochlorite are some of the concerns in the treatment using E33 media. Comparing the matrices for the two treatment alternatives, it was found that the E33 treatment media has a higher rating of 42.08% against 36.46% for the AA media.

Table 20.12 Social sustainability evaluation matrix for the treatment of water with two adsorbents: activated alumina and bayoxide E33 (Reddy et al. 2014).

Social Sustainability Evaluation Matrix (SSEM), V1.0

Krishna R. Reddy (kreddy@uic.edu), University of Illinois at Chicago

Option		Specify for particular action					
		Score					Total score
		Positive impact		No impact or not applicable	Negative impact		
		Ideal	Improved		Diminished	Unacceptable	
Dimension	Key theme area	2	1	0	−1	−2	
Social	Overall public health and happiness						
	Justice and equality						
	Care for the elderly						
	Care for those with special needs						
	Degree to which postremediation project will result in learning opportunities and skills development for community						
	Enhancement of community/civic pride resulting remediation and postremediation project						
	Degree to which tangible community needs are incorporated into remediation design						
	Transformation of perceptions of project and environs within greater community						
Socioinstitutional	Enhancement to the architecture/aesthetics of built environment						
	Enhancement and participation of government institutions (i.e. new facilities) in community						
Socioeconomic	Employment opportunities during construction/remediation						
	Employment opportunities postconstruction/remediation						

Socioenvironmental	Remediation of naturally-occurring contaminants (i.e. naturally occurring asbestos, radon)	
	Degree of protection afforded to remediation workers by proposed remediation	
	Degree of disruption (noise, truck traffic) from proposed remedial method to the surrounding neighborhoods	
	Degree of contaminant removal/destruction vs. in-place capping or immobilization	
	Restoration or impact to productive surface water or groundwater use	
	Degree proposed remediation will affect other media (i.e. emissions/air pollution resulting from soil or groundwater remediation)	
	Grand Score	

Table 20.13 Social sustainability evaluation matrix for the use of activated alumina and bayoxide E33 in the treatment of water.

	Social sustainability matrix		
	No remedy	Activated alumina	AdEdge E33 bayoxide
Social	−5	8	9
Socioinstitutional	−1	0	0
Socioeconomic	0	2	4
Socioenvironmental	−4	3	3
Grand score	−10	13	16

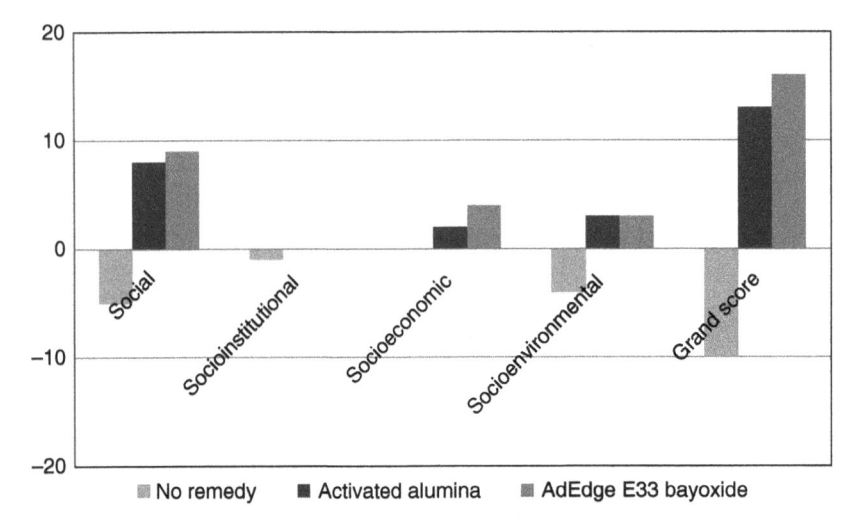

Figure 20.17 Comparison of the social impacts of the adsorbents (activated alumina and bayoxide E33) in the water treatment.

20.3.7 Envision

The results from the Envision assessment for this project are shown in Figure 20.19 for AA and Figure 20.20 for E33. Water treatment using E33 media scores higher on the Quality of Life, Leadership, Natural World, and Climate and Risk categories. Treatment using AA media has a higher score than E33 media in the Resource Allocation category. E33 media is more specifically designed for the removal of arsenic, while in AA media arsenic removal occurs alongside competing fluoride and silicate compositions. While the selectivity for arsenic is improved by lowering the pH to 5.5 in AA treatment, using an adsorbent engineered for a specific purpose at competing price has a favorable impact in terms of reassurance to the investors and the society. Since E33 is transported over 400 miles to the treatment project site, the environmental costs

20.3 Adsorbent for the Removal of Arsenic from Groundwater

	Materials choice	Energy use	Solid residue	Liquid residue	Gaseous residue	Worker safety	Row scores
Chemical transport	2.0	2.0	0.5	0.5	0.5	2.0	7.5
Chemical Storage	2.0	2.0	0.5	0.5	0.0	0.5	5.5
Process	3.0	2.5	0.5	3.0	0.0	3.0	12.0
Backwash	3.0	2.0	1.5	3.0	0.0	2.5	12.0
Regeneration	3.0	2.0	0.0	3.0	0.0	2.0	10.0
Waste disposal	1.5	0.5	1.0	0.5	0.0	2.0	5.5
Column scores	14.5	11.0	4.0	10.5	0.5	12.0	52.5/144 36.46%

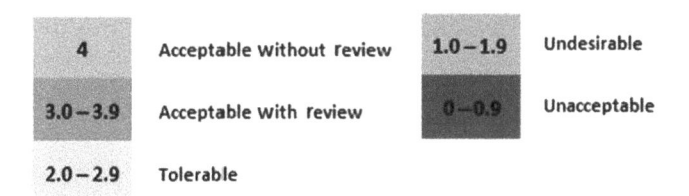

4	Acceptable without review		1.0–1.9	Undesirable
3.0–3.9	Acceptable with review		0–0.9	Unacceptable
2.0–2.9	Tolerable			

(a)

	Materials choice	Energy use	Solid residue	Liquid residue	Gaseous residue	Worker safety	Row scores
Chemical transport	2.0	3.0	0.5	0.5	0.5	2.0	8.5
Chemical Storage	2.0	2.0	0.5	0.5	0.5	2.0	7.5
Process	3.0	2.5	0.5	3.0	3.0	3.0	15.0
Backwash	3.0	2.0	1.5	3.0	0.0	3.0	12.5
Waste disposal	1.5	1.0	1.0	0.5	0.0	3.0	7.0
Column scores	11.5	10.5	4.0	7.5	4.0	13.0	52.5/120 42.08%

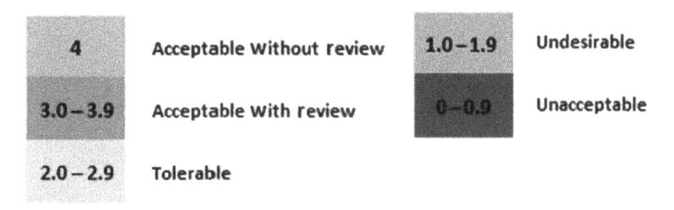

4	Acceptable without review		1.0–1.9	Undesirable
3.0–3.9	Acceptable with review		0–0.9	Unacceptable
2.0–2.9	Tolerable			

(b)

Figure 20.18 Streamlined life cycle analysis for (a) activated alumina and (b) bayoxide E33.

in terms of greater fuel usage and the need to discard the adsorbent after its useful life with no regeneration option affect its scores in the Resource Allocation category.

20.3.8 Conclusions

The results showed that AA caused a higher environmental (negative) impact, for each category under TRACI 2008 LCIA method, and it was attributed to the use of large quantities of hazardous chemicals (sulfuric acid and sodium hydroxide). The most significant impacts of bayoxide E33 were related to the transportation of the absorbent over 400 miles to the project site and the need for the disposal of the spent media. Social and economic analysis also resulted in bayoxide E33 media to be more sustainable. Further, an additional sustainability assessment was carried out using SLCA and Envision, and these results also show that bayoxide E33 media to be the more sustainable adsorbent media.

Credit category	Applicable points	Earned points	Innovation points	Total points pursued	Percentage of applicable points(%)
Quality of life	57	17	5	22	39
Leadership	88	35	4	39	44
Resource allocation	124	71	8	79	64
Natural world	38	17	4	21	55
Climate and risk	40	10	1	11	28
Total project points	347	150	22	172	50

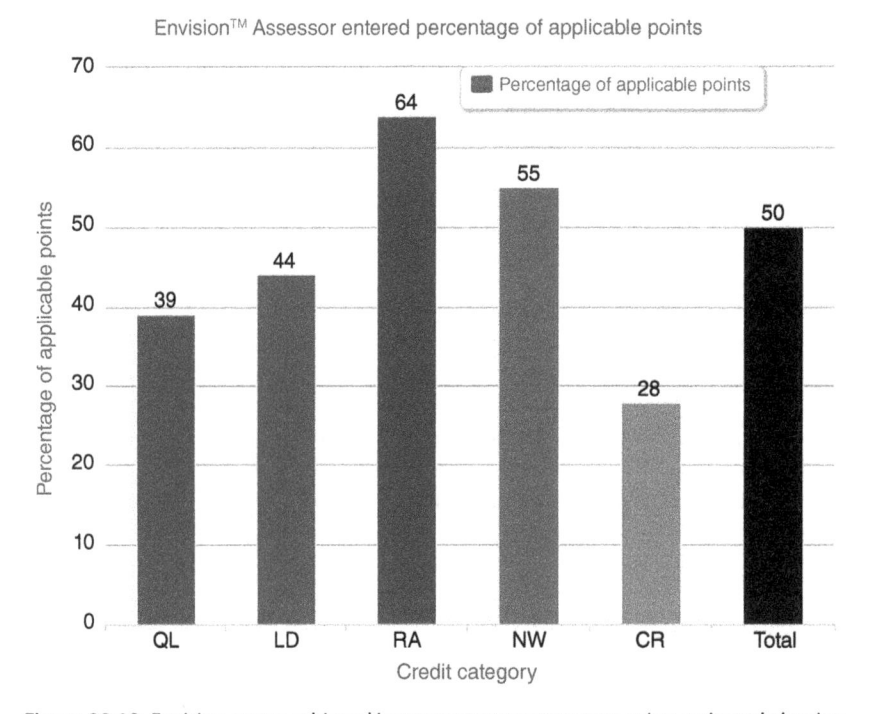

Figure 20.19 Envision scores achieved in water treatment process using activated alumina.

Credit category	Applicable points	Earned points	Innovation points	Total points pursued	Percentage of applicable points(%)
Quality of life	57	49	7	56	98
Leadership	88	38	5	43	49
Resource allocation	124	54	8	62	50
Natural world	38	23	7	30	79
Climate and risk	40	10	3	13	33
Total project points	347	174	30	204	59

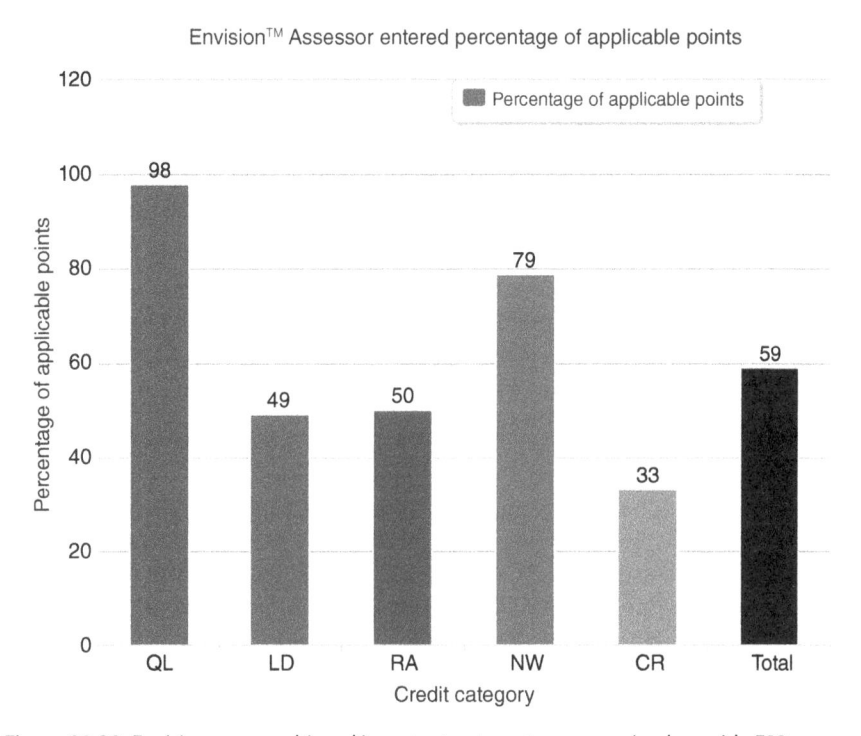

Figure 20.20 Envision scores achieved in water treatment process using bayoxide E33.

20.4 Conventional Versus Biocover Landfill Cover System

Landfill gas (LFG) emissions from MSW landfills are considered as one of the major sources of anthropogenic greenhouse gasses (CH_4 and CO_2) leading to global warming. The final soil cover constructed on the landfill acts as a physical barrier to prevent the infiltration of water from precipitation into the waste mass, thus reducing production of leachate in the landfills. Moreover, the final soil cover also prevents the escape of LFG, generated as a result of biodegradation of organic matter in the MSW. Interestingly, cover soil that contains microbes (especially methanotrophs) utilize methane and oxidize it to carbon dioxide, reducing methane emissions (which has higher GWP) into the atmosphere. However, the performance and efficiency of the cover retaining and reducing gas emissions largely depend on the components of the cover system and its design. Moreover, the microbial oxidation of methane in the final cover is limited due to inadequate microbial population in the soil cover resulting from unsuitable environmental conditions for the microbes to grow and survive.

In this study, an alternate landfill cover consisting of biochar and steel slag amendments to cover soil is proposed to reduce LFG emissions. The study is conducted to analyze the sustainability from a triple bottom line standpoint of the proposed alternate biocover (ALC) in comparison to the conventional RCRA Subtitle D conventional cover (CLC). The technical design of both the systems was carried out based on the hydraulic assessment using the Hydrologic Evaluation of Landfill Performance (HELP) software program. The environmental sustainability assessment was performed using LCA methodology. The economic sustainability assessment was performed by comparing the cost of materials and transportation involved in the construction of each type of cover. Social impacts were assessed with the help of Social Sustainability Evaluation Matrix (SSEM) developed by Reddy et al. (2014). A sensitivity analysis was performed to identify the sources of major environmental impacts and to recommend several ways to make the alternate biocover system a sustainable option.

20.4.1 Background

The anaerobic decomposition of MSW generates leachate and biogas, which are the major sources of environmental pollution from landfill if they are not properly managed. LFG contains about 55–65% CH_4 and 45–50% CO_2, along with trace amounts of ammonia, hydrogen sulfide, and other gasses. The GWP as well as the foul smell of some of these emissions (e.g. hydrogen sulfide) are challenges for both a local and global scenario.

RCRA created a framework for management of hazardous and nonhazardous waste in the United States and required certain minimum requirements for the construction of the components of landfill systems – bottom liner, cover system, gas and leachate collection – of the landfill. The cover systems, especially the final cover that is placed over the waste, after the landfill has reached its capacity, are meant to address the issues of LFG emissions from the landfills. However, the technical design requirements for the construction of a final cover is more focused on the reduction of infiltration of precipitation through the cover system and minimize the leachate levels within the landfill. The growing concerns about the global warming and greenhouse gas emission has led to the investigation of alternative cover systems that help mitigate the LFG emissions.

The LFG emissions into the atmosphere can be significantly reduced with landfill covers specially engineered for sequestering both methane and carbon dioxide. The landfill cover soils may contain microorganisms called methanotrophs that can microbially oxidize CH_4 to CO_2, thereby reducing CH_4 emissions into the atmosphere. In a conventional landfill cover, the proliferation of this bacteria community is limited due to the absence of essential nutrients required for its growth. Recent investigation with the use of biochar as an organic amendment to cover soil is showing great promise in alleviating CH_4 emissions (Yargicoglu and Reddy 2017a). Biochar is a product obtained from gasification or pyrolysis of wood and other biomass under anoxic conditions. The higher surface area, stability, and nutrient holding capacity of biochar aid in creating a better environment for the growth and proliferation of the methanotrophs in the landfill cover, thus increasing the methane oxidation capacity of the cover soil.

Carbon dioxide emissions in a landfill can be reduced by direct engineered capture methods using several sorbent materials such as calcium hydroxide. These sorbent materials are either synthesized or come from parent materials such as fly ash and steel slag. From a sustainability point of view, the use of industrial by-products, such as steel slag, is an attractive alternative since it can mitigate greenhouse gas emissions by sequestering CO_2 emissions into stable compounds while providing a reuse option for steel slag waste. There have been several studies (e.g. Sarperi et al. 2014, Pan et al. 2016 and Reddy et al. 2019) that proved the utility of steel slag in CO_2 capture and storage through mineral carbonation methods. The high quantity of CaO and

MgO present in steel slag absorbs carbon dioxide from atmosphere in the presence of moisture to convert it into more stable carbonates.

The carbon capture and storage (CCS) capacity of steel slag along with its high strength properties makes steel slag a desirable material for its use in landfill covers for CO_2 sequestration alongside biochar and soil. This combination in the landfill cover can be used as a promising alternative cover system to effectively mitigate both the CH_4 and CO_2 emissions from landfills.

20.4.2 Methodology

20.4.2.1 Goal and Scope

CLC provides a physical barrier between the waste mass in the landfill and reduce the infiltration of rainwater but does not provide a significant mitigation of the GHG emissions into the atmosphere (CO_2 and CH_4). The alternative landfill cover ALC or biocover combines layers of topsoil, biochar-amended soil, and steel slag to minimize the LFG emissions into the atmosphere. The sustainability of the two covers will be determined by conducting a LCA, economic impact assessment based on the costs of construction, and a social impact assessment.

20.4.2.2 Landfill Location

The landfill site for the purpose of the study was assumed to be the Zion Landfill in Illinois, and the Indiana Harbor East Steel Plant was assumed to be the source of steel slag. The vendors for other materials were chosen such that it was closest to the Zion Landfill to keep the impacts due to transportation low. Transportation of top soil and barrier soil was ignored as they were assumed to be acquired on site or from a location near the landfill.

20.4.2.3 Technical Design of Landfill Covers

The technical design of both CLC and ALC was developed using a landfill modeling software called HELP version 3.0 (USEPA 2017b) that determines the moisture infiltration through the various layers of cover to check the design criteria of maximum leachate level in the liner system. The basic profile and minimum thickness of each landfill layer were fixed based on the RCRA Subtitle D requirements for the conventional layer. The height of each layer of the landfill cover was further fixed after meeting the maximum equivalent vertical hydraulic conductivity of 10^{-5} cm/s for each profile. Each of the profiles were individually analyzed by HELP software to assure comparable performance. The soil properties used for CLC were obtained from default material property sets in HELP, while the properties such as permeability of biochar-amended soil and steel slag were obtained from Yargicoglu and Reddy (2017a) laboratory test data. The low permeability soil layer was replaced by a 10% (w/w) biochar-amended soil layer and the drainage layer by steel slag in ALC. The profiles of both ALC and CLC are shown in Figure 20.21. Tables 20.14 and 20.15 show the input soil properties and the performance obtained for each landfill cover from the HELP analysis.

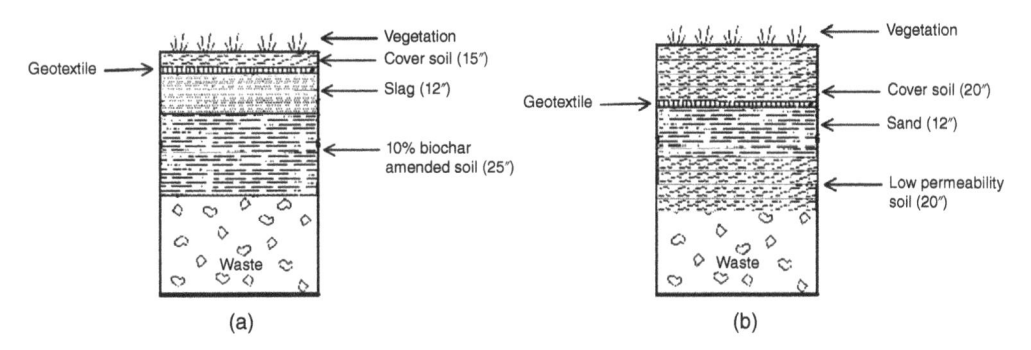

Figure 20.21 Design profile of (a) biocover and (b) conventional cover.

Table 20.14 Characteristic properties of cover materials in conventional cover (CLC).

Layer #	Soil texture	Layer	Height (in.)	Porosity (V/V)	Field capacity (V/V)	Wilting point (V/V)	Drainage length (ft)	Drain slope (%)	K (cm/s)	Equivalent vertical K (cm/s)
1	6	Cover soil	20	0.453	0.19	0.085	1000	33.3	7.2×10^{-4}	2.59×10^{-7}
2	1	Drainage layer/sand	12	0.417	0.045	0.018	1000	33.3	10^{-2}	
3	16	Low permeability layer	20	0.427	0.418	0.367	1000	33.3	10^{-7}	
Total thickness			52				Annual average percolation			1.89%

Table 20.15 Characteristic properties of cover materials in biocover (ALC).

Layer #	Soil texture	Layer	Height (in)	Porosity (V/V)	Field capacity (V/V)	Wilting point (V/V)	Drainage length (ft)	Drain slope (%)	K (cm/s)	Equivalent vertical K (cm/s)
1	24	Cover soil	15	0.365	0.305	0.202	1000	33.3	2.7×10^{-6}	1.96×10^{-6}
2	–	Drainage layer/steel slag	12	0.473	0.222	0.104	1000	33.3	4.2×10^{-4}	
3	–	Low K layer/biochar-amended layer	25	0.445	0.393	0.277	1000	33.3	1.2×10^{-6}	
Total thickness			52				Annual average percolation			1.83%

20.4.3 Environmental Sustainability

The LCA of CLC and ALC was carried out considering the raw material acquisition, transportation of materials, and construction of the landfill covers. Thus, the cradle-to-gate scope of LCA is adopted in this study. The end-use and disposal are not within the scope and applicability of this study. The goal of the LCA study was to evaluate the impacts of both the conventional and the biocover. The functional unit for this LCA was considered to be 1 acre of landfill cover.

LCA for both the landfill covers under consideration was carried out using the LCA software SimaPro 8. The study also makes use of TRACI to carry out initial environmental impact assessment. The uncertainty and sensitivity analysis were also performed by changing the parameters that affected environment negatively in the initial impact assessment. Table 20.16 summarizes the assumptions and conditions used in the study.

The inventory of materials quantity and materials transportation is shown in Table 20.17, and this inventory data was used as input to SimaPro for the LCA. Environmental Impact Assessment was carried out using the TRACI LCIA method by incorporating the inventory data from Table 20.17. The materials, construction equipment, and transportation mode were selected from the available database in the SimaPro 8.0. The steel slag was considered as a dummy material since steel slag is an unused product at the steel mills and does not require separate production. Equipment use, such as excavators and loaders, was also considered.

Table 20.16 Assumptions and conditions for the sustainability assessment of landfill covers.

Item	Input for assessment
Functional unit	One acre of landfill cover
Location of landfill	Zion Landfill, IL, USA
Original steel slag supplier	Indiana Harbor East Steel Mills
Other vendors (biochar, geotextile, etc.)	Closest to the site selected (distances in Table 20.17)
Dummy materials	Steel slag, barrier soil/clay
Database used in SimaPro	All databases were selected due to limited material availability
Method used in SimaPro	TRACI (US–Canada)

Table 20.17 Inventory of materials and transportation.

Cover design	Material	(t/acre)	Material	Distance (km)	t-km
CLC	Cover soil	3283	Sand	12.47	26183
	Sand	2100	Steel slag	123.6	269120
	Barrier soil	3439	Biochar	12.4	4340
	Geotextile	2.42	Geotextile	55.2	134
ALC	Cover soil ALC	3812			
	Steel slag	2177			
	Barrier soil	3500			
	Biochar	350			
	Geotextile	2.42			

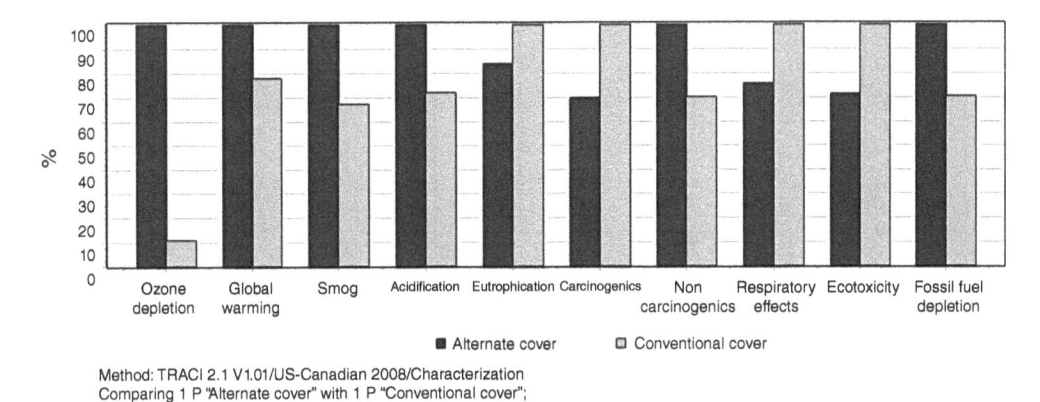

Figure 20.22 Life cycle assessment of biocover (alternative cover) and conventional landfill cover.

The environmental impacts from SimaPro for ALC (dark gray bars) and CLC (light gray bars) are illustrated in Figure 20.22. The values are normalized such that higher values correspond to higher negative impacts to the environment. Figure 20.22 illustrates that ALC has more negative effects to the environment than the CLC in some of the environmental impact categories.

Figure 20.23a shows the breakdown of environmental impacts in each category for the CLC. It also gives insight into what material or activity is contributing the most to the environmental impacts. It can be observed that the major contributors to the negative impacts are the hydraulic loader and spreader as well as sand. The contribution due to sand is due to the energy that goes into the mining and processing of the raw material by the manufacturers of sand. Figure 20.23b shows the breakdown of environmental impacts from different processes involved in ALC life cycle, and it can be seen that the hydraulic loader and spreader and the transportation of slag are the highest contributors to the environmental impacts of ALC.

The observations from the initial analysis were used for the sensitivity analysis where the major contributors were replaced and minimized to reduce the environmental impacts in both the covers. In this part of analysis, efforts were made to curb the impacts due to transportation and hydraulic loader and spreader. For the second analysis, the equipment hydraulic loader and spreader, along with the excavator, were replaced with skid-steer loader and excavator, which are capable of housing several attachments and thus, serving multiple purposes on site such as excavation, spreading, and compaction (Table 20.18). This change was made in both the cover systems for unbiased comparison.

Figure 20.24 shows the environmental impacts after the change in equipment was made in the analysis. It shows that the conventional cover still remained more sustainable compared to biocover. The alternate cover resulted to be a major contributor of ozone depletion and global warming, which was contrary to expectations. The characterization of impact due to conventional cover and the biocover from the second analysis shows that the replacement by skid-steer loader greatly reduced the negative impact due to the construction equipment. However, it was clear that transportation of steel slag was the main cause of environmental unsustainability of biocover (Figure 20.25). Thus, for subsequent analysis, reduced steel slag transportation distances were considered.

For the third sensitivity analysis, the transportation distance for slag was reduced to 20 km after observing that the impact due to transportation over a distance of about 50 km still leads to unsustainable results for the biocover. Figure 20.26a shows the results from this analysis, and it was found that the biocover was more sustainable compared to the conventional cover. The only category in which biocover did not appear sustainable was ozone depletion. Hence,

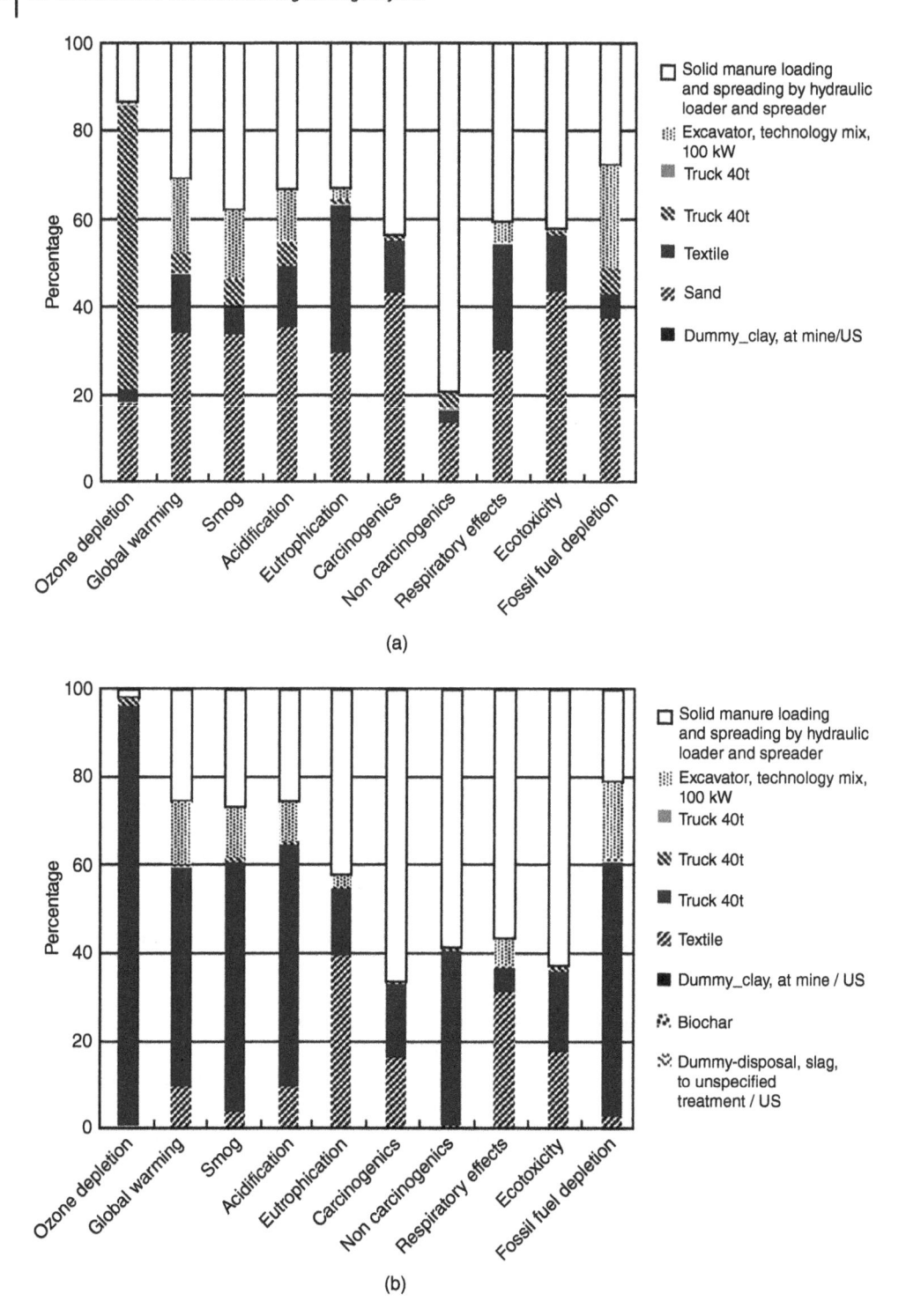

Figure 20.23 Characterization of environmental impacts due to (a) conventional cover (CLC) and (b) biocover (ALC).

Table 20.18 Sensitivity analysis.

Second analysis data	
Materials	Table 20.17
Transportation	Table 20.17
Construction equipment	Skid-steer loader (52 acre-in volume)
Third analysis data	
Materials	Table 20.17
Transportation	Table 20.17 (except steel slag transport – 20 km)
Construction equipment	Skid-steer loader (52 acre-in volume)
Fourth analysis data	
Materials	Table 20.17
Transportation	Table 20.17 (except steel slag transport – 5 km)
Construction equipment	Skid-steer loader (52 acre-in volume)

Modifications on the landfill design and construction to reduce the negative environmental impacts.

Method: TRACI 2.1 V1.01/US-Canadian 2008/Characterization
Comparing 1 p "Alternate cover" with 1 p "Conventional cover";

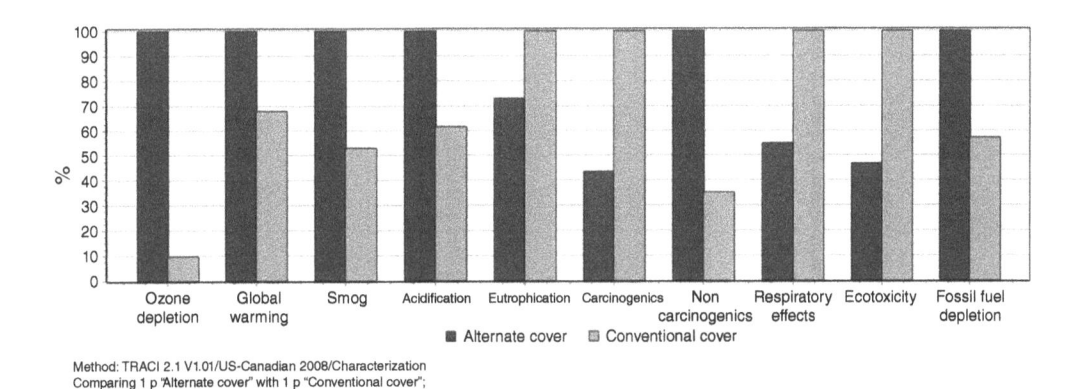

Figure 20.24 LCA of landfill covers with the second analysis data with skid steer loader and excavator.

as a final attempt (fourth analysis) at studying the optimum distance, transportation of steel slag was reduced to 5 km, and the results are represented as shown in Figure 20.26b. It was concluded that the optimum transportation distance for slag would be 5 km, so as to make biocover more environmentally sustainable than the conventional cover in terms of every impact category under consideration.

Overall, the results from the LCA showed that biocover was more sustainable in every impact category if the transportation distance of slag was kept within 5 km of the landfill. This may not be feasible, since not every landfill would be close to a steel plant. However, biocover has the capacity to capture the LFG emissions which could make up for the ozone depletion caused by transportation of materials. Eloneva et al. (2012) reported the carbon dioxide emission data for the 40-ton truck that was used in SimaPro 8 for the transportation of steel slag. The carbon dioxide emission when the 40-ton truck runs empty is 895 g/km and that when full is 1335 g/km CO_2. Hence, for a total of 92 trips required to transport the approximately 3700 tons of steel slag, the total CO_2 emission due to transportation would be 25 316 kg of CO_2. On the other hand, the CO_2 uptake capacity by the basic oxygen furnace steel slag was 63 g CO_2/kg as per

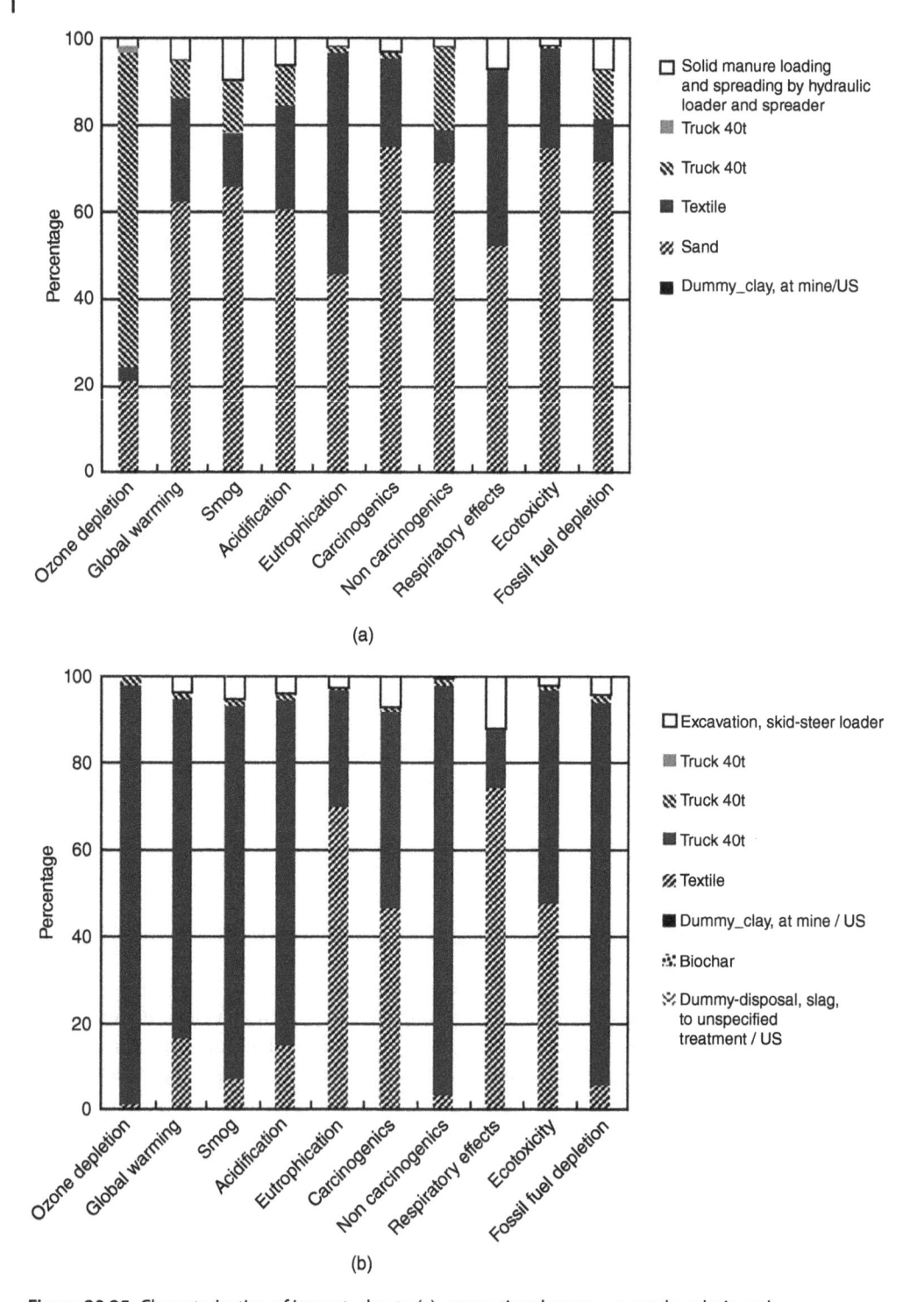

Figure 20.25 Characterization of impacts due to (a) conventional cover – second analysis and (b) biocover – second analysis.

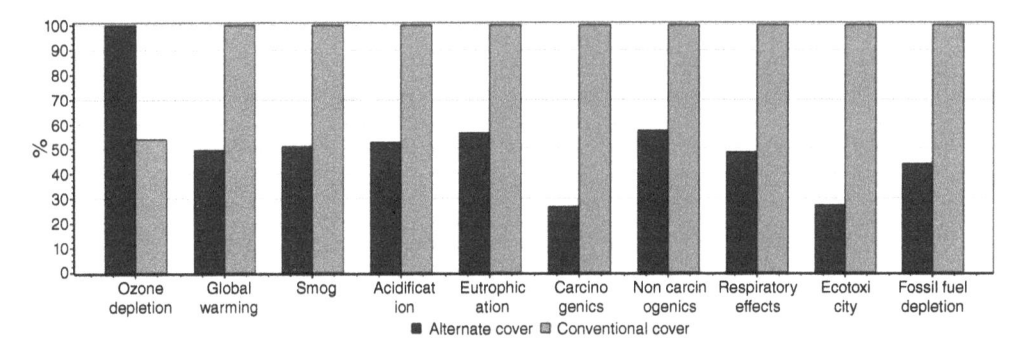

Figure 20.26 LCA of landfill covers with the third analysis data with (a) 20 km transportation distance for steel slag and (b) 5 km transportation distance for steel slag.

Sarperi et al. (2014). The CO_2 capture, therefore, with the same quantity of steel slag is nine times that of the emission due to its transportation over the 123 km. The quantitative gas emission and capture study, along with the fact that the use of steel slag in biocover would provide a market, was considered a valid reason to conclude that the biocover is an environmentally sustainable option when compared to the conventional cover, even under the circumstances where steel slag may be required to be hauled over a long distance.

20.4.4 Economic Sustainability

The economic sustainability assessment of the cover systems was carried out by considering the cost incurred during the construction of each cover and their comparison as shown in Table 20.19. In this assessment, only the materials and transportation costs were compared. The construction methods for both landfill covers were assumed to be the same and thus eliminated the need to take the labor and equipment cost into consideration. It was also assumed that the excavated soil/clay from the landfill site itself was used in the construction of the low permeability layer and top soil layer. Hence, the material and transportation cost of the soil/clay was neglected while assessing the total cost. The cost of each material was taken as an average cost of the many options available in the market as obtained from several market rate reports

Table 20.19 Economic assessment of the alternate cover systems.

		Cover system		Unit cost		Total cover cost ($)	
		CLC	ALC	Cost	Unit	CLC	ALC
Material	Soil/clay	X	X				
	Sand	X		29.45	$/CY	47,512	
	Steel Slag		X	3.75	$/ton		8,165
	Biochar		X	200–500*	$/ton		70,000 to 175,000
	Geotextile	X	X	1.82	$/SY	8,809	8,809
Transport	Soil/clay						
	Sand	X		5.80	$/CY	9,357	
	Steel Slag		X	5.80	$/CY		9,357
	Biochar		X	Included in material cost			
	Geotextile	X	X	Included in material cost			
Total cost						65,678	96,331 to 201,331
Cost benefit of methane mitigation (in 2 years)	Social cost of methane			1000**	$/ton	−389,221	−625,821
Net effective cost or benefit						−323,543	−424,590 to −529,490
Net benefit compared to CLC						0	100, 947 to 205, 947

Note:
a) Methane oxidation rates (Yargicoglu and Reddy 2017) for soil only = 179 and 10% biochar amended soil = 270 µg CH_4/d/g soil
b) Total amount of CH_4 mitigated in two years by soil = 389.2 ton and biochar amended soil = 625.8 ton CH_4
c) *Kulyk (2012)
d) **3% Discount rate (USEPA 2017c)
e) The net cost saving due to the use of biocover will increase in the long run as the methane production continues for many years of waste placement in a landfill.

available online. The average costs of biochar and steel slag were obtained with reference to the Jirka and Tomlinson (2014), Kulyk (2012) and Van Oss (2004).

Following the economic assessment results (Table 20.19), it was concluded that the total cost of conventional cover was significantly lower than that of biocover, thus making it more economically sustainable. It was observed that the high unit cost of biochar escalated the total cost of biocover, which provided opportunities to consider a biocover with much lower proportion of biochar amended to the soil. The cost was also recalculated based on the low-end cost of biochar as obtained from Jirka and Tomlinson (2014) and Kulyk (2012) for the same quantity. This reduced the cost of biocover significantly to $96331 per acre, but this was still more expensive than the conventional cover. However, when benefits of methane mitigation and associated social costs are taken into account, biocover proves to be economically advantageous (Table 20.19). It should be noted that economic benefits are even higher if long-term methane mitigation (by biochar-amended soil) and simultaneous carbon dioxide sequestration (by steel slag) are accounted.

20.4.5 Social Sustainability

The SSEM developed by Reddy et al. (2014) was used. The key areas for evaluation were changed, recognizing the relevance and applicability to this project. The scoring ranges from −2 to 2, with 2 being the highest positive impact and −2 being the highest negative impact.

The results from the SSEM are tabulated in Table 20.20 and graphically represented in Figure 20.27. Constructing no landfill cover evidently has numerous negative impacts in all the dimensions evaluated in the SSEM, whereas construction of landfill cover is much more socially acceptable and sustainable. The results show that biocover is the most socially sustainable option compared to conventional landfill cover.

Table 20.20 Social sustainability final score.

Criteria	No landfill cover	Biocover	Conventional cover
Socialindividual	−14	18	16
Socioinstitutional	−16	30	25
Socioeconomic	−4	15	14
Socioenvironmental	−14	18	9
Grand score	−48	81	64

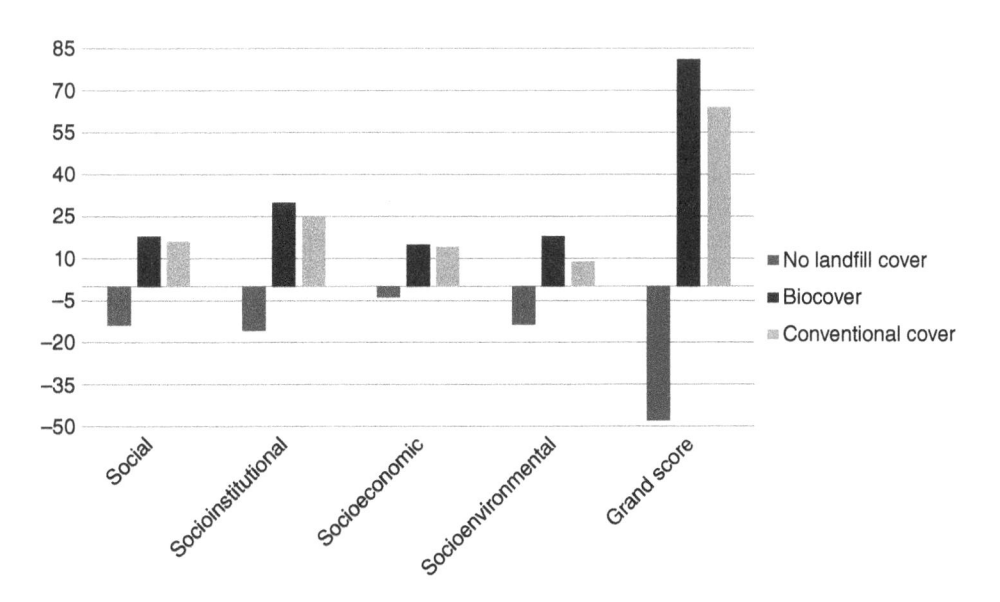

Figure 20.27 Social sustainability of CLC and ALC.

20.4.6 Conclusions

The environmental sustainability assessment based on LCA using SimaPro resulted in negative results for the proposed biocover, though it was clear that this was due to the long transportation distance for steel slag and the construction equipment used. The sensitivity analysis was thus used to assess the biocover after reducing these impacts. Further, the environmental benefits of the biocover for LFG mitigation were analyzed by estimating the emissions due to transportation and carbon dioxide uptake capacity of steel slag. The uptake capacity of the same amount of steel slag was calculated to be approximately equal to nine times the emission due to its transportation from the steel slag source to the landfill. Hence, the biocover was determined to be an environmentally sustainable option, owing to the benefits of LFG capture capacity. The use of biocover is further promoted due to the scope of recycling huge masses of stockpiled steel slag at steel-making plants.

The biocover was assessed as more expensive compared to the conventional cover due to the high cost of biochar. This cost could be reduced by decreasing the amount of biochar used. Moreover, it should also be noted that biochar is a better organic amendment than other materials, such as compost, due to the stability and other unique characteristics of the biochar discussed earlier. Considering the social cost savings due to mitigation of methane and carbon dioxide, biocover is economically preferable. Furthermore, the sustainability of biocover is well established from the social sustainability score of the biocover. The analysis of the key areas affecting the society was useful in assessing the overall social sustainability of the biocover.

Hence, from the overall performance and sustainability assessment, the proposed biocover with biochar and steel slag is concluded as a sustainable alternative to the conventional landfill cover. The conventional landfill cover, which has only very limited capacity to capture LFG, can be replaced with the proposed biocover to reduce these emissions resulting in a positive impact on the environment and society.

20.5 Algae Biomass Deep Well Reactors Versus Open Pond Systems

The production of algae biomass has been studied as a sustainable source of biofuels in an attempt to cover the growing demand of fossil fuels for energy and transportation. Various photobioreactors have been designed and tested for the production of algae biomass. This study deals with the sustainability assessment of two bioreactor systems: a single deep well photobioreactor (DWP) utilizing Neochloris patented technology, and conventional concrete runway ponds considering environmental, economic, and social impacts. The sustainability study focuses on the construction of the two bioreactors; energy and materials required for the bioreactor operation are not considered in this study. Thus, the environmental impacts were assessed at three life cycle stages: raw materials acquisition, material transportation from suppliers to site, and construction.

20.5.1 Background

Energy demand is growing rapidly, and it is expected to reach 800 quadrillion BTU by 2040, or about 25% from the current global demand. The use of fossil fuels to cover the present and future demand is an unattractive option considering the huge amount of CO_2 emissions from the combustion of these fuels, which is responsible for global warming and climate change. In order to meet growing demand, it is necessary to embrace new renewable energy sources. A possible solution to this potential crisis is the production and utilization of algae biofuel.

20.5 Algae Biomass Deep Well Reactors Versus Open Pond Systems

According to the United States Department of Energy, "most algae convert sunlight into energy in a similar manner as plants; however, the genetic diversity of the many different kinds of algae gives researchers an incredible number of unique opportunities that can be harnessed to develop promising algal biofuel technologies (Barry et al. 2016)." In addition, the energy conversion performed by algae organisms can be accelerated by introducing CO_2 from the sources, which would otherwise release it into the atmosphere, thus adding credits of CO_2 to the environment. Therefore, by creating the appropriate environment, algae convert light into energy (Figure 20.28). The objective of this study is to assess sustainability of the construction of two types of bioreactors to produce algae biomass.

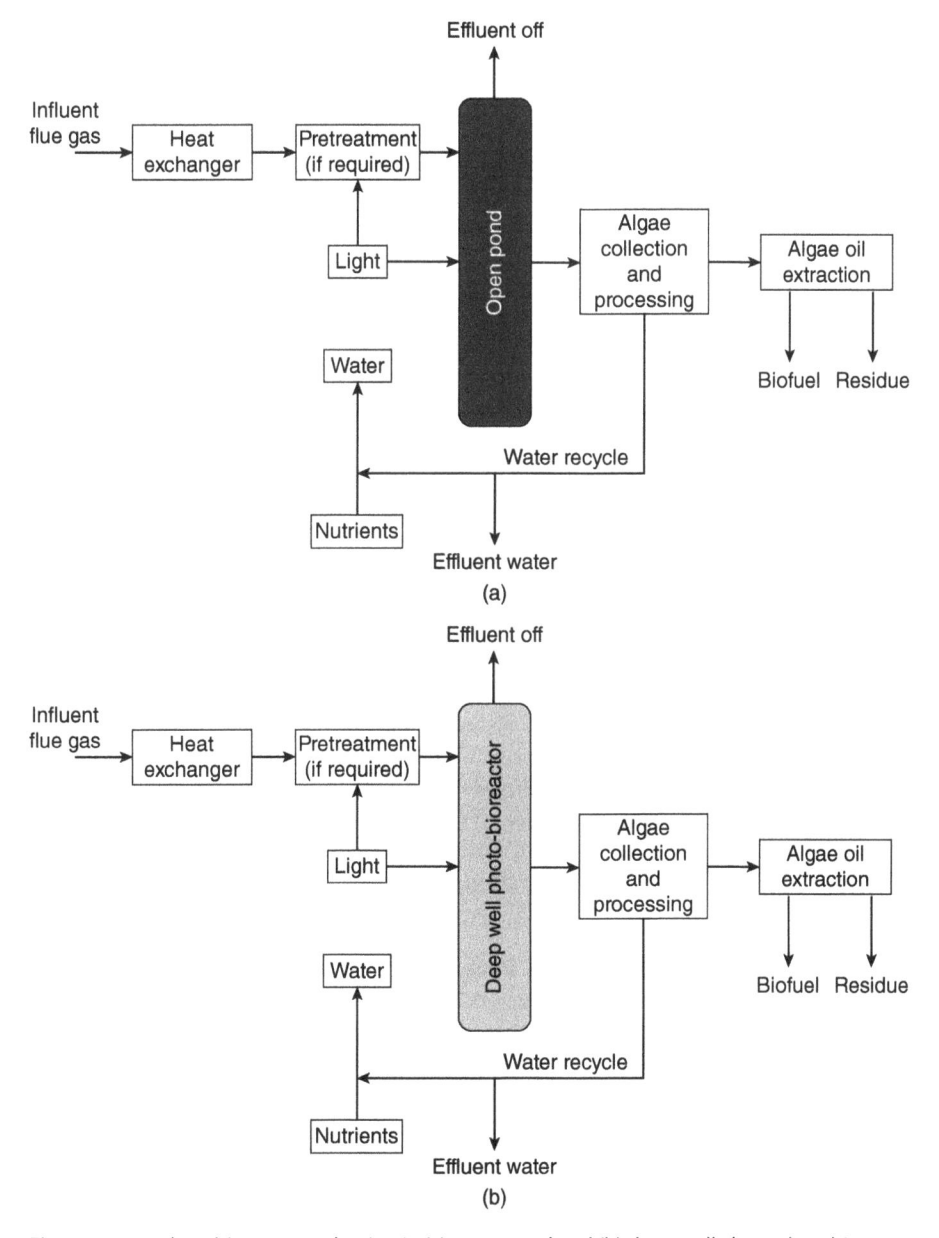

Figure 20.28 Algae biomass production in (a) open pond and (b) deep well algae photobioreactor.

20.5.2 Methodology

20.5.2.1 Goal and Scope
The scope of this study is the comparison of two bioreactor systems for the production of algae biomass: the classical open pond system and the DWP (Figure 20.28). Each option was designed so that the algae biomass production is equal. Based on the observed laboratory and field data and some extrapolation, it was estimated that the open pond system requires a total of 150 "dual ponds" for the production of same amount of biomass with one DWP.

20.5.2.2 Site Location
The project location selected was Casa Grande, AZ. This location averages nearly 200 clear-sky days annually, highest among all US states, and provides accommodating temperatures for the open pond system. The landscape of both properties used were assumed to be level with no obstructions on site. A 22-acre plot was used for the open pond system, and 2.3 acres for the DWP. Municipal infrastructure and utilities were assumed to be adequate enough to accommodate the proposed bioreactors. The soil profile at the site is entirely composed of clay loam with bedrock encountered at a depth of 400 ft below the ground surface.

20.5.2.3 Technical Design
The DWP (Neochloris photobioreactor) was designed for a relatively small footprint while achieving the desired volume by constructing the reactor below ground level (Figure 20.29). The DWP is a 25-ft-diameter, 404-ft-long shaft that uses its depth as an advantage to place algae under pressure to enhance the algae growth rate. The toe of DWP was filled with 4 in. of concrete in height. The steel wall thickness was sized as per the soil pressure acting against the circular outer walls. The length of shaft was divided into four segments increasing in wall thickness with depth (Table 20.21). The entire interior wall was layered with a high density polyethylene (HDPE) liner to prevent contamination.

The open pond system was designed for the same algae biomass production of the DWP. Given the DWP produces 171 ton/1000 ft^3/yr and knowing open ponds provide 0.398 tons/1000 ft^3/yr (Jorquera et al. 2010), for a fluid depth of 0.98 ft, the pond surface area required would be 435 675 ft^2. So, 300 ponds with a surface area of 1456 ft^2 were required. In order to save constructing materials and space, 150 dual ponds with an area of 2913 ft^2 are proposed. Prior to installation, the site was layered with compacted gravel. Each pond consisted of poured in place reinforced concrete slab with 8 and 6-in. walls. Minimum steel reinforcement was used to account for shrinkage. The slab consisted of 6 × 6 W1.4 × W1.4 wire mesh, weighing approximately 21 lb/100 ft^2. The walls incorporated #3 dowels at 24-in. spacing, combined with #3 bars top and bottom. In addition, the interior surface area was entirely lined with HDPE liner to contain the fluid and prevent site contamination. A total of 150 dual ponds were designed similarly, separated 15 ft edge-to-edge to allow operational access for vehicles (Figure 20.30).

20.5.2.4 Sustainability Assessment
The environmental sustainability of the two bioreactors was determined by conducting LCA using SimaPro 8.0, following the procedure described in ISO 14040:2006. The system boundaries include three stages: raw material acquisition, transportation of raw materials, and construction phase. The operational technology, such as mixers, lighting, tubing, nutrient delivery, and collection/processing are excluded. Materials, fuel consumption to power material transportation, as well as operating diesel-driven construction vehicles are included. The functional unit to compare the two bioreactors was based on the biomass production: tons of biomass/1000 ft^3/yr. The raw materials were quantified to be locally available and transported within a 100 km radius (Figure 20.31). The LCA inventory analysis is shown in Figures 20.32 and 20.33 for the dual open pond and the DWP, respectively.

Figure 20.29 Deep shaft profile (not to scale).

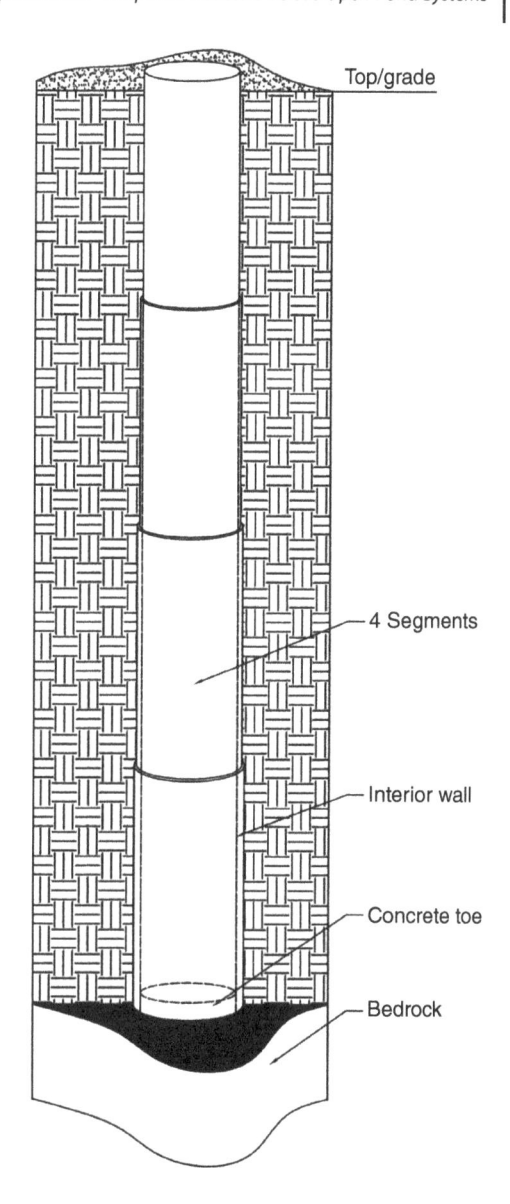

Table 20.21 Wall thickness for the deep well photobioreactor.

Specific weight of loam clay (pcf)	Depth (ft)	Effective stress (psi)	$P_b = \dfrac{2E_p}{(1-\mu_p^2)(D_m/\delta)^3}$ (psi)	$E \times 10^5$ (psi)	Poisson ratio	Diameter (in.)	Wall thickness (in.)
80	100	55.56	55.68	290	0.27	300	3.25
	200	111.11	124.51				4.25
	300	166.67	173.83				4.75
	400	222.22	234.70				5.25

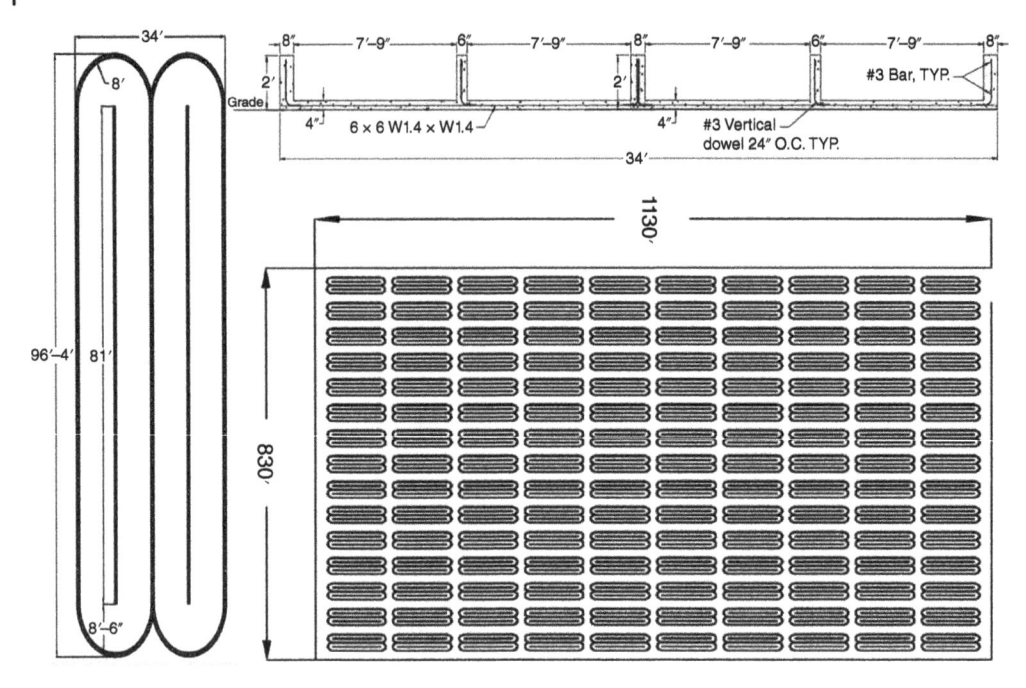

Figure 20.30 Dual pond design.

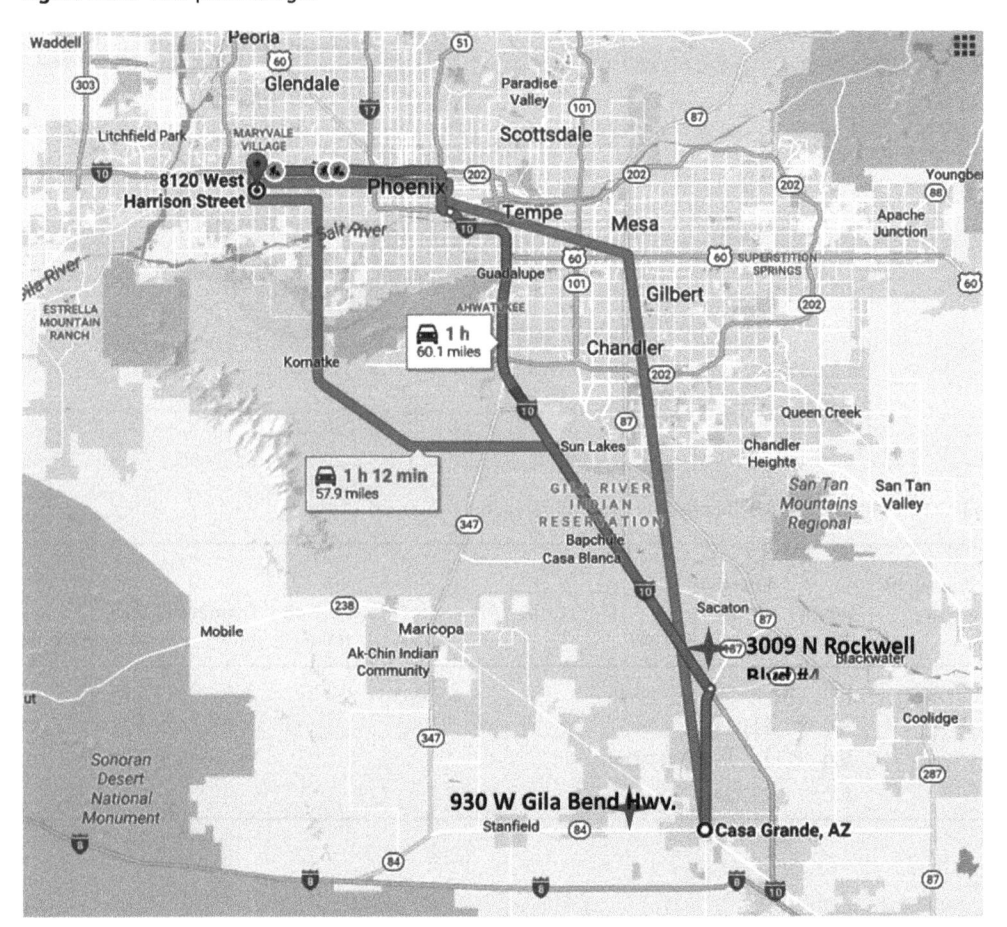

Figure 20.31 Raw material acquisition for conventional open ponds and DWP (Neochloris photobioreactor).

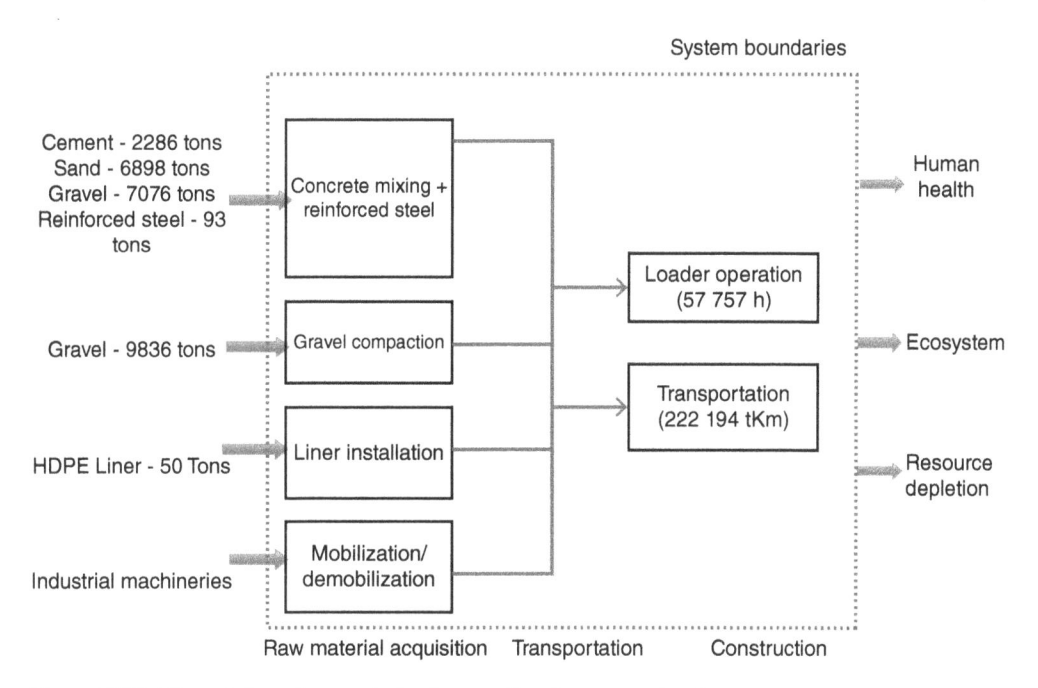

Figure 20.32 Life cycle inventory analysis of conventional open pond system.

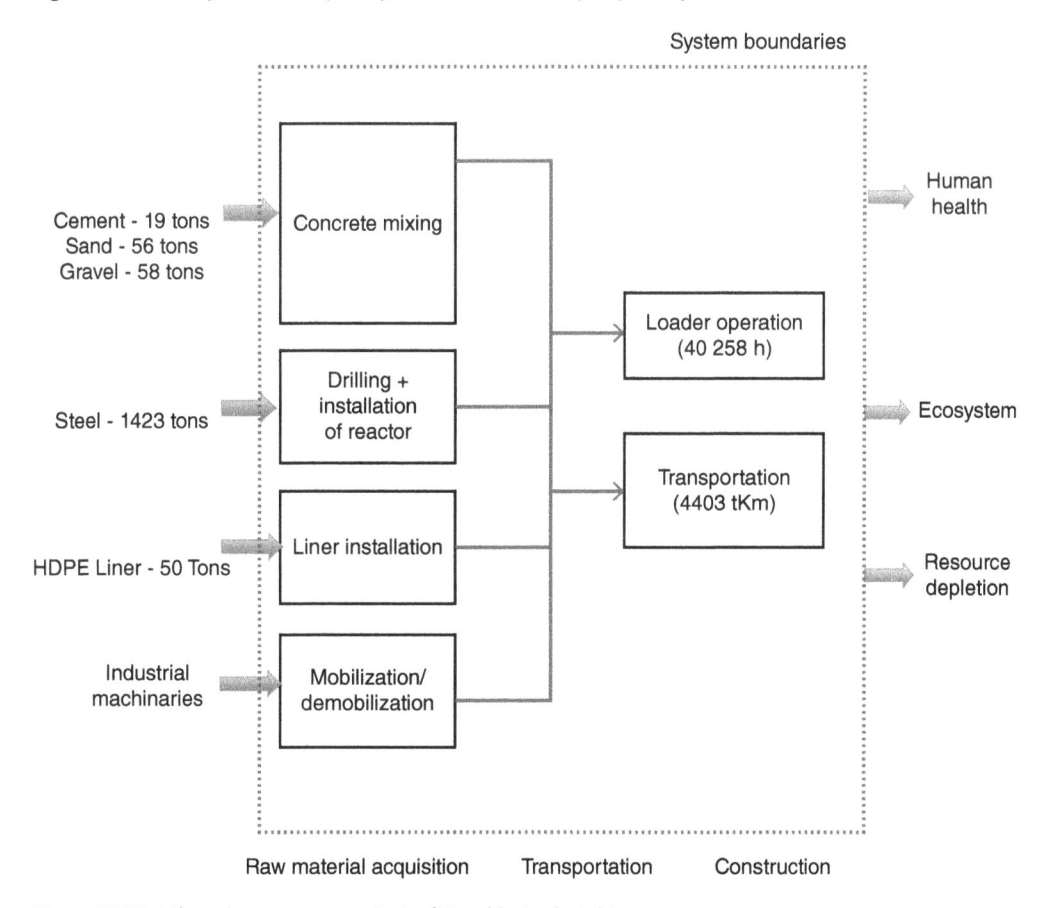

Figure 20.33 Life cycle inventory analysis of Neochloris photobioreactor.

The environmental impact assessment was conducted using TRACI and ReCiPe endpoint (H) method. ReCiPe endpoint (H) comprises two sets of impact categories with associated sets of characterization factors, of which 18 impact categories are addressed at mid-point level, which are further converted and aggregated into three end-point categories (Heijungs et al. 2003). This project focuses on three end-point levels that determine the overall damage to human health, eco-toxicity and depletion of resources.

In the economic sustainability, a local construction company provided all the material cost and construction cost (labor charges, heavy machinery charges, and service charge) involved in the two projects. The transportation cost was calculated based on the fuel consumption of the truck used and the distance covered; other transportation costs (industrial machineries to site and indirect cost) were not considered. The cost of the fabricated steel reactor was not considered but the cost of the steel based on the quantity required in fabricating the steel was included.

The SSEM tool developed by Reddy et al. (2014) was adopted to conduct a social sustainability evaluation for this study. For this project, all the activities under each dimension were identified and a final rating was set.

20.5.3 Environmental Sustainability

The LCAs for DWP and open ponds were carried out for raw material acquisition, transportation, and construction (Table 20.22). The results for raw material acquisition are shown in Figure 20.34. For open ponds, the use of Portland cement (95%) showed major contribution to environmental damage. The major contribution was seen toward global warming, noncarcinogenic effects, and eco-toxicity. For DWP, the use of steel (98%) had major impacts on the environment. The major contribution being toward eco-toxicity, noncarcinogenic effects, and fossil fuel depletion.

Table 20.22 Materials and transportation for the open ponds and DWP.

Open Ponds

Design Components (Materials)	Portland Cement	Fine Aggregate	Coarse Aggregate	HDPE Liners	Reinforced Steel
	2286 tons	6898 tons	16912 tons	49.9 tons	92.86 tons
Material Transportation	8.3 km	8.3 km	8.3 km	96.5 km	8.3 km
	Truck 40 ton	Truck 40 ton	Truck 40 ton	Truck 40 ton	Truck 40 ton

Neochloris Photobioreactor (DWP)

Design Components (Materials)	Portland Cement	Fine Aggregate	Coarse Aggregate	HDPE Liners	Steel
	18.79 tons	56 tons	58 tons	0.8 tons	1422.6 tons
Material Transportation	8.3 km	8.3 km	8.3 km	96.5 km	2.25 km
	Truck 40 ton	Truck 40 ton	Truck 40 ton	Truck 40 ton	Single Unit Diesel Truck, US

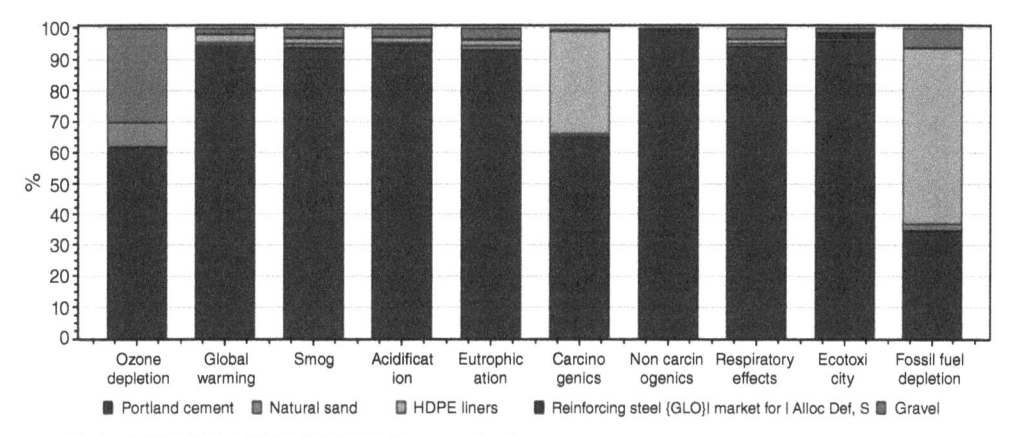

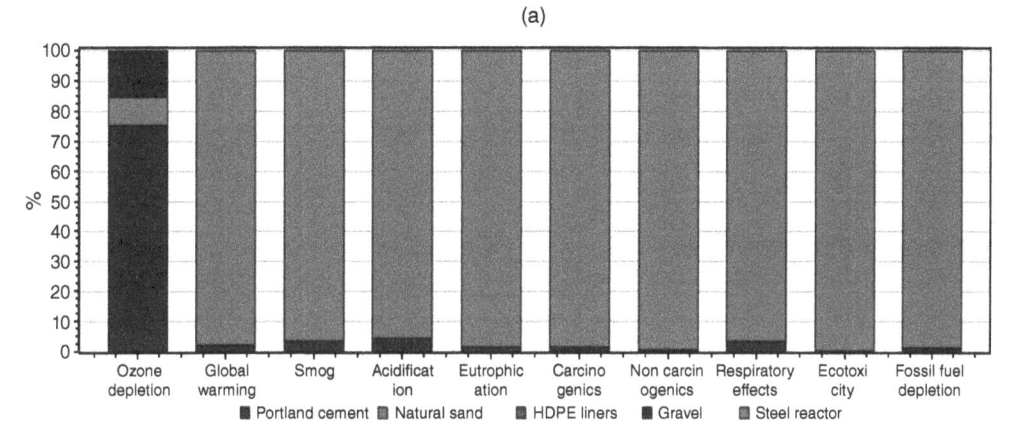

20.5 Algae Biomass Deep Well Reactors Versus Open Pond Systems

Figure 20.34 Environmental impact – raw material acquisition for (a) conventional open ponds and (b) DWP.

Figure 20.35 shows the overall impacts for both open pond system and photobioreactor based on TRACI. Conventional open ponds had major contribution to environment under each impact category when compared to DWP. Eco-toxicity, carcinogenic/noncarcinogenic effects, and resource depletion were the major impact categories for both systems.

Similarly, the ReCiPe endpoint (H) evaluation is shown in Figure 20.36. Open ponds show major environmental damage toward eco-toxicity, carcinogenic/noncarcinogenic effects, and resource depletion along with global warming and ozone depletion. These results were in-line with what was obtained using TRACI method. Results confirm that conventional open ponds are prone to be less sustainable in terms of environmental damage when compared to DWP.

The ReCiPe endpoint (H) has a feature of representing their results considering three main impact category groups that focus on overall human health, eco-toxicity (land conversion, acidification, eutrophication), and resource depletion (fossil fuel and metal/mineral depletion). Based on the above categories, results in Figure 20.37 show that open ponds have major effects on all the three categories when compared to DWP.

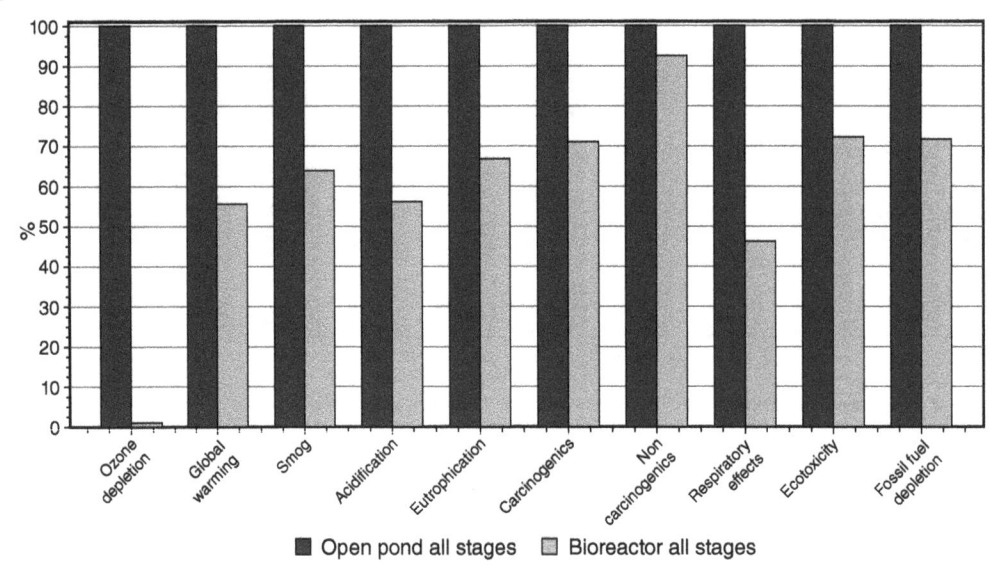

Figure 20.35 Life cycle assessment of DWP versus conventional open ponds – TRACI.

20.5.4 Economic Sustainability

The overall cost of the two projects was estimated and the results are shown in Table 20.23. Conventional open ponds are more cost-effective than Neochloris photobioreactor (DWP). This is due to the cost of steel, which is significantly more expensive than the materials used in open ponds. Moreover, the soil drilling with heavy industrial machineries increases the overall construction cost of the DWP. On the other hand, the cost of land is more demanding for open ponds. The cost of land for constructing 150 open dual ponds (22 acres) was 12 times the cost of the land for one Neochloris photobioreactor (only 2.3 acres). In order to make the DWP more economically feasible, it is recommended to optimize the design constructing smaller reactors.

20.5.5 Social Sustainability

The SSEM tool (Reddy et al. 2014) was used for rating the social impacts (Figure 20.38). Neochloris photobioreactors are more favorable and acceptable over the conventional open pond in the categories socioinstitutional and socioindividual. This is because the large area of land used for cultivating algae biomass, which minimizes community expansion, and the postoperation phase leading to odor problems affecting the quality of life living in the community. Under the socioenvironmental category, both bioreactors show negative environmental impacts in all life cycle stages, but open pond shows more negative impact due to the construction operation over 22 acres of land (compared with only 2.3 acres for DWP) affecting the esthetics and quality of air. The drilling activities for DWP also involve land disturbance and heavy noise pollution, but the latest technologies largely reduce those impacts although with increased cost. Under the socioeconomic category, both technologies generate revenue in the form of local business; although more employment opportunities can be foreseen for open ponds when compared to DWP. Overall, the grand score for Neochloris photobioreactor is high and is acceptable over conventional open ponds.

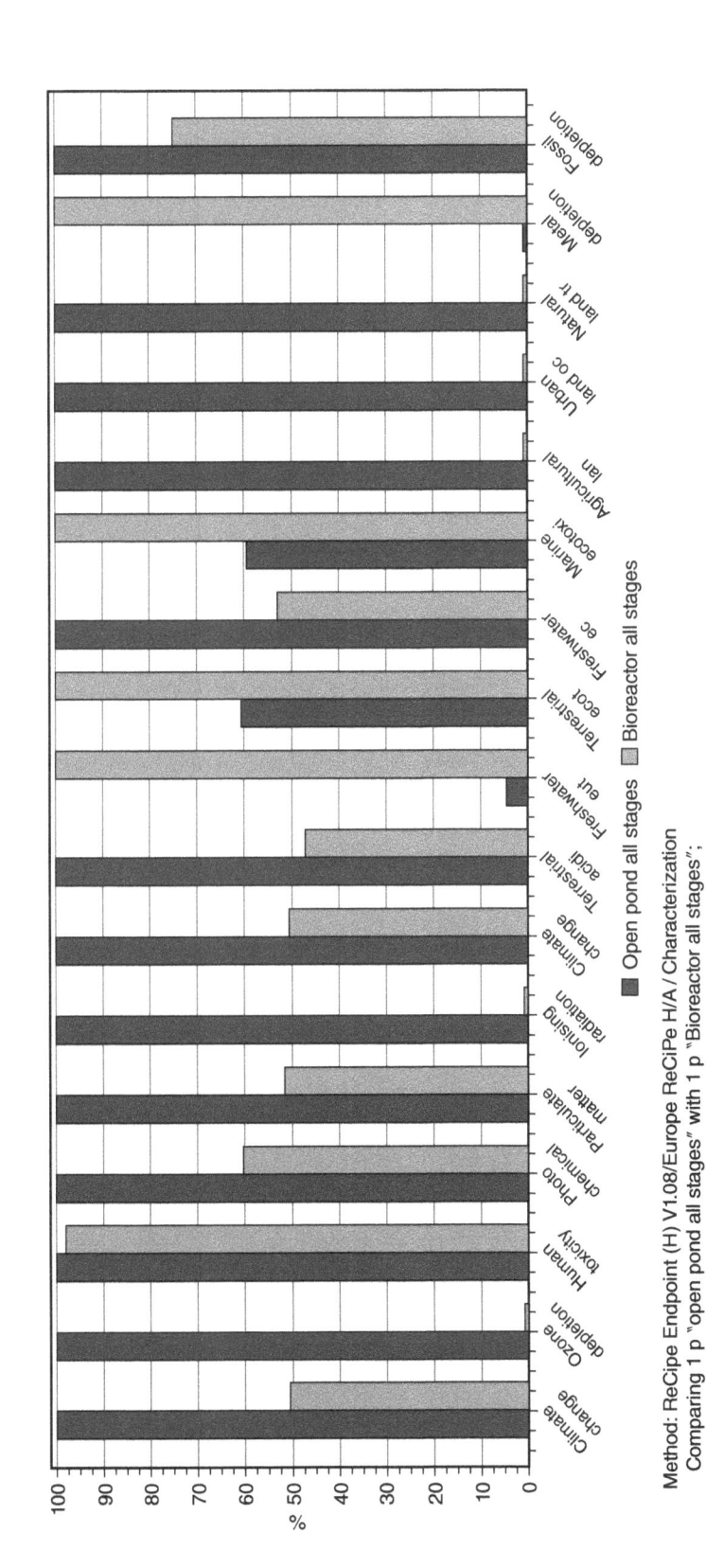

Figure 20.36 Life cycle assessment of DWP versus conventional open pond system – ReCiPe endpoint (H).

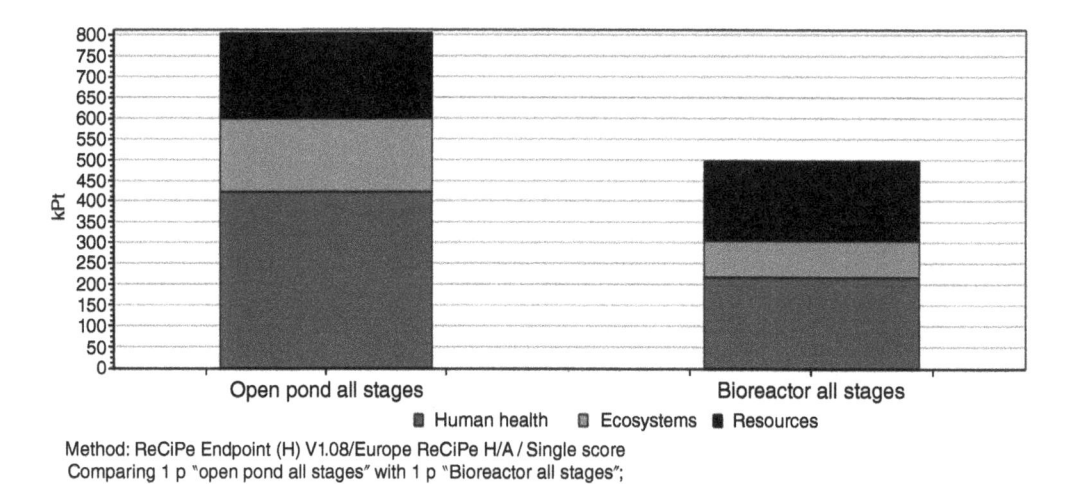

Figure 20.37 Damage assessment of DWP and conventional open pond system – ReCiPe endpoint (H).

Table 20.23 Cost evaluation of conventional open ponds and Neochloris photobioreactor.

	Conventional open ponds ($)	Neochloris photobioreactor ($)
Labor cost	4 674 460	4 838 400
Equipment cost	1 429 200	1 545 600
Raw material cost	4 588 890	5 342 276
Transportation cost (fuel cost)	22 000	10 000
Cost of land	770 000	60 000
Total cost ($)	11 484 550	11 796 276

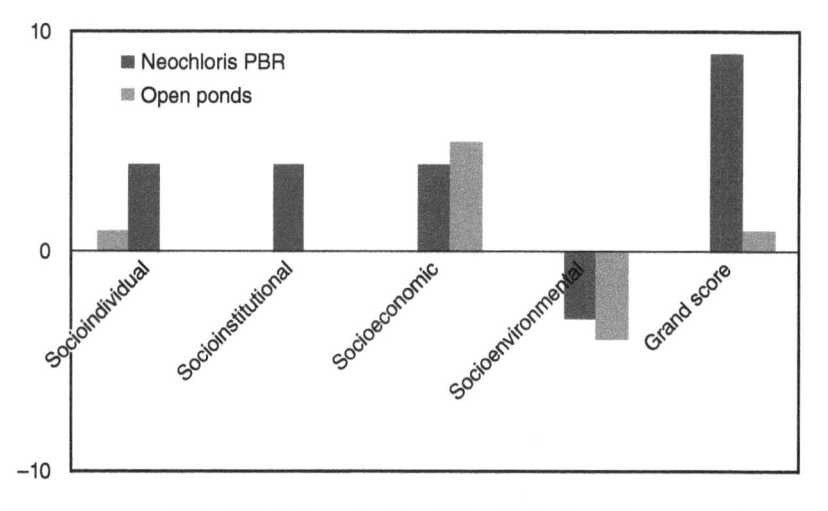

Figure 20.38 Social sustainability evaluation of Neochloris photobioreactor and conventional open pond system – SSEM Tool.

20.5.6 Conclusions

Results demonstrate that Neochloris photobioreactors had low environmental impact when compared to conventional open ponds. The raw material acquisition, transportation, and constructional activities of conventional open ponds contributed major environmental impact over Neochloris photobioreactors. Neochloris photobioreactor had a favorable societal acceptance over the open pond system. The reasons being less land required for constructing the Neochloris photobioreactor over conventional open pond system, which would impact the society in community expansion. In addition, open ponds construction and operation may release obnoxious gasses that may not be accepted by the community living in that area. Under economic assessment, Neochloris photobioreactor showed slightly higher investment when compared to open ponds and this was mainly due to heavy drilling activities; however, the land requirement is very less (2.3 acres) when compared to open ponds that needs large land area (22 acres) and could turn out to be very expensive based on site-specific locations. Overall, this study concludes that in spite of major investment involved in Neochloris photobioreactor, positive environmental and social impacts were observed. Thus, Neochloris photobioreactor is acceptable over conventional open pond systems. It is suggested that a change in the design (multiple small-size reactors) may reduce the overall cost of the project and be economically feasible over conventional open pond systems.

20.6 Remedial Alternatives for PCB- and Pesticide-Contaminated Sediment

Cedar Lake is an approximately 150-acre lake located north of downtown Cedar Rapids, Iowa. Cedar Lake was used as a source and discharge point of cooling water for a power plant and other nearby industries. The lake sediment has historically been contaminated with polychlorinated biphenyls (PCBs) and pesticides. Recent site investigations have shown that the concentrations of contaminants have dropped by an order of magnitude in the past 20 years. The lake has three portions: North Lake, South Lake, and West Lake. The contaminations in North Lake and West Lake are found to be below regulatory levels. About 90 000 ft^2 of sediment in South Lake is reported to have exceeded probable effect concentrations (PECs) for PCBs and pesticides. In this study, four remediation technologies: monitored natural attenuation (MNA), dredging and disposal, conventional capping, and modified capping were considered for the remediation of the contaminated sediment. A sustainability assessment based on triple bottom-line approach was conducted to evaluate the preferable option among the four selected remedial alternatives.

20.6.1 Background

In 1983, the Iowa Department of Natural Resources (Iowa DNR), with the financial support of USEPA, sampled and analyzed fish tissue for the presence of PCBs, pesticides, and metals at 23 lakes across Iowa, including Cedar Lake. Two species, catfish and trout, were selected for the analysis because they were very appropriate for the detection of contaminants in the benthic zone (sediments) due to their feeding habits and potential for contaminant biomagnification. The analysis found elevated levels of PCBs and certain pesticides, including DDT and chlordane, in both catfish and trout tissues. At that time, a fish consumption advisory was issued for Cedar Lake, which entered into a biannual fish tissue sampling program. A preliminary site investigation (Lubben 1994) found elevated concentrations of both PCBs and pesticides throughout North Lake and South Lake with concentrations up to 1 ppm. Over the last 35 years, concentrations of PCBs and pesticides within fish have dropped below advisory levels and the fish consumption advisory was lifted in 2014. In the winter of 2016–2017, the Iowa DNR conducted a phase II site assessment at Cedar Lake to further define the extent of contamination. The study concluded that sediments in a large area in the South Lake exceeded the regulatory levels for either pesticides or PCBs and required remediation.

Sediment remediation is often difficult due to the unique environmental conditions, as well as the continued use of waterways for various purposes. Sediment remediation was often accomplished through dredging and disposal, or placement of a sand and gravel cap. However, innovative technologies (e.g. reactive caps) may result in faster and more effective remediation. These caps often consist of activated carbon interwoven between two pieces of geotextile fabric. The activated carbon adsorbs contamination as it moves to the surface of the sediment.

20.6.2 Methodology

20.6.2.1 Goal and Scope

The goal of this study is to evaluate the sustainability of various sediment remediation technologies to determine which technology is the most sustainable for the remediation of the PCB- and pesticides-contaminated sediments.

20.6.2.2 Study Area

Cedar Lake is divided into two main sections roughly bisected by a railroad causeway. The north portion of the causeway includes North Lake, which is approximately 80 acres in size, and West Lake, which is approximately 10 acres in size. The south portion of the causeway includes South Lake, which is approximately 60 acres in size. The geology of the site consists primarily of quiet-water sediments within North Lake. South Lake is dominated by the delta deposit near the Kenwood Ditch Outfall, which consists of coarser-grained sand and gravel deposits near the outfall, which grade finer with distance.

Currently, there is no established risk levels for sediments within the Iowa Brownfield Assessment program. In lieu of official standards, contaminant concentration in the sediments were compared to the National Oceanic and Atmospheric Administration (NOAA) PEC, NOAA Threshold Effects Concentration (TEC), and the EPA Region III Sediment Benchmarks. The TEC is the concentration at which a negative environmental or ecological impact may be observed, while the PEC is the concentration at which a negative environmental or ecological impact is likely to be observed. For the risk assessment, the PEC was used as the screening level for defining the areas that require remediation. As shown in Figure 20.39, multiple samples located within South Lake by the Kenwood Ditch Outfall exceeded the PEC for either pesticides or PCBs. The final area was determined by using the halfway rule. Half the distance between a sample exceeding the PEC and a sample below the PEC was determined to be the extent of the remediation footprint. A total of 90 000 ft^2 of sediment exceeds the PEC for at least one contaminant of concern and will require further remediation.

20.6.2.3 Technical Design

Alternative 1: Monitored Natural Attenuation MNA assumes that remediation of contaminants will occur through natural processes, including degradation of contaminants through microbial activity as well as natural burial of contaminated sediment. The studies from 1994 to 2017 on Cedar Lake showed a decrease in concentrations of contaminants as high as 1 order of magnitude. This indicates that natural attenuation is occurring and will likely continue to occur.

Figure 20.39 Results of sampling events at Cedar Lake and location of area requiring further remediation.

For the protection of human health and the environment, the benthic zone, which is the top 12 in. (30.48 cm) of sediment, must have contaminant concentrations below the PEC. Given the sediment deposition rate (approximately 1.5 cm/year), the contaminated sediments would be sufficiently buried within 20 years, assuming no other degradation mechanism. Therefore, a monitoring program would need to be in place for at least that length of time.

The proposed MNA will consist of annual sampling of the top foot of sediment in 20 locations across the affected area. The samples will be shipped in coolers from Cedar Lake to an analytical laboratory in Lenexa, Kansas. It is assumed that the sampling team will come from 50 miles away and the samples will be collected using a direct-push rig with disposal acetate liners pushed 1 ft into sediment. Waste disposal from each event will consist of the acetate liners and related sampling materials, such as gloves. It is assumed that all soil collected will be shipped to the laboratory for analysis. Waste generated from sediment sampling was considered negligible.

Alternative 2: Dredging of Contaminated Sediment The target area in the lake with PCBs and pesticides over the regulatory limit was determined to be 90 000 ft^2, but the studies did not fully evaluate the depth of the contamination. The contamination depth can be calculated considering that PCBs were used from 1929 (when PCBs were first synthetized and commercialized) to 1970s when their use was discontinued. Chlorinated pesticides were first used in the 1940s. Assuming a 1.5 cm/year deposition rate over 89 years (from 1929 to 2018), approximately 135 cm (4.5 ft) of sediment has been deposited in South Lake. A conservative thickness of 5 ft was used for the design of dredging. A total volume of 450 000 ft^3 will need to be removed from the targeted area. Assuming a sediment density of 169 lb/ft^3, a total of 38 025 tons will be dredged and will require disposal at a special waste landfill. The nearest landfill to the site is located 9 miles south of the site and all dredged material was assumed to be transported there in 20-ton trucks. The remediation would be complete in 48 days (4 trips per day with 10 trucks). Due to the shallow water depth across the affected area (less than 1 ft), dredging can be completed on land using clam shell buckets to avoid potential resuspension of contaminated material.

Alternative 3: Conventional Capping A conventional cap generally consists of 12 in. of sand over the contaminated sediment to isolate the contaminants and create a new benthic zone for aquatic organisms. An additional 3 in. of sand is placed to minimize the exposure potential. On top of the sand, 4 in. of angular gravel will be placed to keep the sand in place. Similar to the sand, an additional 3 in. of gravel would be placed as an over allowance. An armored cap will need to be installed adjacent to the Kenwood Ditch Outfall to accommodate effects of periodic high flows. This armoring consists of a 4-in. layer of angular stones ranging in size from 3 to 4 in. (riprap). The target area is located near shore and has shallow water depths, so sediments can be placed with the use of a backhoe from the shore. All the cap materials were assumed to be sourced from a quarry located approximately 12 miles from Cedar Lake. Figure 20.40a shows a schematic of the conventional capping system to be implemented at Cedar Lake.

Cedar Lake is also used as a detention basin during storm events and high flow conditions. Since 195 500 ft^3 of sediment is being introduced into South Lake, the same volume must be removed from a different portion of the lake to maintain the same storage volume. It is proposed to remove 1 ft of sediment across 195 500 ft^2 of North Lake leaving 1 ft of clean benthic zone

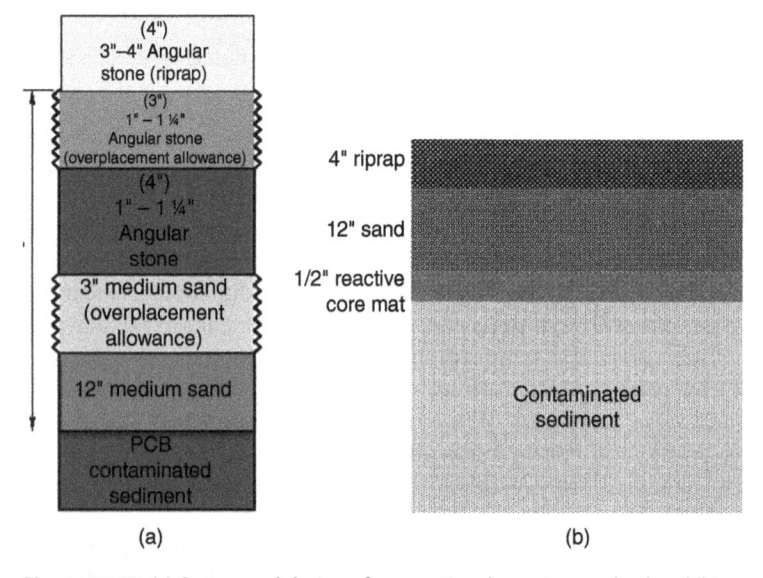

Figure 20.40 (a) Conceptual design of conventional capping method and (b) conceptual design of modified capping method.

based on the contamination study of 2016. The sediments in North Lake are generally finer grained silts and clays and are therefore not suitable for use in the cap. These sediments must therefore be disposed of or reused at a different site. For the purposes of this evaluation, the sediments to be dredged to maintain storage volume will also be disposed of as special waste due to its potential for historic contamination.

Alternative 4: Modified Cap Employing a Reactive Core Mat The modified cap includes a reactive core mat made of granular activated carbon in between two pieces of geotextile fabric. The reactive core mat will be placed directly overlying the contaminated sediment. Then, it is covered with sand and angular stones (riprap) similar to the conventional cap (Figure 20.40b). The reactive core mat contains 0.4 lb/ft^2 of granular activated carbon that acts to adsorb any potential contamination, thereby limiting the exposure risk from the contaminants in the sediment. Since Cedar Lake is located within a floodplain and the net change in storage must be zero, sediments from North Lake must be excavated to account for the volume occupied by the modified cap in South Lake. A total volume of 123 750 ft^3 of sediment is being added to South Lake; therefore, a 1-ft thickness will be removed across 123 750 ft^3 of North Lake and disposed of as special waste due to its potential for historic contamination.

20.6.2.4 Sustainability Assessment Methodology

The environmental impacts were assessed using SimaPro v8.5. In this study, the major life cycle stages involved were material acquisition, operation, use and maintenance, and waste management. The environmental impacts were assessed for each remediation option, and a comparison was made between the alternatives on the basis of the degree of environmental impacts obtained from each life cycle stage.

The functional unit for performing LCA on this project was the total square feet of contaminated sediment requiring remediation, which was 90 000 ft^2. The material quantities inputs used for the LCA are shown in Table 20.24. Table 20.25 shows the location and distance of the site to the quarries and the disposal site. The impact assessment was performed using TRACI and Building for Environmental and Economic Sustainability (BEES) methods. The inventory used in performing LCA was adopted from the database in the LCA software, SimaPro v8.5. For the materials not listed in the database, a dummy material that closely represented the properties of the original material was created for the use in the analysis.

Table 20.24 Input material quantities associated with each remediation option used in LCA.

Material	MNA	Dredging and disposal	Conventional capping	Modified capping
Dredged sediment (dummy-material-clay) (ton)	–	38 025	16 520	10 457
Sand (ton)	–	–	6188	4950
Gravel (1″–1$^{1/4″}$) (ton)	–	–	3019	–
Gravel (3″–4″) (ton)	–	–	1725	1725
Geotextile fabric	–	–	–	21.2
Granular activated carbon (GAC) (Charcoal) (ton)	–	–	–	1.8
Sampling event (km)	1600	–	–	–

Table 20.25 Location of raw material sources and disposal sites.

Source/location	Material	Distance (km)
Covanta Environmental Solutions in Cedar Rapids, IA	Dredged sediment	10.6
Martin Marietta Quarry – Cedar Rapids, IA	Sand Gravel-(1″–1 ¼″) Gravel-(3″–4″)	16.25
CETCO, Arlington Heights, IL	Reactive core mat (RCM)	396
Sediment testing laboratory, Lenexa, Kansas	Contaminated sediment samples	482

The economic assessment was based on the direct cost associated with the materials and the processes involved in each remediation option. The direct cost includes the cost of the materials used in the capping, transportation costs, disposal costs, sampling and laboratory analysis costs in MNA, dredging costs, and labor costs. Indirect costs associated with the remediation options are also included. For example, the social cost of greenhouse gasses such as CO_2 and CH_4 as well as other emissions engendering environmental impacts. Total direct and indirect costs were determined for each remediation option and a comparison was made to make a decision about the most preferred remediation option among all four options considered.

The social impact of each remediation option was assessed by conducting an online survey of indicators describing the impact of each of the remediation options on social aspects at the individual, community, economic, and environmental levels. Most of the indicators were chosen from the SSEM tool developed by Reddy et al. (2014). Table 20.26 shows the various indicators applied for social sustainability assessment.

20.6.3 Environmental Sustainability

A comparison of the environmental impacts of each remediation option is given in Figure 20.41 showing the impact assessment by the TRACI and the BEES methods. The impacts are normalized with respect to the highest contributor in each impact category and expressed in terms of percentage. On analyzing the results of both the TRACI and BEES methods, the conventional capping option appears to have the highest environmental impact in most of the impact categories, while MNA has the lowest impacts in total. This is because MNA does not involve any material acquisition or equipment. In addition, MNA includes significantly less transportation than the other remedial options. Conventional capping seems to be the least sustainable option among the four remediation alternatives.

In the environmental impact assessment of various stages involved in each remediation option, the transportation stage generated most of the negative impacts. To eliminate the dominance of transportation impact, an LCA was performed for all the options with the transportation distances limited to 5 km. Figure 20.42 shows the environmental impact assessment of all four remediation options using 5 km transportation distance and the TRACI method. The results show that even when the transportation distance is minimal, conventional capping remains the highest contributor toward negative environmental impacts.

SEFA was also used to analyze the environmental impact of the various remedial alternatives. This is a tool specifically developed for assessing the sustainability of remediation projects.

Table 20.26 Social sustainability assessment indicators (Reddy et al. 2014).

Criteria	Indicators
Socioindividual	Overall health and happiness
	Income generating activities
	Contaminant exposure (trespasser, worker)
	Accident risk-injury
	Recreational activity
Sociocommunity	Appropriateness of future land use with respect to the community environment
	Enhancement of commercial/income-generating land uses
	Enhancement of recreational facilities
	Degree of "grass-roots" community outreach and involvement
	Time for completion of remediation and opening of park to public
	Degree of improvement in esthetic value
Socioeconomic	Economic impacts of project on community
	Damage to property
	Effect on tourism
	Disruption of businesses and local economy during construction/remediation
	Employment opportunities during construction/remediation
	Impact on fishing activities
Socioenvironmental	Impact on aquatic habitat
	Degree of consumption of natural resources
	Degree to which proposed project will affect other media (i.e. emissions/ air pollution resulting from soil or groundwater remediation)
	Effects of anthropogenic contaminants at "chronic" concentrations
	Effects of anthropogenic contaminants at "acute" concentrations

Since the dredging option required the disposal for the highest quantity of material, this option had the greatest negative environmental impact. MNA showed minor environmental impacts in all the categories: energy consumption, GHG, NO_x, SO_x, particulate matter, and Hazardous Air Pollutants (HAPs) emission. Conventional and modified capping showed intermediate values for all the categories, except for HAP emission where modified capping showed the highest impact.

20.6.4 Economic Sustainability

Table 20.27 summarizes the direct cost for each remediation option. Costs were estimated using an online inventory of construction cost data (available at www.allcostdata.com). MNA is the least expensive option, followed by modified capping, then conventional capping, and finally dredging, which is the most expensive option. The indirect cost, also presented in Table 20.27, was calculated using the USEPA cost of carbon calculator based on the numerical results of carbon dioxide and nitrous oxide compound emissions evaluated from environmental sustainable assessment (LCA). The indirect cost followed the same trend as the direct cost.

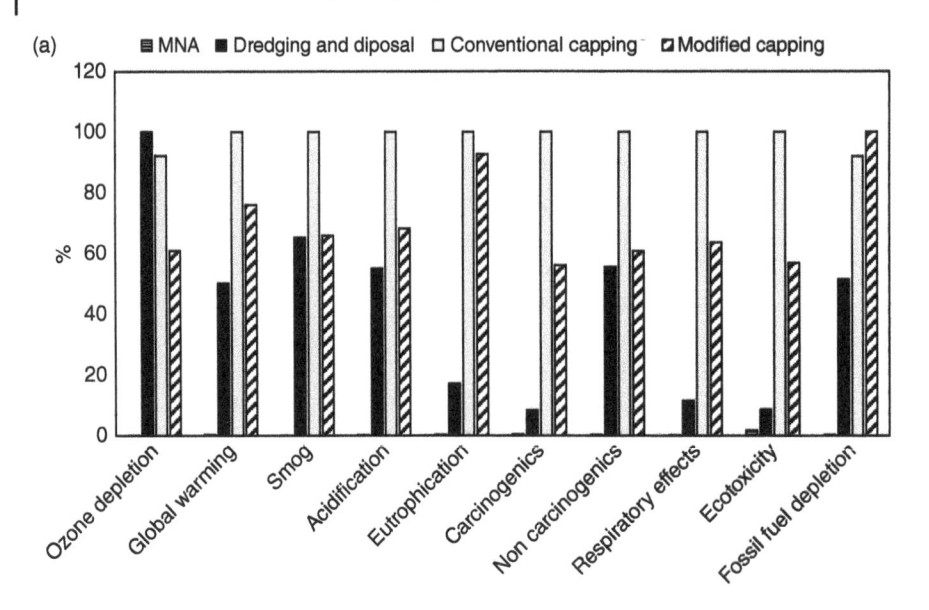

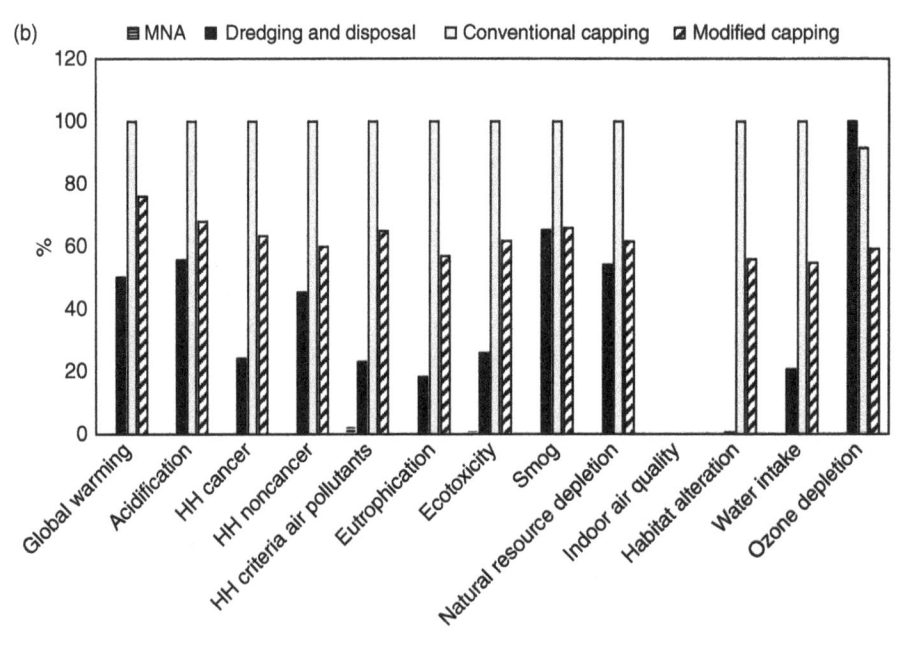

Figure 20.41 Environmental impact assessment of the four remediation options using (a) TRACI and (b) BEES methods.

20.6.5 Social Sustainability

Social impacts were assessed based on the responses obtained through an online survey of respondents with an interest in sustainability. Each remediation option was ranked in a scale of 1–4 (1 being the best and 4 being the worst) for each indicator that best describes the impacts of remediation activity at individual, community, economic, and environmental level. The overall score based on the summation of the survey results is summarized in Table 20.28. Each indicator was ranked 1–4 and a final ranking was determined based on the average score from the

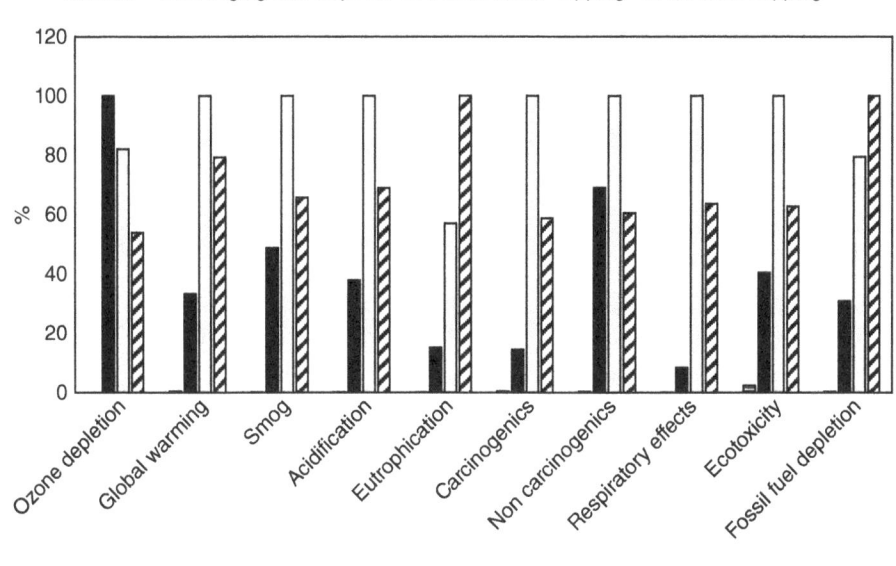

20.6 Remedial Alternatives for PCB- and Pesticide-Contaminated Sediment | 413

Figure 20.42 Environmental impact assessment of the four remediation options using TRACI method limiting the transportation of materials to 5 km.

Table 20.27 Direct and indirect costs for remedial options.

Remedial option	MNA	Dredging and disposal	Conventional capping	Modified capping
Direct cost	$545 100	$1 447 101	$806 404	$751 731
Cost of CO_2	$1498	$18 151	$6779	$11 156
Cost of NO_x	$39.06	$543.90	$200.97	$191.10
Total cost	$546 637	$1 465 796	$813 384	$763 078

Table 20.28 Overall score obtained by the four remediation alternatives in social sustainability assessment.

Social sustainability matrix				
	Monitored natural attenuation	Dredging of contaminated sediments	Conventional capping	Modified capping
Socioindividual	17	12	13	8
Sociocommunity	24	14	15	7
Socioeconomic	15	18	17	9
Socioenvironmental	15	15	12	8
Grand score	71	59	57	32

respondents. The scores from each category (socioindividual, sociocommunity, socioeconomic, and socioenvironmental) were then summed for a total score for each category. Among the four remediation options, MNA received the highest score indicating that it has the greatest negative social impact. Modified capping received the lowest score making it the preferred choice. These results were from responses obtained from total respondents. Since social sustainability is a subjective field, the results may vary with a larger sample size or with respondents outside the engineering field.

20.6.6 Overall Sustainability

The overall sustainability of the four remediation options was assessed with the help of Integrated Value Model for Sustainability Assessment (MIVES) method involving value functions. MIVES method involves various steps: (i) defining the problem, (ii) defining the variables in terms of criteria/indicators that best describe the impacts due to the project/activity, (iii) establish value functions that convert all the qualitative and quantitative variables into a set of variables with same units and scales, (iv) define the weightages to be assigned to each criterion/indicator used in the analysis, (v) evaluate the scores obtained in each criterion/indicator, and (vi) assessment of the results and decision making based on the scores obtained (Pons et al. 2016).

Various criteria and their indicators were defined for the environmental, economic, and social aspects with an objective to determine a final value for the overall sustainability. For the environmental pillar, the indicators were the impact categories obtained from the LCA; for the economic pillar, the indicators were the various direct and indirect costs associated with the project; and for the social pillar, the indicators were the ones defined for the survey. The value function analysis was performed following the methodology developed by Josa and Alavedra (2006). This method involves establishing representative value functions that convert all the quantitative and qualitative variables into a set of variables with same units or scales. Each indicator is assigned a weight (in terms of percentage) depending upon the preference of the stakeholder. The value obtained for each indicator using the value function ($V_{indicator}$) is multiplied by the respective weightage ($W_{indicator}$) as-signed to the indicator. Each of the indicator can be grouped into a criterion. The sum of the products of the $V_{indicator}$ and $W_{indicator}$ of the indicators given under a criterion becomes the value of the criterion ($V_{criterion}$). Further, each of the value obtained for criterion are multiplied by their corresponding weightages assigned for each criterion to obtain the value for each sustainability requirement ($V_{requirement}$) (e.g. environmental, economic, and social). The final value (V_{final}) for each sustainability pillar is obtained by taking the sum of the product of $V_{requirement}$ and the weightage ($W_{requirement}$) assigned to each of the main requirement. The sum of the V_{final} values of the environmental, economic, and social pillars gives the final sustainability index for a remedial. The final score ranges from 0 to 1.

Four cases (a, b, c, and d) were evaluated by assigning different weights to the three pillars of sustainability to examine how the sustainability decision is affected by the relative importance given to the three pillars. The results are shown in Figure 20.43. MNA resulted to be the most sustainable option followed by modified capping for case a (equal weights), b (heaviest weight to economic pillar), c (heaviest weight to environmental pillar), and d (heaviest weight to social pillar). Figure 20.43d shows that the MNA scored less than modified capping when the social pillar was assigned the highest weightage. In this weighting scenario, modified capping appeared as the most sustainable option. Thus, overall sustainability is subjected to the importance given to each assessment pillar. The weights are assigned to the three pillars depending upon the stakeholder's preference for the relative importance of the three pillars.

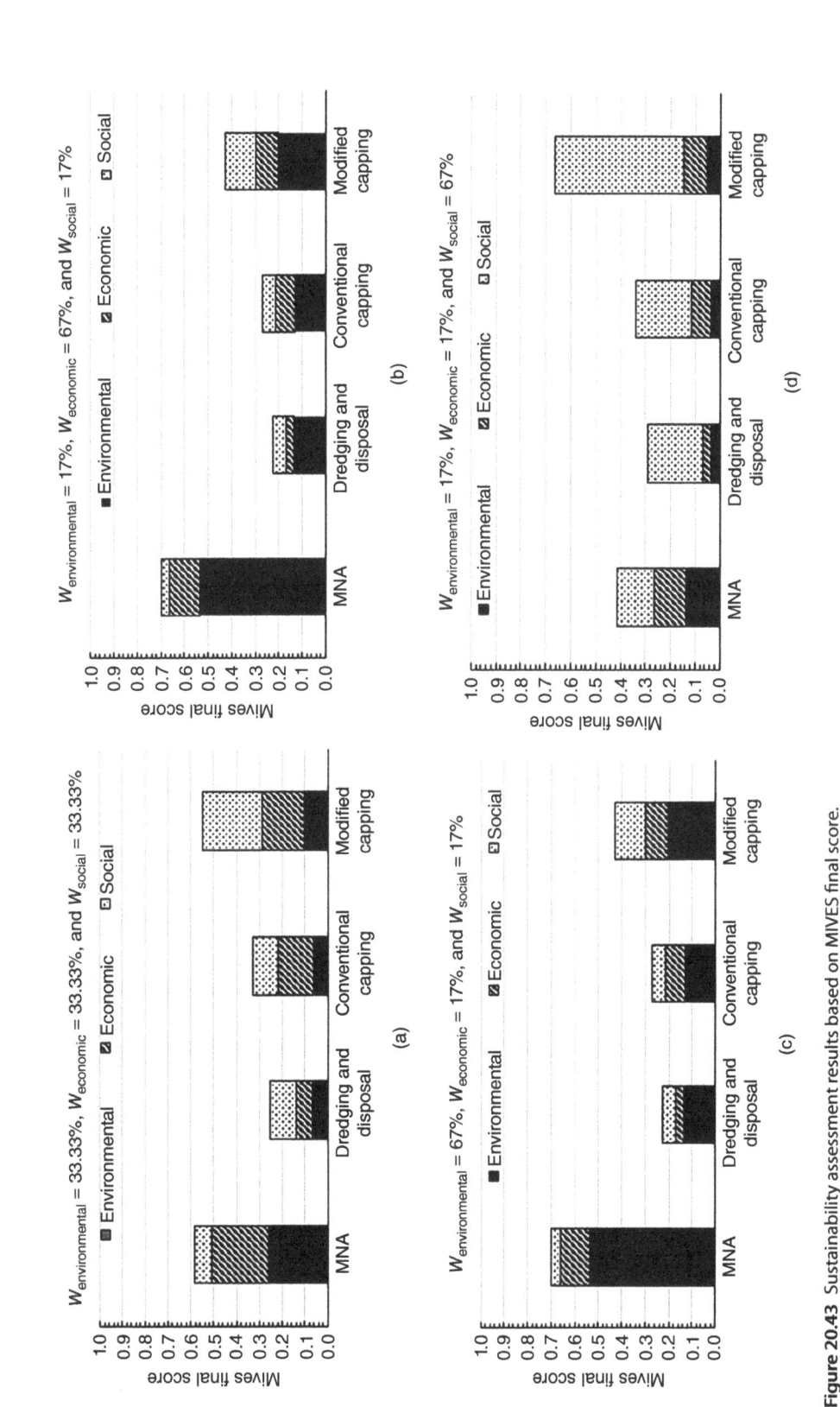

Figure 20.43 Sustainability assessment results based on MIVES final score.

Decision making solely depends upon the importance of the project at environmental, economic, and social level, and it depends upon the preference of stakeholders. For example – in this study, though the MNA option scored highest in the MIVES scale for most of the weighting scenarios, it might not be considered as the most preferred option. Here, the decision could be based on the case where social pillar was given the highest weightage. MNA had highest negative impact at social level as it is associated with the risk of exposing the community as well as the surrounding environment to the contamination for a longer duration of time. Hence, it can be deduced from the results of the various weighting preferences that the overall sustainability depends upon the weighting preferences given to each sustainability pillar.

20.6.7 Conclusions

LCIA carried out using TRACI and BEES methods showed that conventional capping had the highest negative environmental impacts. MNA appeared to have the least negative impact which is justifiable, since it does not require any dredging or transportation activities. However, it poses a risk to the benthic organisms and the surrounding environment. The overall sustainability was found to be a function of the preferences given to the three sustainability pillars. The weights assigned to the three pillars depend upon the relative importance of the pillars for the project. Based on the overall sustainability assessment, MNA turned out to be the most preferred choice for most of the cases (varying weights) studied. However, for a case when social pillar was given the highest preference, modified capping turned out to be the most preferred choice due to the reduced risk to the community from the presence of the contaminants in the lake while MNA poses unacceptable risk due to exposure to contaminated sediments for long periods of time.

20.7 Summary

Comprehensive sustainability assessments incorporating environmental, economic, and social aspects can be applied to a range of projects and products. This chapter demonstrated detailed quantitative sustainability assessments using different quantitative and qualitative tools of typical environmental and chemical engineering projects. As demonstrated, system boundaries should be carefully drawn to incorporate desired LCA activities. Further, the assessments can include specific weighting or incorporate sensitivity analyses to incorporate certain aspects of concern for a project or a product. As seen LCA is commonly accepted for the assessment of environmental sustainability, but it does not address the economic and social dimensions of sustainability. Standardized methods to assess economic sustainability and social sustainability are not yet developed; therefore, these aspects are assessed by different methods. Efforts are being made by several sustainability professionals to develop new economic and social sustainability assessment methods consistent with the same standardized environmental LCA framework. Currently, integrating environmental, economic, and social assessments into overall sustainability requires weighing between these three sustainability dimensions (triple bottom line), which maybe subjective. There is a need for development of a standardized sustainability life cycle assessment method that can integrate three pillars of sustainability(environment, economy, and society). This is also true for the example projects described in Chapters 21–22.

References

ArcGIS (2017). Mapping without limits. www.arcgis.com (accessed 15 September 2017).

ASCE (2018). Envision. American Society of Civil Engineers. https://www.asce.org/envision/ (accessed September 2018).

Barry, A., Wolfe, A., English, C., Ruddick, C., and Lambert, D. (2016). 2016 National Algal Biofuels Technology Review (No. DOE/EE--1409). USDOE Office of Energy Efficiency and Renewable Energy (EERE), Bioenergy Technologies Office (EE-3B).

Chandler, K., Norton, P., and Clark, N. (2001). *Waste Management's LNG Truck Fleet: Final Results*. US Department of Energy.

Eloneva, S., Said, A., Fogelholm, C.J., and Zevenhoven, R. (2012). Preliminary assessment of a method utilizing carbon dioxide and steelmaking slags to produce precipitated calcium carbonate. *Applied Energy* 90 (1): 329–334.

Heijungs, R., Goedkoop, M., Struijs, J., Efftıng, S., Sevenster, M., & Huppes, G. (2003). Towards a life cycle impact assessment method which comprises category indicators at the midpoint and the endpoint level. *Report of the first project phase: design of the new method VROM report*.

Illinois Recycling Association (2009). *Illinois Commodity/Waste Generation and Characterization Study*. Report, vol. 323.

Jirka, S. and Tomlinson, T. (2014). *State of the Biochar Industry*. International Biochar Initiative http://www.biochar-international.org/State_of_industry_2013 (accessed 4 May 2017.

Jorquera, O., Kiperstok, A., Sales, E.A. et al. (2010). Comparative energy life-cycle analyses of microalgal biomass production in open ponds and photobioreactors. *Bioresource Technology* 101 (4): 1406–1413.

Josa, A. and Alavedra, P. (2006). El concepto de sostenibilidad. In: *La Medida de la Sostenibilidad en Edificación industrial: MIVES* (ed. R. Losada, E. Rojí and J. Cuadrado), 59–70. Bilbao, Spain: Universidad Politécnica de Valencia, Universitat Politècnica de Catalunya and Labein-Tecnalia (in Spanish).

Kulyk, N. (2012). Cost-benefit analysis of the biochar application in the US cereal crop cultivation. School of PublicPolicy. *University of Massachusetts Amherst*.

Lubben, D.R. (1994). Assessment of urban lake contamination: a diagnostic/feasibility study of Cedar Lake, Iowa. PhD thesis. University of Iowa.

Nordhaus, W.D. (2007). A review of the stern review on the economics of climate change. *Journal of Economic Literature* 45 (3): 686–702.

Pan, S.Y., Adhikari, R., Chen, Y.H. et al. (2016). Integrated and innovative steel slag utilization for iron reclamation, green material production and CO_2 fixation via accelerated carbonation. *Journal of Cleaner Production* 137: 617–631.

Pons, O., de la Fuente, A., and Aguado, A. (2016). The use of MIVES as a sustainability assessment MCDM method for architecture and civil engineering applications. *Sustainability* 8 (5): 460.

Reddy, K. R., Gopakumar, A., Chetri, K. J., Kumar, G., and Grubb, G. D. (2018) Sequestration of landfill gas emissions using basic oxygen furnace slag: Effects of moisture content and humid gas flow conditions. *Journal of Environmental Engineering* (in press)

Reddy, K.R., Sadasivam, B.Y., and Adams, J.A. (2014). Social sustainability evaluation matrix (SSEM) to quantify social aspects of sustainable remediation. In: *ICSI 2014: Creating Infrastructure for a Sustainable World* (eds. J. Crittenden, C. Hendrickson and B. Wallace). Proceedings of the 2014 International Conference on Sustainable Infrastructure. 831–841. Reston, VA: ASCE. DOI:10.1061/9780784478745.

Sarperi, L., Surbrenat, A., Kerihuel, A., and Chazarenc, F. (2014). The use of an industrial by-product as a sorbent to remove CO_2 and H_2S from biogas. *Journal of Environmental Chemical Engineering* 2 (2): 1207–1213.

USEPA (2017a). Advancing Sustainable Materials Management: Facts and Figures. www.epa.gov (accessed 15 September 2017).

USEPA (2017b). *Hydrologic Evaluation of Landfill Performance (HELP) Model*. US Environmental Protection Agency https://www.epa.gov/land-research/hydrologic-evaluation-landfill-performance-help-model accessed 15 May 2017.

Van Oss, H.G. (2004). *Slag-Iron and Steel. Minerals Yearbook – 2004*. U.S. Geological Survey https://minerals.usgs.gov/minerals/pubs/commodity/iron_&_steel_slag/islagmyb04.pdf.

Yargicoglu, E.N. and Reddy, K.R. (2017). Effects of biochar and wood pellets amendmentsadded to landfill cover soil on microbial methane oxidation: A laboratorycolumn study. *Journal of Environmental Management* 193, 19–31.

Yargicoglu, E.N. and Reddy, K.R. (2017b). Biochar-Amended Soil Cover for Microbial Methane Oxidation: Effect of Biochar Amendment Ratio and Cover Profile. *Journal of Geotechnical and Geoenvironmental Engineering* 144 (3): 04017123. DOI:10.1061/(ASCE)GT.1943-5606.0001845

21

Civil and Materials Engineering Sustainability Projects

21.1 Introduction

Several detailed sustainability assessments of potential environmental and chemical engineering projects were presented in the previous chapter. Comprehensive sustainability assessments incorporating environmental, economic, and social aspects may also be applied to civil and materials engineering-related projects. This chapter presents several example studies of sustainability assessments applied to these fields.

21.2 Sustainable Translucent Composite Panels

In this example study, the use of translucent composite panels is considered for the construction of the external façade of a warehouse building. The main benefit of the composite panels is the energy saving because of the natural light that the translucent panels let into the building. This project analyzed the environmental, economic, and social benefits of the translucent composite panel versus the conventional masonry wall for the construction of the warehouse façade.

21.2.1 Background

Concrete has been a key material for many structural applications, from residential homes to multistory buildings and bridges. Since its first application more than 2000 years ago, its primary use has slightly altered over the years with improved applications in structural and architectural elements alike.

Buildings are the largest contributors to CO_2 gases in the atmosphere, taking into account its entire life cycle from material acquisition, construction, use and maintenance, to demolition. On a global scale, the building sector, in general, is responsible for the consumption of 30% of the energy use around the world, increasing to 39% when only considering the United States (USDOE 2012). The commercial sector is a great contributor to the total building carbon footprint in the United States, and the greenhouse gas (GHG) emission rate from commercial buildings has steadily increased 2.5% per year since the 1980s (USDOE 2012). Such emissions are related to the energy consumption of commercial facilities that occupy large square footage buildings. Engineers and architects are increasingly searching for "greener" building solutions for many different construction applications.

This study considers the implementation of new materials and designs in building practices. The change specifically being evaluated is the introduction of composite materials that can offer alternatives to traditional material uses for nonstructural elements. For example, commercial warehouse facilities often include a primary structural system that will resist gravity and lateral

Sustainable Engineering: Drivers, Metrics, Tools, and Applications, First Edition.
Krishna R. Reddy, Claudio Cameselle, and Jeffrey A. Adams.
© 2019 John Wiley & Sons, Inc. Published 2019 by John Wiley & Sons, Inc.

loads and have an exterior facade that is usually a concrete-based material, such as a cement masonry unit (CMU) or a clay brick veneer. They are used widely around the world because they are easily available, cost-effective, offer satisfactory insulation properties, and can hold up well against the atmospheric conditions. However, their raw material consumption accounts for a significant contribution to the total amount of GHG emissions. As an alternative, new composite materials with a smaller carbon footprint can be used as a substitute to the traditional materials.

21.2.2 Methodology

21.2.2.1 Goal and Scope

A warehouse that is under construction in the city of Ukiah, California, is considered for this study. The actual building design will be compared with a new scenario in which composite translucent panels will be used for the exterior façade but keeping the modifications in the design and structure to a minimum. As shown in Figure 21.1, lighting is responsible for over 25% of commercial energy use. Translucent panels can increase natural light and decrease energy use but also can have positive impacts on occupants' well-being and productivity. The benefits of the new construction scenario will be assessed considering social, environmental, and economical aspects.

21.2.2.2 Technical Design

Scenario 1. Conventional design: The proposed warehouse consists of a 148 000-ft^2 floor plan. The building has structural steel as its primary support system. The current design includes a CMU façade, using $8'' \times 8'' \times 16''$ CMU blocks. A cross section taken from the structural plans for the facility shown in Figure 21.2 further provides the detailing of the CMU construction for the exterior façade. The roofing system consists of open web steel trusses, also known as bar joists, and a galvanized steel decking platform, commonly used for concrete pouring in composite construction.

A total of 32 431 pieces of hollowed celled masonry unit are needed for the construction of the exterior façade. This number was calculated by using the total linear footage of wall (1548 ft) and the average wall elevation of 30 ft, determined considering the variations on the parapet wall. Moreover, 216 000 lb of type M cement mortar mixture commonly used in exterior

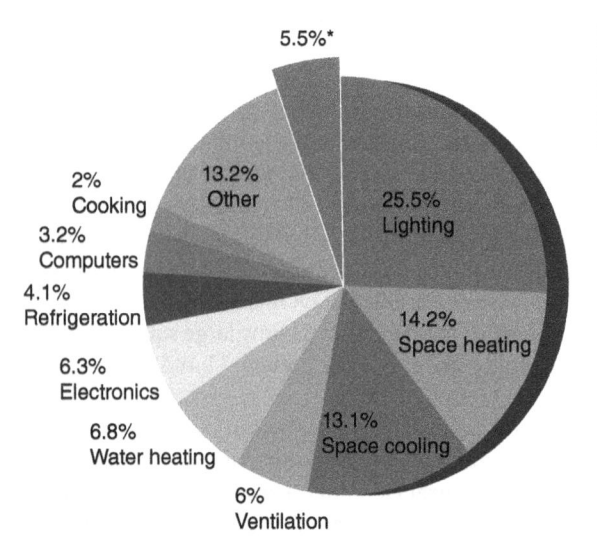

Figure 21.1 Commercial primary energy end-use splits in 2005 (USDOE 2008). *Energy Information Administration (EIA) adjustment factor that accounts for incomplete data in EIA's sampling and survey methodology.

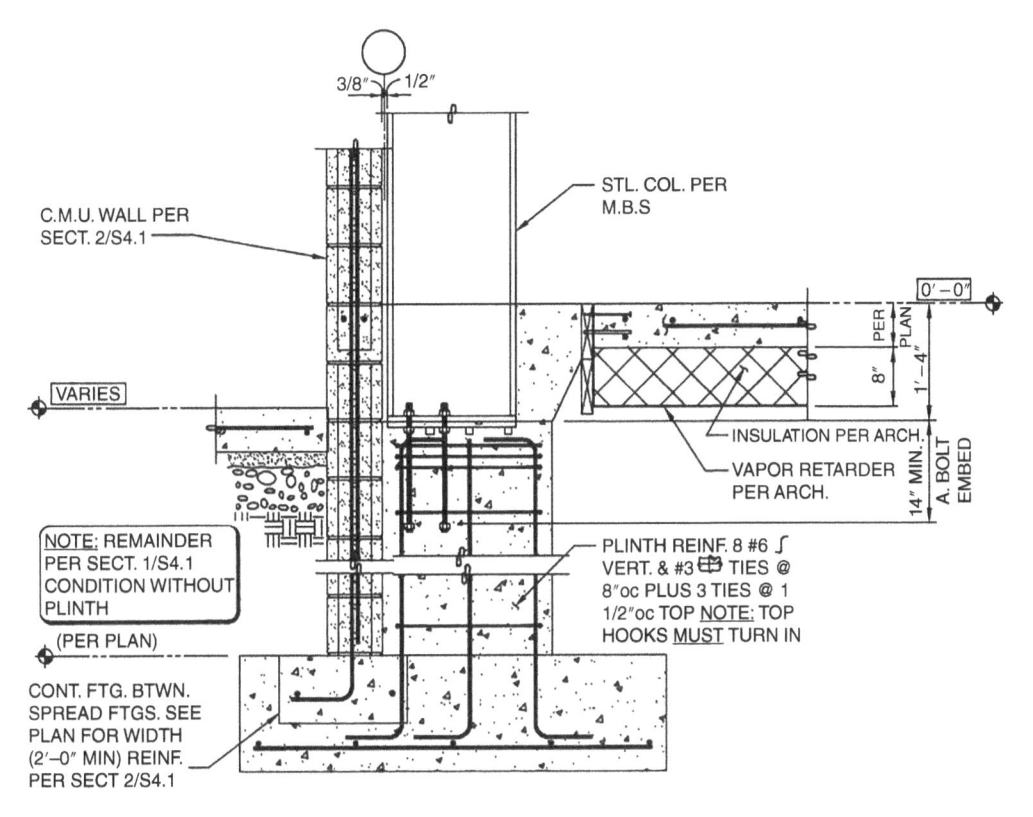

Figure 21.2 Wall cross section per plan elevation.

nonstructural CMU walls will be required to bond masonry units together. In addition, 116.5 ft^3 of grout will be needed to fill in the cell for vertical steel reinforcement anchoring of the wall. The grout is also a cement-based material but with a different cement ratio than mortar.

Scenario 2. Translucent composite panels: The use of the CMU blocks as a veneer (nonstructural component) is a common construction standard that can be substituted with other, more sustainable materials. Translucent composite panels in the façade may be a sustainable alternative for the warehouse. Currently, there are manufacturers who produce reinforced fiberglass sandwich panels that provide daylighting transmissions between 5% and 60% while providing good insulating qualities. This study focuses on the energy saving by providing daylight into the building, but the energy savings resulting from insulating properties are not considered.

The translucent panels will be supplied by a manufacturer of composite walls located in New Hampshire. The selected panel for this specific design has a light transmittance of 20% natural lighting due to its white, nonclear exterior color (Figure 21.3). The wall panels will be manufactured to the maximum allowed width of 5 ft and height of 30 ft.

A schematic of the panel is shown in Figure 21.4. The composite panel is made of a structural fiberglass exterior and interior layer with an aerogel midlayer. The panels themselves only weigh approximately 3 lb/ft^2, which is significantly less when compared to the CMU wall (about 90 lb/ft^2). The major cost parameters include transportation and erection costs and fuel consumption. The raw materials used for panel manufacturing include 20% of recycled materials. A total of 310 panels are required to cover the wall perimeter of 1548 ft. The panels will be transported via rail from the manufacturer location to the construction site.

Figure 21.3 Composite translucent panels in exterior façade. Source: Kalwall Inc. Manchester, NH.

Panel anatomy

KWS weatherable coating technology

Bondline ('shoji' grid pattern shown)

Exterior translucent FRP face sheet

Translucent thermal insulation

Interior translucent FRP face sheet

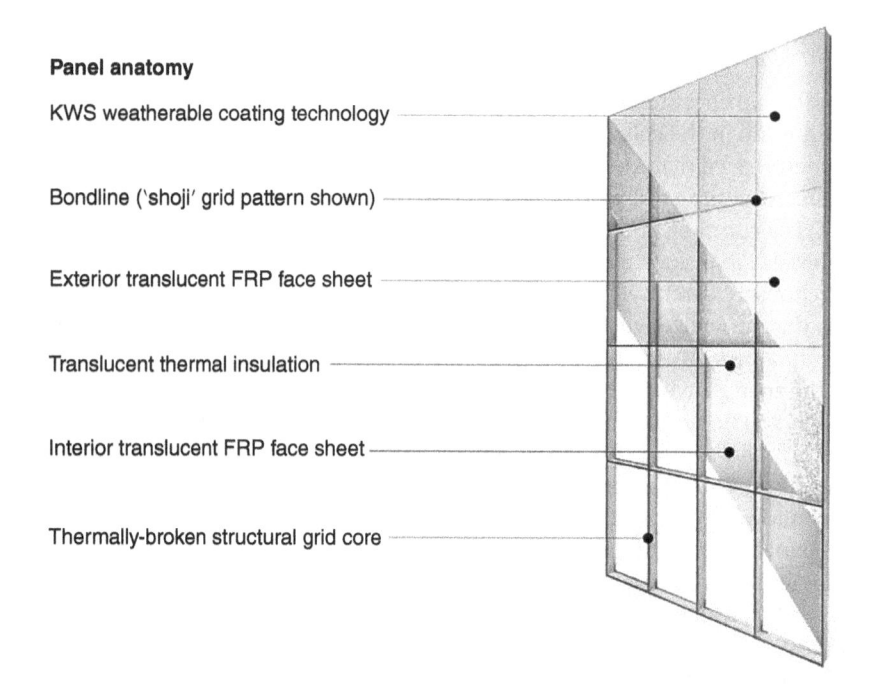

Thermally-broken structural grid core

Figure 21.4 Structure of a composite translucent panel. Source: Kalwall Inc. Manchester, NH.

21.2.3 Environmental Sustainability

The environmental impact of the two scenarios, CMU wall and the translucent composite panels, has been analyzed by using SimaPro™ LCA software. The raw material quantities and transportation were considered for the two scenarios (Figure 21.5). The greatest impact for CMU wall is from the transportation of the cement-based products to the facility. For this specific warehouse location in Ukiah, California, there are a few suppliers located within a 30-mi radius of the facility. The reduction of the transportation distance will decrease the total impact. The impacts from the materials itself is only evident in the category "ozone depletion," and this result is related to the energy consumption in the fabrication of CMU and mortar.

The greatest environmental impacts in each category for the composite panel are related to the fiberglass layer. Aluminum alloy used in the panel framing is the second element in the environmental impacts. The impact due to transportation is minimal, even though the travel distance was approximately 3200 mi from the panel manufacturing facility to the building location, but the characteristic lightweight of the panels (only 3 lb/ft^2) justifies the low impact in transport. When comparing both the scenarios, the CMU wall and translucent wall panels (Figure 21.6), the composite panel's environmental effects are smaller than the impact of cement-based materials. In every category, the CMU wall impacts are much higher than that of the composite panels.

The energy consumption in the warehouse was simulated with "eQuest 3-65" software. This software simulates the anticipated energy consumption for any facility, residential or commercial, using site-specific characteristics including the position of the building in relation to the sun. This software permits to simulate the daylight available to determine the energy requirements of the building for the two scenarios: CMU wall and composite panel wall.

The total energy consumption of the current design using CMU construction (Figure 21.7) resulted in 1.5 million kWh of electricity, of which 922 000 kWh were directly for facility lighting. The energy required for lighting is more than half of the total energy consumption. The carbon emission resulting from the consumption of 922 000 kWh is approximately 512 000 lb CO$_2$e (eGrid2010, Version 1.1, GHG annual emission rates for region-specific sites). In comparison, in Figure 21.8, the resulting consumption of energy from using the translucent panels was around 1.13 million kWh, which is approximately 430 000 kWh less than that of the use of CMU construction. This result proves that the emission rates can be shortened by just changing the façade design. The total consumption for lighting using the translucent panels was 540 300 kWh/year. With this result, the carbon emissions for the composite panels would be 369 700 lb of CO$_2$e opposed to 511 860 lb of CO$_2$e with the CMU wall. This is a reduction of 38% in the overall carbon emissions. It should be noted that these results are site-specific and can change from location to location.

21.2.4 Economic Sustainability

The total cost of the CMU wall was estimated with a calculator provided by "homewise.com" and the results are reported in Table 21.1. This calculator conservatively approximates the cost for various masonry construction applications; in this case, it focused on the erection of the CMU walls. The only cost parameter not defined in the calculator was the grouting of the cells. This cost can be approximated by increasing the overall cost by 10%, as found for similar wall erections.

The cost of the wall with translucent composite panels is much higher due to the fabrication costs of the panels, the lack of suppliers, and transportation over long distance. Additionally, skilled personnel from the panel manufacturer is recommended for the installation of the panels

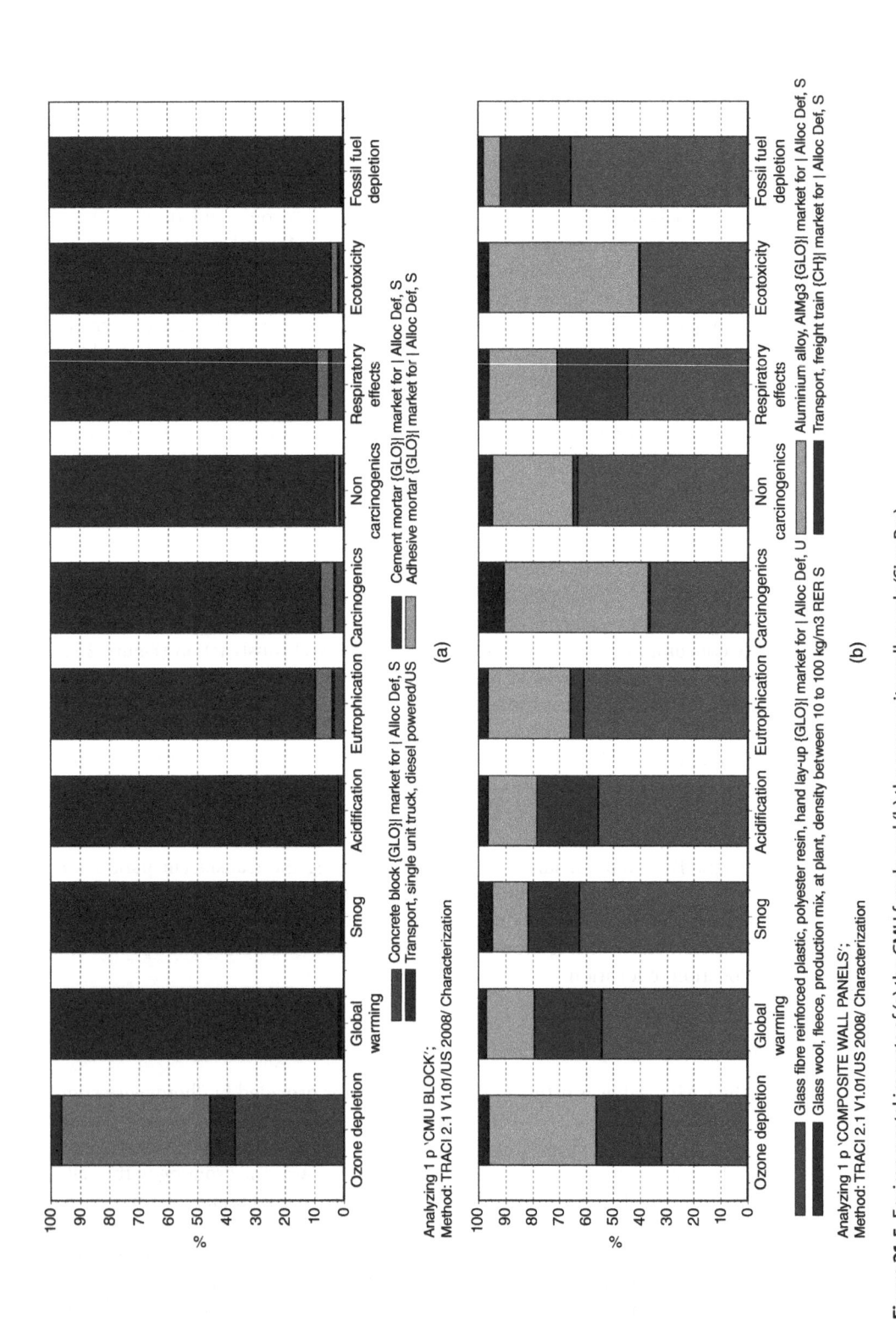

Figure 21.5 Environmental impacts of (a) the CMU façade and (b) the composite wall panels (SimaPro).

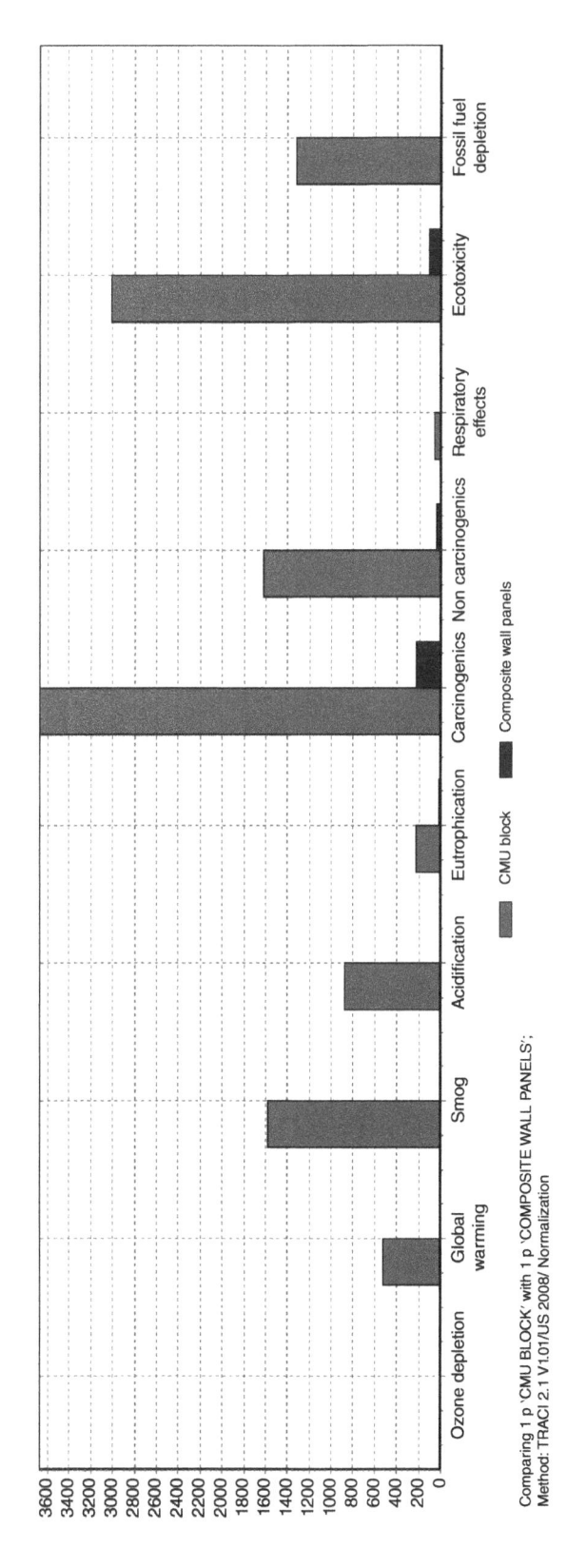

Figure 21.6 Comparison of environmental impacts of the CMU façade and the composite wall panels (SimaPro).

Electric consumption (kWh × 000)

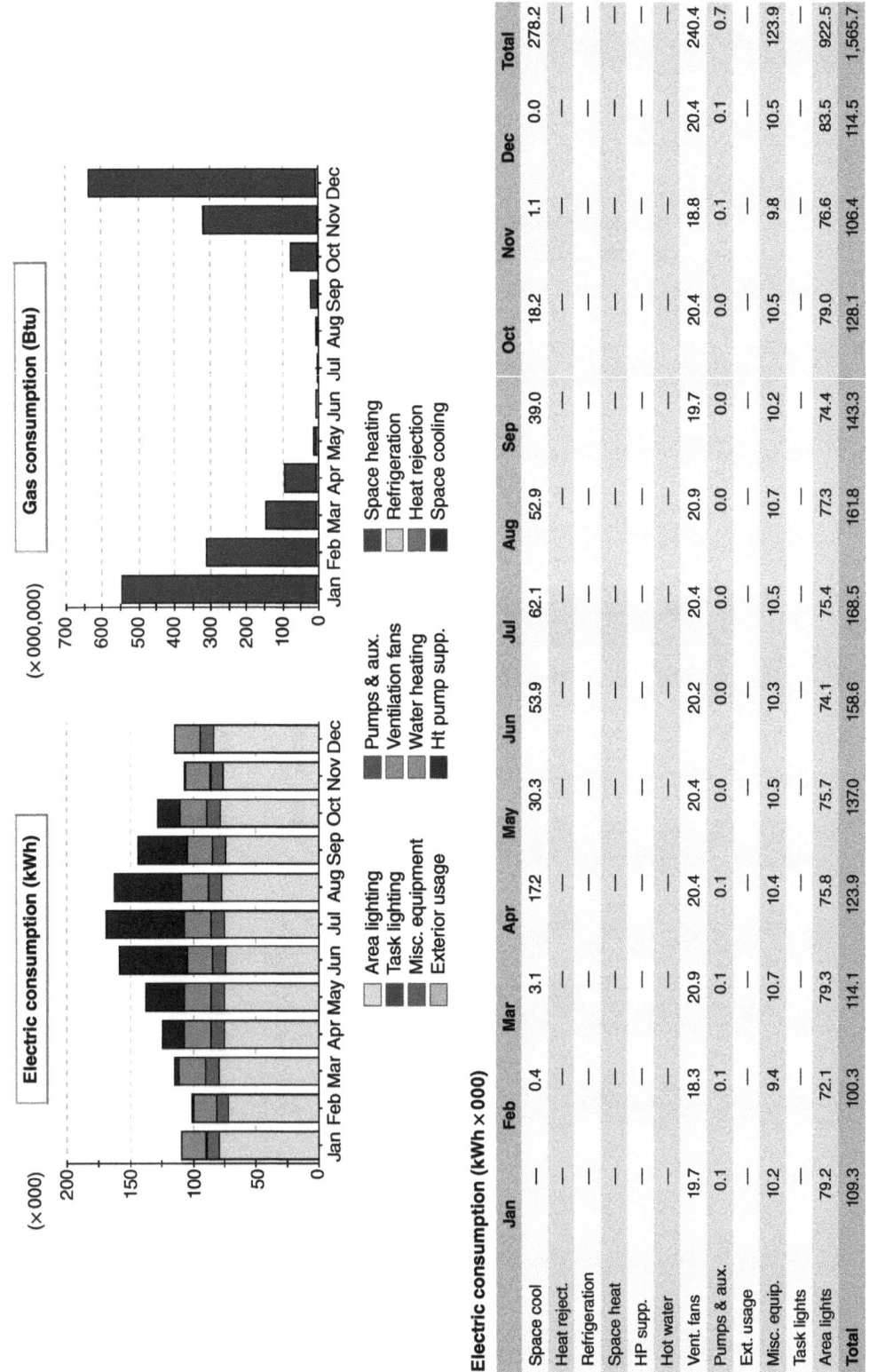

	Jan	Feb	Mar	Apr	May	Jun	Jul	Aug	Sep	Oct	Nov	Dec	Total
Space cool	—	0.4	3.1	17.2	30.3	53.9	62.1	52.9	39.0	18.2	1.1	0.0	278.2
Heat reject.	—	—	—	—	—	—	—	—	—	—	—	—	—
Refrigeration	—	—	—	—	—	—	—	—	—	—	—	—	—
Space heat	—	—	—	—	—	—	—	—	—	—	—	—	—
HP supp.	—	—	—	—	—	—	—	—	—	—	—	—	—
Hot water	—	—	—	—	—	—	—	—	—	—	—	—	—
Vent. fans	19.7	18.3	20.9	20.4	20.4	20.2	20.4	20.9	19.7	20.4	18.8	20.4	240.4
Pumps & aux.	0.1	0.1	0.1	0.1	0.0	0.0	0.0	0.0	0.0	0.0	0.1	0.1	0.7
Ext. usage	—	—	—	—	—	—	—	—	—	—	—	—	—
Misc. equip.	10.2	9.4	10.7	10.4	10.5	10.3	10.5	10.7	10.2	10.5	9.8	10.5	123.9
Task lights	—	—	—	—	—	—	—	—	—	—	—	—	—
Area lights	79.2	72.1	79.3	75.8	75.7	74.1	75.4	77.3	74.4	79.0	76.6	83.5	922.5
Total	109.3	100.3	114.1	123.9	137.0	158.6	168.5	161.8	143.3	128.1	106.4	114.5	1,565.7

Figure 21.7 End-user energy consumption for the CMU wall scenario (eQuest 3-65 software).

Annual energy consumption by enduse

	Electricity (kWH)	Natural gas (MBtu)	Steam (Btu)	Chilled water (Btu)
Space cool	213.9	—	—	—
Heat reject.	—	—	—	—
Refrigeration	—	—	—	—
Space heat	—	886.84	—	—
HP supp.	—	—	—	—
Hot water	—	—	—	—
Vent. fans	252.7	—	—	—
Pumps & aux.	0.2	—	—	—
Ext. usage	—	—	—	—
Misc. equip.	123.9	—	—	—
Task lights	—	—	—	—
Area lights	−540.3	—	—	—
Total	**1 130.9**	**886.84**	**—**	**—**

- Misc. equipment
- Space cooling
- Space heating
- Ventilation fans
- Area lighting

Electricity: 48%, 19%, 22%, 11%

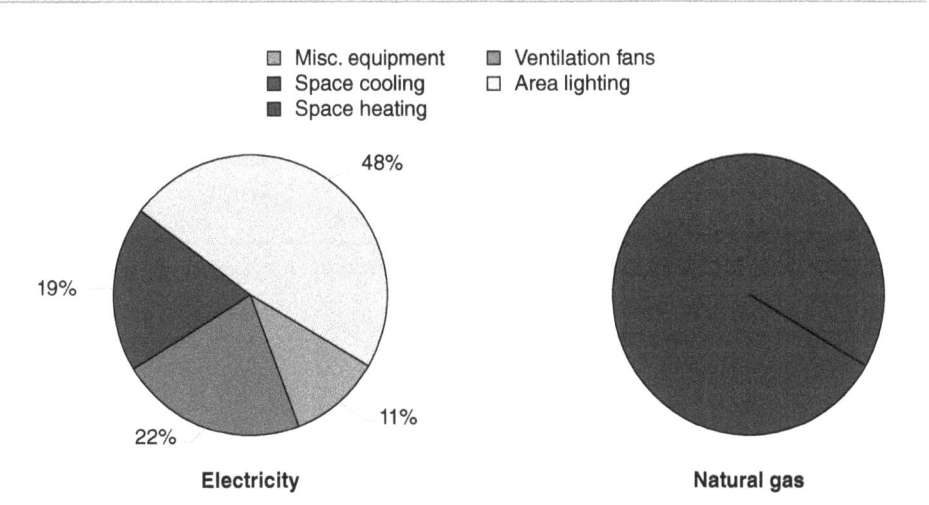

Natural gas

Figure 21.8 End-user energy consumption for the translucent panel wall scenario (eQuest 3-65 software).

due to the limited practice from local contractors. The installation cost of the panels is approximately $80 per sq. ft., of which nearly half is due to the labor itself. The construction time is about 60% less than the CMU wall, since the prefabricated panels could be readily delivered to the site. Table 21.2 shows the breakdown of the direct costs for panel wall erection including materials. The total cost for this design is approximately $2.1 million.

In Table 21.3, a side-by-side cost comparison is presented for each alternative scenario. The energy savings were estimated to be $78 000 per year on electricity for lighting when using translucent panels. The initial investment is four times higher for the composite panels. The large cost difference between the two scenarios can be explained by the proprietary nature of the composite panels, the lack of manufacturers, and skilled laborers available for this application.

21.2.5 Social Sustainability

The social assessment was performed using two different tools for estimating social impacts: ENVISION and the social sustainability evaluation matrix (SSEM) matrix (Reddy et al. 2014). The social assessment with Envision incorporated various effects that may be foreseen on social

Table 21.1 Cost to install a CMU block wall – 2014 cost calculator.

Item	Quantity	Low	High
Block wall cost	49 576 ft^2	$83 039.36	$93 883.10
Nondiscounted retail cost for common, midgrade block wall. Quantity includes typical installation waste, fabrication overage, material for future repairs, and delivery within 25 miles			
Block wall labor	7 421.5 h	$427 797.88	$487 988.58
Direct labor expenses to install CMU block wall			
Block wall job materials and supplies	46 440 ft^2	$9 413.39	$10 193.58
Cost of supplies that may be required to install CMU block wall including cutting and grinding materials, mortar, and reinforcement			
Block wall equipment allowance		$56.25	$78.75
Job-related costs of specialty equipment used for job quality and efficiency, including 115 V wet masonry saw, 5 ft^3 mortar box, and small plate compactor			
Totals – cost to install CMU block wall	46 440 ft^2	$520 306.89	$592 144.02
	Average cost per square foot	$11.20	$12.75

Table 21.2 Composite wall cost.

Wall: 4″ thick/20% transmittance	
Materials	$1 673 480
Transportation from New Hampshire (distance 3200 mi.)	$25 600
Labor costs, panels and sealants, includes crane use for lifting panels	$434 520
Total cost	$2 133 600

Table 21.3 Cost comparison for CMU versus translucent composite panels.

8″ cmu wall (annual cost)		4″ translucent panel (annual cost)	
Energy consumption	Cost	Energy consumption	Cost
Total consumption 1 565 700 kWh	$281 826 (0.18 cent/kWh)	Total consumption 1 130 900 kWh	$203 562 (0.18 cent/kWh)
Lighting usage 922 000 kWh	$165 960 (0.18 cent/kWh)	Lighting usage 540 300 kWh	$97 254 (0.18 cent/kWh)
Cost savings for using translucent panels		$78 264 (annual savings)	

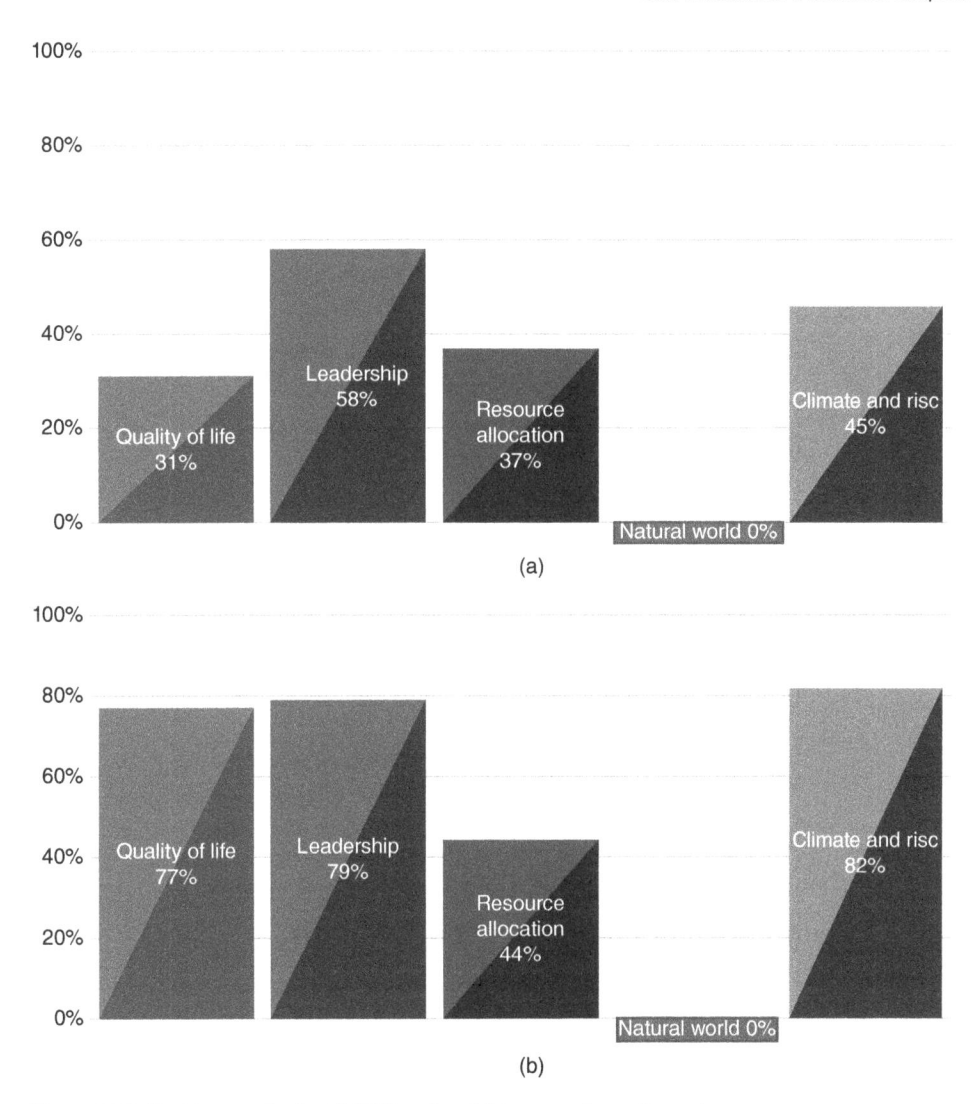

Figure 21.9 Envision results for (a) CMU wall and (b) composite wall panels.

level. The results are shown in Figure 21.9 for the two scenarios. The use of the composite panels had a greater positive effect on the people as per the score in the categories of quality of life, leadership, and climate and risk.

The SSEM method developed by Reddy et al. (2014) was also used for the social sustainability evaluation. Figure 21.10 shows the results of this specific analysis, which resulted in favor of the translucent wall panels. In every category, the composite panel outweighed the use of the CMU wall. A significant element in favor of the translucent panel is the possibility of using more natural light and save electricity. Studies have shown great impact of natural light on people's health and social well-being. The use of natural light provides a better atmosphere from the day-to-day activities, increase people's productivity, and reduce depression and number of sick days.

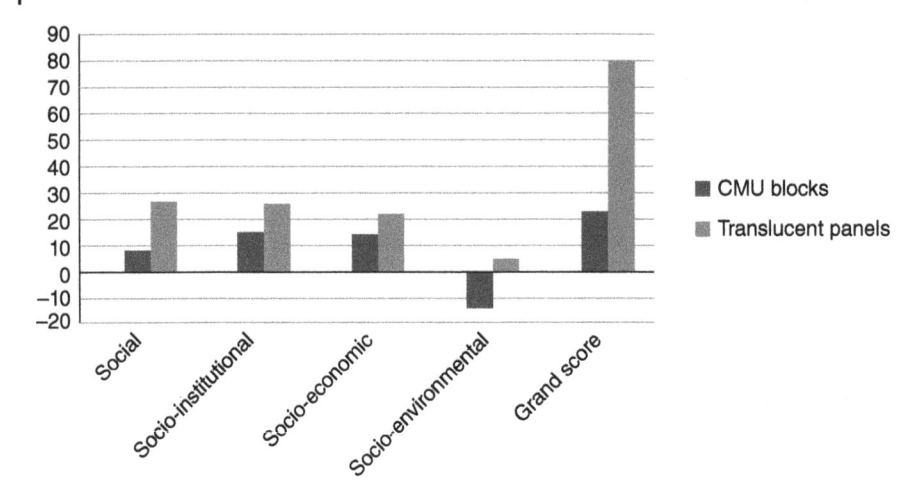

Figure 21.10 Social sustainability results from SSEM.

21.2.6 Conclusions

The use of translucent composite panels as an alternative construction material was found to be more sustainable than the conventional cement-based materials, as it can be concluded from the sustainability assessment results in this study. Although the use of composite panels is currently more expensive, additional competition would contribute to reducing total panel cost. The environmental and social factors clearly demonstrate that the translucent composite panel is a much more sustainable alternative.

21.3 Sustainability Assessment of Concrete Mixtures for Pavements and Bridge Decks

This project considers the use of recycled materials such as limestone waste, fly ash, and steel slag as additives in the concrete. The use of recyclable materials in concrete may help in the reduction of the environmental and economic impacts of the concrete itself and the constructions where it is used. In this project, 12 concrete mixtures designed specifically for pavements and bridge decks were selected. The possible benefits of the 12 concrete mixtures for the rehabilitation of Morgan Street Bridge (Chicago, Illinois) were studied by considering the three pillars of sustainability. Finally, the most sustainable mix design that can be used for the rehabilitation of the bridge was identified based on life cycle sustainability assessment.

21.3.1 Background

The primary environmental impacts associated with concrete production for civil infrastructure stem from the high carbon footprint of concrete. Cement and lime production for concrete represent the fifth and sixth largest contributors to US GHG emissions in 2012 (Figure 21.11). Among all the materials and stages required for concrete production, cement production represents the largest contributor to the overall GHG emissions associated with concrete. On an average, each ton of cement produced accounts for 0.92 tons of CO_2 emissions, 60% of which can be attributed to the calcination process used to process cement (Marceau et al. 2006).

21.3 Sustainability Assessment of Concrete Mixtures for Pavements and Bridge Decks | 431

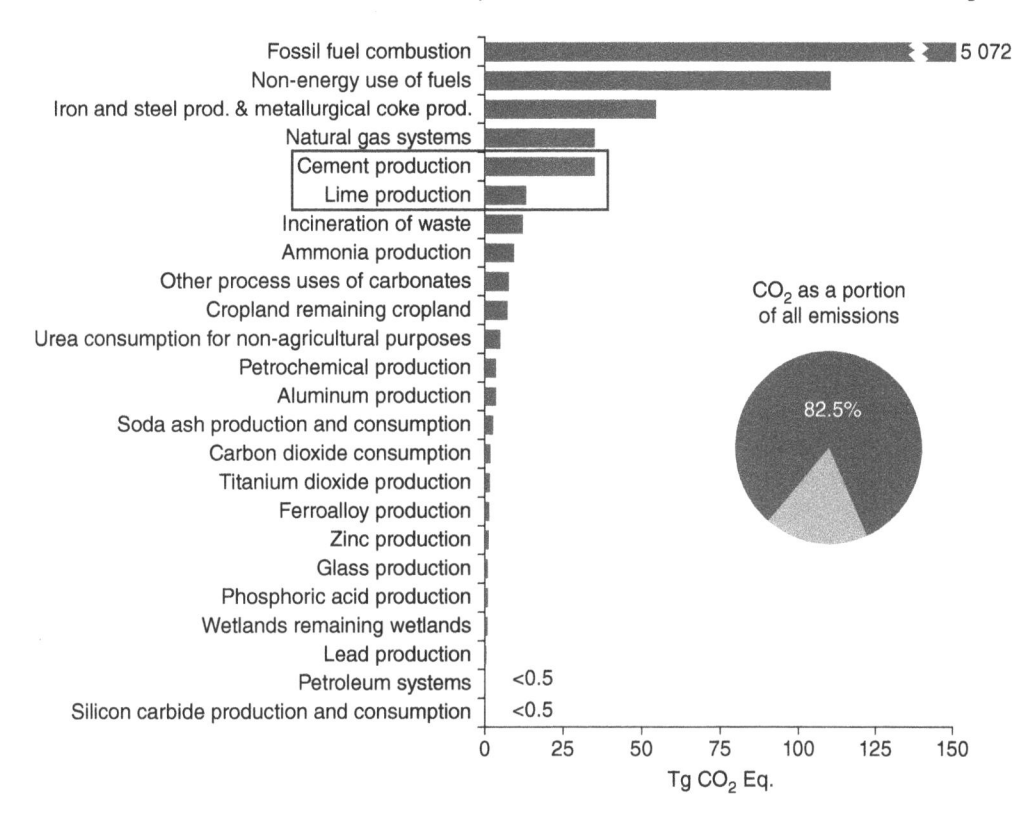

Figure 21.11 Anthropogenic sources of greenhouse gas emissions in the US in 2012. Source: USEPA (2014).

The remaining CO_2 emissions are generated during fossil fuel consumption by machinery for grinding and heating of the raw cementitious materials. Increased industrialization and demand for cement for infrastructure have led to a progressive increase in cement production from 1998 to 2012.

The increase in the cement and concrete production has pushed many organizations and agencies to seek alternative solutions that can help to reduce the GHG emission without negatively affecting cost, strength, durability, and performance. The need to produce sustainable concrete with sustainable cement has pushed the Canadian Standard Association (CSA-A3001-08) to reduce the GHG emission in their cement production. This was facilitated by allowing cement producers to replace up to 15% of their typical Portland cement (PC) mix with limestone and produce what is known as Portland limestone cement (PLC). The advantage of using PLC is the reduced energy and cost required to burn the raw materials for cement. Supplementary cementitious materials (SCMs) such as ground granulated blast furnace slag (slag) and fly ash were adopted for use to replace the cement while batching with concrete by more than 20% by weight.

This study evaluates the sustainability of concrete mixtures used in Thomas et al. (2010). The modified mixtures with PLC and SCMs will be evaluated in order to assess their impact on the overall sustainability of concrete for pavements. This assessment selected a ready-mix concrete plant (RMCP) located in Chicago, Illinois. In order to produce reliable metrics and indices for sustainability measures between the different concrete mixtures, all the materials selected for concrete proportioning are assumed to be procured to the RMCP from locally available cement plants, SCM suppliers, and quarries in Illinois.

21.3.2 Methodology

21.3.2.1 Goal and Scope

The goal of this study is to determine and compare the sustainability of concrete made with typical PC with that of alternative concrete mixes incorporating various types and amounts of SCMs. The different concrete mix designs will be used specifically for pavements and bridge decks in the city of Chicago. The concrete mixtures were made with PC and PLC and batched with either no SCMs or fly ash or slag. The concrete mix designs were evaluated for strength and durability by Thomas et al. (2010). All of the mixtures have shown similar performance without any notable difference between the PC and PLC and the PLC with different SCM combinations.

21.3.2.2 Materials

Limestone Waste is a by-product obtained as a leftover from the process of crushing limestone rocks to produce crushed limestone aggregates for concrete (Figure 21.12). Using this waste in building materials such as in cement can alleviate some environmental problems and reduce the depletion of natural resources. Limestone has been used in combination with natural sand as a fine aggregate material for concrete.

The process of adding limestone to cement is illustrated in Figure 21.13. The cement manufacturing requires proportioning different raw materials (limestone, silica, alumina, and iron) to achieve the desired chemical composition. Once the materials are proportioned, they are blended together by grinding into a powder, after which they are delivered to a rotary kiln where they get burned at a temperature between 2500 and 2800 °F to form cement clinkers (roughly ¾ in. in diameter). The clinker is then ground with gypsum to produce the PC. The limestone is added by either grinding or blending it in the cement. The process of intergrinding limestone takes place with the clinker and gypsum in the grinding mill, while blending takes place after grinding (ASTM C150 requires limestone to be ground with cement).

Fly Ash is defined by the ACI Committee 116 as "the finely divided residue resulting from the combustion of ground or powdered coal, which is transported from the firebox through the boiler by flue gases" (Figure 21.12). It is a by-product of the burnt coal required to generate electricity in coal-fired power plants. Fly ash has been shown to possess pozzolanic properties attributed to its large quantities of reactive silicates, which can help improve the durability, strength, and long-term performance of fly ash-amended concrete (O'Brien et al. 2009). For example, it has been found that fly ash in concrete can reduce thermal cracking by lowering the heat of hydration (Malhotra 2002). Additionally, fly ash concrete also tends to have low permeability and thus can better resist chemical weathering due to chlorides, sulfates, and carbonates in rainwater (Dhir 2016). Moreover, the sustainability of concrete can be improved since the incorporation of fly ash can reduce the amount of PC needed for a given volume of concrete, thus reducing the overall carbon footprint of the fly-ash-amended cement. An added benefit is

Figure 21.12 Materials: (a) excess crushed limestone usable for concrete, (b) fly ash, (c) ground slag byproduct.

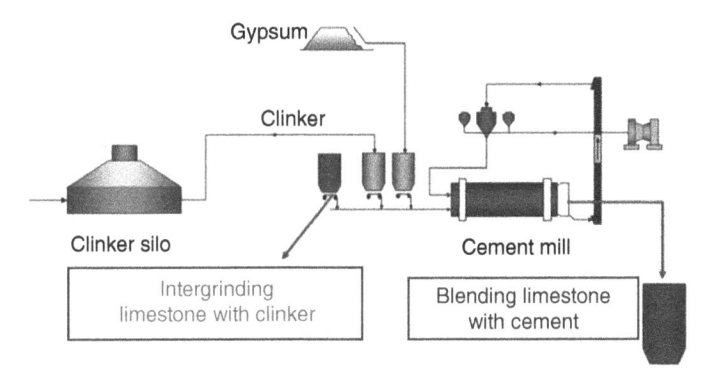

Figure 21.13 Process of blending or intergrinding limestone with cement.

that the fly ash that would have required disposal in a sanitary landfill now is diverted for use in a relatively safe and stable form once incorporated into concrete.

Ground granulated blast furnace slag is an industrial by-product that has been used as a SCM in concrete. Iron blast furnace slag – or simply slag – is formed during smelting process when iron ore, coke, and flux are melted together at a temperature of about 1500 °C. The molten slag is cooled by quenching the material in water, then grinding them to give the final form of ground granulated blast furnace slag (Figure 21.12). Similar to fly ash, studies have shown added advantage to concrete performance made with slag such as improved durability characteristics and long-term strength properties (Kosmatka and Wilson 2011).

21.3.2.3 Technical Design

Twelve concrete mixtures designed specifically for pavements and bridge decks were selected from Thomas et al. (2010) for the sustainability assessment. Thomas et al. (2010) examined the strength and durability of concrete mixtures made with PLC (12% limestone) in comparison with PC (3–4% limestone). The total cementitious content in the batched concrete mixes was around 600 lb/yd^3 upon which the cement (PC or PLC) was prepared with no SCMs, 20% fly ash replacement, and 35% slag replacement. The chemical constituents of the 12 mixtures are presented in Table 21.4.

Figure 21.14 shows the cementitious combinations and aggregate types for the 12 concrete mixes. Natural sand and combined sand were selected as fine aggregate and crushed limestone was selected as coarse aggregate for batching with concrete. These aggregates are the most commonly available materials for concrete in the state of Illinois. The combined sand is produced

Table 21.4 Chemical constituents of cementing materials.

	\multicolumn{9}{c	}{Chemical constituents (%)}	Blaine (m^2/Kg)	Limestone (%)
	SiO$_2$	Al$_2$O$_3$	Fe$_2$O$_3$	CaO	MgO	Na$_2$O	K$_2$O	SO$_3$	LOI		
PC	20.34	5.30	1.94	62.50	1.72	0.11	1.14	4.14	2.65	366	4
PLC	19.75	5.19	1.88	61.25	1.66	0.10	1.18	3.90	4.83	520	12
Fly ash	53.98	23.52	3.82	11.66	1.27	3.08	0.69	0.22	0.89		
Slag	36.84	10.15	0.53	36.41	12.92	0.42	0.62	3.63	−1.27		

LOI: loss on ignition.

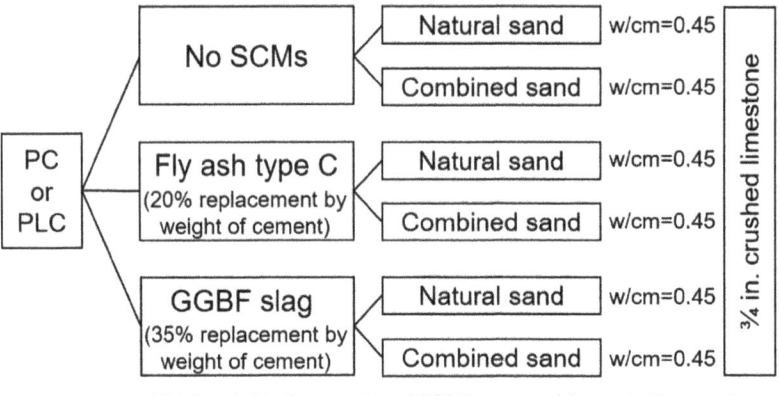

434 | *21 Civil and Materials Engineering Sustainability Projects*

Figure 21.14 Concrete mix combinations.

Table 21.5 Mix designation and design per 1 cubic yard (yd³) of concrete.

Mix designation		Cement type PC	Cement type PLC	SCM type (lb) Slag	SCM type (lb) Fly ash	Fine aggregate Natural sand	Fine aggregate Combined sand	Coarse aggregate (crushed limestone) (lb)
Mixtures batched with natural sand	PC–NS	600	—	—	—	1205	—	1823
	PC–S–NS	390	—	210	—	1187	—	1823
	PC–F–NS	480	—	—	120	1183	—	1823
	PLC–NS	—	600	—	—	1205	—	1823
	PLC–S–NS	—	390	210	—	1187	—	1823
	PLC–F–NS	—	480	—	120	1183	—	1823
Mixtures batched with combined sand	PC–CS	600	—	—	—	—	1205	1823
	PC–S–CS	390	—	210	—	—	1187	1823
	PC–F–CS	480	—	—	120	—	1183	1823
	PLC–CS	—	600	—	—	—	1205	1823
	PLC–S–CS	—	390	210	—	—	1187	1823
	PLC–F–CS	—	480	—	120	—	1183	1823

PC: portland cement; PLC: portland limestone cement; S: slag; F: fly ash; NS: natural sand; CS: combined sand.

by mixing half of the natural sand with leftovers of crushed limestone collected as waste from the limestone quarries.

The mix designations and proportions per one cubic yard of concrete are presented in Table 21.5. PC–NS and PC–CS represents the conventional concrete mixtures made with only PC and batched with natural sand and combined sand.

Functional unit: A concrete bridge deck located in Chicago, the Morgan Street Bridge, was selected as the project site and served as a functional unit for the life cycle analysis (Figure 21.15). The Morgan Street Bridge is part of the Circle Interchange located just west of downtown Chicago and has been recently reconstructed.

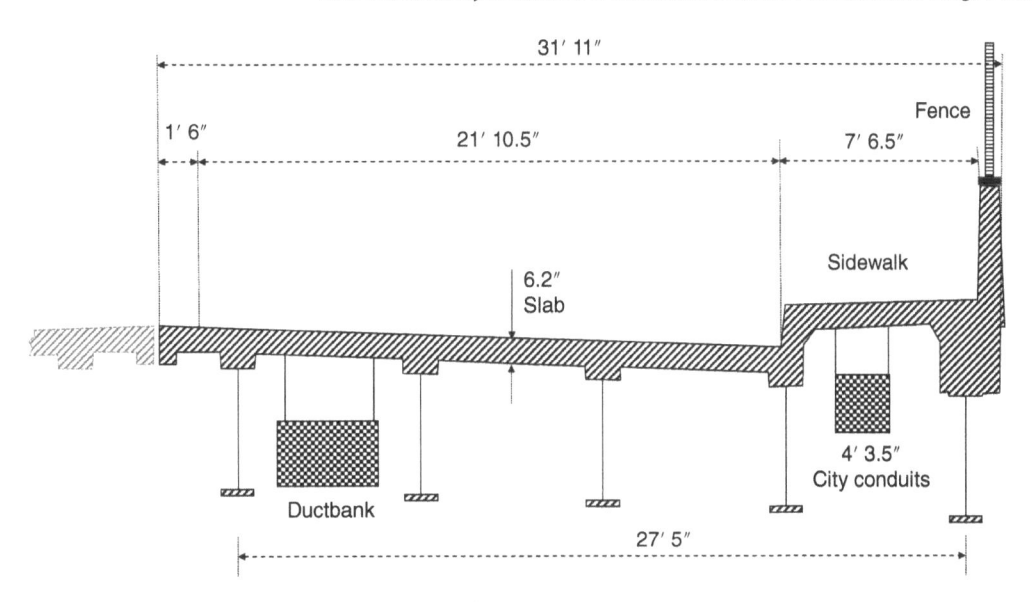

Figure 21.15 Cross-section of Morgan Street Bridge Deck.

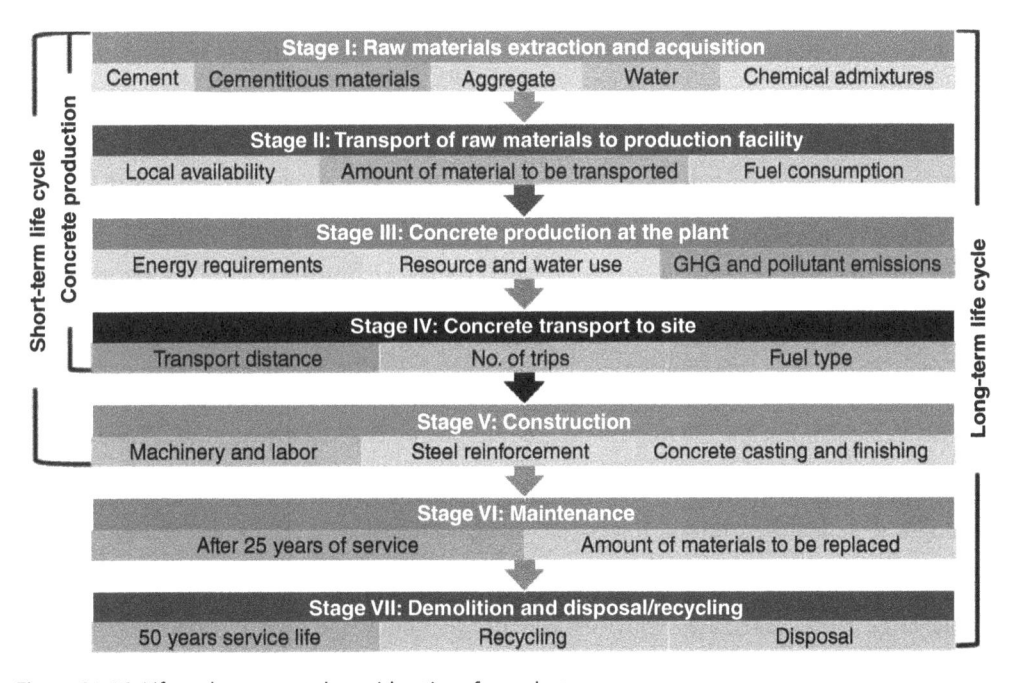

Figure 21.16 Life cycle stages and considerations for each stage.

System boundaries: The entire life cycle of the bridge deck was considered across seven stages as shown in Figure 21.16. The system boundaries were thus defined by a bridge deck design life of 50 years, with maintenance (partial bridge deck replacement, 25% of concrete by volume) occurring at 25 years. Material quantities of steel and concrete required were calculated based on the bridge deck design dimensions (Table 21.6) and a correction factor of 1.25 for concrete was added to account for losses during transport and construction.

Table 21.6 Concrete and steel requirements for the bridge deck (Functional unit).

Structure	Volume concrete required (yd^3)	Amount of steel required (lb)
Bridge deck	401.8	125 020
Sidewalks	103.4	
Parapets	69.3	
Total	574.5	125 020
Total with correction factor	718.1	

The location of the raw materials and the required quantities of each source per concrete mix are given in Table 21.7 for mixes batched with natural sand and Table 21.8 for mixes batched with combined sand. Transport distances were determined from locations of raw material vendors relative to the selected ready-mix plant, located 2.3 mi from the construction site. The water consumption per one cubic yard of concrete is 270 lb/yd^3. However, additional water is needed to wash the concrete truck mixer. This is estimated by assuming an additional 270 lb of water for every 8 yd^3 of concrete (concrete truck mixer capacity). Thus, the total amount of water required is 304 lb (99 metric tons) and 218 302 lb (71 092 metric tons) per cubic yard of concrete and functional unit, respectively.

21.3.2.4 Sustainability Assessment

Sustainability was evaluated based on environmental impacts, economic viability, and social sustainability. Each of these categories was examined separately across the life cycle of the various concrete mixtures used for the bridge deck. The overall sustainability of each mix is a function of all three of these aspects.

Environmental assessment: SimaPro v.8.1 was used to quantify and compare environmental impacts among the 12 mixes considered across each of the life cycle stages. The software used the EcoInvent™ 3.0 database (Weidema et al. 2013) for energy and resource inputs and outputs associated with different materials and processes within the life cycle of concrete production and bridge deck construction, maintenance, and demolition and disposal. For impact assessment calculations, the TRACI method developed by the USEPA was used. The material and energy resource input assumptions for LCA are summarized in Table 21.9.

Economic assessment: A life cycle cost assessment (LCCA) was performed based on the LCA stages to assess the overall project cost using each concrete mix. Local vendors for the raw materials required for concrete production were selected as shown in Tables 21.7 and 21.8. Cost of raw materials and transport, as well as cost of bridge over its entire life cycle, is reported in Table 21.10. Costs associated with concrete production at the RMCP (excluding costs for raw materials), as well as bridge deck construction, maintenance, and demolition costs were assumed to be equivalent across the 12 mixes considered.

Social impact assessment: Social sustainability of the bridge deck rehabilitation project was assessed on an overall basis and compared to the current scenario without any rehabilitation ("No Rehab" scenario). This approach was taken because the social impacts of using alternate concrete mixes do not differ significantly, other than improved public perception reaped from the use of waste materials in PLC, fly ash, and/or slag-amended concrete mixes. Social impacts were quantified using SSEM tool developed by Reddy et al. (2014).

21.3 Sustainability Assessment of Concrete Mixtures for Pavements and Bridge Decks

Table 21.7 Materials' quantities, locations, and distances from RMCP for mixes with natural sand (NS).

Mix designation		Material		Distance to RMCP (km)	Bulk density (kg/m³)	Weight (ton) per yd³ of concrete	Weight (ton) per functional unit	Bulk volume (m)³	Ton km
		Type	Supplier						
Mixes batched with natural sand	PC–NS	PC	Illinois cement Plant	94.4	1506	0.272	195	130	18440
		Natural sand	Bluff city Materials	95	1762	0.547	392	223	37269
		Crushed Limestone	Hanson material Services	25	1602	0.827	594	371	14838
	PC–S–NS	PC	Illinois cement Plant	94.4	1506	0.177	127	84	11986
		Slag	Holcim US, Inc.	11.8	961	0.095	68	71	807
		Natural sand	Bluff city Materials	95	1762	0.538	386	219	36712
		Crushed Limestone	Hanson material Services	25	1602	0.827	594	371	14838
	PC–F–NS	PC	Illinois cement Plant	94.4	1506	0.218	156	104	14752
		Fly ash	Lafarge North America	46.2	961	0.054	39	41	1805
		Natural sand	Bluff city Materials	95	1762	0.537	385	219	36589
		Crushed Limestone	Hanson material Services	25	1602	0.827	594	371	14838
	PLC–NS	PLC	Illinois cement Plant	94.4	1506	0.272	195	130	18440
		Natural sand	Bluff city Materials	95	1762	0.547	392	223	37269
		Crushed Limestone	Hanson material Services	25	1602	0.827	594	371	14838
	PLC–S–NS	PLC	Illinois cement Plant	94.4	1506	0.177	127	84	11986
		Slag	Holcim US, Inc.	11.8	961	0.095	68	71	807
		Natural sand	Bluff city Materials	95	1762	0.538	386	219	36712
		Crushed Limestone	Hanson material Services	25	1602	0.827	594	371	14838
	PLC–F–NS	PLC	Illinois cement Plant	94.4	1506	0.218	156	104	14752
		Fly ash	Lafarge north America	46.2	961	0.054	39	41	1805
		Natural sand	Bluff city Materials	95	1762	0.537	385	219	36589
		Crushed Limestone	Hanson material Services	25	1602	0.827	594	371	14838

21 Civil and Materials Engineering Sustainability Projects

Table 21.8 Materials' quantities, locations, and distances from RMCP for mixes with combined sand (CS).

Mix designation	Material Type	Supplier	Distance to RMCP (km)	Bulk density (kg/m³)	Weight (ton) per yd³ of concrete	per functional unit	Bulk volume (m³)	Ton km
PC–CS	PC	Illinois cement plant	94.4	1 506	0.272	195	130	18 440
	Combined sand	Hanson material Services	30.3	1 762	0.547	392	223	11 887
	Crushed Limestone	Hanson material Services	25	1 602	0.827	594	371	14 838
PC–S–CS	PC	Illinois cement plant	94.4	1 506	0.177	127	84	11 986
	Slag	Holcim US, Inc.	11.8	961	0.095	68	71	807
	Combined sand	Hanson material Services	30.3	1 762	0.538	386	219	11 709
	Crushed Limestone	Hanson material Services	25	1 602	0.827	594	371	14 838
PC–F–CS	PC	Illinois cement plant	94.4	1 506	0.218	156	104	14 752
	Fly ash	Lafarge North America	46.2	961	0.054	39	41	1 805
	Combined sand	Hanson material Services	30.3	1 762	0.537	385	219	11 670
	Crushed Limestone	Hanson material Services	25	1 602	0.827	594	371	14 838
PLC–CS	PLC	Illinois cement plant	94.4	1 506	0.272	195	130	18 440
	Combined sand	Hanson material Services	30.3	1 762	0.547	392	223	11 887
	Crushed Limestone	Hanson material Services	25	1 602	0.827	594	371	14 838
PLC–S–CS	PLC	Illinois cement plant	94.4	1 506	0.177	127	84	11 986
	Slag	Holcim US, Inc.	11.8	961	0.095	68	71	807
	Combined sand	Hanson material Services	30.3	1 762	0.538	386	219	11 709
	Crushed Limestone	Hanson material Services	25	1 602	0.827	594	371	14 838
PLC–F–CS	PLC	Illinois cement plant	94.4	1 506	0.218	156	104	14 752
	Fly ash	Lafarge North America	46.2	961	0.054	39	41	1 805
	Combined sand	Hanson material Services	30.3	1 762	0.537	385	219	11 670
	Crushed Limestone	Hanson material Services	25	1 602	0.827	594	371	14 838

Mixes batched with combined sand

Table 21.9 Material and energy resource input used for the LCA of various concrete mixtures.

Stage I	• PC: 100% Class I cement (Portland cement) (SimaPro™) • PLC: 94% Class I cement (Portland cement) and 6% crushed stone (SimaPro™) • Slag: Ground granulated blast furnace slag (SimaPro™) • Fly ash: Fly ash and scrubber sludge (SimaPro™) ○ Assuming inter-grinding and blending are similar in terms of energy requirements • Crushed limestone aggregate: Crushed stone (SimaPro™) • Natural sand: sand (SimaPro™) • Combined sand: 50% natural sand (normal sand) and 50% crushed stone (SimaPro™)

Stage II	• Same transit truck used for all materials (aggregate, cementitious materials)
	  ○ Single diesel powered lorry truck (SimaPro™)
	  ○ Capacity of aggregate truck: 12 yd^3
	  ○ Capacity of Cement truck: 8 yd^3

Stage III	• Ready mix concrete plant
	  ○ Amount of water used (SimaPro™)
	  ○ Concrete production plant (SimaPro™)

Stage IV	• Concrete truck mixer
	  ○ Single diesel powered lorry truck (SimaPro™)
	  ○ Capacity of truck: 8 yd^3

Stage V	• Bridge Deck construction
	  ○ Steel reinforcement (SimaPro™)
	  ○ Building hall per m^2 (SimaPro™)

Stage VI	• Maintenance requirements are equivalent among all 12 mixes (i.e. there are no special maintenance requirements for fly ash or slag-amended concrete) • 25% of Stages I–V

Stage VII	• Bridge Deck demolition (SimaPro™)
	  ○ Case 1: Disposal
	    – Treatment of waste reinforced concrete, collection for final disposal
	    – Replacement with equivalent amount of raw materials
	  ○ Case 2: Recycling:
	    – Treatment of waste reinforced concrete, recycling

21.3.3 Environmental Sustainability

Results of the environmental impact assessment using the TRACI method in SimaPro are shown in Figures 21.17–21.22. Differences in the environmental impact among the 12 concrete mixes are observed in Stages I and II, while the remaining stages (Stages III–VII) have no known differences, with the exception of Stage VI (Maintenance, partial bridge deck replacement) since it requires additional concrete for the reconstruction of 25% of the existing bridge deck.

The main differences among concrete mixes are incurred at Stage I (Figure 21.17). Slag amendments and fly ash amendments both reduced the environmental impacts related to fossil fuel depletion and global warming by way of reducing the amount of raw materials required to produce cement. However, fly ash apparently had a greater positive impact than slag. This may be due to the fact that slag requires additional grinding prior to use as an SCM in cement, adding to the total energy requirements for slag-amended concrete. Fly ash, on the other hand, is already a fine, powdery substance and requires little modification prior to incorporation with other cementitious materials. PLC (with 6% more of limestone than PC) showed minor influence in the environmental impact. The addition of 6% limestone to cement exhibited, on average, less than 5% positive environmental impact and reduced the carbon footprint by 5.7%.

21 Civil and Materials Engineering Sustainability Projects

Table 21.10 Unit cost per item and capacity of transporting trucks.

Item	Unit cost ($)	Capacity (yd^3)
Materials		
Type I Portland cement (metric ton)	75	—
Portland limestone cement (metric ton)	70.5	—
Class C fly ash (metric ton)	40	—
Grade 100 slag (metric ton)	40	—
Natural sand (metric ton)	7	—
Combined sand (metric ton)	8	—
Crushed limestone (metric ton)	8	—
Chemical admixtures (metric ton)	3000	—
Transport type		
Aggregate dump truck, per mile	2	12.0
Cement silo truck, per mile	3	8.0
Concrete mixing truck, per mile	5	8.0
Bridge deck		
Bridge deck casting, per yd^2	900	12.0
Bridge deck removal, per yd^3	400	8.0
Reinforcement bars, epoxy coated, lb	2	8.0

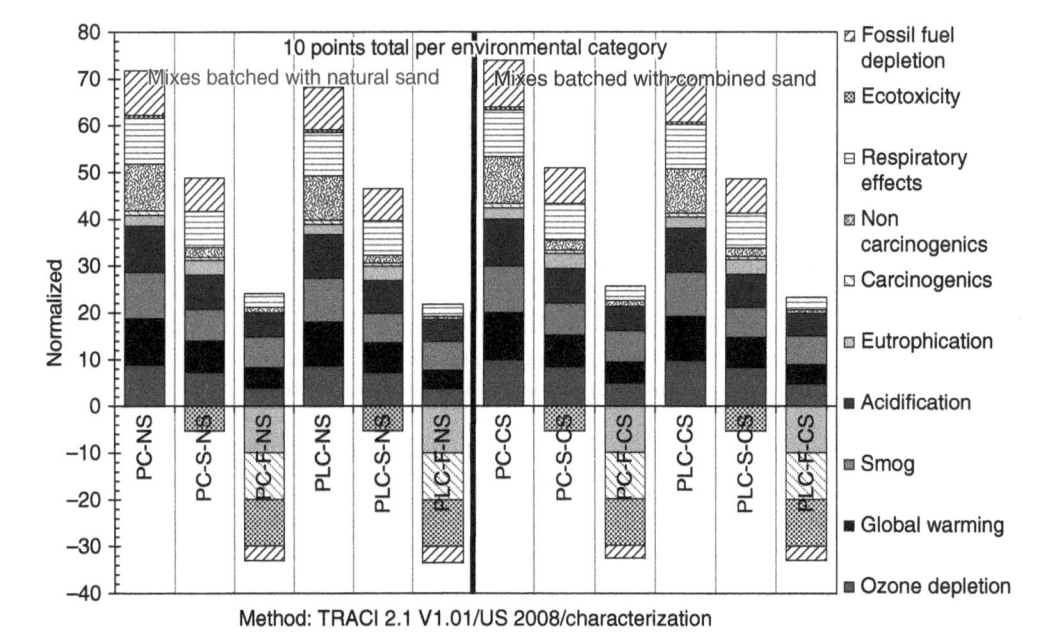

Figure 21.17 Environmental impact on Stage I (raw material acquisition).

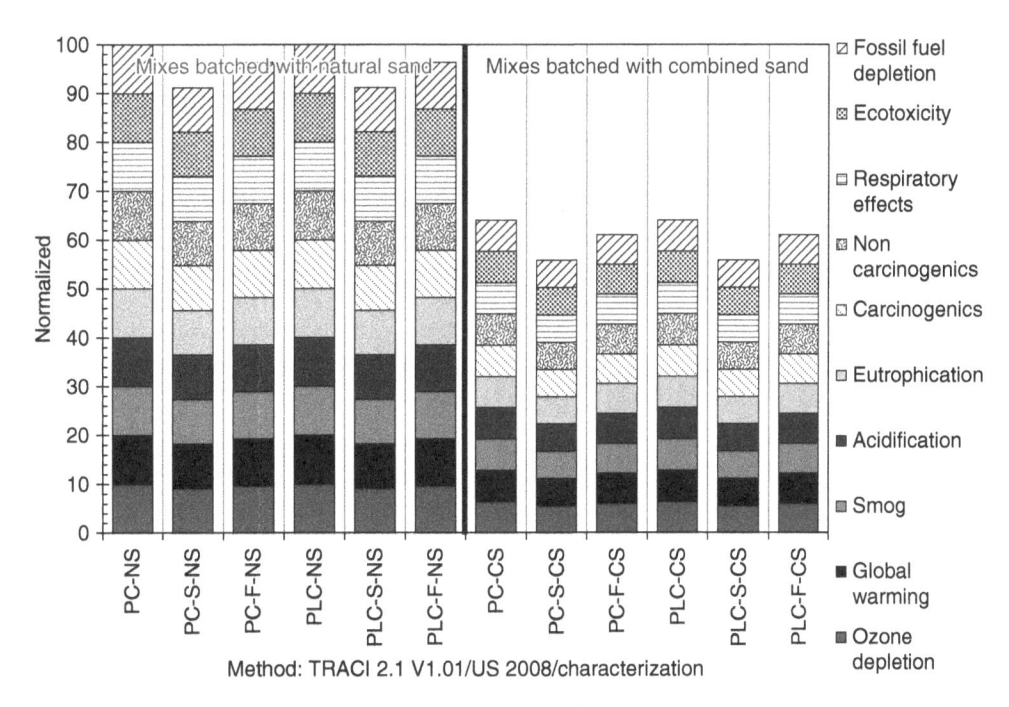

21.3 Sustainability Assessment of Concrete Mixtures for Pavements and Bridge Decks | 441

Figure 21.18 Environmental impact on Stage II (transport of raw materials to RMCP).

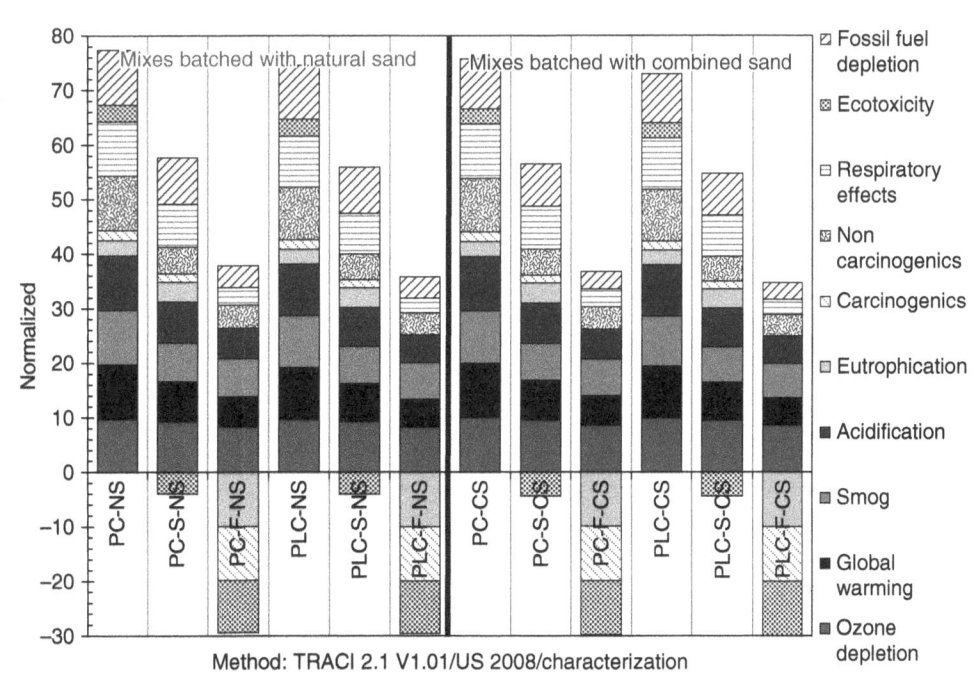

Figure 21.19 Environmental impact on Stages I–IV (concrete production).

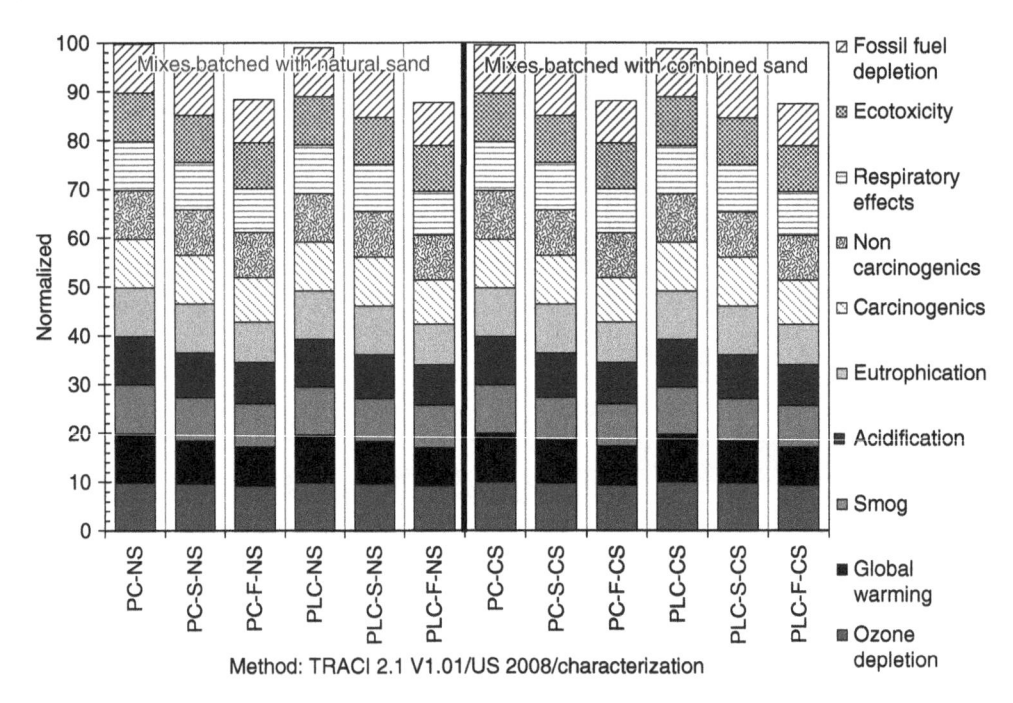

442 | *21 Civil and Materials Engineering Sustainability Projects*

Figure 21.20 Environmental impact on Stages I–V (Bridge Deck construction).

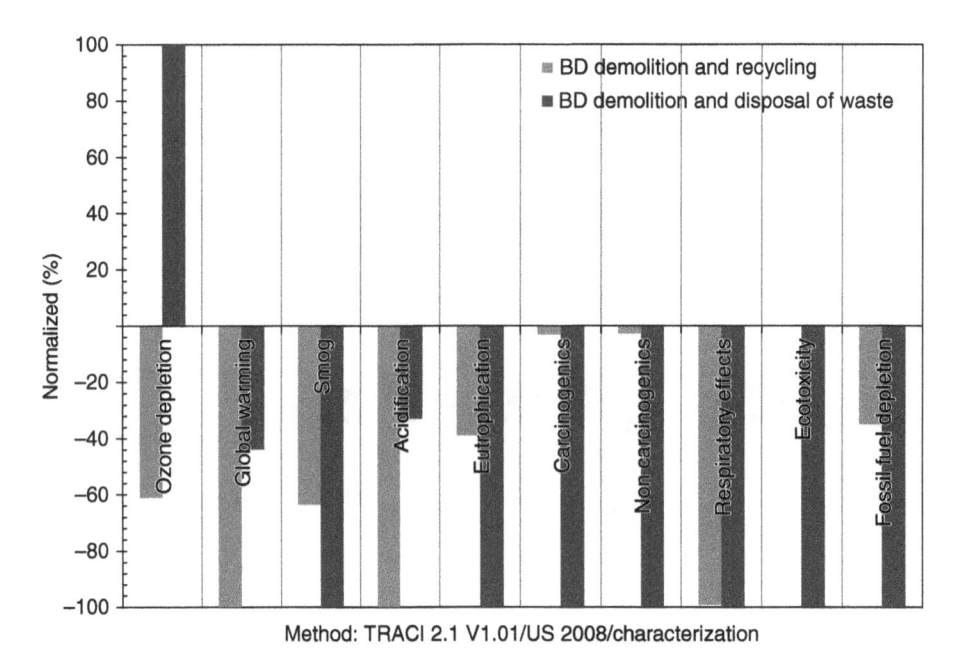

Figure 21.21 Environmental impact of Bridge Deck disposal versus recycling after demolition.

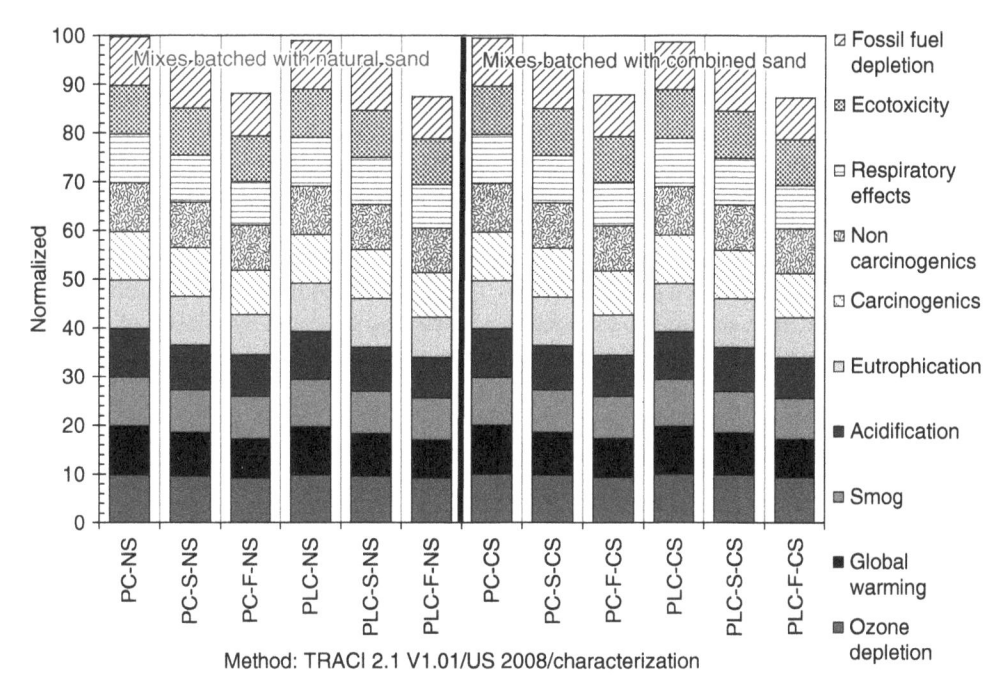

Figure 21.22 LCA of bridge deck from Stage I to VII (cradle to grave).

It seems there is a linear relationship between the limestone addition and the reduction in carbon footprint. It is noteworthy to mention that concrete mixtures batched with combined sand did not show any notable environmental impact when compared to natural sand, which indicates that the cement production is the main contributor to the negative environmental consequences in Stage I.

Figure 21.18 shows that the differences in impacts in Stage II (transport of raw materials to RMCP) are primarily seen among mixes made with natural sand versus those made with combined sand given the greater transport distance required for natural sand (95 mi versus approximately 30 mi from RMCP). Thus, using locally available raw materials is key to achieving environmental sustainability, as transporting materials accounts for a significant portion of the negative environmental impacts associated with raw concrete mix materials. Differences among mixes were relatively minor in terms of the relative impact of Stage II to the overall impact across the entire project life cycle, which contributed only approximately 1.5% to the total environmental impact.

Figure 21.19 shows the environmental impacts accrued from concrete production (Stages I–IV). Stage III (concrete production at the RMCP) took into consideration the energy required to produce the concrete at the RMCP and the total amount of water used. Stage IV considered the fuel consumption for transporting the concrete to the site. The effect of slag and fly ash amendments is also prevalent in reducing the environmental impacts for producing the concrete required for the bridge deck, while being more pronounced for fly ash, as it was explained in Figure 21.17. The primary contributor to the negative environmental consequences is attributed to the cement production. This explains the reduction in CO_2 emissions across Stages I through IV in response to partial replacement of a portion of the cement with slag or fly ash.

Figure 21.20 illustrates the environmental impacts of each concrete mix across the short-term life cycle of the bridge deck (Stages I–V). The bridge deck construction (Stage V) is the biggest contributor to the overall environmental impact. Stage V took into consideration energy required to manufacture the steel reinforcements, install them, and the energy required from the machineries to cast the bridge deck. The total contribution of the bridge deck to the carbon footprint is 759 300 equivalent tons of CO_2 in the baseline concrete mix (PC–NS). The construction stage (Stage V) is responsible for more than 55% of the total carbon footprint, while Stage I contributed to 36.6% of the total carbon footprint. In terms of the effect of the material proportion of each concrete mix, it is observed that mixes made with fly ash have the best influence on all the environmental categories, followed by mixes batched with slag, while similar to the findings in the previous stages, the PLC showed minor contribution to the overall environmental assessment. Mixes batched with natural sand and fly ash (PC–F–NS) or slag (PC–S–NS) were able to reduce the carbon footprint by 20.4% and 12%, respectively, for the short-term life cycle.

Two scenarios following demolition were modeled using the LCA software SimaPro: recycling of construction and demolition debris or disposal of waste concrete in a landfill. The recycling scenario includes materials and energy required for deconstruction of the bridge deck and sorting of C&D wastes at the recycling plant, in addition to any processing required prior to reuse. The waste disposal scenario includes materials and energy required for bridge deck demolition and transport of the deconstructed materials to a landfill. Figure 21.21 shows that the recycling scenario has a greater positive environmental impact than disposal for the following impact categories: ozone depletion, global warming, and acidification. All other categories, the negative impacts associated with recycling were greater than for disposal. These differences are attributed to the energy needed to sort the reinforced concrete materials as well as the crushing process required to break the concrete, in addition to the reuse of potentially toxic waste by-products, which may increase the risk of human exposure to these toxic substances, especially for workers dealing directly with the recycled waste materials.

The overall LCA results for each concrete mix across the long-term life cycle of the bridge deck are shown in Figure 21.22. A summary of the overall environmental impacts for the LCA per category for concrete mixes batched with natural sand is presented in Table 21.11. In terms of the effect of the material proportion of each concrete mix, it is observed that mixes made with fly ash has overall the best influence on all the environmental categories, this is followed

Table 21.11 Summary of bridge deck LCA results per impact category for mixes batched with natural sand (NS).

Impact category	Unit	Maximum value	\multicolumn{6}{c}{Percentage with respect to baseline mix (PC–NS) (%)}
			PC–NS	PC–S–NS	PC–F–NS	PLC–NS	PLC–S–NS	PLC–F–NS
Ozone depletion	kg CFC-11 eq	0.044 41	100	98.0	93.6	99.8	97.9	93.4
Global warming	kg CO_2 eq	942 458	100	87.9	79.4	97.9	86.5	77.8
Smog	kg O_3 eq	63 676	100	88.1	87.8	98.0	86.8	86.2
Acidification	kg SO_2 eq	5 653	100	91.6	84.8	98.2	90.4	83.4
Eutrophication	kg N eq	1 625.7	100	101.1	83.7	99.9	101.0	83.6
Carcinogenics	CTUh	0.133 48	100	99.8	90.5	100.0	99.7	90.4
Noncarcinogenics	CTUh	0.371 60	100	93.6	92.9	99.5	93.3	92.5
Respiratory effects	kg PM2.5 eq	697	100	96.6	89.3	99.2	96.1	88.7
Ecotoxicity	CTUe	8 488 627	100	96.5	93.7	100.0	96.5	93.7
Fossil fuel depletion	MJ surplus	519 662	100	97.1	87.6	99.6	96.9	87.3

Table 21.12 Contribution of Stages I and V to the full LCA for the baseline mix (PC–NS) in %.

Item category	PC–NS Stage I	PC–NS Stage V
Ozone depletion	9.2	44.3
Global warming	29.5	44.4
Smog	28.4	49.9
Acidification	25.2	53.5
Eutrophication	3.3	76.7
Carcinogenics	0.7	78.9
Noncarcinogenics	6.4	70.0
Respiratory effects	12.2	72.7
Ecotoxicity	0.3	78.8
Fossil fuel depletion	7.3	65.7

by mixes batched with slag, while similar to the findings in the previous stages, the PLC showed minor contribution to the overall environmental assessment. Table 21.11 shows that replacing the cement by 20% of fly ash (PC–F–NS) can reduce the overall carbon footprint by 20% while 35% replacement of slag to cement (PC–S–NS) reduced the carbon footprint by 12%. The addition of limestone to cement only resulted in marginal reduction, approximately 2%, in the carbon footprint.

The long-term life cycle takes into account Stages I–V besides the maintenance process required for the bridge deck and the demolition considerations after the end of the service life of the bridge. The bridge deck was assumed to be maintained after 25 years in service by assuming 25% of the bridge deck materials to be replaced. The Full LCA results showed that Stage V (bridge construction) and Stage I (materials) are the biggest contributors to negative environmental impacts (Table 21.12). Thus, Stage V exhibited the highest negative environmental impacts with contribution ranging between 44.3% for ozone depletion and 78.9% for carcinogens.

21.3.4 Economic Sustainability

The results of the LCCA are summarized in Figures 21.23–21.25. Cost estimates for Stage VI (Maintenance) and Stage VII (Demolition) are given in Table 21.13. The main cost is the bridge deck construction (Stage V), which is estimated to contribute nearly 55% ($900 000) to the total project cost (approximately $1.63 million USD). Maintenance (Stage VI) and demolition and disposal (Stage VII) are estimated to contribute about 20% each to the total cost ($329 000 and $342 400, respectively). Only 5% of the total project cost is related to raw material acquisition and concrete production (Stage I–III). There are no cost differences among the 12 concrete mixes in terms of deck construction, although small differences can be seen in Stages I and II based on current market prices and transport distances to the RMCP for the raw materials. Some cost savings are realized by the reuse of industrial by-products such as slag and fly ash, as these materials tend to be relatively inexpensive and also reduce the cost of virgin cement production. If Stage I alone is considered, the cost savings by using fly ash and crushed limestone can be up to 10% as compared to conventional Portland cement without any amendments

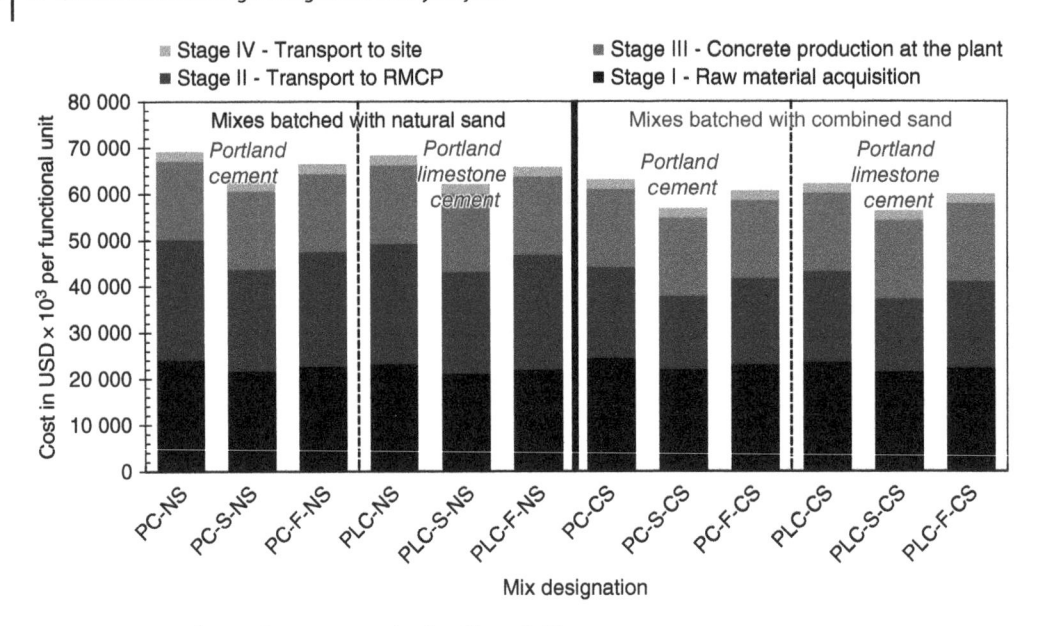

Figure 21.23 Total cost of concrete production (Stage I–IV).

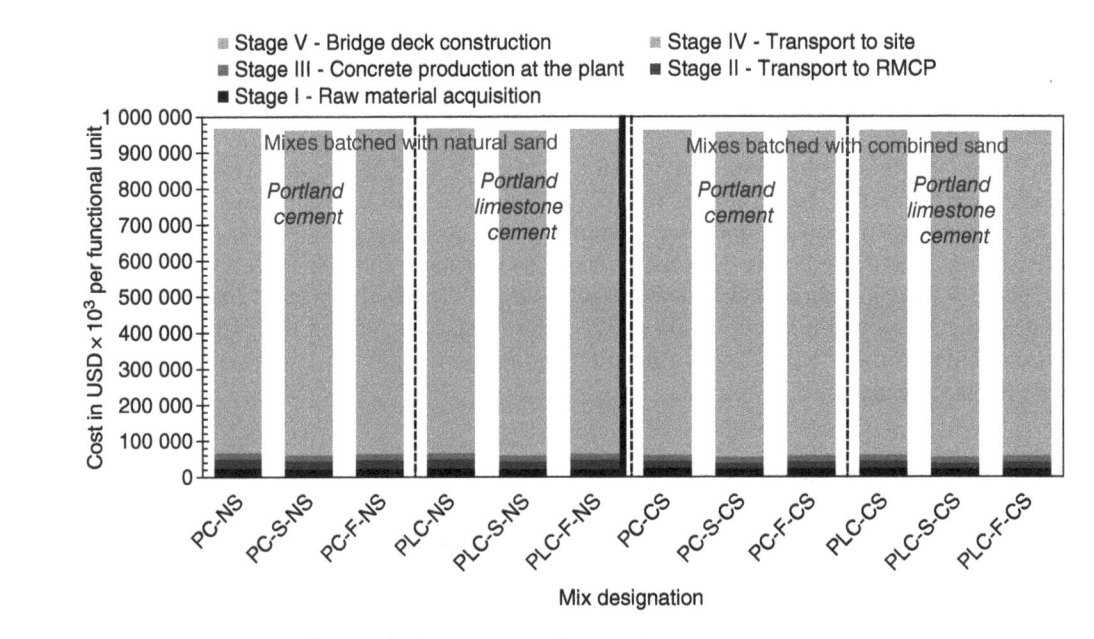

Figure 21.24 Total cost of bridge deck construction (Stage I–V).

(PC–NS). However, as is shown in Figure 21.24, the differences in cost due to the use of slag and fly ash are minor in the context of the entire project. Overall, cost considerations are not strong drivers for selecting the modified concrete mixes but do provide an added benefit for their use.

Indirect economic benefits from the bridge construction project as a whole are also important to consider for the overall project sustainability. These include the jobs created during each stage of the project life cycle, as well as the resultant impact on local businesses during and after construction. During construction, some economic disruption is anticipated due to detours and

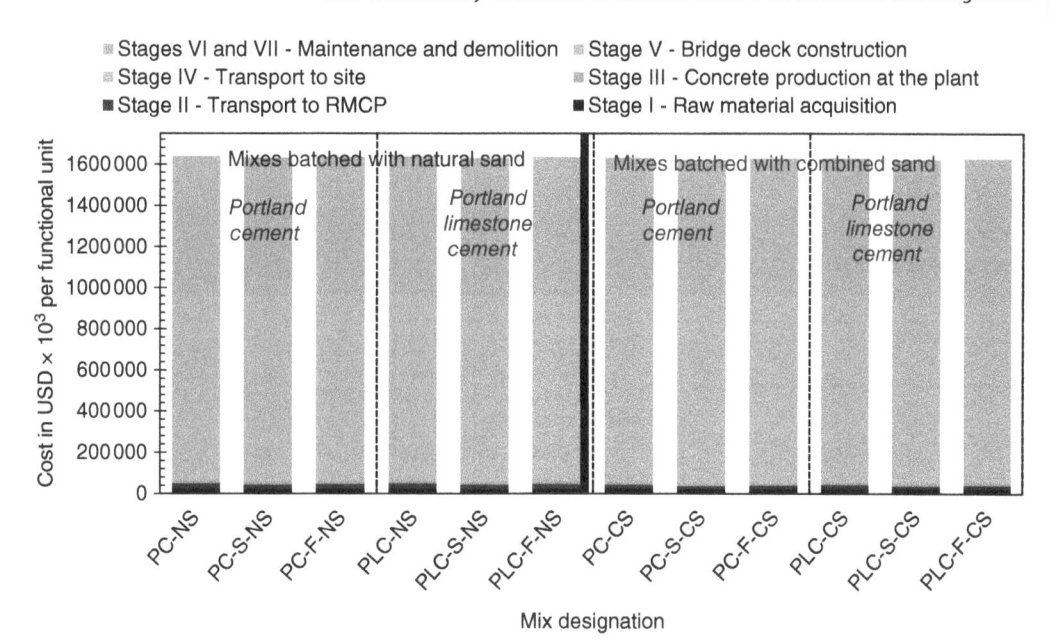

21.3 Sustainability Assessment of Concrete Mixtures for Pavements and Bridge Decks | **447**

Figure 21.25 Total cost of the bridge deck after 50 years in service (Stage I–VII).

Table 21.13 Cost assessment of Stages VI and VII.

Item	Quantity		Unit cost ($)	Cost ($)
Stage VI – bridge deck Maintenance – after 25 years in service				
Protective shield	1 344	sq. yd	40	53 760
Bridge deck grooving	1 155	sq. yd	6	6 930
Deck slab repair (partial 15% of total deck)	203	sq. yd	300	60 900
Deck slab repair (full depth, type II) (10% of total deck)	136	sq. yd	500	68 000
Bridge deck scarification 3/4"	1 155	sq. yd	20	23 100
Bridge deck concrete overlay	1 155	sq. yd	100	115 500
Total				328 190
Stage VII – bridge deck demolition – after 50 years in service				
Protective shield	1 344	sq. yd	40	53 760
Concrete removal	722	cu. yd	400	288 640
Total				342 400

traffic delays. However, once construction is complete, business activity is expected to increase due to improved traffic flow and better access to public transit (i.e. improved connectivity).

21.3.5 Social Sustainability

Social benefits were evaluated from the project as whole, as major differences in terms of quality of life indices are not expected to arise due to differences in the concrete mixes used. Thus, the social impacts from the bridge deck rehabilitation project were identified independently and

Table 21.14 Social sustainability evaluation matrix (SSEM) summary.

	No rehabilitation	Bridge rehabilitation
Socio-individual	−7	20
Socio-community	−7	24
Socio-economic	−5	12
Socio-environmental	0	3
Grand score	−19	59

Socio-individual
- Bike lane access
- Improved pedestrian access (e.g. widened sidewalks)
- Improved disability access to public transit

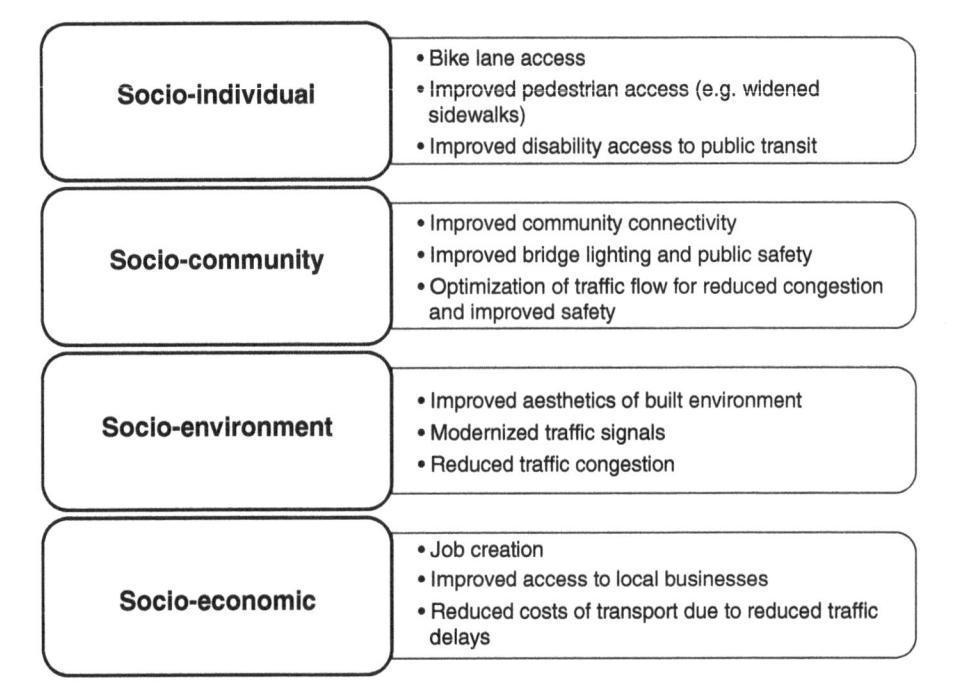

Socio-community
- Improved community connectivity
- Improved bridge lighting and public safety
- Optimization of traffic flow for reduced congestion and improved safety

Socio-environment
- Improved aesthetics of built environment
- Modernized traffic signals
- Reduced traffic congestion

Socio-economic
- Job creation
- Improved access to local businesses
- Reduced costs of transport due to reduced traffic delays

Figure 21.26 Positive social benefits from the Morgan street bridge deck rehabilitation project.

then compared to the current scenario, without any bridge rehabilitation. The social impacts are summarized in Table 21.14 and Figure 21.26. The categories that contribute the most to the social acceptance of the bridge rehabilitation are those at individual and community level, and in second term at economic level due to the job creation during the rehabilitation works.

21.3.6 Conclusions

The environmental LCA analyses reveal that the modified concrete mixtures have significant positive environmental impacts relative to the conventional concrete mixtures. These impacts were primarily related to energy and fuel savings resulting from a reduction in the amount of virgin cement required for a given volume of concrete when using SCMs such as slag, fly ash, or limestone by-product. These reductions in fuel consumption and raw material use also translated to significant cost savings for the raw concrete mix materials (Stage I). However, these savings are minor in the context of the entire project, accounting for 1.3–1.5% of the total project cost, depending on the mix selected. Therefore, environmental impacts and improved public perception are stronger drivers for the use of modified concrete mixtures in civil infrastructure than economic considerations. These benefits include a lower carbon footprint of modified concrete relative to conventional mixtures due to lower virgin cement production requirements; reduced air pollutant emissions associated with transport of raw materials and cement calcination processes; and diversion of industrial waste by-products (slag, fly ash, and excess crushed limestone) from landfill disposal.

21.4 Sustainability Assessment of Parking Lot Design Alternatives

The purpose of this study is to compare the sustainability of several design alternatives for a parking lot, known as Parking Lot 1, at the University of Illinois at Chicago (UIC). The constructed design made use of concrete pavement and permeable pavers to drain rainwater runoff and store it in a layer of coarse aggregate beneath the pavement. In accordance with the City of Chicago stormwater standards, a technical design of the parking lot was completed proposing the use of traditional drainage structures, RCP pipe, and a StormTrap® Detention Vault to store runoff. A sustainability assessment was conducted to compare the constructed design with the proposed design for the drainage structures, while also considering design alternatives using concrete, asphalt, and brick pavers for the proposed parking lot pavement. The assessment used various metrics to compare the environmental, economic, and social aspects of each design alternative.

21.4.1 Background

The key water resources design components for urban parking lots are the proper drainage of stormwater from the lot itself and the limiting of flow rates at the lot's downstream outfall. The drainage of the lot includes the placement of water collection structures at the low spots and the design of a stormwater network, which usually takes the form of storm sewers in urban environments, which can accommodate a designed flow rate. The 10-year storm is the standard required in the City of Chicago. The limitation of flow rates is used to ensure that flow leaving the site is not compromising the capacity of the existing storm sewer. Stormwater detention is often used to temporarily store water and release it at a slower rate. This storage can take the form of basins or underground storage vaults. The City of Chicago has specific release rate standards for each neighborhood, where the discharge from 100-year, 24-hour storm must be monitored (Powers and Emanuel 2016).

The site being analyzed is the UIC Parking Lot 1. It covers approximately 5.5 acres of land and holds 778 parking stalls. This site drains into the combined sewer system of Chicago. The site consists of mostly impervious area and contains clay soil beneath the parking lot (Soil Survey Staff 2018). Due to these conditions, a very high percentage of rainfall is expected to become runoff. The State of Illinois uses the document Bulletin 70 as a standard for rainfall data specific to each location throughout the state (Huff and Angel 1989).

The use of best management practices (BMPs) is a common focus for many sustainability assessments involving stormwater. BMPs essentially treat and improve the quality and slow the flow of stormwater runoff onsite before it discharges downstream. The location of this project made the use of BMPs infeasible, as the flow discharges into a combined sewer system. In this system, the stormwater will be treated at a wastewater treatment plant, negating any benefits of improved quality onsite. While BMPs can provide benefits by reducing flow rates and therefore reducing occurrences of combined sewer overflows, this aspect of the design was mainly accounted for in the detention component of this project.

21.4.2 Methodology

21.4.2.1 Goal and Scope

The objective of this study was to compare the constructed design with the proposed design for the drainage structures of the UIC Parking Lot 1. The sustainability assessment also included the analysis of design alternatives using concrete, asphalt, and brick pavers for the proposed parking lot pavement. The goal of the study was to evaluate the most sustainable design alternative among the aforementioned choices.

21.4.2.2 Study Area

In order to reconstruct the parking lot, the design and construction process was split into two phases, with Phase 1 comprising the design of the East half of the lot, and Phase 2 comprising the West half. This was done to ensure partial functionality and use of the lot throughout the construction process.

Figure 21.27 shows the as-built typical cross section information. Even though there is a greater expense, underground storage was deemed necessary for this project. This was due to the fact that the construction of an above ground detention basin would be impractical due to land space constraints in the heavily urban location.

The majority of the constructed lot used Portland cement concrete (PCC) pavement. In the sections where the central parking stalls were located, Filtercrete Porous concrete was installed. The parking lot was graded so that permeable concrete sections were located at the low spots of the pavement. This allowed stormwater runoff to flow through the parking lot below the pavement. An 8-in. depth was used for both the traditional and permeable concrete. Beneath the pavement, an 8-in. layer of coarse aggregate was installed to allow for stormwater storage within its voids while also providing structural support for the pavement (Figure 21.27). The specific aggregate used was CA-7, which has a void storage ratio of 0.36. This layer of aggregate spans the entire parking lot, resulting in approximately 1.27 acre-feet of stormwater detention. Perforated PVC underdrains were also used to collect the water from the bottom of the aggregate layer. These underdrains direct the flow toward a restrictor manhole, which uses an orifice in a steel plate to control the release rate of the water. In each phase of the design, flow from these restricted manholes discharges into a 24-inch-diameter brick combined sewer owned by the City of Chicago.

21.4.2.3 Technical Design

The areas of Phase 1 and Phase 2 were hydraulically independent, and therefore, their designs were completed separately. A selection of precast concrete manholes, catch basins, and inlets were used to construct a storm sewer network, and the inlets and catch basins were placed at low spots to capture runoff in the parking lot. It was assumed that the grading would remain the same among each proposed alternative. The specific structures and their frame and lids were selected from the list of Illinois Department of Transportation (IDOT) standard pay items (Illinois Department of Transportation 2018).

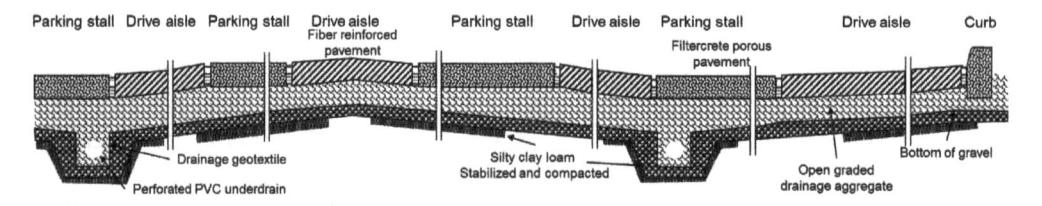

Figure 21.27 Typical cross-section of as-built design.

A hydrologic analysis was then conducted for the parking lot. The relevant parameters were determined for each drainage structure; these included the tributary area, curve number, maximum flow path, and slope. Curve number methodology was used to calculate expected runoff for each structure. The higher curve number values correspond to a higher imperviousness. Equations (21.1) and (21.2) were used in this calculation, which determine the amount of runoff from a given rainfall event.

$$S = 1000/CN - 10 \tag{21.1}$$
$$R = (P - 0.2S)^2/(P + 0.8S) \tag{21.2}$$

where CN is the Curve Number, P is the rainfall depth, and R is the resultant flow depth.

The next step in the design was the calculation of the allowable release rates and required detention. The Chicago Stormwater Ordinance Manual dictates allowable release rate ratios for specific neighborhoods. The allowable release rate ratio for the Taylor Neighborhood, in which the project is located, is 0.68 cfs per impervious-acre. This enabled the calculation of the allowable release rates.

As an alternative, the StormTrap Single Cell detention vault product was chosen to store stormwater beneath the parking lot. This product consists of individual concrete cells, which can be combined to form a sizable vault with approximately 80% of its total volume available for water storage. The Single Cell product was chosen with a 2-ft depth due to cover constraints and invert elevation of the downstream combined sewer that the drainage network needed for discharge. For each phase, this vault stores water while flow is restricted by an orifice in a steel plate located in a proposed manhole (StormTrap LLC 2016).

In order to determine the proper size of the vaults and diameters of the orifices, EPA SWMM software was used to model the 100-year, 24-hour rainfall event. The physical and hydrologic parameters were input to construct a model for both phases. The models were calibrated to ensure that the flow rates through the orifices were within the allowable limits and that the vaults provided adequate storage. A summary of design values can be found in Table 21.15.

The final design can be seen as a hydraulic improvement to the as-built design, as release rates are lower and greater storage is provided. Finally, these technical design components were combined with the three choices of pavement (i.e. concrete, asphalt, and brick pavers) to complete the proposed design alternatives.

21.4.2.4 Sustainability Assessment

The environmental impact assessment was performed using SimaPro software. The inventory of materials and their amount was determined based on the parking lot size and requirements

Table 21.15 Summary of technical design of parking lot 1: phase 1 and phase 2.

Design criteria	Phase 1	Phase 2
Impervious area (ac)	3.23	2.96
Allowable release rate (cfs)	2.20	2.01
Restrictor diameter (in)	7	7
Design release rate (cfs)	1.68	1.61
Volume required (ac-ft)	0.54	0.54
Volume provided (ac-ft)	0.72	0.67

for its reconstruction. The environmental impact assessment was performed using the TRACI method (USEPA 2016).

The economic assessment used the IDOT Manual to determine the standard pay items and descriptions used for the cost analysis based on today's industry standards. Date from USEPA has been used to determine the cost of GHG emissions based on the results from SimaPro software.

The social impacts for each design alternatives were quantified through a survey. This survey was distributed to UIC faculty and students, as they are among the primary shareholders of the parking lot analyzed in this study. For the survey, 25 key indicators that were considered relevant to the parking lot and university projects were identified. Most of the indicators referenced the pavement options, as they were thought to have greater variance in social impacts than the underground detention. Furthermore, these indicators were categorized into four separate categories to help better assess the specific social impacts, that is, Socioindividual, Sociocommunity, Socioeconomic, and Socioenvironmental.

The survey format consisted of rating each indicator for each proposed alternative. The ratings were conducted on a scale from High Negative Impact to High Positive Impact, which was converted to a numeric scale of 1–5. A not applicable (N/A) option was also included. Weights varied only slightly between indicators, as ones that had a higher proportion of respondents answer "N/A" were treated as less relevant to the project and therefore given a lower weight. Average values for each of the proposed alternatives were then tabulated.

21.4.3 Environmental Sustainability

Figure 21.28 shows the comparison of the environmental impacts associated with the production of each material considered in this study. A higher positive percentage represents more adverse environmental impacts. The asphalt pavement is the largest contributor to the environmental impacts. This is due to the energy used in the process of making the asphalt itself. The

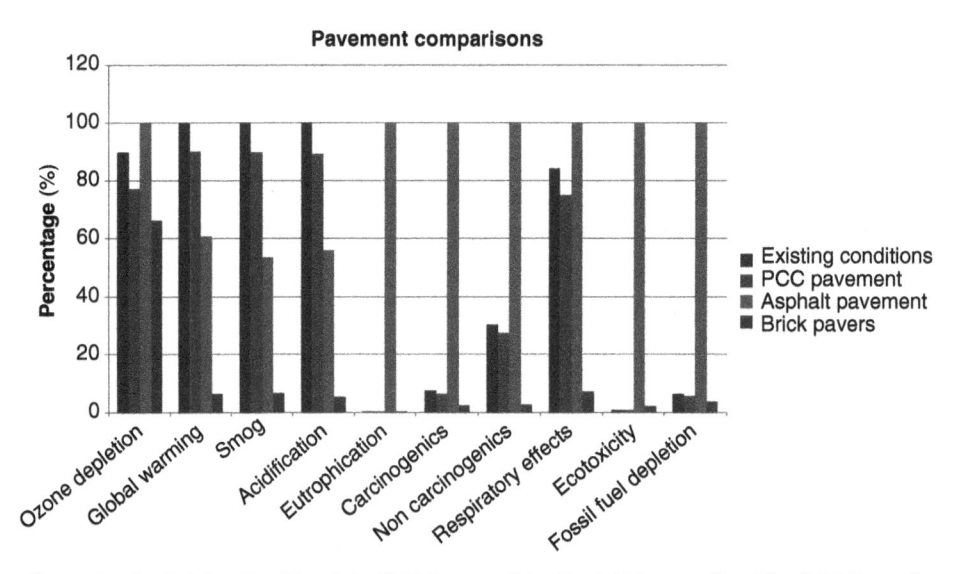

Figure 21.28 Emissions of pavement alternatives.

21.4 Sustainability Assessment of Parking Lot Design Alternatives | 453

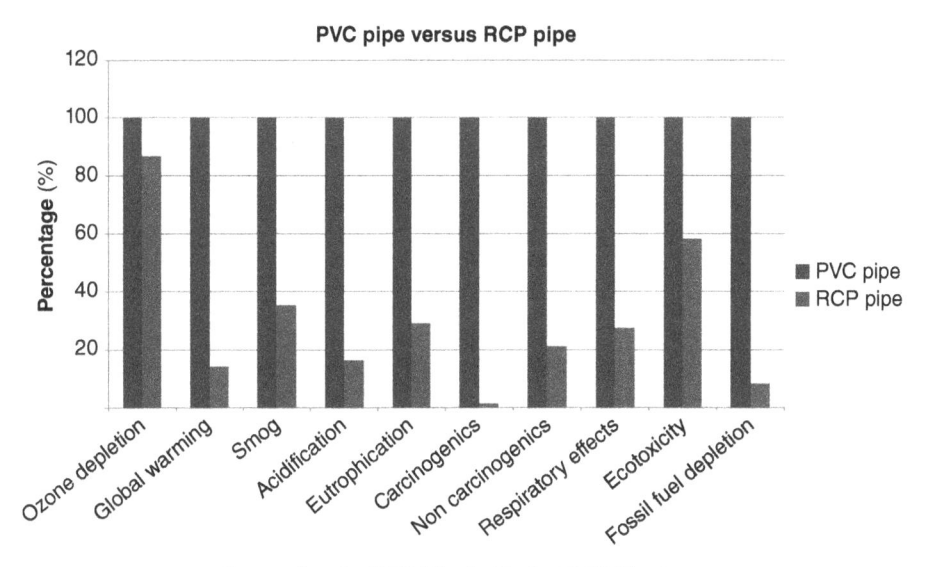

PVC pipe versus RCP pipe

Comparing 1 p 'PVC Pipe' with 1 p 'RCP Pipe';
Method: TRACI 2.1 V1.01/US 2008/Characterization/Excluding infrastructure processes

Figure 21.29 Emissions of PVC and RCP pipes.

second largest contributor is the existing conditions in the parking lot. This is due to the energy used in the process of making the PCC and permeable concrete pavement. The third largest contributor is the PCC Pavement. This is largely because of the PCC. The lowest contributor is the brick pavers. Out of all the four options, brick pavers had the least overall environmental impact, in terms of manufacturing and transportation to the site.

Figure 21.29 shows the comparison between the two different pipe materials, polyvinyl chloride (PVC) and reinforced concrete pipe (RCP). Quantities between both pipes are fairly similar. Based on the results, PVC pipe produces the greater emissions compared to RCP pipe. This is primarily due to the PVC pipe being made up of different thermoplastic polymers that are used in the production of the pipe. As shown in Figure 21.29, RCP pipe outperforms the PVC pipe in every emissions category. Only in two emission categories, ozone depletion and ecotoxicity, RCP impact is over 50%. PVC is the most common plastic used in today's construction due to its abundance and relatively easy manufacturing process.

Figure 21.30 shows the comparison between the two different types of storage options, stone void storage and StormTrap Single Cell detention vault. Stone void storage consists of holding stormwater runoff beneath the pavement surface in the stone media (Figure 21.27). The StormTrap Single Cell detention vault is a series of prefabricated reinforced concrete walls that hold and store water beneath a parking lot surface. Based on the results, the stone void storage produced the greater set of emissions compared to the StormTrap system. This is in part because of the production of the stone material used for the site. The high emissions of the StormTrap were largely due to the transportation and fabrication of the storm vault itself (StormTrap system is located and manufactured 55 mi away). If the StormTrap system were manufactured and transported only 5 mi away from the site, the total emissions would be significantly reduced (Figure 21.31). For example, ozone depletion and ecotoxicity were reduced by more than 50%.

Figure 21.32 shows the pavement surface comparisons of two "green" options, which include the permeable concrete pavement (existing conditions) and brick pavers. Permeable concrete pavement is pavement that has controlled amounts of cementitious materials and water to form

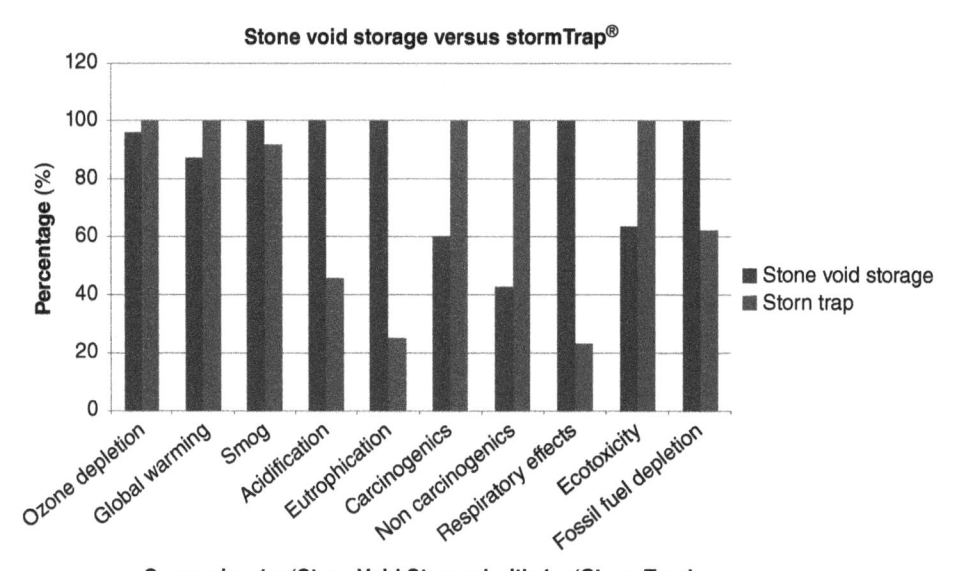

454 | 21 Civil and Materials Engineering Sustainability Projects

Figure 21.30 Emissions of the storage alternatives (storm trap and stone void storage).

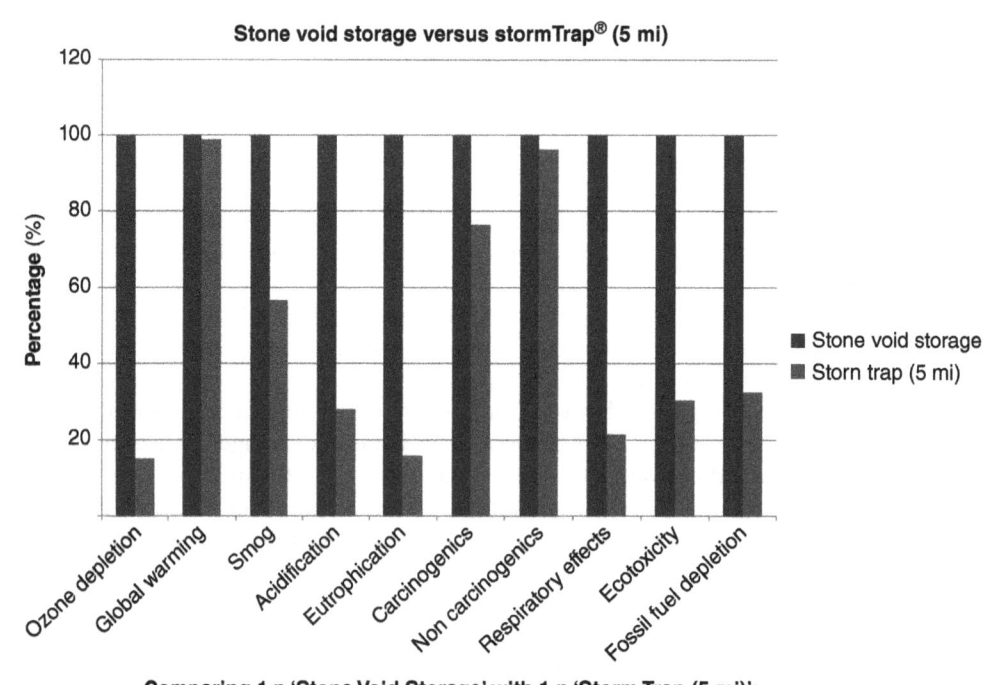

Figure 21.31 Emissions of the storage alternatives considering the StormTrap® system located within 5 mi.

21.4 Sustainability Assessment of Parking Lot Design Alternatives | 455

Existing conditions versus brick pavers

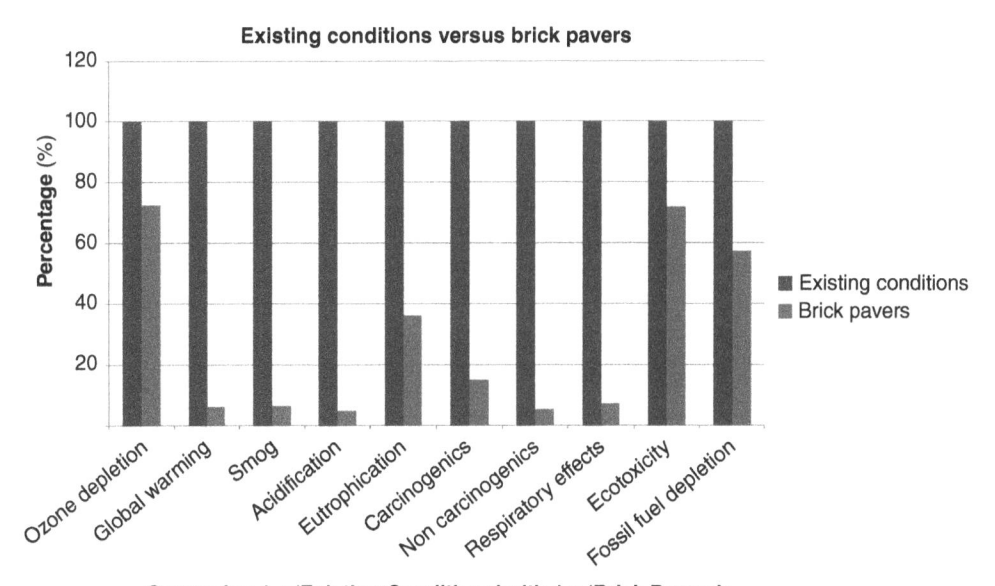

Comparing 1 p 'Existing Conditions' with 1 p 'Brick Pavers';
Method: TRACI 2.1 V1.01 / US 2008 / Characterization / Excluding infrastructure processes

Figure 21.32 Emissions of the "green" brick pavers compared to the existing conditions.

a so-called paste around the larger aggregates. Since little to no fine aggregates are used in this mix, it creates a void space large enough for stormwater runoff to seep through it. The brick pavers selected are made from a concrete mix that interlock with each other and allow stormwater runoff to flow through the sides of the brick. After analyzing both options, it is shown that the existing condition pavement produces the greatest emissions due largely in part to the manufacturing of both the concrete pavement and permeable concrete pavement. Meanwhile, the brick pavers, specifically the brick themselves, produce the least emissions in all categories.

Overall, the selection of pavement surface and water resource components greatly influences the emissions in all categories of the TRACI method. As shown in Figures 21.28–21.32, brick pavers are shown to have the least emissions in all categories compared to the asphalt, brick, and concrete pavements. As per the water resources components, the StormTrap system fared far better than the stone void storage.

21.4.4 Economic Sustainability

Table 21.16 shows the comparison of the economic assessment of the as-built design and proposed alternatives. Using industry standards and prices, and based upon local sources, the unit cost was determined per each item or component. The different excavation depths needed for each material and design was considered in the evaluation of the cost. The cost analysis includes the construction cost, the maintenance cost (considering an approximate lifespan of 25 years), and the return cost for each scenario. For determining return cost, an assumption was made that the parking lot was one-third full during the 16-week periods of both the spring and fall school session. Since the UIC Pavilion also holds major events, it is assumed that 52 days a year the parking lot will be full. Based on the economic assessment, asphalt pavement proved to be the most economical alternative. The maintenance cost of each option was incorporated in the overall return of each scenario. As shown in Table 21.16, the existing conditions pavement and brick pavers have the highest maintenance costs. This is largely due to the cleaning of the

Table 21.16 Cost analysis per parking lot type.

Parking lot type	Cost	Maintenance cost/yr	Return (yrs)
Existing conditions	$2 300 000	$65 000	2.7
PCC pavement	$1 661 000	$30 000	1.5
Asphalt pavement	$1 562 000	$25 000	1.4
Brick pavers	$1 949 000	$60 000	1.8

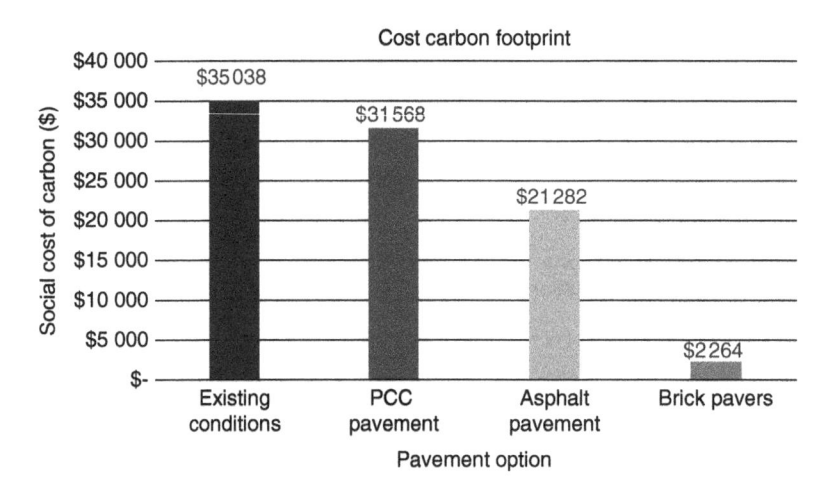

Figure 21.33 Social cost of CO_2 emissions.

pavement surface. For these two options to work as intended, the parking lot needs to be well maintained, especially considering the climatic conditions of Chicago. The lowest maintenance cost corresponds to asphalt pavement.

Figure 21.33 shows how the as-built design and proposed designs fared against each other in terms of the social cost of carbon emissions, using the data from the SimaPro and the estimated cost of GHG emissions as per USEPA. The existing conditions have the greatest carbon footprint, followed by the PCC pavement, asphalt, and finally the brick pavers. Moreover, the carbon footprint was compared to the carbon sequestered to sustain a medium growth coniferous tree for a lifespan of 10 years. Based on the results, brick pavers only need two trees to eliminate the carbon footprint emitted, followed by asphalt pavement with 15 trees, the PCC pavement with 23 trees, and finally the existing conditions that requires 25 trees.

21.4.5 Social Sustainability

The results of the social sustainability survey are shown in Table 21.17. All the categories are given equal weight. The overall score for existing conditions, proposed concrete, and proposed asphalt design show fairly similar impacts. The brick pavers option had substantially the highest overall score, but also in each individual category. This can likely be attributed to the aesthetics and its least harmful environmental impacts. Many of the concerns of higher cost and maintenance for brick pavers were not typically manifested in social impacts perceived by students and faculty.

Table 21.17 Social impact survey average responses.

Criteria	Existing conditions	PCC pavement	Asphalt pavement	Brick pavers
Socio-individual	2.74	2.86	3.05	3.38
Socio-community	3.43	3.33	3.18	4.06
Socio-economic	3.32	3.38	3.34	3.48
Socio-environmental	2.36	2.64	2.50	3.14
Overall average	2.96	3.05	3.02	3.52

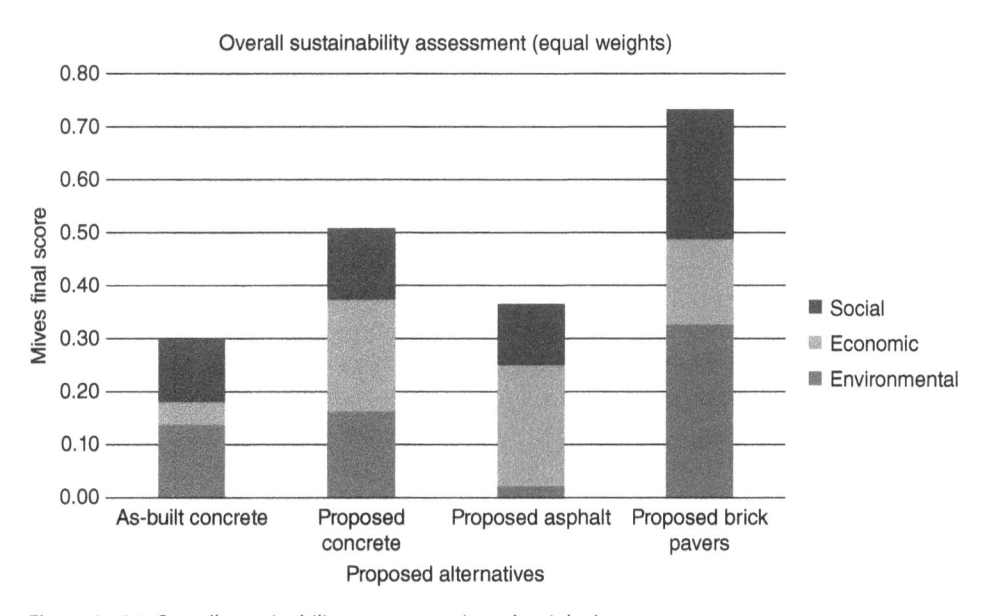

Figure 21.34 Overall sustainability assessment (equal weights).

21.4.6 Overall Sustainability

A value function methodology was applied to the results in an effort to combine each of the three assessments: environmental, economic, and social. The incorporation of a value function was important as many of assessments' results had different units. The value function used the MIVES tool in order to assign each of the indicator results to a scale from 0 to 1 (Alarcon et al. 2010). Figure 21.34 shows the overall assessment considering equal weights for each pillar of sustainability. A higher final MIVES score for a design alternative corresponds with better sustainability. Brick pavers scored the highest, as it was expected from the individual assessment results. The benefits in social and environmental assessment of the brick pavers may compensate the higher economic cost. Thus, brick pavers can be considered the most sustainable alternative.

21.4.7 Conclusions

The detention storage components proved to have roughly equivalent impacts. The asphalt pavement option had substantially the worst environmental impacts, and impacts from brick pavers were determined to be the most positive. The economic impacts followed the common trend of brick pavers, concrete, and asphalt being respectively the most expensive to build and maintain. The permeable pavement storage also was found to perform worse on the economic assessment than the more traditional detention vault storage. Finally, the brick pavers option was found to have the greatest social benefits, while the other three alternatives were roughly average.

The economics of this comparison made overall sustainability assessment challenging as the asphalt option was clearly the cheapest to construct and maintain but was particularly harmful to the environment. The overall assessment concluded that when equal weightings were used among the three categories, brick pavers alternative was the most sustainable option as its positive environmental and social impacts offset its economic deficiencies. However, depending on the owner's budget and the financial climate during the project, an alteration to the weightings can make the concrete and asphalt options more viable for the parking lot.

21.5 Summary

This chapter presents several detailed sustainability assessments of potential civil and materials engineering projects. As was evident in environmental and chemical engineering applications, system boundaries should be carefully drawn to incorporate desired life cycle assessment activities. Further, the assessments can include specific weighting or include sensitivity analyses to incorporate uncertainty aspects in civil and materials engineering projects. It should be mentioned that the sustainability assessment results are project-specific, so the results cannot be generalized. The same methodologies presented in this chapter could be adopted for evaluating various design options and selecting the most sustainable option and achieve sustainability at the project level.

References

Alarcon, B., Aguado, A., Manga, R., and Josa, A. (2010). A value function for assessing sustainability: application to industrial buildings. *Sustainability* 3 (1): 35–50.

Powers T.H. and Emanuel R. (2016). City of Chicago. Stormwater Management Ordinance Manual. City of Chicago. IL https://www.cityofchicago.org/content/dam/city/depts/water/general/Engineering/SewerConstStormReq/2016StormwaterManual.pdf.

Dhir, R. (2006). The potential of fly ash: the future looks bright. *Concrete* 40 (6): 68–70.

Huff, F.A. and Angel, J.R. (1989). *Rainfall Distributions and Hydroclimate Characteristics of Heavy Rainstorms in Illinois (Bulletin 70)*. Illinois State Water Survey.

Illinois Department of Transportation (2018). Coded Pay Items (retrieved 30 April 2018).

Kosmatka, S.H. and Wilson, M.L. (2011). *Design and Control of Concrete Mixes Portland. EB001*, 15e. Skokie, IL: Portland Cement Association.

Malhotra, V.M. (2002). High-performance high-volume fly ash concrete. *Concrete International* 24 (7): 30–34.

Marceau, M., Nisbet, M.A., and Van Geem, M.G. (2006). *Life Cycle Inventory of Portland Cement Manufacture*. Skokie, IL: Portland Cement Association.

O, Brien, K., Ménaché, J., and O, Moore, L. (2009). Impact of fly ash content and fly ash transportation distance on embodied greenhouse gas emissions and water consumption in concrete. *The International Journal of Life Cycle Assessment* 14 (7): 621–629.

Reddy, K.R., Sadasivam, B.Y., and Adams, J.A. (2014). Social sustainability evaluation matrix (SSEM) to quantify social aspects of sustainable remediation. In: *ICSI 2014: Creating Infrastructure for a Sustainable World* (eds. J. Crittenden, C. Hendrickson and B. Wallace). Proceedings of the 2014 International Conference on Sustainable Infrastructure. Reston, VA: ASCE. 831–841. DOI: 10.1061/9780784478745.

Soil Survey Staff (2018). Natural Resources Conservation Service, United States Department of Agriculture. Web Soil Survey. https://websoilsurvey.sc.egov.usda.gov/ (accessed 15 March 2018).

StormTrap LLC (2016). Stormwater detention systems – underground stormwater detention. http://stormtrap.com/products/singletrap/ (retrieved 15 March 2018).

Thomas, M.D.A., Cail, K., Blair, B. et al. (2010). Equivalent performance with half the clinker content using PLC and SCM. In: *Proc., NRMCA Concrete Sustainability Conference*, Tempe, AZ (13–15 April 2010).

USDOE (2008). Energy efficiency trends in residential and commercial buildings. U.S. Department of Energy. Office of Energy Efficiency and Renewable Energy.

USDOE (2012). 2011 Buildings Energy Data Book. U.S. Department of Energy. Buildings Technologies Program, Energy Efficiency and Renewable Energy. D&R International, Ltd.

USEPA (2014). Inventory of U.S. greenhouse gas emissions and sinks: 1990 – 2012. US Environmental Protection Agency. *Report number EPA 430-R-14-003*.

USEPA (2016). Tool for Reduction and Assessment of Chemicals and Other Environmental Impacts (TRACI). https://www.epa.gov/chemical-research/tool-reduction-and-assessment-chemicals-and-other-environmental-impacts-traci (accessed September 2018).

Weidema, Bo Pedersen; Bauer, C.; Hischier, R.; Mutel, C.; Nemecek, T.; Reinhard, J.; Vadenbo, C. O.; Wernet, G. (2013). Overview and Methodology: Data Quality Guideline for the Ecoinvent Database Version 3. Ecoinvent Report Series. Swiss Centre for Life Cycle Inventories.

22

Infrastructure Engineering Sustainability Projects

22.1 Introduction

Chapter 20 provided sustainability assessment of example environmental and chemical engineering projects, and Chapter 21 presented sustainability assessment of example civil and materials engineering projects. These sustainability assessments incorporate environmental, economic, and social aspects (often known as triple bottom line aspects) in quantitative and/or qualitative terms. Moreover, integrated sustainability assessment can combine the three triple bottom line aspects to express overall sustainability index. Such assessment results are helpful to compare various design alternatives or optimize the components of a project design to maximize sustainability. This chapter presents examples of sustainability assessments applied to infrastructure engineering-related projects.

22.2 Comparison of Two Building Designs for an Electric Bus Substation

As a first example, a sustainability analysis, using a triple bottom line-based approach, is presented for two different building types under consideration to house an electric substation for an electric bus service. The substation would be located in an existing park area, allowing the bus system to serve a nearby tourist attraction.

22.2.1 Background

The purpose of the proposed building is to increase the support of electric bus service in order to decrease air pollution near a popular tourist area. The characteristics used for the design of the two buildings are as follows:

Type 1: Foundation – Reinforced concrete spread footing. Wall – 7-inch-thick, cast-in-place reinforced concrete. Roof – 18-gauge steel deck. Earth retaining system (ERS) – Soldier piles with timber lagging.

Type 2: Foundation – Friction piles, reinforced concrete pile caps, and grade beams. Wall – 14-in.-thick masonry wall with brick veneer (facade). Roof – Modified bituminous roof system. ERS – None; a sloped soil excavation is used as an alternative.

Emphasis was also given to the site preparation before the construction of the building. The designed perimeter of the substation was minimized to account for unforeseen circumstances; such as right of way restrictions imposed by the city or existing infrastructure close to the construction zone.

Sustainable Engineering: Drivers, Metrics, Tools, and Applications, First Edition.
Krishna R. Reddy, Claudio Cameselle, and Jeffrey A. Adams.
© 2019 John Wiley & Sons, Inc. Published 2019 by John Wiley & Sons, Inc.

22.2.2 Methodology

22.2.2.1 Goal and Scope
The building designs were compared using the same functional unit. The footprint area of the structure was used as the normalizing factor, which was set to 1320 sq. ft. Identical system boundaries were also established, and environmental, economic, and social impacts were considered for materials transportation, building construction, and waste management.

22.2.2.2 Subsurface Soil Profile and Design Requirements
A baseline soil profile was used for design purposes for both the building types: sandy loose fill, sand, soft clay, and hard clay. Two different foundation types that meet all minimum load requirements for buildings in accordance with ASCE 7, ACI 318, and ACI 530 were considered for both building types (ACI 2013, 2014; ASCE 2014). Standard reinforced concrete, masonry, and steel deck details were implemented and designed to resist lateral wind and seismic loads.

22.2.2.3 Technical Design
Type 1 building – spread footing: The building is designed with a shallow spread footing foundation (Figure 22.1a), and a bearing pressure of 2000 lb per square foot (psf) was used for foundation dimension design. To achieve the design bearing pressure, soil to a depth of 3 ft below the bottom of the design footing depth was removed and replaced with engineered fill. The wall system of this building includes 7-in.-thick, cast-in-place concrete walls. Lastly, the roof system of this building is composed of nine 12 × 26 wide flange beams that carry an

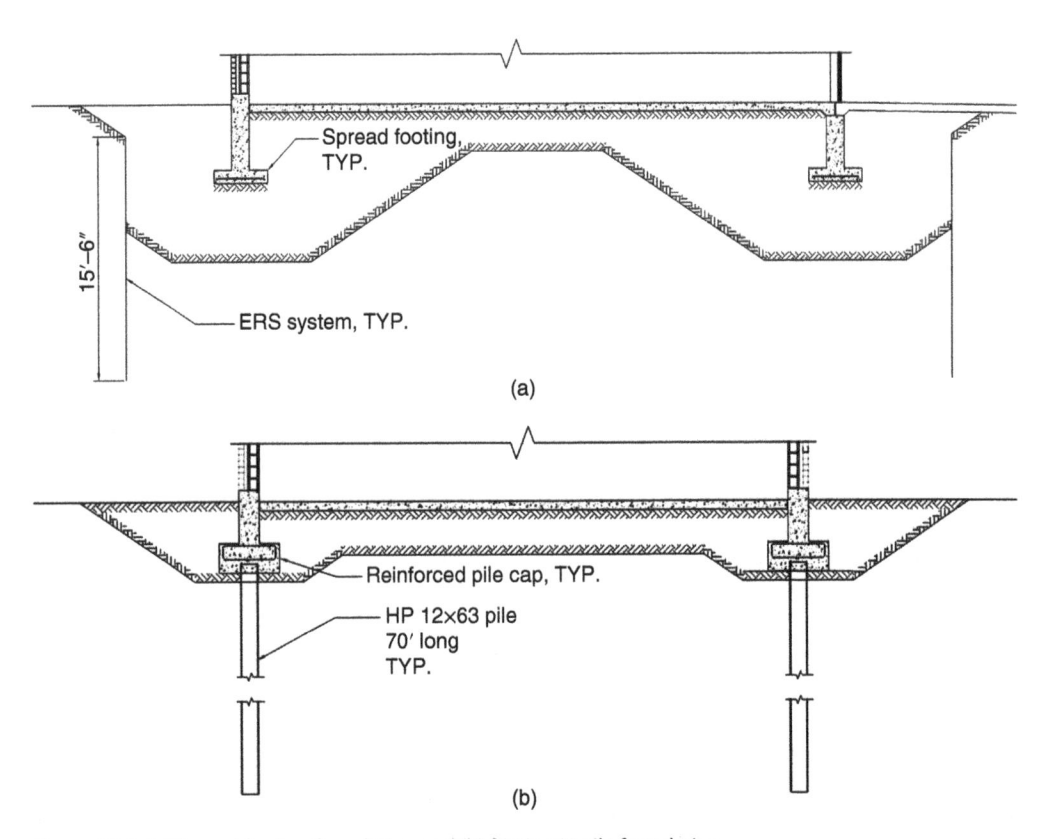

Figure 22.1 (a) Spread footing foundation and (b) friction H-pile foundation.

18-gauge steel deck. The design assumes that the steel roof deck is capable of withstanding the gravity and uplift loads.

An analysis of the ERS was performed to confirm soil would be properly retained during construction activities, meeting OSHA safety standards. Coulomb's earth pressure theory was used in the development of the ERS, and a factor of safety of 3 was assumed for lateral resistance. The ERS design utilized a soldier pile and timber lagging system and FHWA-RD-75-128 was used to design the timber lagging based on clear span between piles and the type of soil retained (FHWA 1976).

Type 2 building – friction piles: The same design principles as the Type 1 building were used to develop the foundation structure. However, revisions in design of the foundation, wall, and roof elements were implemented to achieve a more aesthetically pleasing structure. The foundation system for this building consists of a deep foundation friction pile system with reinforced concrete caps and grade beams (Figure 22.1b). Given the same soil profile that was assumed for the Type 1 building, 70-foot-long H-Piles are required at each corner of the building to resist the imposed gravity loads. The wall system is comprised of reinforced concrete masonry units and a brick veneer facade, requiring a 14-in.-thick wall. The roof system consists of nine 12 × 26 wide flange beams that carry a modified bituminous roof system. It is assumed that the roof system is adequate to carry the uplift and gravity forces.

This type of foundation does not require extensive soil excavation since all loads are transferred to the friction piles and the surrounding soil by skin friction. Thus, the ERS for this building is solely comprised of sloped soil excavation meeting OSHA safety standards.

22.2.2.4 Sustainability Assessment

The system boundaries were established to include the following activities for analysis: transport of construction materials to the site, building construction, and waste management of the building materials. The transport of materials to the site also included the amount of energy required to manufacture the materials. The inventory analysis included the locations of material suppliers, landfills, recycling centers, equipment rental areas in relation to the site where the building construction was planned to take place, the schedule to complete each portion of the building, the equipment needed to complete each portion of the building, the labor needed to complete each portion of the building, and the quantity and types of materials required. The economic analysis includes the materials and transportation costs as well as the construction costs (labor and machinery). The social aspects of the two building designs were assessed with the Social Sustainability Evaluation Matrix (SSEM) developed by Reddy et al. (2014).

22.2.3 Environmental Sustainability

The environmental assessment was performed using SimaPro v8 software for the two building designs, considering the impacts of each stage of the building construction life cycle. Figures 22.2–22.4 show the environmental impacts of the three stages or portions considered for the construction of both buildings. These stages include materials transport (Figures 22.2), construction (Figures 22.3), and waste management (Figures 22.4). These figures show the processes of each project stage as individual blocks that are added together into a tree diagram for each life cycle stage of each building design, revealing the main impacts for each stage of the project. The environmental impact of each phase occurring within the stage is proportional to the thickness of the connecting arrow between blocks. As an example, in considering the transportation of materials (Figure 22.2), most environmental impacts are occurring as a result of the transportation of concrete and steel to the site.

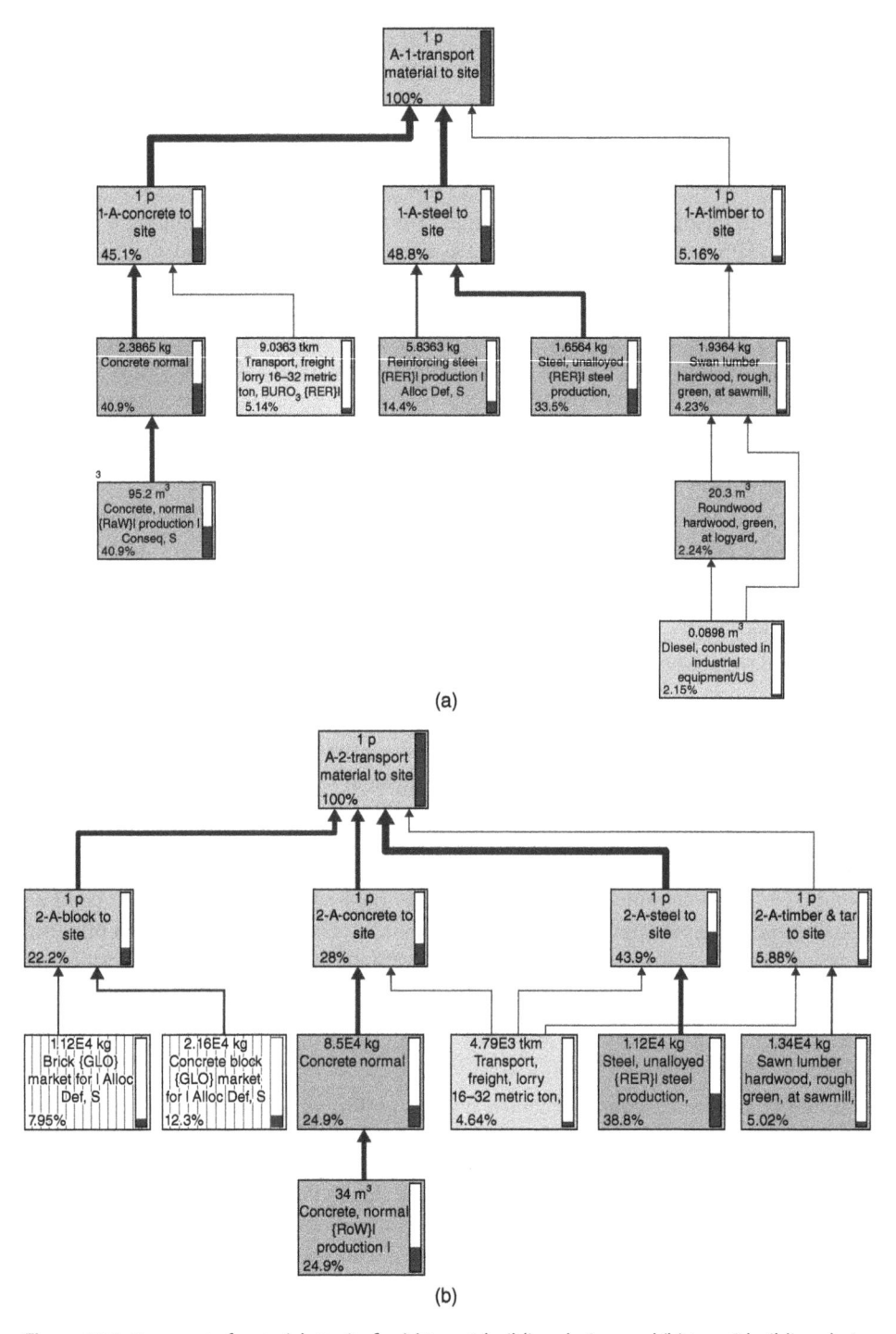

Figure 22.2 Transport of materials to site for (a) type 1 building design, and (b) type 2 building design.

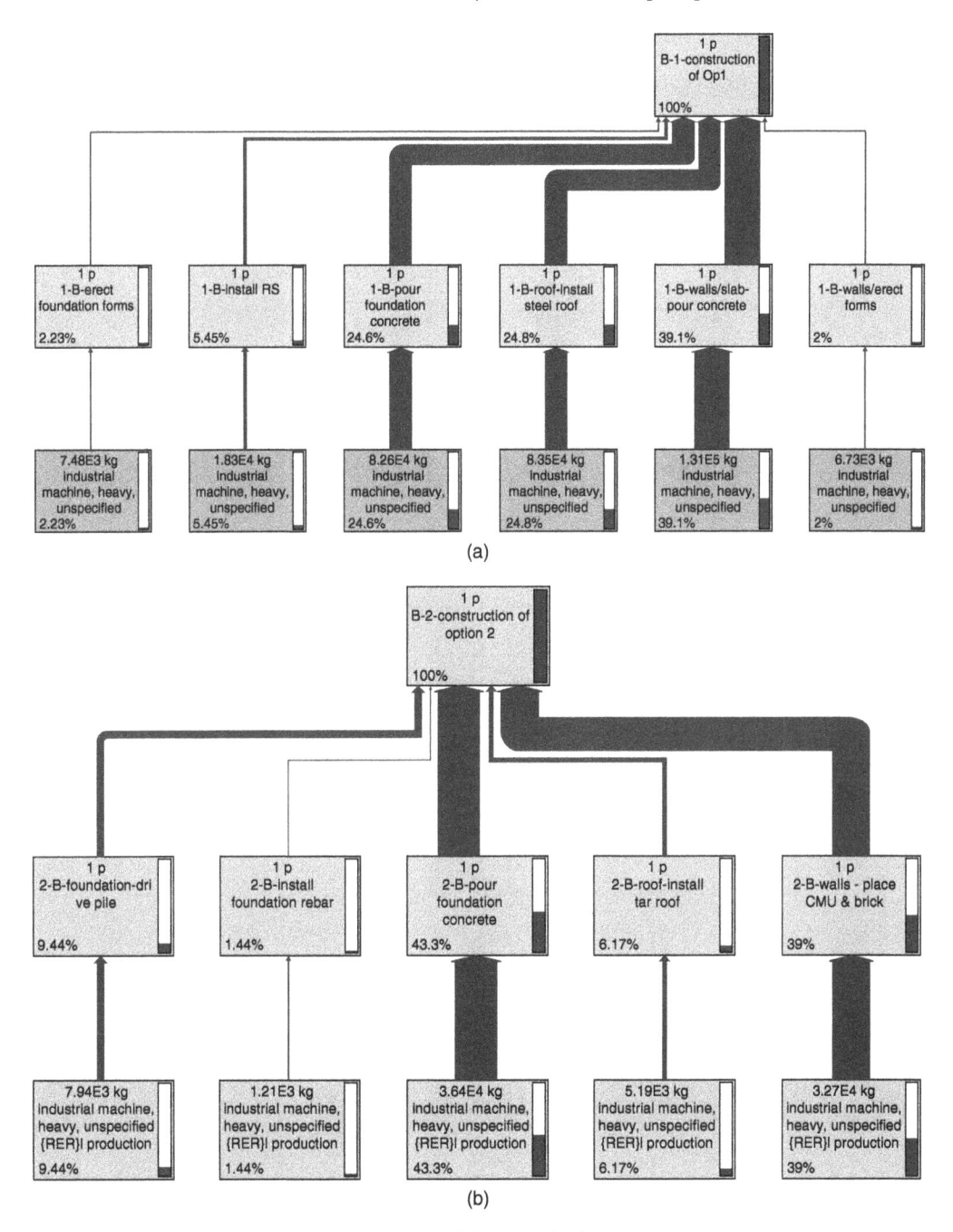

22.2 Comparison of Two Building Designs for an Electric Bus Substation

Figure 22.3 Construction of (a) Type 1 building, and (b) Type 2 building.

In the construction of the buildings (Figure 22.3), most of the impacts are shown to result from the pouring of concrete for footings, slabs, and walls. Since the Type 1 building had almost five times as much concrete as the Type 2 building, the results of the life cycle assessment (LCA) shows that the impacts to the environment of the Type 1 building are much higher than those of the Type 2 building.

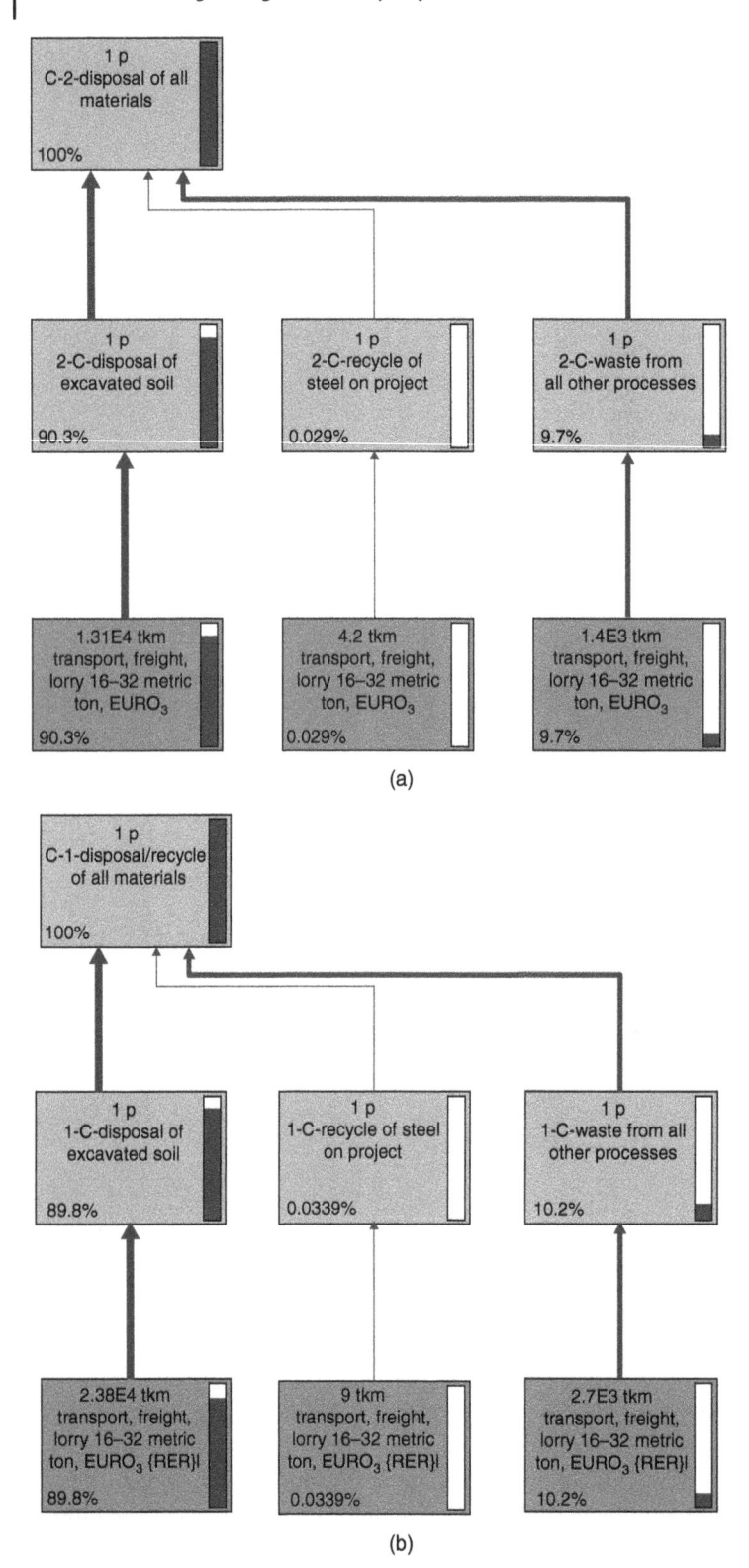

Figure 22.4 Waste management of (a) Type 1 building design materials and (b) Type 2 building design materials.

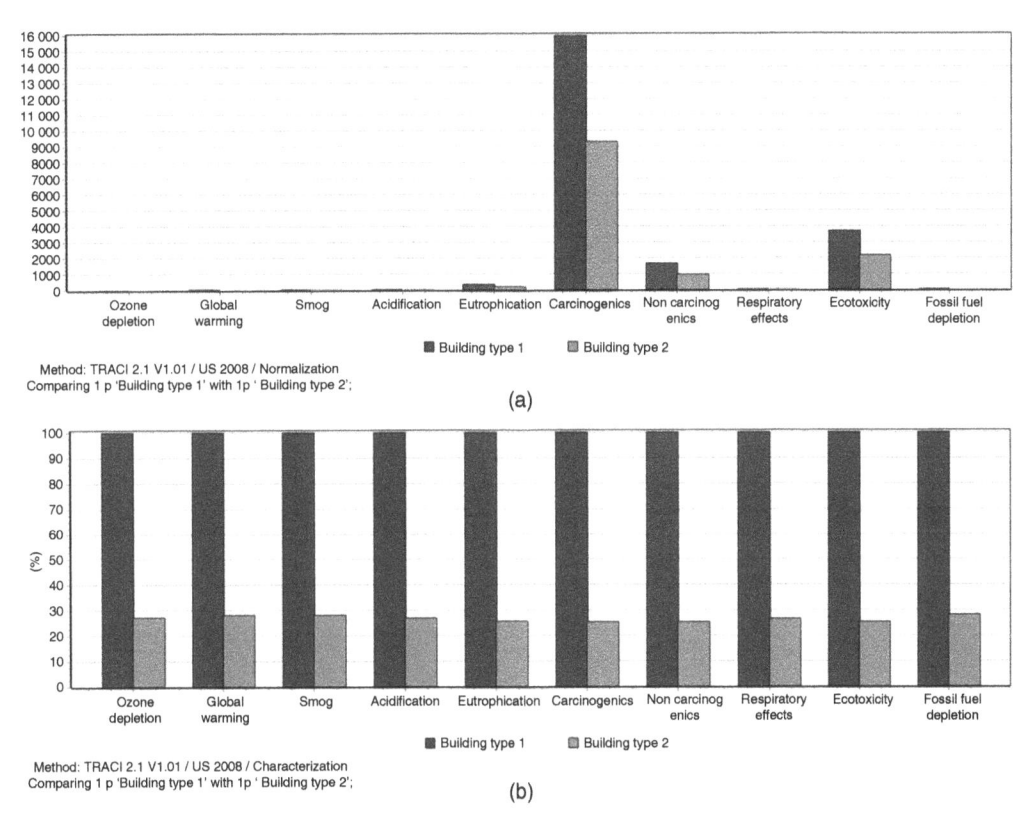

Figure 22.5 Environmental impacts form the LCA with SimaPro 8. (a) Comparison between the outputs of the two building designs, (b) comparison between the outputs of the two building designs as a percentage.

In the waste management stage, the main environmental impact is related to the excavation and transportation of soil for both building designs (Figure 22.4). However, Type 1 building design had almost two times as much soil being excavated and transported to a landfill located 104 km away from the site. Thus, the environmental impact associated with the Type 1 building was more than double for the Type 2 building.

The environmental impacts for each building design for the three stages added together are shown in Figure 22.5. The Type 1 building design has a greater environmental impact than the Type 2 building design, especially in the category of carcinogenic emissions. In all of the categories, the impact of the Type 1 building is four times greater than the Type 2 building.

22.2.4 Economic Sustainability

The economic aspects of building construction include the materials, transportation, labor, and construction costs. The costs of these project elements were obtained from several suppliers, national average material databases and national wage construction databases. Overall costs for the Type 1 and Type 2 buildings are shown in Table 22.1. Assumptions for the economic analysis include an eight-hour work day, five-day work week, and transportation costs are assumed to be 10% of the per unit material cost per mile. The economic assessment results shows that the Type 1 building is cost-effective than the Type 2 building.

Table 22.1 Building costs.

	Type 1 building			
	Unit cost	Unit	Hours	Cost ($)
Excavator	111.64 $/h	5 d	40	$4 465.60
Dump truck	28.17 $/h	5 d	40	$1 162.80
35 ton crane	93.79 $/h	11.5 d	92	$8 628.68
Concrete pump	104.88 $/h	5 d	40	$4 195.20
30 CY dumpster	76.88 $/h	5 d	40	$3 075.20
Pile driver	60.63 $/h	4 d	32	$1 940.16
Laborers	15 $/h	51 d	408	$6 120.00
Operators	38 $/h	28.5 d	228	$8 664.00
Carpenters	32 $/h	28 d	224	$7 168.00
Ironworkers	24 $/h	26.5 d	212	$5 088.00
Steel	$400.00 per ton	18.22 tons	—	$7 287.80
Wood	$50.00 per ton	21.43 tons	—	$3 398.94
Concrete	$90.00 per yd^3	117.37 yd^3	—	$10 563.33
Rebar	$400.00 per ton	6.60 tons	—	$3 111.85
Steel deck	$4.50 per ft^2	1 153.00 ft^2		$5 188.50
Clear land	$0.80 per ft^2	2 673.00 ft^2	—	$2 138.40
Soil removal	$12.00 per yd^3	566.00 yd^3	—	$6 792.00
Soil placement	$6.00 per yd^3	520.70 yd^3	—	$3 124.22
Transportation	10% of material cost per mile			$3 249.00
			Total cost:	$95 325.68

	Type 2 building			
	Unit cost	Unit	Hours	Cost ($)
Excavator	111.64 $/h	3.5 d	28	$3 125.92
Dump truck	28.17 $/h	5 d	40	$1 126.80
35 ton crane	93.79 $/h	20.5 d	164	$15 381.56
Concrete pump	104.88 $/h	2 d	16	$1 678.08
30 CY dumpster	76.88 $/h	5 d	40	$3 075.20
Pile driver	60.63 $/h	2 d	16	$970.08
Laborers	15 $/h	31.5 d	252	$3 780.00
Operators	38 $/h	28 d	224	$8 512.00
Carpenters	32 $/h	3 d	24	$768.00
Ironworkers	24 $/h	22 d	176	$4 224.00
Masons	19 $/h	80 d	640	$12 160.00

(Continued)

Table 22.1 (Continued)

	Type 1 building			
	Unit cost	Unit	Hours	Cost ($)
Steel	$400.00 per ton	12.33 tons	—	$4 932.00
Rebar	$400.00 per ton	1.90 tons	—	$1 231.73
Concrete	$90.00 per yd^3	41.30 yd^3	—	$3 717.34
Wood	$50.00 per ton	9.14 tons	—	$1 371.56
Brick	$3.00 per ft^2	4 879.00 ft^2	—	$24 565.00
Tar bitumen	$3.00 per ft^2	1 153.00 ft^2	—	$3 459.00
Clear land	$ 0.80 per ft^2	2 673.00 ft^2	—	$2 138.40
Soil removal	$12.00 per yd^3	2 448.40 yd^3	—	$3 720.00
Soil placement	$6.00 per yd^3	290.03 yd^3	—	$1 740.18
Transportation	10% of material cost per mile			$5 113.61
		Total cost:		$106 790.46

22.2.5 Social Sustainability

The social analysis attempts to quantify the social impacts that the construction of the substation would have on the community or the society around the site. Since the two building designs were similar with respect to their size and building type, not many factors differed with respect to their social sustainability. As a result, a comparison was performed considering a "building the substation" and a "not building the substation" alternative to determine if it would be more harmful to the environment and community to build the electric substation instead of choosing to follow the "do nothing" approach.

The method used to compare the two options discussed above was the SSEM developed by Reddy et al. (2014). The matrix considered four different areas of social interaction with the building (or lack-thereof) and the surrounding environment: Social, Social-Institutional, Social-Economic and Social-Environmental (Table 22.2).

(a) *Social dimension* – This considered whether or not it is more advantageous to build the substation versus not building the substation in the area at an individual person level.
(b) *Socio-institutional dimension* – Since the proposed substation is sited within an existing park area, the interaction between the park community and the building was an important aspect. This analysis resulted in a neutral conclusion as to whether or not to build the substation building, since the existing park community provided a strong institutional community for the area and removing it could endanger its ability to assist the community. Further, outreach and educational efforts would be beneficial to demonstrate the advantage and importance of having the presence of park areas and "green" transportation within the city.
(c) *Socio-economical dimension* – A major aspect of choosing whether or not to build the building was its economic importance to the area and/or if it would benefit the economy. One consideration is to add a coffee shop or some other type of economic and social stimulus to the substation in order to assist with building up the economy within the area. This would assist the area not only by generating revenue but also by increasing the amount of people visiting the area, which would in turn assist with generating more revenue for the tourist area.

Table 22.2 Social aspects of deciding to build a structure.

Dimension	Key theme area	Score					Total score
		Positive impact		No impact or not applicable	Negative impact		
		Ideal	Improved		Diminished	Unacceptable	
		2	1	0	−1	−2	
Social	Effect of proposed remediation on quality-of-life issues during and postconstruction						
	Cultural identity and promotion						
	Degree to which postconstruction project will result in learning opportunities and skills development for community						
	Degree to which postconstruction project will result in leadership development/empowerment opportunities						
	Enhancement of community/civic pride resulting construction and postconstruction project						
	Degree to which tangible community needs are incorporated into building design						
	Transformation of perceptions of project and environs within greater community						
	Enabling knowledge management (including access to E-knowledge)						
Socio-institutional	Appropriateness of future land use with respect to the community environment						
	Degree of land use planning fostered by proposed construction						
	Involvement of community in land use planning decisions						
	Enhancement of commercial/income-generating land uses						
	Enhancement to the architecture/aesthetics of built environment						
	Degree of "grass-roots" community outreach and involvement						
	Involvement of community organizations pre- and postconstruction						

(Continued)

Table 22.2 (Continued)

Dimension	Key theme area	Score					Total score
		Positive impact		No impact or not applicable	Negative impact		
		Ideal	Improved		Diminished	Unacceptable	
		2	1	0	−1	−2	
	Enhancement of cultural heritage institutions within community						
	Incorporation of green and sustainable infrastructure into construction						
	Enhancement of transportation system improvements						
Socio-economic	Disruption of businesses and local economy during construction						
	Employment opportunities during construction						
	Employment opportunities post-construction						
	Relative degree of increased tax revenue from Site Reuse						
	Degree to which green/sustainable or other "new economy" businesses may be created						
	Degree of stimulated informal activities/economy						
	Degree of anticipated partnership and collaboration with outside investors/institutions						
Socio-environmental	Degree of disruption (noise, truck traffic) from proposed construction method to the surrounding neighborhoods						
	Degree of future characterization/construction required by rezoning or altered land use						
	"Greenness"/sustainability of proposed construction action						
	Incorporation of green energy sources into construction activity						
	Potential of future environmental impact (i.e. diesel exhaust from trucks) that resulted from construction and allowable land reuse						

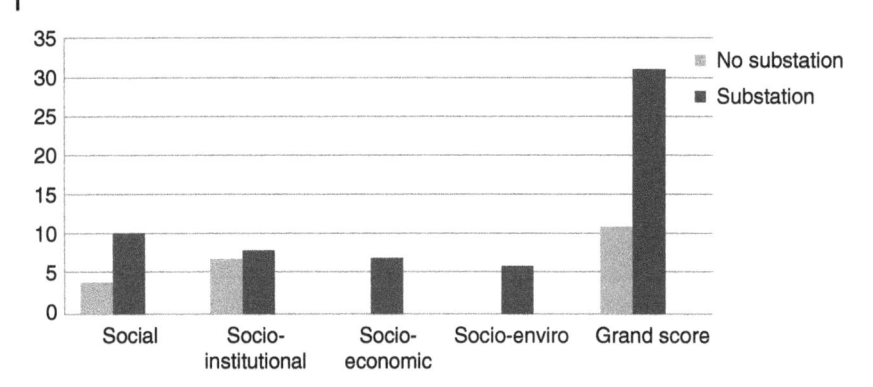

Figure 22.6 Comparison of the social aspects of "to build" or "not to build" options.

(d) *Socio-environmental* – This portion of analysis looked at areas of interest such as if the construction of the building would increase or decrease emissions in the area, the "greenness/sustainability" of a proposed construction of the building, potential of future environmental impact resulting from construction, and the degree of noise and traffic derived from the proposed construction. Some attributes of the building design that made construction sustainable were the addition of solar panels atop the building roof and taking into consideration that the construction of the substation building would lead to future savings in resulting emissions directly entering the air by supporting the development of electric buses for the city. While it can be argued that it takes a power plant to generate the electricity, and this may produce even more carbon emissions, a counterargument would be that the emissions would be generated at a common point, which could more easily be controlled at the power generating plant.

The overall social analysis (Figure 22.6) shows the global results from the four categories of the SSEM and the grand score for the two alternative scenarios (building an electric substation or the "do nothing" approach). The social analysis shows that it is better to build the substation building as opposed to not building it.

22.2.6 Conclusion

The Type 2 building design was demonstrated to be the more sustainable choice. While the cost of the Type 2 building design was higher than the Type 1 design, the Type 2 building had significantly lower environmental impacts, and its exterior aesthetics allowed it to better blend in with the surrounding park area. These two aspects were of greater importance in the overall analysis since they justify the increase in cost in order to support the placement of the building within the park area. The social analysis of the project showed that the overall impact of building versus not building the structure was more advantageous. The substation will assist with supporting mass transportation run by clean energy and thus assist the overall advancement of the quality of life in the city.

22.3 Prefabricated Cantilever Retaining Wall versus Conventional Cantilever Cast-in Place Retaining Wall

The construction of a retaining wall for highway application in Chicago has been considered in this study. The purpose is to conduct triple bottom line analysis for two alternatives of retaining wall types. The first option is the cast-in-place retaining wall which is the conventional type used. The second alternative is using precast concrete retaining wall.

22.3.1 Background

Retaining wall systems constitute a crucial part of any highway construction project. There are several alternatives available for retaining walls construction projects such as conventional cantilever, counterfort retaining walls, MSE walls (mechanically stabilized earth), and anchored walls. It is a common practice to use conventional, cast-in-place techniques for any of the selected wall types. However, cast-in-place is associated with several environmental and social impacts due to the time-intensive-related site works. As a result, greater attention has been focused toward achieving a constructible system that would combine the efficiency of a retaining wall system, speed of constructability, lower cost, and safety. This combination may be best utilized through the use of precast concrete.

The use of precast concrete elements for highway construction is considered efficient, as it requires less time of operation than the cast-in-place procedure. Additionally, precasting can eliminate several issues associated with cast-in-place approaches, including:

- Site preparation procedures such as installation of formwork, casting, curing of concrete.
- Traffic detouring and lane closures causing traffic congestion.
- Construction works leading to labor exposure to active traffic.
- Finishing works that require skilled workmanship.

22.3.2 Methodology

22.3.2.1 Goal and Scope

The goal of this study is to evaluate the sustainability of different construction practices used specifically for retaining walls. Thus, the cast-in-place technique is compared to retaining walls constructed using a precast concrete technique. The Circle Interchange Project in Chicago was selected as the site to evaluate the two retaining wall construction procedures. The main objective of this study is to identify the environmental, economic, and social impacts associated with the life cycle of the retaining walls constructed using cast-in-place concrete versus precast concrete. The functional unit for this study was taken as 100-ft-wide and 20-ft-high concrete retaining wall units.

22.3.2.2 Study Area

The Circle Interchange project is located within Chicago. The study is focused on the new flyover ramp from inbound Interstate 90/94 Dan Ryan Expressway to outbound Interstate 290 Eisenhower Expressway (IDOT 2017). The purpose of the construction project is to provide an improved transportation facility at the Circle Interchange by addressing the existing and 2040 transportation needs. Sustainable design concepts for the Circle Interchange project are wide reaching and include consideration for both construction and maintenance. The most noticeable improvement for the motoring public and communities that surround the Circle Interchange is the complete redesign of the area within the interchange itself. The geotechnical site data were obtained from the geotechnical report associated with the construction (Table 22.3).

22.3.2.3 Technical Design

Scenario 1: Precast concrete products: Precast concrete products are fabricated in a precast manufacturing plant and transported to the site where they are erected and assembled. These procedures are preceded by the required excavation and surveying actions.

The precast cantilever type walls consisted of 7000 psi precast footing and stem segments. The length of each segment is limited to 12 ft to facilitate transportation and erection as shown in Figure 22.7a. Full moment connection is provided between the stem and the footing through

Table 22.3 Geotechnical soil data for the project site.

Soil type	Cohesion (psi)	Friction angle (°)	Unit weight (pcf)
Foundation	3.62	30.00	117.7
Backfill soil	0.00	34.00	121.00
Ultimate bearing capacity (ksf)		24.31	
Allowable bearing capacity (ksf)		8.1	

Figure 22.7 (a) Precast concrete retaining wall system; (b) cast-in place concrete wall base slab; and (c) cast-in place concrete wall stem.

grout-filled mechanical splicers. Vertical joints between precast elements utilized shear keys, which are filled with nonshrink high strength grout. Retaining wall heights ranged from 4 ft to a maximum of 26 ft. The processes flow involved with the precast retaining wall option is represented in Figure 22.8a.

The dimensions of the retaining wall were calculated based on the existing soil conditions. The technical design covers the need to satisfy the AASHTO LRFD requirements for strength and stability. The required stability checks according to the provisions of AASHTO LRFD are the following:

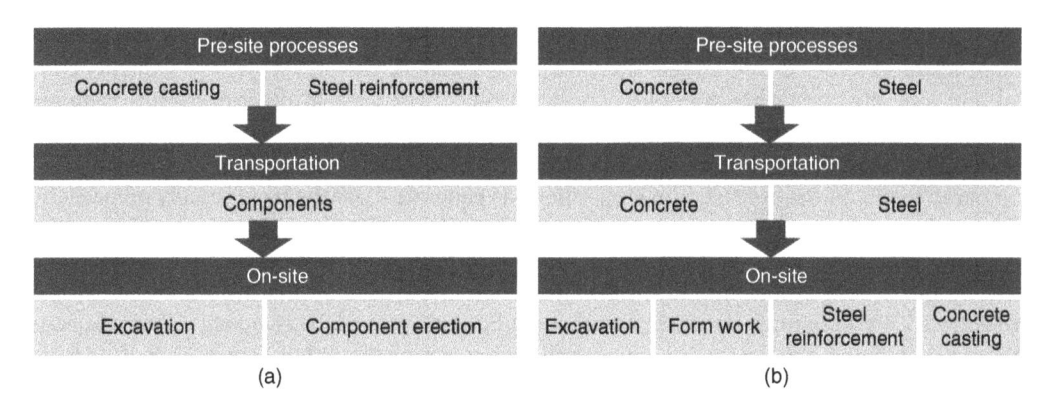

22.3 Prefabricated Cantilever Retaining Wall versus Conventional Cantilever Cast-in Place Retaining Wall

Figure 22.8 (a) Processes flow of precast concrete retaining wall. (b) Processes flow of precast concrete retaining wall.

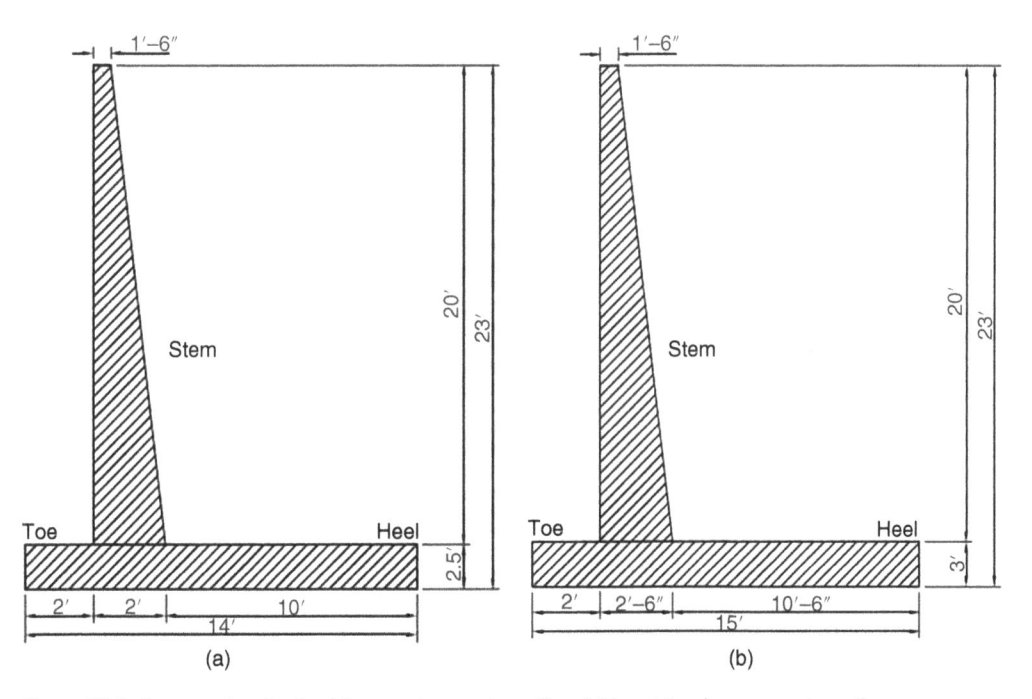

Figure 22.9 Cross section for the (a) precast concrete wall and (b) cast-in-place concrete wall.

- Bearing capacity pressure;
- Sliding of the system with factor of safety (F.O.S.) of 1.5;
- Overturning with F.O.S. of 2;
- Eccentricity limits is confined within the middle two-thirds of the base slab.

High-quality control standards are applied to precast concrete products to produce high-performance materials. The resulting material is durable and meets the strength requirements. Concrete compressive strength of 7000 psi is used in the design for the precast components. Figure 22.9a shows the cross section and the dimensions of the precast retaining wall system.

Scenario 2: Cast-in-place concrete products: Cast-in-place concrete products require several procedures to be carried out on-site. Site preparation, formwork, steel works, quality control, and concrete casting are examples of on-site procedures. Cast-in-place procedure requires several additional processes that are performed to limit the environmental, social, and economic impacts of the project. For instance, lane closures are required for longer periods of times compared to that of the precast. The longer periods generate from the need to fully prepare the construction site prior to concrete casting as shown in Figure 22.7b,c.

Full moment connection is provided between the stem and the footing by extending the main reinforcement of the stem toward the base slab at the time of pouting the slab as shown in Figure 22.7b. Retaining walls are assumed to be with a fixed 20-ft height. The processes flow associated with the precast retaining wall option is represented in Figure 22.8b. Concrete compressive strength of 3500 psi is used in the design for the cast in place retaining wall components. Figure 22.9b shows the cross section and the dimensions of the cast-in-place retaining wall system.

22.3.2.4 Sustainability Assessment

The environmental impact assessment was determined with SimaPro software. The impact assessment methodology selected is the USEPA TRACI method.

In the life-cycle assessment, each structural element (material, product, or process) was modelled independently. These elements were then combined to comprise a complete building subassembly (e.g. precast concrete wall subassembly, excavation subassembly, and steel subassembly). Each of these subassemblies was combined to model the complete building structure and envelope as fabricated or constructed on-site. Maintenance, replacement, end-of-life demolitions, recycling, and landfilling were not considered as they are similarly shared for both types of retaining walls. Table 22.4 summarizes the materials and transportation distances from the source materials to the construction site.

The economic impacts of the two scenarios were determined by performing the life-cycle cost assessment (LCCA) including direct and indirect costs. Direct costs include materials, transportation, and processes involved in construction. Transportation distances were determined based on the availability of the concrete producers within a reasonable distance from the project site (Table 22.4). Indirect costs were based on the social cost of carbon and other gaseous emissions at an appropriate discount rate.

Social impacts of the two scenarios were determined using three different tools or methods: SSEM; streamlined life cycle assessment (SLCA) health and safety matrix (Table 22.5); and Envision rating checklist.

Table 22.4 Sources of materials needed for the projects.

Type	Source	Location	Travel distance (mi)	Travel time (min)
Precast concrete plant	Dukane Precast, Inc.	Naperville, IL	38	45
Ready mix plant	Ozinga Ready Mix Concrete, Inc.	Chicago, IL	3	10
Rebar	Elston Materials, LLC	Chicago, IL	3	10

Table 22.5 Scoring system for streamlined life cycle assessment health and safety matrix.

0.0–0.9	Unacceptable
1.0–1.9	Undesirable
2.0–2.9	Tolerable
3.0–3.9	Acceptable with review
4.0	Acceptable without review

22.3.3 Environmental Sustainability

The environmental impacts calculated with SimaPro for the two scenarios: cast-in-place and precast retaining walls, are shown in Figures 22.10–22.12. In the cast-in-place system, concrete and steel production are responsible for all of the impacts (Figure 22.10a). Construction works do not show any contribution to the environmental impacts. Even the contribution of transportation of materials is negligible considering the short distance of the material suppliers. As an example, concrete and steel production are responsible for 50% each on the global warming category (Figure 22.11a). In the precast system, the highest impacts were attributed to the concrete production and transportation (Figure 22.10b), followed by the construction works on site. It is important to note that the predefined precast system in SimaPro includes all the activities of concrete production with all the processes associated to 0.5% steel reinforcements. The actual designed reinforcement is calculated to be 0.63%. As a result, additional 0.13% steel production was added to the processes. The concrete production, steel reinforcement, and transportation contribute to 94.3% of the overall global warming (Figure 22.11b). Figure 22.12 shows a comparison of the environmental impacts of the two scenarios. A precast concrete system shows higher environmental impacts in all the categories when compared to the cast-in-place system. The CO_2 emissions (Table 22.6) are also higher for the precast system. This is due to the longer transportation distances that contributes not only to the CO_2 emissions but most of the environmental impacts categories. Thus, for shorter transportation distances of the precast system, this option may become more environmentally friendly than the cast-in-place option.

22.3.4 Economic Sustainability

Table 22.7 summarizes the required amount of steel and concrete for both alternatives. The total costs of concrete and steel (including the installation of the steel reinforcement) is presented in Table 22.8. The transportation of materials is calculated based on the transportation distance and the capacity of the transportation vehicles. For the cast-in-place option, the capacity of the mixing truck is around 9 cubic yards. This will require 35 trips from the plant to the construction site to deliver the mixed concrete. Similarly, precast components are made with 13-ft interval lengths. The base and the stem are transported separately. This required 16 trips of the transportation truck (Table 22.9).

The form work is calculated based on the required covered area (Table 22.10). The wood is assumed to be reusable every 20 ft. The labor cost calculation for each case is based on the number of working hours. For a cast in place retaining wall, it is assumed that four days of work are needed to completely build, cast, and remove forms for a 20-ft-long retaining wall.

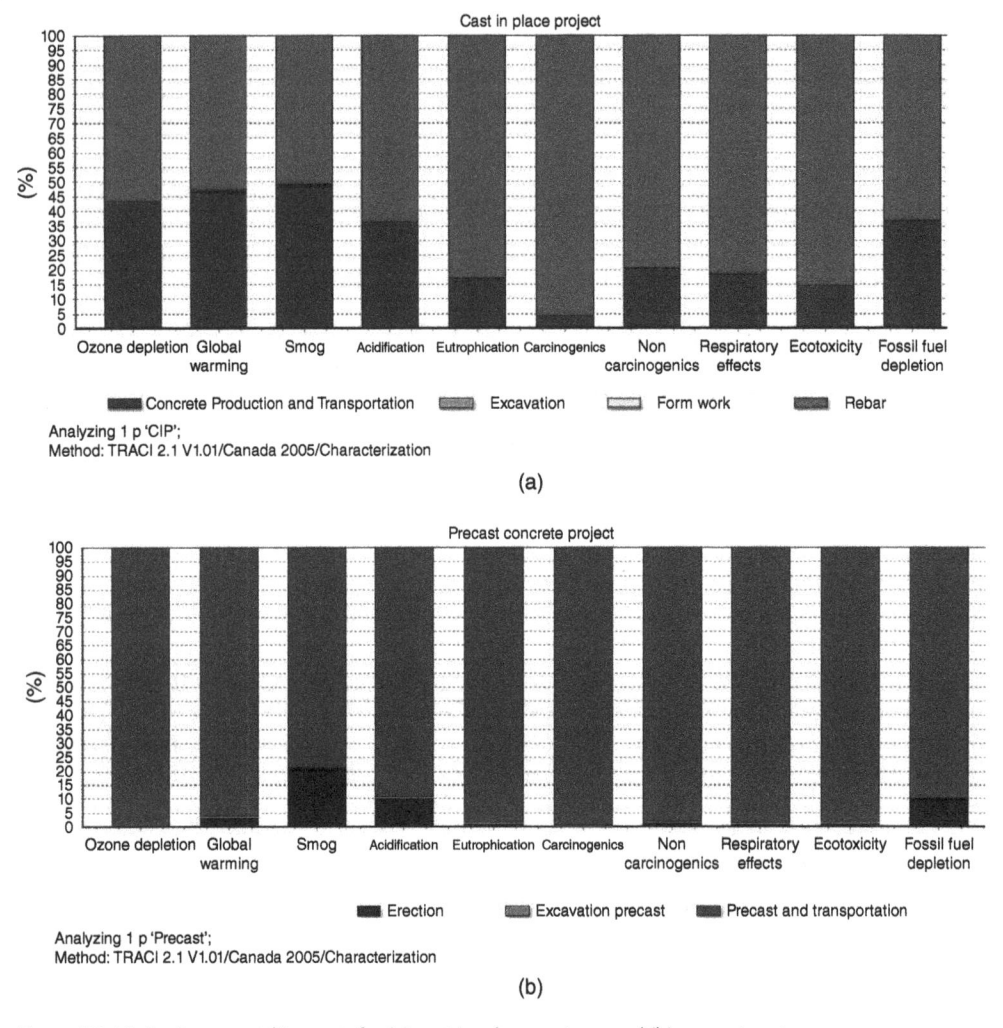

Figure 22.10 Environmental impacts for (a) cast-in-place system, and (b) precast system.

Eight-hour work days were considered for estimating the cost of labor and equipment. In the precast concrete wall, the duration is shorter because no curing is required.

The final cost estimates per functional unit is in Table 22.11. The cost for both alternatives is similar. The cast-in-place option shows higher cost for on-site procedures. However, the precast shows higher cost for transportation. Thus, the precast option would be more economically sustainable for shorter transportation distances.

22.3.5 Social Sustainability

The social impact assessment with SSEM is shown in Figure 22.13. The results show that the precast retaining wall attains better social acceptance when compared to the cast-in-place system. The precast wall received better scoring in all the categories, but the biggest difference is in the socio-environmental dimension, where the cast-in-place system receives a negative score (negative impact or social acceptability). This is mainly due to the safety factor associated with

22.3 Prefabricated Cantilever Retaining Wall versus Conventional Cantilever Cast-in Place Retaining Wall

(a)

(b)

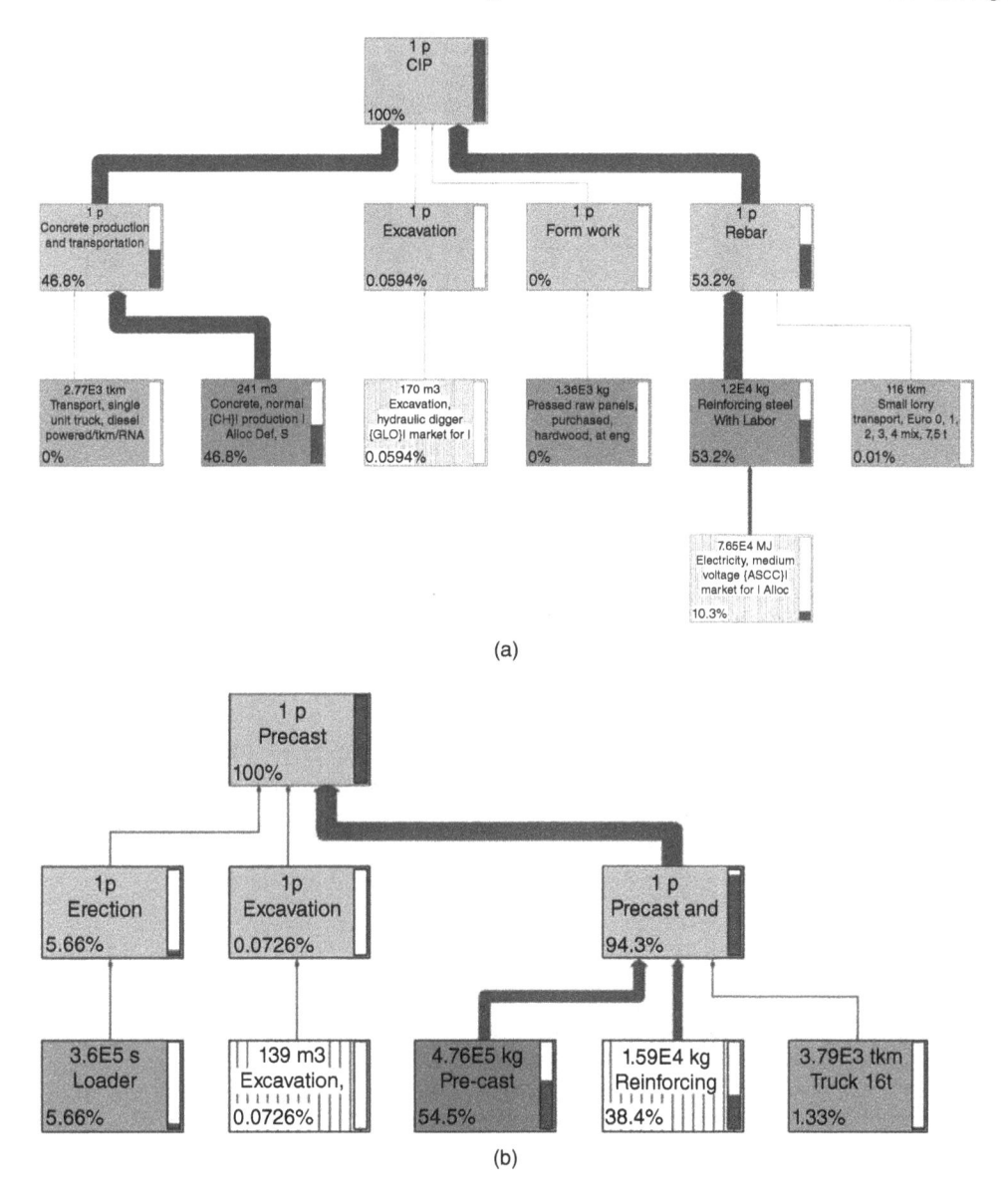

Figure 22.11 Detailed network for process contribution to global warming for (a) cast-in-place and (b) precast system

the precast concrete erection procedure. The precast construction system is characterized by less exposure to traffic and less exposure of people with the construction works.

The social assessment with the SLCA health and safety matrix is shown in Table 22.12 for cast-in-place and precast retaining wall scenarios. In this rating method, the higher the score, the higher the safety and social acceptability of the project. The results show 83.6% social acceptance for the precast concrete alternative and 76.2% for the cast-in-place alternative. This indicates that the precast concrete provides enhanced working environment which is characterized by safety, less exposure to weather conditions, less delays due to weather and traffic, and less

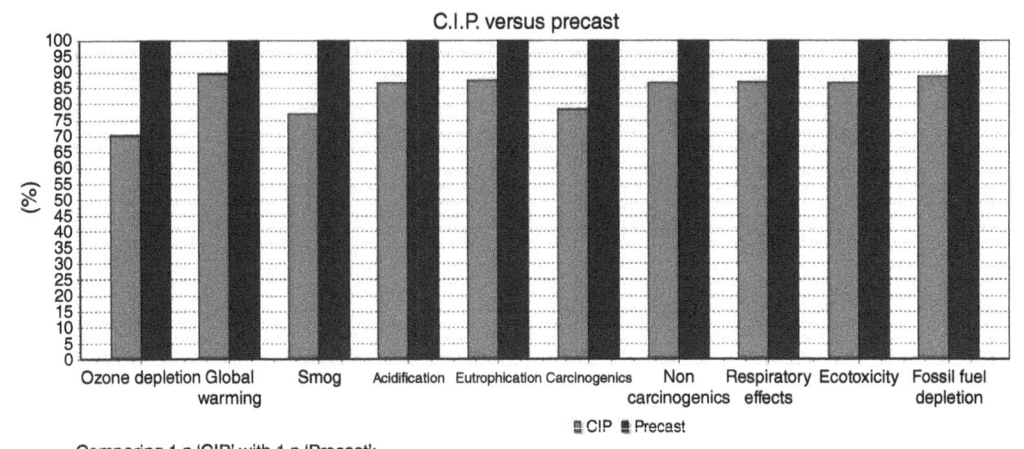

22 Infrastructure Engineering Sustainability Projects

C.I.P. versus precast

Comparing 1 p 'CIP' with 1 p 'Precast';
Method: TRACI 2.1 V1.01/Canada 2005/Characterization

Figure 22.12 Comparison of environmental impacts from precast versus cast-in-place systems.

Table 22.6 Comparison of carbon dioxide emissions between cast-in-place and precast concrete.

Carbon dioxide type	Cast-in-place	Precast
Biogenic (kg)	1 894.9	763.1
Fossil (kg)	148 481.7	107 261.9
In air (kg)	943.0	899.9
Land transformation (kg)	138.2	56 263.4
Total (kg)	151 457.9	165 188.3

Table 22.7 Volume of concrete and amount of steel reinforcement required.

Retaining wall	Volume of concrete required (yd³/ft)	Average amount of steel required (lb/ft)
CIP	3.15	265.50
Precast concrete	2.6	170.12

Table 22.8 Total costs for steel and concrete required for both projects.

	Concrete and steel production cost				
	Cost of concrete production		Cost of steel (with installment)		Total cost ($)
Wall type	per cubic yard of concrete	per long 1 ft	Average per lb	per long 1 ft	
CIP	85	268	2	531	798.75
Precast	85	221	2	340	561.24

Table 22.9 Concrete transportation to the construction site.

Wall type	Concrete transport					Functional unit	
	Distance from concrete plant to site (mi)	Truck characteristics				Number of trucks required	Cost of transport
		Truck type	Transport cost (USD/mi)	Capacity yd³ (or piece)			
CIP	3	Concrete mixer	5.00	8	35	525
Precast	38	Long truck	2.00	1	16	1216

Table 22.10 Form work and labor costs for cast-in-place (CIP) and precast concrete.

Wall type	No.	Item	Unit	Quantity	Unit Cost	Cost
CIP	1	Form work	sq yd	61	$3	$183
	2	Labor	hours/4 d	24	$30	$11 520
			per 100 ft	96	4 employees	
	Total					$298.20
Precast	1	In-plant form work	sq yd	61	$1	$61
	2	Labor	hours/4 d	14	$25	$5 600
			per 100 ft	56	4 employees	
	Total					$117.00

Table 22.11 Total cost per functional unit.

	Total cost/100 long ft
Retaining wall	Total cost
CIP	$162 195
Precast concrete	$189 424

risk of shocks due to moving traffic. In addition, it is also characterized by less noise due to accelerated construction works.

The Envision rating checklist assesses the social impact of a project over several aspects: quality of life (QL), leadership (LD), resource allocations (RA), natural world (NW), and climate and risk (CR). The results of the Envision checklist for cast-in-place concrete and precast concrete are represented in Figure 22.14. It can be concluded that precast system leads the cast-in-place concrete system in all the categories.

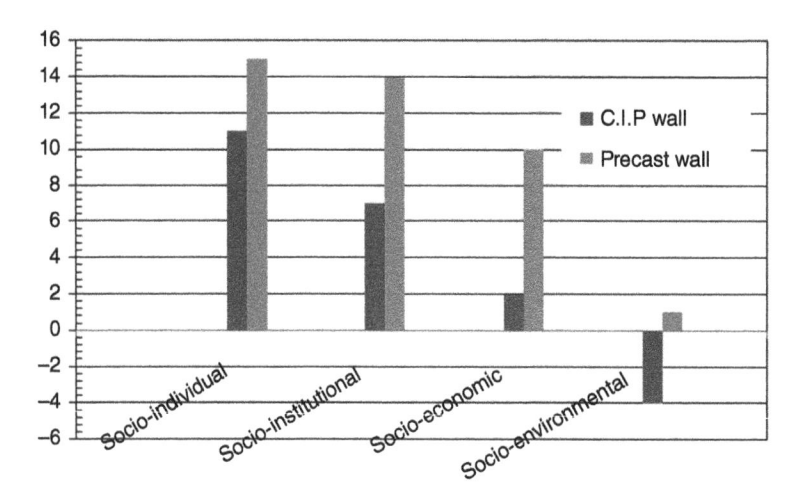

Figure 22.13 Social Impact assessment using SSEM tool for CIP (cast-in-place) and precast walls.

Table 22.12 SLCA Health and safety matrix for cast in place and precast concrete.

	Cast in place wall: SLCA health and safety matrix					
	Physical hazard	Chemical hazard	Shock hazard	Ergonomic hazard	Noise hazard	Row score
Site excavation	2.5	4.0	3.8	2.5	1.5	14.3
Concrete Manufacture	3.0	2.0	3.5	2.9	1.5	12.9
Product delivery	2.9	4.0	4.0	2.5	2.5	15.9
Installation/construction	2.0	3.8	4.0	1.5	2.1	13.4
Field service	4.0	4.0	4.0	4.0	4.0	20.0
End of life	3.5	4.0	4.0	3.0	0.5	15.0
Column score	17.9	21.8	23.3	16.4	12.1	91.5/120 76.2%
	Precast concrete wall: SLCA Health And Safety Matrix					
	Physical hazard	Chemical hazard	Shock hazard	Ergonomic hazard	Noise hazard	Row Score
Material Excavation	2.5	4.0	3.8	2.5	1.5	14.3
Product Manufacture	3.5	3.0	3.5	4.0	4.0	18.0
Product delivery	1.5	4.0	4.0	3.2	3.5	16.2
Installation/construction	3.5	4.0	4.0	3.9	1.5	16.9
Field service	4.0	4.0	4.0	4.0	4.0	20.0
End of life	3.5	4.0	4.0	3.0	0.5	15.0
Column score	18.5	23.0	23.3	20.6	15.0	100.4/120 83.6%

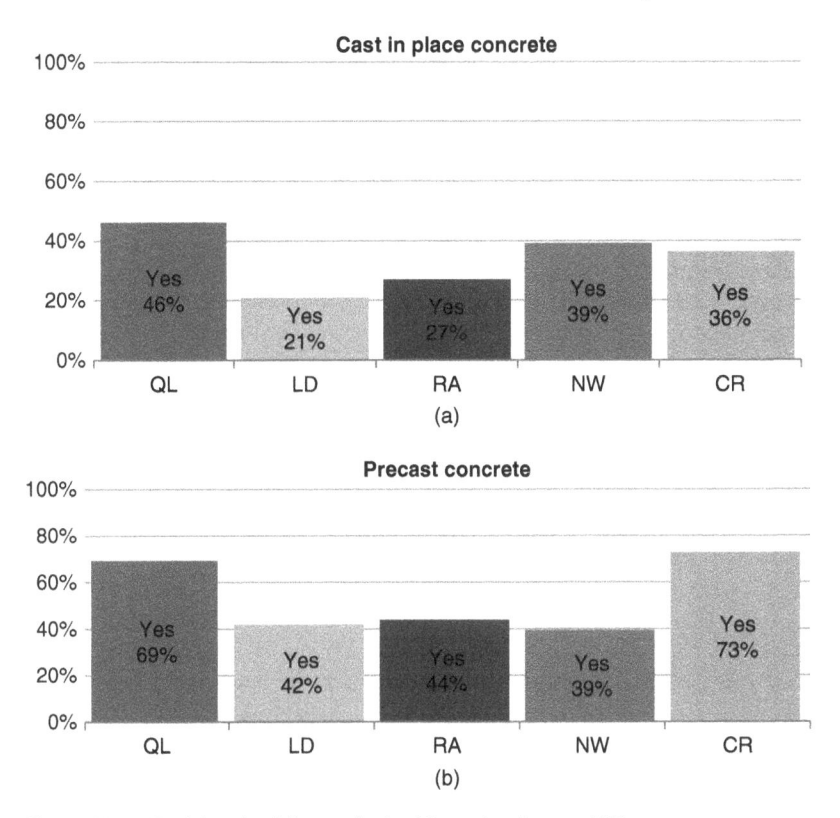

Figure 22.14 Envision checklist results for (a) cast-in-place and (b) precast concrete.

22.3.6 Conclusion

The sustainability assessment of the two construction alternatives identified similar impacts considering the environmental and economic assessment. The main environmental differencing factor between the two alternatives is transportation for the precast wall and site works for the cast-in-place wall. Precast wall shows better social acceptability because of shorter contraction time and less noise. However, the results are site-specific, and the precast option could be the more sustainable alternative for shorter transportation distance of the precast concrete to the construction site.

22.4 Sustainability Assessment of Two Alternate Water Pipelines

This study presents the sustainability assessment of two alternate water pipelines: (a) conventional ductile iron with cement-mortar lining and (b) clean energy producing (CEP) steel with cement-mortar lining, over their respective life cycle period. The environmental impact assessment is performed by considering life cycle stages from raw material acquisition, manufacturing/construction, use/maintenance, and disposal.

22.4.1 Background

Gravity-fed water systems rely solely on pressure-reducing valves (PRVs) to release the buildup of excess pressure in their pipes. New systems, like Lucid Energy, plan to insert turbines into replacement pipes to convert this excess head into electricity. This new system has the advantage of allowing for clean and renewable energy generation with an innovative idea. It will also reduce the wear on existing PRVs, thus extending their design life.

22.4.2 Methodology

22.4.2.1 Goal and Scope

In this study, two alternate water pipeline systems, the conventional pipe and the CEP pipe, were assessed for their overall sustainability by considering their different life cycle stages that includes raw material acquisition, manufacturing, construction, transportation, use and maintenance, and final disposal and recycling. Environmental impacts were investigated by LCA by SimaPro 8.0 software for the entire life cycle. Economic and social sustainability of both systems were also studied in detail.

22.4.2.2 Site Background

The project site is a residential area located in Portland, Oregon. The site was selected because the community's water system utilized a gravity-fed system and a specific length of pipe was due for replacement.

22.4.2.3 Technical Design

The functional unit for the project is a 52.5-ft pipe length with a 42-in. diameter. The life-span of the replacement pipes was set at 50 years. In addition, the project considered the output of 900 MWh of energy per year over that same 50-year period. Assumptions and estimations were made for the installation phase.

Scenario 1: Conventional pipe design: The pipe will be made of ductile iron and lined with cement-mortar. The ductile iron will be 5/8 in. thick, while the cement-mortar will only be 1/8 in. thick. The installation of the pipe will occur in three stages. First, the old pipe will be excavated and removed. The pipe will require an excavation pit 10 ft wide and 55 ft long. The walls will be sloped to a 2-ft-wide base, 55 ft long, and an overall depth of 7 ft. The new cement structures, six in total, will then be set into the ground. The new pipe will be placed, and the pit will be refilled with the previously excavated dirt. Figure 22.15a shows the three stages of installation as well as a rendering of the project below ground.

Scenario 2: CEP pipe design: The pipe will be made of steel and lined with cement-mortar. The steel will be slightly thinner than the ductile iron with a thickness of only 3/8 in. However, the lined cement-mortar will have a thickness of ½ in. The installation of the pipe will occur in four stages. First, the old pipe will be excavated and removed. The pipe will require an excavation pit 14 ft wide and 55 ft long. The walls will be slightly angled to maintain the 10-ft-wide base, again 55 ft long, with an overall depth of 12 ft. Then, a 10-ft by 55-ft cement base, with a thickness of 10 in., will be placed to act as the floor for the Clean Energy pipe. After the floor is set, the four walls will be placed, and then the six structures to support the pipe will be placed. The new pipe and all its components will then be inserted and connected. Finally, the structure will be capped with a 10-ft by 55-ft cement top. Figure 22.15b shows the four stages of installation as well as a rendering of the project.

22.4.2.4 Sustainability Assessment

The environmental sustainability was studied by conducting a comprehensive LCA using SimaPro 8.0 and related databases and impact assessment methods (ISO 2006; Goedkoop et al. 2010). In this study, the goal of the LCA was to model the life cycle stages of material acquisition, manufacturing, construction, use and maintenance, and disposal of two alternate

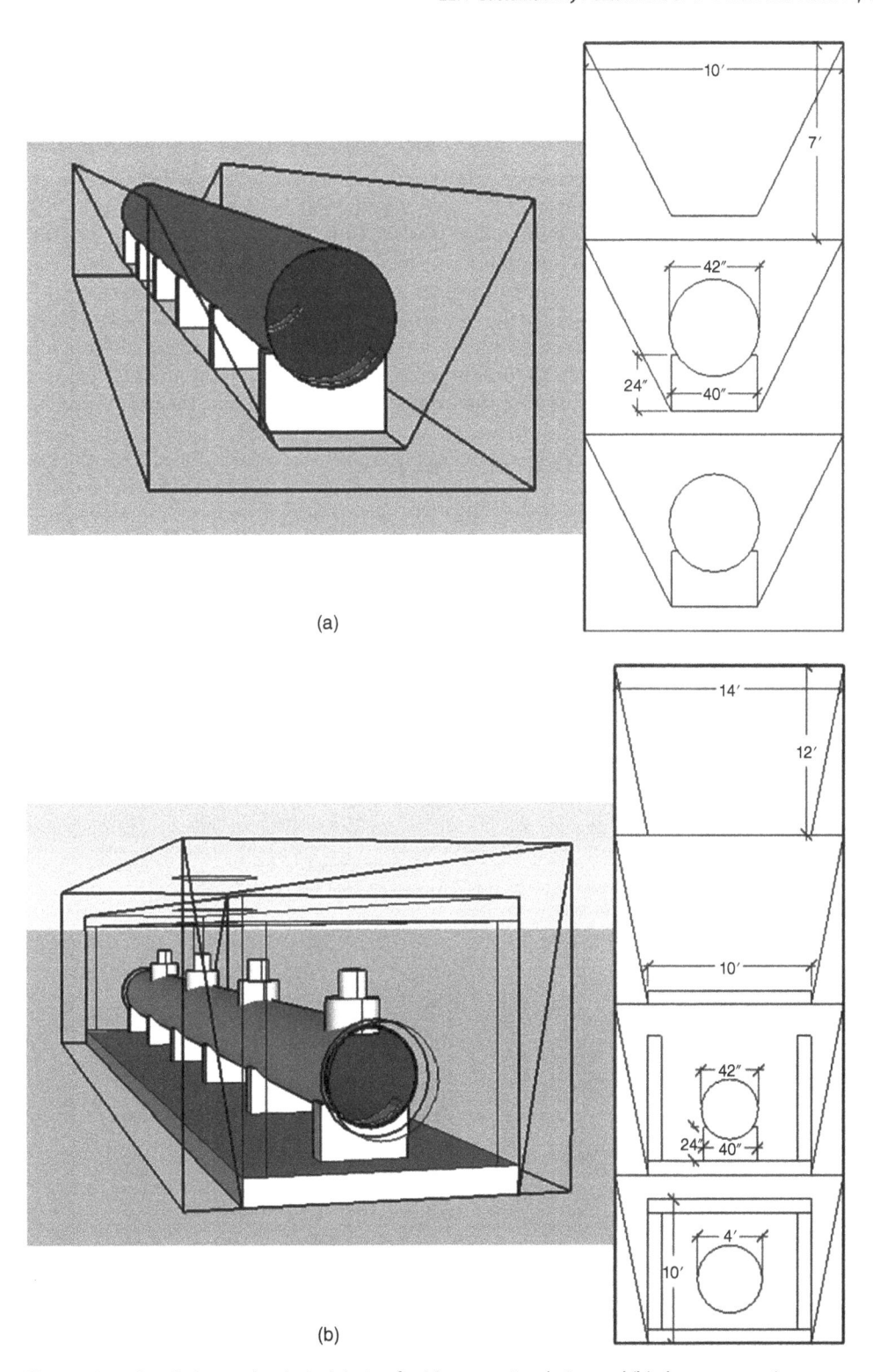

Figure 22.15 Rendering and technical design for (a) conventional pipe and (b) clean energy pipe.

water pipeline systems to evaluate environmental impacts and investigate which system is more environmentally sustainable over the systems' design life.

22.4.3 Environmental Sustainability

The results of the environmental assessment of the two alternative pipelines using SimaPro are shown in Figures 22.16 and 22.17. Table 22.13 shows the different weights and amounts used as input in the LCA program. Because limited data was available for the turbines and generators used for the CEP pipe, simplifying assumptions were incorporated. These values were then used with varying distances and manufacturing processes found within the SimaPro database.

Figure 22.16 highlights the difference in the resource acquisition stage between the two pipes, with the CEP pipe being more energy and resource-intensive. The increased amount of concrete required for the CEP pipe is the main factor for the large difference in environmental impacts. The resource acquisition of concrete is more environmentally intensive than the manufacturing and installation.

The use and maintenance stage is where the largest impact discrepancy occurs for the two scenarios. In this stage, the emissions for the energy demand of 900 MWh for 50 years are accounted in the environmental assessment. The CEP pipe uses clean, renewable energy; therefore, its total emissions are zero. However, the conventional pipe relies on coal to generate that same amount of energy for 50 years. Thus, the yearly output of global warming potential (GWP)

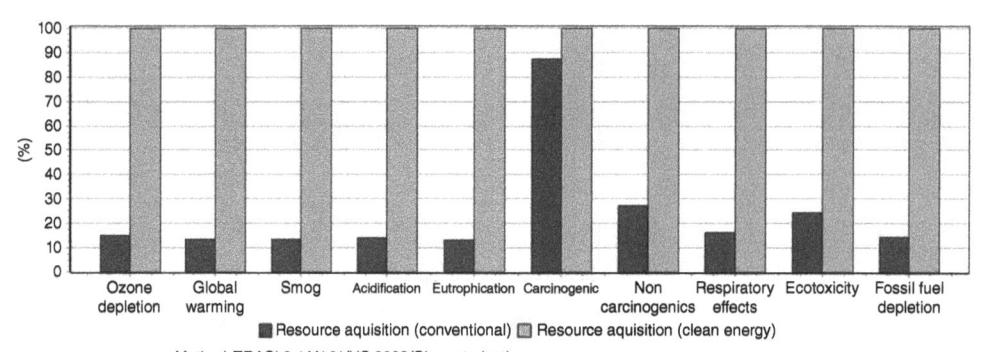

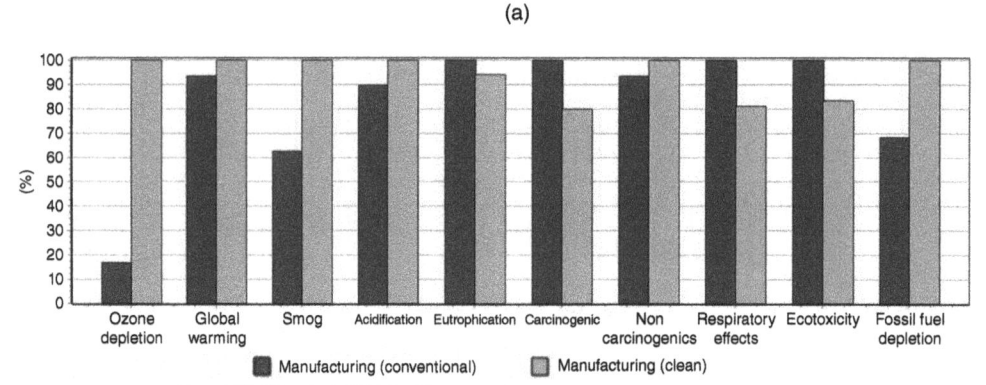

Figure 22.16 Environmental impacts of (a) resource acquisition and (b) manufacturing for the two pipelines.

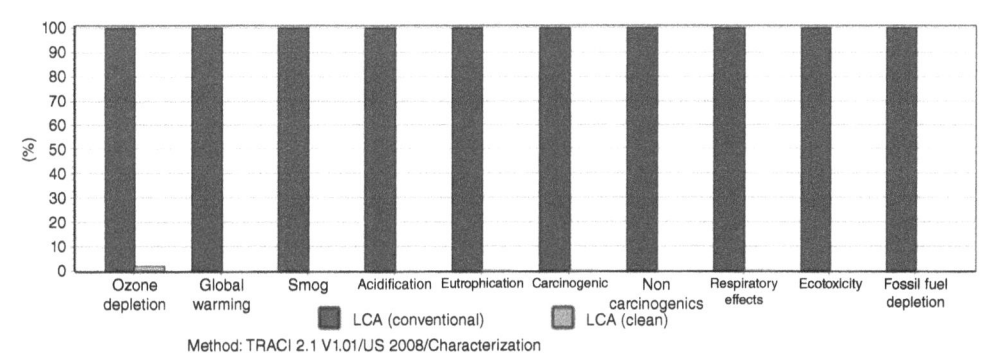

Figure 22.17 Comparison of global environmental impacts of the CEP and conventional pipeline (including use and maintenance stage).

Table 22.13 Materials data for SimaPro.

Materials	Amount for 1 unit	
Conventional pipe		
Ductile iron	12 064.5	lb
Cement-mortar lining (for iron)	935.55	lb
Soil	2660	ft^3
Concrete	270.75	ft^3
Asphalt	15.4	tons
Clean energy pipe		
Stainless steel	8 760.15	lb
Cement-mortar lining (for steel)	3 756.9	lb
Turbine (×4)	400	lb
Generator (×2)	400	lb
Soil	10 650	ft^3
Concrete	2 250	ft^3
Asphalt	15.4	tons

for coal generation is roughly equal to the GWP of CEP pipe for the entire process (Table 22.14). Figure 22.17 shows the overall impact difference between the two pipelines. In all the categories, the CEP pipe shows minor impacts compared to the conventional pipe. These results are due to the coal emissions for energy generation.

22.4.4 Economic Sustainability

The economic assessment of the two water pipelines include the total cost of installing the pipeline in place (materials, transportation, labor, construction costs, and annual costs and benefits). The total costs were estimated from various sources or estimated from a similar project. Capital costs are shown in Table 22.15, and annual costs and benefits are shown in Table 22.16. The annual costs and benefits were calculated and produced based on the current

Table 22.14 LCA results for the CEP and conventional pipeline.

LCA results	Total global warming (kgCO$_2$ eq) Conventional	Total global warming (kgCO$_2$ eq) Clean energy	Total ozone depletion (kg CFC-11 eq) Conventional	Total ozone depletion (kg CFC-11 eq) Clean energy	Total fossil fuel depletion (MJ energy) Conventional	Total fossil fuel depletion (MJ energy) Clean energy
Resource acquisition	136 735.47	1 013 729.80	0.004 7	0.031 8	96 691.28	676 211.36
Manu/Installation	13 457.44	14 384.76	0.001 3	0.007 4	9 886.42	14 452.46
Use	54 706 800.49	0.00	1.297 6	0.000 0	10 157 648.94	0.00
Disposal	65.90	69.69	0.000 1	0.000 1	125.18	132.38
Total	54 857 059.30	1 028 184.24	1.303 7	0.039 4	10 264 351.82	690 796.20
Total w/o use	150 258.81	1 028 184.24	0.006 1	0.039 4	106 702.88	690 796.20

Table 22.15 Pipeline capital cost analysis.

	Conventional pipe	Clean energy pipe
Total capital cost	$66 150	$1 700 000
Total labor cost	$4 604	$9 208
Equipment cost	$5 180	$9 930
Materials (raw, manufacturing, transport, monitoring equipment)	$56 366	$1 680 862

Table 22.16 Annual cost and benefit analysis.

	Conventional pipe		Clean energy pipe	
Electricity	0.037 26 $/kWh	Cost to produce	0.111 1 $/kWh	Price purchased
Annual cost	33 534	900 MWh/year	6 000	(Estimated maintenance $)
Over 50 years	1 676 700		300 000	
Annual profit	0		100 000	900 MWh/year
Over 50 years	0		5 000 000	
Total value	−1 676 700		4 700 000	

power purchasing agreement (PPA) for alternative energy in place for the electricity produced by the CEP pipeline and the cost to produce the electricity based on the rates for the region.

Results show that despite the greater initial cost for the clean energy pipeline, the benefits of the production of hydropower and its sale will make up its value and more over the estimated 50-year life span. The total value of the clean energy pipe comes out to be $3 million and the conventional pipe is −$1.7 million. Even excluding the cost of energy for the conventional pipe, the CEP will be more cost-effective by selling the electricity produced from it.

22.4.5 Social Sustainability

Social sustainability was evaluated in this project by a combination of the Benoît-Norris et al. (2011) framework and the SSEM tool (Reddy et al. 2014) as shown in Table 22.17. The social

Table 22.17 Social aspects criteria use in the social assessment.

Jobs/income	Creates new jobs (monitoring system)
	Generator and turbine maintenance
	More income related to energy-generation in coal power plants
	More parts required for manufacturing
Health and safety	Monitor drinking water quality
	Monitor pipe pressure
	More resources = more possibilities for health and safety risks
	More intricate parts and construction = more possibilities to health and safety risks
Environmental impact	Recycling of pipe materials, transportation of soil, energy generation
Value added	Provide green energy that does not generate emissions
	Generates electricity in new source
	Does not harm ecosystem like typical hydroelectric sources such as dams
Social responsibility	Community is committed to reducing carbon emissions and alternative energy sources
Community development	Knowing that the community is supporting a greener environment
Cultural aspects	Promotes cultural values, making the earth and greener, healthier environment
Quality of life	Air quality is better
	Better living environment
Product value	Clean energy is generating value (electricity) where before there was none

matrix used in this study demonstrates an approach to assess the social sustainability of both projects. Based on the results (Table 22.18), the CEP was found to be more socially viable than conventional water pipeline, with a positive impact score of 10, mainly coming from the use life stage and the production of electricity.

22.4.6 Conclusion

Results demonstrated that the CEP pipeline was more environmentally friendly than the conventional pipeline over the estimated lifespan of the two systems. The largest factor of environmental impact was the use/maintenance cycle for the conventional pipe system. The environmental impact of providing 900 MWh/year of electricity from conventional sources to make up for the electricity produced by clean hydropower is higher than the global impacts of the clean energy pipeline. The social sustainability assessment also identified the CEP pipeline as the most sustainable alternative. The economical assessment, based on capital and annual costs and benefits, showed that the CEP pipeline has greater upfront costs but returns value over the 50-year lifespan of the two systems because the CEP pipeline will produce approximately 900 MWh/year of clean electricity, which would be sold back to the power grid. Overall, this study concluded that the CEP pipeline is more sustainable than the conventional pipeline in terms of environmental, economic, and social aspects over its entire life cycle for the specific conditions considered.

Table 22.18 Social sustainability assessment.

Life stage		Jobs/ Income	Health and safety	Environment impact	Value added	Social liability	Community development	Cultural aspects	Quality of life	Product value	Total
Resource Acquisition	Conventional	0	−1	−1	0	0	0	0	0	0	−2
	Clean energy	1	−2	−2	0	0	0	0	0	0	−3
Manufacturing	Conventional	0	−1	−1	0	0	0	0	0	0	−2
	Clean energy	1	−2	−2	1	1	0	0	0	0	−1
Use/Maintenance	Conventional	2	0	−2	0	0	0	0	0	0	0
	Clean energy	1	2	2	2	2	2	1	2	2	16
Disposal	Conventional	0	0	1	0	0	0	0	0	0	1
	Clean energy	1	−1	−1	0	−1	0	0	0	0	−2
Total	Conventional	2	−2	−3	0	0	0	0	0	0	−3
	Clean energy	4	−3	−3	3	2	2	1	2	2	10

Scale

−2	−1	0	1	2
Negative impact	Slight negative impact	No impact	Slight positive impact	Positive impact

22.5 Sustainable Rural Electrification

Paisley is a small rural town in Oregon. Due to its location, far from major road service, Paisley is unable to receive electrical infrastructure support from the neighboring villages. The problem of electricity supply to the population of Paisley can be solved with various off-grid solutions, including a photovoltaic system with lead-acid or lithium battery storage or a diesel power generation system.

22.5.1 Background

The electricity supply to rural areas is often difficult and expensive, especially when the nearest grid connection is located at far distances. A possible solution is the design and installation of off-grid power generation systems using fossil fuels or other renewable energy sources. Photovoltaic and diesel power generation systems are available in the market, and they can be used to generate *in situ* electricity for rural communities. Renewable energy sources may be a sustainable solution because of the positive environmental and social impacts compared to the conventional use of fossil fuels.

Photovoltaic systems are power systems designed to supply power using solar panels that absorb and convert sunlight into usable electricity. Solar photovoltaic (PV) systems include mounting equipment, cabling, electronic devices such as charge battery controller, and an inverter to change the electrical current from DC to AC. Solar PV systems often include advanced tracking systems that can adjust the tilt of the solar panel in the line of the most direct sunlight to improve panel performance. Most solar PV systems are grid-connected, but they can also operate as a stand-alone or off-grid system if the excess of energy produced during the day is stored in rechargeable batteries to be used during the night. Diesel power generation systems contain a diesel engine connected to an electric generator. Usually, diesel power generators are implemented as emergency power-supply in case of grid failure, although it can be used also in areas without connection to a power grid.

22.5.2 Methodology

22.5.2.1 Goal and scope

The aim of this study is to determine the sustainability of two alternative systems for the electrification of Paisley, Oregon. A diesel power generation system and a solar photovoltaic (PV) plant were considered in this study. The two alternative systems were analyzed technically and economically. The cost analysis includes installation, maintenance, and operation costs. Social and environmental impact assessment were also considered for the final recommendation to implement a sustainable electrification system in Paisley.

22.5.2.2 Study Area

The town of Paisley is a small rural town in Oregon with a population of 254 people occupying 156 homes. Paisley has no highway or close major road service. The closest interstate highway to Paisley is 206 mi away. Therefore, Paisley is unable to receive electrical infrastructure support from neighboring areas.

22.5.2.3 Technical Design
Scenario 1: The solar PV system to be considered for installation in Paisley will contain a PV panel array, an inverter, a battery bank for energy storage, and a panel board for energy distribution to the community (Figure 22.18). Lead-acid and lithium ion batteries were considered for electricity storage.

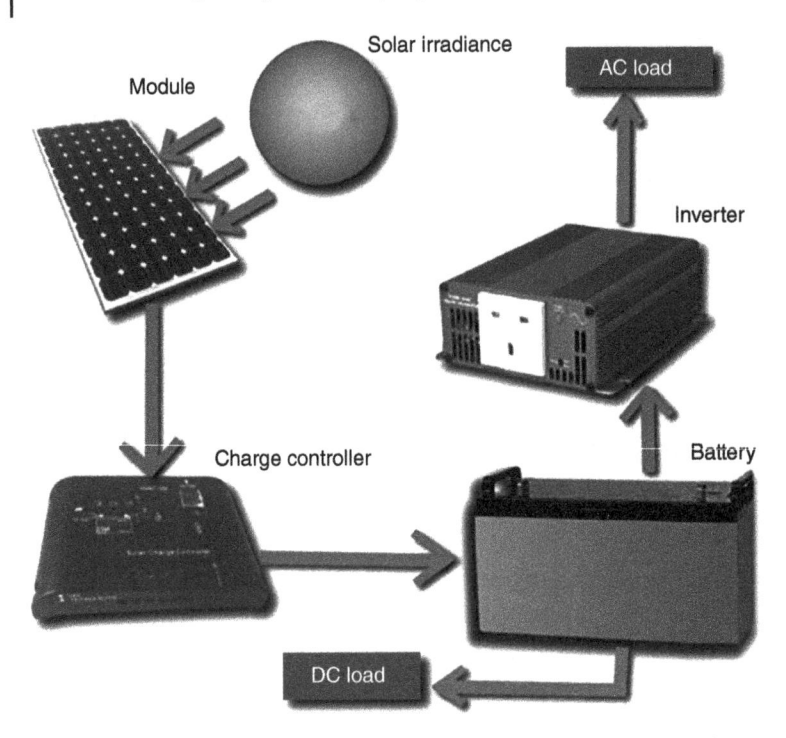

Figure 22.18 Process flow diagram for stand-alone solar photovoltaic system.

Scenario 2: A diesel power generation plant of 3000 kW (Caterpillar 3000 kW CAT C175) is proposed to cover the electric demand of the town of Paisley. The system includes the electric generator, diesel engine, pumps, compressors, a digital volt regulator, and a 5000-gallon diesel fuel tank above ground.

22.5.2.4 Sustainability Assessment

The environmental impacts of the two scenarios for the electrification of Paisley was determined with SimaPro software, TRACI, and BEES methods. Two stages were considered in the life cycle analysis: system installation and operation. The environmental impacts of the solar PV systems could be analyzed upstream from the purchase of materials for solar panel manufacturing. The materials and manufacturing of other components (inverter, battery pack, and charge controller) could have been included in the study. However, only the installation and operation phases of both scenarios were considered for comparison purposes with the economic study that only include installation and operation stages.

LCA data for solar PV system: The installation of a 570-kWp open ground, multi-Silicon solar PV system includes the component modules, mounting system, electric installation, inverter, and fence. It also includes the energy and labor for mounting, transport of materials, as well as the disposal at the end of the life. The life-cycle analysis for installation assumes that 2% of the modules purchased will be damaged and 1% of the modules are lost during handling and transport to the site. The solar panel modules used are 14.9% efficient in the solar cells and have a panel efficiency of 13.6%. The transportation of plant parts to the construction site is assumed to be a delivery of 100 km. The photovoltaic modules are delivered from a site 500 km away. The lifetime of the inverter is 15 years (it must be changed once during the lifetime of the solar PV plant, estimated to be 30 years). The operation stage of the solar PV system includes energy output of the plant, waste heat and sewage sludge produced, and tap water for maintenance and cleaning of solar panels.

LCA data for diesel generator system: The installation life cycle of the diesel system includes the purchase and transport of the diesel engine to the site and the fuel tank. The operation stage includes the transport of diesel to the site (accounted for a year of operation) as well as amount of diesel burned in MJ over one year of operation. In the calculations stage this was scaled to account for 30 years of solar PV operation in order to compare life cycles on a yearly basis.

22.5.3 Environmental Sustainability

Figures 22.19–22.21 show the results of the LCA for the installation and operation stages of the two scenarios with BEES and TRACI. The results show that installation of diesel power generation impacts the environment much less than solar PV system (Figure 22.19). This is related to the large area occupied by the solar plant that may disrupt natural habitats. The results from BEES and TRACI are very similar, except for the carcinogenic emissions that are higher for the diesel system with TRACI, due to the transportation of the diesel engine to the site. The main impact of the solar PV installation stage is related to the water intake (Figure 22.19). This is because the open ground mounting process for solar PV systems requires large volumes of water for soil stabilization. The environmental impacts for the operation stage are negligible for the solar PV system compared to the diesel engine (Figure 22.20). There are almost zero emissions for the PV systems during the operation stage, whereas the transportation and consumption of fuel are the responsible for the emissions of the diesel system. The actual impact values from Figure 22.21 show that the operation stage of the diesel system is the major responsible for the environmental impacts.

22.5.4 Economic Sustainability

22.5.4.1 Solar PV Power Generation System Proposal (CAPEX Costs)

The total load requirement for the Paisley community was estimated to be 1873.51 kW, and daily energy usage was estimated to be 6819.16 kWh/day. The total load requirement is increased by a pre-factor of 1.5 as recommended by the Texas State Energy Conservation Office, resulting in 2810.25 kW total load. This load increase is necessary to account for various inefficiencies or "ghost loads" in the grid and appliances. For a similar reason, the daily energy usage is multiplied by a pre-factor of 1.2. The daily energy usage estimated for Paisley is 8182.99 kWh/day.

The amount of equivalent full-sun hours is 2.9 hours per day in Paisley. This value describes the amount of time it would take to supply the daily average energy given that the sun is directly overhead the panel array on a cloudless day where the collector is directed at a perpendicular tilt to the sun's rays. Thus, the solar panels must provide 2821.72 kW to meet the daily energy usage estimated for Paisley.

The design of the solar PV system was based on raw data from a 570-kWp solar photovoltaic plant in Switzerland (Edisun Power Corporation). The 570 kWp plant has a solar array area of 4400 m^2, which was scaled to 22 000 m^2 to fit the solar array requirement for the plant in Paisley. The total cost for the solar array ($400/$m^2$) is $8.8 million.

The maximum number of continuous cloudy days in Paisley is four days. The battery bank is designed to hold reserves for four days. Moreover, batteries will have a longer life if they are only discharged 33%. This allows 67% of charge stored for emergency uses. The total power requirement of the battery bank will be 48 853.71 kWh.

The total cost of the battery back using lithium-ion batteries ($500/kWh) is $24.4 million. This cost may be significantly reduced with the use of lead-acid batteries ($65/kWh). The cost of the

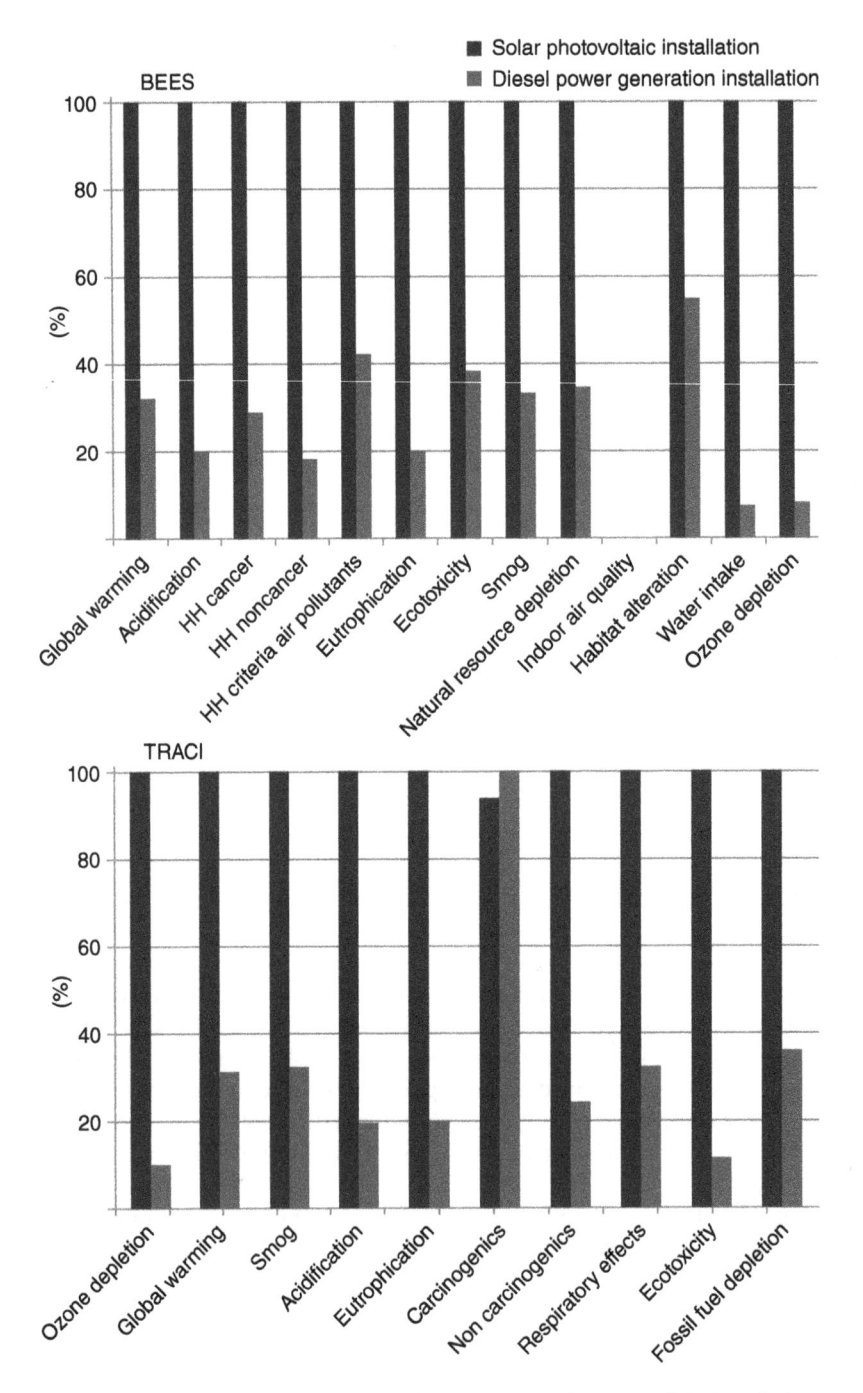

494 | *22 Infrastructure Engineering Sustainability Projects*

Figure 22.19 BEES and TRACI results for the environmental impacts of the installation stage of solar PV plant and diesel power generation.

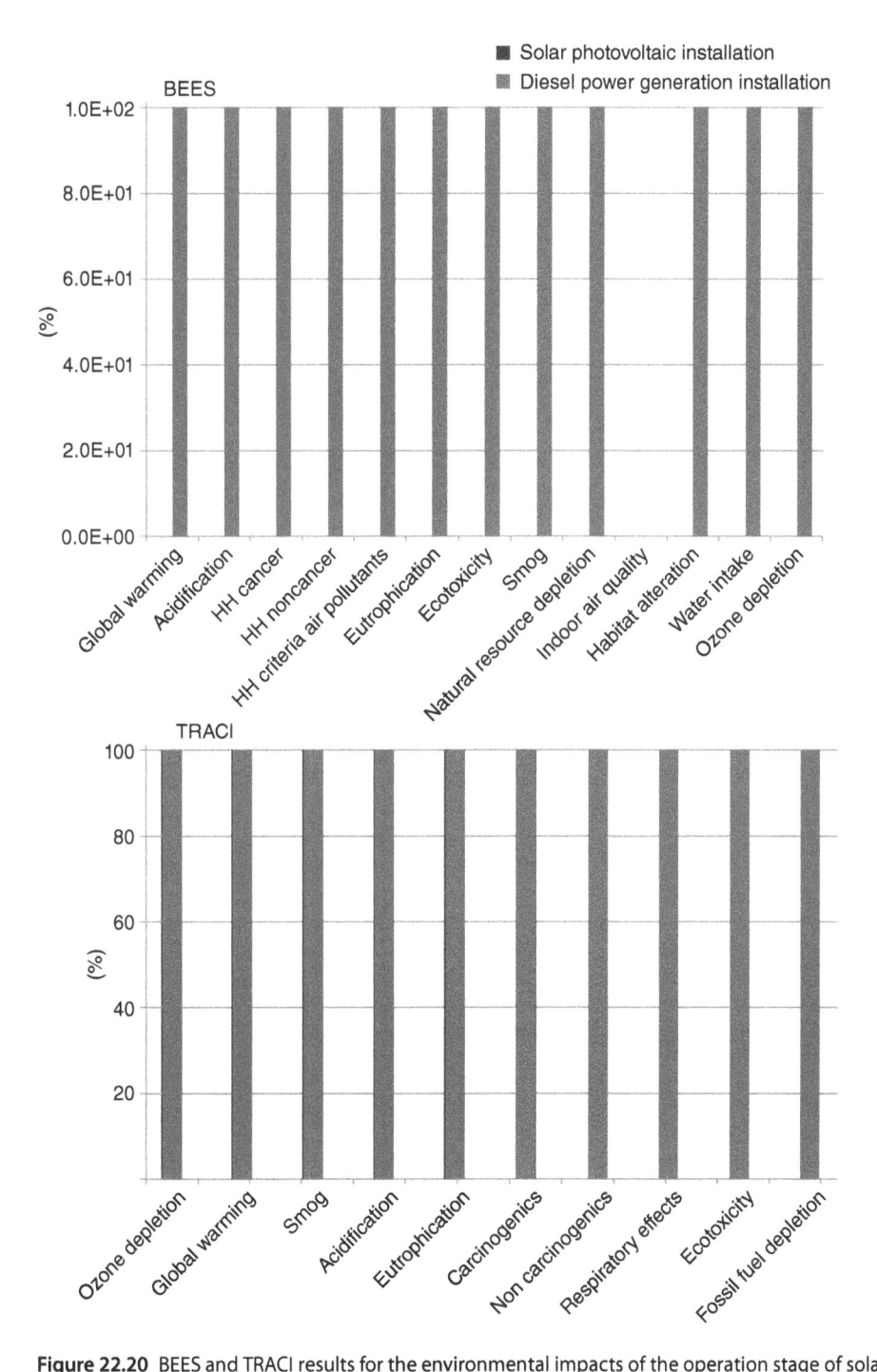

Figure 22.20 BEES and TRACI results for the environmental impacts of the operation stage of solar PV plant and diesel power generation system. The impacts of installation the solar plant are negligible compared to the diesel system.

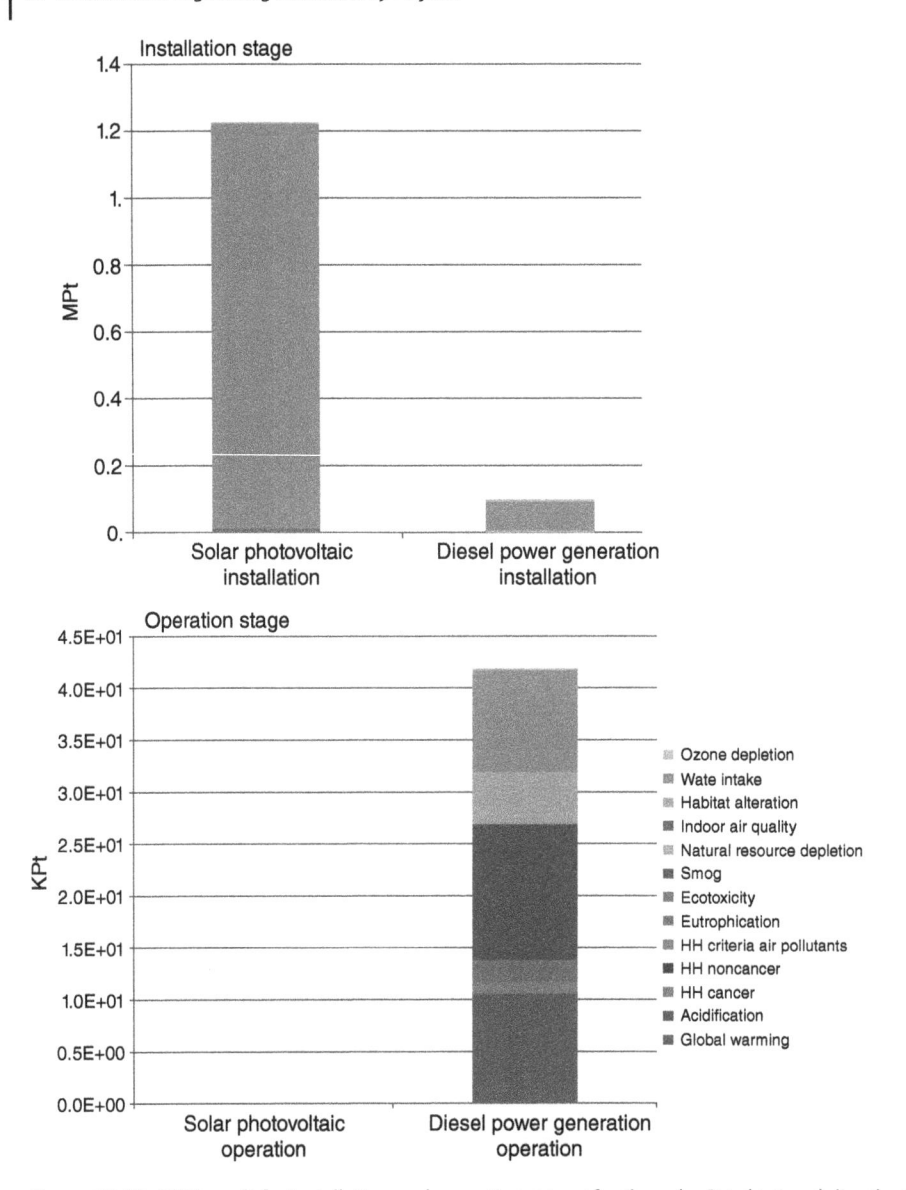

496 | *22 Infrastructure Engineering Sustainability Projects*

Figure 22.21 BEES result for installation and operation stages for the solar PV plant and diesel power generation system.

inverter, according to the Solar Market Research and Analysis group SolarBuzz, is $400/kW. Thus, the cost of an inverter for the solar PV system in Paisley is $1.12 million.

The last step in the economic analysis for the solar PV system is to determine the upfront investment or life-cycle cost of the system. The total cost that must be provided to Paisley to start-up the solar PV system must be equal to the life-cycle cost for the system. This includes replacement of solar panels as well as the battery bank. According to Edisun Corporation, the life cycle of the 570-kWp solar panels is 30 years. The life cycle of the lithium-ion batteries is approximately seven years. Thus, the life-cycle cost for the batteries is $104.69 million. Assuming that the costs for engineering, equipment procurement, maintenance, and operation are 25% of the upfront investment cost, the total upfront cost on a 30-year basis for a solar PV

Table 22.19 Investment cost for solar PV systems with varying battery pack and peak power.

Solar photovoltaic system	Total upfront investment cost for life-cycle of solar panel including CAPEX ($M)
Solar PV system 570 kWp (scaled, Li-Ion)	118.78
Solar PV system 570 kWp (scaled, Li-Ion, 1.2 prefactor)	143.26
Solar PV system 570 kWp (scaled, Pb-acid)	34.54
Solar PV system 570 kWp (scaled, Pb-acid, 1.2 prefactor)	42.17
Solar PV system 3.5 MWp (scaled, Li-Ion)	124.78
Solar PV system 3.5 MWp (scaled, Pb-acid)	40.54

Table 22.20 Investment cost for diesel power generation system.

System type	Initial cost of startup (1 year) ($M)	Total cost of diesel power generation (OPEX + CAPEX) ($M)
3000 kW diesel power generation system	3.72	125.64

system is $143.26 million. Table 22.19 shows the total upfront cost for various PV systems using Li-ion or lead-acid batteries.

22.5.4.2 Diesel Power Generation System Proposal (OPEX and CAPEX Costs)

A conventional diesel power generation plant, Caterpillar 3000 kW CAT C175, was selected to meet the total load requirement for the community. The cost of the diesel engine is $500 000. This cost includes the generator, diesel engine, pumps, compressors, and digital volt regulator.

The cost of a 5000-gallon diesel fuel tank (above ground) is $7000. The cost of diesel fuel is estimated to be $2.50/gallon after signing a contract with a company for the diesel delivery to the plant. The fuel requirement is calculated considering 75% consumption load with an average 153.2 gallons/hour, operating 24 hours per day and 350 days/year (additional days are used in maintenance operations). Thus, the cost of the first year including the equipment is estimated to be $3.72 million. In a period of 30 years, the total cost of the system is estimated to be $125.645 million. This cost includes the equipment, fuel consumption, maintenance, and substitution of the equipment (Table 22.20).

22.5.5 Social Sustainability

The social impacts were assessed with SSEM matrix (Reddy et al. 2014), and the results are shown in Figure 22.22 for the solar PV system, the diesel generator and the "do-nothing" approach. The solar PV system scored higher in all the categories, but the socioenvironmental. The large area occupied by the solar plant and its effect on natural habitats is the reason for the negative score in the socio-environmental category. Socio-economically, the solar PV system has ranked higher than the diesel power generation because of government support to projects that implement green and sustainable technologies. It has been shown that tax relief and financial investments to new small businesses would be granted by the government if sustainable technologies were incorporated in rural areas. In socioinstitutional

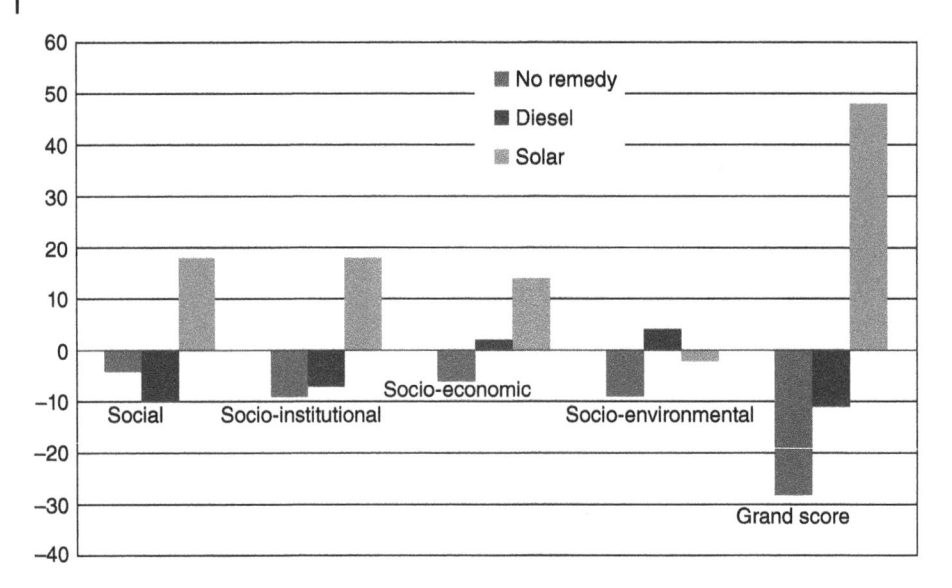

Figure 22.22 Social sustainability assessment with the SSEM.

and social-individual categories, the solar PV system ranks higher for the good perception of renewable energy versus fossil fuel combustion. The diesel generation system ranked higher than the no-remedy approach because the diesel generation system is a step in the right direction toward sustainable rural electrification.

The Envision checklist is an alternative method for assessing the sustainability of infrastructure (Figure 22.23). The solar PV system is rated higher than the diesel generation system in all the categories but natural world. This is due to the fact that the solar PV system takes up significantly more physical space than the diesel operation. This does not preserve or protect the habitat and has a negative effect on species and biodiversity in the area. On the other hand, diesel power generation shows also negative impact in natural world due to the production of GHG gases and air pollutants.

22.5.6 Conclusion

The upfront investment cost, at present value, for a solar PV system to be installed in Paisley, Oregon, on a 30-year basis, is significantly more expensive than installing a diesel power generation system. Most of the expenses for the solar PV system are due to the batteries that are about 90% of the total cost. The only way that a solar PV system could become a viable option is if the battery bank becomes cheaper or the lifetime of the batteries increases. The results from the environmental analysis and social analysis (SSEM and Envision) shows that the solar PV system is more sustainable for the entire life cycle of the project. Moving forward, it is proposed as a solution to combine the results of the economic, environmental and social analysis. The alternative solution is the implementation of a photovoltaic-diesel hybrid system. These hybrid systems consist of a photovoltaic system, a diesel generator, and controls that will determine which supply of energy is used at a certain time. The photovoltaic system can be used to supply energy when there is an average load and the diesel generator can relieve the system at peak loads. The battery storage system can store the excess of energy produced by the solar system. This energy can be used as power supply in the nighttime. This solution requires lower upfront investment but reduces also the cost of fuel consumption and GHG emissions. Thus, this can be considered an environmentally friendly and cost-effective solution.

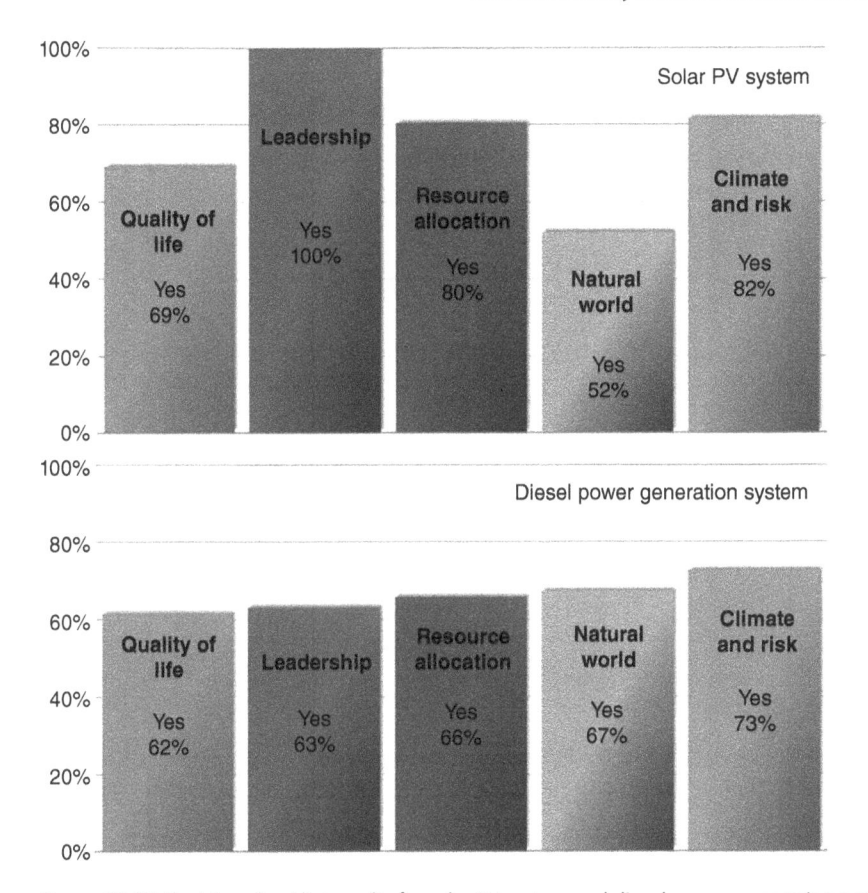

Figure 22.23 Envision checklist results for solar PV system and diesel power generation system.

22.6 Sustainability Assessment of Shear Wall Retrofitting Techniques

Masonry shear walls are widely used in medium- to high-rise buildings in order to resist lateral loads arising from high winds and earthquakes. There is a need to upgrade the lateral performance of existing walls to meet newly refined standards. For this reason, several different techniques for retrofitting shear walls have been developed in recent decades. However, many of these processes are fairly new, and little has been done in the way of analyzing their various impacts. This project presents a sustainability assessment of two such retrofitting techniques: steel bracings and fiber-reinforced polymer laminates. The advantages and disadvantages of each process are thoroughly discussed, analyzing the impacts of each technique through a typical life cycle. For a simplified analysis, a specific site renovation is used as an example in order to analyze possible environmental, economic, and social impacts of each retrofitting technique. These impacts are then compiled, weighed, averaged and normalized in order to provide a balanced comparison of each proposal.

22.6.1 Background

The O'Rourke Building, constructed in 1917 (Figure 22.24), is a six-story plus basement concrete-framed structure located in East Moline, Illinois. The building is to be remodeled

Figure 22.24 Existing condition of the O'Rourke building.

into a hotel and the new Quad-Cities Amtrak Multimodal Rail Station. The proposed Multimodal Station is located adjacent to the existing Centre Station and is currently an urban Brownfield site.

An inspection performed by a professional engineering consulting company in 2015, identified a need to strengthen the 20 shear walls along the north and south side of the building in order to be compliant with current code provisions. Each shear wall is 10 ft × 15 ft, with a thickness of 8 in. The strength of each masonry wall was calculated based on ACI 530-13 code, using 8-inch concrete masonry units, with a nominal strength of 2000 psi and Type S PCL mortar. The maximum shear strength of the walls was calculated and compared with required capacity and shown to be inadequate. The largest discrepancy between provided and required shear strength occurs within the top-most walls, which exhibit a disparity of 21.85 kips. Thus, designing for the worst-case scenario, each shear wall is in need of supplemental reinforcement capable of supplying an additional 21.85 kips of lateral strength.

Several retrofitting techniques have been developed in recent years to efficiently strengthen existing masonry walls. These processes are quite varied, implementing a wide range of different materials, from standards like steel and concrete, to fiber-reinforced polymer laminates and shape-memory alloys. Steel bracing is the most commonly used practice today, often chosen for its ease of application and inexpensive materials. However, in recent years, there has been a noticeable increase in the use of Fiber-Reinforced Polymer Laminates (FRPLs). FRPLs offer the same amount of lateral support but requires far less material.

22.6.2 Methodology

22.6.2.1 Goal and Scope

The objective of this study is to determine the sustainability of two retrofitting techniques to strengthen existing masonry walls: Steel bracing and Fiber-Reinforced Polymer Laminates. The

Table 22.21 Assessment approach.

Assessment	Weightage (%)	Assessment components
Environmental assessment	50	- SimaPro analysis of material production and transport
		- Analysis and quantification of disposal options
Economic assessment	30	- Direct costs of material, transport, application, and disposal
		- Social cost of life cycle emissions
Social assessment	20	- Survey of potential social impact indicators

life-cycle stages of the two options include transport, application, use, and disposal of all materials involved. The only stage of either product's life cycle excluded from this analysis is the extraction of precursory material. A use period of 40 years is assumed for estimation purposes, though the impact of this stage is minimal for each proposed retrofit. A general weighting among the three parts of the assessment is provided in Table 22.21, along with a description of the metrics and tools used to quantify the three types of impact.

22.6.2.2 Technical Design

Scenario 1: Steel bracing: The process of applying steel bracing to shear walls in need of supplemental reinforcement is common place, though not always applicable. The steel must be attached to the existing vertical reinforcement within the wall, and thus is only applicable if the vertical reinforcement is accessible. Fortunately, the concrete frame of the O'Rourke Building is easily accessed and still retains its structural integrity. Thus, the application of steel bracing is as simple as bolting a set of four gusset plates along the corners of each wall, securing them to the building's existing frame; opposite gusset plates are then connected with standard steel braces, which are then themselves bolted to the masonry wall itself. In certain conditions, additional vertical reinforcement is required along the perimeter of the masonry wall, though this is typically used in cases where the shear deficit is much higher than that seen in the O'Rourke Building (Galal and El-Sokkary 2008).

A high-strength low-alloy steel is selected for the steel bracing. Low-alloy steel contains small amounts of chromium and other alloys that provide the steel with exceptional resistance to corrosion and rust. Having chosen a material, a standard procedure is taken to determine the dimensions of required steel based on the 14th edition AISC Steel Construction Manual and the load discrepancy provided by WJE and the results are outlined in Figure 22.25a and Table 22.22.

Scenario 2: Carbon-Based fiber-reinforced polymer laminate: FRPLs are a composite material formed by weaving together fine strands of high-strength fibers such as carbon, glass,

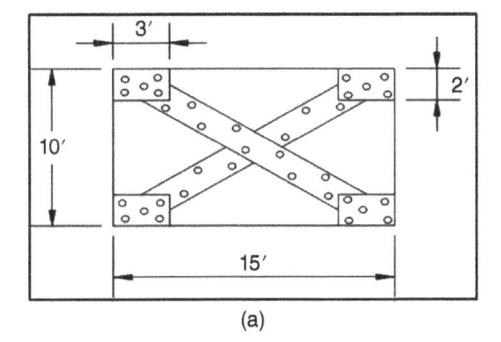

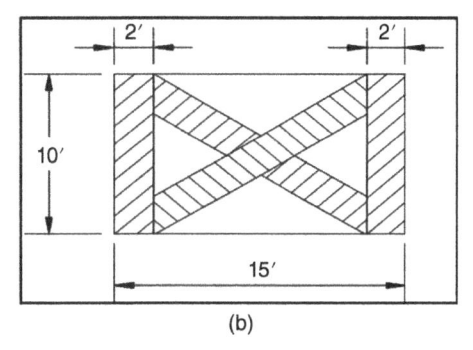

Figure 22.25 (a) Required dimensions of steel bracing. (b) Proposed configuration of FRP laminate.

Table 22.22 Required component dimensions of steel bracing.

Component	Dimensions	Number required per wall
Gusset plates	2′ × 3′ × 3/8″	4
Brace plates	17′6″ × 1′ × 1/4″	2
Low strength bolts	1″ Diameter, Grade A325	12
High strength bolts	2″ Diameter, Grade A325	24

Table 22.23 Summary of selected FRP laminate.

Laminate	Total area per wall	Provided strength	Composition
Carbon fiber 514-RL	112 sq. ft	23 kips	95% carbon, 5% polyester

or aramid along with more easily produced and accessible polymer fibers, such as polyester or epoxy. The resulting laminates exhibit incredibly high levels of compressive and tensile strengths, especially when compared to their low weight and ease of transport.

Structural applications of FRPLs are a recent development, first making an appearance in 1991 as a method for providing supplemental shear strength to beams. By 1997, more than 1500 concrete structures around the world had been retrofitted with FRPLs. The use and popularity of this technique has grown quickly in a short time, yet its potential applications still remain a great source of excitement for the industry. One of the most recently developed applications is the retrofitting of shear walls. This is done simply by attaching the required amounts of FRPL to the wall using a high-strength epoxy resin. Typically, carbon-based FRPLs are used for structural applications as they provide the highest levels of strength (Alferjani et al. 2013).

There is little literature concerning the strength provided by an arbitrary configuration of FRPL, though a method has been proposed by Triantafillou (1998), which provides a fair estimate for the lateral strength of a given configuration of a FRPL. Using these equations, an analysis of the lateral strength for various FRPLs is conducted using the configuration shown in Figure 22.25b.

The resulting strengths are compared to the required 21.85 kips, and the laminate which best meets these requirements is selected. As expected, a high-strength carbon-based FRPL with exceptionally high capacity is shown to be the most applicable. Details of the selected FRPL and configuration are provided in Table 22.23.

22.6.2.3 Sustainability Assessment

The environmental sustainability assessment of the two scenarios was calculated with SimaPro software using the TRACI method. The summary of the required materials for each proposal (Table 22.24) were determined based on the technical design for each scenario. The amount of required epoxy resin is estimated based on industry standards. Material distributors were chosen based on proximity to the site. A summary of material transport between chosen distributors and the site is given in Table 22.25. Data in Tables 22.24 and 22.25 are used in SimaPro to determine the environmental impacts of production and transport of base materials. In addition to environmental sustainability assessment, the economic (including the direct and indirect costs) and social sustainability was evaluated for the different alternatives.

Table 22.24 Total amount of required materials.

High-strength low-alloy steel	Carbon-based FRPL
80 2′ × 3′ × 3/8″ gusset plates	2240 ft² of laminate
40 17′6″ × 1′ × 1/4″ brace plates	264 lb of carbon fiber
240 1″ diameter bolts	14 lb of polyester fiber
320 2″ diameter bolts	394 lb of polyacrylonitrile precursor
Total: 8984 lb of steel	20 gallons of epoxy resin

Table 22.25 Total transport of required materials.

	Steel (8984 lb)	FRPL (280 lb)
Distributor:	East moline metal products	Simpson strong tie
Distance to site:	3 mi	117 mi
Energy expenditure:	13.5 tmi	16.38 tmi

22.6.3 Environmental Sustainability

The environmental impacts of the competing proposals are shown in Figures 22.26 and 22.27. Table 22.26 and Figure 22.28 compare the total impacts of the two proposals.

Figures 22.26 and 22.27 clearly indicate an almost negligible contribution from the material transport stage for both proposals. This may be considered as a general result, as FRPL is an incredibly light material, and steel a readily available one. However, it should be noted that the scope of this assessment does not include the manufacturer's process of gathering precursor, which could potentially involve transport of heavy materials over large distances, especially concerning the manufacturing of steel.

More interesting is the significant contribution of the epoxy resin used to apply the FRPL. The epoxy resin (Figure 22.26a) is shown to produce high negative impacts in terms of carcinogenics, ecotoxicity, respiratory effects, and eutrophication; all strong indicators of environmental sustainability. This is possibly the most surprising result of the analysis, and one that is not discussed frequently in the current literature. However, these impacts are still minimal when compared to those of the steel bracing.

The comparison of the two proposals (Table 22.26 and Figure 22.28) clearly shows that the impacts of steel production far outweigh those of the carbon FRPL across all considered indicators, most notably in terms of carcinogen emissions. However, this analysis is not quite complete. Producing carbon fibers requires an additional expenditure of energy in order to convert polyacrylonitrile ($[C_3H_3N]_n$) fibers into carbon fibers through pyrolysis. SimaPro does not account for this process. The energy expenditure is estimated using data taken from Park and Heo (2015) to be roughly 100 MJ. The derived impact is then accounted for in the final assessment of this study along with all environmental impacts evaluated by SimaPro.

The greatest advantage of steel as a structural medium is its recyclability. In fact, it has often been termed "infinitely recyclable." This is because 90% of structural steel is removed during demolition, taken off site, and melted down in order to make new steel of equal quality to the original. There is some energy expenditure involved in this process, though the costs are minimal considering the benefit of a waste-free product. Fiber-Reinforced Polymer Laminates, however, are completely nonrecyclable. So far, no technology has been developed in order to

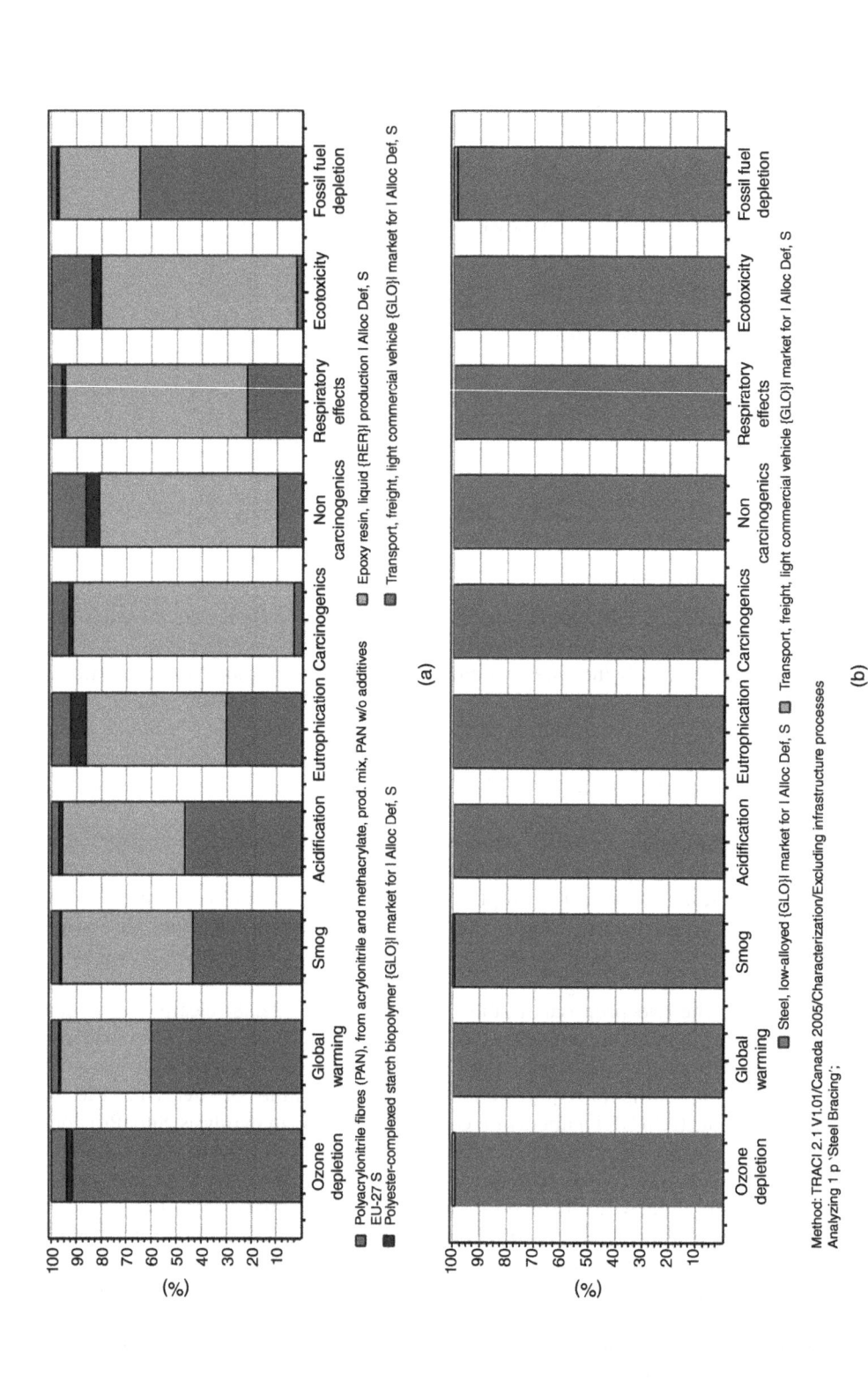

Figure 22.26 Environmental Impact distribution for (a) carbon-based FRP laminate and (b) steel bracing.

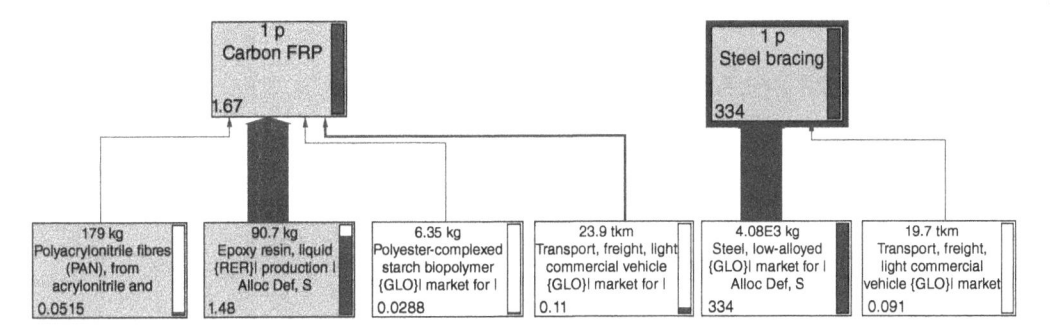

22.6 Sustainability Assessment of Shear Wall Retrofitting Techniques | 505

Figure 22.27 Distribution of carcinogenic impact for the two proposals (carbon-based FRP laminate and steel bracing).

Table 22.26 Summarized environmental impact comparison of the two proposals.

Impact category	Unit	Carbon FRP	Steel bracing	Carbon FRP normalized	Steel bracing normalized
Ozone depletion	kg CFC-11 eq	6.80E−05	3.48E−04	0.017	0.087
Global warming	kg CO_2 eq	1.69E+03	8.18E+03	0.073	0.354
Smog	kg O_3 eq	1.50E+02	5.29E+02	0.081	0.286
Acidification	kg SO_2 eq	8.13E+00	4.07E+01	0.061	0.306
Eutrophication	kg N eq	1.06E+00	4.06E+01	0.081	3.134
Carcinogens	CTUh	3.89E−05	7.75E−03	1.673	333.744
Noncarcinogens	CTUh	1.34E−04	9.25E−03	0.150	10.302
Respiratory effects	kg PM2.5 eq	8.31E−01	1.37E+01	0.010	0.172
Ecotoxicity	CTUe	2.92E+03	2.40E+05	0.280	23.011
Fossil fuel depletion	MJ surplus	4.27E+03	4.73E+03	0.128	0.142

reuse structural FRPLs. The current practice is to send FRPLs to a landfill. It is difficult to quantify this impact, and thus it is not included in the final metric. However, given the length of the FRPL's life cycle, it is reasonable to assume that a technique for the recycling of FRPLs will be developed in that time, leading to negligible contributions associated with this stage for both proposals.

22.6.4 Economic Sustainability

Table 22.27 provides a summary of direct costs for each proposal. Material prices are taken directly from their respective distributors. Cost of labor is estimated assuming a three-man crew, average cost of labor being $40/man-hour. Based on common practice, steel bracings require roughly one hour to affix for a three-man crew, while retrofitting with FRPLs requires half that. However, application of FRPLs requires specialized equipment, which leads to a higher cost of labor, estimated at $60/man-hour. Transport costs were estimated at $1.50 per mile, given industry standards.

The emissions quantified using SimaPro are converted into their corresponding social costs using EPA standards given for the year of 2040. This is done to account for an expected life cycle of 40 years. The results are summarized in Table 22.28.

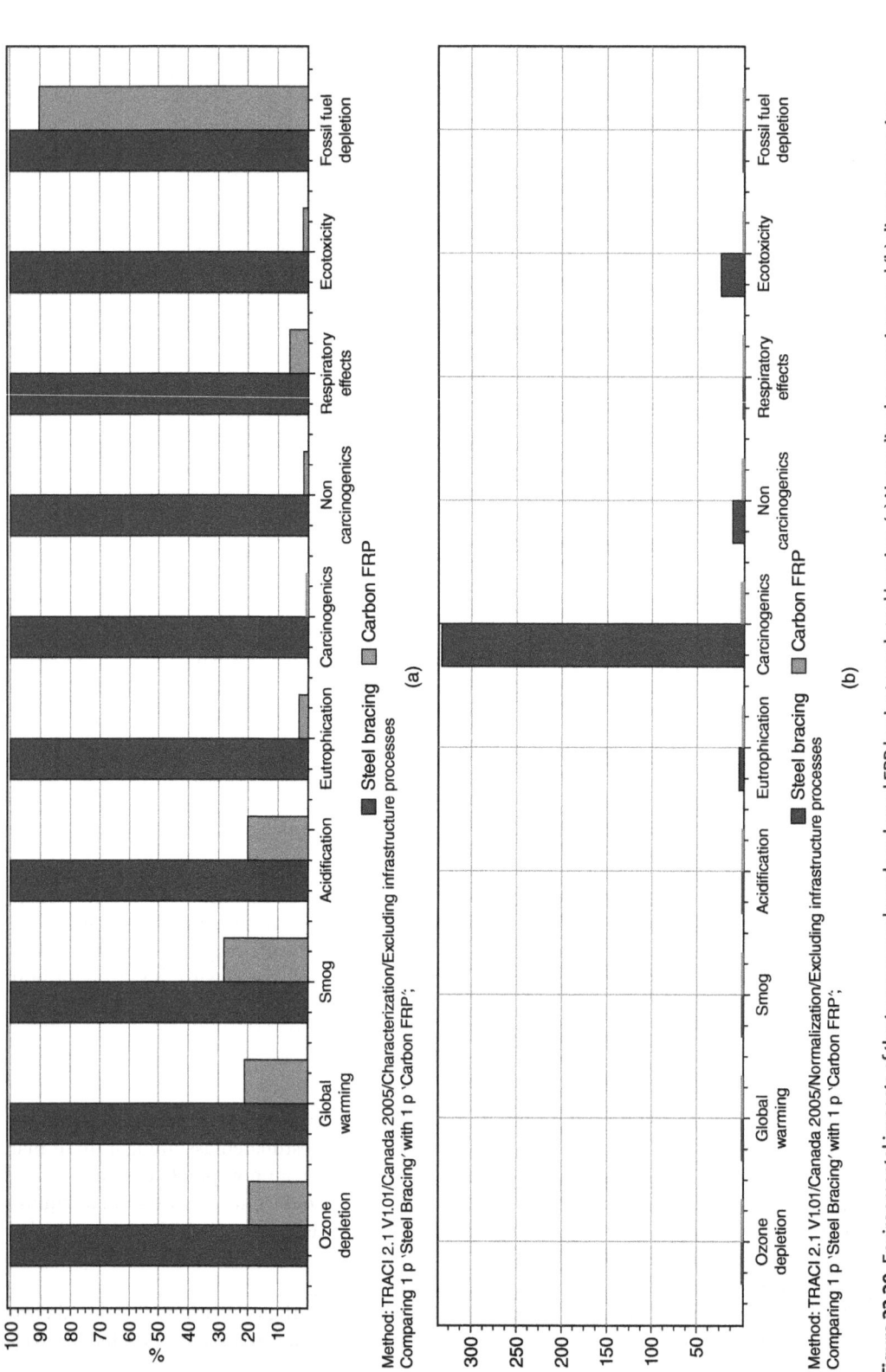

Figure 22.28 Environmental impacts of the two proposals: carbon-based FRP laminate and steel bracing. (a) Normalized comparison and (b) direct comparison (TRACI method).

Table 22.27 Direct costs of the two wall retrofitting alternatives.

	Materials/Transport	Application	Disposal	Total
Steel	80 gussets – $1200 40 steel plates – $1000 240 1″ diameter bolts – $200 320 2″ diameter bolts – $320	Labor (60 manhrs) - $2400	Labor (30 manhrs) - $1200	$6320
Carbon FRP	12 rolls laminate – $2952 20 gallons epoxy – $2000 Transport – $175	Labor (30 manhrs) - $1800	None	$6927

Table 22.28 Social costs of the two wall retrofitting alternatives.

	Emissions → Cost	Total cost	Percent comparison to direct cost (%)
Steel	CO_2 – 203 kg → $13.40 CH_4 – 24 kg → $52.80 N_2O – 21 kg → $531.3 PM – 9.3 kg → $232.5	$950.60	15
Carbon FRP	CO_2 – 930 kg → $61.40 CH_4 – 5.5 kg → $12.10 N_2O – 6 kg → $151.80 PM – 0.4 kg → $10	$235.3	3.35

Both the direct and indirect costs are compiled, weighted, averaged, and normalized in the final assessment of this study. The weighting were given to each indicator so as to be proportional to relative differences between the proposals in terms of the given indicator. The economic assessment as a whole is given little weighting, as the results are quite similar for each proposal.

22.6.5 Social Sustainability

In order to quantify the social impact of each proposal, a list of indicators is developed and categorized into four subsections: socio-individual, socio-communal, socio-economic, and socio-environmental impacts. The selected indicators are summarized in Table 22.29.

These indicators are then sent out to a small community of individuals who understand the entire scope of each proposed design's life cycle. They are asked to rank each proposal in terms of each indicator, providing a quantification of each proposal's social impact. An additional question on the survey asks the individuals to rank the proposals for each subsection as a whole. The results of the survey are given in Figure 22.29. The results clearly show the carbon FRPL outranking the steel bracing in nearly every impact category; however, in many cases, this is only true by a small margin.

22.6.6 Overall Sustainability

The environmental, economic, and social assessment results are compiled, weighted, averaged, and normalized into a single number from zero to one, representing the "sustainability score"

Table 22.29 Social impact indicators.

Socio-individual impact	Socio-communal impact	Socio-economic impact	Socio-environmental impact
– Noise pollution – Traffic congestion – Overall health and happiness – Higher contaminating effects – Life disturbance – Accident risk – Skills development – Future land use	– Appropriateness of future land use – Involvement of community in land use planning – Enhancement of recreational facilities – Involvement of community organizations – Trust and voluntary organizations – Enhancements to the architecture and aesthetics of environment – Employment opportunity for locals	– Effect on tourism – Employment opportunity for locals – Use of site for landscaping – Trust and voluntary organizations – Degree of increased tax revenue from nearby properties – Degree of increased tax revenue from site reuse – New business opportunities	– Water generation – Reduction, reuse, and treatment – Impact on biodiversity – Potential for future environmental impact – Degree to which external media are affected – Emissions

for each proposal. Individual impact indicators are given weightings proportional to the relative difference of the impact between each proposal. This is done by taking the difference in results for said indicator and dividing by the total difference of results for said category. These categories are then themselves given weights. Averaging over each category results in a number from zero to one, which represents the relative benefit of each proposal in terms of its environmental, economic, and social impacts. As impact indicators used to measure environmental and economic impacts were negative in scope, the score is simply subtracted from one. Then these three scores are averaged using the weighting given in Table 22.21, producing a final score for each proposal from zero to one which represents the sustainability of the solution across all three impact assessments. A higher score implies the solution is more sustainable than the other. Results are outlined in Table 22.30, providing the relative benefit of each proposal.

22.6.7 Conclusion

The sustainability assessment concludes with a final relative score of 0.37 for the steel bracing and 0.63 for the Carbon FRPL. The Carbon FRPL is more than 1.5 times more sustainable than the steel bracing. The bulk of this difference appears to be due to the massive environmental impact associated with the production of steel. The steel bracing option requires significantly more material that manufacturing emissions overwhelm almost all other environmental impact indicators. Further, the economic assessment shows how the additional labor requirements for steel bracing nearly balance the relatively high material costs of the FRPL. The increased emissions of the steel solution are also shown to produce significant social costs. The social assessment similarly provides strong support for the FRPL, exhibiting exactly three times more favorability toward the FRPL solution.

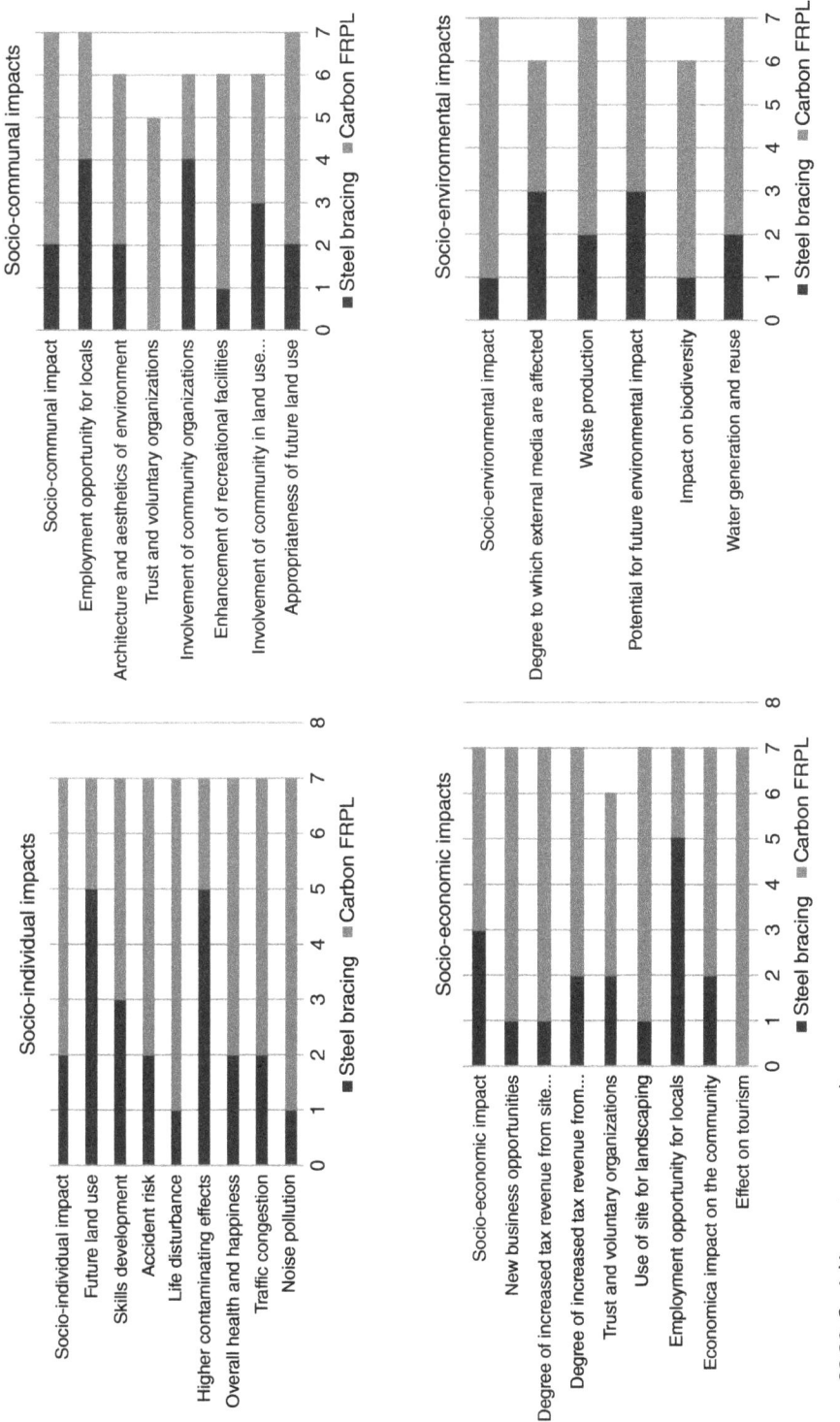

Figure 22.29 Social impact survey results.

Table 22.30 Global sustainability assessment of the two wall retrofitting techniques.

	Steel bracing	Carbon FRPL
Environmental impact	0.35	0.65
Economic impact	0.49	0.51
Social impact:	0.25	0.75
Final score	0.37	0.63

22.7 Summary

This chapter presented various infrastructure sustainability projects. As it was presented in previous chapters, the most sustainable alternative is the result of the overall score of the three pillars of sustainability: environment, economy, and society. Often, infrastructure proposals that involve the use of new materials or renewable energy show better environmental impacts and social acceptability, but the implementation of such projects may be limited by economic factors like the upfront investment. Even in these conditions, the low rating in the economic pillar can be compensated with higher rating in social and environmental sustainability. Furthermore, the reduced operation/maintenance costs of innovative infrastructure solutions may result in the selection of such alternatives as the most sustainable.

References

ACI (American Concrete Institute) (2013). *Building Code Requirements and Specification for Masonry Structures and Companion Commentaries*. Standard. ACI 530-13. First Printing. Reston, VA: ACI.

ACI (American Concrete Institute) (2014). *Building Code Requirements for Structural Concrete and Commentary*. Standard. ACI 318-14. First Printing. Reston, VA: ACI.

Alferjani, M.B.S., Samad, A.A., Elrawaff, B.S. et al. (2013). Use of carbon fiber reinforced polymer laminate for strengthening reinforced concrete beams in shear: a review. *International Refereed Journal of Engineering and Science (IRJES)* 2 (2): 45–53.

ASCE (American Society of Civil Engineers) (2014). *Minimum Design Loads for Buildings and Other Structures*, Standard ASCE/SEI 7-10. Third printing. Reston, VA: ASCE.

Benoît-Norris, C., Vickery-Niederman, G., Valdivia, S. et al. (2011). Introducing the UNEP/SETAC methodological sheets for subcategories of social LCA. *The International Journal of Life Cycle Assessment* 16 (7): 682–690.

FHWA (Federal Highway Administration) (1976). *Lateral Support Systems and Underpinnings*. Guide. First Printing. FHWA-RD-75-128. Washington, DC: FHWA.

Galal, K. and El-Sokkary, H. (2008). Recent advancements in retrofit of RC shear walls. In: *The 14th World Conference on Earthquake Engineering October*, 12–17.

Goedkoop, M., Schryver, A.D., Oele, M. et al. (2010). *Introduction to LCA with SimaPro 7*. Netherlands: Pré Consultants.

IDT (2017). Circle Interchange. Illinois Department of Transportation. http://www.circleinterchange.org/ (accessed May 2017).

ISO (2006). *ISO 14040:2006 Environmental Management – Life Cycle Assessment – Principles and Framework*. Geneva: International Organization for Standardization.

Park, S.J. and Heo, G.Y. (2015). Precursors and manufacturing of carbon fibers. In: *Carbon Fibers*, 31–66. Dordrecht: Springer.

Reddy, K.R., Sadasivam, B.Y., and Adams, J.A. (2014). Social sustainability evaluation matrix (SSEM) to quantify social aspects of sustainable remediation. In: *ICSI 2014: Creating Infrastructure for a Sustainable World* (eds. J. Crittenden, C. Hendrickson and B. Wallace). Proceedings of the 2014 International Conference on Sustainable Infrastructure. 831–841. Reston, VA: ASCE. DOI: 10.1061/9780784478745.

Triantafillou, T.C. (1998). Strengthening of masonry structures using epoxy-bonded FRP laminates. *Journal of Composites for Construction* 2 (2): 96–104.

Index

a

acidification impact category   137
acid rain   47–48
activated alumina (AA)
   Envision scores   378
   properties of   369
   SLCA   373, 377
aesthetic degradation   61
air emissions   244–247
air pollution   44–46, 268
air sparging   318
albedo   338
ALC. *see* alternate biocover (ALC)
algae biomass deepwell reactors *vs.* open pond systems
   background   392–393
   economic sustainability   400, 402
   environmental sustainability   396–400
   methodology
      goal and scope   393, 394
      site location   394
      technical design   394, 395–396
   social sustainability   400, 402, 403
   sustainability assessment   394, 396, 397–398
alternate biocover (ALC)
   characteristic properties of cover materials   382, 384
   design profile of   382
   environmental impacts   382, 385, 387
   LCA of   382
   social sustainability of   391
   technical design of   381
American Council of Engineering Companies (ACEC)   300
American Public Works Association (APWA)   300
American Society of Civil Engineers (ASCE)   112, 298, 299
American Society of Testing and Materials (ASTM) framework   323, 324
analytic hierarchy process (AHP)   238
anthropogenic carbon dioxide emissions   245
anthropogenic/natural flows for elements   141–142
ArcGIS   356
arsenic removal in groundwater
   arsenic concentrations   367, 368
   background   366–367
   economic sustainability   371, 373
   environmental sustainability   370–371
   Envision   376, 378–379
   methodology, goal and scope   367–368
   social sustainability   373, 374–376
   technical design
      activated alumina (AA)   369
      bayoxide E33 (E33)   370
   well water characteristics   368
ash   263
atom economy   270, 271
Attest Engagements on GHG Information   163

b

background consumption   98
bayoxide E33
   Envision scores   373, 379
   physical and chemical properties of   370
   SLCA   373, 377
benzene   45
best management practices (BMPs)   300, 322, 447
biodegradable wastes   273
biodiversity   41–42
   loss of   14
biogas   55
biological oxygen demand (BOD)   50
biomass energy   260–263
bioremediation   317

Sustainable Engineering: Drivers, Metrics, Tools, and Applications, First Edition.
Krishna R. Reddy, Claudio Cameselle, and Jeffrey A. Adams.
© 2019 John Wiley & Sons, Inc. Published 2019 by John Wiley & Sons, Inc.

bioswales   307, 309
BREEAM tool   232
budget
   construction of a   143
   definition   141
   national   207
   of natural resources   141–143
Building Research Establishment Environmental Assessment Method (BREEAM)   291

C

California Green Remediation Evaluation Matrix (GREM)   325
Campus Carbon Calculator™   165
carbon capture and sequestration   250
carbon capture and storage (CCS)   336, 381
carbon capture and underground storage   251
carbon dioxide (CO_2)   353
carbon dioxide emissions   331, 333, 381
carbon dioxide equivalent (CO_2e)   362
carbon dioxide removal (CDR) methods
   direct engineered capture   337, 339
   marine organism sequestration   336–339
   subsurface sequestration   334–336
   surface sequestration   336
carbon footprint   138, 232
   emission measurement   159–162
   GHG emission mitigation   169–170
   GHG inventory   163–164
   global warming potential   157–159
   independent verification   162
   off-set unavoidable emissions   162
   scope of   159
   standards for calculating   162–163
   targets and strategies   162
   tools for GHG inventory   164–165
   University of Illinois at Chicago   165–169
carbon monoxide   45
cast-in-place retaining wall
   background   471
   carbon dioxide emissions   475, 478
   cross section for   473, 474
   environmental impacts   475, 476, 478
   Envision checklist results   479, 481
   global warming   475, 477
   methodology, goal and scope   471
   SLCA health and safety matrix   477, 480
   social impact assessment   476, 480
   work and labor costs for   475, 479
CDR methods. *see* carbon dioxide removal (CDR) methods

Cedar Lake   403
cement masonry unit (CMU)
   conventional design   418
   cost to install   426
   end-user energy consumption for   424
   environmental impact   421, 422
   envision results for   427
   non-structural component   419
   social sustainability results   427, 428
   total cost of   421
   vs. translucent composite panels   426
Champaign County in Illinois   353, 354
chemical oxidation technology   317
chemicals of potential concerns (COPCs)   223
Chicago Stormwater Ordinance Manual   449
Chile's wildfires by drought and record heat 2016, 106
chlorofluorocarbons (CFCs)   43
chromium cycle   143–144
Circle Interchange project   471
circular economy   280–281
civil infrastructure
   "D+" grade   298, 299
   Envision rating system   300–302
   report card   298, 299
clay brick veneer   418
CLC. *see* conventional cover (CLC)
Clean Air Act   45
Clean Air and Clean Water Acts   57
clean energy   250–252
Clean Energy pipe   482
clean energy producing (CEP) steel with cement-mortar lining
   annual cost and benefit analysis.   485, 486
   background   481–482
   economic sustainability   485–486
   environmental impacts of   484
   global environmental impacts of   484, 485
   LCA results for   485, 486
   methodology
      goal and scope   482
      rendering and technical design for   482, 483
      site background   482
      sustainability assessment   482–484
   pipeline capital cost analysis   485, 486
   social sustainability   486–488
climate and risk   376
climate change   10, 32–39, 104–105
Climate Change Action Plan   162
climate geoengineering
   CDR methods

applicability of   342–343
direct engineered capture   337, 339
marine organism sequestration   336–339
subsurface sequestration   334–336
surface sequestration   336
risks and challenges   343–345
SRM methods
applicability of   342–343
reflectors and mirrors   341–342
sulfur injection   340–341
theoretical framework   343
climate leadership in parks tool (CLIP)   164
climate resilience evaluation and awareness tool (CREAT) 3.0   116
climate "tipping point"   339
Clinton Landfill with no landfill gas (LFG) recovery   356, 357
CMU. *see* cement masonry unit (CMU)
CO_2 emissions   36–37
Committee of Institutional Means for Assessment of Risks to Public Health   222
community and institutional structures   74
community resources   74
comparable emissions database (CEDB)   164
composite panel wall   421
compost system analysis   356
Comprehensive Environmental Response, Compensation, and Liability Act (CERCLA)   58, 268
concrete mixtures for pavements and bridge decks
background   428–429
economic sustainability   443–445
environmental sustainability   437–443
methodology
cementitious combinations   431
concrete mix combinations   431, 432
economic assessment   434–436, 438
environmental assessment   434, 437
fly ash   430–431
functional unit   432, 433
goal and scope   430
ground granulated blast furnace slag   430, 431
limestone waste   430, 431
social impact assessment   434
system boundaries   433–436
social sustainability   445–446
constructed wetlands   308–310
contaminated site remediation
case studies   326–327
challenges and opportunities   327–328

economic sustainability indicators   324–325
environmental indicators   324
green and sustainable remediation technologies   316–321
qualitative tools   325
semiquantitative assessment tools   325
social sustainability   325
subsurface contamination sources   313, 314
sustainable remediation framework   321–324
systematic approach   315–316
in United States   313, 314
conventional cover (CLC)
characteristic properties of cover materials   382, 383
design profile of   382
environmental impacts   382, 385, 387
LCA of   382
social sustainability of   391
technical design of   381
conventional ductile iron with cement-mortar lining
annual cost and benefit analysis.   485, 486
background   481–482
economic sustainability   485–486
environmental impacts of   484, 485
global environmental impacts of   484, 485
LCA results for   485, 486
methodology
goal and scope   482
rendering and technical design for   482, 484
site background   482
sustainability assessment   482–484
pipeline capital cost analysis   485, 486
social sustainability   486–488
conventional LCA   214–216
conventional pipe
conventional *versus* biocover landfill cover system
background   380–381
economic sustainability   388, 390, 391
environmental sustainability   382, 384–390
methodology
goal and scope   381
landfill location   381
technical design of landfill covers   381–384
social sustainability   391
copper use   145
cost–benefit analysis   183, 235
"cradle to grave"   268
criteria maximum concentration (CMC)   51
curve number   449
cycle   142

d

Danville Landfill with LFG recovery for   356, 357
data gap analysis   179
decline of ecosystems   13
deep well algae photo-bioreactor   393
deep well photo-bioreactor (DWP)
   economic sustainability   400
   environmental impact   396, 399
   goal and scope   392, 393
   LCA inventory analysis   396
   life cycle assessment of   400
   materials and transportation   398
   raw material acquisition for   396, 397
   site location   393
   social sustainability   400, 402
   technical design   393
deforestation   40–41
democracy and governance   72
desertification   40
developed and developing societies   70–71
Development, and Demonstration programs (DOE 2018)   336
diesel power generation system
   BEES and TRACI   491–493
   economic sustainability   491, 494–495
   goal and scope   489
   LCA data for   491
   OPEX and CAPEX costs   495
direct engineered capture   337, 339
disaster resiliency
   challenges   126
   2016 Chile's wildfires by drought and record heat   106
   climate change and extreme events   104–105
   2012 Hurricane Sandy in New York City   105–106
   overview of   103
   resiliency (*see* resiliency)
   2017 worst South Asian monsoon floods   106
diversity   72
DOE's Clean Coal Research   336
domestic hidden flows (DHF)   150
dose–response assessment   223–225
dry lands   40
dual-phase extraction   318
ductile iron with cement-mortar lining
   annual cost and benefit analysis.   485, 486
   background   481–482
   economic sustainability   485–486
   environmental impacts of   484, 485
   global environmental impacts of   484, 485
   LCA results for   485, 486
   methodology
      goal and scope   482
      rendering and technical design for   482, 484
      site background   482
      sustainability assessment   482–484
   pipeline capital cost analysis   485, 486
   social sustainability   486–488
durable goods calculator (DGC)   164
DWP. *see* deep well photo-bioreactor (DWP)

e

Earth retaining system (ERS)   459
Easter Island   4–5
ecological resiliency   106
ecological risk assessment   228–229
economic assessment framework   78–79
economic assessment tools   233–234
   cost–benefit analysis   235
   life-cycle costing   234
economic input–output (EIO) analysis   138
economic input-output life-cycle analysis (EIO-LCA)
   available models   210
   boundary definition and circularity effects   209
   vs. conventional LCA   214–216
   environment sector   209
   example of   212–214
   industry transactions   210
   interpretation of results   211
   online tool   210
   other issues and considerations   212
   uncertainty   211–212
economic input–output (EIO) model
   cause-and-effect relationship   209
   constant technical coefficients   209
   GDPs   207
   input–output transaction table   207–208
   limitations   209
   mass balance markets   209
   national budgets   207
   NFAs   207
   vs. physical input–output (PIO) analysis   216–219
   two-sector economy   208
economic issues   77–78
   economic assessment framework   78–79
   life cycle costing   79
   true-cost accounting   79–80

Index

economic resiliency   106
ecosystem capital   13
ecosystems, decline of   13
ecosystem services valuation tools   235–236
eco-toxicity impact category   137
ecotoxicity potentials (ETPs)   371
Edisun Corporation   494
effective concentration (EC)   224
electric bus service
  background   459
  economic sustainability   465, 466–467
  environmental sustainability   461–465
  methodology
    design requirements   460
    friction piles   460, 461
    goal and scope   460
    spread footing foundation   460–461
    subsurface soil profile   460
    sustainability assessment   461
  social sustainability
    social dimension   467, 468
    socio-economical dimension   467, 469
    socio-environmental   469, 470
    socio-institutional dimension   467, 468
electrokinetics   317
electronic waste (e-waste)   57
elemental cycles   91–94
"elephants and scorpions" analogy   151
employee commute emissions calculator   164
endpoint analysis   184
end-use sector emissions   245–246
energy budget, atmosphere   331, 332
energy efficiency and reduced consumption   250
energy efficient buildings   251
energy infrastructure   297
energy-intensive aggressive technologies   321
energy recovery   274, 275
engineering resiliency   107
environmental accounting   78
Environmental and Social Impact Assessment (ESIA)   74
environmental assessment tools/indicators   231–233
environmental CBA (eCBA)   234
environmental concerns   31
  acid rain   47–48
  aesthetic degradation   61
  air pollution   44–46
  deforestation   40–41
  desertification   40
  eutrophication   51–52
  global warming and climate change   32–39
  land contamination   59–60
  land use patterns   61–62
  loss of habitat and biodiversity   41–42
  noise pollution   62
  odors   61
  ozone layer depletion   43–44
  salinity   52–53
  smog   47
  thermal pollution   62
  visibility   60–61
  waste generation and disposal   53–59
  water usage and pollution   48–51
environmental health risk assessment
  ecological risk assessment   228–229
  emergence of the risk era   221–222
  risk assessment and management   222–228
environmental impact assessment (EIA)   73, 183
environmental impacts
  current products and activities   178
  of energy generation
    air emissions   244–247
    land resource use   248–249
    solid waste generation   248
    water resource use   248
  matrix for accounting the   192–193
  relative   192, 194
  when selecting between several product and activity options   178–179
environmentalism   18, 69
Environmental Justice Screening and Mapping Tool (EJSCREEN)   236
Environmental Justice Strategic Enforcement Assessment Tool (EJSEAT)   236
environmental justice tools   236–237
environmental LCA (eLCA)   234
environmental LCC (eLCC)   234
environmentally responsible product matrix (ERP matrix)   192
environmental pollution   6–7
environmental remediation techniques   8
Envision   300–302
Envision Sustainable Infrastructure Rating System   364, 365, 366
EPA's electronic greenhouse gas reporting tool (e-GGRT)   164
EPA SWMM software   449
EPAWARM model   357, 358, 363
"eQuest 3-65" software   421
equity   72

E33 media 370
euryhalines 52
eutrophication 51–52
eutrophication impact category 137
events emissions calculator 164
excessive nutrient loading 50
expanded SLCA 195–198
exposure assessment 225–226
ex-situ remediation method 316
extreme events 104–105

f

Federal Water Pollution Control Act of 1948 51
fiber-reinforced polymer laminates (FRPLs)
  carcinogenic impact for 500, 503
  direct costs of 503, 505
  environmental impact 501, 503, 505
  environmental impact distribution for 501, 502
  global sustainability assessment of 506, 508
  social costs of 503, 505
  social impact survey results 505, 507
  technical design 499–500
  total amount of 500, 501
  total transport of 500, 501
financial CBA (fCBA) 234
financial LCC (fLCC) 234
fleet emissions calculator 165
flooding/drought 38
fly ash 430–431
food scrap landfilling *vs.* composting
  background 351
  economic sustainability
    cost of transportation 357
    indirect costs for landfilling *vs.* composting 362–363
    operation and maintenance costs 357, 362
  environmental sustainability 356–358
  ENVISION™ 363–366
  life cycle assessment 357, 359, 360, 361
  methodology
    functional unit 356
    goal and scope 353
    study area 353, 354
    system boundaries 353–356
    transportation 356
  social sustainability 363, 364
foreign hidden flows (FHF) 150
fossil fuels
  coal 85–86
  combustion 245
  natural gas 86–88
  oil shales and tar sands 86
  petroleum 86–87
Fourth National Climate Assessment (NCA4) 105
fresh water 90
FRPLs. *see* fiber-reinforced polymer laminates (FRPLs)
full environmental LCC (feLCC) 234
Fuzzy Evaluation for Life Cycle Integrated Sustainability Assessment (FELICITA) 237–238
fuzzy logic 237
FY2008 Consolidated Appropriations Act 164

g

geothermal energy 257–260
GHG inventory 138
GHG protocol calculation tools and guidance 165
global carbon cycle 36–37, 91–92, 331, 332
Global Change Research Act of 1990 105
global climate change impact category 137
global warming 32–39
global warming potential (GWP) 32, 34–36, 157–159, 353
green accounting 78
green and sustainable remediation (GSR) 313
green building
  average savings of 286
  components of
    features of 289–291
    geothermal energy system 289
    hot water generation with solar energy and circulation system 289
    solar power 288
    stormwater harvesting 288
  concepts 287
  design and operation 287
  history 286
  impact of 286
  LEED 291–295
Green Building Certification Institute (GBCI) 293
green chemistry 270–272
green engineering (GE) 21
Greener Cleanup Matrix 326
greenhouse effect 10, 32
greenhouse gases (GHGs) 10–13, 157, 268, 313
  CDR 334
  CO_2e 362

commercial buildings   417
developments in the United States   163–164
global warming potential   157–158
mitigation   169–170
SRM methods   334
UNFCCC   345
WARM model   356–357
greenhouse gas reporting program (GHGRP)   164
green infrastructure   303
green remediation   313
green roofs   304, 305
green stormwater infrastructure
  bioswales   307, 309
  constructed wetlands   308–310
  green roofs   304, 305
  permeable pavements   304–305, 306
  rain gardens and planter boxes   307, 308
  rainwater harvesting   305, 307–308
  tree canopies   310
green vehicles   251–252
greenwashing   327
GREET model   165
gross domestic products (GDPs)   207
groundwater plumes   320
Groundwater Quality Database (GWQDB)   367
groundwater remediation   317–318

h

hard infrastructure   297
hazard identification   223
hazard index (HI)   227
hazardous air pollutants (HAPs) emission   410–411
hazardous waste
  integrated waste management   276–278
  landfill   279
  RCRA regulations   268
high density polyethylene (HDPE)   394
high-level waste (HLW)   87
"homewise.com"   421
human health cancer impact category   137
human health noncancer impact category   137
human health particulate matter (PM) impact category   137
human population   7–8
Hurricane Sandy in New York City   2012, 105–106
hydrofluorocarbons (HFCs)   158
Hydrologic Evaluation of Landfill Performance (HELP) software program   380, 382

hypoxia   51

i

Illinois Department of Transportation (IDOT)   448
Illinois EPA   326
Illinois Green Business Association (IGBA)   353
Illinois State Water Survey (ISWS) Public Service Laboratory   366
IMPACT 2002+database   357
Indian Ridge Marsh (IRM)   326
industrial smog   47
industrial solid wastes   268
industrial waste   146
infrastructure. see civil infrastructure; sustainable infrastructure
infrastructure engineering sustainability projects
  electric bus service
    background   459
    economic sustainability   465, 466–467
    environmental sustainability   461–465
    methodology   459–461
    social sustainability   467–470
  retaining wall systems
    background   471
    environmental sustainability   475–478
    methodology   471–475
    social sustainability   476–477, 479, 480, 481
in-situ remediation method   316
in-situ soil flushing   317
institute for sustainable infrastructure (ISI)   300
integrated sustainability assessment tools   237–239
Integrated Value Model for Sustainability Assessment (MIVES)   238, 412
integrated waste management   276–279
Intergovernmental Forum on Forests (IFF)   41
Intergovernmental Panel on Climate Change (IPCC)   32, 111–112
Intergovernmental Panel on Forests (IPF)   41
International standard on Assurance Engagements   163
Interstate Technology and Regulatory Council (ITRC) framework   322–323
Iowa Department of Natural Resources (Iowa DNR)   403
IPAT equation   17
iron fertilization   336–339
ISO 14040:2006   177
ISO 14044:2006   177

l

land contamination   59–60
land degradation   40
landfill gases   263
landfill-induced groundwater contamination   10
landfills   9
   hazardous waste   279
   nonhazardous waste   274, 276
land resource use   248–249
landscape recycling center (LRC)   353
land use patterns   61–62
large-scale sulfur dioxide emissions   45
LCA. *see* life cycle analysis (LCA)
leachate   54
leachate-based groundwater pollution   54
leached solid wastes   9
lead   45
leadership category   378
Leadership in Energy and Environmental Design (LEED)   139, 231–232
   certification an attractive benefit   293
   credentials   293
   energy and atmosphere   292
   GBCI   293
   indoor environmental quality   292
   innovation   292
   integrative process   291
   life-cycle aspects   292, 293
   locations and transportation   291
   materials and resources   292
   regional priority   292
   scoring system
      certification level based   292, 295
      for new constructions   292, 294
   sustainable sites   291
   water efficiency   292
LEED. *see* Leadership in Energy and Environmental Design (LEED)
legal issues   80–81
less/least developed countries (LDCs)   70–71
lethal concentration (LC)   224
LFG emissions   380, 381
life cycle analysis (LCA)   25, 79, 280, 323
   biocover and conventional landfill cover   382, 385, 388
   concrete mixtures   434, 437
   contaminated sediment   407
   definition and objective   174
   "food waste" product   357, 359
   framework and guidance   177
   goal and scope definition   178–179
   history   176–177
   interpretation   186–187
   ISO standards   177–178
   LCI   179–182
   LCIA   182–186
   overview   173–174
   procedure   174–176
   tools and applications   187–188
life-cycle assessment (LCA)   138
life cycle cost assessment (LCCA)   434
life-cycle costing (LCC)   79, 234
life cycle impact assessment (LCIA)   182–186, 370, 371
life cycle inventory (LCI)   371
   feedback loop   181
   format   182
   material flows within systems   180–181
   outputs from raw materials   180
   specific stages for the life cycle   181
   system boundary   179–180
   treatment of data collection   181
   unit of function   181
   unresolved problems   182
life cycle stages of food   354
limestone waste   430, 431
London Fog   47
loss of habitat   41–42
lowest observable effect concentration (LOEC)   224
low impact development (LID)   303
low/intermediate level radioactive waste (LILW)   87

m

management of waste
   energy recovery   274, 275
   green chemistry   270–272
   hierarchy of   269
   integrated   276–279
   landfills   274
   minimize and/or reduce waste generation   272
   pollution prevention   270
   reuse/recycling   272–274
March Point Landfill site   119
marine organism sequestration   336–339
master equation   17
material flow analysis (MFA)
   chromium cycle   143–144
   copper use   145
   wastes   146–148
material flow and budget analysis   137–138

Index 521

material recovery facility (MRF)   273
material recycling   147
maturity   72
maximum allowable toxicant concentration (MATC)   224
measured mineral resource   88
medical waste   57
Medical Waste Tracking Act (MWTA)   57
mental and physical health   73
metallic ores consumption   5–6
metals   273
methane (CH$_4$)   353
midpoint effect   184
Millennium Ecosystem Assessment   13, 42
mineral resource   88–89
Minnesota Pollution Control Agency   326
misconsumption   98
MIVES tool   455
money economy   78
monitored natural attenuation (MNA)   326, 405–406, 409– 411, 413
monsoon floods   106
more/most developed countries (MDCs)   70
Morgan Street Bridge   432, 433
MOtor Vehicle Emission Simulator (MOVES)   165
multi-criteria decision analysis (MCDA)   238
municipal solid waste (MSW)   9, 53–54, 260, 263, 268
  anaerobic decomposition of   380
  organic fraction of   353
  in US   351, 352

n

National Ambient Air Quality Standards (NAAQS)   46
national budgets   207
National Emission Standards for Hazardous Air Pollutants (NESHAPs)   46
National Financial Accounts (NFAs)   207
national material account (NMA)   149–153
National Oceanic and Atmospheric Administration (NOAA)   404
National Pollution Discharge Elimination System (NPDES)   51, 303
National Priority List (NPL)   268
National recycling coalition's (NRC) environmental benefits calculator   165
national stormwater calculator with climate assessment tool   116
natural attenuation (MNA)   403
natural eutrophication   52

natural habitats   304
natural resources
  budget of   141–143
  Easter Island example   4–5
  elemental cycles   91–94
  fossil fuels   85–88
  metallic ores consumption example   5–6
  mineral resource   88–89
  radioactive fuels   87–88
  resource depletion   94–99
  water resources   89–91
Natural World category   376, 378
Neochloris patented technology   392
Neochloris photobioreactor
  economic sustainability   400
  life cycle inventory analysis of   398
  raw material acquisition for   397
  social sustainability evaluation of   400, 402
nitrogen oxide emissions   245
nitrous oxide (N$_2$O)   45, 157
NOAA sea level rise viewer   116
NOAA Surging Seas   116
noise pollution   62
nonhazardous waste
  landfill for   274, 276
  municipal solid waste   268
  RCRA regulations   268
no observed adverse effects level (NOAEL)   224
northeast recycling council's environmental benefits calculator   165
noxious by-product chemicals   6
nuclear energy   249–250

o

odors   61
office footprint calculator™   165
off-set unavoidable emissions   162
open pond systems
  cost evaluation of   402
  damage assessment of   402
  economic sustainability   400
  environmental impact   396, 399
  goal and scope   393, 394
  LCA inventory analysis   396, 397
  life cycle assessment of   400, 401
  materials and transportation for   398
  raw material acquisition for   397
  site location   394
  social sustainability   400
  social sustainability evaluation of   402

522 | Index

open pond systems (*contd.*)
  technical design   394
  TRACI   399, 400
optimization model for reducing emissions of greenhouse gases from automobiles (OMEGA)   165
organic wastes   50
O'Rourke Building   497, 498
overconsumption   98
overshooting   98
ozone-depleting substances (ODS)   43
ozone depletion   43–44, 421
  impact category   137

p

paper   273
paper use emissions calculator   165
parking lots
  economic sustainability   453–454
  environmental sustainability   450–453
  methodology
    goal and scope   448
    study area   448
    sustainability assessment   449–450
    technical design   448–449
  overall sustainability   455
  social sustainability   454, 455
particulate matter   45
passive solar heating   253–254
perfluorocarbons (PFCs)   158
permeable pavements   304–305, 306
permeable reactive barriers (PRBs)   318
persistent organic pollutants (POPs)   60
personal GHG emissions calculator   165
photobioreactor
  economic sustainability   400
  life cycle inventory analysis of   398
  raw material acquisition for   397
  social sustainability evaluation of   400, 402
photochemical smog   47
  formation impact category   137
photovoltaic cells   252
physical economy   78
physical input–output (PIO) analysis   216–219
phytoremediation technology   278
plasma arc torch methods   278
plastics   273
political and social resources   74
pollution controls   250

pollution prevention (P2)   270
polychlorinated biphenyls (PCBs) and pesticides
  background   403–404
  environmental sustainability   409–411
  goal and scope   404
  overall sustainability   412–414
  social sustainability   411–412
  study area   404
  sustainability assessment methodology   407–409
  technical design
    conventional capping   406, 407
    dredging of contaminated sediment   406
    modified capping method   407
    monitored natural attenuation   405–406
polynucleic land use approach   16
population characteristics   74
Portland cement (PC)   396, 429
Portland limestone cement (PLC)   429
poverty   72
power purchasing agreement (PPA)   486
precast concrete retaining wall
  carbon dioxide emissions   475, 478
  cross section for   473, 474
  environmental impacts   475, 476, 478
  Envision checklist results   479, 481
  global warming   475, 477
  methodology
    goal and scope   471
    study area   471
  processes flow of   472–473, 474
  SLCA health and safety matrix   477, 480
  social impact assessment   480
  technical design   471, 472–473
  work and labor costs for   475, 479
precipitation   38
pressure-reducing valves (PRVs)   481
primary pollutants   45
probable effect concentrations (PECs)   403
product life cycle   173, 192–193
projections of Earth's surface temperatures   331, 334
public policies and legislative actions   179
public water system (PWS)   303

q

quality of life   72, 376, 378
Quantitative Assessment of Life Cycle Sustainability (QUALICS)   238

quantitative environmental health risk assessment   138

r

radioactive fuels   87–88
radioactive waste   57
rainwater harvesting   305, 307, 308
RCRA corrective measure study   316
RCRA Subtitle C regulations (USEPA 2017b)   268
RCRA Subtitle D regulations (USEPA 2017b)   268
RCRA Subtitle D requirements   382
ready-mix concrete plant (RMCP)   429
ReCiPe endpoint   396, 399
recycled content (ReCon) tool   165
recycling   8, 272–274, 276, 277
reference dose (RfD)   225
remedial technology
   bioremediation   321
   cap, vertical barrier and bottom barrier   318, 319
   GSR criteria   320
   incineration   320
   pump-and-treat operations   320
   pumping well systems   318, 319
   saturated zone/groundwater   317–318
   subsurface drain system   318, 319
   vadose zone soil remedial methods   316–317
renewable energy   250
   biomass energy   260–263
   geothermal energy   257–260
   solar energy   252–254
   water energy   255–257
   wind energy   253, 255–256
renewable feedstock   270, 271
resiliency
   ecological   106
   economic   106
   engineering   107
   framework   112–115
   infrastructure   115–116
   initiatives and policies on   109–112
   resilient environmental remediation   119–126
   San Francisco firehouse resilient design   117
   San Francisco resilient CSD design   117–121
   social   106
   vs. sustainability   108–109
resilient environmental remediation   119–126
Resource Allocation category   378, 379
Resource Conservation and Recovery Act (RCRA)   8, 56, 267, 351, 353
resource depletion
   causes of   95
   description   94
   effects of   95–98
   overshooting   98
   urban metabolism   98–99
retaining wall. *see* cast-in-place retaining wall; precast concrete retaining wall
retrofitting   270
reuse   8
risk assessment   183
risk assessment and management
   dose–response assessment   223–225
   exposure assessment   225–226
   hazard identification   223
   risk characterization   226–228
   schematic   222
risk characterization   226–228
Rockefeller Foundation's 100 Resilient Cities Initiative   109–110

s

Safe Drinking Water Act (SDWA)   51
salinity   52–53
San Francisco firehouse resilient design   117
San Francisco Public Works (SFPW)   117
San Francisco resilient CSD design   117–121
sanitary landfills   274
SBTool^{PT} 232
SBTool^{PT}-UP   232
sea level change curve calculator tool   116
Seattle City Light   362
secondary and/or refined/synthesized products   13
secondary pollutants   45
sediment runoff   6
Sertraline   24
sewage-borne pathogens   51
Sewer System Improvement Program (SSIP)   117
shear wall retrofitting techniques
   background   497–498
   economic sustainability   503–505
   methodology
      environmental sustainability   501–503
      goal and scope   498–499
      sustainability assessment   500, 501
      technical design   499–500
   overall sustainability   505–506, 508
   social sustainability   505, 506
shipping emissions calculator   165

SimaPro   174
  arsenic adsorption   370–371
  food waste   357, 359, 360
SimaPro™ LCA software   421
SiteWise™   326
SLCA. *see* streamlined life-cycle assessment (SLCA)
"SMART"   324
smog   47
snow cover   38
social CBA (sCBA)   234
social cohesion/interconnection   72
social impact assessment (SIA)   73–77
social indicators   72–73
social injustice   14–16
social issues
  developed and developing societies   70–71
  social impact assessment   73–77
  social indicators   72–73
  social sustainability implementation   77
  society   69–70
  sustainability concept   71–72
social LCA (sLCA)   234
social resiliency   106
social sustainability concept   71–72
social sustainability evaluation matrix (SSEM)   325, 380, 461
social sustainability implementation   77
societal LCC (sLCC)   234
society   69–70
soft infrastructure   297
soil stabilization   317
soil vapor extraction (SVE)   316
solar energy   252–254
solar photovoltaic (PV) system
  BEES and TRACI   491–493
  CAPEX costs   491, 494–495
  Envision checklist results for   496, 497
  goal and scope   489
  LCA data for   490–491
  process flow diagram for   489, 490
solar power   288
solar radiation management (SRM) methods
  reflectors and mirrors   341–342
  sulfur injection   340–341
solar thermal technology   252–253
solid waste   263, 274
Solid Waste Disposal Act   56
solid waste generation   248
South Asian monsoon floods   2017, 106
spread footing foundation   460–461

spreadsheets for environmental footprint analysis (SEFA)   407, 410
steel bracing
  carcinogenic impact for   500, 503
  direct costs of   503, 505
  environmental impact   501, 503, 505
  environmental impact distribution for   501, 502
  global sustainability assessment of   506, 508
  social costs of   503, 505
  social impact survey results   505, 507
  technical design   499–500
  total amount of   500, 501
  total transport of   500, 501
StormTrap Single Cell   449
stormwater harvesting   288
stormwater infrastructure   303. *see also* green stormwater infrastructure
stormwater runoff   303
streamlined life-cycle assessment (SLCA)   138, 373, 377
  alternative graphical representation   194–195
  applications of   200–204
  environmental concerns   192–193
  expanded   195–198
  graphical representation   194
  life cycle stages   192
  matrix for accounting the environmental impacts   192–193
  overview   191
  product-related   192
  relative environmental impact   192, 194
  simple example   198–200
subsurface sequestration   334–336
sulfur dioxide emissions   245
sulfur hexafluoride (SF6)   158
sulfur injection   340–341
Superfund Amendments and Reauthorization Act (SARA)   58
Superfund feasibility study   316
SuperstormSandy   105–106
supplementary cementitious materials (SCMs)   429
surface capping   318
surface reflectivity   338
surface temperature   331, 333
sustainability   17–20
sustainability assessment tools   137–139
sustainability evaluation matrix (SSEM) matrix   425, 427
sustainability indicators

actionable/achievable   134
economic   135
key indicators   135–136
measurable   134
relevant   134
specific   133
timely   134
UN   134–135
sustainability metrics   136–137
sustainable energy engineering
energy consumption   243, 245
energy sources in the United States   243–244
environmental impacts of energy generation   244–249
fuel mix for electricity generation   243–244
nuclear energy   249–250
renewable energy   252–263
strategies for clean energy   250–252
sustainable engineering (SE)   21–25
sustainable infrastructure
principles of   298
of water infrastructure
green stormwater infrastructure   304–306
groundwater   303
public drinking water system   303
stormwater runoff   303
Sustainable Remediation Forum (SURF) framework   322
sustainable remediation framework
ASTM framework   322–323
ITRC framework   322–323
SURF framework   322
USEPA framework   321
sustainable remediation tool (SRT)   326
sustainable rural electrification
background   489
economic sustainability   491, 494–495
environmental sustainability   491–494
methodology
goal and scope   489
study area   489
sustainability assessment   490–491
technical design   489–490
social sustainability   495–497
sustainable waste management   279–280

t

tailings   248
telecommunications infrastructure   297
Texas State Energy Conservation Office   491
thermally enhanced remediation   321
thermal pollution   62
Threshold Effects Concentration (TEC)   404
Tool for Reduction and Assessment of Chemicals and Other Environmental Impacts (TRACI)   137, 187–188, 370, 371–372, 396
Toolkit for Greener Practices   326
Toxicity Characteristic Leaching Procedure (TCLP) test   370
translucent composite panels
background   417–418
economic sustainability   421
environmental sustainability   421, 422–426
methodology
conventional design   418–419
in exterior façade   419, 420
goal and scope   418
structure of   419, 420
social sustainability   425, 427–428
transportation infrastructure   297
tree canopy   310
tropical storms   38–39
true-cost accounting (TCA)   79–80
two-sector EIO model   212–213

u

UIC Climate Action Plan   168
UIC Parking Lot 1   447, 448
uncontrolled urban sprawl   16
UN Convention to Combat Desertification (UNCCD)   40
underground geologic carbon sequestration   334–336
underground storage tanks (USTs)   313
The United Nations (UN)   110–111
United Nations Convention on Biological Diversity (UNCBD)   42
United Nations Framework Convention on Climate Change (UNFCCC)   345
United States Environmental Protection Agency (USEPA)   313, 316
University of Illinois at Chicago (UIC)   165–169, 447
urban metabolism   98–99
urban sprawl   16–17
US Department of Defense (DOD)   112
USEPA   164
USEPA Environmental Footprint Analysis Tool (SEFA)   326
USEPA green remediation framework.   321
USEPA's Food Recovery Hierarchy   351

US Presidential Policy Directives (PPDs)   109
US raw nonfuel minerals   141–142
US Resource Conservation and Recovery Act (RCRA)   351, 352

v
vacuum-enhanced recovery   318
vadose zone soil remediation technologies   316–317
vehicular GHG emissions   159
visibility   60–61
vitrification (ISV)   317
volatile organic compounds (VOCs)   6

w
waste   146–148
  circular economy   280–281
  definition   267
  effects and impacts of   268
  gasification   278
  generation   8–10
  generation and disposal   53–59
  generation flow   146
  hazardous waste   268
  nonhazardous waste   268
  streams   287
  sustainable waste management   279–280
  waste management
    energy recovery   274, 275
    green chemistry   270–272
    hierarchy of   269
    integrated   276–279
    landfills   274
    minimize and/or reduce waste generation   272
    pollution prevention   270
    reuse/recycling   272–274
waste reduction model (WARM)   164, 356–357
waste-to-energy incinerators   274, 275
waste to energy (WTE) process   56
water and environment infrastructure   297
water energy   255–257
water footprint   91
water infrastructure
  green stormwater infrastructure   304–306
  groundwater   303
  public drinking water system   303
  stormwater runoff   303
water pipeline systems
  background   481–482
  economic sustainability   485–486
  environmental sustainability   484–485
  methodology
    CEP pipe design   482, 484
    conventional pipe design   482
    goal and scope   482
    site background   482
    sustainability assessment   482–484
  social sustainability   486–488
water pollution   6, 51
water resources   89–91
water resource use   248
water supply wells   303
water treatment   378
water usage and pollution   48–51
Western Australia Council of Social Services (WACOSS)   72
wetlands, constructed   308–310
Whitmarsh Landfill   119
wildfires   106
wind energy   253, 255–256
World Commission on Environment and Development   107

z
Zion Landfill in Illinois   381
Zofnass Program for Sustainable Infrastructure   300, 363